Also available from McGraw-Hill

Schaum's Outline Series in Mechanical Engineering

Most Outlines include basic theory, definitions, hundreds of example problems solved in step-by-step detail, and supplementary problems with answers.

Related titles on the current list include:

Acoustics
Continuum Mechanics
Elementary Statics & Strength of Materials
Engineering Economics
Engineering Mechanics
Engineering Thermodynamics
Fluid Dynamics
Fluid Mechanics & Hydraulics
Heat Transfer
Lagrangian Dynamics
Machine Design
Mathematical Handbook of Formulas & Tables
Mechanical Vibrations
Operations Research
Statics & Mechanics of Materials
Strength of Materials
Theoretical Mechanics
Thermodynamics with Chemical Applications

Schaum's Solved Problems Books

Each title in this series is a complete and expert source of solved problems with solutions worked out in step-by-step detail.

Related titles on the current list include:

3000 Solved Problems in Calculus
2500 Solved Problems in Differential Equations
2500 Solved Problems in Fluid Mechanics & Hydraulics
1000 Solved Problems in Heat Transfer
3000 Solved Problems in Linear Algebra
2000 Solved Problems in Mechanical Engineering Thermodynamics
2000 Solved Problems in Numerical Analysis
700 Solved Problems in Vector Mechanics for Engineers: Dynamics
800 Solved Problems in Vector Mechanics for Engineers: Statics

Available at most college bookstores, or for a complete list of titles and prices, write to: Schaum Division
McGraw-Hill, Inc.
Princeton Road, S-1
Hightstown, NJ 08520

Also Available from McGraw-Hill

Schaum's Outline Series in Mechanical Engineering

Schaum's Solved Problems Books

FUNDAMENTALS OF FLUID FILM LUBRICATION

Bernard J. Hamrock

The Ohio State University
Columbus, Ohio

McGraw-Hill, Inc.

New York St. Louis San Francisco Auckland Bogotá
Caracas Lisbon London Madrid Mexico City Milan
Montreal New Delhi San Juan Singapore Sydney Tokyo Toronto

This book was set in Times Roman by Science Typographers, Inc.
The editors were John J. Corrigan and Jack Maisel;
the production supervisor was Louise Karam.
The cover was designed by Joseph Gillians.
R. R. Donnelley & Sons Company was printer and binder.

FUNDAMENTALS OF FLUID FILM LUBRICATION

This book is printed on recycled, acid-free paper containing a minimum of 50% total recycled fiber with 10% postconsumer de-inked fiber.

1 2 3 4 5 6 7 8 9 0 DOC DOC 9 0 9 8 7 6 5 4 3

ISBN 0-07-025956-9

Library of Congress Cataloging-in-Publication Data

Hamrock, Bernard J.
Fundamentals of fluid film lubrication/Bernard J. Hamrock.
p. cm. — (McGraw-Hill series in mechanical engineering)
ISBN 0-07-025956-9
1. Fluid-film bearings. I. Title. II. Series
TJ1063.H35 1994
621.8'9—dc20 93-32107

ABOUT THE AUTHOR

Bernard J. Hamrock joined The Ohio State University as a professor in the department of mechanical engineering in September 1985. Prior to coming to The Ohio State University, he spent over 18 years as a research consultant in the tribology branch of NASA Lewis Research Center in Cleveland, Ohio. He received his Ph.D. degree from the University of Leeds, Leeds, England, and has done postdoctoral research work at the University of Luleå, Luleå, Sweden. Professor Hamrock has done research work for the past 32 years in the field of lubrication. Primary areas of research have been elastohydrodynamic lubrication as related to the lubrication of rolling-element bearings and gears and hydrodynamic lubrication as related to fluid film journal and thrust bearings, seals, and fluid dampers. During the 32 years Professor Hamrock has worked in the lubrication field, he has published a book with Duncan Dowson entitled *Ball Bearing Lubrication*, three separate chapters for handbooks, and over 100 technical publications. His awards include the 1976 Melville Medal from ASME, the best paper of the year for 1975 and 1976 from the lubrication division of ASME, the NASA Lewis Research Center Best Paper of the Year (1977), and the NASA Exceptional Achievement Medal in 1984. Professor Hamrock is a Fellow of ASME.

To Rosemary

CONTENTS

PREFACE

The title of this book was specifically chosen as *Fundamentals of Fluid Film Lubrication*, rather than the more general title of *Tribology*, since fluid film lubrication will be the book's primary emphasis. Fluid film lubrication occurs when opposing bearing surfaces are completely separated by a lubricant film. Hydrodynamic and elastohydrodynamic lubrication are modes of fluid film lubrication and are emphasized in this text, whereas boundary lubrication is given only a cursory treatment. The reason for this slant of the book is that fluid film lubrication has been the focal point of my research throughout my professional career.

The organization of the text is such that it is divided into three parts. The first part covers the fundamentals required in understanding fluid film lubrication. That is, an understanding of surface characterization (Chapter 3), lubricant properties (Chapter 4), bearing materials (Chapter 5), viscous flow (Chapter 6), and the Reynolds equation (Chapter 7) is important in understanding fluid film lubrication. The second part of the book then covers hydrodynamic lubrication (Chapters 8 to 18), and the third part covers elastohydrodynamic lubrication (Chapters 19 to 27).

Hydrodynamic lubrication can be achieved by sliding motion (as discussed in Chapters 8 to 12), by squeeze motion (as discussed in Chapter 13), and by external pressurization (as discussed in Chapter 14). Generally, in hydrodynamic lubrication oil is the lubricant. However, as discussed in Chapters 16 and 17, gas can be an effective lubricant in certain applications.

The treatment of elastohydrodynamic lubrication begins with the consideration of elasticity effects in Chapters 19 and 20. Elastohydrodynamic lubrication of rectangular conjunctions is considered in Chapter 21 and of elliptical conjunctions in Chapter 22. Film thicknesses for different fluid film lubrication regimes are presented in Chapter 23. In Chapters 24 and 25 the theory of elastohydrodynamic lubrication is applied to a range of lubricated conjunctions: in roller and ball bearings; between a ball and a flat plate; between concave and convex surfaces; in power transmission devices such as involute gears and variable-speed drives; between a railway wheel and wet or oily rails; and finally, in synovial joints.

The material presented through Chapter 25 with the exception of Section 4.15 assumes the fluid behavior to be Newtonian. Non-Newtonian fluid rheology is applied to elastohydrodynamic line contacts in Chapters 26 and 27. Both steady-state and transient effects are considered in Chapter 26, and in Chapter 27 the added consideration of thermal effects is presented.

Throughout the book emphasis is given to deriving formulas from basic theory and providing physical understanding of these formulas. Although at times this proves to be lengthy, I feel it is important that the reader develop a firm understanding of how information provided in design charts has been obtained. Also the importance and influence of the assumptions made in all derivations based on the theory are discussed. The assumptions emphasize the limits to which the results of the derivations are valid and applicable. The application of the theory to the design of machine elements that use fluid film lubrication helps the development of the material of the text. It is, however, not intended to consider all types of machine element and all types of bearing within this text. Rather I hope that the understanding gained from this book will enable the reader to properly analyze any machine element that uses fluid film lubrication. The material in this book in its entirety is best suited for a one-semester course (15 weeks of 3 hours of lecture per week), but a somewhat shortened version can be given in a one-quarter course (10 weeks of 3 hours of lecture per week). The book was written for senior undergraduate and graduate engineering students. Engineers who encounter machine elements that use fluid film lubrication should also find this book useful.

ACKNOWLEDGMENTS

During my career I have had the opportunity to work with a number of outstanding tribologists who have had a great effect on this book. Frank Archibald (deceased) at Northrop Nortronics, William Anderson at NASA Lewis, and Professor Duncan Dowson at Leeds University all played a significant role in the development of the material presented here.

A number of other coworkers not only advised me but also spent a great deal of time reviewing the text, for which I am ever so grateful. These include Dr. John Tripp and Dr. Bo Jacobson of SKF-ERC; Professor Erik Höglund of the University of Luleå; David Brewe and Dr. David Fleming of NASA Lewis; Dr. Shigeaki Kuroda of the University of Electro-Communications; Dr. Yeau-Ren Jeng of General Motors Research Laboratories; Professor Francis Kennedy of Dartmouth College; Professor Jack Booker of Cornell University; and Dr. Jorgen Lund of the Technical University of Denmark. The following McGraw-Hill reviewers also gave many thoughtful critiques in their reading of the manuscript: Itzhak Green, Georgia Tech; Ing T. Hong, Oklahoma State University; Michael M. Khonsari, University of Pittsburgh; H. G. Rylander, University of Texas at Austin; Andras Z. Szeri, University of Pittsburgh; and John A. Tichy, Rensselaer Polytechnic Institute.

In addition, a number of my graduate students have extensively reviewed the text, for which I am extremely grateful. They are Dr. Rong-Tsong Lee, presently at the National Sun Yat-Sen University, Taiwan; Claus Marner Myllerup of the Technical University of Denmark; Hannu Iivonen of Tampere University, Finland; and Hsing-Sen Hsiao, Jinn-An Shieh, Mohsen Esfahanian, Juan Carlos Gonzalez, and Abdalla Elsharkawy, all of The Ohio State University. Lastly, but most importantly, I would like to acknowledge Carol Vidoli of NASA Lewis Research Center for her careful editing and the publications staff at the NASA Lewis Research Center, who prepared the photographs and illustrations, especially Kathleen Malzi, who typed the various drafts of the book and the final NASA version. I acknowledge the use of tables and illustrations from the following publishers: American Society of Mechanical Engineers, American Society for Testing and Materials, BHRA Fluid Engineering, Buttersworths, Elsevier Science Publishing Company, Engineering Sciences Data Unit, Ltd., Heinemann (London), Hemisphere Publishing Corporation, MacMillan Publishing Company, Inc., McGraw-Hill Book Company, Mechanical Technology Incorporated, Non-Ferrous Founders Society, Oxford University Press, Inc., Penton Publishing, Inc., Society of Automotive Engineers, Society of Tribologists and Lubrication Engineers, VCH Publishers, John Wiley & Sons, and Wykeham Publications (London), Ltd. The specific sources are identified in the text.

The first draft of this book appeared as *NASA Reference Publication 1255*. Two entire new chapters plus significant revisions to a number of other chapters, as well as many more problems, complete the present text. Of the chapters with revisions, the major revisions occur in Chapters 11 and 12. The development of expressing the dynamic stability of journal bearings in terms of stiffness and damping coefficients is a major change.

Bernard J. Hamrock

LIST OF SYMBOLS

A	area, m^2
A_a	dimensionless amplitude of asperity, $a_a R_x / b^2$
A_{av}	average failure risk that is only partially accessible
A_c	slide-roll ratio, $2(u_b - u_a)/(u_b + u_a)$
A_p	projected pad area, m^2
A_r	recess area, m^2
A_s	sill area, m^2
A_1	defined in Eq. (18.23)
A^*, B^*	constants
$\tilde{A}$	integration constant
$\mathscr{A}$	dimensionless shear strcss, τ/τ_L
a	radius of pipe, m
a_a	amplitude of asperity, m
a_b	bearing pad load coefficient
a_1	$-s_1/(s_1+1)$ [Eq. (18.6)]
$\bar{a}$	linear thermal expansion coefficient, K^{-1}
$\tilde{a}$	semilength in x direction of uniform pressure, m
B	total conformity ratio
B_2	defined in Eq. (18.11)
B_3	defined in Eq. (18.12)
$B_{xx}, B_{xz}, B_{zx}, B_{zz}$	dimensionless damping coefficients [Eq. (11.65)]
$\tilde{B}$	integration constant
$\mathscr{B}$	dimensionless shear strain rate, $\eta s/\tau_L$
b	width of bearing; dimension in side-leakage direction, m
$b_{xx}, b_{xz}, b_{zx}, b_{zz}$	damping coefficients [Eq. (11.39)], N · s/m
b^*	contact semiwidth, m
$\bar{b}$	intercept on z axis, m
$\tilde{b}$	semiwidth in y direction of uniform pressure, m
b_0	characteristic length in y direction, m

b_1 groove width, m
b_2 half-width of uniform pressure $\tilde{p}_c$, m
C dimensionless constant used in pressure-density formula [Eq. (4.23)]
C_j weighting factors given in Eq. (21.32)
C_p specific heat of material at constant pressure, J/(kg · °C) [or J/(N · °C)]
C_r surface roughness correction factor
C_s volumetric specific heat, $\rho^* C_p$, kJ/(m^3 · °C) [or N/(m^2 · °C)]
C_ℓ thermal correction factor, $h_{\text{thermal}}/h_{\text{isothermal}}$
C_v specific heat of material at constant volume, J/(kg · K)
$\bar{C}$ specific dynamic capacity or load rating of bearing, N
$\tilde{C}$ integration constant
$\tilde{C}_0$ specific static capacity of bearing, N
c radial clearance, m
c_b bearing clearance at pad minimum film thickness (Fig. 12.11), m
c_d diametral clearance, m
c_e free endplay, m
c_f correction factor [Eq. (6.74)]
c_p constant used in Eqs. (4.10) and (27.9), Pa (or N/m^2)
c_r distance between race curvature centers, m
c' pivot circle clearance, m
$\tilde{c}$ number of divisions in semidiameter of contact ellipse in y direction
$\tilde{c}_d$ orifice discharge coefficient
D pipe diameter, m
$D_{i,j}$ influence coefficients
D_x diameter of contact ellipse along x axis, m
D_y diameter of contact ellipse along y axis, m
D_1 percentage difference between $H_{e,\min}$ and $\tilde{H}_{e,\min}$
D_2 parameter used in viscosity equation [Eq. (4.16)]
D^* influence coefficient defined in Eqs. (20.68) and (22.8)
$\bar{D}$ material factor
$\tilde{D}$ integration constant
d diameter of rolling element, m
d_a overall diameter of rolling-element bearing, m
d_b bore diameter of rolling-element bearing, m
$d_b N_a$ ball bearing bore diameter in millimeters times rotational speed in revolutions per minute
d_c diameter of capillary tube, m

d_e pitch diameter, m
d_i, d_0 inner- and outer-race diameters, m
d_p pipe inside diameter, m
$\bar{d}$ difference in two ways of evaluating z
$\tilde{d}$ number of divisions in semidiameter of contact ellipse in x direction
$\tilde{d}_o$ orifice diameter, m
E modulus of elasticity, Pa (or N/m^2)
E_{cv} stored energy of a control volume, J
E' effective elastic modulus, $2\left(\dfrac{1-\nu_a^2}{E_a}+\dfrac{1-\nu_b^2}{E_b}\right)^{-1}$, Pa (or N/m^2)
$\overline{\mathrm{E}}$ metallurgical processing factor
$\tilde{E}$ integration constant
$\mathscr{E}$ complete elliptic integral of second kind
$\bar{\mathscr{E}}$ complete elliptic integral of second kind obtained from approximate formula
e eccentricity of journal bearing, m
e_r percentage of error [e.g., $(\bar{k}-k)100/\mathrm{k}$]
e_1 constant
e' pivot circle eccentricity of journal bearing, m
$\tilde{e}$ plane strain component in a solid
$\bar{e}$ dimensionless variable given in Eq. (26.34)
F dimensionless shear force, N
F_{KE} dimensionless kinetic energy parameter, $\rho_0(\bar{u}_{ab})^3 D_x/2K_0(T_0-T_R)$
F_{WK} dimensionless flow work parameter, $E'\bar{u}_{ab}D_x/2K_0(T_0-T_R)$
F^* Fourier coefficient
$\bar{F}$ cumulative distribution of all-ordinate distribution curve
$\bar{F}_\ell$ lubrication factor
$\tilde{F}$ integration constant
$\mathscr{F}$ complete elliptic integral of first kind
$\bar{\mathscr{F}}$ complete elliptic integral of first kind obtained from approximate formula
f tangential (friction) force, N
f_c coefficient dependent on material and bearing type
f' tangential (friction) force per unit width, N/m
G dimensionless materials parameter, $\xi E'$
G_f groove factor
G_s shear modulus of elasticity, Pa (or N/m^2)
G_0 dimensionless constant used in Eq. (4.13)

$\overline{G}$ speed effect factor
$\tilde{G}$ integration constant
g acceleration due to gravity (9.8 m/s^2)
g_E dimensionless elasticity parameter, $W^{8/3}/U^2$
g_V dimensionless viscosity parameter, GW^3/U^2
H dimensionless film thickness
H_a film thickness ratio, h_s/h_r
H_b bearing pad power coefficient
H_e dimensionless film thickness for elliptical conjunctions, h/R_x
$H_{e,c}$ dimensionless central film thickness for elliptical conjunctions, h_c/R_x
$H_{e,c,s}$ dimensionless central film thickness for elliptical conjunctions under starved conditions, $h_{c,s}/R_x$
$H_{e,\min}$ dimensionless minimum film thickness for elliptical conjunctions, $h_{\min}/R_x$
$H_{e,\min,s}$ dimensionless minimum film thickness for elliptical conjunctions under starved conditions, $h_{\min,s}/R_x$
H_i inlet film thickness ratio, h_i/h_o
H_{in} dimensionless fluid inlet level, h_{in}/R_x
H_m dimensionless film thickness where $dP/dX = 0$
$H_{\max}$ dimensionless maximum film thickness
$H_{\min}$ dimensionless minimum film thickness
$H_{\min,s}$ dimensionless minimum film thickness under starved conditions
H_o outlet film thickness ratio, h_o/s_h
H_p dimensionless total power loss
H_r dimensionless film thickness for rectangular conjunctions [Eq. (21.7)]
$H_{r,a}$ dimensionless film thickness for rectangular conjunctions at inflection point, $d^2P_r^*/dX_r^2 = 0$
$H_{r,m}$ dimensionless film thickness for rectangular conjunctions when $dP/dX = 0$
$H_{r,o}$ dimensionless film constant for rectangular conjunctions
H_t total power loss, Pa · s (or N · s/m^2)
H_v power loss due to viscous dissipation, Pa · s (or N · s/m^2)
H_{in}^* dimensionless fluid inlet level where starvation first occurs, h_{in}^*/R_x
$\overline{H}$ Saybolt universal viscosity of reference oil with high viscosity index
$\overline{H}_m$ misalignment factor
$\overline{H}_p$ power loss due to pumping loss, Pa · s (or N · s/m^2)

$\tilde{H}$	integration constant
$\tilde{H}_{e,c}$	dimensionless central film thickness for elliptical conjunctions obtained from curve-fitting results, $\tilde{h}_c/R_x$
$\tilde{H}_{e,\min}$	dimensionless minimum film thickness for elliptical conjunctions obtained from curve-fitting results, $\tilde{h}_{\min}/R_x$
$\hat{H}$	dimensionless film thickness parameter, $H_e(W/U)^2$
$\hat{H}_c$	dimensionless central film thickness parameter, $H_{e,c}(W/U)^2$
$\hat{H}_{\min}$	dimensionless minimum film thickness parameter, $H_{e,\min}(W/U)^2$
h	film thickness, m
h_c	central film thickness, m
h_i	inlet film thickness, m
h_{in}	fluid inlet level, m
h_m	film thickness where $dp/dx = 0$, m
$h_{\min}$	minimum film thickness, m
h_o	outlet film thickness, m
h_p	film thickness at pivot, m
h_r	film thickness in ridge region, m
h_s	film thickness in step or groove region, m
h_t	height of reservoir (Fig. 6.10), m
h_0	central film thickness, m
h_{01}	outlet film thickness at time t_1, m
h_{02}	outlet film thickness at time t_2, m
h_{in}^*	fluid inlet level where starvation first occurs, m
$\bar{h}_t$	height from which sphere falls, m
$\tilde{h}_c$	central film thickness obtained from curve-fitting results, m
$\tilde{h}_{\min}$	minimum film thickness obtained from curve-fitting results, m
$\hbar_p$	rate of working against viscous stresses (power loss), hp
I	integral defined in Eq. (20.42)
i	number of rows of rolling elements
J	Joule's mechanical equivalent of heat, $N \cdot m/J$
$\tilde{J}$	number of stress cycles
$\check{K}_a$	dimensionless interface conductivity parameter at interface a, $K_0R/2\pi K_aD_x$
$\check{K}_b$	dimensionless interface conductivity parameter at interface b, $K_0R/2\pi K_bD_x$
K_f	lubricant thermal conductivity, $W/(m \cdot °C)$
$\bar{K}_f$	dimensionless prezothermal conductivity of lubricant
K_g	dimensionless stiffness
k_o	lubricant thermal conductivity at atmospheric conditions, $W/(m \cdot °C)$

K_∞	dimensionless stiffness, defined in Eq. (16.64)
K_1	constant defined in Eq. (24.34)
$K_{1.5}$	constant defined in Eq. (24.33)
$K_{xx}, K_{xz}, K_{zx}, K_{zz}$	dimensionless stiffness coefficients [Eq. (11.64)]
$\bar{K}$	constant defined in Eq. (21.9), $3\pi^2 U/4(W')^2$
$\mathscr{K}$	Knudsen number, $\lambda_m/h_{\min}$
k	ellipticity parameter, D_y/D_x
k_c	flow rate constant of capillary tube, $\mathrm{m}^4/(\mathrm{s}\cdot\mathrm{N}^{1/2})$
k_o	flow rate constant of orifice, $\mathrm{m}^4/(\mathrm{s}\cdot\mathrm{N}^{1/2})$
$k_{xx}, k_{xz}, k_{zx}, k_{zz}$	stiffness coefficients, N/m
$\bar{k}$	ellipticity parameter from approximate formula, $\alpha_r^{2/\pi}$
L	hydrodynamic lift
L_A	life adjustment
L^*	defined in Eq. (23.23)
$\bar{L}$	Saybolt universal viscosity of reference oil having low viscosity index
$\tilde{L}$	fatigue life
$\tilde{L}_{10}$	fatigue life for probability of survival of 0.90
ℓ	length in x direction, m
ℓ_a	length of annulus, m
ℓ_c	length of capillary tube, m
ℓ_e	length of roller land, m
ℓ_g	length of groove, m
ℓ_ℓ	roller effective length, m
ℓ_r	length of ridge, m
ℓ_s	length of step or groove, m
ℓ_t	roller length, m
ℓ_v	length dimension in stressed volume, m
ℓ_0	characteristic length in x direction, m
$\bar{\ell}_t$	length of capillary tube, m
$\tilde{\ell}$	constant used to determine side-leakage region
$(M_a)_{\mathrm{cr}}$	dimensionless critical mass, $(cm_a\omega_b^2/w_r)(\bar{\Omega}_v)^2_{\mathrm{cr}}$.
M_n	nth moment of distribution curve $\psi(z)$ about mean
$\bar{M}$	dimensionless stability parameter, $m_a p_a h_r^5/2r^5 b\eta_0^2$
$\tilde{M}$	probability of failure
$\dot{M}^*$	dimensionless rate of total mass flow passing a section at $x = x_i$, $\bar{\rho}_i^* H_i$
$\dot{M}_{\mathrm{end}}$	dimensionless rate of total mass flow passing the Reynolds boundary, $\bar{\rho}^*_{T,\mathrm{end}} H_{\mathrm{end}}$
m_a	mass of body, kg
m_k	load-life exponent
m_p	preload factor

m_W fully flooded–starved boundary as obtained from Wedeven

m'_a mass of body per unit width, kg/m

m^* fully flooded–starved boundary

$\overline{m}$ slope

$\hat{m}$ constant used to determine length of inlet region

$\dot{m}$ rate of mass flow per unit length passing surface of control volume, kg/(m · s)

$\dot{m}^*$ dimensionless rate of mass flow passing surface of control volume, $2m/\rho_0\overline{u}_{ab}D_x$

N number of measurements or number of modes

N_a rotational speed, r/min

N_c cycles to failure

N_{max} maximum number of nodes

N_0 number of pads or grooves

$\overline{N}$ number of grooves

n number of rolling elements per row

n_s location of parallel step or pivot from inlet

n_2 asymptote, $(\text{GPa})^{-2}$

$\overline{n}$ polytropic gas-expansion exponent

$\tilde{n}$ dimensionless constant used to determine length of outlet region

$O_{a1i,k}$ influence coefficients, heat flux on temperature rise at interface a, first kind

$O_{a2i,k}$ influence coefficients, heat flux on temperature rise at interface a, second kind

$O_{b1i,k}$ influence coefficients, heat flux on temperature rise at interface b, first kind

$O_{b2i,k}$ influence coefficients, heat flux on temperature rise at interface b, second kind

O_{ai} thermal influence variable of interface a, $O_{a2i,i} + O_{a1i,i+1} - O_{a2i,i+1}$

O_{bi} thermal influence variable of interface b, $O_{b2i,i} + O_{b1i,i+1} - O_{b2i,i+1}$

P dimension pressure

PV limit of operation, where P is load in pounds force per square inch and V is surface speed in feet per minute

P_a dimensionless pressure at inlet, $p_a h_0^2/\eta_0 \ell u_b$

P_e dimensionless pressure, p/E'

P_{ef} Peclet number of lubricant, $\overline{u}_{ab}D_x/2\lambda_0 = \rho_0 c\overline{u}_{ab}D_x/2K_0$

$P_{e,s}$ dimensionless pressure spike, p_{sk}/E'

P_h	homogeneous solution to dimensionless pressure, $2p_h R/\eta_0(u_a + u_b)$
P_i	dimensionless inlet pressure
P_m	dimensionless pressure where $dP/dX = 0$
P_o	dimensionless outlet pressure
P_p	particular solution to dimensionless pressure, $2p_p R_x/\eta_0(u_a + u_b)$
P_r	dimensionless pressure in ridge region
P_s	dimensionless pressure in step region
P^*	dimensionless reduced pressure, $p^*/p_{\max}$
P_r^*	dimensionless reduced pressure for rectangular contact
$\tilde{P}_{e,s}$	curve-fit dimensionless pressure spike, $\bar{p}_{sk}/E'$
$\hat{P}_h$	$P_h/4\varphi$
p	pressure, Pa (or N/m^2)
p_a	ambient pressure, Pa (or N/m^2)
p_c	pressure in capillary, Pa (or N/m^2)
p_H	maximum Hertzian pressure, Pa (or N/m^2)
p_h	homogeneous solution of pressure, Pa (or N/m^2)
p_i	inlet pressure, Pa (or N/m^2)
$p_{iv,as}$	isoviscous asymptotic pressure, Pa (or N/m^2)
p_ℓ	lift pressure, Pa (or N/m^2)
p_m	pressure where $dp/dx = 0$, Pa (or N/m^2)
$p_{\max}$	maximum pressure, Pa (or N/m^2)
p_{mean}	mean pressure, Pa (or N/m^2)
p_o	outlet pressure, Pa (or N/m^2)
p_p	particular solution of pressure, Pa (or N/m^2)
p_r	recess pressure, Pa (or N/m^2)
p_R	pole pressure, 1.96×10^8 Pa (or N/m^2)
p_s	supply pressure, Pa (or N/m^2)
$p_{\mathfrak{s}}$	solidification pressure, Pa (or N/m^2)
p_{sk}	pressure spike amplitude, Pa (or N/m^2)
p_x	pressure gradient in x direction for steady-state condition, $(\partial p/\partial x)_0$, N/m^3
p_z	pressure gradient in z direction for steady-state condition, $(\partial p/\partial z)_0$, N/m^3
$p_{\dot{z}}$	pressure gradient in z direction for dynamic condition, $(\partial p/\partial \dot{z})_0$, $N \cdot s/m^3$
$p_{\dot{x}}$	pressure gradient in x direction for dynamic condition, $(\partial p/\partial \dot{x})_0$, $N \cdot s/m^3$
p_0	steady-state pressure, Pa (or N/m^2)
$p_1, p_2, \ldots$	first-, second-,... order perturbation pressure, Pa (or N/m^2)

p^* reduced pressure [see Eq. (21.5)], $(1 - e^{-\xi p})/\xi$, Pa (or N/m^2)
$\bar{p}_i$ initial pressure, Pa (or N/m^2)
$\tilde{p}_c$ constant uniform pressure, Pa (or N/m^2)
$\tilde{p}_s$ curve-fit dimensional pressure, Pa (or N/m^2)
$\tilde{p}_{sk}$ pressure spike amplitude from curve fitting, Pa (or N/m^2)
Q dimensionless volume flow rate, $2q/bu_b s_h$
Q_{cv} heat conducted into a control volume, J
Q_m dimensionless mass flow
Q_∞ dimensionless mass flow rate, $3\eta_0 q/\pi p_a h_r^3$
$\tilde{Q}_m$ curve-fit dimensionless mass flow rate
$\hat{Q}$ quantity of heat, J
q volumetric flow rate, m^3/s
q_a constant, $(\pi/2) - 1$
q_b bearing pad flow coefficient
q_c volumetric flow rate of (laminar flow through) capillary tube, m^3/s
q_m mass flow rate, Pa · s (or $N \cdot s/m^2$)
q_o volumetric flow rate of (laminar flow through) orifice, m^3/s
q_s side volume flow, m^3/s
q' volumetric flow rate per unit width, m^2/s
q'_r radial volumetric flow rate per circumference, m^2/s
q'_x volumetric flow rate per unit width in sliding direction, m^2/s
q'_y volumetric flow rate per unit width in transverse direction, m^2/s
q'_ϕ volumetric flow rate per unit width in ϕ direction, m^2/s
$\bar{q}$ dimensionless volumetric flow rate, $2\pi q/r_b c b \omega_b$
$\dot{q}$ heat flux, $[2K_0(t_{m,0} - t_{m,R})R/D_x^2]\dot{q}^*$, W/m^2
$\dot{q}^*$ dimensionless heat flux
R curvature sum [Eq. (19.2)], m
R_a centerline average or arithmetic average, m
R_b dimensionless radius of asperity, r_a/b
R_c decrease in hardness
R_d curvature difference [Eq. (19.3)]
R_e radius ratio defined in Eq. (22.21)
R_g groove length fraction, $(r_0 - r_m)/(r_0 - r_i)$
R_q root-mean-square (rms) surface roughness, m
R_r race conformity, $r/2r_r$
R_s radius ratio defined in Eq. (22.21)
R_t maximum peak-to-valley height, m

R_x, R_y	effective radii in x and y directions, respectively, m
$\bar{R}$	gas constant
$\mathscr{R}$	Reynolds number, $\rho_0 u_0 \ell_0/\eta_0$
$\mathscr{R}_x$	modified Reynolds number in x direction, $\rho_0 u_0 h_0^2/\eta_0 \ell_0$
$\mathscr{R}_y$	modified Reynolds number in y direction, $\rho_0 v_0 h_0^2/\eta_0 b_0$
$\mathscr{R}_z$	modified Reynolds number in z direction, $\rho_0 w_0 h_0/\eta_0$
r	radius, m
r, θ, z	cylindrical polar coordinates
r, θ, ϕ	spherical polar coordinates
r_a	radius of asperity, m
r_b	radius of journal bearing, m
r_c	corner radius, m
r_i	inner radius, m
r_m	inner radius of extent of spiral groove (see Fig. 16.6)
r_o	outer radius, m
r_r	crown radius, m
$S(x, y)$	separation due to geometry of solids, m
S_c	dimensionless variable described in Eq. (26.17), $\eta\tilde{u}/\tau_L h$
S_0	dimensionless constant used in Eq. (4.13)
S^*	inverse of dimensionless circular shear strain rate coefficient, $\pi U\bar{\eta}^*/8WH\hat{\tau}_L$
$\bar{S}$	distance from pitch line, m
$\tilde{S}$	probability of survival
s	shear rate, s^{-1}
s_h	shoulder height, m
s_0	dimensionless viscosity-temperature index
s_1	r/h_0
$\bar{s}$	distance from origin, m
T	dimensionless time, t/t_0
T_q	dimensionless torque [defined in Eq. (16.66)]
T^*	dimensionless field temperature, $(t_m - t_{m,0})/(t_{m,0} - t_{m,R})$
$\tilde{T}$	dimensionless variable, $\dfrac{h}{\tau_L}\dfrac{dp}{dx} = \dfrac{2WH_r}{\pi\bar{\tau}_L}\dfrac{dP_r}{dX_r}$
$\check{T}$	difference of dimensionless shear stresses, $\tau_b^* - \tau_a^* q = \dfrac{2WH(dP/dX)}{\pi\hat{\tau}_L}$
$\mathbf{T}$	stress tensor
t	time, s
t_i	inlet temperature, °C
t_m	temperature, °C
$t_{m,0}$	ambient temperature, °C

$t_{m,R}$ reference temperature, $-273°$ C
$t_{m,\infty}$ extreme temperature, °C
Δt_m temperature change, °C
t_0 outlet temperature, °C
t_q frictional torque, N · m
t_0 characteristic time, s
t' start transient time period used to calculate interface temperature, s
t^* auxiliary parameter; dimensionless time variable for Gauss-Legendre integration
$\bar{t}^*$ approximate value of t^* [see Eq. (19.36)]
U dimensionless speed parameter, $\eta_0\tilde{u}/E'R_x$
U_V dimensionless speed parameter, $\eta_0 V/E'R_x$
u velocity in x direction, m/s
u_a surface a velocity, m/s
u_b surface b velocity, m/s
u_0 characteristic velocity in x direction, m/s
$\bar{u}$ dimensionless velocity in x direction, u/u_0
$\tilde{u}$ mean surface velocity in x direction, $(u_a + u_b)/2$, m/s
u^* dimensionless x component of field velocity, $u/\tilde{u}$
$\bar{u}^*_{\mathrm{fab}}$ dimensionless average lubricant velocity at interface, $[u^*(Z = 0) + u^*(Z = 1)]/2$
$\check{u}$ specific internal energy, $c_v(t_m - t_{m,0})$
V velocity vector, $(\tilde{u}^2 + \tilde{v}^2)^{1/2}$, m/s
VI viscosity index
V_r relative volume
V_0 average volume, m^3
V^* dimensionless magnitude of field velocity, $V/\bar{u}_{ab}$
$\tilde{V}$ elementary volume, m^3
$\mathbf{V}$ field velocity vector, m/s
v velocity in y direction, m/s
v_i, v_o linear velocities of inner and outer contacts, m/s
v_r, v_z, v_θ velocities in r, z, and θ directions of cylindrical polar coordinates, m/s
v_r, v_θ, v_ϕ velocities in r, θ, and ϕ directions of spherical polar coordinates, m/s
v_0 characteristic velocity in y direction, m/s
$\bar{v}$ dimensionless velocity in y direction, v/v_0
$\tilde{v}$ mean surface velocity in y direction, $(v_a + v_b)/2$, m/s
W dimensionless load parameter, $w'_z/E'R_x$
W_{cv} flow work (pressure work and shear energy) done by control volume

W_K dimensionless load predicted by Kapitza

W_r dimensionless resultant load in a journal bearing, $\frac{w_r}{\eta_0 \omega_b r_b b}\left(\frac{c}{r_b}\right)^2$

W_x dimensionless tangential load, $\frac{w_x}{\eta_0 u_b}\left(\frac{s_h}{\ell}\right)$

W_z dimensionless normal load, $\frac{w'_z}{\eta_0 u_b}\left(\frac{s_h}{\ell}\right)^2$

W_∞ dimensionless load for spiral-groove thrust bearing [Eq. (16.63)]

W^* $w/2r_b b$ [Eq. (12.4)], MPa

W' dimensionless load for rectangular contact, $w'_z/E'R_x$

W'_r dimensionless resultant load in a journal bearing when side leakage is neglected, $\frac{w'_r}{\eta_0 \omega_b r_b b}\left(\frac{c}{r_b}\right)^2$

$\overline{W}_r$ dimensionless resultant first-order-perturbation load for gas-lubricated bearings, $\bar{w}_r/\pi p_a rb\varepsilon$

W_z^* dimensionless normal load, $w'_z/2rp_a$

w velocity of fluid in z direction, m/s

w_a, w_b velocities of fluid in z direction (squeeze velocity) acting at surfaces a and b, respectively, m/s

w_0 characteristic velocity in z direction, m/s

w^* dimensionless z component of field velocity, $w/\tilde{u}$

$\bar{w}$ dimensionless velocity in z direction, w/w_0

w resultant load, $(w_x^2 + w_z^2)^{1/2}$, N

w_c pad load component along line of centers (Fig. 17.3), N

w_e bearing equivalent load, N

w_r resultant load in a journal bearing, N

w_s pad load normal to line of centers (Fig. 17.3), N

w_t total thrust load of bearing, N

w_x tangential load, N

w_{xb}, w_{xb} tangential load components, N

w_z normal load component, N

w' load per unit width, N/m

w'_a, w'_b resultant loads per unit width acting on surfaces a and b, respectively, N/m

w'_{xa}, w'_{xb} tangential loads per unit width acting on surfaces a and b, respectively, N/m

w'_r resultant load per unit length in a journal bearing, N/m

w'_z normal load per unit width, N/m

w'_{za}, w'_{zb} normal loads per unit width acting on surfaces a and b, respectively, N/m

$\bar{w}_r$ resultant first-order-perturbation load, $(\bar{w}_x^2 + \bar{w}_z^2)^{1/2}$, N

$\overline{w}_x$	first-order-perturbation tangential load component, N
$\overline{w}_z$	first-order-perturbation normal load component, N
$\hat{w}$	load ratio, $w_{\text{finite}}/w_{\text{infinite}}$
X, Y, Z	dimensionless Cartesian coordinate system
X_a, Y_a, Z_a	body forces in Cartesian coordinates, m/s^2
X_c, Y_c, Z_c	body forces in cylindrical polar coordinates, m/s^2
X_{cp}	dimensionless center of pressure, x_{cp}/ℓ
X_{cr}	dimensionless critical value of X
X_E, Y_E	dimensionless values of X and Y at edge of computational region (Fig. 18.7)
X_e	dimensionless x coordinate, x/R_x
$X_{e,\text{cp}}$	dimensionless location of center of pressure, x_{cp}/R_x
$X_{e,\min}$	dimensionless location of minimum film thickness, $x_{\min}/R_x$
$X_{e,r}$	dimensionless location of film rupture, x_r/R_x
$X_{e,s}$	dimensionless pressure spike location, x_s/R_x
X_m	dimensionless value of X when $dP/dX = 0$
X_r	dimensionless x coordinate for rectangular contact [Eq. (21.7)]
$X_{r,\text{cp}}$	dimensionless center of pressure, x_{cp}/D_x
$X_{r,\text{end}}$	dimensionless x coordinate for rectangular contact at outlet meniscus
X_s, Y_s, Z_s	body forces in spherical polar coordinates, m/s^2
X^*	dimensionless coordinate variable for Gauss-Legendre integration
$\overline{X}$	$X/(2H_0)^{1/2}$ [Eq. (18.67)]
$\tilde{X}, \tilde{Y}$	factors for calculating equivalent load
$\tilde{X}_{e,\text{cp}}$	dimensionless location of center of pressure obtained from curve-fitting results, $\tilde{x}_{\text{cp}}/R_x$
$\tilde{X}_{e,\min}$	dimensionless location of minimum film thickness obtained from curve-fitting results, $\tilde{x}_{\min}/R_x$
$\tilde{X}_{e,s}$	dimensionless pressure spike location obtained from curve-fitting results, $\tilde{x}_s/R_x$
x	Cartesian coordinate in direction of sliding, m
Δx	change in x direction, m
$\Delta \dot{x}$	change in $\dot{x}$, m/s
x_{cp}	center of pressure, m
x_{end}	value of x at exit, m
x_m	value of x when $dp/dx = 0$, m
$x_{\min}$	value of x at minimum film thickness, m
x_r	location of film rupture in x coordinate, m
x_s	location of pressure spike, m
$\bar{x}$	Saybolt universal viscosity of unknown oil [Eq. (4.18)]

$\bar{x}_{cp}$ curve-fitting results for center of pressure, m
$\tilde{x}_{min}$ curve-fitting results for location of minimum film thickness, m
$\tilde{x}_s$ location of pressure spike from curve fitting, m
Y_ℓ fraction of clearance width in y direction occupied by lubricant
$\bar{Y}$ $Y/(2\alpha_r H_0)^{1/2}$ [eq. (18.68)]
y Cartesian coordinate in side-leakage direction, m
Z_{cr} critical dimensionless Z where $u/u_b = 0$
Z_w constant used in Eq. (24.46)
Z_1 viscosity-pressure index, a dimensionless constant
$\bar{Z}$ dimensionless coordinate, $\bar{z}/h$
$\tilde{Z}, \tilde{z}$ dimensions in continuously variable-speed drive (Fig. 25.3)
z Cartesian coordinate in direction of film, m
Δz change in direction, m
$\Delta \dot{z}$ change in $\dot{z}$, m/s
z_0 depth of maximum shear stress, m
z' stress-weighted mean depth, m
z^* distance of mean value of z from value chosen as origin [Eq. (3.10)], m
$\bar{z}$ coordinate measured from lower surface (see Fig. 26.2), $z - h_a$, m
α cone angle, deg
α_a offset factor
α_b groove width ratio, $\ell_s/(\ell_r + \ell_s)$
α_h ratio of radius to central film thickness, r/h_0
α_p angular extent of pivoted pad, deg
α_r radius ratio, R_y/R_x
α_t thermal diffusivity, m^2/s
$\alpha_{\kappa 1}$ piezothermal-conductivity constant, first, 1.73×10^{-9} m^2/N
$\alpha_{\kappa 2}$ piezothermal-conductivity constant, second, 6.91×10^{-10} m^2/N
$\alpha_{\rho 1}$ piezodensity constant, first, 0.6×10^{-9} m^2/N
$\alpha_{\rho 2}$ piezodensity constant, second, 1.7×10^{-9} m^2/N
α_ω ratio of squeeze to entraining velocity, $w_a/\tilde{u}$
α' angle described in Fig. 11.1
$\bar{\alpha}$ skewness
$\bar{\alpha}_r$ radius ratio of thrust bearing, r_i/r_o
$\tilde{\alpha}$ dimensionless viscosity index, $G(W/2\pi)^{1/2}$
β contact angle, deg
β_a groove angle, deg

β_f free contact angle, deg
β_g groove width ratio, $(\ell_s + \ell_r)/(\ell_s + \ell_r + \ell_g)$
β_p angle between load direction and pivot, deg
β_s film thickness reduction factor due to starvation, $H_{\min, s}/H_{\min}$
β_t dynamic load ratio, $W/(W)_{\beta_\omega = 0}$
β_η temperature-viscosity coefficient, K^{-1}
β_ρ dimensionless thermodensity constant
β_{τ_L} temperature-limiting-shear-strength coefficient, K
β_ω dimensionless normal velocity parameter, $\alpha_\omega/(2H_{\min})^{1/2}$
β^* angle between line of centers and x axis, deg
$\bar{\beta}$ kurtosis
Γ $PH^{3/2}$ [Eq. (18.44)]
γ Sommerfeld variable
γ_a auxiliary angle, deg
γ_c surface pattern parameter [see Eq. (23.25)]
γ_g groove width ratio, b_1/b
γ_n defined in Table 24.10
γ^* limiting-shear-strength proportionality constant, $\partial\tau_L/\partial p$
$\bar{\gamma}$ angle between load components [see Eq. (21.41)], deg
$\dot{\gamma}^*$ dimensionless field shear strain rate, $du^*/dZ = \dot{\gamma}h/\tilde{u}$
Δ uniform interval length, m
δ elastic deformation, m
δ_m elastic deformation where $dp/dx = 0$, m
$\delta_{\max}$ interference, or total elastic compression on load line, m
δ_t deflection due to thrust load, m
δ_ψ elastic compression of ball, m
$\bar{\delta}$ dimensionless elastic deformation
$\bar{\delta}_H$ maximum dimensionless elastic deformation when pressure distribution is assumed to be Hertzian
ε eccentricity ratio, e/c
ε_0 steady-state eccentricity ratio
ε_1 strain in axial direction
ε_2 strain in transverse direction
ζ coordinate defined in Eq. (17.1); coordinate perpendicular to surface of semi-infinite solid
ζ_1 dimensionless boundary shear stress variable, first kind, $\sin^{-1}\tau_b^* - \sin^{-1}\tau_a^* - \tau_b^*\sqrt{1-\tau_a^{*2}} + \tau_a^*\sqrt{1-\tau_b^{*2}}$
ζ_2 dimensionless boundary shear stress variable, second kind, $\sin^{-1}(\tau_a^* + \check{T}) - \sin^{-1}\tau_a^* + (\tau_a^* - \check{T})\sqrt{1-(\tau_a^* + \check{T})^2}$

ζ_3 dimensionless boundary shear stress variable, third kind, $\sin^{-1}\tau_b^* - \sin^{-1}(\tau_b^* - \check{T}) - (\tau_b^* + \check{T})\sqrt{1 - (\tau_b^* - \check{T})^2}$
ζ_5 dimensionless boundary shear stress variable, fifth kind, $-1/\sin^{-1}\tau_a^*$
ζ_p angle between pad leading edge and pivot, deg
ζ^* defined in Eq. (19.26)
η absolute viscosity, Pa · s (or N · s/m^2)
η_v dimensionless variable described in Eq. (26.36)
η_e effective viscosity, τ/s, Pa · s (or N · s/m^2)
η_k kinematic viscosity, m^2/s
η_p ambient piezoviscosity, $\eta_0\eta_p^*$, Pa · s (or N · s/m^2)
η_R pole viscosity, 6.315×10^{-5} Pa · s (or N · s/m^2)
η_{T,p_0} atmospheric thermoviscosity, $\eta_0\eta_T^*$, Pa · s (or N · s/m^2)
η_0 absolute viscosity at $p = 0$ and constant temperature, Pa · s (or N · s/m^2)
η_∞ constant used in Eqs. (4.10) and (27.9), Pa · s (or N · s/m^2)
$\overline{\eta}$ dimensionless absolute viscosity, η/η_0
$\overline{\eta}_{avg}$ average dimensionless viscosity across film thickness, $\overline{\eta}_p\overline{\eta}_{T,avg}$
$\overline{\eta}_p$ dimensionless piezoviscosity
$\overline{\eta}_T$ dimensionless thermoviscosity
$\overline{\eta}_{T,avg}$ average dimensionless thermoviscosity across film thickness
$\hat{\eta}_1$ dimensionless non-Newtonian viscosity variable, first kind, $\check{T}^3\overline{\eta}_{avg}/6\zeta_1$
$\hat{\eta}_2$ dimensionless non-Newtonian viscosity variable, second kind, $\check{T}^3\overline{\eta}_{avg}/6\zeta_2$
$\hat{\eta}_3$ dimensionless non-Newtonian viscosity variable, third kind, $\check{T}^3\overline{\eta}_{avg}/6\zeta_3$
θ coordinate in cylindrical polar coordinates; slide-roll ratio, $\tan^{-1}(\tilde{v}/\tilde{u})$
θ' angle described in Fig. 11.1, deg
θ_g angle extended by groove region, deg
θ_r angle extended by ridge region, deg
θ_s angle used to define shoulder height, deg
Λ dimensionless film parameter, $h_{min}/(R_{q,a}^2 + R_{q,b}^2)^{1/2}$
Λ_a dimensionless bearing number, $6\eta_0 u_b b/p_a h_r^2$
Λ_b function of Λ_j defined in Eq. (17.25)
Λ_c function of Λ_j defined in Eq. (17.26)
Λ_g dimensionless bearing number, $6\eta_0 u_b \ell/p_a h_{min}^2$
Λ_j dimensionless bearing number for journal bearings, $6\eta_0\omega r^2/p_a c^2$

Λ_s dimensionless bearing number, $3\eta_0\omega(r_o^2 - r_i^2)/p_a h_r^2$
λ length-to-width ratio, ℓ/b; thermal diffusivity of incompressible substance, $\kappa/\rho c$, m^2/s
λ_0 thermal diffusivity of lubricant at atmospheric conditions, $\kappa_0/\rho_0 c_v$, m^2/s
λ_a second coefficient of viscosity, Pa · s (or N · s/m^2); thermal diffusivity of solid a, $\kappa_a/\rho_a c_a$, m^2/s
λ_b thermal diffusivity of solid b, $\kappa_b/\rho_b c_b$, m^2/s
λ_f dimensionless friction force
λ_h dimensionless elastic energy lost from hysteresis
λ_j width-to-diameter ratio, $b/2r$
λ_k diameter-to-width ratio, $2r_b/b$
λ_m mean free molecular path of gas, m
λ_x autocorrelation length in x direction
λ_y autocorrelation length in y direction
$\bar{\lambda}$ dimensionless velocity parameter, $3\pi^2 U/4W^2$
μ coefficient of sliding friction
μ_r coefficient of rolling friction
ν Poisson's ratio
ξ pressure-viscosity coefficient, m^2/N
ξ_a dilatation, $\partial u/\partial x + \partial v/\partial y + \partial w/\partial z$, s^{-1}
ξ_p angle between line of centers and pad leading edge, deg
ξ_s dynamic peak pressure ratio, $P_m/(P_m)_{\beta_\omega=0}$
ξ^* defined in Eq. (19.27)
ρ force density of lubricant, N · s^2/m^4
ρ_f force density of sphere, N · s^2/m^4
ρ_m force density where $dp/dx = 0$, N · s^2/m^4
ρ_s force density of sphere, N · s^2/m^4
ρ_0 force density at $p = 0$, N · s^2/m^4
ρ^* mass density, $(kg)_{mass}/m^3$
$\bar{\rho}_{avg}$ average dimensionless density of lubricant across film thickness, $\bar{\rho}_p \bar{\rho}_{T,avg}$
$\bar{\rho}_p$ dimensionless piezodensity of lubricant
$\bar{\rho}_T$ dimensionless thermodensity of lubricant
$\bar{\rho}_{T,avg}$ average dimensionless thermodensity of lubricant across film thickness
$\bar{\rho}$ dimensionless force density, ρ/ρ_0
$\bar{\rho}_k$ autocorrelation
σ normal stress, Pa (or N/m^2)
σ_g dimensionless squeeze number, $12\eta_0\omega\ell^2/p_a h_{min}^2$

σ_i magnitude of normal stress
σ_r stress function
σ_s squeeze number, $\rho_0 h_0^2/\eta_0 t_0$
$\sigma_x, \sigma_y, \sigma_z$ normal stress components, Pa (or N/m^2)
σ_1 axial stress, Pa (or N/m^2)
$\bar{\sigma}$ standard deviation, m
τ shear stress, Pa (or N/m^2)
τ_a shear stress at surface a, Pa (or N/m^2)
τ_E shear stress at which fluid first starts to behave nonlinearly when stress is plotted against shear strain rate, Pa (or N/m^2)
τ_L limiting shear stress, Pa (or N/m^2)
τ_{L0} limiting shear strength at atmospheric conditions, Pa (or N/m^2)
τ_u endurance strength or threshold stress, Pa (or N/m^2)
τ_0 shear stress at zero pressure, Pa (or N/m^2)
$\bar{\tau}$ dimensionless shear stress, τ/τ_L
$\bar{\tau}_a$ dimensionless shear stress at surface a, τ_a/τ_L
$\tau_{L,\text{avg}}$ average limiting shear strength across film thickness, $\tau_{L0}\bar{\tau}_L$, N/m^2; dimensionless limiting shear strength, τ_L/E'
$\bar{\tau}_{L,\text{avg}}$ average dimensionless limiting shear strength across film thickness, $\bar{\tau}_{Lp}\bar{\tau}_{LT,\text{avg}}$
$\bar{\tau}_{Lp}$ dimensionless-pressure limiting shear strength
$\bar{\tau}_{LT}$ dimensionless-temperature limiting shear strength
$\bar{\tau}_{LT,\text{avg}}$ average dimensionless-temperature limiting shear strength
$\hat{\tau}_L$ average dimensionless limiting shear strength across film thickness, $\bar{\tau}_L/E'$
Φ attitude angle; location of minimum film thickness, deg
Φ_L angle between y axis and resultant load (Fig. 11.1), deg
$\bar{\Phi}$ dimensionless stream function, $\phi_s/u_b s_h$
ϕ angle in spherical polar coordinates, deg
ϕ_a auxiliary angle, deg
ϕ_b Boussinesq stress function, N
ϕ_m location of maximum pressure, deg
ϕ_p angular extent from inlet to pivot location, deg
ϕ_s angle locating ball-spin vector, deg
ϕ_0 location of terminating pressure, deg
ϕ^* angle of ϕ when $p = 0$ and $dp/dx = 0$
$\bar{\phi}$ dimensional stream function, m^2/s

$\tilde{\phi}$ resultant load angle (see Fig. 8.4)

φ side-leakage factor [see Eq. (18.64)], $(1 + 2/3\alpha_r)^{-1}$

ψ angle to load line, deg

ψ_a groove width ratio, θ_r/θ_g

ψ_g step location parameter, $\ell_s/(\ell_s + \ell_r + \ell_g)$

ψ_ℓ angular extent of bearing loading, deg

$\bar{\psi}$ pressure angle in mating gear teeth, deg

$\tilde{\psi}$ probability density function, m^{-1}

Ω_d real component of angular speed, rad/s

Ω_v complex component of angular speed, rad/s

$(\Omega_v)_{cr}$ dimensionless critical speed, $(\Omega_v/\omega_v)_{cr}$

ω angular velocity, $2\pi N_a$, rad/s

ω_b angular velocity of surface b or rolling element about its own axis, rad/s

ω_c angular velocity of separator or ball set, rad/s

ω_i, ω_o angular velocity of inner and outer races, rad/s

ω_ℓ rotational speed of load vector, rad/s

ω_r angular velocity of race, rad/s

ω_s angular velocity due to spinning, rad/s

ω_t rotational speed of load vector, rad/s

$\bar{\omega}_d$ whirl frequency ratio

Subscripts:

a solid a

b solid b

cr critical

cs control surfaces

cv control volumes

EHL elastohydrodynamic lubrication

g feed groove

HEHL hard EHL

HL hydrodynamic lubrication

IE isoviscous-elastic regime

IR isoviscous-rigid regime

i inner; inlet

max maximum

min minimum

o outer; outlet

R	rectangular
r	ridge
s	step
sk	spike
st	starved
VE	viscous-elastic regime
VR	viscous-rigid regime
x, y, z	coordinates

CHAPTER 1

INTRODUCTION

In 1966 with the publication in England of the "Department of Education and Science Report," sometimes known as the "Jost Report," the word "tribology" was introduced and defined as the science and technology of interacting surfaces in relative motion and of the practices related thereto. A better definition might be the lubrication, friction, and wear of moving or stationary parts. The "Department of Education and Science Report" (1966) also claimed that industry could save considerable money by improving their lubrication, friction, and wear practices.

This book focuses on the fundamentals of fluid film lubrication. Fluid film lubrication occurs when opposing bearing surfaces are completely separated by a lubricant film. The applied load is carried by pressure generated within the fluid, and frictional resistance to motion arises entirely from the shearing of the viscous fluid. The performance of fluid film bearings can be determined by applying well-established principles of fluid mechanics, usually in terms of slow viscous flow.

Boundary lubrication, where considerable contact between the surfaces occurs, is defined in this book but only presented in a cursory way. See either Rabinowicz (1965), Bowden and Tabor (1973), or Hutchings (1992) for a discussion of boundary lubrication.

1.1 CONFORMAL AND NONCONFORMAL SURFACES

Conformal surfaces fit snugly into each other with a high degree of geometrical comformity so that the load is carried over a relatively large area. For example, the lubrication area of a journal bearing would be 2π times the radius times the

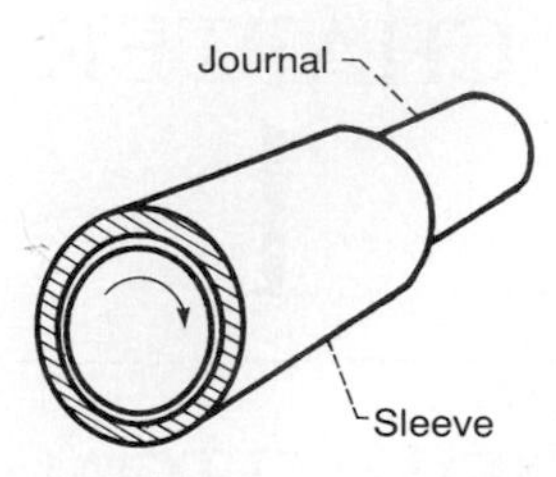

FIGURE 1.1
Conformal surfaces. [*From Hamrock and Anderson (1983).*]

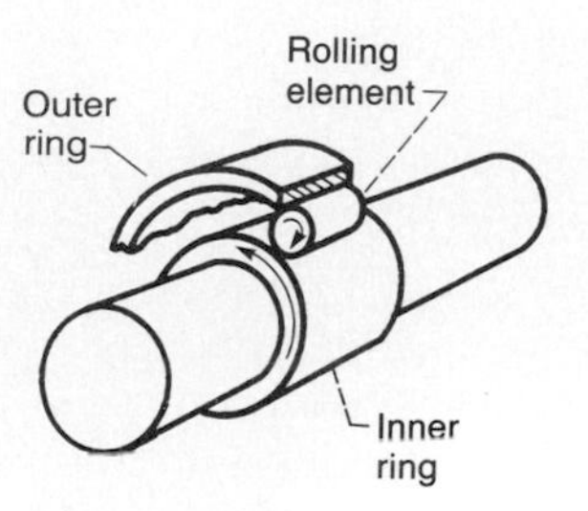

FIGURE 1.2
Nonconformal surfaces. [*From Hamrock and Anderson (1983).*]

length. The load-carrying surface area remains essentially constant while the load is increased. Fluid film journal bearings (Fig. 1.1) and slider bearings have conformal surfaces. In journal bearings the radial clearance between the journal and the sleeve is typically one-thousandth of the journal diameter; in slider bearings the inclination of the bearing surface to the runner is typically one part in a thousand.

Many machine elements that are fluid-film-lubricated have surfaces that do not conform to each other well. The full burden of the load must then be carried by a small lubrication area. The lubrication area of a nonconformal conjunction is typically three orders of magnitude less than that of a conformal conjunction. In general, the lubrication area between nonconformal surfaces enlarges considerably with increasing load, but it is still smaller than the lubrication area between conformal surfaces. Some examples of nonconformal surfaces are mating gear teeth, cams and followers, and rolling-element bearings (Fig. 1.2).

1.2 LUBRICATION REGIMES

A lubricant is any substance that reduces friction and wear and provides smooth running and a satisfactory life for machine elements. Most lubricants are liquids (such as mineral oils, synthetic esters, silicone fluids, and water), but they may be solids (such as polytetrafluoroethylene, or PTFE) for use in dry bearings, greases for use in rolling-element bearings, or gases (such as air) for use in gas bearings. The physical and chemical interactions between the lubricant and the lubricating surfaces must be understood in order to provide the machine elements with satisfactory life. As an aid in understanding the features that distinguish the four lubrication regimes from one another, a short historical perspective is given, followed by a description of each regime.

1.2.1 HISTORICAL PERSPECTIVE. By the middle of this century two distinct lubrication regimes were generally recognized: hydrodynamic lubrication and boundary lubrication. The understanding of hydrodynamic lubrication began with the classical experiments of Tower (1885), in which the existence of a film was detected from measurements of pressure within the lubricant, and of Petrov (1883), who reached he same conclusion from friction measurements. This work was closely followed by Reynolds' (1886) celebrated analytical paper in which he used a reduced form of the Navier-Stokes equations in association with the continuity equation to generate a second-order differential equation for the pressure in the narrow, converging gap between bearing surfaces. This pressure enables a load to be transmitted between the surfaces with extremely low friction, since the surfaces are completely separated by a fluid film. In such a situation the physical properties of the lubricant, notably the dynamic viscosity, dictate the behavior in the conjunction.

The understanding of boundary lubrication is normally attributed to Hardy and Doubleday (1922a, b), who found that extremely thin films adhering to surfaces were often sufficient to assist relative sliding. They concluded that under such circumstances the chemical composition of the fluid is important, and they introduced the term "boundary lubrication." Boundary lubrication is at the opposite end of the lubrication spectrum from hydrodynamic lubrication. In boundary lubrication the physical and chemical properties of thin films of molecular proportions and the surfaces to which they are attached determine contact behavior. The lubricant viscosity is not an influential parameter.

In the last 40 years, research has been devoted to a better understanding and more precise definition of other lubrication regimes between these extremes. One such lubrication regime occurs between nonconformal surfaces, where the pressures are high and the surfaces deform elastically. In this situation the viscosity of the lubricant may rise considerably, and this further assists the formation of an effective fluid film. A lubricated conjunction in which such effects are found is said to be operating "elastohydrodynamically." Significant progress has been made in understanding the mechanism of elastohydrodynamic lubrication, generally viewed as reaching maturity.

Since 1970 it has been recognized that between fluid film and boundary lubrication some combined mode of action can occur. This mode is generally termed "partial lubrication" or is sometimes referred to as "mixed lubrication." To date, most of the scientific unknowns lie in this lubrication regime. An interdisciplinary approach will be needed to gain an understanding of this important lubrication mechanism. Between conformal surfaces, where hydrodynamic lubrication occurs if the film gets too thin, the mode of lubrication goes directly from hydrodynamic to partial. For nonconformal surfaces, where elastohydrodynamic lubrication occurs if the film gets too thin, the mode of lubrication goes from elastohydrodynamic to partial. A more in-depth historical development of lubrication, or tribology in general, can be obtained from Dowson (1979).

1.2.2 HYDRODYNAMIC LUBRICATION. Hydrodynamic lubrication (HL) is generally characterized by conformal surfaces. A positive pressure develops in a hydrodynamically lubricated journal or thrust bearing because the bearing surfaces converge and the relative motion and the viscosity of the fluid separate the surfaces. The existence of this positive pressure implies that a normal applied load may be supported. The magnitude of the pressure developed (usually less than 5 MPa) is not generally large enough to cause significant elastic deformation of the surfaces. It is shown later that the minimum film thickness in a hydrodynamically lubricated bearing is a function of normal applied load w_z, velocity u_b of the lower surface, lubricant viscosity η_0, and geometry (R_x and R_y). Figure 1.3 shows some of these characteristics of hydrodynamic lubrication. Minimum film thickness h_{min} as a function of u_b and w_z for sliding motion is given as

$$(h_{min})_{HL} \propto \left(\frac{u_b}{w_z} \right)^{1/2} \tag{1.1}$$

The minimum film thickness normally exceeds 1 μm.

In hydrodynamic lubrication the films are generally thick so that opposing solid surfaces are prevented from coming into contact. This condition is often referred to as "the ideal form of lubrication," since it provides low friction and high resistance to wear. The lubrication of the solid surfaces is governed by the bulk physical properties of the lubricant, notably the viscosity, and the frictional characteristics arise purely from the shearing of the viscous lubricant.

For a normal load to be supported by a bearing, positive-pressure profiles must be developed over the bearing length. Figure 1.4 illustrates three ways of developing positive pressure in hydrodynamically lubricated bearings. For a positive pressure to be developed in a slider bearing [Fig. 1.4(a)] the lubricant film thickness must be decreasing in the sliding direction. In a squeeze film bearing [Fig. 1.4(b)] the squeeze action with squeeze velocity w_a has the bearing

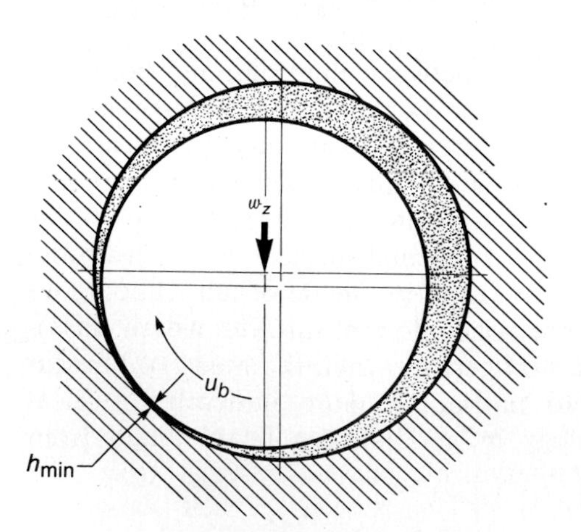

FIGURE 1.3
Characteristics of hydrodynamic lubrication.

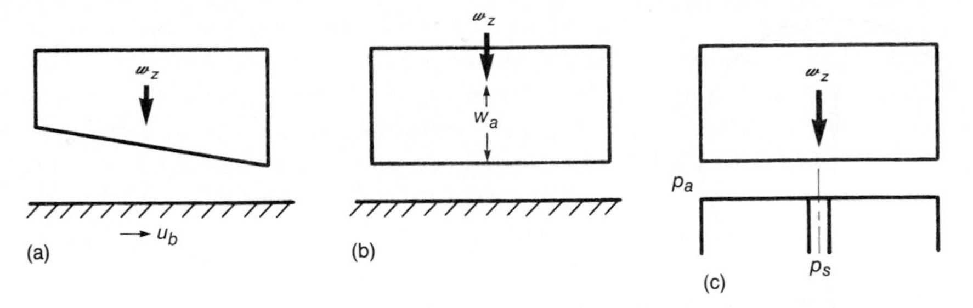

FIGURE 1.4
Mechanism of pressure development for hydrodynamic lubrication. (a) Slider bearing; (b) squeeze film bearing; (c) externally pressurized bearing.

surfaces approach each other. The squeeze mechanism of pressure generation provides a valuable cushioning effect when the bearing surfaces approach each other. Positive pressures will be generated only when the film thickness is diminishing. In an externally pressurized bearing, sometimes referred to as "a hydrostatic bearing" [Fig. 1.4(c)], the pressure drop across the bearing supports the load. The load-carrying capacity is independent of bearing motion and lubricant viscosity. There is no surface contact wear at starting and stopping as there is with the slider bearing.

1.2.3 ELASTOHYDRODYNAMIC LUBRICATION. Elastohydrodynamic lubrication (EHL) is a form of hydrodynamic lubrication where elastic deformation of the lubricated surfaces becomes significant. The features important in a hydrodynamically lubricated slider bearing [Fig. 1.4(a)]—converging film thickness, sliding motion, and a viscous fluid between the surfaces—are also important here. Elastohydrodynamic lubrication is normally associated with nonconformal surfaces. There are two distinct forms of EHL.

1.2.3.1 Hard EHL. Hard EHL relates to materials of high elastic modulus such as metals. In this form of lubrication the elastic deformation and the pressure-viscosity effects are equally important. Figure 1.5 gives the characteristics of hard elastohydrodynamically lubricated conjunctions. The maximum pressure is typically between 0.5 and 3 GPa; the minimum film thickness normally exceeds 0.1 μm. These conditions are dramatically different from those found in a hydrodynamically lubricated conjunction (Fig. 1.3). At loads normally experienced in nonconformal machine elements the elastic deformations are several orders of magnitude larger than the minimum film thickness. Furthermore, the lubricant viscosity can vary by as much as 10 orders of magnitude within the lubricating conjunction. The minimum film thickness is a function of the same parameters as for hydrodynamic lubrication (Fig. 1.3) but with the additions of the effective elastic modulus E' and the pressure-viscosity coefficient ξ of the

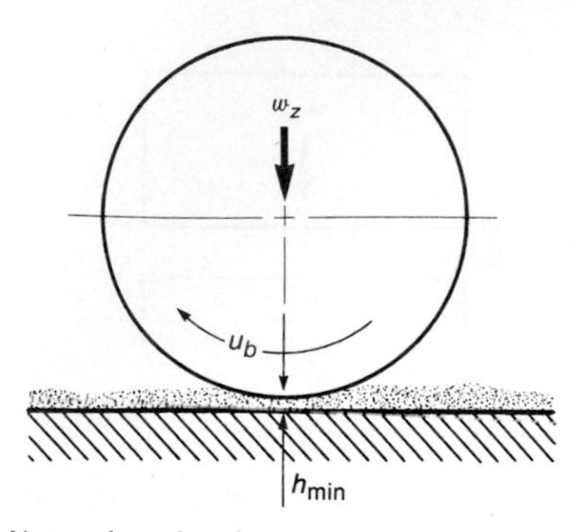

Nonconformal surfaces
High-elastic-modulus material (e.g., steel)
$p_{max} \approx 1$ GPa
$h_{min} = f(w_z, u_b, \eta_0, R_x, R_y, E', \xi) > 0.1\ \mu$m
Elastic and viscous effects both important

FIGURE 1.5
Characteristics of hard elastohydrodynamic lubrication.

lubricant.

$$E' = \frac{2}{\dfrac{1-\nu_a^2}{E_a} + \dfrac{1-\nu_b^2}{E_b}} \tag{1.2}$$

The relationships between the minimum film thickness and the normal applied load and speed for hard EHL as obtained from Hamrock and Dowson (1977) are

$$(h_{min})_{HEHL} \propto w_z^{-0.073} \tag{1.3}$$

$$(h_{min})_{HEHL} \propto u_b^{0.68} \tag{1.4}$$

Comparing the results for hard EHL [Eqs. (1.3) and (1.4)] with those for hydrodynamic lubrication [Eq. (1.1)] yielded the following conclusions:

1. The exponent on the normal applied load is nearly seven times larger for hydrodynamic lubrication than for hard EHL. This implies that the film thickness is only slightly affected by load for hard EHL but significantly affected for hydrodynamic lubrication.
2. The exponent on mean velocity is slightly higher for hard EHL than for hydrodynamic lubrication.

Some of the important results are presented in this chapter and substantiated in subsequent chapters. Engineering applications in which elastohydrodynamic lubrication is important for high-elastic-modulus materials include gears, rolling-element bearings, and cams.

1.2.3.2 Soft EHL. Soft EHL relates to materials of low elastic modulus such as rubber. Figure 1.6 shows the characteristics of soft-EHL materials. In soft EHL the elastic distortions are large, even with light loads. The maximum pressure

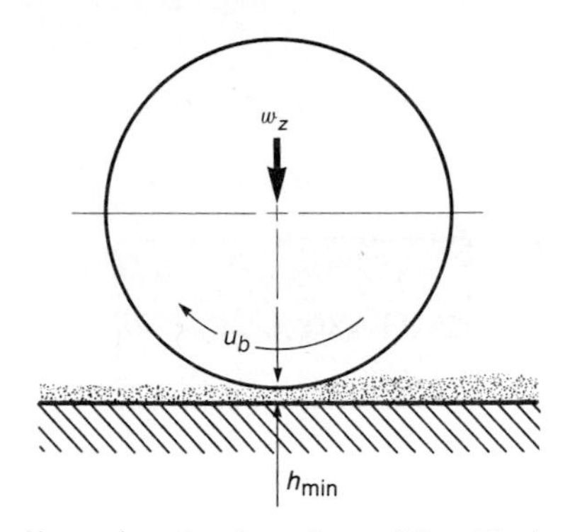

FIGURE 1.6
Characteristics of soft elastohydrodynamic lubrication.

for soft EHL is typically 1 MPa, in contrast to 1 GPa for hard EHL (Fig. 1.5). This low pressure has a negligible effect on the viscosity variation throughout the conjunction. The minimum film thickness is a function of the same parameters as in hydrodynamic lubrication with the addition of the effective elastic modulus. The minimum film thickness for soft EHL is typically 1 μm. Engineering applications in which elastohydrodynamic lubrication is important for low-elastic-modulus materials include seals, human joints, tires, and a number of lubricated machine elements that use rubber as a material. The common features of hard and soft EHL are that the local elastic deformation of the solids provides coherent fluid films and that asperity interaction is largely prevented. This implies that the frictional resistance to motion is due to lubricant shearing.

1.2.4 BOUNDARY LUBRICATION. Because in boundary lubrication the solids are not separated by the lubricant, fluid film effects are negligible and there is considerable asperity contact. The contact lubrication mechanism is governed by the physical and chemical properties of thin surface films of molecular proportions. The properties of the bulk lubricant are of minor importance, and the friction coefficient is essentially independent of fluid viscosity. The frictional characteristics are determined by the properties of the solids and the lubricant film at the common interfaces. The surface films vary in thickness from 1 to 10 nm—depending on the molecular size.

Figure 1.7 illustrates the film conditions existing in fluid film and boundary lubrication. The surface slopes in this figure are greatly distorted for purposes of illustration. To scale, real surfaces would appear as gently rolling hills rather than sharp peaks. The surface asperities are not in contact for fluid film lubrication but are in contact for boundary lubrication.

Figure 1.8 shows the behavior of the friction coefficient in the different lubrication regimes. In boundary lubrication, although the friction is much higher than in the hydrodynamic regime, it is still much lower than for

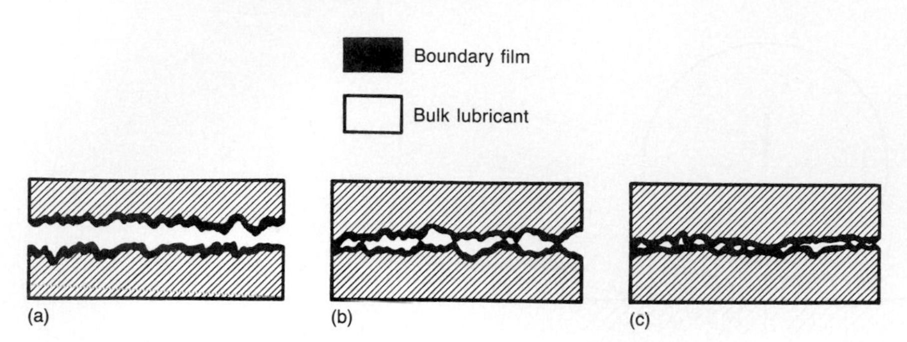

FIGURE 1.7
Film conditions of lubrication regimes. (a) Fluid film lubrication—surfaces separated by bulk lubricant film; (b) partial lubrication—both bulk lubricant and boundary film play a role; (c) boundary lubrication—performance depends essentially on boundary film.

unlubricated surfaces. The mean friction coefficient increases a total of three orders of magnitude in going from the hydrodynamic to the elastohydrodynamic to the boundary to the unlubricated regime.

Figure 1.9 shows the wear rate in the various lubrication regimes as determined by the operating load. In the hydrodynamic and elastohydrodynamic regimes there is little or no wear, since there is no asperity contact. In the

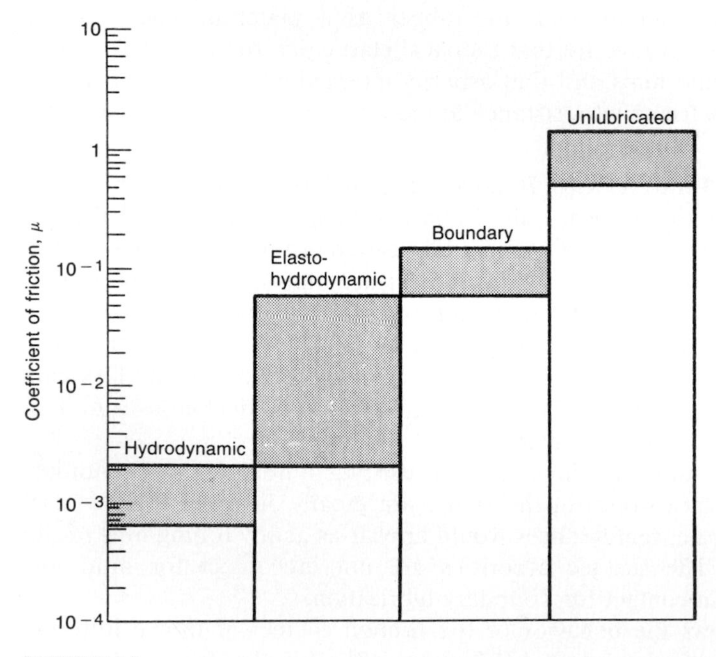

FIGURE 1.8
Bar diagram showing friction coefficient for various lubrication conditions.

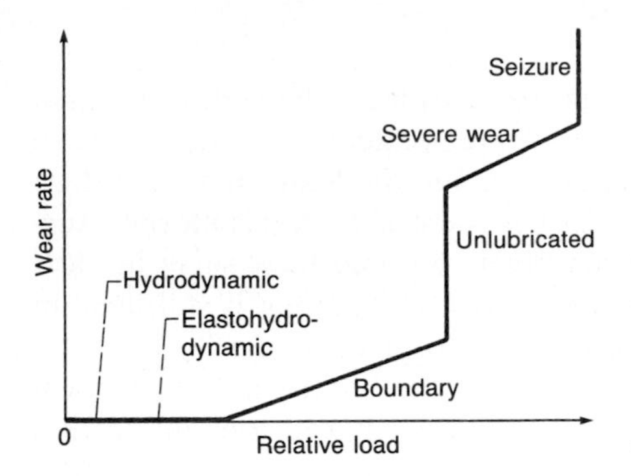

FIGURE 1.9
Wear rate for various lubrication regimes. *From Beerbower (1972).*

boundary lubrication regime the degree of asperity interaction and the wear rate increase as the load increases. The transition from boundary lubrication to an unlubricated condition is marked by a drastic change in wear rate. As the relative load is increased in the unlubricated regime, the wear rate increases until scoring or seizure occurs and the machine element can no longer operate successfully. Most machine elements cannot operate long with unlubricated surfaces. Together Figs. 1.8 and 1.9 show that the friction and wear of unlubricated surfaces can be greatly decreased by providing boundary lubrication.

Boundary lubrication is used for heavy loads and low running speeds, where fluid film lubrication is difficult to attain. Mechanisms such as door hinges operate under conditions of boundary lubrication. Other applications where low cost is of primary importance use boundary lubrication in rubbing sleeve bearings.

1.2.5 PARTIAL LUBRICATION. If the pressures in elastohydrodynamically lubricated machine elements are too high or the running speeds are too low, the lubricant film will be penetrated. Some contact will take place between the asperities, and partial lubrication (sometimes referred to as "mixed lubrication") will occur. The behavior of the conjunction in a partial lubrication regime is governed by a combination of boundary and fluid film effects. Interaction takes place between one or more molecular layers of boundary lubricating films. A partial fluid film lubrication action develops in the bulk of the space between the solids. The average film thickness in a partial lubrication conjunction is less than 1 μm and greater than 0.01 μm.

It is important to recognize that the transition from elastohydrodynamic to partial lubrication does not take place instantaneously as the severity of loading is increased, but rather a decreasing proportion of the load is carried by pressures within the fluid that fills the space between the opposing solids. As the load increases, a larger part of the load is supported by the contact pressure between the asperities of the solids. Furthermore, for conformal surfaces the regime of lubrication goes directly from hydrodynamic to partial lubrication.

1.3 CLOSURE

In this chapter conformal and nonconformal surfaces were defined. Conformal surfaces fit snugly into each other with a high degree of geometric conformity so that the load is carried over a relatively large area and the load-carrying surface area remains essentially constant as the load is increased. Nonconformal surfaces do not geometrically conform to each other well and have small lubrication areas. The lubrication area enlarges with increasing load but is still small in comparison with the lubrication area of conformal surfaces.

The development of understanding of a lubricant's physical and chemical action within a lubricated conjunction was briefly traced, and four lubrication regimes were described: hydrodynamic, elastohydrodynamic, partial, and boundary. Hydrodynamic lubrication is characterized by conformal surfaces. The lubricating film is thick enough to prevent the opposing solids from coming into contact. Friction arises only from the shearing of the viscous lubricant. The pressures developed in hydrodynamic lubrication are low (usually less than 5 MPa) so that the surfaces may generally be considered rigid and the pressure-viscosity effects are small. Three modes of pressure development within hydrodynamic lubrication were presented: slider, squeeze, and external pressurization. For hydrodynamic lubrication with sliding motion the minimum film thickness is quite sensitive to load, being inversely proportional to the square root of the normal applied load.

Elastohydrodynamic lubrication is characterized by nonconformal surfaces, and again there is no asperity contact of the solid surfaces. Two modes of elastohydrodynamic lubrication exist: hard and soft. Hard EHL is characterized by metallic surfaces, and soft EHL by surfaces made of elastomeric materials. The pressures developed in hard EHL are high (typically between 0.5 and 3 GPa) so that elastic deformation of the solid surfaces becomes important as do the pressure-viscosity effects of the lubricant. As with hydrodynamic lubrication, friction is due to the shearing of the viscous lubricant. The minimum film thickness for hard EHL is relatively insensitive to load because the contact area increases with increasing load, thereby providing a larger lubrication area to support the load. For soft EHL the elastic distortions are large, even for light loads, and the viscosity within the conjunction varies little with pressure because the pressures are relatively low and the elastic effect predominates. Both hydrodynamic and elastohydrodynamic lubrication are fluid film lubrication phenomena in that the film is thick enough to prevent opposing solid surfaces from coming into contact.

In boundary lubrication considerable asperity contact occurs, and the lubrication mechanism is governed by the physical and chemical properties of thin surface films that are of molecular proportion (from 1 to 10 nm). The frictional characteristics are determined by the properties of the solids and the lubricant film at the common interfaces. Partial lubrication (sometimes referred to as "mixed lubrication") is governed by a mixture of boundary and fluid film effects. Most of the scientific unknowns lie in this lubrication regime.

1.4 PROBLEMS

1.4.1 Describe at least three applications for each of the four lubrication regimes.

1.4.2 Describe the differences between comformal and nonconformal surfaces.

1.5 REFERENCES

Beerbower, A. (1972): Boundary Lubrication. GRU.IGBEN.72, Report on Scientific and Technical Application Forecasts (Avail. NTIS, AD–747336).

Bowden, F. P., and Tabor, D. (1973): *Friction—An Introduction to Tribology*. Anchor Press/ Doubleday, New York.

Department of Education and Science, Great Britain (1966): *Lubrication (Tribology), Education and Research; A Report on the Present Position and Industry's Needs*. HMSO, London.

Dowson, D. (1979): *History of Tribology*. Longman, London and New York.

Hamrock, B. J., and Anderson, W. J. (1983): Rolling-Element Bearings. *NASA Ref. Publ.* 1105.

Hamrock, B. J., and Dowson, D. (1977): Isothermal Elastohydrodynamic Lubrication of Point Contacts, Part III—Fully Flooded Results, *J. Lubr. Technol.*, vol. 99, no. 2, pp. 264–276.

Hardy, W. B., and Doubleday, I. (1922a): Boundary Lubrication—The Paraffin Series. *Proc. R. Soc. London Ser. A*. vol. 100, Mar. 1, pp. 25–39.

Hardy, W. B., and Doubleday, I. (1922b): Boundary Lubrication—The Temperature Coefficient. *Proc. R. Soc. London Ser. A*, vol. 101, Sept. 1, pp. 487–492.

Hutchings, I. M. (1992): *Tribology-Friction and Wear of Engineering Materials*. Edward Arnold, London.

Petrov, N. P. (1883): Friction in Machines and the Effect of the Lubricant. *Inzh. Zh. St. Petersburg*, vol. 1, pp. 71–140; vol. 2, pp. 227–279; vol. 3, pp. 377–436; vol. 4, pp. 535–564.

Rabinowicz, E. (1965): *Friction and Wear of Materials*. Wiley, New York.

Reynolds, O. (1886): On the Theory of Lubrication and Its Application to Mr. Beauchamp Tower's Experiments, Including an Experimental Determination of Viscosity of Olive Oil. *Philos. Trans. R. Soc. London Ser. A*, vol. 177, pp. 157–234.

Tower, B. (1885): Second Report on Friction Experiments (Experiments on the Oil Pressure in a Bearing). *Proc. Inst. Mech. Eng.*, pp. 58–70.

CHAPTER 2

BEARING CLASSIFICATION AND SELECTION

Design is a creative process aimed at finding a solution to a particular problem. In all forms of design a particular problem may have many different solutions, mainly because design requirements can be interpreted in many ways. For example, it may be desirable to produce:

- The cheapest design
- Or the easiest to build with available materials
- Or the most reliable
- Or the one that takes up the smallest space
- Or the one that is lightest in weight
- Or the best from any of a whole variety of possible standpoints

The task of the designer is therefore not clear-cut, because he or she has to choose a reasonable compromise between these various requirements and then has to decide to adopt one of the possible designs that could meet this compromise.

The process of bearing selection and design usually involves these steps:

1. Selecting a suitable type of bearing
2. Estimating a bearing size that is likely to be satisfactory
3. Analyzing bearing performance to see if it meets the requirements
4. And then modifying the design and the dimensions until the performance is near to whichever optimum is considered the most important

The last two steps in the process can be handled fairly easily by someone who is trained in analytical methods and understands the fundamental principles of the subject. The first two steps, however, require some creative decisions to be made and for many people represent the most difficult part of the design process.

2.1 BEARING CLASSIFICATION

A bearing is a support or guide that locates one machine component with respect to others in such a way that prescribed relative motion can occur while the forces associated with machine operation are transmitted smoothly and efficiently. Bearings can be classified in several ways: according to the basic mode of operation (rubbing, hydrodynamic, hydrostatic, or rolling element), according to the direction and nature of the applied load (thrust or journal), or according to geometric form (tapered land, stepped parallel surface, or tilting pad). There is much to be said for classification according to the basic mode of operation, with subdivisions to account for different geometric forms and loading conditions. That classification is used in this book.

2.1.1 DRY RUBBING BEARINGS. In dry rubbing bearings the two bearing surfaces rub together in rolling or sliding motion, or both, and are lubricated by boundary lubrication. Examples of dry rubbing bearings are unlubricated journals made from materials such as nylon, polytetrafluoroethylene, and carbon and diamond pivots used in instruments. The load-carrying and frictional characteristics of this class of bearings can be related directly to the basic contact properties of the bearing materials.

2.1.2 IMPREGNATED BEARINGS. In this type of bearing a porous material (usually metal) is impregnated with a lubricant, thus giving a self-lubricating effect. The porous metal is usually made by sintering (heating to create a coherent mass without melting) a compressed metal powder (e.g., sintered iron or bronze). The pores serve as reservoirs for the lubricant. The load-carrying and frictional characteristics of the bearing depend on the properties of the solid matrix and the lubricant in conjunction with the opposing solid. The lubricant may be a liquid or a grease.

In general, the application of impregnated bearings is restricted to low sliding speeds (usually less than 1 or 1.5 m/s), but they can carry high mean pressures (often up to 7 to 15 MPa). A great advantage of these bearings is that they are simple and cheap, just like rubbing bearings, and they are frequently used in low-speed or intermittent-motion situations such as automobile chassis, cams, and oscillating mechanisms.

Impregnated separators for small ball bearings such as those used in precision instruments are sometimes used as lubricant reservoirs for the rolling elements when a minimum amount of lubricant is required. In this case the porous material is generally a plastic (e.g., nylon).

It is usually doubtful that the impregnated bearing operates in true hydrodynamic fashion owing to the small amount of lubricant that is present. The behavior can be described as partial hydrodynamic lubrication, therefore implying partial lubrication. The bearing's hydrodynamic performance can be analyzed by assuming that a full film exists in the clearance space and that the lubricant flow within the porous material is covered by Darcy's law as pointed out, for example, in Cameron (1976). Darcy's simple formula for porous bearings relates the pressure gradient to the flow within a porous material while neglecting inertial effects and assuming that there is no relative surface velocity. A simultaneous solution of the Reynolds equation and the flow equation for a porous matrix yields flow patterns, pressure distributions, and load-carrying capacities that can be used to construct design charts. Satisfactory design procedures, however, usually embody a considerable amount of experimental information and operating experience to supplement the hydrodynamic analysis. Difficulty in qualifying the separate actions that govern bearing behavior reflects the partial hydrodynamic operation of many bearings in this class.

2.1.3 CONFORMAL FLUID FILM BEARINGS. The opposing surfaces of hydrodynamic fluid film bearings are completely separated by a lubricant film. The lubricant may be a liquid or a gas, and the load-carrying capacity derived from the pressure within the lubricating film may be generated by the motion of the machine elements (self-acting or hydrodynamic bearings) or by external pressurization (hydrostatic) or hydrodynamic squeeze motion, or by a combination of these actions. In all cases the frictional characteristics of the bearings are governed by the laws of viscous flow. The load-carrying capacities are similarly dictated by hydrodynamic action, but the properties of the bearing materials have to be considered (e.g., the fatigue life or low friction properties) at extremely low speeds.

The methods of feeding lubricant to a conformal fluid film bearing vary considerably. At low speeds and modest loads a simple ring-oiler that draws oil up to the bearing from a reservoir by means of viscous lifting might suffice, but in many modern machines the oil is supplied to the bearing under pressure to ensure adequate filling of the clearance space. Externally pressurized, or hydrostatic, bearings require elaborate lubricant supply systems, and the lubricant enters the bearing under a pressure of the order of a megapascal. This type of bearing is particularly useful at high loads and low speeds or when film stiffness perpendicular to surface motion is important.

A simple subdivision of conformal fluid film bearings that accounts for the nature of the lubricant, the mode of operation, the direction of motion, the nature of the load, and the geometric form of the bearing is shown in Fig. 2.1.

2.1.4 ROLLING-ELEMENT BEARINGS. The machine elements in rolling-element bearings are separated by elements in predominately rolling motion. Figure 2.2 shows the subgrouping of rolling-element bearings. The rolling elements might be balls, rollers, or needles (rollers with large width-to-diameter ratios). Relative motion between the machine elements is permitted by replac-

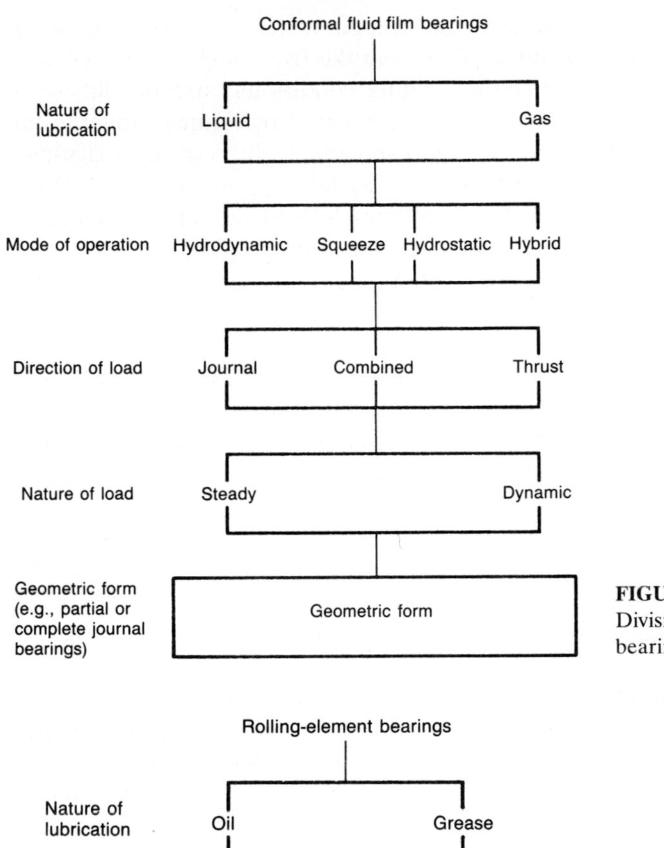

FIGURE 2.1
Divisions of conformal fluid film bearings.

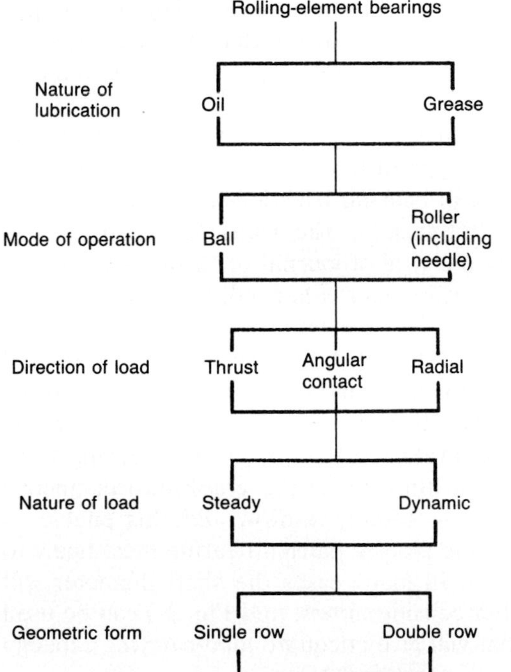

FIGURE 2.2
Divisions of rolling-element bearings.

ing the sliding action with a motion that is mainly rolling. Normally some slipping, sliding, or spinning also takes place, and the friction characteristics are determined by the relative motion, the loading conditions, and the lubricant properties. Rolling-element bearings may be lubricated by liquids (mineral oils or synthetic lubricants) or greases. The lubricant (normally a grease) is sometimes sealed into the bearing assembly, or it may be applied in a mist of fine droplets. There are innumerable types of rolling-element bearings designed to meet the varied operating conditions encountered in industry.

2.2 BEARING SELECTION

The designer is often confronted with decisions on whether a rolling-element or hydrodynamic bearing should be used in a particular application. The following characteristics make rolling-element bearings *more desirable* than hydrodynamic (conformal fluid film) bearings in many situations: (1) low starting and good operating friction, (2) the ability to support combined radial and thrust loads, (3) less sensitivity to interruptions in lubrication, (4) no self-excited instabilities, (5) good low-temperature starting, and (6) the ability to seal the lubricant within the bearing. Within reasonable limits, changes in load, speed, and operating temperature have but little effect on the satisfactory performance of rolling-element bearings.

The following characteristics make rolling-element bearings *less desirable* than hydrodynamic (conformal fluid film) bearings: (1) finite fatigue life subject to wide fluctuations, (2) larger space required in the radial direction, (3) low damping capacity, (4) higher noise level, (5) more severe alignment requirements, and (6) higher cost.

Each type of bearing has its particular strong points, and care should be taken in choosing the most appropriate type of bearing for a given application. Useful guidance on the important issue of bearing selection has been presented by the Engineering Sciences Data Unit (ESDU). The ESDU documents (1965, 1967) are excellent guides to selecting the type of journal or thrust bearing that is most likely to give the required performance when considering the load, speed, and geometry of the bearing.

Figure 2.3, reproduced from ESDU (1965), shows the typical maximum load that can be carried at various speed, for a nominal life of 10,000 h at room temperature, by various types of journal bearings on shafts of the diameters quoted. The heavy curves indicate the preferred type of journal bearing for a particular load, speed, and diameter and thus divide the graph into distinctive regions. The applied load and speed are usually known, and this enables a preliminary assessment to be made of the type of journal bearing most likely to be suitable for a particular application. In many cases the shaft diameter will already have been determined by other considerations, and Fig. 2.3 can be used to find the type of journal bearing that will give adequate load-carrying capacity at the required speed.

These curves are based on good engineering practice and commercially available parts. Higher loads and speeds or smaller shaft diameters are possible

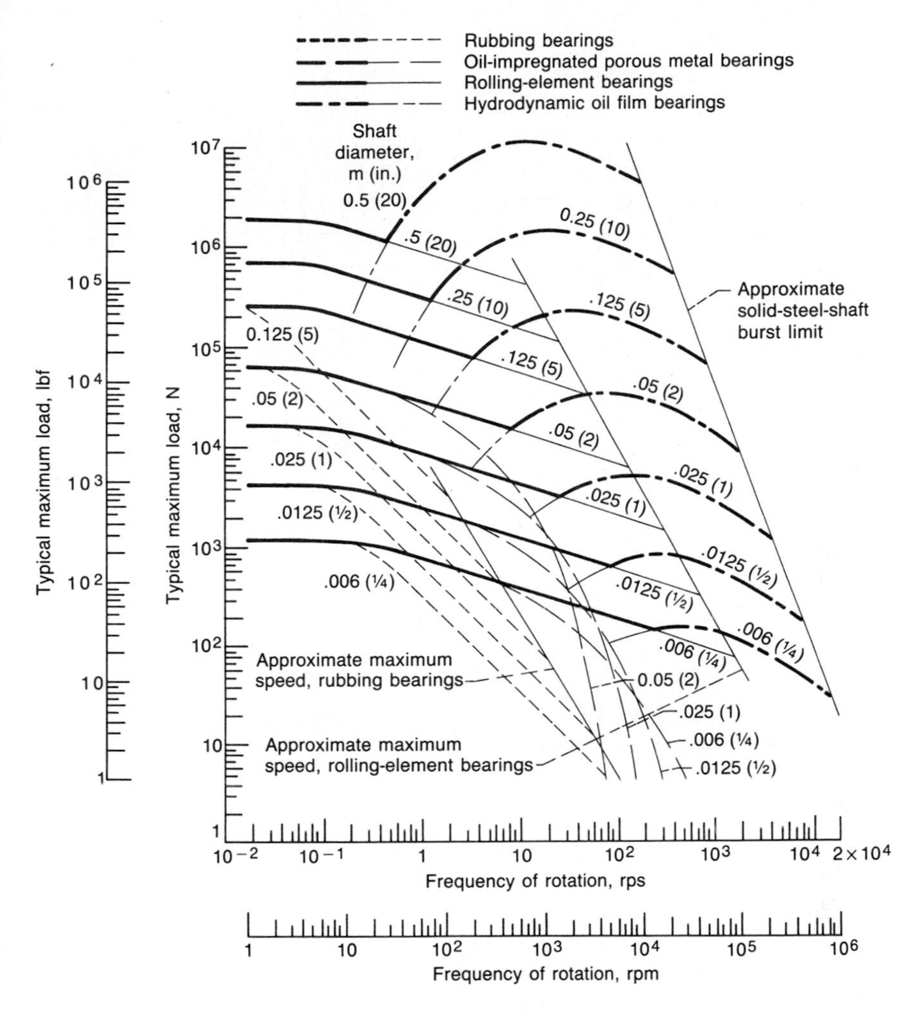

FIGURE 2.3
General guide to journal bearing type. Except for rolling-element bearings, curves are drawn for bearings with width equal to diameter. A medium-viscosity mineral oil is assumed for hydrodynamic bearings. [*From* ESDU *(1965).*]

with exceptionally high engineering standards or specially produced materials. Except for rolling-element bearings the curves are drawn for bearings with widths equal to their diameters. A medium-viscosity mineral oil lubricant is assumed for the hydrodynamic bearings.

Considerations other than the load and speed may often have an overriding importance in bearing selection. Table 2.1 gives the advantages and limitations of various bearings in relation to environmental conditions and particular requirements. It is emphasized that Fig. 2.3 and Table 2.1 are only intended as guides.

TABLE 2.1 Advantages and limitations of journal bearings

[From ESDU 1965)]

Condition	General comments	Journal bearing type						
		Rubbing bearings	Oil-impregnated porous metal bearings	Rolling-element bearings	Hydrodynamic fluid film bearings	Hydrostatic fluid film bearings	Self-acting gas bearings	Externally pressurized gas bearings
High temperature	Attention to differential expansions and their effect on fits and clearances is necessary	Normally satisfactory depending on material	Attention to oxidation resistance of lubricant is necessary	Up to 100 °C no limitations; from 100 to 250 °C stabilized bearings and special lubrication procedures are probably required	Attention to oxidation resistance of lubricant is necessary		Excellent	Excellent
Low temperature	Attention to differential expansions and starting torques is necessary		Lubricant may impose limitations; consideration of starting torque is necessary	Below −30 °C special lubricants are required; consideration of starting torque is necessary	Lubricant may impose limitations; consideration of starting torque is necessary	Lubricant may impose limitations	Excellent; thorough drying of gas is necessary	
External vibration	Attention to the possibility of fretting damage is necessary (except for hydrostatic bearings)	Normally satisfactory except when peak of impact load exceeds load-carrying capacity		May impose limitation; consult manufacturer	Satisfactory	Excellent	Normally satisfactory	Excellent
Space requirements		Small radial extent		Bearings of many different proportions; small axial extent	Small radial extent but total space requirement depends on the lubrication feed system		Small radial extent	Small radial extent, but total space requirement depends on the gas feed system
Dirt or dust		Normally satisfactory; sealing is advantageous	Sealing is important		Satisfactory; filtration of lubricant is important		Sealing important	Satisfactory
Vacuum		Excellent	Lubricant may impose limitations				Not normally applicable	Not applicable when vacuum has to be maintained

TABLE 2.1 *Continued*

<table>
<tr><th rowspan="2">Condition</th><th rowspan="2">General comments</th><th colspan="7">Journal bearing type</th></tr>
<tr><th>Rubbing bearings</th><th>Oil-impregnated porous metal bearings</th><th>Rolling-element bearings</th><th>Hydrodynamic fluid film bearings</th><th>Hydrostatic fluid film bearings</th><th>Self-acting gas bearings</th><th>Externally pressurized gas bearings</th></tr>
<tr><td>Simplicity of lubrication</td><td></td><td colspan="2">Excellent</td><td>Excellent with self-contained grease or oil lubrication</td><td>Self-contained assemblies can be used with certain limits of load, speed, and diameter; beyond this, oil circulation is necessary</td><td>Auxiliary high pressure is necessary</td><td>Excellent</td><td>Pressurized supply of dry, clean gas is necessary</td></tr>
<tr><td>Availability of standard parts</td><td></td><td>Good to excellent depending on type</td><td colspan="2">Excellent</td><td>Good</td><td colspan="3">Not available</td></tr>
<tr><td>Prevention of contamination product and surroundings</td><td></td><td>Improved performance can be obtained by allowing a process liquid to lubricate and cool the bearing, but wear debris may impose limitations</td><td colspan="4">Normally satisfactory, but attention to sealing is necessary, except where a process liquid can be used as a lubricant</td><td colspan="2">Excellent</td></tr>
<tr><td>Frequent stop-starts</td><td rowspan="2"></td><td rowspan="2">Excellent</td><td>Good</td><td rowspan="2">Excellent</td><td>Good</td><td rowspan="2">Excellent</td><td rowspan="2">Poor</td><td rowspan="2">Excellent</td></tr>
<tr><td>Frequent change of rotating direction</td><td>Generally good</td><td>Generally good</td></tr>
<tr><td>Running costs</td><td></td><td colspan="3">Very low</td><td>Depends on complexity of lubrication system</td><td>Cost of lubricant supply has to be considered</td><td>Nil</td><td>Cost of gas supply has to be considered</td></tr>
</table>

TABLE 2.1 *Concluded*

Condition	General comments	Journal bearing type						
		Rubbing bearings	Oil-impregnated porous metal bearings	Rolling-element bearings	Hydrodynamic fluid film bearings	Hydrostatic fluid film bearings	Self-acting gas bearings	Externally pressurized gas bearings
Wetness and humidity	Attention to possibility of metallic corrosion is necessary	Normally satisfactory depending on material	Normally satisfactory; sealing advantageous	Normally satisfactory, but special attention to sealing may be necessary	Satisfactory		Satisfactory	
Radiation		Satisfactory	Lubricant may impose limitations				Excellent	
Low starting torque		Not normally recommended	Satisfactory	Good	Satisfactory	Excellent	Satisfactory	Excellent
Low running torque								
Accuracy of radial location		Poor	Good			Excellent	Good	Excellent
Life		Finite but predictable			Theoretically infinite but affected by infinite filtration and number of stops and starts	Theoretically infinite	Theoretically infinite but affected by number of stops and starts	Theoretically infinite
Combination of axial and load-carrying capacity		A thrust face must be provided to to carry the axial loads		Most types capable of dual duty	A thrust face must be provided to carry the axial loads			
Silent running		Good for steady loading	Excellent	Usually satisfactory; consult manufacturer	Excellent	Excellent except for possible pump noise	Excellent	Excellent except for possible compressor noise

TABLE 2.2 Advantages and limitations of thrust bearings

[From ESDU (1967)]

Condition	General comments	Thrust bearing type						
		Rubbing bearings	Oil-impregnated porous metal bearings	Rolling-element bearings	Hydrodynamic fluid film bearings	Hydrostatic fluid film bearings	Self-acting gas bearings	Externally pressurized gas bearings
High temperature	Attention to differential expansions and their effect upon axial clearance is necessary	Normally satisfactory depending on material	Attention to oxidation resistance of lubricant is necessary	Up to 100 °C no limitations; from 100 to 250 °C stabilized bearings and special lubrication procedures are probably required	Attention to oxidation resistance of lubricant is necessary		Excellent	
Low temperature	Attention to differential expansions and starting torques is necessary		Lubricant may impose limitations; consideration of starting torque is necessary	Below −30 °C special lubricants are required; consideration of starting torque is necessary	Lubricant may impose limitations; consideration of starting torque is necessary	Lubricant may impose limitations	Excellent; thorough drying of gas is necessary	
External vibration	Attention to the possibility of fretting damage is necessary (except for hydrostatic bearings)	Normally satisfactory except when peak of impact load exceeds load-carrying capacity		May impose limitations; consult manufacturer	Satisfactory	Excellent	Normally satisfactory	Excellent
Space requirements		Small radial extent		Bearings of many different proportions are available	Small radial extent but total space requirement depends on the lubrication feed system		Small radial extent	Small radial total space requirement depends on gas feed system
Dirt or dust		Normally satisfactory; sealing advantageous		Sealing is important	Satisfactory; filtration of lubricant is important		Sealing important	Satisfactory
Vacuum		Excellent	Lubricant may impose limitations				Not normally applicable	Not applicable when vacuum has to be maintained
Wetness and humidity	Attention to possibility of metallic corrosion is necessary	Normally satisfactory depending on material	Normally satisfactory; sealing advantageous	Normally satisfactory, but special attention to sealing is perhaps necessary	Satisfactory			
Radiation		Satisfactory	Lubricant may impose limitations				Excellent	

TABLE 2.2 *Continued*

Condition	General comments	Thrust bearing type						
		Rubbing bearings	Oil-impregnated porous metal bearings	Rolling-element bearings	Hydrodynamic fluid film bearings	Hydrostatic fluid film bearings	Self-acting gas bearings	Externally pressurized gas bearings
Low starting torque		Not normally recommended	Satisfactory	Good	Satisfactory	Excellent	Satisfactory	Excellent
Low running torque					Satisfactory		Excellent	
Accuracy of radial location		Good				Excellent	Good	Excellent
Life		Finite but can be estimated			Theoretically infinite but affected by filtration and number of stops and starts	Theoretically infinite	Theoretically infinite but affected by number of stops and starts	Theoretically infinite
Combination of axial and load-carrying capacity		A journal bearing surface must be provided to carry the radial loads		Some types capable of dual duty	A journal bearing surface must be provided to carry the radial loads			
Silent running		Good for steady loading	Excellent	Usually satisfactory; consult manufacturer	Excellent	Excellent, except for possible pump noise	Excellent	Excellent, except for possible compressor noise
Simplicity of lubrication		Excellent		Excellent with self-contained grease lubrication; with large sizes or high speeds, oil lubrication might be necessary	Self-contained assemblies can be used with certain limits of load, speed, and diameter; beyond this, oil circulation is necessary	Auxiliary high pressure is necessary,	Excellent	Pressurized supply of dry, clean gas is necessary

TABLE 2.2 *Concluded*

Condition		General comments	Thrust bearing type						
			Rubbing bearings	Oil-impregnated porous metal bearings	Rolling-element bearings	Hydrodynamic fluid film bearings	Hydrostatic fluid film bearings	Self-acting gas bearings	Externally pressurized gas bearings
Availability of standard parts			Good to excellent depending on type	Excellent		Good	Poor		
Prevention of contamination of product and surroundings			Performance can be improved by allowing a process liquid to lubricate and cool the bearing, but wear debris may impose limitations	Normally satisfactory, but attention to sealing is necessary, except where a process liquid can be used as a lubricant				Excellent	
Tolerance to manufacturing and assembly inaccuracies			Good		Satisfactory	Poor	Satisfactory	Poor	Satisfactory
Type of motion	Frequent start-stops		Excellent			Good	Excellent		Excellent
	Unidirectional		Suitable						
	Bidirectional		Suitable			Some types are suitable	Suitable	Some types are suitable	Suitable
	Oscillatory					Unsuitable		Unsuitable	
Running costs			Very low			Depends on complexity of lubrication system	Cost of lubricant supply has to be considered	Nil	Cost of gas supply has to be considered

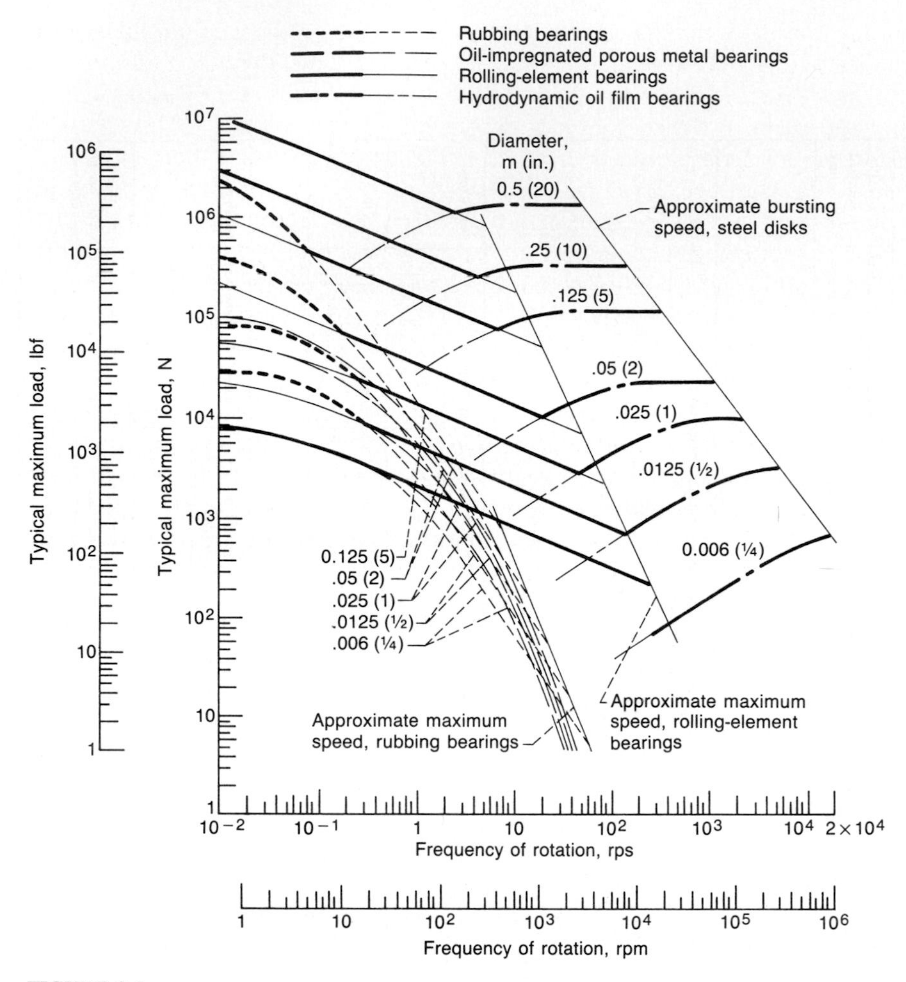

FIGURE 2.4
General guide to thrust bearing type. Except for rolling-element bearings, curves are drawn for typical ratios of inside diameter to outside diameter. A medium-viscosity mineral oil is assumed for hydrodynamic bearings. [*From* ESDU *(1967)*.]

Similarly, Fig. 2.4, reproduced from ESDU (1967), shows the typical maximum load that can be carried at various speeds, for a nominal life of 10,000 h at room temperature, by various types of thrust bearings on shafts of the diameters quoted. The heavy curves again indicate the preferred type of bearing for a particular load, speed, and diameter and thus divide the graph into major regions. Considerations other than load and speed are given in Table 2.2 for thrust bearings.

2.3 CLOSURE

This chapter began with a general discussion of the process of bearing design. The four primary steps used in bearing design are selecting a suitable type, estimating bearing size, analyzing performance, and modifying or fine-tuning. It was pointed out that the first two steps are the most difficult and require creative decisions, whereas the last two steps can be handled fairly easily by a person trained in analytical methods. After considering several options, it was decided that bearings would best be classified by considering their modes of operation. The four primary classes of bearing that were considered were dry or rubbing bearings, which use boundary lubrication if any; impregnated bearings, which use partial lubrication; rolling-element bearings, which use elastohydrodynamic lubrication; and hydrodynamic fluid film bearings, which use hydrodynamic lubrication. The Engineering Sciences Data Unit documents can be used as guides in selecting the type of journal or thrust bearing most likely to give the required performance when considering its load, speed, and geometry.

Considerations other than load and speed are important in bearing selection. Thus, tables are presented that give the advantages and limitations of various bearings in relation to environmental conditions and particular requirements. It should be recognized that information on bearing selections given in this chapter is intended to be a guide to selecting a suitable type of bearing and to estimating a bearing size that is likely to be satisfactory.

2.4 PROBLEMS

2.4.1 Figures 2.3 and 2.4 show the relationship between load and speed for four different types of bearing. How would you use these figures to help you select the appropriate bearing for your particular application?

2.4.2 Suggest suitable types of bearing to meet the following situations:
(*a*) High load, very low speed, very low friction
(*b*) Light load, very high speed, no liquid lubricant
(*c*) Light load, low speed, no liquid lubricant

2.4.3 Explain why gas-lubricated bearings are appealing. Describe the limiting features of this type of bearing.

2.5 REFERENCES

Cameron, A. (1976): *Basic Lubrication Theory*, 2d ed. Ellis Horwood Limited, Chichester, England.

Engineering Sciences Data Unit (ESDU) (1965): *General Guide to the Choice of Journal Bearing Type*. Item 65007, Institution of Mechanical Engineers, London.

Engineering Sciences Data Unit (ESDU) (1967): *General Guide to the Choice of Thrust Bearing Type*. Item 67033, Institution of Mechanical Engineers, London.

CHAPTER 3

SURFACE TOPOGRAPHY

Increasing production speeds and new cutting methods, such as plasma cutting, spark erosion, and laser cutting, change the characteristics of machined surfaces. Requirements with regard to surface accuracy and surface refinement have also greatly increased. The importance of a fine-scale surface description is well-demonstrated in tribology. The breakdown of lubrication layers of oil in engine cylinders or in bearings can be caused by improper microlevel surface shape.

Although bearing design theory relies heavily on fluid mechanics and kinematics, it is still ultimately a problem of two surfaces that are either in contact or separated by a thin fluid film. In either case the surface texture can be important in ensuring proper lubrication.

The first step in gaining insight into the lubrication of solid surfaces is to examine the surface profile, or topography. Smooth surfaces are not flat on an atomic scale. The roughness of manufactured surfaces used in lubrication is between 0.01 and 10 μm, whereas typical atomic diameters are between 0.0001 and 0.001 μm. Even a highly polished surface, when examined microscopically or with a profilometer, has an irregular nature. The surface consists of high and low spots. The high spots, or protuberances, are called "asperities."

3.1 GEOMETRIC CHARACTERISTICS OF SURFACES

The geometric characteristics, or texture, of surfaces as shown in Fig. 3.1 may conveniently be divided into three main categories:

1. *Error of form.* The surface deviates from a well-defined pattern because of errors inherent in the manufacturing process.

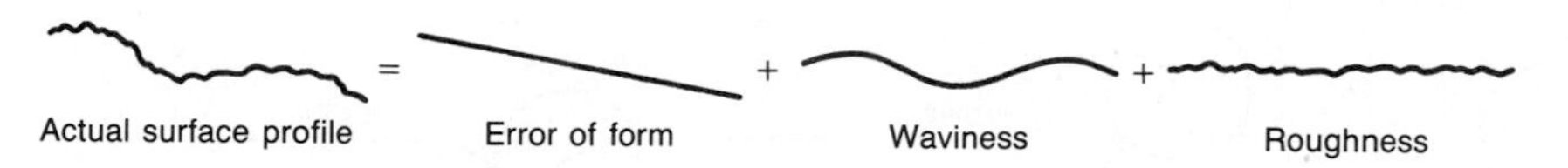

FIGURE 3.1
Geometric characteristics of solid surfaces. [*From Halling (1976).*]

2. *Waviness*. Relatively long waves in a surface profile are often associated with unwanted vibrations that always occur in machine tool systems.
3. *Roughness*. Irregularities, excluding waviness and error of form, are inherent in the cutting and polishing process during production.

In the study of lubricated surfaces, roughness is the geometric variation that is generally of interest. Although often no sharp distinction can be drawn between these categories, roughness simply concerns the horizontal spacing (wavelength) of the surface features. From a practical point of view, in characterizing surfaces used in tribology both the vertical direction (or amplitude parameter) and the horizontal direction (or wavelength) are important.

3.2 STYLUS MEASUREMENTS

Two general classes of hardware are commonly used for measuring surface finish: contacting methods using stylus techniques and noncontacting methods. Stylus measurements are discussed in this section. Stylus measurements are based on transforming the vertical motion of the stylus tip as it traverses a surface into an electrical analog voltage. This voltage is then processed either by using analog circuitry or by converting it to digital information for processing. The method was introduced by Abbot and Firestone in 1933.

The stylus is normally made of diamond and has a tip radius of 2 μm and a static load of less than 0.0007 N (0.00256 oz). The tip radius is relatively large in comparison with the typical roughness. Therefore, it is often difficult to

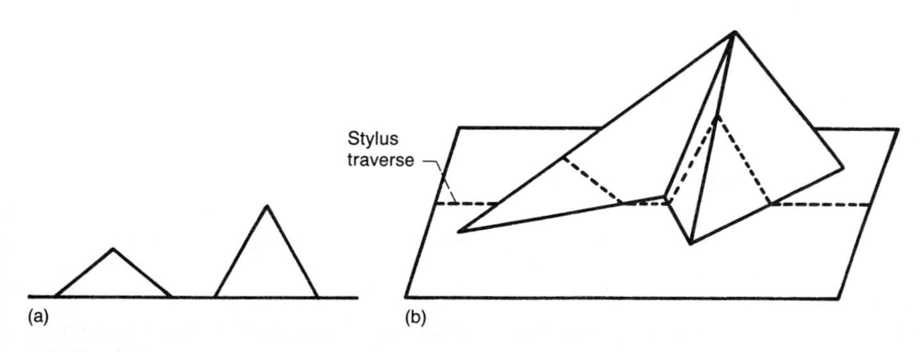

FIGURE 3.2
Difficulty in interpreting profilometer traces. (a) Surface profile; (b) surface asperity.

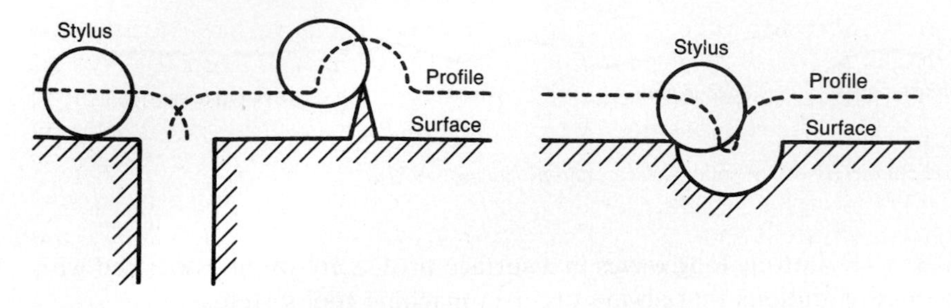

FIGURE 3.3
Error due to stylus radius.

obtain a true picture of a surface from a stylus measurement. Features that appear to be asperity peaks on a single profile may in fact be local ridges on the flank of a true summit, as shown in Fig. 3.2. Moreover, many real surfaces of practical interest to tribologists have highly anisotropic surface textures. Thus, profiles taken in different traverse directions look quite different.

However, the main limitation of stylus measurements is the finite size of the stylus tip, which distorts the surface profile, as shown in Fig. 3.3, broadening

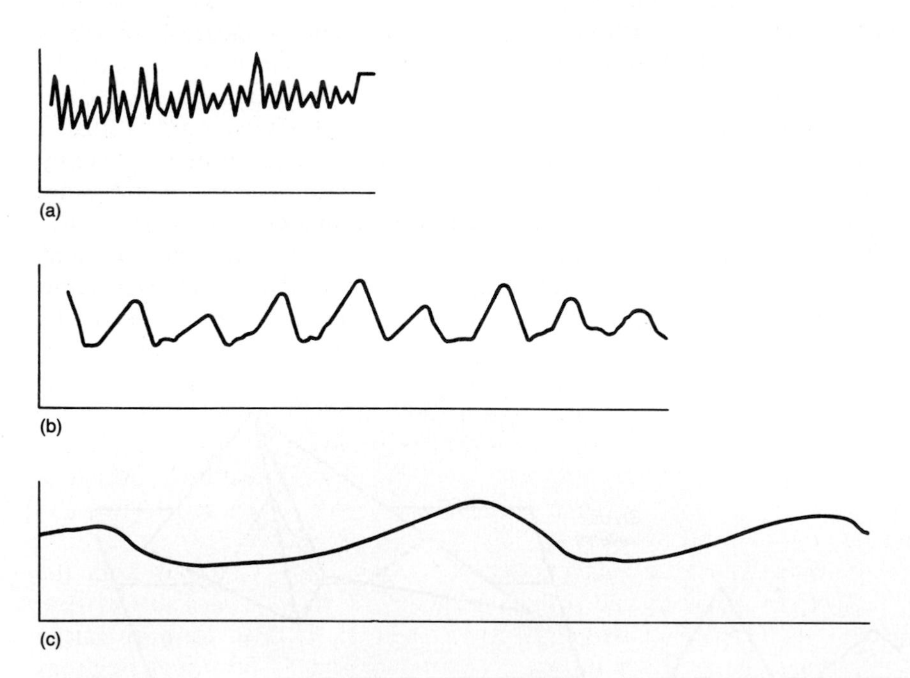

FIGURE 3.4
Misleading impression from stylus-records-shaped surface at vertical magnification of 200 and three horizontal magnifications. Centerline average, 10 μm. (a) Horizontal magnification, 8; (b) horizontal magnification, 40; (c) horizontal magnification, 200 (true form). [*From Barwell (1979).*]

peaks and narrowing valleys. The magnification in the vertical direction is generally 100 to 100,000; that in the horizontal direction is 10 to 5000. A typical ratio of vertical to horizontal magnification is 50 : 1. It is therefore important to take into account this difference in magnification. This is often not done, and thus a false impression of the nature of surfaces has become prevalent. Figure 3.4 attempts to illustrate this. Figure 3.4(a) shows a typical profilometry trace at a horizontal magnification of 8, Fig. 3.4(b) shows the trace at a horizontal magnification of 40, and Fig. 3.4(c) shows a small portion of the trace at a magnification of 200 based on equal horizontal and vertical scales.

3.3 SOME NONCONTACTING MEASUREMENT DEVICES

The noncontacting methods for determining the characteristics of surfaces use a number of different measurement principles and devices:

1. *Pneumatic devices.* This method is based on the measurement of air leakage. A flat and finely finished measurement head has an orifice of air. Measuring the airflow when the head is placed on the surface indicates changes in surface roughness. Pneumatic devices provide a simple, inexpensive, portable, quick, and robust way of assessing surface roughness, well-suited for use in determining the quality of surfaces on the shop floor.
2. *Optical devices.* The intensity of reflected laser light defines the surface texture, producing an average over a surface area. This method is not well established in terms of the number of people who use it routinely.
3. *Electron microscope devices.* This method offers better resolution and depth of field than optical methods, chiefly because of the extremely short wavelength of electron microscope beams in comparison with that of light. Sherrington and Smith (1988) point out the two types of electron microscopy:
   a. Transmission electron microscopy (TEM). Electrons are incident on a thin specimen (less than 1 μm thick) that deflects and scatters the electrons as they pass through it. A lens system magnifies and focuses scattered electrons to form an image on a screen or photographic film. To examine the surface of a metal component by TEM, it is usually necessary to fabricate a replica of the specimen surface. TEM is typically able to resolve features down to a separation of around 0.3 nm and has been used to study the changes in surface structure during wear.
   b. Reflection electron microscopy (REM). Electrons are scattered from the surface of the specimen and strike a collector, which generates an electrical signal. This signal is subsequently processed and used to form an image representing the specimen surface on a monitor screen. Scattered electrons are produced by a beam of finely focused electrons that scans the specimen in a raster pattern, a process called "scanning electron microscopy" (SEM). Scanning electron microscopes can be adjusted to have a maximum resolu-

TABLE 3.1
Summary off typical specifications of devices used in laboratory measurement of surface topography
[From Sherrington and Smith (1988)]

Device	Resolution				Vertical measurement range or depth of field		Measurable area
	Lateral		Vertical				
	Lowest	Highest	Lowest	Highest	At lowest resolution	At highest resolution	
Stylus instrument	(a)	(a)	0.5 μm	0.00025 μm	500 μm	0.25 μm	Depends on traverse length; typically a few millimeters
Optical light microscope	2.5 μm	0.1 μm	(b)	(b)	42 μm	0.04 μm	Depends on magnification
Transmission electron microscope	2.5 nm	0.5 nm	(c)	(c)	400 nm	80 nm	Depends on magnification
Scanning electron microscope	5 μm	10 nm	(d)	(d)	1 mm	2 μm	Depends on magnification

[a]Not easily defined.
[b]Not applicable.
[c]Approximately the same as the lateral resolution.
[d]Not available.

tion of about 10 nm, a little less than that available in TEM. However, this disadvantage is compensated for by the fact that specimen preparation is considerably easier.

Besides these noncontacting devices, scanning tunneling microscopy is also available but beyond the scope of this book. Sherrington and Smith (1988) give an excellent description of this device as well as describing in general the modern measurement techniques used in determining the characteristics of tribological surfaces.

Several methods of measuring surface topography have been described in this section as well as in the preceding section. Of these methods the stylus measurement device is the most widely used. Table 3.1, obtained from Sherrington and Smith (1988), summarizes the specifications of a number of the devices described in this section as well as in the preceding section. From this table the resolution, the depth of field, and the measurable area are described for four different devices.

3.4 REFERENCE LINES

In computing the parameters that define the surface texture all height measurements are made from some defined reference line. Several methods have been used. They are summarized below and shown graphically in Fig. 3.5.

3.4.1 MEAN, OR M SYSTEM. The mean, or M system, method is based on selecting the mean line as the centroid of the profile. Thus, the areas above and below this line are equal. If for discrete profiles the area of each profile is a rectangle, this method turns out to be simply finding the average of the measured heights. This method gives a "horizontal" reference that does not compensate for errors of form, or tilt.

3.4.2 TEN-POINT AVERAGE. The ten-point-average method is based on finding the five highest peaks and the five lowest valleys. The average of these 10 points gives the reference line. For deeply pitted surfaces this method can lead to a reference line that is below the major surface features.

3.4.3 LEAST SQUARES. The least-squares method is based on postulating a sloping reference line instead of a horizontal line as is the case for the M system. Therefore, the major advantage of this approach is that it can be used to compensate for the linear error of form, or tilt. The following will attempt to describe mathematically the least-squares reference lines. From the equation of a line,

$$z = \overline{m}x + \overline{b}$$

where $\overline{m}$ = slope
$\overline{b}$ = intercept on z axis

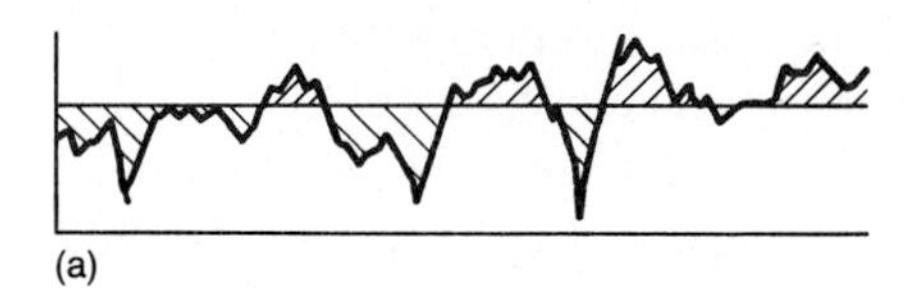

(a)

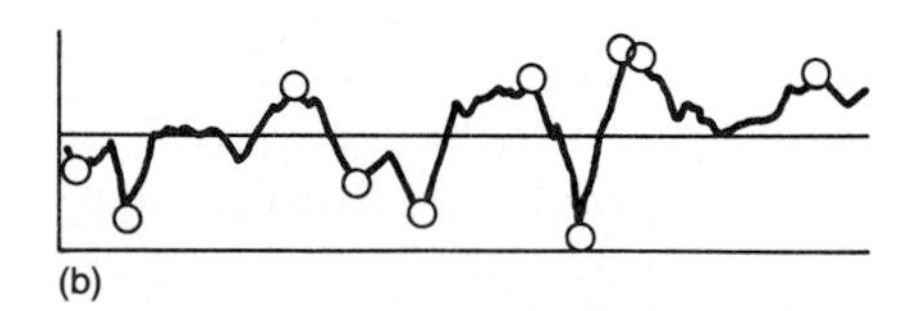

(b)

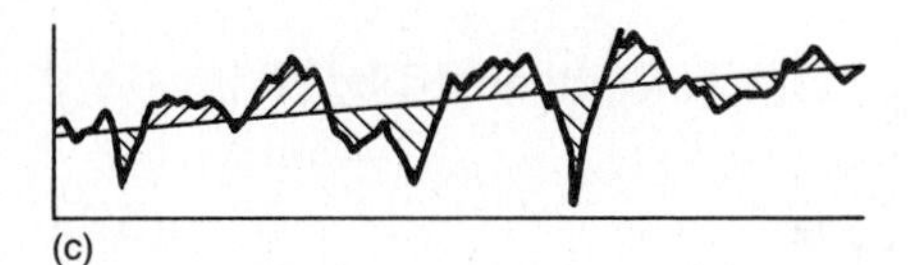

(c)

FIGURE 3.5
Comparison of three types of reference line: (a) M system; (b) ten-point average; (c) least squares.

Given a set of points $P_1(x_1, z_1), P_2(x_2, z_2), \ldots, P_n(x_n, z_n)$ corresponding to each value of x, consider two values of z: (1) z measured and (2) z obtained from $\overline{m}x + \overline{b}$. Call the difference $\overline{d}$ such that

$$\overline{d}_1 = \left[z_1 - (\overline{m}x_1 + \overline{b})\right], \ldots, \overline{d}_n = \left[z_n - (\overline{m}x_n + \overline{b})\right]$$

The set of all deviations gives a picture of how well the observed data fit a line. If $\sum_{i=1}^{n} \overline{d}_i^2 = 0$, the fit is perfect. This never occurs in real situations, and that is where the method of least squares comes in.

Not only is it important to establish a reference line or mean line in the z direction, it is also important to establish a sampling length or distance in the x direction that adequately differentiates between roughness and waviness.

$$\bar{f}(\overline{m}, \overline{b}) = \sum_{i=1}^{n} \overline{d}_i^2$$

or

$$\bar{f}(\overline{m}, \overline{b}) = \left(z_1 - \overline{m}x_1 - \overline{b}\right)^2 + \left(z_2 - \overline{m}x_2 - \overline{b}\right)^2 + \cdots + \left(z_n - \overline{m}x_n - \overline{b}\right)^2$$

Finding the values of $\overline{m}$ and $\overline{b}$ that give the smallest $\bar{f}(\overline{m}, \overline{b})$ requires finding values of $\overline{m}$ and $\overline{b}$ that satisfy

$$\frac{\partial \bar{f}}{\partial \overline{m}} = \frac{\partial \bar{f}}{\partial \overline{b}} = 0$$

Solving these two equations for the two unknowns produces the reference line that gives the slope-intercept form.

Once the method of selecting a reference line is agreed upon, all surface height measurements are made relative to it. The centroid (or M system) and the least-squares reference line methods are the easiest to deal with analytically, because the sum of all the profile height deviations is always zero.

3.5 COMPUTATION OF SURFACE PARAMETERS

Measurements made with a stylus instrument are assumed. The reference line used is obtained by either the M system or the least-squares method so that the average of z_i is zero. Samples taken at uniform length intervals Δ are assumed. The Δ is defined as small. Resulting discretized height values are denoted by z_i, $i = 1, 2, \ldots, N$.

Three different surface parameters may be computed:

1. Centerline average (CLA) or arithmetic average (AA), denoted by R_a,

$$R_a = \frac{1}{N} \sum_{i=1}^{N} |z_i| \tag{3.1}$$

From $\partial \bar{f}/\partial \bar{b} = 0$ it immediately follows that $\Sigma \bar{d}_i = 0$ so $z_i' = z_i - \bar{m}x_i - \bar{b}$ has zero mean.

2. Root mean square (rms), denoted by R_q,

$$R_q = \left(\frac{1}{N} \sum_{i=1}^{N} z_i^2 \right)^{1/2} \tag{3.2}$$

If a Gaussian height distribution is assumed, the R_q has the advantage of being the standard deviation of the profile.

3. Maximum peak-to-valley height, denoted by R_t,

$$R_t = \max(z) - \min(z) \tag{3.3}$$

In general,

$$R_a \le R_q \le R_t \tag{3.4}$$

Also for a simple sine distribution the ratio of R_q to R_a is

$$\frac{R_q}{R_a} = \frac{\pi}{2\sqrt{2}} = 1.11 \tag{3.5}$$

Terms such as "rough," "fine," "smooth," and "supersmooth" should be avoided in describing surface topography, since the meaning is relative to the application being used. For example, a surface with an $R_q = 40$ nm is "very rough" to people working in the field of optics, whereas it is considered "very smooth" for a machined surface.

Table 3.2 gives typical values of the arithmetic mean R_a for various processes and components. Figure 3.6 shows six different surface profiles, all with the same R_a, or arithmetic average roughness. The R_a is thus an ambiguous parameter, since it does not indicate whether the R_a value is the mean of many small deviations from the mean value or of a few large ones. For this reason tribologists seek additional surface parameters that are rather more informative.

Thus far, the discussion of surface roughness has been on defining the profile in the z direction. To incorporate the length of the asperities, the bearing length is introduced. The profile bearing length is obtained by cutting the profile peaks by a line parallel to the mean line within the sampling length at a given section level. Figure 3.7 is helpful in describing the bearing length ℓ^*. Mathematically, it is defined as

$$\ell^* = \ell_1^* + \ell_2^* + \cdots + \ell_n^* \tag{3.6}$$

The profile bearing length ratio t_p is defined as

$$t_p = \frac{\ell^*}{\ell} \tag{3.7}$$

where ℓ = sampling length, m.

TABLE 3.2 Typical arithmetic averages for various processes and components

	Arithmetic average, R_a	
	μm	μin.
Processes		
Sand casting; hot rolling	12.5-25	500-1000
Sawing	3.2-25	128-1000
Planing and shaping	.8-25	32-1000
Forging	3.2-12.5	128-500
Drilling	1.6-6.3	64-250
Milling	.8-6.3	32-250
Boring; turning	.4-6.3	16-250
Broaching; reaming; cold rolling; drawing	.8-3.2	32-128
Die casting	.8-1.6	32-64
Grinding, coarse	.4-1.6	16-120
Grinding, fine	.1-.4	4-16
Honing	.1-.8	4-32
Polishing	.05-.4	2-16
Lapping	.025-.4	1-16
Components		
Gears	0.25-10	10-400
Plain bearings—journal (runner)	.12-.5	5-20
Plain bearings—bearing (pad)	.25-1.2	10-50
Rolling bearings—rolling elements	.025-.12	1-5
Rolling bearings—tracks	.1-.3	4-12

FIGURE 3.6
Geometric profiles having same values of arithmetic average. [*From Halling (1976).*]

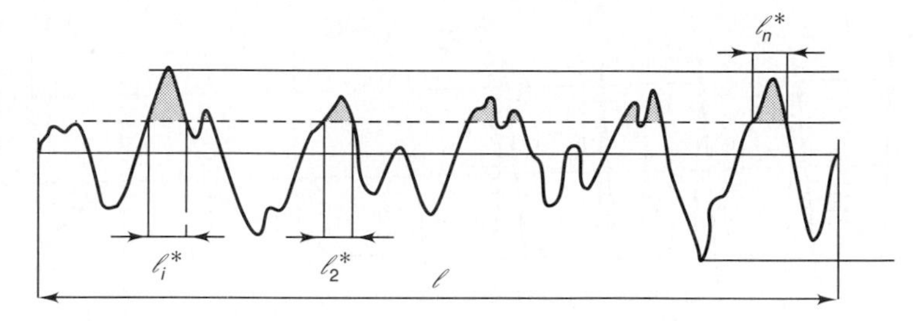

FIGURE 3.7
Surface profile showing bearing length. [*From Persson (1992).*]

The Abbot curve (also known as the "bearing length curve") is shown in Fig. 3.8. The vertical axis is the bearing length ratio multiplied by 100. These curves describe how much of the profile protrudes a given distance above the surface. During running-in, a certain amount of the surface will be removed by plastic deformation. These curvcs also calculate the peakedness and the amount of material between given heights. In the top profile the asperities have steeper peaks than in the bottom profile.

The texture of a surface can be described in terms of the distribution function of its profile heights. In statistical terms the cumulative distribution of the all-ordinate distribution curve can be written as

$$\bar{F}(z) = \int_{-\infty}^{z} \tilde{\psi}\, dz$$

where z refers to the profile height and $\tilde{\psi}$ is the probability density function of the distribution of these heights. The probability density function may be viewed as the fraction of heights in a given interval. Therefore, the practical derivation of such a distribution curve involves taking measurements of z_1, z_2, etc., at some discrete interval and summing the number of ordinates at any given height level. Figure 3.9 illustrates the method used in obtaining the all-ordinate distribution. The distribution curve in Fig. 3.9 is the smoothest curve that can be drawn through the histogram produced by such a sampling procedure. This smooth curve for many surfaces tends to exhibit a Gaussian distribution of surface texture heights.

If these ideas of height distribution are introduced, the empirical probability density, or histogram, can be used to evaluate R_a and R_q. If the fraction of heights in an interval $z_j - \Delta \leq z_j \leq z_j + \Delta$ is denoted by $\tilde{\psi}_j$, for $j = -L, \ldots, 0, \ldots, L$, then approximately the same values as those given in Eqs. (3.1) and (3.2) can be expressed as

$$R_a = \int_{-L}^{L} |z| \tilde{\psi}\, dz \tag{3.8}$$

$$R_q = \left(\int_{-L}^{L} z^2 \tilde{\psi}\, dz \right)^{1/2} \tag{3.9}$$

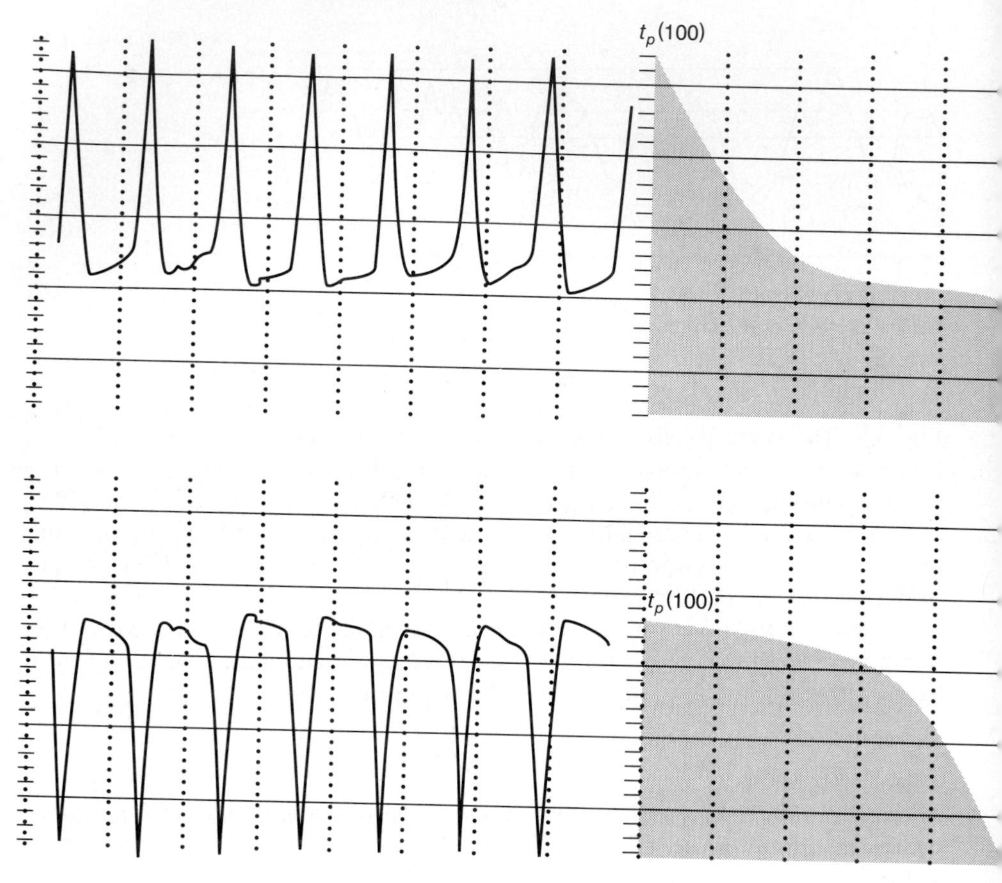

FIGURE 3.8
Abbot curves for two different profiles. [*From Persson (1992).*]

From the Gauss-Laplace law a general expression for the probability density function while assuming a Gaussian distribution is

$$\tilde{\psi} = \frac{1}{\bar{\sigma}(2\pi)^{1/2}} \exp \frac{-(z - z^*)^2}{2\bar{\sigma}^2} \tag{3.10}$$

where $\bar{\sigma}$ is the standard deviation and z^* is the distance of the mean from the value chosen as the origin. The values of the ordinate $\tilde{\psi}$ of the Gaussian distribution curve for $z^* = 0$ are found in most books on statistics. The form of the Gaussian distribution necessitates a spread of $-\infty$ to ∞, which cannot happen with practical surfaces. In practice the distribution curve is truncated to

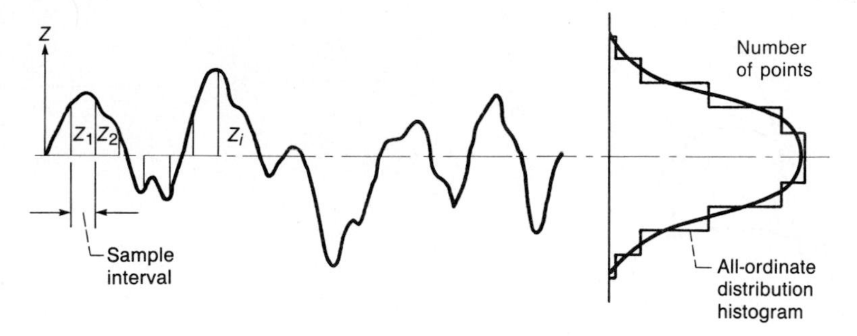

FIGURE 3.9
Method of deriving all-ordinate distribution. [*From Halling (1975).*]

$\pm 3\bar{\sigma}$. Because approximately 99.9 percent of all events occur within this region, the truncation leads to negligible error while providing useful simplification.

The nth moment of the distribution curve $\tilde{\psi}\,dz$ about the mean is defined as

$$M_n = \int_{-\infty}^{\infty} z^n \tilde{\psi}\,dz \qquad (3.11)$$

It can be observed that twice the first moment of half $\tilde{\psi}\,dz$ is equivalent to the centerline average R_a defined in Eq. (3.8) or

$$R_a = 2\int_0^{\infty} z\tilde{\psi}\,dz = \text{twice the first moment of half } \tilde{\psi}\,dz$$

The first moment of the whole $\tilde{\psi}\,dz$ about the mean reference line is zero. Likewise, comparing the second moment of $\tilde{\psi}\,dz$ with the root mean square R_q expressed in Eq. (3.9) gives

$$R_q = \bar{\sigma} = \left(\int_{-\infty}^{\infty} z^2 \tilde{\psi}\,dz\right)^{1/2} = (\text{second moment of } \tilde{\psi}\,dz)^{1/2}$$

The third moment of $\tilde{\psi}\,dz$ relates to the skewness of a curve, or the departure of a curve from symmetry. The mathematical expression for the normalized skewness is

$$\bar{\alpha} = \frac{1}{R_q^3}\int_{-\infty}^{\infty} z^3 \tilde{\psi}\,dz \qquad (3.12)$$

If the peaks and valleys deviate from the reference line by approximately the same amount, the skewness will be zero. If, on the other hand, the surfaces are deeply pitted, the skewness will be some negative value.

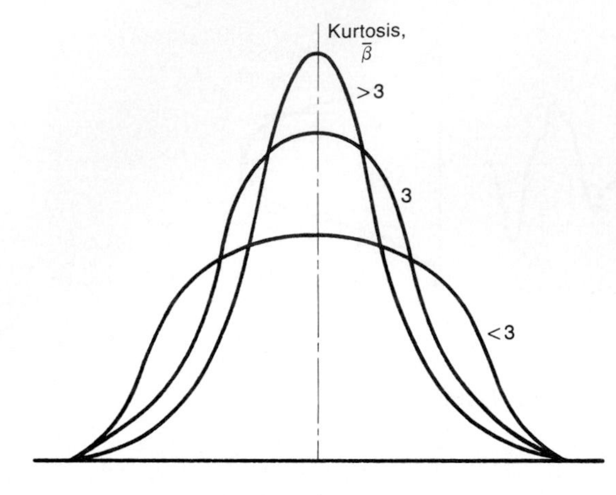

FIGURE 3.10
Illustration of three different kurtosis values. [*From Halling (1975).*]

The fourth moment of $\tilde{\psi}\,dz$ relates to the peakedness, or kurtosis, of the curve and is expressed as

$$\bar{\beta} = \frac{1}{R_q^4}\int_{-\infty}^{\infty} z^4 \tilde{\psi}\,dz \tag{3.13}$$

The kurtosis $\bar{\beta}$ always has a positive value and measures the sharpness of a symmetric distribution. For a Gaussian distribution the curve has a kurtosis of 3. When most of the profile heights are close to the reference line, $\bar{\beta}$ is quite large; a relatively flat height distribution will have a $\bar{\beta}$ near zero. Curves with values of $\bar{\beta}$ less than 3 are called "platykurtic," and those with $\bar{\beta}$ greater than 3 are called "leptokurtic." Figure 3.10 illustrates the various types of kurtosis curves.

For a Gaussian distribution where the curve is represented by $\tilde{\psi}$, as expressed in Eq. (3.10), the general expression for the nth moment is given by

$$M_n = \frac{1}{\bar{\sigma}(2\pi)^{1/2}}\int_{-\infty}^{\infty} z^n \exp\frac{-z^2}{2\sigma^2}\,dz \tag{3.14}$$

where the standard deviation $\bar{\sigma} = R_q$. From Eq. (3.14) it is observed that, if n is odd, M_n vanishes and the curve must be symmetrical. If n is even, then

$$M_n = \frac{n!}{2^{n/2}(n/2)!}\bar{\sigma}^n \tag{3.15}$$

Note that the second moment becomes $\bar{\sigma}^2$, the variance.

$$\therefore M_2 = \bar{\sigma}^2 = R_q^2 = \text{variance} \tag{3.16}$$

Some additional parameters used to define surfaces utilized in fluid film lubrication as well as the range of values normally found are listed below:

Density of asperities, 10^2 to 10^6 peaks/mm^2
Asperity spacing, 1 to 75 μm
Asperity slopes, 0 to 25°, but mainly 5 to 10°
Radii of peaks, mostly 10 to 30 μm

These additional surface definitions better define the surface texture used in fluid film lubrication.

3.6 AUTOCORRELATION PARAMETER

The previously discussed parameters (R_a, R_q, and R_t) depend only on the profile heights, not on the spacing between heights. As a result they do not truly reflect the shape of the profile, as illustrated in Fig. 3.9. The autocorrelation does incorporate the spacing between heights and is obtained by multiplying each profile height by the height of the point at some fixed horizontal distance farther along the profile. After averaging the product over a representative profile length and normalizing by the variance R_q^2, the expression for the autocorrelation is

$$\tilde{\rho}_k = \frac{1}{R_q^2(N-k)} \sum_{i=1}^{N-k} z_i z_{i+k} \tag{3.17}$$

The autocorrelation depends on k and is a measure of the similarity of heights separated by the distance $k\Delta$ (where Δ is the sample interval, assumed to be constant). In machining processes such as turning or milling, where there are pronounced feed marks, the autocorrelation will have a maximum where $k\Delta$ is some integer multiple of the linear feed. This distance is often called the "characteristic wavelength."

Typical plots of the autocorrelation function for two different profiles are shown in Fig. 3.11. The variance is defined as $\bar{\sigma}^2$. The shape of this function is

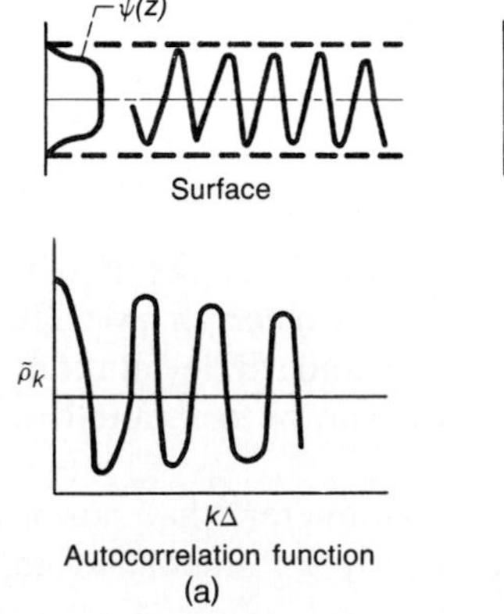

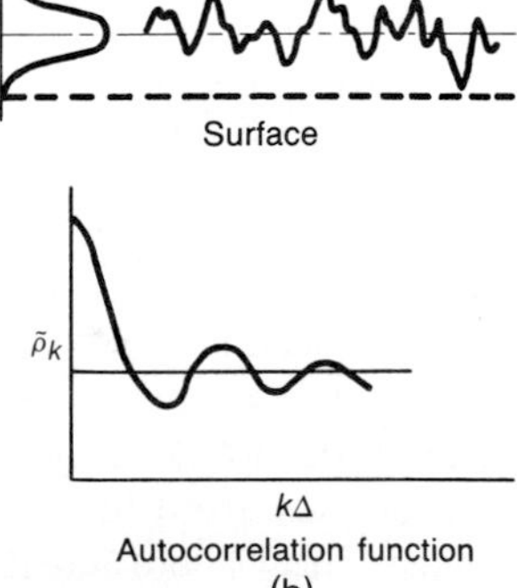

FIGURE 3.11
Two different surfaces and resulting autocorrelation functions. (*a*) Periodicity profile; (*b*) decay profile. [*From Halling (1975).*]

most useful in revealing some of the characteristics of the profile. The general decay of the function is the random component of the surface profile and indicates a decrease in correlation as $k\Delta$ increases, as shown in Fig. 3.11(b). The oscillatory component of the function indicates any inherent periodicity of the profile, as shown in Fig. 3.11(a).

3.7 DISTRIBUTION OF SLOPE AND CURVATURE

A basic geometric description of a profile can be given, considering the profile as a random function, with first and second derivatives. Each of the derivatives also has a statistical distribution associated with it. The most straightforward way of estimating this information is to use differences in the profile to represent the derivatives, or

$$\left(\frac{dz}{dx}\right)_i \approx \frac{z_{i+1} - z_i}{\Delta} \tag{3.18}$$

$$\left(\frac{d^2z}{dx^2}\right)_i \approx \frac{z_{i+1} - 2z_i + z_{i-1}}{\Delta^2} \tag{3.19}$$

Equation (3.18) defines the slope of the profile and Eq. (3.19) defines its curvature. However, because the profile is a random function with "noise," differencing is not the best way, since it makes this noise greater. Some sort of smoothing of the data should be done to minimize this effect. One way of doing this is to use Eqs. (3.18) and (3.19) with samples taken at intervals of Δ, $2\Delta, 3\Delta, \ldots$ and then to make weighted averages of these expressions. For example, by choosing weights of 1.5, -0.6, and 0.1 for derivatives formed at Δ, 2Δ, and 3Δ, respectively, the following expressions are obtained:

$$\left(\frac{dz}{dx}\right)_i = \frac{1}{60\Delta}\left[45(z_{i+1} - z_{i-1}) - 9(z_{i+2} - z_{i-2}) + (z_{i+3} - z_{i-3})\right] \tag{3.20}$$

$$\left(\frac{d^2z}{dx^2}\right)_i = \frac{1}{180\Delta^2}\left[-490z_i + 270(z_{i+1} - z_{i-1}) - 27(z_{i+2} - z_{i-2}) + 2(z_{i+3} - z_{i-3})\right] \tag{3.21}$$

A slope and curvature profile can be generated in this way from the height profile, and by using the same relations as for the heights, a mean or average slope and curvature can be found as well as the respective standard deviation of the distributions. Likewise, the skewness and the kurtosis can be computed to characterize the shape of these distributions.

The previously defined properties are a few of the parameters that can be found from analyzing a profile. These characteristics may be important in relating function to manufactured surfaces or in analyzing how a surface interacts with another surface. They illustrate some of the efforts at extracting more information from the surface measurement than simply the height-sensitive parameters.

Finally, the relation between function and surface geometry has to be more fully understood. This will require closer interaction between the manufacturing engineer who makes the surface and the designer who specifies the surface.

3.8 FILM PARAMETERS FOR DIFFERENT LUBRICATION REGIMES

If a machine element is adequately designed and fluid film lubricated, the lubricated surfaces are completely separated by a lubricant film. Endurance testing of ball bearings, for example, as reported by Tallian et al. (1967) has demonstrated that when the lubricant film is thick enough to separate the contacting bodies, the fatigue life of the bearing is greatly extended. Conversely, when the film is not thick enough to provide full separation between the asperities in the contact zone, the life of the bearing is adversely affected by the high shear resulting from direct metal-to-metal contact.

The four lubrication regimes were defined in Sec. 1.2, and calculation methods for determining the rms surface finish were introduced in Sec. 3.5. This section introduces a film paramcter and describes the range of values for the four lubrication regimes. The relationship between the dimensionless film parameter Λ and the minimum film thickness $h_{\min}$ is

$$\Lambda = \frac{h_{\min}}{\left(R_{q,a}^2 + R_{q,b}^2\right)^{1/2}} \tag{3.22}$$

where $R_{q,a}$ = rms surface finish of surface a
$R_{q,b}$ = rms surface finish of surface b

The film parameter is used to define the four important lubrication regimes. The range of Λ for these four regimes is

1. Hydrodynamic lubrication, $5 \le \Lambda < 100$
2. Elastohydrodynamic lubrication, $3 \le \Lambda < 10$
3. Partial lubrication, $1 \le \Lambda < 5$
4. Boundary lubrication, $\Lambda < 1$

These values are rough estimates. The great differences in geometric conformity in going from hydrodynamically lubricated conjunctions to elastohydrodynamically lubricated conjunctions make it difficult for clear distinctions to be made.

3.9 TRANSITION BETWEEN LUBRICATION REGIMES

As the severity of loading is decreased, there is no sharp transition from boundary to fluid film lubrication; rather an increasing proportion of the load is carried by pressures within the film that fills most of the space between the opposing solids. Indeed, it is often difficult to eliminate fluid film lubrication effects so that true boundary lubrication can occur, and there is evidence to suggest that micro-fluid-film lubrication formed by surface irregularities is an important cffcct.

The variation of the friction coefficient μ with the film parameter Λ is shown in Fig. 3.12. The friction coefficient is defined as

$$\mu = \frac{f}{w_z} \tag{3.23}$$

where f is the tangential (friction) force and w_z is the normal applied load. In Fig. 3.12 the approximate locations of the various lubrication regimes discussed in Sec. 3.8 are shown. This figure shows that as the film parameter Λ increases, the friction coefficient initially decreases in the elastohydrodynamic regime and then increases in the hydrodynamic regime. In explaining this phenomenon let us assume the surface roughness is the same in both lubrication regimes.

In hydrodynamic lubrication of conformal surfaces as found in journal and thrust bearings, $w_z \propto 1/h^2$. In elastohydrodynamic lubrication of nonconformal surfaces the normal applied load has little effect on the film thickness. Therefore, w_z is essentially proportional to a constant. In both hydrodynamic and elastohydrodynamic lubrication the frictional force is due to lubricant shearing and can be expressed in both lubrication regimes as $f \propto 1/h$. Making use of

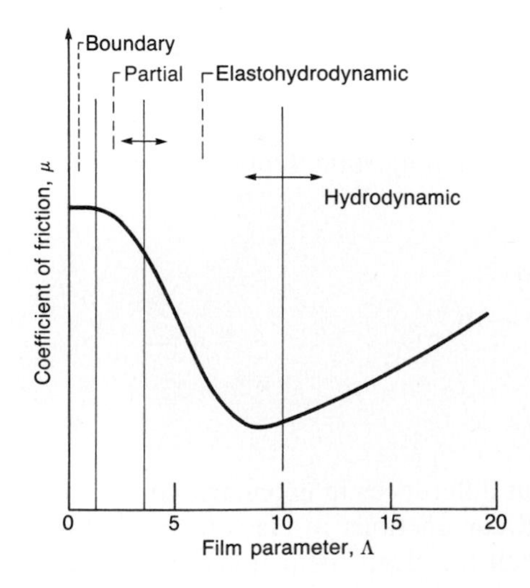

FIGURE 3.12
Variation of friction coefficient with film parameter. [*From Hamrock and Dowson (1981).*]

this,

$$\mu_{\mathrm{HL}} \propto \frac{1/h}{(1/h)^2} \propto h \tag{3.24}$$

$$\mu_{\mathrm{EHL}} \propto \frac{1/h}{\text{constant}} \propto \frac{1}{h} \tag{3.25}$$

This then explains the reversing of the slope of the friction coefficient in Fig. 3.12.

3.10 CLOSURE

Since fluid film lubrication is concerned with the lubrication between solids separated by small film thicknesses, it is essential that the physical nature of the solid's surface topography be understood. To obtain this understanding, this chapter investigated surface measurement hardware. The contact method of stylus measurement is based on transforming the vertical motion of the stylus tip as it traverses a surface into an electrical analog voltage. The main limitation of this approach is the finite size of the stylus tip, which distorts the surface profile broadens peaks and narrows valleys. The following noncontacting measurement devices were presented.

1. Pneumatic devices measure the airflow when a finely finished head is placed on the surface and indicate changes in surface roughness.
2. Optical devices use the intensity of reflected laser light to define the surface texture.
3. Electron microscopic devices produce images when electrons are incident on the specimen.

In computing the parameters that define the surface texture, all height measurements were made from some defined reference line. Two ways of defining the reference line are

1. Mean, or M system, based on selecting the mean line as the centroid of the profile. The area above this line equals the area below this line.
2. Least squares, based on postulating a slope-intercept form for the reference line. This can compensate for errors of form, or tilt.

Ways of computing surface parameters were also discussed. The following two important parameters were presented:

1. Centerline average (CLA) or arithmetic average (AA), denoted by R_a,

$$R_a = \frac{1}{N} \sum_{i=1}^{N} |z_i| \tag{3.1}$$

2. Root-mean-square (rms) roughness, denoted by

$$R_q = \left(\frac{1}{N} \sum_{i=1}^{N} z_i^2 \right)^{1/2} \tag{3.2}$$

In general, $R_a \le R_q$.

The film parameter Λ was defined as the ratio of the minimum film thickness to the composite surface roughness. The film parameter was used to define four important lubrication regimes: hydrodynamic, elastohydrodynamic, partial, and boundary.

3.11 PROBLEMS

3.11.1 Show that for a Gaussian distribution and a zero mean ($z^* = 0$) from the reference line (determined by the M system) the theoretical skewness is zero and the kurtosis is 3.

3.11.2 Show that the skewness for a deeply pitted surface is less than zero ($\bar{\alpha} < 0$) and that the kurtosis for a relatively flat height distribution approaches zero ($\bar{\beta} \to 0$).

3.11.3 Find R_q/R_a for a Gaussian distribution with zero mean ($z^* = 0$).

3.11.4 Prove that $R_a \le R_q$.

3.11.5 What profile would produce an $R_a = R_q$?

3.11.6 Prove that the kurtosis is greater than or equal to 1.

3.11.7 What is the skewness-kurtosis inequality? That is, when plotting skewness versus kurtosis, describe the critical curve that separates where the results are allowed and where they are forbidden.

3.11.8 Show that for a Gaussian distribution with a nonzero mean ($z^* \neq 0$) from the reference line (determined by the M system), the theoretical skewness is 0 and the kurtosis is 3.

3.12 REFERENCES

Abbot, E. J., and Firestone, F. A. (1933): Specifying Surface Quality. *J. Mech. Eng.*, vol. 55, no. 9, pp. 569–572.

Barwell, F. T. (1979): *Bearing System: Principles and Practice*, Oxford University Press, Oxford.

Halling, J. (ed.) (1975): *Principles of Tribology*. Macmillan Press, London and Basingstoke.

Halling, J. (1976): *Introduction to Tribology*. Wykeham Publications, London.

Hamrock, B. J., and Dowson, D. (1981): *Ball Bearing Lubrication—The Elastohydrodynamics of Elliptical Contacts*. Wiley-Interscience, New York.

Persson, U. (1992): "Surface Topography—Speckle Technique and Image Analysis Applied to Surface Roughness Measurement on Machined Surfaces." Ph.D. Thesis. Instrumentation Laboratory, Royal Institute of Technology, Stockholm, TRITA-ILA 92.06

Sherrington, I., and Smith, E. H. (1988): Modern Measurement Techniques in Surface Metrology: Part I: Stylus Instruments, Electron Microscopy and Non-Optical Comparators. *Wear*, vol. 125, pp. 271–288.

Tallian, T. E., et al. (1967): On Computing Failure Modes in Rolling Contacts. *ASLE Trans.*, vol. 10, no. 4, pp. 418–435.

CHAPTER 4

LUBRICANT PROPERTIES

The primary function of a lubricant is to control friction and wear. Liquid lubricants, however, also have desirable secondary properties and characteristics:

1. They can be drawn between moving parts by hydraulic action.
2. They have relatively high heat-sink capacity to cool the contacting parts.
3. They are easily mixed with chemicals to give a variety of properties such as corrosion resistance, detergency, or surface-active layers.
4. They can remove wear particles.

Lubricants can be divided into those of petroleum origin, known as "mineral oils," and those of animal or vegetable origin, known as "fatty oils." Synthetic oils are often grouped with the latter. In order for a lubricant to be effective, it must be viscous enough to maintain a lubricant film under operating conditions but should be as fluid as possible to remove heat and to avoid power loss due to viscous drag. A lubricant should also be stable under thermal and oxidation stresses, have low volatility, and possess some ability to control friction and wear by itself. As can be seen, quite a bit is expected from a lubricant, and to better understand the role of a lubricant, some basic chemistry is needed.

4.1 BASIC CHEMISTRY

Since some lubricants are derived from petroleum, which consists of compounds of carbon and hydrogen, a brief discussion of hydrocarbon chemistry is needed. This discussion can also serve as a basis for the study of alcohols, fatty acids, and cyclic hydrocarbons. Much of this section was obtained from two sources, Pugh (1970) and Hess (1981).

4.1.1 HYDROCARBONS. Hydrocarbons are compounds of carbon and hydrogen. Carbon has a valency, or chemical bonding power, of 4; hydrogen has a valency of 1. The simplest hydrocarbon can be represented diagrammatically as

```
     H
     |
  H—C—H
     |
     H
```

and is named methane with a chemical formula of CH_4. Carbon atoms have the unique property of being able to link together, and each can join further hydrogen atoms as in ethane (C_2H_6), given as

```
     H   H
     |   |
  H—C—C—H
     |   |
     H   H
```

Note that the chain is symmetrical.

By varying the number of carbon atoms in the molecule, it is possible to present "straight-chain hydrocarbons." This family is known as "alkanes" or "paraffins" and has the general formula C_nH_{2n+2}, where n is the total number of carbon atoms present in the molecule. This forms a series in which such physical characteristics as boiling point and specific gravity increase as the value of n increases. Table 4.1, which lists the first 10 members of the series, shows that an increase in the number of carbon atoms increases the boiling point of the compounds, thus reducing their chemical activity. Such families of hydrocarbons are known as "homologous series," and all the members have formulas that fit the general formula of the series. The principal homologous series of hydrocarbons is summarized in Table 4.2.

Much of the remainder of this section on basic chemistry concerns more details of the members of the homologous series given in Table 4.2. The main difference in the series is the type of bonding. A single covalent bond consists of a single pair of electrons shared between two atoms, as in methane. Double bonds involve two shared pairs of electrons, and triple bonds involve three shared pairs. The bonding in organic compounds is usually indicated by dashes, as shown for the following examples:

TABLE 4.1 Straight-chain paraffins
[From Pugh (1970)]

Number of carbon atoms	Name	Formula	Boiling point, °C	Specific gravity	Physical state at NTP[a]
1	Methane	CH_4	−161.5	-----	Gas
2	Ethane	C_2H_6	−88.3	-----	↓
3	Propane	C_3H_8	−44.5	-----	↓
4	Butane	C_4H_{10}	−.5	-----	↓
5	Pentane	C_5H_{12}	36.2	0.626	Liquid
6	Hexane	C_6H_{14}	69	.660	↓
7	Heptane	C_7H_{16}	98.4	.684	↓
8	Octane	C_8H_{18}	125.8	.704	↓
9	Nonane	C_9H_{20}	150.6	.718	↓
10	Decane	$C_{10}H_{22}$	174	.730	↓

[a]Normal temperature and pressure.

1. Single (methane, or CH_4): H—C—H (with H above and below C)

2. Double (ethylene, or C_2H_4): $H_2C{=}CH_2$ (each C bonded to two H)

3. Triple (acetylene, or C_2H_2): H—C≡C—H

TABLE 4.2 Homologous series of hydrocarbons
[From Hess (1981)]

Name	Formula	Sample of familiar member
Alkane or paraffin	C_nH_{2n+2}	Methane (CH_4)
Olefin or alkene	C_nH_{2n}	Ethylene (C_2H_4)
Acetylene or alkyne	C_nH_{2n-2}	Acetylene (C_2H_2)
Cycloparaffin or naphthene	C_nH_{2n}	Cyclopentane (C_5H_{10})
Aromatic	C_nH_{2n-6}	Benzene (C_6H_6)

TABLE 4.3 Petroleum products with boiling point range and number of carbon atoms present

Petroleum product	Boiling point range, °C	Number of carbon atoms present
Natural gas	<32	1–4
Gasoline	40–200	4–12
Naphtha (benzine)	50–200	7–12
Kerosene	175–275	12–15
Fuel oil	200–300	15–18
Lubricating oil	>300	16–20
Wax	>300	20–34
Asphalt	Residue	Large

Although the olefin or alkene and the cycloparaffin or naphthene series in Table 4.2 have the same formula, they behave quite differently. In the alkene series the double bond present between two carbon atoms greatly increases the chemical reactivity of these hydrocarbons. In the cycloparaffin series the carbon atoms are joined to each other, forming a ring. Such a structure is relatively inert.

In the refining of petroleum many useful mixtures of hydrocarbons are made available. The basic process in refining is fractional distillation, which separates the various products according to ranges of boiling points. The products formed by the initial distillation of crude petroleum are known as "straight-run products." Table 4.3 gives the main products formed, together with other pertinent information about them.

4.1.2 ALCOHOLS. If in the structure of methane one hydrogen atom is replaced by the monovalent hydroxyl group OH, one gets methanol, or methyl alcohol,

```
      H
      |
  H — C — OH
      |
      H
```

with a chemical formula of CH_3OH. This volatile liquid is also known as "wood alcohol" because it is an important byproduct from the destructive distillation of wood to produce charcoal. Methanol is an important solvent and is used as an antifreeze and as a denaturant for ethyl alcohol. It is the starting point in the production of many synthetic chemicals.

The structure of ethane can be modified in the same way as was done for methane to give ethanol, or ethyl alcohol (C_2H_5OH),

```
     H   H
     |   |
  H—C—C—OH
     |   |
     H   H
```

Ethanol is also known as "grain alcohol." It is produced from fermentation of carbohydrate compounds contained in molasses, corn, rye, barley, and potatoes. Enzymes in yeast cause the fermentation. Ethanol has the property of absorbing moisture from the atmosphere until it has a composition of 95 percent alcohol and 5 percent water. It is used extensively in industries preparing drugs, medicinals, and cosmetics. Ethanol is present in a variety of beverages, causing them to be intoxicating.

From observing the structures of methanol and ethanol, the family compounds of alcohol may be written in a general formula as

$$C_nH_{2n+1}OH$$

Each is named after the hydrocarbon from which it is derived, the terminal "e" being replaced by "ol."

4.1.3 FATTY ACIDS. The fatty acids may be considered from the appearance of their molecular structure to be derived from the paraffins by replacing an end methyl (CH_3) group with a carboxyl (CO_2H) group. The corresponding acid is named after the root hydrocarbon, and the terminal "e" in the name is changed to "oic." For example, hexanoic acid contains six carbon atoms and its structure is

```
     H   H   H   H   H    O
     |   |   |   |   |   //
  H—C—C—C—C—C—C
     |   |   |   |   |   \
     H   H   H   H   H    O—H
```

This could be expressed as $CH_3(CH_2)_4CO_2H$. Table 4.4 lists the straight-chain fatty acids that result when the number of carbon atoms present in the molecule is varied from 1 to 20. Note that the molecular structures of the fatty acids are no longer symmetrical as was true for the hydrocarbons, since in all cases one end methyl group has been replaced by the CO_2H, or carboxyl, group.

If a straight-chain hydrocarbon has a formula suggesting that it is short two hydrogen atoms, it is called an "olefin," as pointed out in Table 4.2. The olefins form a family series of unsaturated hydrocarbons. They are called

TABLE 4.4 Formulas for straight-chain hydrocarbons and fatty acids
[From Pugh (1970)]

Number of carbon atoms in molecule	Hydrocarbon		Fatty acid		
	Formula	Name	Formula	Chemical name	Commo name
1	$H.CH_3$ or CH_4	Methane	$H.CO_2H$	Methanoic	Formic
2	$H.(CH_2).CH_3$ or C_2H_6	Ethane	$CH_3.CO_2H$	Ethanoic	Acetic
3	$CH_3.(CH_2)CH_3$	Propane	$CH_3.CH_2.CO_2H$	Propanoic	Propioni
4	$CH_3.(CH_2)_2.CH_3$	Butane	$CH_3.(CH_2)_2.CO_2H$	Butanoic	Butyric
6	$CH_3.(CH_2)_4.CH_3$	Hexane	$CH_3.(CH_2)_4.CO_2H$	Hexanoic	Caproic
8	$CH_3.(CH_2)_6.CH_3$	Octane	$CH_3.(CH_2)_6.CO_2H$	Octanoic	Caprylic
10	$CH_3.(CH_2)_8.CH_3$	Decane	$CH_3.(CH_2)_8.CO_2H$	Decanoic	Capric
12	$CH_3.(CH_2)_{10}.CH_3$	Dodecane	$CH_3.(CH_2)_{10}.CO_2H$	Dodecanoic	Lauric
14	$CH_3.(CH_2)_{12}.CH_3$	Tetradecane	$CH_3.(CH_2)_{12}.CO_2H$	Tetradecanoic	Myristic
16	$CH_3.(CH_2)_{14}.CH_3$	Hexadecane	$CH_3.(CH_2)_{14}.CO_2H$	Hexadecanoic	Palmitic
18	$CH_3.(CH_2)_{16}.CH_3$	Octadecane	$CH_3.(CH_2)_{16}.CO_2H$	Octadecanoic	Stearic
20	$CH_3.(CH_2)_{18}.CH_3$	Eicosane	$CH_3.(CH_2)_{18}.CO_2H$	Eicosanoic	Arachidi

unsaturated because the valencies of some of the carbon atoms in the molecule are not completely satisfied by the hydrogen atoms, and consequently the compound is particularly chemically active. The main feature of the olefin family is the double bonding of the carbon atom, which is easily broken and is a source of weakness in the molecule. The family, whose formula is C_nH_{2n}, starts with ethylene.

As pointed out by Pugh (1970), the olefins are not of direct interest from the lubrication point of view, although their reactivity is exploited in the manufacture of synthetic lubricants, but unsaturation in fatty acids is of definite interest, since a number of fatty acids present in animal and vegetable oils have some degree of unsaturation. The unsaturated fatty acids of primary interest in the study of lubricating oils are listed in Table 4.5.

TABLE 4.5 Formulas for some unsaturated fatty acids
[From Pugh (1970)]

Number of carbon atoms in molecule	Common name	Chemical name	Formula
16	Palmitoleic	Hexadec-9-enoic	$CH_3.(CH_2)_5.CH:CH.(CH_2)_7.CO_2H$
18	Oleic	Octadec-9-enoic	$CH_3.(CH_2)_7.CH:CH.(CH_2)_7.CO_2H$
↓	Ricinoleic	12-Hydroxyoctadec-9-enoic	$CH_3.(CH_2)_5.CH(OH).CH_2.CH:CH.(CH_2)_7.CC$
	Linoleic	Octadeca-9:12-dienoic	$CH_3.(CH_2)_4(CH:CH.CH_2)_2.(CH_2)_6.CO_2H$
	Linolenic	Octadeca-9:12:15-trienoic	$CH_3.CH_2.(CH:CH.CH_2)_3.(CH_2)_6.CO_2H$
20	Arachidonic	Eicosa-5:8:11:14-tetraenoic	$CH_3.(CH_2)_4.(CH:CH.CH_2)_4.(CH_2)_2.CO_2H$

4.1.4 CYCLIC HYDROCARBONS. Cyclic hydrocarbons form two distinct groups, the cycloparaffins, or naphthenes, and the aromatics. An example of a naphthene is cyclohexane, which has the structure shown below:

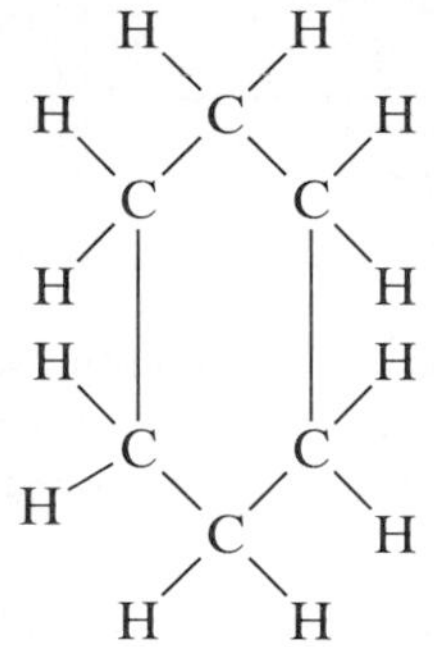

```
          H     H
           \   /
     H       C       H
      \    /   \    /
        C         C
      /  |        | \
    H    |        |   H
    H    |        |   H
      \  |        | /
        C         C
      /    \    /   \
    H        C        H
           /   \
          H     H
```

It is a saturated compound, and the molecule contains 6 carbon atoms and 12 hydrogen atoms. The naphthene family formula is C_nH_{2n}. Recall that the olefins have the same formula but that their structure differs considerably from the naphthene family's. One difference is that the olefins have double-bonded carbon atoms, whereas the naphthenes have single-bonded carbon atoms.

The cycloparaffins are named after the corresponding straight-chain paraffins containing the same number of carbon atoms, but since it takes at least three carbon atoms to form a ring, the series starts with cyclopropane. The physical properties of the cycloparaffins are different from those of the corresponding straight-chain paraffins, but they are chemically quite similar.

The last member of the homologous series given in Table 4.2 is the aromatics. They form a separate, distinct, and peculiar group of compounds, and much time has been spent studying their general structure. Aromatics have a basic ring structure containing three double bonds per molecule. A member of this family is benzene, which is represented structurally below:

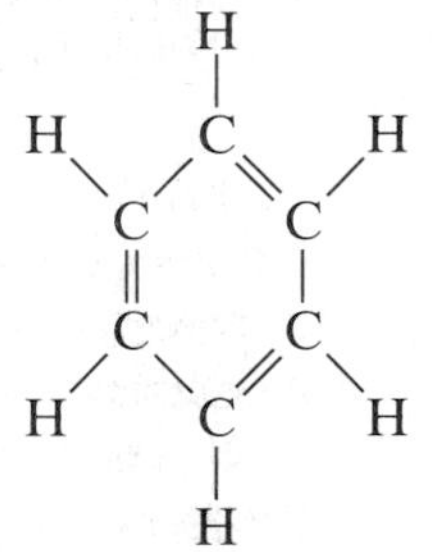

```
             H
             |
     H       C       H
      \    /   \\   /
        C         C
        ||        |
        C         C
      /   \    //   \
    H        C        H
             |
             H
```

The other aromatics can be derived from this basic structure by replacing one or more of the hydrogen atoms with a CH_3 or more complex group of atoms. The ring system is exceptionally stable. Observe from the preceding discussion that there is an appreciable difference between a lubricating oil made from a naphthenic base and one made from an aromatic base.

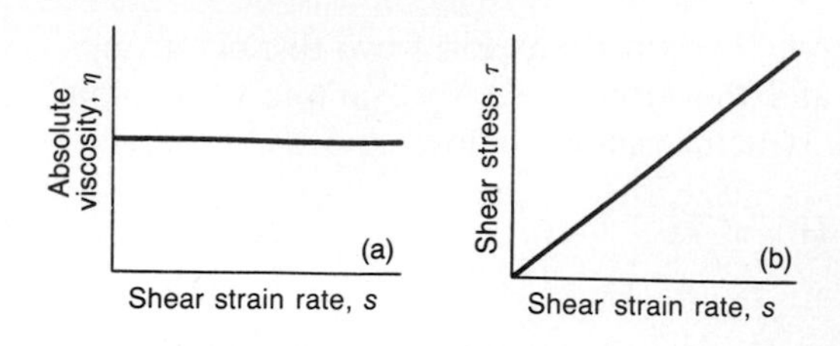

FIGURE 4.1
Properties of a Newtonian fluid. (a) Effect of viscosity on shear strain rate; (b) effect of shear stress on shear strain rate.

4.2 NEWTONIAN FLUIDS

The friction between surfaces that are completely separated (no asperity contact) is due solely to the internal friction of the liquid, namely, its viscosity. Newton in 1687 found that the absolute viscosity of a liquid η can be defined as

$$\eta = \frac{\tau}{s} \tag{4.1}$$

where τ = shear stress, N/m^2
s = shear strain rate, du/dz, s^{-1}
η = absolute viscosity, $N \cdot s/m^2$

The viscosity of a fluid may be associated with its resistance to flow, that is, with the resistance arising from intermolecular forces and internal friction as the molecules move past each other. Thick fluids such as molasses have relatively high viscosity; they do not flow easily. Thinner fluids such as water have lower viscosity; they flow vcry casily. The flow characteristics of Newtonian liquids as a function of shear strain rate are shown in Fig. 4.1.

4.3 NEWTON'S POSTULATE

Let us relook at Eq. (4.1) in the way Newton did, while making use of the sketch shown in Fig. 4.2. Oil molecules were visualized as small balls that roll along in layers between flat planes. Since the oil will "wet" and cling to the two surfaces, the bottommost layer will not move at all, the uppermost layer will move with a velocity equal to the velocity of the upper plane, and the layer in between will move with a velocity directly proportional to the distance between the two

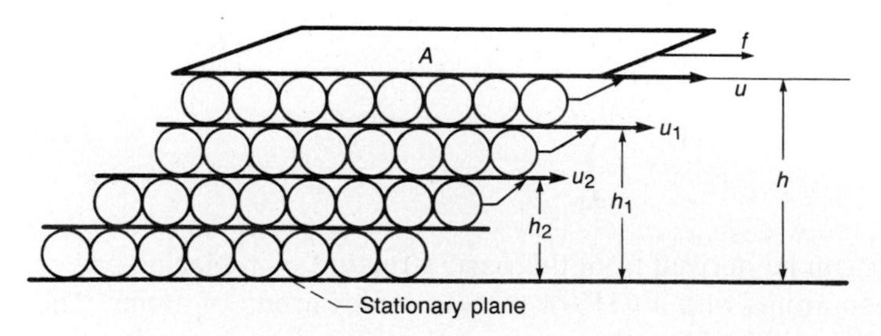

FIGURE 4.2
Physical illustration of Newton's postulate, where f = friction force, N; A = area, m^2; u = velocity, m/s; h = film thickness, m.

planes. This type of orderly movement in parallel layers is known as "streamline," "laminar," or "viscous" flow.

Newton went on to define

$$s = \text{shear strain rate} = \frac{u}{h} = \frac{u_1}{h_1} = \frac{u_2}{h_2} = \cdots$$

Newton correctly deduced that the force required to maintain a constant velocity u of the upper layer was proportional to the area A and the velocity gradient or shear strain rate ($s = u/h$).

$$f = \eta A \frac{u}{h} \tag{4.2}$$

where η is the proportionality constant, the viscosity coefficient, or the absolute viscosity. Viscosity, absolute viscosity, and dynamic viscosity all mean the same as defined above. By rearranging Eq. (4.2), absolute viscosity can be written as

$$\eta = \frac{f/A}{u/h} = \frac{\text{shear stress}}{\text{shear strain rate}} \tag{4.3}$$

4.4 UNITS OF ABSOLUTE VISCOSITY

It follows from Eq. (4.3) that the units of viscosity must be the units of shear stress divided by the units of shear rate. The units of viscosity η for three different systems are

1. SI system: $N \cdot s/m^2$ or, since a newton per square meter is also called a pascal, $Pa \cdot s$
2. cgs system: $dyn \cdot s/cm^2$ (dyne-second per square centimeter) or poise, where $1\ cP = 10^{-2}\ P$
3. English system: $lbf \cdot s/in^2$ (pound-force-second per square inch), called a reyn in honor of Osborne Reynolds

Conversion of absolute viscosity from one system to another can be facilitated by Table 4.6. To convert from a unit in the column on the left side of the table to a unit at the top of the table, multiply by the corresponding value given in the table.

Sample problem 4.1

$$\eta = 0.04\ N \cdot s/m^2 = (0.04)1.45 \times 10^{-4}\ lbf \cdot s/in^2 = 5.8 \times 10^{-6}\ lbf \cdot s/in^2$$

Note also that

$$\eta = 0.04\ N \cdot s/m^2 = 0.04\ Pa \cdot s = 5.8 \times 10^{-6}\ \text{reyn}$$

and

$$\eta = 0.04\ N \cdot s/m^2 = (0.04)(10^3)cP = 40\ cP = 0.4\ P$$

TABLE 4.6 Viscosity conversion factors

To convert from–	To–			
	cP	kgf s/m^2	N s/m^2	lbf s/in.^{2}
	Multiply by–			
cP	1	1.02×10^{-4}	10^{-3}	1.45×10^{-7}
kgf s/m^2	9.807×10^{3}	1	9.807	1.422×10^{-3}
N s/m^2	10^{3}	1.02×10^{-1}	1	1.45×10^{-4}
lbf s/in.^{2}	6.9×10^{2}	7.03×10^{2}	6.9×10^{3}	1

4.5 KINEMATIC VISCOSITY

In many situations it is convenient to use kinematic viscosity rather than absolute viscosity. The kinematic viscosity η_k is defined as

$$\eta_k = \frac{\text{absolute viscosity}}{\text{force density}} = \frac{\eta}{\rho} = \frac{\mathrm{N\cdot s/m^2}}{\mathrm{N\cdot s^2/m^4}} = \mathrm{m^2/s} \tag{4.4}$$

The mass density ρ^* of a fluid is the mass of a unit volume of the fluid. The SI unit of mass density is kilogram mass per cubic meter $[(\mathrm{kg})_{\mathrm{mass}}/\mathrm{m}^3]$. A kilogram mass is equal to a kilogram force divided by gravitational acceleration. We can relate kilogram force to a newton so that the force density ρ is represented in SI units as newton-seconds squared per meter to the fourth power ($\mathrm{N\cdot s^2/m^4}$). More specifically,

$$\frac{(\mathrm{kg})_{\mathrm{mass}}}{\mathrm{m}^3} = \frac{\mathrm{N\cdot s^2}}{\mathrm{m}^4}$$

$$\therefore \rho^* = \rho \tag{4.5}$$

where ρ^* = mass density, $(\mathrm{kg})_{\mathrm{mass}}/\mathrm{m}^3$
ρ = force density, $\mathrm{N\cdot s^2/m^4}$

The ratio given in Eq. (4.4) is literally kinematic, all trace of force or mass canceling out. The units of kinematic viscosity are

1. SI units: square meters per second ($\mathrm{m^2/s}$)
2. cgs units: square centimeters per second ($\mathrm{cm^2/s}$), called a stoke (St)
3. English units: square inches per second ($\mathrm{in^2/s}$)

Table 4.7 shows the difference between absolute and kinematic viscosity with increasing temperature for two types of lubricating oil. Because of the decrease in density, the difference between absolute and kinematic viscosity increases with increasing temperature. Absolute viscosity is required for calculating elastohydrodynamic lubrication as related to rolling-element bearings and gears. However, kinematic viscosity can be determined experimentally more easily and with great precision and is therefore preferred for characterizing lubricants.

TABLE 4.7 Divergence between kinematic and absolute viscosity data with increasing temperature
[From Klaman (1984)]

Temperature t_m, C	Paraffinic base oil			Naphthenic base oil		
	Kinematic viscosity, η_k, mm^2/s	Absolute viscosity, η, mPa s	Viscosity difference, Δ, percent	Kinematic viscosity, η_k, mm^2/s	Absolute viscosity, η, mPa s	Viscosity difference, Δ, percent
0	287	253	13.4	1330	1245	6.8
20	78.4	68	15.3	218	201.0	8.5
40	30.2	25.8	17.1	60.5	55.0	10.0
60	14.7	12.33	19.2	23.6	21.2	11.3
80	8.33	6.91	20.5	11.6	16.2	13.7
100	5.3	4.32	22.7	6.66	5.80	14.8
120	3.65	2.93	24.6	4.27	3.66	16.7
150	2.33	1.83	27.3	2.53	2.12	19.3

4.6 VISCOSITY GRADE SYSTEM

The International Organization for Standardization (ISO) system is based on the kinematic viscosity of the oil, in centistokes, at 40°C. Each viscosity grade is numerically equal to the kinematic viscosity at the midpoint of the range. To fall into a given viscosity grade, the oil must be within 10 percent of the midpoint kinematic viscosity.

Figure 4.3 shows the various viscosity grades. The intent of this figure is to show how the various designations and the grades within them compare. The lowest grades are not presented because of space limitations. The Saybolt universal viscosity unit given in this figure is a commercial measure of kinematic viscosity expressed as the time in seconds required for 60 cm^3 of a fluid to flow through the orifice of the standard Saybolt universal viscometer at a temperature of 38°C.

Herschel in 1918 found that the kinematic viscosity in centistokes can be expressed in terms of a Saybolt second by the general formula

$$\eta_k = \mathscr{A} t - \frac{\mathscr{B}}{t}$$

where $\mathscr{A}$ and $\mathscr{B}$ are experimentally determined constants having units of centimeters squared per second squared and centimeters squared, respectively. In this expression t is expressed in Saybolt seconds. Values of $\mathscr{A}$ and $\mathscr{B}$ were obtained from experiments, and the preceding equation became

$$\eta_k = 0.22t - \frac{180}{t} \tag{4.6}$$

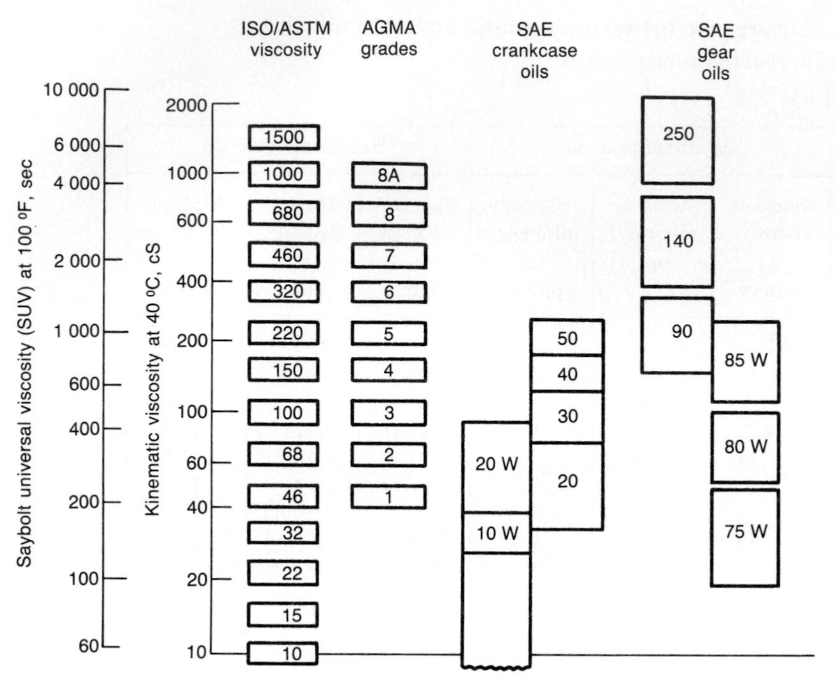

FIGURE 4.3
Viscosity grade comparisons. [*From Litt (1986).*]

This equation is helpful in converting Saybolt seconds into kinematic viscosity expressed in centistokes. The appropriate expression, if we know the kinematic viscosity and wish to know what the Saybolt seconds would be, can be determined from the equation

$$t = 2.27\left[\eta_k + \left(\eta_k^2 + 158.4\right)^{1/2}\right] \tag{4.7}$$

Equations (4.6) and (4.7) are only valid when the kinematic viscosity η_k is expressed in terms of centistokes and the time t in terms of Saybolt seconds.

Viscosity is the most important property of the lubricants employed in hydrodynamic and elastohydrodynamic lubrication. In general, however, a lubricant does not simply assume a uniform viscosity in a given bearing. This results from the nonuniformity of the pressure and/or the temperature prevailing in the lubricant film. Indeed, many elastohydrodynamically lubricated machine elements operate over ranges of pressure and/or temperature so extensive that the consequent variations in the viscosity of the lubricant may become substantial and, in turn, may dominate the operating characteristics of the machine element. Consequently, an adequate knowledge of the viscosity-pressure and viscosity-pressure-temperature relationships of lubricants is indispensable. The next three sections deal with such relationships.

4.7 VISCOSITY-PRESSURE EFFECTS

As long ago as 1893, Barus proposed the following formula for the isothermal viscosity-pressure dependence of liquids:

$$\ln \frac{\eta}{\eta_0} = \xi p \tag{4.8}$$

where ln = natural, or Napierian, logarithm, $\log_e$
η_0 = absolute viscosity at $p = 0$ and at a constant temperature, $\mathrm{N \cdot s/m^2}$
ξ = pressure-viscosity coefficient of the lubricant dependent on temperature, $\mathrm{m^2/N}$
p = pressure, $\mathrm{N/m^2}$

Table 4.8 lists the kinematic viscosities in square meters per second and absolute viscosities in centipoise of 11 lubricants at zero pressure and three temperatures. These values of the absolute viscosity correspond to η_0 in Eq. (4.8) for the particular fluid and temperature used. The manufacturer and the manufacturer's designation for these 11 lubricants are given in Table 4.9 and the pressure-viscosity coefficients ξ, expressed in square meters per newton, in Table 4.10. The values correspond to ξ used in Eq. (4.8).

Although Eq. (4.8) is extensively used, it is not generally applicable and is valid as a reasonable approximation only at moderate pressures. Because of this

TABLE 4.8 Absolute and kinematic viscosities of fluids at atmospheric pressure and three temperatures

[From Jones et al. (1975)]

Fluid	Temperature, t_m, °C					
	38	99	149	38	99	149
	Absolute viscosity at $p = 0$, η_0, cP			Kinematic viscosity at $p = 0$, η_k, m²/s		
Advanced ester	25.3	4.75	2.06	2.58×10^{-5}	0.51×10^{-5}	0.23×10^{-5}
Formulated advanced ester	27.6	4.96	2.15	2.82	.53	.24
Polyalkyl aromatic	25.5	4.08	1.80	3.0	.50	.23
Synthetic paraffinic oil (lot 3)	414	34.3	10.9	49.3	4.26	1.4
Synthetic paraffinic oil (lot 4)	375	34.7	10.1	44.7	4.04	1.3
Synthetic paraffinic oil (lot 2) plus antiwear additive	370	32.0	9.93	44.2	4.0	1.29
Synthetic paraffinic oil (lot 4) plus antiwear additive	375	34.7	10.1	44.7	4.04	1.3
C-ether	29.5	4.67	2.20	2.5	.41	.20
Superrefined naphthenic mineral oil	68.1	6.86	2.74	7.8	.82	.33
Synthetic hydrocarbon (traction fluid)	34.3	3.53	1.62	3.72	.40	.19
Fluorinated polyether	181	20.2	6.68	9.66	1.15	.4

TABLE 4.9 Fluids with manufacturer and manufacturer's designation
[From Jones et al. (1975)]

Fluid	Manufacturer	Designation
Advanced ester	Shell Oil Co.	Aeroshell turbine oil 555 (base oil)
Formulated advanced ester	Shell Oil Co.	Aeroshell turbine oil 555 (WRGL-358)
Polyalkyl aromatic	Continental Oil Co.	DN-600
Synthetic paraffinic oil (lot 3)	Mobil Oil Co.	XRM 109-F3
Synthetic paraffinic oil (lot 4)	↓	XRM 109-F4
Synthetic paraffinic oil (lot 2) plus antiwear additive	↓	XRM 177-F2
Synthetic paraffinic oil (lot 4) plus antiwear additive	↓	XRM 177-F4
C-ether	Monsanto Co.	MCS-418
Superrefined naphthenic mineral oil	Humble Oil and Refining Co.	FN 2961
Synthetic hydrocarbon (traction fluid)	Monsanto Co.	MCS-460
Fluorinated polyether	DuPont Co.	PR 143 AB (lot 10)

shortcoming of Eq. (4.8), several isothermal viscosity-pressure formulas have been proposed that usually fit experimental data better than that suggested by Barus (1893). One of these approaches, which is used in this book, was developed by Roelands (1966), who undertook a wide-ranging study of the effect of pressure on lubricant viscosity. For isothermal conditions the Roelands (1966) formula can be written as

$$\log \eta + 1.200 = (\log \eta_0 + 1.200)\left(1 + \frac{p}{2000}\right)^{Z_1} \tag{4.9}$$

where log = common, or Briggsian, logarithm, $\log_{10}$

η = absolute viscosity, cP

η_0 = absolute viscosity at $p = 0$ and a constant temperature, cP

p = gage pressure, $(\mathrm{kg})_{\mathrm{force}}/\mathrm{cm}^2$

Z_1 = viscosity-pressure index, a dimensionless constant

Taking the antilog of both sides of Eq. (4.9) and rearranging terms gives

$$\bar{\eta} = \frac{\eta}{\eta_0} = 10^{-(1.2+\log \eta_0)[1-(1+p/2000)^{Z_1}]}$$

TABLE 4.10 Pressure-viscosity coefficients for fluids at three temperatures
[From Jones et al. (1975)]

Fluid	Temperature, t_m, °C		
	38	99	149
	Pressure-viscosity coefficient, ξ, m^2/N		
Advanced ester	1.28×10^{-8}	0.987×10^{-8}	0.851×10^{-8}
Formulated advanced ester	1.37	1.00	.874
Polyalkyl aromatic	1.58	1.25	1.01
Synthetic paraffinic oil (lot 3)	1.77	1.51	1.09
Synthetic paraffinic oil (lot 4)	1.99	1.51	1.29
Synthetic paraffinic oil (lot 2) plus antiwear additive	1.81	1.37	1.13
Synthetic paraffinic oil (lot 4) plus antiwear additive	1.96	1.55	1.25
C-ether	1.80	.980	.795
Superrefined naphthenic mineral oil	2.51	1.54	1.27
Synthetic hydrocarbon (traction fluid)	3.12	1.71	.939
Fluorinated polyether	4.17	3.24	3.02

Rearranging this equation gives

$$\bar{\eta} = \frac{\eta}{\eta_0} = \left(\frac{\eta_\infty}{\eta_0}\right)^{1-(1+p/c_p)^{Z_1}} \tag{4.10}$$

where $\eta_\infty = 6.31 \times 10^{-5}\ \mathrm{N \cdot s/m^2}\ (9.15 \times 10^{-9}\ \mathrm{lbf \cdot s/in^2})$

$c_p = 1.96 \times 10^8\ \mathrm{N/m^2}\ (28{,}440\ \mathrm{lbf/in^2})$

In Eq. (4.10) care must be taken to use the same dimensions in defining the constants η_∞ and c_p as are used for η_0 and p.

Figure 4.4 compares the Barus equation [Eq. (4.8)] with the Roelands equation [Eq. (4.10)] for three different fluids presented in Tables 4.8 to 4.10. The temperature is fixed at 38°C. This figure indicates that for pressures normally experienced in elastohydrodynamically lubricated conjunctions ($\sim$ 1 GPa), the difference between the two formulas is appreciable. Figure 4.4 shows that the viscosity rises more rapidly for the Barus formula than for the Roelands formula for oils 1 and 2, but for oil 3 the Roelands formula rises more rapidly than the Barus formula. The difference between the formulas varies with the lubricant, the difference being less for the traction fluid. When the pressure approaches zero, the two formulas produce the same viscosity.

Table 4.11 gives values of the viscosity-pressure index Z_1 as obtained from Jones et al. (1975) for the same lubricants considered in Tables 4.8 to 4.10. Roelands (1966) found that for most fluids Z_1 is usually constant over a wide temperature range. This is confirmed in Table 4.11, the only exceptions being the synthetic hydrocarbon (traction fluid) and the C-ether.

Blok (1965) arrived at the important conclusion that all elastohydrodynamic lubrication results achieved from the Barus formulas [Eq. (4.8)] can be generalized, to a fair approximation, simply by substituting the reciprocal of the asymptotic isoviscous pressure $1/p_{\text{iv, as}}$ for the pressure-viscosity coefficient ξ occurring in those results. This implies

$$\xi \approx \frac{1}{p_{\text{iv, as}}}$$

where

$$p_{\text{iv, as}} = \eta_0 \int_0^\infty \frac{dp}{\eta} \tag{4.11}$$

Substituting Eq. (4.8) into (4.11) quickly proves that $\xi = 1/p_{\text{iv, as}}$.

The preceding can be used to find a tie between the pressure-viscosity coefficient ξ used in the Barus formula [Eq. (4.8)] and the viscosity-pressure index Z_1 used in the Roelands formula [Eq. (4.10)]. Making use of the fact that the two different formulas approach each other as the pressure approaches zero gives

$$\frac{\partial}{\partial p}(\ln \eta|_{\text{Roelands}})_{p \to 0} \equiv \frac{\partial}{\partial p}(\ln \eta|_{\text{Barus}})_{p \to 0} = \xi$$

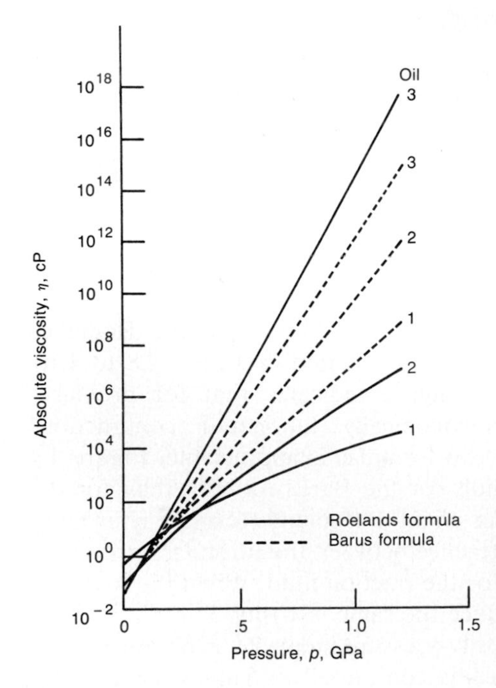

FIGURE 4.4
Comparison of absolute viscosity obtained from Barus' and Roelands' formulas for a wide range of pressure. Results are shown for three different lubricants at 38°C: oil 1—synthetic paraffinic oil (lot 3); oil 2—superrefined naphthenic mineral oil; oil 3—synthetic hydrocarbon (traction fluid).

TABLE 4.11 Viscosity-pressure index for fluids at three temperatures

Fluid	Temperature, t_m, °C					
	38	99	149	38	99	149
	Dimensionless viscosity-pressure index, Z_1,					
	From Jones et al. (1975)			From equation (4–12)		
Advanced ester	0.48	0.48	0.48	0.42	0.45	0.48
Formulated advanced ester	.49	.47	.49	.44	.45	.49
Polyalkyl aromatic	.55	.54	.55	.52	.59	.59
Synthetic paraffinic oil (lot 3)	.43	.44	.39	.40	.47	.42
Synthetic paraffinic oil (lot 4)	.44	.46	.47	.45	.47	.50
Synthetic paraffinic oil (lot 2) plus antiwear additive	.43	.44	.43	.41	.43	.44
Synthetic paraffinic oil (lot 4) plus antiwear additive	.44	.46	.46	.44	.48	.48
C-ether	.72	.50	.50	.57	.45	.44
Superrefined naphthenic mineral oil	.67	.67	.64	.71	.64	.66
Synthetic hydrocarbon (traction fluid)	1.06	.85	.69	.97	.83	.57
Fluorinated polyether	.77	.79	.80	1.03	1.10	1.27

Making use of Eq. (4.10) gives

$$Z_1 = \frac{\xi}{(1/c_p)(\ln \eta_0 - \ln \eta_\infty)} \tag{4.12}$$

But

$$\frac{1}{c_p} = \frac{1}{1.96 \times 10^8} \text{ m}^2/\text{N} = 5.1 \times 10^{-9} \text{ m}^2/\text{N}$$

$$\ln \eta_\infty = \ln\left(6.31 \times 10^{-5} \text{ N} \cdot \text{s/m}^2\right) = -9.67$$

$$\therefore Z_1 = \frac{\xi}{5.1 \times 10^{-9}(\ln \eta_0 + 9.67)}$$

This equation is unit sensitive and only applicable for SI units. Equation (4.12), which is general, thus enables one to quickly determine the viscosity-pressure index directly from knowing the pressure-viscosity coefficient, or vice versa.

Table 4.11 also compares the viscosity-pressure indexes as obtained from Jones et al. (1975) and from Eq. (4.12) for the same fluids considered in Tables 4.8 to 4.10. The agreement between the two methods of obtaining Z_1 is quite good, the exception being the fluorinated polyether. Thus, Eq. (4.12) can be

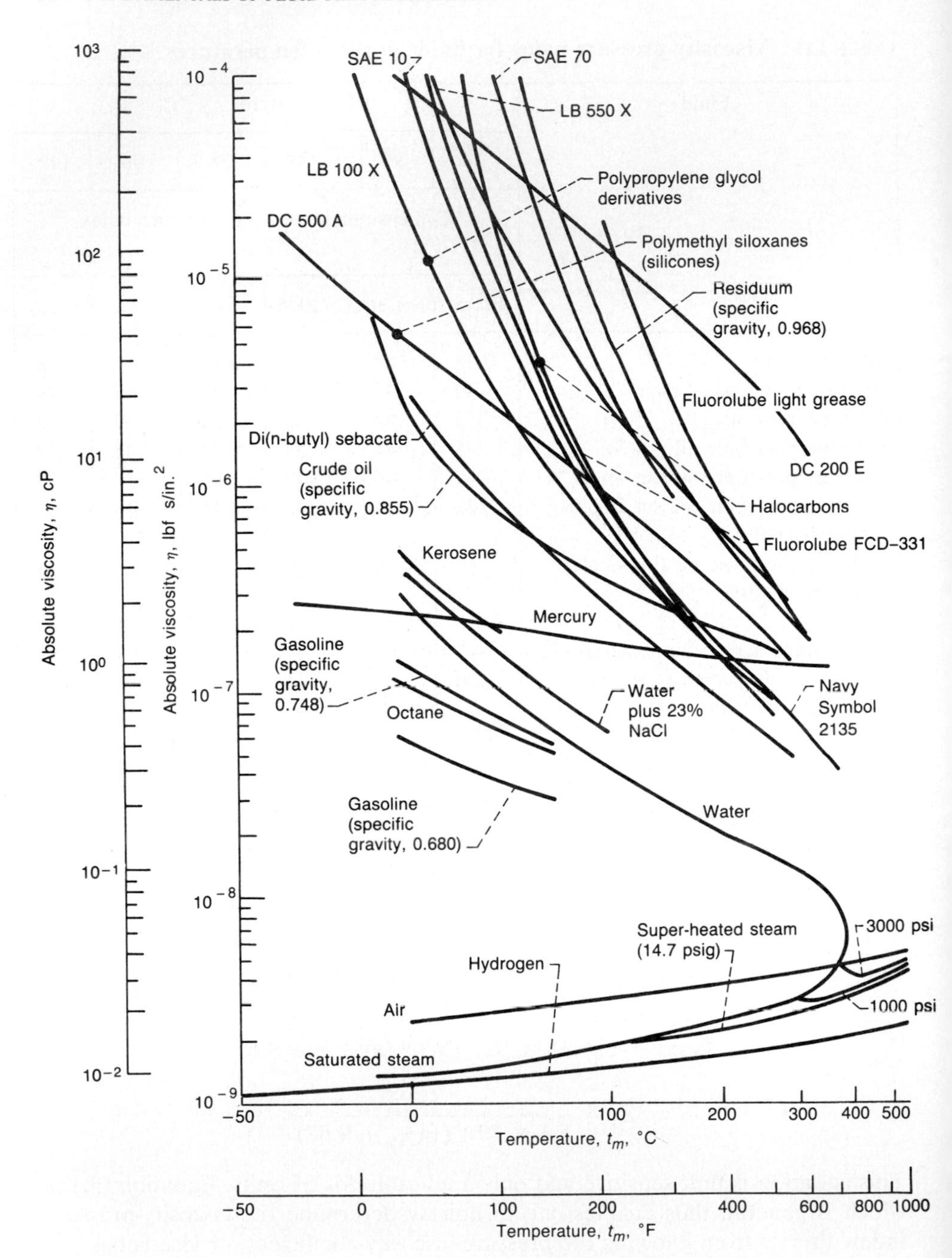

FIGURE 4.5
Absolute viscosities of a number of fluids for a wide range of temperatures.

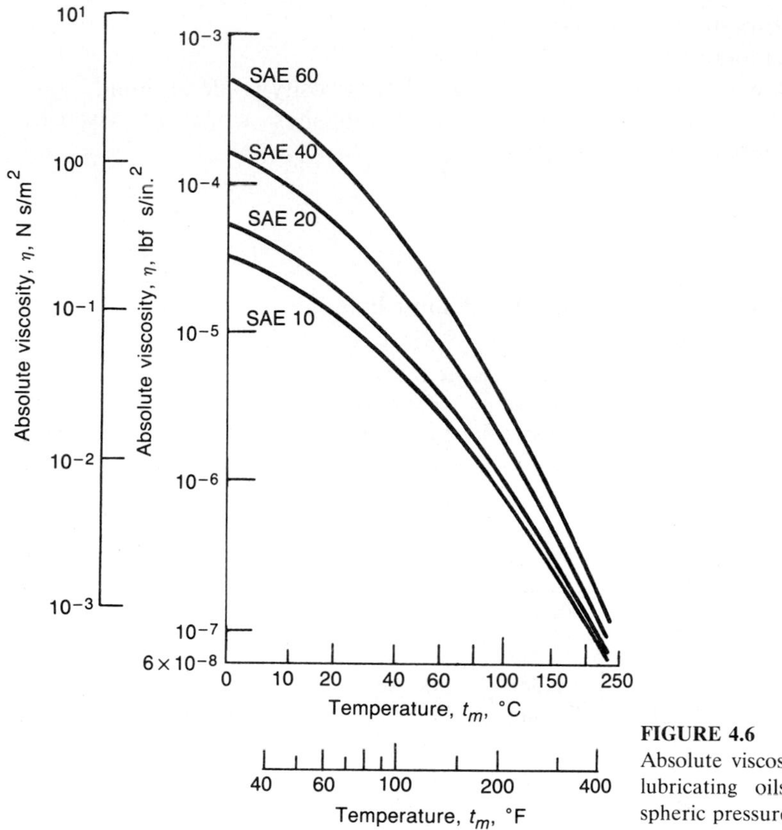

FIGURE 4.6
Absolute viscosities of SAE lubricating oils at atmospheric pressure.

useful for relating the viscosity as obtained from the Roelands and Barus formulas.

4.8 VISCOSITY-TEMPERATURE EFFECTS

The viscosity of mineral and synthetic oils decreases with increasing temperature. Therefore, the temperature at which the viscosity was measured must be quoted with every viscosity reported. Figures 4.5 and 4.6 show how absolute viscosity varies with temperature. The absolute viscosity of a number of different fluids for a wide range of temperatures is presented in Fig. 4.5. The interesting point of this figure is how drastically the slope and the level of viscosity change for different fluids. The viscosity varies five orders of magnitude, with the slope being highly negative for the SAE oils and positive for gases. Figure 4.6 gives the viscosity-temperature effect for the SAE oils.

The expression used by Roelands (1966) to describe the effect of temperature on viscosity is given as

$$\log(\log \eta + 1.200) = -S_0 \log\left(1 + \frac{t_m}{135}\right) + \log G_0 \tag{4.13}$$

where η = absolute viscosity, cP
t_m = temperature, °C
G_0 = dimensionless constant indicative of viscosity grade of liquid
S_0 = dimensionless constant that establishes slope of viscosity-temperature relationship

TABLE 4.12 Typical thermal properties of some liquids
[From Winer and Cheng (1980)]

Temperature, t_m, °C	Mass density, ρ^*, kg/m³	Specific heat, C_p, kJ/kg °C	Kinematic viscosity, $\eta_k = \eta/\rho$, m²/s	Thermal conductivity, K_f, W/m °C	Thermal diffusivity, $\alpha_t = K_f/\rho^* C_p$, m²/s
Glycerin ($C_3H_5(OH)_3$)					
0	1 276	2.261	0.00831	0.282	0.983×10^{-7}
10	1 270	2.319	.00300	.284	.965
20	1 264	2.386	.00118	.286	.947
30	1 258	2.445	.00050	.286	.929
40	1 252	2.512	.00022	.286	.914
50	1 244	2.583	.00015	.287	.893
Ethylene glycol ($C_2H_4(OH)_2$)					
0	1 130	2.294	57.53×10^{-6}	0.242	0.934×10^{-7}
20	1 116	2.382	19.18	.249	.939
40	1 101	2.474	8.69	.256	.939
60	1 087	2.562	4.75	.260	.932
80	1 077	2.650	2.98	.261	.921
100	1 058	2.742	2.03	.263	.908
Engine oil (unused)[a]					
0	899	1.796	0.00428	0.147	0.911×10^{-7}
20	888	1.880	.00090	.145	.872
40	876	1.964	.00024	.144	.834
60	864	2.047	$.839 \times 10^{-4}$	.140	.800
80	852	2.131	.375	.138	.769
100	840	2.219	.203	.137	.738
120	828	2.307	.124	.135	.710
140	816	2.395	.080	.133	.686
160	805	2.483	.056	.132	.663

TABLE 4.12 *Concluded*

Temperature, t_m, °C	Mass density, ρ^*, kg/m^3	Specific heat, C_p, kJ/kg °C	Kinematic viscosity, $\eta_k = \eta/\rho$, m^2/s	Thermal conductivity, K_f, W/m °C	Thermal diffusivity, $\alpha_t = K_f/\rho^* C_p$, m^2/s
Mercury (Hg)					
0	13 628	0.1403	0.1240×10^{-6}	8.20	42.99×10^{-7}
20	13 579	.1394	.1140	8.69	46.06
50	13 505	.1386	.1040	9.40	50.22
100	13 384	.1373	.0928	10.51	57.16
150	13 264	.1365	.0853	11.49	63.54
200	13 144	.1570	.0802	12.34	69.08
250	13 025	.1357	.0765	13.07	74.06
315.5	12 847	.1340	.0673	14.02	81.50
Diester					
30	910	1.93	-------------	0.151	0.860×10^{-7}
Phosphate ester					
30	1 060	1.76	-------------	0.125	0.670×10^{-7}
Polyglycol					
30	1 000	1.97	-------------	0.152	0.772×10^{-7}
Polyphenylether					
30	1 180	1.80	-------------	0.132	0.621×10^{-7}
Dimethyl silicone					
30	970	1.42	-------------	0.142	1.03×10^{-7}
Chlorofluorocarbon					
30	1 900	1.22	-------------	0.069	0.298×10^{-7}
Fluorinated polyether					
30	1 870	0.96	-------------	0.093	0.518×10^{-7}

[a]The viscosity values should only be used if no other information on the particular lubricant is available. The thermal properties ($\rho^*, C_p, K_f, \alpha_t$) should be representative of most mineral oils.

Taking the antilog of Eq. (4.13) gives

$$\log \eta + 1.2 = G_0 \times 10^{-S_0 \log(1+t_m/135)} \tag{4.14}$$

Equation (4.14) can be expressed in dimensionless terms as

$$\bar{\eta} = \frac{\eta}{\eta_0} = \left(\frac{\eta_\infty}{\eta_0}\right) 10^{G_0(1+t_m/135)^{-S_0}} \tag{4.15}$$

where

$$\eta_\infty = 6.31 \times 10^{-5}\ \mathrm{N \cdot s/m^2}\ (9.15 \times 10^{-9}\ \mathrm{lbf \cdot s/in^2})$$

Besides the variations of viscosity with temperature, a number of other thermal properties of fluids are important in lubrication. Some of these are specific heat, volumetric specific heat, thermal conductivity, and thermal diffusivity. The volumetric specific heat C_s is defined as the mass density multiplied by the specific heat ($C_s = \rho^* C_p$) and therefore has the units of kilojoules per degree Celsius per cubic meter ($\mathrm{kJ/°C \cdot m^3}$). Winer and Cheng (1980) point out that C_s is relatively constant for classes of fluids based on chemical composition.

Winer and Cheng (1980) also point out that the thermal conductivity K_f, like the volumetric specific heat, is relatively constant for classes of lubricants based on chemical composition. For mineral oils the thermal conductivity is between 0.12 and 0.15 W/m · °C. Recall that a watt is equivalent to a joule per second (W = J/s), a joule is equivalent to a newton-meter (J = N · m), and a newton is equivalent to a kilogram-meter per second squared ($\mathrm{N = kg \cdot m/s^2}$). Typical values of thermal properties of other fluids are shown in Table 4.12. In this table the thermal diffusivity is defined as $K_f/\rho^* C_p$.

4.9 VISCOSITY-PRESSURE-TEMPERATURE EFFECTS

Viscosity is extremely sensitive to both pressure and temperature. This extreme sensitivity forms a considerable obstacle to the analytical description of the consequent viscosity changes. Roelands (1966) noted that at constant pressure the viscosity increases more or less exponentially with the reciprocal of absolute temperature. Similarly, at constant temperature the viscosity increases more or less exponentially with pressure as shown earlier in this chapter. In general, however, the relevant exponential relationships constitute only first approximations and may be resorted to only in moderate-temperature ranges.

From Roelands (1966) the viscosity-temperature-pressure equation can be written as

$$\log \eta + 1.200 = G_0 \frac{(1 + p/2000)^{-C_2 \log(1+t_m/135)+D_2}}{(1 + t_m/135)^{S_0}} \tag{4.16}$$

Comparing Eqs. (4.9) and (4.14) with the preceding equation reveals the contribution of the pressure and temperature effects that are contained in Eq.

(4.16). Equation (4.16) can also be expressed as

$$\bar{\eta} = \frac{\eta}{\eta_0} = \frac{\eta_\infty}{\eta_0} 10^{G_0(1+t_m/135)^{-S_0}(1+p/2000)^{D_2-C_2 \log(1+t_m/135)}} \tag{4.17}$$

According to Eq. (4.17), four parameters (G_0, S_0, C_2, and D_2) that are determined experimentally are sufficient to enable the viscosity η to be expressed in centipoise as a function of temperature t_m in degrees Celsius and gage pressure p in kilograms per square centimeter.

4.10 VISCOSITY–SHEAR RATE EFFECTS

Liquids whose viscosities are independent of the shear rate are known as "Newtonian." Liquids whose viscosities vary with shear rate are known as "non-Newtonian." Figure 4.7 shows viscosity and shear stress versus shear rate for Newtonian and non-Newtonian fluids. In Fig. 4.7(a) the pseudoplastic fluids show a decrease in viscosity with increasing shear rate. This behavior may be restricted to ranges of shear rate. The dilatant fluid shows an increase in viscosity with increasing shear rate.

In Fig. 4.7(b) the shear stress is shown as a function of shear rate for a Newtonian, a pseudoplastic, and a dilatant fluid and a Bingham solid. A Bingham solid is a plastic solid such as a grease that flows only above a certain yield shear stress. The pseudoplastic fluids are characterized by linearity at extremely low and extremely high shear rates. The dilatant fluid exhibits an increase in apparent viscosity with increasing shear rate. An explanation of this behavior in the case of suspensions is that particles in a concentrated suspension will be oriented at rest so that the void space is a minimum. The suspending liquid is just sufficient to fill the voids in this state. The increase in voidage caused by shearing a dilatant material means that the space between particles becomes incompletely filled with liquid. Under these conditions of

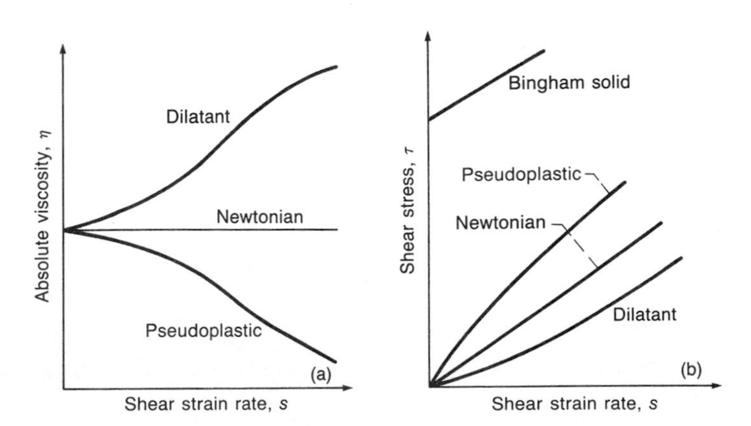

FIGURE 4.7
Characteristics of different fluids as a function of shear rate. (a) Viscosity curve; (b) flow curve.

inadequate lubrication the surfaces of adjacent particles come into direct contact, causing an increase in apparent viscosity with increasing shear rate.

4.11 VISCOSITY INDEX

To better evaluate the relationship between viscosity and temperature, Dean and Davis (1929) developed an arbitrary system of comparison called a "viscosity index" (VI). The standard of comparison is based on Pennsylvania oils refined by the sulfuric acid method and on Texas or California oils refined by the same process. The Pennsylvania oils thinned (became less viscous) less rapidly than other types of mineral oils when subjected to increases in temperature. Those oils were rated 100 for this property. The Texas naphthenic oils were rated 0.

Although oils have changed considerably since 1929, the interest in expressing the relative change of viscosity with temperature has not. As far as evaluating between two limits is concerned, there is no question as to the sense and meaning of the viscosity index. A viscosity index of 75 means that the quality of an oil with regard to viscosity has reached 75 percent of the range from the poorest naphthenic oil (VI = 0) to the best Pennsylvania oil (VI = 100). When the limiting oils were selected, no allowance was made for development in manufacturing methods and the introduction of synthetic lubricants. As a result, lubricants are produced today with viscosity indices that are considerably higher than 100. It is nevertheless customary to compute the viscosity index

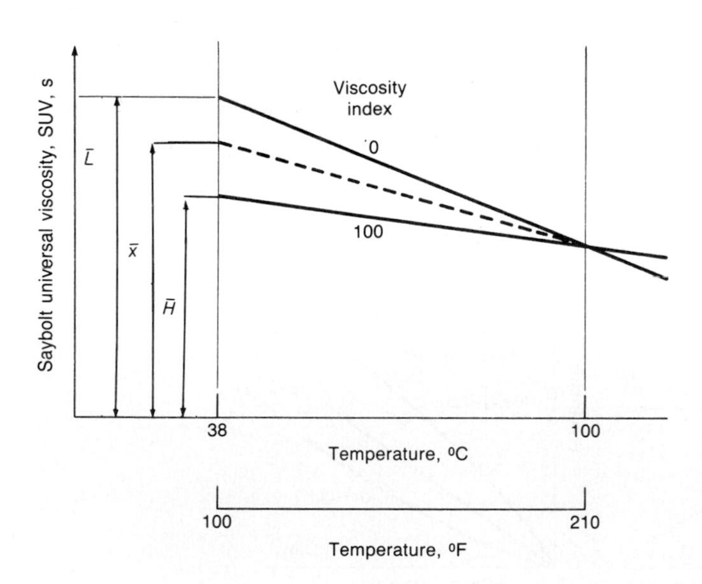

FIGURE 4.8
Graphical explanation of viscosity index where $\overline{L}$ = low VI oil, $\bar{x}$ = unknown oil, and $\overline{H}$ = high VI oil.

TABLE 4.13 Viscosity-index data to be used in Eq. (4.18)

[An abridgement from ASTM D567, "Standard Method for Calculating Viscosity Index"]

SUV at 100 °C, $\bar{x}$	SUV at 38 °C		SUV at 100 °C, $\bar{x}$	SUV at 38 °C		SUV at 100 °C, $\bar{x}$	SUV at 38 °C	
	$\bar{H}$	$\bar{L}$		$\bar{H}$	$\bar{L}$		$\bar{H}$	$\bar{L}$
40	93	107	---	----	----	---	----	----
41	109	137	81	810	1674	121	1643	3902
42	124	167	82	829	1721	122	1665	3966
43	140	197	83	849	1769	123	1688	4031
44	157	228	84	868	1817	124	1710	4097
45	173	261	85	888	1865	125	1733	4163
46	189	291	86	907	1914	126	1756	4229
47	205	325	87	927	1964	127	1779	4296
48	222	356	88	947	2014	128	1802	4363
49	238	389	89	966	2064	129	1825	4430
50	255	422	90	986	2115	130	1848	4498
51	272	456	91	1006	2166	131	1871	4567
52	288	491	92	1026	2217	132	1894	4636
53	305	525	93	1046	2270	133	1918	4705
54	322	561	94	1066	2322	134	1941	4775
55	339	596	95	1087	2375	135	1965	4845
56	356	632	96	1107	2428	136	1988	4915
57	374	669	97	1128	2481	137	2012	4986
58	391	706	98	1148	2536	138	2036	5058
59	408	743	99	1168	2591	139	2060	5130
60	426	481	100	1189	2646	140	2084	5202
61	443	819	101	1210	2701	141	2108	5275
62	461	857	102	1231	2757	142	2132	5348
63	478	897	103	1252	2814	143	2156	5422
64	496	936	104	1273	2870	144	2180	5496
65	514	976	105	1294	2928	145	2205	5570
66	532	1016	106	1315	2985	146	2229	5645
67	550	1057	107	1337	3043	147	2254	5721
68	568	1098	108	1358	3102	148	2278	5796
69	586	1140	109	1379	3161	149	2303	5873
70	604	1182	110	1401	3220	150	2328	5949
71	623	1225	111	1422	3280	151	2353	6026
72	641	1268	112	1444	3340	152	2378	6104
73	660	1311	113	1466	3400	153	2403	6182
74	678	1355	114	1488	3462	154	2428	6260
75	697	1399	115	1510	3524	155	2453	6339
76	716	1444	116	1532	3585	156	2478	6418
77	734	1489	117	1554	3648	157	2503	6498
78	753	1534	118	1576	3711	158	2529	6578
79	772	1580	119	1598	3774	159	2554	6659
80	791	1627	120	1620	3838	160	2580	6740

according to the formula

$$\text{VI}\% = \left(\frac{\overline{L} - \bar{x}}{\overline{L} - \overline{H}}\right) 100 \tag{4.18}$$

where $\overline{L}$ = Saybolt universal viscosity (SUV) of reference oil of low VI
$\overline{H}$ = SUV of reference oil of high VI
$\bar{x}$ = SUV of unknown oil

Note that the values of $\overline{L}$, $\overline{H}$, and $\bar{x}$ are assumed to be equal at 100°C. Equation (4.18) is expressed graphically in Fig. 4.8.

Figure 4.8 shows how the viscosity index is obtained. The figure shows that the two reference oils and the unknown oil have the same Saybolt viscosity at 100°C. At 38°C the Saybolt viscosities are different, and these values of $\overline{H}$, $\overline{L}$, and $\bar{x}$ are then used in Eq. (4.18).

Table 4.13 gives tabular values of $\overline{H}$ and $\overline{L}$ at 38°C for given values of the Saybolt universal viscosity of the unknown oil at 100°C. These values are to be used in Eq. (4.18).

Sample problem 4.2 Suppose that an unknown oil has a Saybolt universal viscosity of 2730 SUV at 38°C and 120 SUV at 100°C. What is the VI of the oil?

Solution. From the first column of Table 4.13, find the two oils that will have a Saybolt universal viscosity of 120 SUV at 100°C. Therefore,

$$\overline{H} = 1620 \text{ SUV at } 38°\text{C}$$

$$\overline{L} = 3838 \text{ SUV at } 38°\text{C}$$

Thus, making use of Eq. (4.18) for $\bar{x}$ = 2730 SUV gives

$$\text{VI} = \left(\frac{3838 - 2730}{3838 - 1620}\right) 100 = 50$$

4.12 OXIDATION STABILITY

The chemical reaction whereby oxygen in air combines with hydrocarbon in an oil is called “oxidation.” Straight mineral oils possess a certain resistance to oxidation during the early stages of service, but subsequently deterioration due to oxidation tends to accelerate. Operating conditions where oxidation can occur are

1. High temperature
2. Presence of metallic wear particles
3. Presence of moisture and other contaminants, such as sludge, dirt, rust, and other corrosion products
4. Churning and agitation

Oxidation of straight mineral oils proceeds very slowly at room temperature, at 140°F (60°C) oxidation is still slow but significant, and above 200°F (93°C) it is greatly accelerated.

Oil oxidation products are undesirable for the following reasons: Insoluble products (sludge) may prevent effective lubrication due to clogging. Soluble products circulating with the oil tend to be acidic and eventually either lead to corrosion or pitting of bearing surfaces or form varnish deposits on parts operating at high temperatures.

4.13 POUR POINT

The pour point is the lowest temperature at which a lubricant can be observed to flow under specified conditions. Pour point is related to viscosity, since its concern is if the oil will flow at low temperatures or just barely flow under prescribed conditions.

Oils used under low-temperature conditions must have low pour points. Oils must have pour points (1) below the minimum operating temperature of the system and (2) below the minimum surrounding temperature to which the oil will be exposed.

4.14 DENSITY

The mass density ρ^* of a fluid is the mass of a unit volume of the fluid. The SI unit of mass density is kilogram mass per cubic meter. The effects of temperature on viscosity were found in Sec. 4.8 to be important. For a comparable change in pressure, temperature, or both, the density change is small relative to the viscosity change. However, extremely high pressure exists in elastohydrodynamic films, and the lubricant can no longer be considered as an incompressible medium. It is therefore necessary to consider the dependence of density on pressure.

The variation of density with pressure is roughly linear at low pressures, but the rate of increase falls off at high pressures. The limit of the compression of mineral oils is only 25 percent, for a maximum density increase of about 33 percent. From Dowson and Higginson (1966) the dimensionless density for mineral oil can be written as

$$\bar{\rho} = \frac{\rho}{\rho_0} = 1 + \frac{0.6p}{1 + 1.7p} \tag{4.19}$$

where ρ_0 = density when $p = 0$, $\mathrm{N \cdot s^2/m^4}$
p = gage pressure, GPa

Therefore, the general expression for the dimensionless density can be written as

$$\bar{\rho} = 1 + \frac{0.6E'P}{1 + 1.7E'P} \tag{4.20}$$

where $E' = 2\left(\dfrac{1 - \nu_a^2}{E_a} + \dfrac{1 - \nu_b^2}{E_b}\right)^{-1}$

E = modulus of elasticity, Pa
ν = Poisson's ratio

and where E' in Eq. (4.20) must be expressed in gigapascals and $P = p/E'$.

TABLE 4.14 Base fluids tested, with corresponding kinematic viscosity and average molecular weight
[From Hamrock et al. (1987)]

Base fluid	Kinematic viscosity, at 40 °C, η_k, mm²/s	Average molecular weight
Naphthenic distillate	26	300
Naphthenic raffinate	23	320
Polypropylene glycol 1	175	2000
Polypropylene glycol 2	80	2000
Ditridecyl adipate	26	510
Poly alpha olefin	450	500

Equation (4.20) states that the variation of density with pressure is roughly linear at low pressures but that its rate of increase falls off at high pressures. The maximum density increase from atmospheric pressure is 35 percent. The data used in obtaining Eq. (4.20) were restricted to relatively low pressures (< 0.4 GPa).

Recently, Hamrock et al. (1987) obtained experimental data showing the effect of pressure on density for a range of pressures from 0.4 to 2.2 GPa and for six base fluids at an assumed constant temperature of 20°C. The six base fluids tested are listed in Table 4.14 along with corresponding kinematic viscosity and average molecular weight. An important parameter used to describe the results is the change of relative volume with change of pressure, dV_r/dp. For pressures less than the solidification pressure ($p < p_s$), a small change in pressure results in a large change in dV_r/dp. For pressures greater than the solidification pressure ($p > p_s$), there is little change in dV_r/dp. Once the molecules of the lubricant become closely packed, increasing the pressure fails to alter the value of dV_r/dp.

In elastohydrodynamic lubrication, which is covered more fully later in the text, the rate of pressure increase is extremely high, typically 10^{13} Pa/s. The lubricant under these conditions will not have time to crystallize but will be compressed to an amorphous solid. Physically, this means that as the lubricant is compressed, the distance between the molecules of the lubricant becomes smaller and smaller. There exists a point where the molecules are not free to move and any further compression will result in deformation of the molecules. The pressure where this first starts to occur is the solidification pressure, which varies considerably for the different lubricants.

Figure 4.9 shows change in relative volume with changing pressure for six base fluids. The solidification pressure varies considerably for the different base

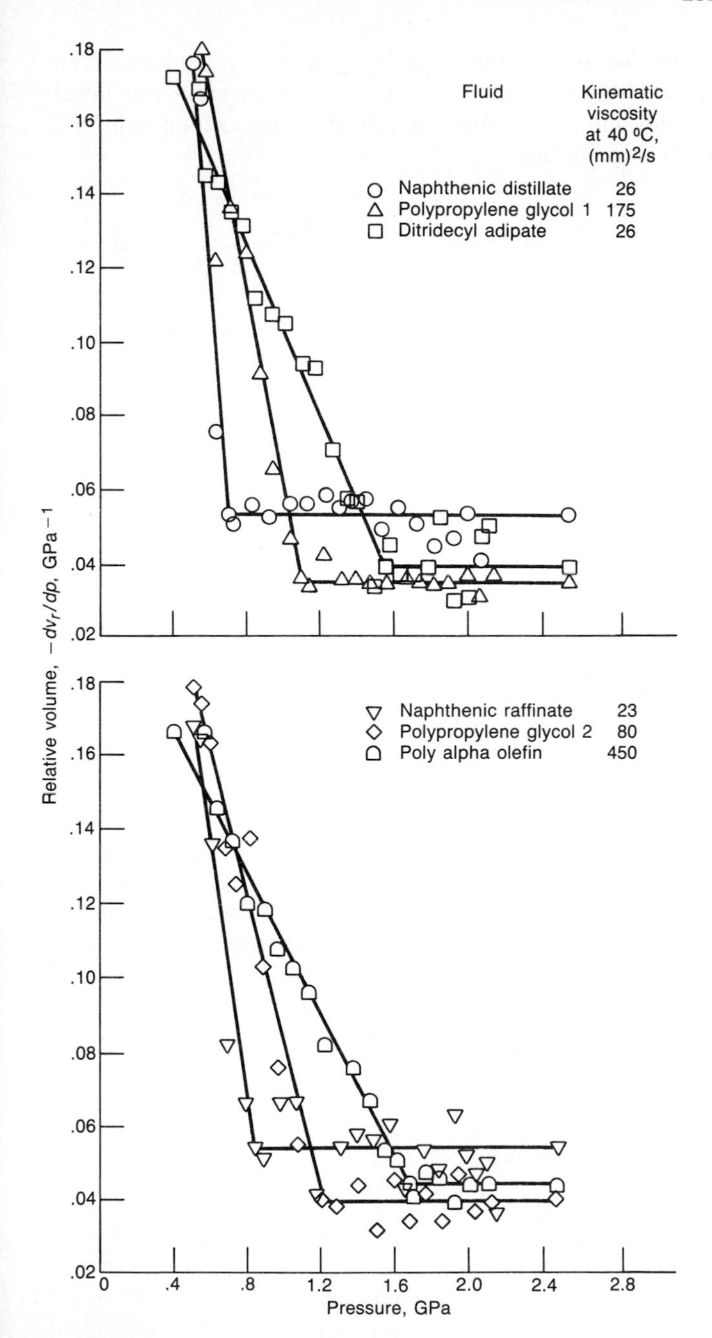

FIGURE 4.9
Effect of pressure on relative volume for six base fluids. Constant temperature of 20°C assumed. [*From Hamrock et al. (1987).*]

fluids tested. Furthermore, for $p < p_s$ the experimental data assume different slopes for the different base fluids. A new pressure-density formula was developed by Hamrock et al. (1987) that describes the effect of pressure on density in terms of four constants. It is given here.

$$\bar{\rho} = \begin{cases} \dfrac{1}{1 - C_1 p^2 - C_2 p} & \text{for } p \le p_s \quad (4.21) \\[2ex] \dfrac{1}{1 - C_3 p + C_4} & \text{for } p > p_s \quad (4.22) \end{cases}$$

$$C_1 = \frac{\bar{m}}{2C} \qquad (\text{GPa})^{-2} \tag{4.23}$$

$$C_2 = \frac{n_2 - \bar{m}p_s}{C} \qquad (\text{GPa})^{-1} \tag{4.24}$$

$$C_3 = \frac{n_2}{C} \qquad (\text{GPa})^{-1} \tag{4.25}$$

$$C_4 = \frac{\bar{m}p_s^2}{2C} \tag{4.26}$$

where

$$C = 1 + \frac{\bar{m}}{2}(\bar{p}_i)^2 + (n_2 - \bar{m}p_s)\bar{p}_i \tag{4.27}$$

and $\bar{p}_i$ is the initial pressure in gigapascals. Experimentally obtained values of $\bar{m}$, n_2, and p_s for each base fluid are given in Table 4.15; Table 4.16 gives corresponding values of C_1, C_2, C_3, and C_4.

Hamrock et al. (1987) used their derived formula in obtaining Fig. 4.10, which shows the effect of pressure on density for two of the fluids. Also in this figure is the pressure-density curve obtained from Dowson and Higginson

TABLE 4.15 Parameters obtained from least-squares fit of experimental data

[From Hamrock et al. (1987)]

Base fluid	Slope, $\bar{m}$, $(\text{GPa})^{-2}$	Asymptote, n_2, $(\text{GPa})^{-1}$	Solidification pressure, p_s, GPa
Naphthenic distillate	−0.626	0.0538	0.706
Naphthenic raffinate	−.336	.0542	.839
Polypropylene glycol 1	−.271	.0360	1.092
Polypropylene glycol 2	−.195	.0395	1.213
Ditridecyl adipate	−.115	.0395	1.561
Poly alpha olefin	−.0958	.0439	1.682

TABLE 4.16
Constants used in defining effect of pressure on density
[From Hamrock et al. (1987)]

Base fluid	Pressure-density constants			
	C_1, $(\mathrm{GPa})^{-2}$	C_2, $(\mathrm{GPa})^{-1}$	C_3, $(\mathrm{GPa})^{-1}$	C_4
Naphthenic distillate	−0.271	0.430	0.0466	−0.135
Naphthenic raffinate	−.151	.302	.0487	−.106
Polypropylene glycol 1	−.121	.297	.0323	−.145
Polypropylene glycol 2	−.0887	.251	.0395	−.131
Ditridecyl adipate	−.0531	.202	.0365	−.129
Poly alpha olefin	−.0444	.190	.0407	−.126

(1966). As the pressures increased above 2 GPa, the deviation between the Dowson and Higginson (1966) formula and the present results also increased. For pressures to 2 GPa for the naphthenic distillate there was good agreement with the Dowson and Higginson (1966) formula. The same cannot bc said for the poly alpha olefin. It is anticipated that these deviations from the normally used Dowson and Higginson pressure-density formula will significantly influence the definition of the pressure profile in elastohydrodynamically lubricated conjunctions.

Furthermore, it can be observed from Fig. 4.10 that for fluids other than mineral oils the Dowson and Higginson (1966) formula is not valid, as demon-

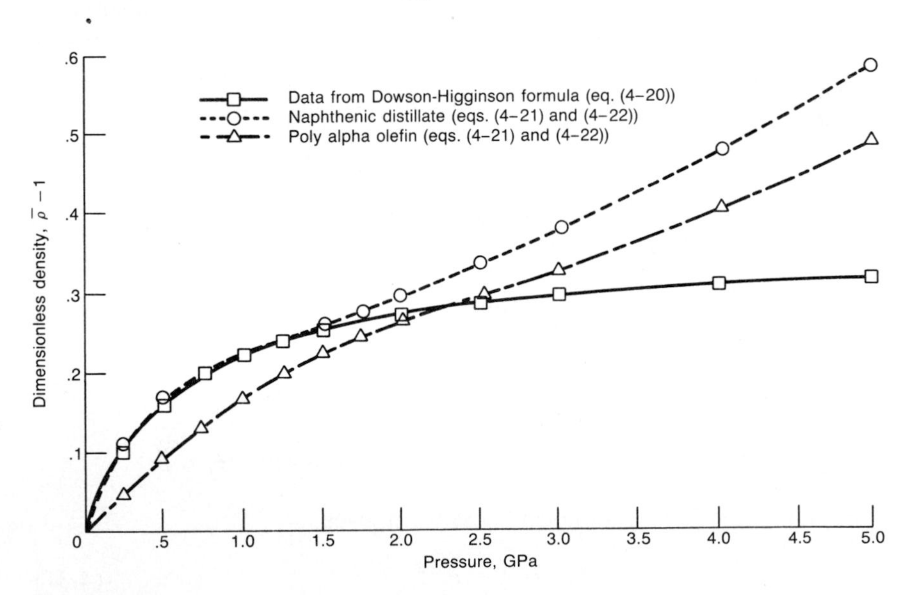

FIGURE 4.10
Effect of pressure on density. [*From Hamrock et al. (1987).*]

strated with poly alpha olefin. Therefore, Hamrock et al. (1987) claim that Eqs. (4.21) and (4.22) are valid for any lubricant as long as there are measured values for the constants C_1, C_2, C_3, and C_4 as well as for $\overline{m}$, n_2, and p_s.

4.15 LIMITING SHEAR STRESS

For most fluid film lubrication analyses the lubricant is assumed to behave in a Newtonian manner. This implies from Eq. (4.1) that the shear stress τ is linearly related to the shear strain rate s, or

$$s = \frac{\tau}{\eta} \tag{4.28}$$

where η is the absolute viscosity or proportionality constant and may vary with both pressure and temperature. However, the lubricant in elastohydrodynamically lubricated conjunctions experiences rapid and extremely large pressure variations, a rapid transit time, possibly large temperature changes, and particularly in sliding contacts, high shear rates. The great severity of these conditions has called into question the normal assumptions of Newtonian behavior of the fluids in elastohydrodynamically lubricated conjunctions. This implies that the lubrication shear stress is still a function of the shear strain rate but that the relationship is no longer linear as shown in Eq. (4.28). For a fluid having non-Newtonian characteristics, the shear rate increases more rapidly than the shear stress.

Figure 4.11 shows the relationship between the dimensionless shear strain rate and the dimensionless shear stress for three nonlinear, viscous rheological fluid models. Isothermal conditions are assumed. The two limiting-shear-stress models shown in Fig. 4.11 were obtained from Bair and Winer (1979) and can be expressed as

$$s = \frac{\tau_L}{\eta} \ln \left(1 - \frac{\tau}{\tau_L}\right)^{-1} \tag{4.29}$$

$$s = \frac{\tau_L}{\eta} \tanh^{-1} \frac{\tau}{\tau_L} \tag{4.30}$$

where τ_L = limiting shear stress, $\tau_0 + \gamma^* p$, Pa
τ_0 = shear stress at zero pressure, Pa
γ^* = limiting-shear-strength proportionality constant, $\partial \tau_L / \partial p$

Rewriting these equations in dimensionless form gives

$$\mathscr{A} = 1 - e^{-\mathscr{B}} \tag{4.31}$$

and

$$\mathscr{A} = \tanh \mathscr{B} \tag{4.32}$$

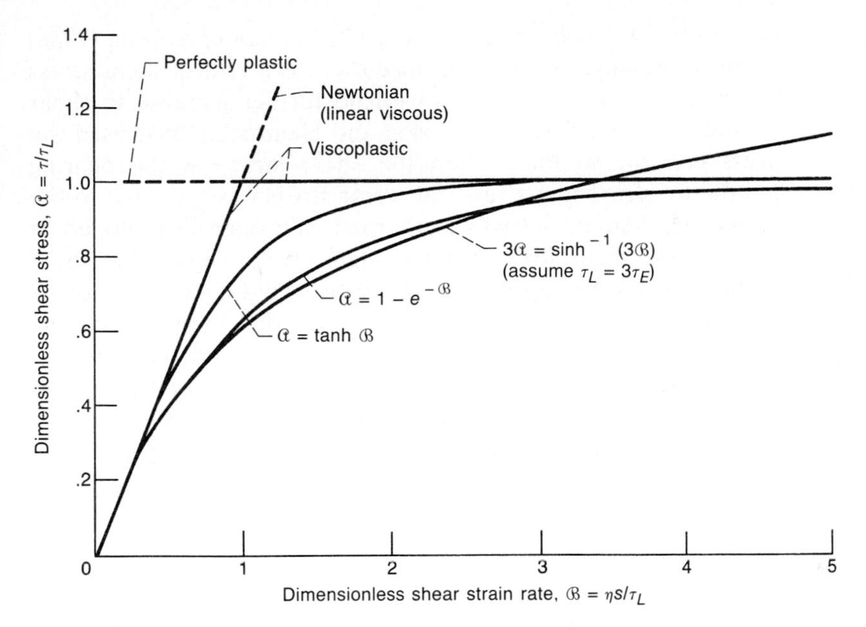

FIGURE 4.11
Comparison of rheological models for isothermal conditions.

where
$$\mathscr{A} = \frac{\tau}{\tau_L} \tag{4.33}$$

$$\mathscr{B} = \frac{\eta s}{\tau_L} \tag{4.34}$$

The other nonlinear viscous model shown in Fig. 4.11 is that attributed to Eyring (1936) and can be expressed as

$$s = \frac{\tau_E}{\eta} \sinh \frac{\tau}{\tau_E} \tag{4.35}$$

where τ_E is the shear stress at which the fluid first starts to behave nonlinearly when stress is plotted against shear strain rate. In comparing this model with the limiting-shear-stress models, the following relationship between τ_E and τ_L will be assumed:

$$\tau_E = \frac{\tau_L}{3} \tag{4.36}$$

Making use of Eq. (4.36) while rewriting Eq. (4.35) in dimensionless form gives

$$3\mathscr{A} = \sinh^{-1}(3\mathscr{B}) \tag{4.37}$$

Figure 4.11 shows that for the Eyring model the shear stress increases monotonically with increasing strain rate.

Bair and Winer (1979), Jacobson (1985), and Höglund and Jacobson (1986) observe that at a given pressure and temperature there is a critical shear stress at which the lubricant will shear plastically with no further increase in shear stress with increasing shear strain rate. Jacobson and Hamrock (1984) used the limiting-shear-stress concept so that, when the shear stress at the bearing surfaces exceeded the limiting shear stress, the shear stress was set equal to the limiting shear stress. As their model was isothermal, slip planes developed at the bearing surfaces if the shear stress was too high. Depending on the shear stress at the bearing surfaces, three different pressure equations were used. As discussed earlier, Houpert and Hamrock (1985) used the Eyring model in developing a new Reynolds equation that shows the nonlinear viscous effects.

The various properties of lubricants having been described in the preceding sections, the next four sections will describe how these properties vary for three different types of lubricant, namely, liquids, greases, and gases. Liquid lubricants will be separated into petroleum or mineral oils and synthetic oils.

4.16 PETROLEUM OR MINERAL OIL BASE STOCKS

Petroleum or mineral oils are generally complex mixtures of hydrocarbons but can roughly be divided, according to the chemical family to which their predominating constituents belong, as paraffins or naphthenes (sometimes referred to as cycloparaffins). Paraffinic oils are characterized by their pour points, usually -17.8 to -6.7°C, and by a moderate change in viscosity with an increase in temperature. In general, their viscosity index will range from 85 to 100. Paraffinic oils have a lower density than naphthenic oils. Naphthenic oils are characterized by pour points from -50 to -12°C and a larger change in viscosity with an increase in temperature. In general, their viscosity index will range from 0 to 60. Both naphthenic and paraffinic oils have a wide range of flash and fire points.

Paraffinic oils are high in paraffin hydrocarbons and contain some wax. Naphthenic oils are high in naphthenic hydrocarbons and contain little wax. In applications that operate over a wide range of temperatures, a naphthenic oil would generally be less suitable than a paraffinic oil. Naphthenic products are usually used in applications exhibiting a limited range of operating temperatures and requiring a relatively low pour point. Also, naphthenic oils tend to swell seal materials more than most paraffinic oils.

4.17 SYNTHETIC OIL BASE STOCKS

Synthetic lubricants have the potential of satisfying a wide range of requirements, since they can be formulated with nearly any desired range of a specific property. However, certain other properties fixed by the chemical structures must be accepted in many cases. Applications must be considered in terms of all properties associated with the proposed synthetic fluid. Choosing the right synthetic fluid can be tricky because to get special characteristics the user usually must trade off some other performance feature. Generally, synthetics have good thermal and oxidation stability, but a common weakness is limited lubricity (the ability of the lubricant to reduce wear and friction other than by its purely viscous properties). In general, synthetic oils cost considerably more per unit volume than the petroleum oils they replace. However, the real value of the lubricant must be calculated on a price-for-performance basis. Much of the remainder of this section summarizes a paper by Hatton (1973).

The field of synthetic lubricants comprises hundreds of organic and semiorganic compounds that cannot be easily classified within the scope of this book. Therefore, synthetic oils will be grouped here by chemical structure. This section briefly summarizes the properties of some of the more popular synthetics (see Table 4.17).

4.17.1 SYNTHETIC HYDROCARBONS. Synthetic hydrocarbons are compounds containing only carbon and hydrogen that are prepared by chemical reactions starting with low-molecular-weight materials. Synthetic hydrocarbons typically possess narrower boiling point ranges for a given viscosity than petroleum oils. In compatibility with other fluids, corrosivity, etc., they are similar to mineral oils.

The primary reasons for preparing synthetic hydrocarbons for use as lubricants are that chemical synthesis provides specific structures and characteristics and that molecular weight can be controlled within narrow ranges. Therefore, properties that are functions of molecular weight, such as vapor pressure, boiling point, viscosity, and low-temperature characteristics, can be controlled within narrow ranges.

4.17.2 ORGANIC ESTERS. The term "organic esters" is applied to those materials that consist of carbon, hydrogen, and oxygen and contain an ester or carboxyl linkage in the molecule. The most widely used as lubricants are those that contain two ester groups and are made from dibasic acids. They are commonly called "diesters." Diesters are the most widely used synthetic lubricants.

The esters have a good overall balance of properties, particularly in liquid range and in viscosity-volatility characteristics. They possess fair lubricity because they respond well to additives. Aliphatic diesters are thermally stable to about 260°C but are exceedingly vulnerable to oxidation above 149°C. This results in increased viscosity and the generation of oil insolubles and large amounts of acid or corrosive material.

Polyol esters were developed to improve upon the thermal stability of the diesters while maintaining other desirable properties. However, they give poorer low-temperature performance. The polyol esters are used in other applications requiring enhanced thermal stability.

TABLE 4.17 Comparative rating of synthetic lubricants

[From Hatton (1973)]

Class	Property[a]												
	Viscosity-temperature relationship	Liquid range	Low-temperature properties	Thermal stability	Oxidative stability	Hydrolytic stability	Fire resistance	Lubricating ability	Bulk modulus	Volatility	Radiation resistance	Density	Handling and storage
Petroleum oils	G	G	G	F	F	E	L	G	A	A	H	L	G
Superrefined petroleum oils	E	G	↓	G	↓	E	↓	↓	↓	L	H	L	↓
Synthetic hydrocarbons	G	G	↓	G	↓	E	↓	↓	↓	L	H	L	↓
Organic esters	G	E	↓	F	↓	F	↓	↓	↓	A	A	A	↓
Polyglycols	G	G	↓	F	↓	G	↓	↓	↓	L	A	A	↓
Polyphenyl esters	P	G	P	E	G	E	↓	↓	H	A	H	H	↓
Phosphate esters, alkyl	G	G	G	F	G	F	H	↓	H	A	L	H	↓
Phosphate esters, aryl	F	P	P	G	G	F	H	↓	H	L	↓	H	↓
Silicate esters and polysiloxanes	E	E	E	↓	F	P	L	F	A	A	↓	A	F
Silicones	E	E	E	↓	G	G	L	P	L	L	↓	A	G
Silanes	G	G	G	↓	F	E	L	F	A	H	↓	L	G
Halogenated polyaryls	G	↓	F	↓	G	E	H	G	H	H	↓	H	G
Fluorocarbons	F	↓	F	↓	G	F	H	P	L	A	↓	H	F
Perfluoropolyglycols	F	↓	G	↓	G	G	H	G	L	A	↓	H	G

[a] Ratings: E = excellent, G = good, F = fair, P = poor, H = high, L = low, A = average.

4.17.3 POLYGLYCOLS. The polyalkylene glycols are the most widely used of this class. They are high-molecular-weight polymers of ethylene or propylene oxide that are available in a wide range of viscosities. Some polymers are completely soluble in water and are often diluted and used as fire-resistant hydraulic fluids or lubricants. Another type of polyglycol is insoluble in water and is used as a lubricant base stock.

The polyglycols are excellent lubricants and respond well to additives. They have high flash points, good viscosity-temperature properties, low wax-free pour points, and shear stability. They have little or no adverse effect on many of the common seal materials but have a strong solvent action on nonresistant paints. Volatility can be a problem, particularly under severe thermal and oxidative conditions. Their stability characteristics, even when improved with appropriate additives, are not outstanding among the synthetics. Their rust-preventing characteristics are generally poor.

Polyglycols are used as industrial lubricants in rubber-processing applications, as machining lubricants, as lubricants for rubber seals, and in heat transfer applications.

4.17.4 PHOSPHATE ESTERS. The phosphate esters are a diverse group of chemical compounds varying widely in physical and chemical characteristics. The oxidative stability of most phosphate esters is good; their thermal stability is excellent at medium temperatures but poorer at higher temperatures. In severe environments extensive thermal, oxidative, or hydrolytic breakdown of the phosphate esters can form acidic substances that may corrode metals. The outstanding properties of phosphate esters are their ability to lubricate moving surfaces and their good fire resistance. They are used as the sole component or as the major component of synthetic lubricants and hydraulic fluids. They are also widely used as an additive in both synthetic lubricants and petroleum oils. The phosphate esters require special consideration with respect to material compatibility—proper material matching is critical to successful performance.

4.17.5 SILICON-CONTAINING COMPOUNDS. One of the more fruitful areas of research in modifying carbon, hydrogen, and oxygen compounds has been the inclusion of silicon in the molecule.

4.17.5.1 Silicate esters. These synthetics have found use as base stocks for wide-temperature-range fluids and lubricants. They have excellent viscosity-temperature characteristics and good lubricating properties.

4.17.5.2 Silicones. Properly called "siloxane polymers" silicones are characterized by the nature of the substituents that are attached directly to the silicon atoms. Dimethyl silicones are characterized by low freezing points and probably the best viscosity-temperature properties of any synthetic lubricant. They have better thermal and oxidative properties than corresponding hydrocarbons, polyglycols, or aliphatic diesters; they can form gels when permitted to degrade excessively, particularly above 200°C. Stabilization is possible, permitting properly inhibited fluids to be used at temperatures as high as 315°C.

The dimethyl silicones are chemically inert, noncorrosive, and inert to most common plastics, elastomers, and paints. They have low surface tension and are shear stable. Their major shortcoming is their lack of steel-on-steel lubricating ability. They show relatively weak response to the usual lubricant additives. The lubricating properties of silicones can be improved by incorporating chlorine or fluorine into the molecule.

4.17.5.3 Silanes. The silanes are compounds that contain only carbon-silicon bonds. These products possess wide liquid ranges and thermal stabilities up to 370°C, but they are poor lubricants for sliding surfaces.

4.17.6 HALOGEN-CONTAINING COMPOUNDS. The incorporation of halogen atoms into organic molecules results in higher densities and reduced flammability relative to the parent compound. Chlorine tends to increase the pour point and the viscosity. Fluorine has little influence on the pour point or the viscosity but greatly decreases surface tension.

Because chlorine-aliphatic carbon bonds are generally weak, chlorinated aliphatic compounds have found little use as synthetic lubricants, per se. They are used as additives in lubricants to provide a source of chlorine that can be reacted with the surface and thus improve boundary lubrication.

Fluorine-containing materials have not found particular application as lubricants because on a price-for-performance basis the desired properties can better be obtained by using other materials.

4.17.7 HALOGENATED POLYARYLS. The chlorine-containing biphenyls and polyphenyls have found some use as lubricants. These products range from mobile liquids to tacky solids. In the past they have found use as lubricant base stocks, heat transfer agents, industrial lubricants, and additives. However, these uses are now prohibited because of potential environmental problems. The common term for these materials is "PCB."

4.17.8 FLUOROCARBONS. Fluorocarbons are compounds containing only fluorine and carbon. Such compounds have been made and proposed as synthetic lubricant base stocks. In general, they are thermally and oxidatively stable and have physical properties quite similar to those of the corresponding hydrocarbons, but they have a higher density and a lower surface tension. They tend to creep over surfaces but do not appear to wet them in terms of boundary lubrication. Such compounds are particularly useful because of their extreme chemical inertness. These compounds are resistant to ignition by any source and represent some of the most fire-resistant organic compounds known.

Recent studies have been concentrated on the tetrafluoroethylene polymers. These have been used as liquid-oxygen-resistant fluids, lubricants, and greases.

Although not strictly fluorocarbons, the chlorotrifluoroethylene polymers have found some applications and are better lubricants than fluorocarbons because chlorine is more reactive with metals than fluorine. These materials generally have properties similar to those of the fluorocarbons but are significantly less stable.

4.17.9 PERFLUOROPOLYGLYCOLS. The perfluoropolyglycols are polyalkylene glycols in which all the hydrogens have been replaced with fluorines. These types of product derived from propylene oxide are under consideration for a number of applications. Their primary advantages are high thermal stability, extreme fire resistance, relatively good liquid range, and moderate lubricating characteristics. However, they lack stability in the presence of certain commonly used high-temperature metals, have high specific gravities, and do not respond to common additives. Studies are under way to improve performance in these areas.

4.18 GREASE BASE STOCKS

A petroleum grease is a lubricating oil to which a thickener has been added, usually a metallic soap. The type of thickener added determines the characteristics of the grease. Greases are preferred to liquid lubricants when the application of a continuous supply of lubricant is impractical. Greases are also preferred when equipment is not readily accessible and when a sufficiently tight enclosure for retaining a liquid lubricant does not exist.

4.18.1 THICKENERS. A major factor influencing the properties of a lubricating grease is the thickener employed in it. Thickeners compose 5 to 17 percent of a simple grease formulation. Ninety percent of all greases sold in the United States are based on what is termed "metallic soap." Soaps utilized in lubricating greases are produced during grease manufacturing by saponifying (neutralizing) fats; the compounds (neutralizers) most commonly used are the hydroxides of lithium, calcium, sodium, barium, and aluminum. The saponifiable compounds include tallow, lard oil, hydrogenated fats and oils, fish oil, fatty acids, and vegetable oils. The two most commonly used fatty acids are stearic and 12-hydroxystearic.

During the grease manufacturing process the oil and the fatty acids are heated to 135 to 150°C, at which time the alkaline compound is added and saponification occurs. The water resulting from the chemical reaction is boiled off.

The amount of fatty acid and metal hydroxide added to the oil determines the amount of soap formed. The soap is the thickener. The resultant thickening action is referred to in the grease industry as "consistency." Consistency is a measure of the hardness or softness of the grease.

The major thickener types and the properties associated with them are detailed in the following text. The properties described for each thickener are typical; however, in some formulations they may be radically altered owing to the influence of other grease components.

4.18.1.1 Water-stabilized calcium soap (cup-type thickener). Typically, this thickener is based on calcium stearate stabilized with water. It yields a buttery grease with excellent water resistance. However, service is limited to about 80°C because at higher temperatures the stabilizing water is lost, causing the soap to separate from the oil. This type of grease is chiefly used for mild service applications.

4.18.1.2 Anhydrous-calcium soap. Typically, this thickener is calcium 12-hydroxystearate. Greases with this thickener are similar to the cup-type products. However, because they do not need water to stabilize the system, they have a higher operating temperature range, typically about 120°C. At about 145°C they melt, often separating into their soap and oil phases. Commonly, such greases are used on rolling-element bearings, for which temperature extremes do not occur.

4.18.1.3 Sodium soap. Sodium soap greases usually employ sodium stearate or similar materials. They tend to be fibrous relative to other types of thickener. Typically, they are usable to about 120°C with melting points in the range 150 to 230°C. Although the thickener provides some inherent rust protection, large amounts of water contamination cause these greases to wash out. Sodium soap greases also generally lack the oxidation resistance of lithium and clay greases. They are typically used in plain journal and slider bearings and gears. Certain short-fiber products may be used in greases for rolling-element bearings.

4.18.1.4 Lithium soap. This thickener is usually lithium 12-hydroxystearate. Lithium soaps are the most versatile and widely used greases. They are buttery and have a melting point of about 195°C. When melted and recooled, they generally return to a grease texture (although the properties of the resulting recooled grease are usually changed from those of the unmelted grease). Lithium soap greases are also resistant to water, oxidation, and mechanical working. Most formulations will operate for long times at 120°C; some will function for extended times to about 165°C. They are widely used as multipurpose greases and are particularly suited for rolling-element bearings. The use of lithium soap greases has substantially reduced the number of complex specialty lubricating greases that would otherwise be required in modern manufacturing plants.

4.18.1.5 Complex soaps. Complex soap thickeners are generally formed by reacting several distinctly different acids with the alkali. For example, calcium–complex soap greases may be formed from calcium 12-hydroxystearate and calcium acetate. Calcium–complex soap, lithium–complex soap, and aluminum–complex soap thickened greases are fairly common, with some other types being used occasionally. The principal advantage of most complex soap thickeners is their high melting point, typically about 260°C or higher. This permits their use in applications in which the temperature may at times exceed the melting point of the simple soap thickeners and thus is the reason why they are much used today. Generally, if these greases are used in sustained service above about 120°C, frequent relubrication is needed unless the product is specially formulated for sustained high-temperature service.

4.18.1.6 Polyureas. Polyureas are nonsoap thickeners that are polymerized substituted ureas. Like the complex soaps they typically melt at about 260°C and are used in similar types of service.

4.18.1.7 Clay thickeners. These thickeners are generally bentonite or hectorite clay that has been chemically treated to make it thicken oil. The chief feature of clay-thickened greases is that the thickener does not melt; hence, these greases can be used in operations in which temperatures occasionally exceed the melting points of other thickeners. Their oxidation stability is generally no better than that of other petroleum products. Therefore, if these greases are used in sustained service at temperatures above about 120°C, frequent relubrication is necessary unless the product has been explicitly formulated for sustained service at higher temperatures.

4.18.2 LUBRICATING OIL. Lubricating oil is the largest single component of a lubricating grease and is the component that provides the grease with its ability to lubricate. Simple greases, only oil and thickener, usually contain 65 to 95 percent oil. Although the retentive properties of grease, as well as its resistance to heat, water, and extreme loads, depend upon the proportion and type of soap, the frictional characteristics of grease are based on its oil content. The more important oil properties affecting overall grease performance are as follows:

1. Viscosity and viscosity-temperature characteristics, which influence the ability of a grease to form a lubricating film in service and also influence its behavior at low temperatures
2. Oxidation resistance and evaporation characteristics, which influence the ability of a grease to lubricate for extended times, especially at higher temperatures
3. Characteristics affecting elastomers, which influence the compatibility of a grease with seal materials used in bearings and other devices

Most greases employ a petroleum-based oil as the lubricating oil, but some use synthetic fluids. Diesters, silicones, polyol esters, polyalkylene glycols, and

TABLE 4.18 Typical characteristics of lubricating greases

Thickener	Grease solid, percent of total	Texture	Dropping point		Water resistant ?	Mechanical stability	Maximum temperature for continuous use		Relative cost[a]
			°F	°C			°F	°C	
Soap base:									
Lithium	59.2	Smooth to buttery	375	190	Yes	Fair to good	250	120	3
Calcium:	17.0								
Hydrated	9.0	Smooth	190	88		Poor to good	150	65	2
Anhydrous	3.8	Smooth	290	143		Fair to good	---	---	3
Complex	3.8	Smooth	500+	260+	↓	Poor to good	300	149	5
Sodium	5.0	Buttery to fibrous	360	182	No	Fair to good	250	120	1
Aluminum:	6.0								
Normal	.4	Smooth	180	87	Yes	Poor to fair	150	65	2
Complex	5.6	Smooth	480	249		Fair to good	300	149	4
Barium	2.6	Buttery to fibrous	400	204		Good	250	120	4
Nonsoap base:									
Clay	2.8	Smooth	500+	260+		Fair to good	300	149	4
Polyurea	2.3	Smooth	470	243		Good	300	149	5
Other	4.9	Smooth	470+	243+	↓	Fair to good	300	149	5

[a]Cost: 1 = low; 5 = high.

fluorosilicones are most commonly used. These fluids offer special characteristics, such as high-temperature performance, chemical resistance, and low-temperature performance, that elude refined petroleum oils. Their cost is substantially higher than that of the refined petroleum oils.

The typical characteristics of some greases are shown in Table 4.18. The knowledge of these characteristics is important in establishing which grease is to be used in a specific application. Within the table the dropping point of the various greases is given. This characteristic is obtained from a test (ASTM D–566 and D–2265, IP–132, and DIN 51801) that indicates the temperature at which the thickener deteriorates (melts, loses water of stabilization, etc.) Greases generally should not be used above their dropping point temperatures, but many greases cannot be used even near them because of limits of base-oil oxidation stability, additive stability, etc. The test designation of ASTM is for tests developed in the United States, IP for Britain, and DIN for Germany.

The preceding three sections (4.16 to 4.18) deal with base stocks as they relate to mineral oils, synthetic oils, and greases. No mention is made of additives, since they are beyond the scope of this book. It suffices to say that additives can considerably enhance the performance of the machine element being lubricated.

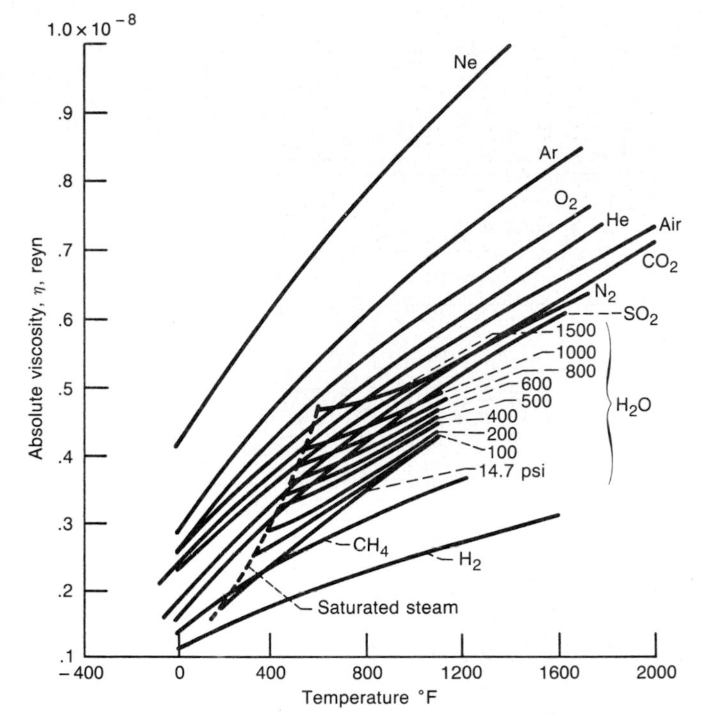

FIGURE 4.12
Viscosity of common gases as function of temperature. [*From Cameron (1976).*]

4.19 GASES

Just as liquids and greases are used as lubricants, so too can gases be used as lubricants, as explained more fully in Chaps. 16 and 17. In this section some of the properties of gases will be described. As shown in Fig. 4.5 a most interesting aspect of gases is their viscosity-temperature relationship. That is, the viscosity of gases increases with temperature and is only moderately affected by changes in temperature and pressure. In contrast to this situation, as mentioned earlier in this chapter, liquids vary inversely with temperature and are strongly sensitive to temperature and pressure variations.

Figure 4.12 shows the viscosity of common gases as a function of temperature. The viscosity of air is midrange of the gases; hydrogen has the lowest and neon the highest value. The information given in Fig. 4.12 is tabulated in Table 4.19 for a complete range of temperatures. Besides the viscosity data, Table 4.19 also gives the boiling temperature and the gas constant. Table 4.20 shows how the properties of gases and liquids differ. Only a small number of quite different liquids and gases are shown to illustrate the differences. Note that Tables 4.19 and 4.20 as well as Fig. 4.12 are in English units rather than SI units. Table 4.6 is helpful in converting to SI.

TABLE 4.19

Viscosity of various gases at 14.7 psia

[From Svehla (1962)]

Temperature		Air	Ar	CO_2	H_2O	He	Kr	N_2	Ne	Xe
°F	°R	Absolute viscosity, η, lbf s/in.^{2} (reyn)								
−280	180	1.07×10^{-9}	1.27×10^{-9}	0.74×10^{-9}	---------	1.45×10^{-9}	---------	1.05×10^{-9}	2.12×10^{-9}	---------
−100	360	1.97	2.41	1.49	---------	2.27	2.50×10^{-9}	1.91	3.48	2.24×10^{-9}
80	540	2.68	3.32	2.20	---------	2.96	3.67	2.57	4.56	3.35
260	720	3.29	4.10	2.84	2.08×10^{-9}	3.56	4.69	3.15	5.50	4.38
440	900	3.84	4.80	3.41	2.59	4.11	5.61	3.66	6.35	5.29
620	1080	4.34	5.44	3.93	3.12	4.62	6.46	4.14	7.16	6.15
800	1260	4.80	6.04	4.41	3.64	5.11	7.23	4.57	7.91	6.93
980	1440	5.24	6.59	4.86	4.17	5.56	7.95	4.99	8.64	7.67
1160	1620	5.64	7.11	5.29	4.69	6.00	8.65	5.39	9.31	8.36
1340	1800	6.05	7.60	5.70	5.20	6.43	9.30	5.75	9.96	9.03
1520	1980	6.44	8.09	6.09	5.70	6.84	9.92	6.13	10.60	9.66
1700	2160	6.80	8.55	6.45	6.19	7.23	10.52	6.49	11.19	10.26
2600	3060	8.55	11.82	8.14	8.41	9.05	13.26	8.14	14.00	13.00
Boiling temperature, °R		---------	147.2	---------	67.2	7.9	219.2	139	48.7	298.4
Gas constant, in.^{2}/(s^2 °R)		2.47×10^5	1.79×10^5	1.63×10^5	39.8×10^5	17.9×10^5	0.85×10^5	2.55×10^5	3.54×10^5	0.55×10^5

TABLE 4.20 Some properties of common liquids and gases at 68°F and 14.7 psia

[From Gross (1980)]

Liquid or gas	Force density, ρ, lbf s^2/in.^{4}	Absolute, viscosity, η, lbf s/in.^{2} (reyn)	Kinematic viscosity, η_k, in.^{2}/s	Specific heat, C_p, in./°F	Thermal conductivity, K_f, lbf/s °F
Liquids					
Carbon tetrachloride	1.48×10^{-4}	1.41×10^{-7}	9.52×10^{-4}	1.87×10^{3}	2.4×10^{-2}
Glycerine	1.18×10^{-4}	1.25×10^{-4}	1.06	5.4×10^{3}	3.54×10^{-2}
Olive oil	8.49×10^{-5}	1.22×10^{-5}	1.44×10^{-1}	4.4×10^{3}	2.10×10^{-2}
Lubricating oil	8.02×10^{-5}	4.44×10^{-5}	5.54	4.7×10^{3}	1.83×10^{-2}
Water	9.33×10^{-5}	1.46×10^{-7}	1.56×10^{-3}	9.32×10^{3}	7.50×10^{-2}
Gases					
Air	1.15×10^{-7}	2.62×10^{-9}	2.28×10^{-2}	2.24×10^{3}	3.22×10^{-3}
Helium	1.61×10^{-8}	2.85×10^{-9}	1.77×10^{-1}	1.17×10^{3}	2.4×10^{-2}
Hydrogen	8.08×10^{-9}	1.31×10^{-9}	1.62×10^{-1}	3.20×10^{4}	2.29×10^{-2}
Nitrogen	1.12×10^{-7}	2.56×10^{-9}	2.28×10^{-2}	2.32×10^{3}	3.11×10^{-3}

4.20 CLOSURE

This chapter described the properties of lubricants that are important in fluid film lubrication. Since lubricating oils are derived from petroleum, which consists of compounds of carbon and hydrogen, basic chemistry has been briefly discussed. In fluid film lubrication the most important physical property of a lubricant is its viscosity. The viscosity of a fluid is associated with its resistance to flow, that is, with the resistance arising from intermolecular forces and internal friction as the molecules move past each other. Newton deduced that the force required to maintain a constant velocity u of an upper plane while the bottom plane is stationary was proportional to the area A and the velocity gradient or rate of shear. Therefore,

$$f = \eta A \frac{u}{h}$$

where η is the proportionality constant or absolute viscosity.

It was also shown in this chapter that the viscosity is greatly affected by temperature, pressure, and shear rate. Appropriate expressions that describe these relationships were presented. Density-pressure effects were also discussed, and the concept of solidification pressure was introduced.

It was found in this chapter that the lubricant in elastohydrodynamically lubricated conjunctions experiences rapid and extremely large pressure variations, a rapid transit time, possibly large temperature changes, and particularly in sliding contacts, high shear rates. The great severity of these conditions has called into question the normal assumption of Newtonian fluid behavior. The concept of a limiting shear stress was introduced, and some non-Newtonian fluid models were presented. The chapter closed with discussions of synthetic oil base stocks, greases, and gases, all of which are used as lubricants in fluid film lubrication.

4.21 PROBLEMS

4.21.1 Given the absolute viscosity of a given fluid at atmospheric conditions to be $6 \times 10^{-3}\ (\text{kg})_{\text{force}} \cdot \text{s/m}^2$, what would this absolute viscosity be in
(*a*) reyn
(*b*) P
(*c*) $\text{lbf} \cdot \text{s/in}^2$
(*d*) $\text{N} \cdot \text{s/m}^2$

4.21.2 A silicon fluid has a Saybolt universal viscosity of 1000 SUV at 38°C and 130 SUV at 100°C. What is the VI of the oil.

4.21.3 Given that a machine element is elastohydrodynamically lubricated with an oil, describe some of the properties you would like to see for that oil. Give an indication whether high or low values are desirable.

4.22 REFERENCES

Bair, S., and Winer, W. O. (1979): Shear Strength Measurements of Lubricants at High Pressure. *J. Lubr. Technol.*, vol. 101, no. 3, pp. 251–257.

Barus, C. (1893): Isothermals, Isopiestics, and Isometrics Relative to Viscosity. *Am. J. Sci.*, vol. 45, pp. 87–96.

Blok, H. (1965): Inverse Problems in Hydrodynamic Lubrication and Design Directions for Lubricated Flexible Surfaces. *Proceedings of International Symposium on Lubrication and Wear*, D. Muster and B. Sternlicht (eds.), McCutchan, Berkeley, pp. 1–151.

Cameron, A. (1976): *Basic Lubrication Theory*, 2d ed. Ellis Harwood Limited, Chichester, England.

Dean, E. W., and Davis, G. H. B. (1929): Viscosity Variations of Oils With Temperature. *J. Chem. Met. Eng.*, vol. 36, no. 10, pp. 618–619.

Dowson, D., and Higginson, G. R. (1966): *Elastohydrodynamic Lubrication: The Fundamentals of Roller and Gear Lubrication*. Pergamon, Oxford.

Eyring, H. (1936): Viscosity, Plasticity, and Diffusion as Examples of Absolute Reaction Rates. *J. Chem. Phys.*, vol. 4, no. 4, pp. 283–291.

Gross, W. A. (1980): *Fluid Film Lubrication*. Wiley-Interscience, New York.

Hamrock, B. J., Jacobson, B. O., and Bergström, S. I. (1987): Measurement of the Density of Base Fluids at Pressures to 2.2 GPa. *ASLE Trans.*, vol. 30, no 2, Apr., pp. 196–202.

Hatton, R. E. (1973): Synthetic Oils. *Interdisciplinary Approach to Liquid Lubricant Technology*. *NASA Spec. Publ.* 318, P. M. Ku (ed.), pp. 101–135.

Herschel, W. H. (1918): Standardization of the Saybolt Universal Viscosimeter. *Tech. Pap.* 112, *Nat. Bur. Stands.*

Hess, F. C. (1981): *Chemistry Made Simple*. Heinemann, London.

Höglund, E., and Jacobson, B. (1986): Experimental Investigations of the Shear Strength of Lubricants Subjected to High Pressure and Temperature. *J. Tribology*, vol. 108, no. 4, pp. 571–578.

Houpert, L. G., and Hamrock, B. J. (1985): Elastohydrodynamic Lubrication Calculations Used as a Tool to Study Scuffing. *Mechanisms and Surface Distress: Global Studies of Mechanisms and Local Analyses of Surface Distress Phenomena*, D. Dowson et al. (eds.), Butterworths, England, pp. 146–162.

Jacobson, B. O. (1985): A High Pressure–Short Time Shear Strength Analyzer for Lubricants. *J. Tribology*, vol. 107, no. 2, pp. 220–223.

Jacobson, B. O., and Hamrock, B. J. (1984): Non-Newtonian Fluid Model Incorporated into Elastohydrodynamic Lubrication of Rectangular Contacts. *J. Tribology*, vol. 106, no. 2, pp. 275–284.

Jeng, Y. R., Hamrock, B. J., and Brewe, D. E. (1987): Piezoviscous Effects in Nonconformal Contacts Lubricated Hydrodynamically. *ASLE Trans.*, vol. 30, no. 4, pp. 452–464.

Jones, W. R., et al. (1975): Pressure-Viscosity Measurements for Several Lubricants to 5.5×10^8 Newtons Per Square Meter (8×10^4 psi) and 149°C (300°F). *ASLE Trans.*, vol. 18, no. 4, pp. 249–262.

Klamann, D. (Killer, A., transl.) (1984): *Lubricants and Related Products*. Verlag Chemie, Weinheim.

Litt, F. A. (1986): Viscosity Index Calculations. *Lubr. Eng.*, vol. 42, no. 12, pp. 752–753.

Newton, I. (1687): *Philosophiae Naturales Principia Mathematica*. Revised and supplied with a historical and explanatory appendix by F. Cajori, edited by R. T. Crawford (1934), and published by the University of California Press, Berkeley and Los Angeles (1966).

Pugh, B. (1970): *Practical Lubrication*. Newnes-Butterworths, London.

Roelands, C. J. A. (1966): *Correlational Aspects of the Viscosity-Temperature-Pressure Relationship of Lubricating Oils*. Druk, V. R. B., Groingen, Netherlands.

Svehla, R. A. (1962): Estimated Viscosities and Thermal Conductivities of Gases at High Temperatures. *NASA Tech. Rep.* R–132.

Winer, W. O., and Cheng, H. S. (1980): Film Thickness, Contact Stress and Surface Temperatures. *Wear Control Handbook*, ASME, New York, pp. 81–141.

CHAPTER 5

BEARING MATERIALS

Another factor that can affect the successful operation of tribological elements is the solid materials used. Bearing materials must have special characteristics if the bearings are to operate successfully. Some desirable characteristics that will be explored in this chapter are compatibility with rubbing counterface materials; embeddability for dirt particles and wear debris; conformability to enable the bearing to accommodate misalignment, geometrical errors, and deflection in the structure; thermal stability; corrosion resistance; and fatigue resistance.

5.1 MATERIAL CHARACTERISTICS

The selection of the bearing material for a particular application depends on (1) the type of bearing (journal, thrust, ball, etc.), (2) the type of lubricant (grease, oil, water, gas, etc.), and (3) the environmental conditions (temperature, pressure, etc.). No single material has been developed that can satisfy all the requirements of a good bearing material. Therefore, the selection must be made on the basis of the characteristics considered of primary importance in the application.

1. *Compatibility.* Although a properly performing hydrodynamic bearing is one in which the shaft and the bearing are separated by a lubricant film, there are times during the operation when the shaft and the bearing come into contact. High spots on the shaft and the bearing rub, localized heating occurs, the high spots can weld, and the microscopic welds can fracture. This sequence of events results in scoring damage to both the shaft and bearing materials. The ability of these material combinations to resist welding and scoring is a measure of their compatibility.

2. *Embeddability.* In the operation of bearings dirt or other foreign debris is carried into the bearing clearance by the lubricant and by the rotation of the shaft. If this dirt cannot be embedded in the bearing material, scoring damage results. The ability to embed or absorb this dirt determines the embeddability characteristic of the bearing material.
3. *Conformability.* As the term implies, conformability is a measure of the ability of the bearing material to conform to misalignment between the shaft and the bearing or to other geometric inaccuracies produced in manufacturing the parts. Usually, bearing materials having a low modulus of elasticity (low E) are readily conformable.
4. *Corrosion resistance.* The bearing material should be resistant to attack by the lubricant or any of the oxidation products produced during lubricant degradation. For example, lubricating oils without oxidation inhibitors produce organic acids, which attack and corrode certain bearing materials. The selection of materials for use with water as the lubricant is of necessity limited to corrosion-resistant materials.
5. *Fatigue resistance.* High fatigue resistance is necessary in applications in which the load changes direction or in which the load intensity varies cyclically. Fatigue failures appear initially as cracks in the bearing surface. These cracks propagate throughout the bearing material, interconnecting with other cracks and resulting in loose pieces of bearing material. Fatigue strength is particularly important where cyclic loading is present.

TABLE 5.1 Properties and characteristics of various conformal bearing metals
[From Clauser (1948)]

Bearing metal	Brinell hardness number		Load-carrying capacity		Maximum operating temperature		Fatigue strength[a]	Antiseizure[a] property	Conformability and embeddability[a]
	Room temperature	149 °C (300 °F)	MPa	psi	°C	°F			
Tin-base babbitt	20–30	6–12	5.5–10.4	800–1500	149	300	3	1	1
Lead-base babbitt	15–20	6–12	5.5–8.3	800–1200	149	300	↓	↓	↓
Alkali-hardened lead	22–26	11–17	8.3–10.4	1200–1500	260	500	↓	↓	↓
Cadmium base	30–40	15	10.4–13.8	1500–2000	260	500	↓	↓	↓
Copper lead	20–30	20–23	10.4–17.2	1500–2500	177	350	2	2	2
Tin bronze	60–80	60–70	>27.6	>4000	260+	500+	1	3	3
Lead bronze	40–70	40–60	20.7–31.1	3000–4500	232–260	450–500	1	3	↓
Phosphor bronze	75–100	65–100	>27.6	>4000	260+	500+	1	3	↓
Aluminum alloy	45–50	40–45	>27.6	>4000	107–149	225–300	2	2	↓
Silver (overplated)	25	25	>27.6	>4000	260+	500+	1	2	↓
Copper-nickel matrix	10	7	13.8	2000	177	350	2	1	2
Trimetal and plated	(b)	(b)	>27.6	>4000	107–149	225–300	↓	↓	↓
Grid type	↓	↓	>27.6	>4000	107–149	225–300	↓	↓	↓
Thin babbitt overlay, 0.051–0.178 mm (0.002–0.007 in.)	↓	↓	13.8	2000	149	300	↓	↓	↓
Conventional babbitt overlays, 0.51 mm (0.020 in.)	↓	↓	10.4	1500	149	300	↓	↓	↓

[a]This is an arbitrary scale with 1 being the highest rating.
[b]Approximately the same as the babbitts.

6. *Dimensional and thermal stability.* The thermal characteristics of the bearing material are important with regard to both heat dissipation and thermal distortion. The thermal conductivity K_f of the bearing material should be high to ensure maximum dissipation of the frictional heat generated if hydrodynamic lubrication conditions cannot be maintained. The linear thermal expansion coefficient $\bar{a}$ should be acceptable within the overall design so that the effects of temperature variation are not detrimental. Values of K_f and $\bar{a}$ are given in Sec. 5.6. Even if a material has these desirable characteristics, additional constraints of acceptable cost and material availability need to be satisfied.

The properties and characteristics of several bearing materials are shown in Table 5.1. Brinell hardness number, load-carrying capacity, and maximum operating temperature are given. Also, in this table ratings of fatigue strength, antiseizure property, conformability, and embeddability are given based on an arbitrary scale, with 1 being the most desirable or best.

5.2 METALLICS

Bearing materials for conformal surfaces fall into two major categories:

1. Metallics: babbitts, bronzes, aluminum alloys, porous metals, and metal overlays such as silver, babbitts, and indium
2. Nonmetallics: plastics, rubber, carbon-graphite, wood, ceramics, cemented carbides, metal oxides (e.g., aluminum oxide), and glass

The principal metallic materials will now be covered in more detail.

5.2.1 TIN- AND LEAD-BASE ALLOYS. The babbitts are among the most widely used materials for hydrodynamically lubricated bearings. Babbitts are either tin- or lead-base alloys having excellent embeddability and conformability characteristics. They are unsurpassed in compatibility and thus prevent shaft scoring.

Tin- and lead-base babbitts have relatively low load-carrying capacity. This capacity is increased by metallurgically bonding these alloys to stronger backing materials such as steel, cast iron, or bronze. Babbitt linings are either still cast or centrifugally cast onto the backing material. Fatigue strength is increased by decreasing the thickness of the babbitt lining. Dowson (1979) points out that at the beginning of the century babbitt linings were rarely less than 3 mm thick and not infrequently at least 6.4 mm thick. The need to provide adequate compressive and fatigue strength gradually brought the thickness of the lining down to 500 μm, at the expense of other desirable features such as embeddability and conformability. The optimum thickness of the bearing layer varies with the application but is generally between 0.02 and 0.12 mm.

Tables 5.2 and 5.3 show the composition and physical properties of some of the tin- and lead-base alloys presently used. Table 5.2 shows the significant

TABLE 5.2 Composition and physical properties of white metal bearing alloys[a]

[From ASTM B23–83. Reprinted by permission of the American Society for Testing and Materials.]

Alloy number[b]	Tin	Antimony	Lead	Copper	Arsenic	Specific gravity[c]	Tin	Antimony	Lead	Copper	20 °C (68 °F)		100 °C (212 °F)	
	Specified nominal composition of alloys, percent						Composition of alloys tested, percent				Yield point[d]			
											MPa	psi	MPa	psi
1	91.0	4.5	--------	4.5	----	7.34	90.9	4.52	None	4.56	30.3	4400	18.3	2650
2	89.0	7.5	--------	3.5	----	7.39	89.2	7.4	0.03	3.1	42.0	6100	20.6	3000
3	84.0	8.0	--------	8.0	----	7.46	83.4	8.2	.03	8.3	45.5	6600	21.7	3150
7	10.0	15.0	Remainder	---	.45	9.73	10.0	14.5	75.0	.11	24.5	3550	11.0	1600
8	5.0	15.0	Remainder	---	.45	10.04	5.2	14.9	79.4	.14	23.4	3400	12.1	1750
15	1.0	16.0	Remainder	---	1.0	10.05	----	----	----	----	----	----	----	----

Alloy number[b]	20 °C (68 °F)		100 °C (212 °F)		20 °C (68 °F)		100 °C (212 °F)		20 °C (68 °F)	100 °C (212 °F)	Melting point		Temperature of complete liquefaction		Proper pouring temperature	
	Johnson's apparent elastic limit[e]				Ultimate strength in compression[f]				Brinell hardness[g]		°C	°F				
	MPa	psi	MPa	psi	MPa	psi	MPa	psi	MPa	psi			°C	°F	°C	°F
1	16.9	2450	7.2	1050	88.6	12 850	47.9	6950	8.0	17.0	223	433	371	700	441	825
2	23.1	3350	7.6	1100	102.7	14 900	60.0	8700	12.0	24.5	241	466	354	669	424	795
3	36.9	5350	9.0	1300	121.3	17 600	68.3	9900	14.5	27.0	240	464	422	792	491	915
7	17.2	2500	9.3	1350	107.9	15 650	42.4	6150	10.5	22.5	240	464	268	514	338	620
8	18.3	2650	8.3	1200	107.6	15 600	42.4	6150	9.5	20.0	237	459	272	522	341	645
15	----	----	----	----	----	------	----	----	13.0	21.0	248	479	281	538	350	662

[a] Compression test specimens were cylinders 1.5 in. (38 mm) in length and 0.5 in. (13 mm) in diameter, machined from chill castings 2 in. (51 mm) in length and 0.75 in. (19 mm) in diameter. The Brinell tests were made on the bottom of parallel machined specimens cast in a mold 2 in. (51 mm) in diameter and 0.625 in. (16 mm) deep at room temperature.

[b] Data not available on alloys 11 and 13.

[c] The specific gravity multiplied by 0.0361 equals the mass density in pounds per cubic inch.

[d] The values for yield point were taken from stress-strain curves at a deformation of 0.125 percent of gage length.

[e] Johnson's apparent elastic limit is taken as the unit stress at the point where the slope of the tangent to the curve is two-thirds its slope at the origin.

[f] The ultimate strength values were taken as the unit load necesary to produce a deformation of 25 percent of the specimen length.

[g] These values are the average Brinell number of three impressions on each alloy, using a 10-mm (0.39-in.) ball and a 500-kg (1102.3 lb) load applied for 30 s.

TABLE 5.3 Chemical composition of alloys in more general use

[From ASTM B23–83. Reprinted by permission of the American Society for Testing and Materials.]

Element	Alloy number[a,b]							
	Tin base				Lead base			
	1	2	3	11	7	8	13	15
	Chemical composition, percent							
Tin	90.0–92.0	88.0–90.0	83.0–85.0	86.0–89.0	9.3–10.7	4.5–5.5	5.5–6.5	0.8–1.2
Antimony	4.0–5.0	7.0–8.0	7.5–8.5	6.0–7.5	14.0–16.0	14.0–16.0	9.5–10.5	14.5–17.5
Lead	.35	.35	.35	.50	Remainder[c]	Remainder	Remainder	Remainder
Copper	4.0–5.0	3.0–4.0	7.5–8.5	5.0–6.5	.50	.50	.50	.6
Iron	.08	.08	.08	.08	.10	.10	.10	.10
Arsenic	.10	.10	.10	.10	.30–.60	.30–.60	.25	.8–1.4
Bismuth	.08	.08	.08	.08	.10	.10	10	.10
Zinc	.005	.005	.005	.005	.005	.005	.005	.005
Aluminum	.005	.005	.005	.005	005	.005	.005	.005
Cadmium	.05	.05	.05	.05	.05	.05	.05	.05
Total named elements, minimum	99.80	99.80	99.80	99.80	---------	--------	--------	

[a] All values not given as ranges are maximum unless shown otherwise.

[b] Alloy 9 was discontinued in 1946 and 4, 5, 6, 10, 11, 12, 16, and 19 were discontinued in 1959. A new number 11, similar to SAE grade 11, was added in 1966.

[c] To be determined by difference.

effect of temperature in decreasing the strength properties of these alloys. The alloys in more general use are shown in Table 5.3. The effect of various percentages of the alloying elements on the mechanical and physical properties of tin- and lead-base alloys can be significant. Increasing the copper or the antimony increases the hardness and the tensile strength and decreases the ductility. However, increases beyond the percentages shown in Table 5.3 can result in decreased fatigue strength.

5.2.2 COPPER-LEAD ALLOYS. Two alloys, one consisting of 60 percent copper and 40 percent lead and the other of 70 percent copper and 30 percent lead, or slight variations, are used as lining materials on steel-backed bearings. These alloys are either strip cast or sintered onto the backing strip, thus providing a bearing with a higher load-carrying capacity than one lined with the babbitt alloys. They also have higher fatigue resistance and can operate at higher temperatures, but they have poor antiseizure properties. They are used in automotive and aircraft internal combustion engines and in diesel engines. Their high lead content provides a good bearing surface but makes them susceptible to corrosion. Their corrosion resistance and antiseizure properties are improved when they are used as trimetal bearings with a lead-tin or lead-indium overlay electrodeposited onto the copper-lead surface.

5.2.3 BRONZES. Several bronze alloys, including lead, tin, and aluminum bronzes, are used extensively as bearing materials. Some are described in Table 5.4. Because of their good structural properties, they can be used as cast bearings without a steel backing. Bearings can also be machined from standard bar stock.

Lead bronzes, which contain up to 25 percent lead, provide higher load-carrying capacity and fatigue resistance and a higher temperature capability than the babbitt alloys. Tin contents up to about 10 percent are used to

TABLE 5.4 Typical bronze and copper alloy bearing materials

[From Booser (1966)]

Designation	Material	Cu	Sn	Pb	Zn	Fe	Al	Brinell hardness number, BNH	Tensile strength		Maximum operating temperature		Maximum load	
		Nominal composition, percent							MPa	ksi	°C	°F	MPa	ksi
SAE 480	Copper lead	65	--	35	--	--	--	25	55.2	8	177	350	13.8	2
AMS 4840	High-lead tin bronze	70	5	25	--	--	--	48	172.5	25	204+	400+	20.7+	3+
SAE 67	Semiplastic bronze	78	6	16	--	--	--	55	207.0	30	232	450	20.7+	3+
SAE 40	Leaded red brass	85	5	5	5	--	--	60	241.5	35	232	450	24.2	3.5
SAE 660	Bronze	83	7	7	3	--	--	60	241.5	35	232+	450+	27.6	4
SAE 64	Phosphor bronze	80	10	10	--	--	--	63	241.5	35	232+	450+	27.6	4
SAE 62	Gunmetal	88	10	--	2	--	--	65	310.5	45	260+	500+	27.6	4
SAE 620	Navy G	88	8	--	4	--	--	68	276.0	40	260	500	27.6+	4+
SAE 63	Leaded gunmetal	88	10	2	--	--	--	70	276.0	45	260	500	27.6+	4+
ASTM B148-52-9c	Aluminum bronze	85	--	--	--	4	11	195	621.0	90	260+	500+	31.1+	4.5+

improve the strength properties. Higher-lead bronze (70 percent copper, 5 percent tin, and 25 percent lead) can be used with soft shafts, but harder shafts (300 BHN) are recommended with the harder lower-lead bronzes, particularly under conditions of sparse lubrication. Lead bronze bearings are used in pumps, diesel engines, railroad cars, home appliances, and many other applications.

Tin bronzes, which contain 9 to 20 percent tin and small quantities of lead (usually < 1 percent), are harder than lead bronzes and are therefore used in heavier-duty applications.

5.3 NONMETALLICS

Although nonmetallic materials such as rubber and graphite have found increasing application, polymeric and plastic materials have had the greatest recent impact in triboelements. These materials fall into two categories: thermosetting and thermoplastic materials. In thermosetting materials the fabrics of nonoriented fibers are generally set in phenolic, or occasionally cresylic, resins.

Of the thermoplastic materials nylon has been recognized as a valuable bearing material as has the remarkable low-friction polymer polytetrafluoroethylene (PTFE). The great merit of these materials is that they can operate effectively without lubricants, although their mechanical properties generally limit their application to lightly loaded conditions and often to low speeds and conforming surfaces.

The limits of applying nonmetallic materials are shown in Table 5.5. The specific limits shown in this table are load-carrying capacity, maximum temperature, maximum speed, and PV limit, where P is the load expressed in pounds-force per square inch and V is the surface speed expressed in feet per minute.

TABLE 5.5 Limits of application of nonmetallic bearing materials
[Revised from O'Conner et al. (1968)]

Material	Load-carrying capacity		Maximum temperature		Maximum speed		PV limit[a]
	MPa	psi	°C	°F	m/s	ft/min	
Carbon graphite	4.1	600	399	750	12.7	2500	15×10^3
Phenolics	41.4	6 000	93	200	12.7	2500	15
Nylon	6.9	1 000	93	200	5.1	1000	3
PTFE (Teflon)	3.4	500	260	500	.51	100	1
Reinforced PTFE	17.2	2 500	260	500	5.1	1000	10
PTFE fabric	414.0	60 000	260	500	.25	50	25
Polycarbonate (Lexan)	6.9	1 000	104	220	5.1	1000	3
Acetal resin (Delrin)	6.9	1 000	82	180	5.1	1000	3
Rubber	.34	50	66	150	7.6	1500	15
Wood	13.8	2 000	66	150	10.2	2000	15

[a] P = load (psi); V = surface speed (ft/min).

5.3.1 CARBON GRAPHITES. In addition to their excellent self-lubricating properties, carbon graphites have several advantages over conventional materials and lubricants. They can withstand temperatures of approximately 370°C in an oxidizing atmosphere such as air and can be used in inert atmospheres to 700°C or at cryogenic temperatures. They can be used in equipment in which lubricant contamination must be prevented, such as textile machinery and food-handling machinery. Carbon graphites are highly resistant to chemical attack and are used in applications where the chemicals attack conventional lubricants. They can be used with low-viscosity lubricants, such as water, gasoline, or air.

Carbon graphites are used for pump shaft bearings, impeller wear rings in centrifugal pumps, and journal and thrust bearings in covered motor pumps and for many other applications. Because of its low expansion coefficient of 2.7×10^{-6} mm/(mm ·°C), a carbon graphite liner is shrunk-fit into a steel sleeve. The steel backing provides mechanical support, improves heat transfer, and helps to maintain shaft clearance. The mating shaft should be made of harder metals. Chromium plates, hardened tool steels, or even some ceramics are used.

A PV value of 15,000 is used when lubricant is present. Depending on the material grade and the application, friction coefficients ranging from 0.04 to 0.25 are obtainable. Absorbed water vapor enhances film formation and reduces the friction and wear of carbon graphite. With no water vapor present (low humidity), wear increases. In general, low speeds and light loads should be used in nonlubricated applications.

5.3.2 PHENOLICS. Among several types of plastic bearings presently in use are the phenolics. These are in the form of laminated phenolics, made by treating sheets of either paper or fabric with phenolic resin, stacking the desired number of sheets, and curing with heat and pressure to bond them together and set the resin. Other filling materials, such as graphite and molybdenum disulfide, are added in powdered form to improve lubrication qualities and strength.

Figure 5.1 shows the various orientations of the phenolic laminates used in bearings. Tubular bearings [Fig. 5.1(a)] are used where complete bushings are required. Bearings in which the load is taken by the edges of the laminations [Fig. 5.1(b) and (c)] are used in light-duty service. Stave bearings [Fig. 5.1(d)] are used mainly for stern-tube and rudder-stock bearings on ships and for guide bearings on vertical waterwheel turbines. Molded bearings [Fig. 5.1(e)] are used for roll-neck bearings in steel mills or for ball-mill bearings. Table 5.6 gives some typical applications of phenolic bearings.

Laminated phenolics operate well with steel or bronze journals when lubricated with oil, water, or other liquids. They have good resistance to seizure. One main disadvantage of these materials is their low thermal conductivity (0.35 W/(m · °C), about 1/150 that of steel), which prevents them from dissipating frictional heat readily and can result in their failure by charring. In large roll-neck bearings the heat is removed by providing a large water flow through the bearing.

Laminated phenolics have good resistance to chemical attack and can be used with water, oil, diluted acid, and alkali solutions. They have good con-

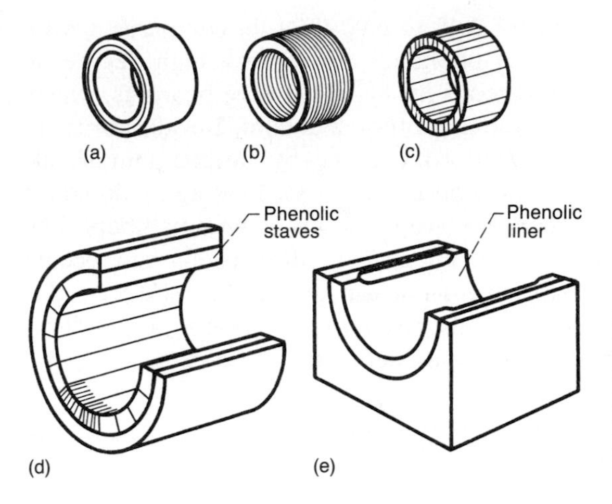

FIGURE 5.1
Phenolic laminate bearings. (a) Tubular bearing; (b) circumferentially lamented bearing; (c) axially laminated bearing; (d) stave bearing; (e) molded bearing. [*From Kaufman (1980).*]

TABLE 5.6 Typical applications of laminated phenolic bearings
[From Kaufman (1980)]

Bearing application	Type[a]	Size range		Fabric weight		Resin, percent	Lubricant	Diametral clearance[b]		Principal reasons for using laminated phenolic bearing material
		mm	in.	g/m	oz/yd			mm	in.	
Roll neck	(e)	76-762	3-30	466-1552	15-50	40-60	Water or emulsion	0-13	0-0.5	Longer life, power saving due to lower friction, lower-cost water lubrication, greater cleanliness of operation, better holding of gage due to less water
Ship, stern tube	(d)	76-660	3-26	248	8	60	Water	0.001/mm diameter over 127 mm	0.001/in. diameter over 5 in.	Longer life, greater ease of handling and installing, higher load-carrying capacity particularly with impact loads, lower friction, greater corrosion and decay resistance, lower journal wear, greater local availability
Rudder, pintle	(a),(d)	76-660	3-26	93-248	3-8	55-60	Grease or water	0.001/mm diameter over 127 mm	0.001/in. diameter over 5 in.	
Small craft, stern tube	(c)	13-76	.5-3	248	8	60	Water	.127	.005	
Centrifugal pump	(a),(b), (c)	13-102	.5-4	93-248	3-8	60	Pumped liquid	.127	.005	Longer life, better lubrication with pumped liquid (water, gasoline, chemical solutions, etc.)
Water wheel turbine, guide bearing	(d)	102-610	4-24	248	8	60	Water	.127	.005	Longer life, lower friction, no decay, less journal wear
Ball mill	(a),(e)	381-1219	15-49	202-466	6.5-15	55-60	Water or emulsion of water and grease	.381-.762	.015-.030	Longer life, higher load-carrying capacity, lower friction, lower lubricant cost
Aircraft, landing gear	(a)	51-381	2-12	93	3	60	Oil	0.001/mm diameter over 127 mm	0.001/in. diameter over 5 in.	Lighter weight, satisfactory dimensional stability and load-carrying capacity
Railway, bolster cup	Molded cone	--------	---	202	6.5	53	Grease	-----------	--------	Longer life, lower noise and vibration transmission

[a]See figure 5-1.
[b]Running clearance, does not include allowance for swelling.

formability, having an elastic modulus of 3.45 to 6.90 GPa, in comparison with about 3.45 GPa for babbitts. Laminated phenolics also have a high degree of embeddability. This property is advantageous in ship stern-tube bearings, which are lubricated by water that contains sand and other sediment. Because of their good resilience, they are highly resistant to damage by fatigue and shock loading. They do not hammer out or extrude under shock loading as do some babbitt alloys. Because laminated phenolics are made up of organic fibers that absorb certain liquids and expand, small changes in dimensions can occur. Water or lubricants containing water have a greater measurable effect on the dimensional stability of phenolics than do oils. Expansion is greater perpendicular to the laminations (2 to 3 percent) than parallel (0 to 0.3 percent).

5.3.3 NYLON. Nylon is one of the classes of thermoplastic materials, as differentiated from the thermosetting plastics, the phenolics. Nylon bearings can be molded, or nylon powders can be sintered in a manner similar to the manufacture of porous metals. Nylon is not affected by petroleum oils and greases, food acids, milk, photographic solutions, etc., and thus can be used in applications where these fluids are handled.

Nylon has good abrasion resistance, a low wear rate, and good embeddability. Like most plastics, it has good antiseizure properties and softens or chars rather than seizing. It has low thermal conductivity [0.24 W/(m · °C)], and failure is usually the result of overheating. Cold flow (creep) under load is one of its main disadvantages. This effect can be minimized by supporting thin nylon liners in metal sleeves. Nylon bearings are used in household applications such as mixers and blenders and for other lightly loaded applications.

5.3.4 TEFLON. Teflon is a thermoplastic material based on the polymer polytetrafluoroethylene (PTFE), which has a low friction coefficient. It has excellent self-lubricating properties and in many applications can be used dry. It is resistant to chemical attack by many solvents and chemicals and can be used in the temperature range −260 to 260°C. Like nylon, it has a tendency to cold-form under loads. Teflon in its unmodified form also has the disadvantages of low stiffness, a high thermal expansion coefficient, low thermal conductivity, and poor wear resistance. These poor properties are greatly improved by adding fibers such as glass, ceramics, metal powders, metal oxides, graphite, or molybdenum disulfide.

5.4 FORM OF BEARING SURFACES

The metallic and nonmetallic materials described in the two preceding sections may be applied to bearing surfaces in several ways, as shown in Fig. 5.2:

1. Solid bearing [Fig. 5.2(a)]. Bearings are machined directly from a single material (cast iron, aluminum alloys, bronzes, porous metals, etc.).

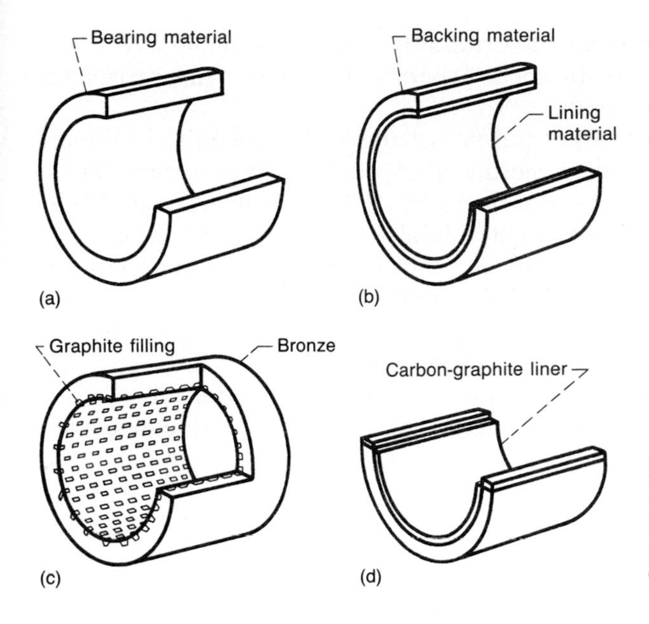

FIGURE 5.2
Different forms of bearing surfaces. (a) Solid bearing; (b) lined bearing; (c) filled bearing; (d) shrink-fit liner bearing.

2. Lined bearing [Fig. 5.2(b)]. Bearing material is bonded to a stronger backing material. The thickness of the bearing lining may range from 0.25 mm to as much as 13 mm. Most modern bonding techniques are metallurgical, although chemical and mechanical methods are also used. The lining material may be cast, sprayed, electrodeposited, or chemically applied.
3. Filled bearing [Fig. 5.2(c)]. A stronger bearing material is impregnated with a bearing material that has better lubricating properties (e.g., graphite impregnated into a bronze backing).
4. Shrink-fit liner bearing [Fig. 5.2(d)]. Carbon-graphite or plastic liners are shrunk into a metal backing sleeve by retaining devices such as setscrews, dowels, and clamping flanges.

5.5 MATERIALS AND MANUFACTURING PROCESSES USED FOR ROLLING-ELEMENT BEARINGS

Nonconformal surfaces such as rolling-element bearings operate under conditions that impose high compressive stresses for millions of stress cycles as the balls or the rollers rotate through the loaded zone of the bearing. For such applications the race and ball materials should be hard and have high fatigue resistance.

Until about 1955 the technology of rolling-element bearing materials did not receive much attention from materials scientists. Bearing materials were restricted to SAE 52100 and some carburizing grades such as AISI 4320 and AISI 9310, which seemed to be adequate for most bearing applications, despite the limitation in temperature of about 176°C for 52100 steel. A minimum acceptable hardness of Rockwell C 58 was specified. Experiments indicated that fatigue life increased with increasing hardness.

The advent of the aircraft gas turbine engine, with its need for advanced rolling-element bearings, provided the major impetus for advancements in the technology of rolling-element bearing materials. Higher temperatures, higher speeds and loads, and the need for greater durability and reliability all served as incentives for developing and evaluating a broad range of new materials and processing methods. The combined research efforts of bearing manufacturers, engine manufacturers, and government agencies over the past three decades have resulted in startling advances in rolling-element bearing life, reliability, and performance. The discussion here is narrow in scope. For a comprehensive treatment of the research status of current bearing technology and current bearing designs, refer to Bamberger et al. (1980).

5.5.1 FERROUS ALLOYS. The need for higher temperature capability led to the evaluation of several available molybdenum and tungsten alloy tool steels as bearing materials. These alloys have excellent high-temperature hardness retention. Such alloys melted and cast in an air environment were, however, generally deficient in fatigue resistance because of nonmetallic inclusions. Vacuum processing techniques can reduce or eliminate these inclusions. Techniques used include vacuum induction melting (VIM) and vacuum arc remelting (VAR). These have been extensively explored, not only with the tool steels now used as bearing materials but with SAE 52100 and some of the carburizing steels as well. Table 5.7 lists a fairly complete array of ferrous alloys, both fully developed and experimental, from which present-day bearings are fabricated. AISI M–50, usually VIM–VAR or consumable electrode vacuum melted (CEVM), has become a widely used quality bearing material. It is usable at temperatures to 315°C, and it is usually assigned a materials life factor of 3 to 5. T–1 tool steel has also come into fairly wide use in bearings, mostly in Europe. Its hot hardness retention is slightly superior to that of M–50 and approximately equal to that of M–1 and M–2. These alloys retain adequate hardness to about 400°C.

Surface-hardened or carburized steels are used in many bearings where, because of shock loads or cyclic bending stresses, the fracture toughness of through-hardened steels is inadequate. Some of the newer materials being developed, such as CBS 1000 and Vasco X–2, have hot hardness retention comparable to that of the tool steels (Fig. 5.3). They too are available as ultraclean, vacuum-processed materials and should offer adequate resistance to fatigue. Carburized steels may become of increasing importance in ultra-high-speed applications. Bearings with through-hardened steel races are currently limited to approximately 2.5 million $d_b N_a$ (where d_b is bore diameter in millimeters and N_a is rotational speed in revolutions per minute). At higher $d_b N_a$ values fatigue cracks propagate through the rotating race as a result of excessive hoop stress (Bamberger et al., 1976).

TABLE 5.7 Typical compositions of selected bearing steels
[From Bamberger et al. (1980)]

Designation	C	P (max)	S (max)	Mn	Si	Cr	V	W	Mo	Co	Cb	Ni
	Alloying element, wt %											
SAE 52100[a]	1.00	0.025	0.025	0.35	0.30	1.45	---	---	---	---	---	---
MHT[b]	1.03	.025	.025	.35	.35	1.50	---	---	---	---	---	---
AISI M-1	.80	.030	.030	.30	.30	4.00	1.00	1.50	8.00	---	---	---
AISI M-2[a]	.83	↓	↓	.30	.30	3.85	1.90	6.15	5.00	---	---	---
AISI M-10	.85	↓	↓	.25	.30	4.00	2.00	---	8.00	---	---	---
AISI M-50[a]	.80	↓	↓	.30	.25	4.00	1.00	---	4.25	---	---	---
T-1 (18-4-1)[a]	.70	↓	↓	.30	.25	4.00	1.00	18.0	---	---	---	---
T15	1.52	.010	.004	.26	.25	4.70	4.90	12.5	.20	5.10	---	---
440C[a]	1.03	.018	.014	.48	.41	17.30	.14	---	.50	---	---	---
AMS 5749	1.15	.012	.004	.50	.30	14.50	1.20	---	4.00	---	---	---
Vasco Matrix II	.53	.014	.013	.12	.21	4.13	1.08	1.40	4.80	7.81	---	0.10
CRB-7	1.10	.016	.003	.43	.31	14.00	1.03	---	2.02	---	0.32	---
AISI 9310[c]	.10	.006	.001	.54	.28	1.18	---	---	.11	---	---	3.15
CBS 600[c]	.19	.007	.014	.61	1.05	1.50	---	---	.94	---	---	.18
CBS 1000M[c]	.14	.018	.019	.48	.43	1.12	---	---	4.77	---	---	2.94
Vasco X-2[c]	.14	.011	.011	.24	.94	4.76	.45	1.40	1.40	.03	---	.10

[a]Balance, iron.
[b]Also contains 1.36% Al.
[c]Carburizing grades.

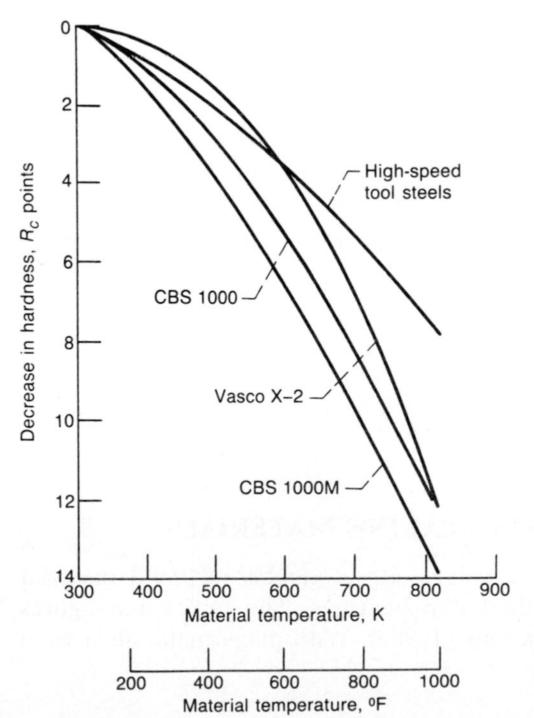

FIGURE 5.3
Hot hardness of CBS 1000, CBS 1000M, Vasco X–2, and high-speed tool steels. [*From Anderson and Zaretsky (1975).*]

In applications where the bearings are not lubricated with conventional oils and protected from corrosion at all times, a corrosion-resistant alloy should be used. Dry-film-lubricated bearings, bearings cooled by liquefied cryogenic gases, and bearings exposed to corrosive environments such as high humidity and salt water are applications where corrosion-resistant alloys should be considered. Of the alloys listed in Table 5.7, both 440C and AMS 5749 are readily available in vacuum-melted heats.

Forging and forming methods that result in improved resistance to fatigue have also been developed. Experiments indicate that fiber or grain flow parallel to the stressed surface is superior to fiber flow that intersects the stressed surface (Bamberger, 1970; Zaretsky and Anderson, 1966). Forming methods that result in more parallel grain flow are now being used in the manufacture of many bearings, especially those for high-load applications.

5.5.2 CERAMICS. Experimental bearings have been made from a variety of ceramics including alumina, silicon carbide, titanium carbide, and silicon nitride. The use of ceramics as bearing materials for specialized applications will probably continue to grow for several reasons. These include

1. High-temperature capability. Because ceramics can exhibit elastic behavior to temperatures beyond 1000°C, they are an obvious choice for extreme-temperature applications.
2. Corrosion resistance. Ceramics are essentially chemically inert and able to function in many environments hostile to ferrous alloys.
3. Low density. This can be translated into improved bearing capacity at high speeds, where centrifugal effects predominate.
4. Low thermal expansion coefficient. Under severe thermal gradients, ceramic bearings exhibit less drastic changes in geometry and internal play than do ferrous alloy bearings.

Silicon nitride has been developed as a bearing material (Sibley, 1982; Cundill and Giordano, 1982). Silicon nitride bearings have exhibited fatigue lives comparable to and in some instances superior to that of high-quality, vacuum-melted M–50 (Sibley, 1982). Two problems remain: (1) quality control and precise nondestructive inspection techniques to determine acceptability and (2) cost. Improved hot isostatic compaction, metrology, and finishing techniques are all being actively pursued.

5.6 PROPERTIES OF COMMON BEARING MATERIALS

This section provides representative values for a number of solid material properties required in evaluating fluid film bearings. The tables and figures presented in this section came from ESDU (1984). With many materials a wide range of property values is attainable by, for example, heat treatment or a small change in composition. The quoted values given in the tables are therefore only typical values likely to be met in fluid film lubrication applications. Unless otherwise stated, all material properties are quoted for room temperature (20°C).

Bearing materials have been conveniently grouped into three basic classifications: metals, ceramics, and polymers. This scheme is based primarily on chemical makeup and atomic structure, and most materials fall into one distinct grouping or another, although there are some intermediates. In addition to these three major classifications, there is one additional group of bearing materials that might be considered, namely, composites. A brief explanation of the material classifications and their representative characteristics is given here:

1. *Metals.* Metallic materials are normally combinations of metallic elements. They have large numbers of nonlocalized electrons; that is, these electrons are not bound to particular atoms. Metals are extremely good conductors of electricity and heat and are not transparent to visible light; a polished metal surface has a lustrous appearance. Furthermore, metals are quite strong yet deformable.
2. *Ceramics.* Ceramics are compounds of metallic and nonmetallic elements; they are most frequently oxides, nitrides, and carbides. The wide range of materials that fall within this classification includes ceramics that are composed of clay, cement, and glass. These materials are typically insulative to the passage of electricity and heat and are more resistant to high temperatures and harsh environments than metals and polymers. With regard to mechanical behavior, ceramics are hard but very brittle.
3. *Polymers.* Polymers include plastic and rubber materials. Many polymers are organic compounds that are chemically based on carbon, hydrogen, and other nonmetallic elements. Furthermore, they have very large molecular structures. These materials typically have low density and may be extremely flexible.
4. *Composites.* Composite materials include more than one material type. Fiberglass is an example, in which glass fibers are embedded within a polymeric material. A composite is designed to display a combination of the best characteristics of each of the component materials. Fiberglass acquires strength from glass and flexibility from the polymer.

Composites are beyond the scope of this book, but the properties of the other three main classifications of materials will be considered along with a number of materials within each classification.

5.6.1 MASS DENSITY. As pointed out in Sec. 4.5, the mass density ρ^* of a solid material is the mass divided by the volume and hence has metric units of kilograms per cubic meter. Typical values lie between 10^3 and 10^4 kg/m^3.

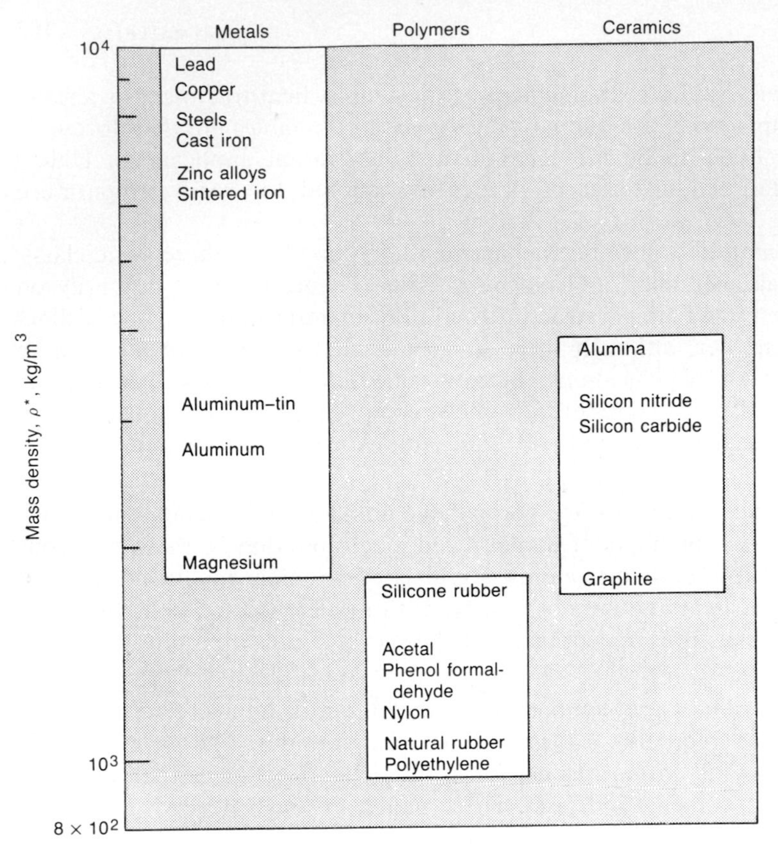

FIGURE 5.4
Illustration of mass density for various metals, polymers, and ceramics at room temperature (20°C; 68°F). [*From* ESDU *(1984).*]

Figure 5.4 illustrates the mass density ordering of various metals, polymers, and ceramics, and Table 5.8 gives quantitative mass density at room temperature (20°C).

Alloying changes the mass density only slightly. To a first approximation, the mass density of an alloy (metallic solid resulting from dissolving two or more molten metals in each other) is given by the "rule of mixtures" (i.e., by a linear interpolation between the mass densities of the alloy components).

5.6.2 MODULUS OF ELASTICITY AND POISSON'S RATIO. A simple tensile load applied to a bar produces a stress σ_1 and a strain ε_1, where

$$\sigma_1 = \frac{\text{load}}{\text{cross-sectional area}} = \text{stress in axial direction}$$

$$\varepsilon_1 = \frac{\text{change in length}}{\text{original length}} = \text{strain in axial direction}$$

The elastic constant, or modulus of elasticity (sometimes referred to as "Young's modulus"), can be written as

TABLE 5.8 Mass densities of various metals, polymers, and ceramics at room temperature (20°C; 68°F)

[From ESDU (1984)]

Material	Mass density, ρ^*	
	kg/m³	lbm/in.³
Metals:		
Aluminum and its alloys[a]	2.7×10^3	0.097
Aluminum tin	3.1	.11
Babbitt, lead-based white metal	10.1	.36
Babbitt, tin-based white metal	7.4	.27
Brasses	8.6	.31
Bronze, aluminum	7.5	.27
Bronze, leaded	8.9	.32
Bronze, phosphor (cast)[b]	8.7	.31
Bronze, porous	6.4	.23
Copper	8.9	.32
Copper lead	9.5	.34
Iron, cast	7.4	.27
Iron, porous	6.1	.22
Iron, wrought	7.8	.28
Magnesium alloys	1.8	.065
Steels[c]	7.8	.28
Zinc alloys	6.7	.24
Polymers:		
Acetal (polyformaldehyde)	1.4	.051
Nylons (polyamides)	1.14	.041
Polyethylene, high density	.95	.034
Phenol formaldehyde	1.3	.047
Rubber, natural[d]	1.0	.036
Rubber, silicone	1.8	.065
Ceramics:		
Alumina (Al_2O_3)	3.9	.14
Graphite, high strength	1.7	.061
Silicon carbide (SiC)	2.9	.10
Silicon nitride (Si_3N_4)	3.2	.12

[a]Structural alloys.
[b]Bar stock typically 8.8×10^3 kg/m³ (0.30 lbm/in.³).
[c]Excluding "refractory" steels.
[d]"Mechanical" rubber.

$$E = \frac{\sigma_1}{\varepsilon_1} \tag{5.1}$$

Although no stress acts transversely to the axial direction, there will nevertheless be dimensional changes in the transverse direction, for as a bar extends axially it contracts transversely. The transverse strains ε_2 are related to the axial strains by Poisson's ratio ν such that

$$\varepsilon_2 = -\nu\varepsilon_1 \tag{5.2}$$

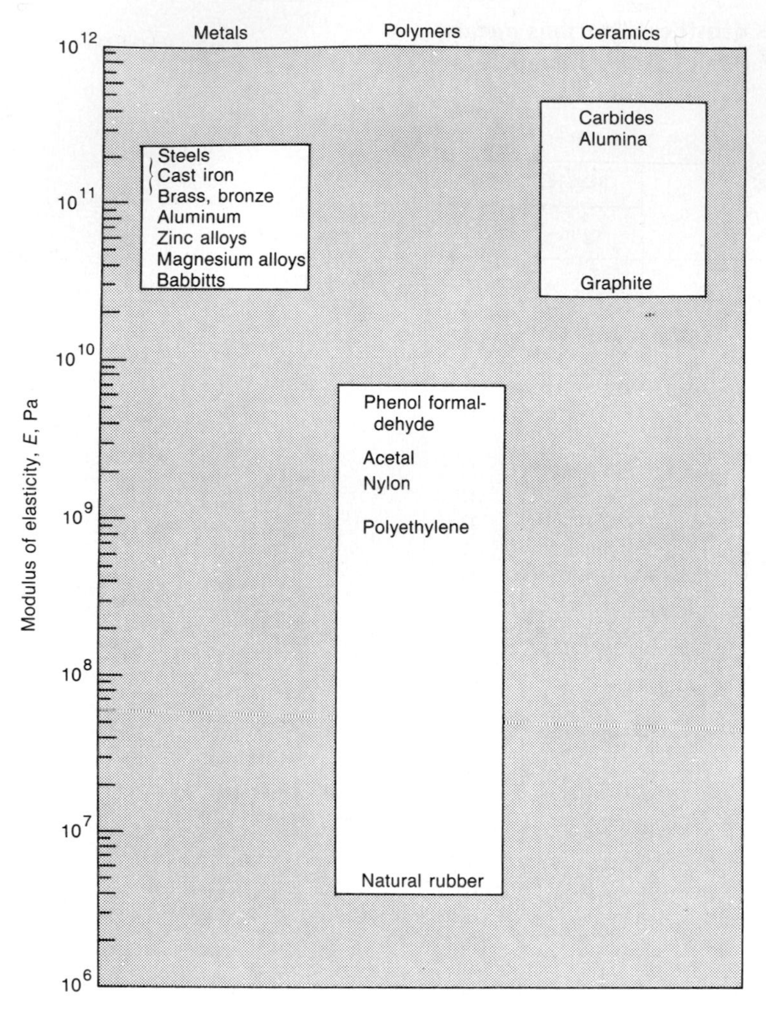

FIGURE 5.5
Illustration of modulus of elasticity for various metals, polymers, and ceramics at room temperature (20°C; 68°F). [*From* ESDU *(1984).*]

where the negative sign simply means that the transverse deformation will be in the opposite sense to the axial deformation. The metric unit of modulus of elasticity is newton per square meter, or pascal, and Poisson's ratio is dimensionless.

Figure 5.5 illustrates values of the modulus of elasticity for various metals, polymers, and ceramics at room temperature (20°C). The moduli of elasticity for metals and ceramics are quite similar, but those for the polymers are considerably lower. Tables 5.9 and 5.10 give quantitative values of the modulus of elasticity and Poisson's ratio, respectively, for various metals, polymers, and ceramics at room temperature.

TABLE 5.9 Modulus of elasticity for various metals, polymers, and ceramics at room temperature (20°C; 68°F)

[From ESDU (1984)]

Material	Modulus of elasticity, E	
	GPa	Mlbf/in.^{2}
Metals:		
Aluminum	62	9.0
Aluminum alloys[a]	70	10.2
Aluminum tin	63	9.1
Babbitt, lead-based white metal	29	4.2
Babbitt, tin-based white metal	52	7.5
Brasses	100	14.5
Bronze, aluminum	117	17.0
Bronze, leaded	97	14.1
Bronze, phosphor	110	16.0
Bronze, porous	60	8.7
Copper	124	18.0
Iron, gray cast	109	15.8
Iron, malleable cast	170	24.7
Iron, spheroidal graphite[b]	159	23.1
Iron, porous	80	11.6
Iron, wrought	170	24.7
Magnesium alloys	41	5.9
Steel, low alloys	196	28.4
Steel, medium and high alloys	200	29.0
Steel, stainless[c]	193	28.0
Steel, high speed	212	30.7
Zinc alloys[d]	50	7.3
Polymers:		
Acetal (polyformaldehyde)	2.7	.39
Nylons (polyamides)	1.9	.28
Polyethylene, high density	.9	.13
Phenol formaldehyde[e]	7.0	1.02
Rubber, natural[f]	.004	.0006
Ceramics:		
Alumina (Al_2O_3)	390	56.6
Graphite	27	3.9
Cemented carbides	450	65.3
Silicon carbide (SiC)	450	65.3
Silicon nitride (Si_3N_4)	314	45.5

[a]Structural alloys.
[b]For bearings.
[c]Precipitation-hardened alloys up to 211 GPa (30 lbf/in.^{2}).
[d]Some alloys up to 96 GPa (14 lbf/in.^{2}).
[e]Filled.
[f]25-Percent-carbon-black "mechanical" rubber.

TABLE 5.10 Poisson's ratio for various metals, polymers, and ceramics at room temperature (20°C; 68°F)

[From ESDU (1984)]

Material	Poisson's ratio, ν
Metals:	
Aluminum and its alloys[a]	0.33
Aluminum tin	----
Babbitt, lead-based white metal	----
Babbitt, tin-based white metal	----
Brasses	.33
Bronze	.33
Bronze, porous	.22
Copper	.33
Copper lead	----
Iron, cast	.26
Iron, porous	.20
Iron, wrought	.30
Magnesium alloys	.33
Steels	.30
Zinc alloys	.27
Polymers:	
Acetal (polyformaldehyde)	----
Nylons (polyamides)	.40
Polyethylene, high density	.35
Phenol formaldehyde	----
Rubber	.50
Ceramics:	
Alumina (Al_2O_3)	.28
Graphite, high strength	----
Cemented carbides	.19
Silicon carbide (SiC)	.19
Silicon nitride (Si_3N_4)	.26

[a]Structural alloys.

5.6.3 LINEAR THERMAL EXPANSION COEFFICIENT. Different materials expand at different rates when heated. A solid object increases in length by a certain fraction for each degree rise in temperature. This result is accurate over a fairly large range of temperatures. It can be used for calculating how much an object will expand for a given change in temperature, once the extent of the material's expansion is measured. This value is given for each material by a number called "linear expansivity" or "linear thermal expansion coefficient" $\bar{a}$. The metric unit of $\bar{a}$ is Kelvin^{-1}.

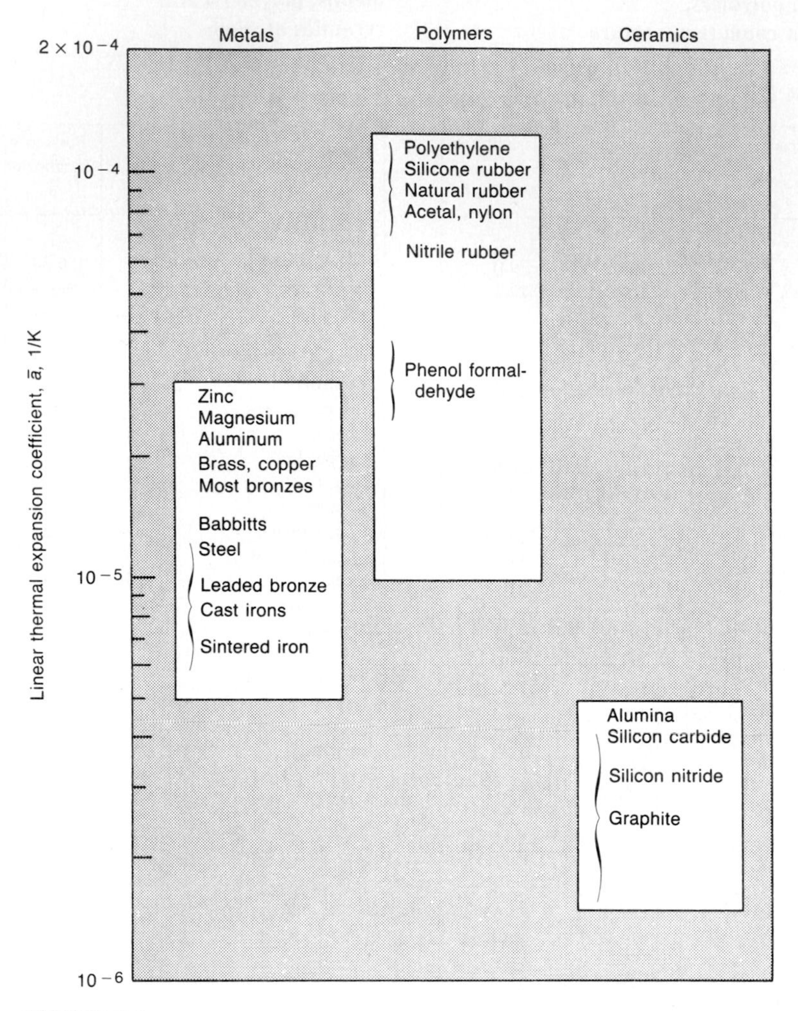

FIGURE 5.6
Illustration of linear thermal expansion coefficient for various metals, polymers, and ceramics applied over temperature range 20 to 200°C (68 to 392°F). [*From* ESDU *(1984)*.]

Figure 5.6 illustrates the linear thermal expansion coefficient for various metals, polymers, and ceramics applied over the temperature range 20 to 200°C. The polymers have the highest value, followed by the metals and then the ceramics. Table 5.11 gives quantitative values of the linear thermal expansion coefficient for various metals, polymers, and ceramics from 20 to 200°C.

5.6.4 THERMAL CONDUCTIVITY. When two bodies at different temperatures are brought together, the faster-moving molecules of the warmer body collide with the slower-moving molecules of the cooler body and transfer some of their

TABLE 5.11 Linear thermal expansion coefficient for various metals, polymers, and ceramics applied over temperature range 20 to 200°C (68 to 392°F)

[From ESDU (1984)]

Material	Linear thermal expansion coefficient, $\bar{a}$	
	1/K	1/°F
Metals:		
Aluminum	23×10^{-6}	12.8×10^{-6}
Aluminum alloys[a]	24	13.3
Aluminum tin	24	13.3
Babbitt, lead-based white metal	20	11
Babbitt, tin-based white metal	23	13
Brasses	19	10.6
Bronzes	18	10.0
Copper	18	10.0
Copper lead	18	10.0
Iron, cast	11	6.1
Iron, porous	12	6.7
Iron, wrought	12	6.7
Magnesium alloys	27	15
Steel, alloy[b]	11	6.1
Steel, stainless	17	9.5
Steel, high speed	11	6.1
Zinc alloys	27	15
Polymers:		
Thermoplastics[c]	$(60\text{–}100) \times 10^{-6}$	$(33\text{–}56) \times 10^{-6}$
Thermosets[d]	$(10\text{–}80) \times 10^{-6}$	$(6\text{–}44) \times 10^{-6}$
Acetal (polyformaldehyde)	90×10^{-6}	50×10^{-6}
Nylons (polyamides)	100	56
Polyethylene, high density	126	70
Phenol formaldehyde[e]	$(25\text{–}40) \times 10^{-6}$	$(14\text{–}22) \times 10^{-6}$
Rubber, natural[f]	$(80\text{–}120) \times 10^{-6}$	$(44\text{–}67) \times 10^{-6}$
Rubber, nitrile[g]	34×10^{-6}	62×10^{-6}
Rubber, silicone	57	103
Ceramics:		
Alumina (Al_2O_3)[h]	5.0	2.8
Graphite, high strength	4.5	.8-2.2
Silicon carbide (SiC)	4.3	2.4
Silicon nitride (Si_3N_4)	3.2	1.8

[a] Structural alloys.
[b] Cast alloys can be up to $15\text{–}10^{-6}$/K.
[c] Typical bearing materials.
[d] 25×10^{-6}/K to 80×10^{-6}/K when reinforced.
[e] Mineral filled.
[f] Fillers can reduce coefficients.
[g] Varies with composition.
[h] 0 to 200°C.

motion to the latter. The warmer object loses energy (drops in temperature) while the cooler one gains energy (rises in temperature). The transfer process stops when the two bodies reach the same temperature. This transfer of molecular motion through a material is called "heat conduction." Materials

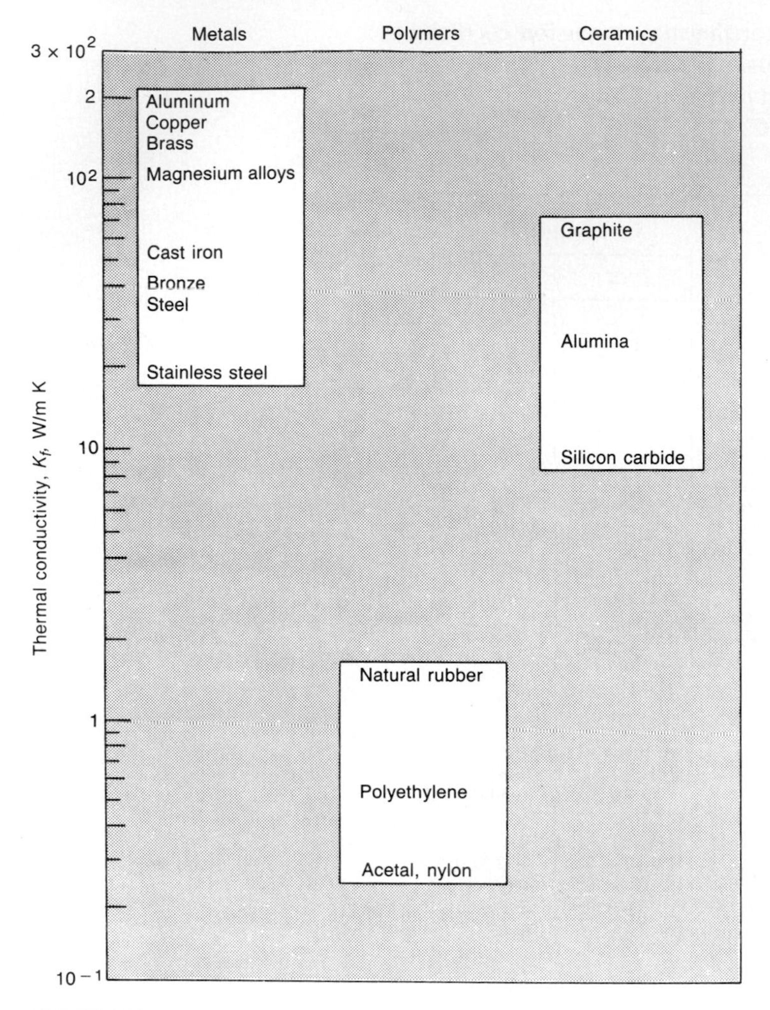

FIGURE 5.7
Illustration of thermal conductivity for various metals, polymers, and ceramics. [*From* ESDU *(1984)*.]

differ in how fast they let this transfer go on. The metric units of thermal conductivity K_f are watts per meter-Kelvin.

Figure 5.7 illustrates the thermal conductivity of various metals, polymers, and ceramics. The metals and ceramics in general are good conductors, and the polymers are good insulators. Table 5.12 quantifies the thermal conductivity results given in Fig. 5.7. In Fig. 5.7 and Table 5.12, unless otherwise stated, the temperature is assumed to be room temperature (20°C; 68°F).

5.6.5 SPECIFIC HEAT CAPACITY. The nature of a material determines the amount of heat transferred to or from a body when its temperature changes by a given amount. Imagine an experiment in which you take a cast iron ball and a

TABLE 5.12 Thermal conductivity for various metals, polymers, and ceramics

Material	Thermal conductivity, K_f	
	W/m K	Btu/ft hr °F
Metals:		
Aluminum	209	120
Aluminum alloys, casting[a]	146	84
Aluminum alloys, silicon[b]	170	98
Aluminum alloys, wrought[c]	151	87
Aluminum tin	180	100
Babbitt, lead-based white metal	24	14
Babbitt, tin-based white metal	56	32
Brasses[a]	120	69
Bronze, aluminum[a]	50	29
Bronze, leaded	47	27
Bronze, phosphor (cast)[d]	50	29
Bronze, porous	30	17
Copper[e]	170	98
Copper lead	30	17
Iron, gray cast	50	29
Iron, spheroidal graphite	30	17
Iron, porous	28	16
Iron, wrought	70	40
Magnesium alloys	110	64
Steel, low alloy[c]	35	20
Steel, medium alloy	30	17
Steel, stainless[f]	15	8.7
Zinc alloys	110	64
Polymers:		
Acetal (polyformaldehyde)	.24	.14
Nylons (polyamides)	.25	.14
Polyethylene, high density	.5	.29
Phenol formaldehyde	------	------
Rubber, natural	1.6	.92
Ceramics:		
Alumina (Al_2O_3)[g]	25	14
Graphite, high strength	125	72
Silicon carbide (SiC)	15	8.6
Silicon nitride (Si_3N_4)	------	------

[a]At 100°C.
[b]At 100°C (~ 150 W/m K at 25°C).
[c]20 to 100°C.
[d]Bar stock typically 69 W/m *K*.
[f]Typically 22 W/m *K* at 200°C.
[g]Typically 12 W/m K at 400°C.

babbitt (lead-based white metal) ball of the same size, heat them both to the temperature of boiling water, and then lay them on a block of wax. You would find that the cast iron ball melts a considerable amount of wax but the babbitt ball, in spite of its greater mass, melts hardly any. It therefore would seem that

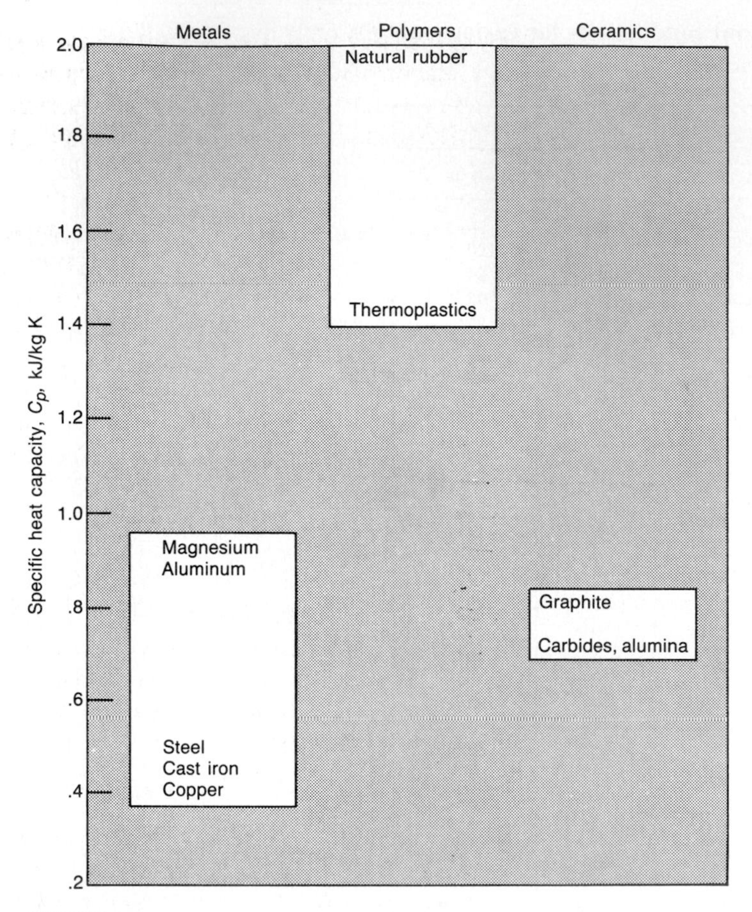

FIGURE 5.8
Illustration of specific heat capacity for various metals, polymers, and ceramics at room temperature (20°C; 68°F). [*From* ESDU *(1984).*]

different materials, in cooling through the same temperature range, give up different amounts of heat.

The quantity of heat energy given up or taken on when a body changes its temperature is proportional to the mass of the object, to the amount that its temperature changes, and to a characteristic number called the "specific heat capacity" of the material the body is made from.

$$\therefore \hat{Q} = C_p m_a (\Delta t_m) \tag{5.3}$$

where $\hat{Q}$ = quantity of heat, J
C_p = specific heat of material, $\mathrm{J/(kg \cdot K)}$
m_a = mass of body, kg
Δt_m = temperature change, K

TABLE 5.13 Specific heat capacity for various metals, polymers, and ceramics at room temperature (20°C; 68°F)

[From ESDU (1984)]

Material	Specific heat capacity, C_p	
	kJ/kg K	Btu/lb °F
Metals:		
Aluminum and its alloys	0.9	0.22
Aluminum tin	.96	.23
Babbitt, lead-based white metal	.15	.036
Babbitt, tin-based white metal	.21	.05
Brasses	.39	.093
Bronzes	.38	.091
Copper[a]	.38	.091
Copper lead	.32	.076
Iron, cast	.42	.10
Iron, porous	.46	.11
Iron, wrought	.46	.11
Magnesium alloys	1.0	.24
Steels[b]	.45	.11
Zinc alloys	.4	.096
Polymers:		
Thermoplastics	1.4	.33
Thermosets	----	----
Rubber, natural	2.0	.48
Ceramics:		
Alumina (Al_2O_3)	----	----
Graphite	.8	.2
Cemented carbides	.7	.17
Silicon carbide (SiC)	----	----
Silicon nitride (Si_3N_4)	----	----

[a]Aluminum bronze up to 0.48 kJ/kg K (0.12 Btu/lb °F).
[b]Rising to 0.55 kJ/kg K (0.13 Btu/lb °F) at 200°C (392°F).

Figure 5.8 illustrates the specific heat capacity of various metals, polymers, and ceramics at room temperature (20°C). Polymers have considerably higher specific heat than metals or ceramics. Table 5.13 quantifies the information presented in Fig. 5.8.

5.7 CLOSURE

In this chapter the general characteristics of bearing materials have been established and discussed. Some desirable characteristics covered in this chapter are compatibility with rubbing counterface materials; embeddability for dirt particles and wear debris; conformability to enable the bearing to accommodate misalignment, geometrical errors, and deflections in structure; strength; corrosion resistance; and fatigue resistance. The various types of bearing material that are now available have been evaluated in terms of these characteristics. These materials consist of metallics (babbitts, bronzes, aluminum alloys, porous metals, and metal overlays such as silver and indium) or nonmetallics (plastics, rubber, carbon-graphite, ceramics, cemented carbides, and metal oxides). Bearing materials applicable to conformal surfaces, where hydrodynamic lubrication occurs, as well as to nonconformal surfaces, where elastohydrodynamic lubrication occurs, were discussed. The stresses acting on conformal and nonconformal surfaces differ considerably, and therefore the solid surface requirements are quite different. The last section listed representative values of a number of solid material properties required in evaluating fluid film bearings. These properties included mass density, modulus of elasticity, Poisson's ratio, linear thermal expansion coefficient, thermal conductivity, and specific heat capacity. Values of these parameters were given for a variety of metals, polymers, and ceramics at room temperature. The material developed in this chapter should prove useful in subsequent chapters.

5.8 PROBLEM

5.8.1 Describe and compare the main types of materials that are available for use as a plain journal bearing when an oil or a grease is undesirable as a lubricant.

5.9 REFERENCES

Anderson, N. E., and Zaretsky, E. V. (1975): Short-Term Hot-Hardness Characteristics of Five Case-Hardened Steels. *NASA Tech. Note* D–8031.

Bamberger, E. N. (1970): Effect of Materials—Metallurgy Viewpoint. *Interdisciplinary Approach to the Lubrication of Concentrated Contacts*, P. M. Ku (ed.), *NASA Spec. Publ.* 237, pp. 409–437.

Bamberger, E. N., Zaretsky, E. V., and Signer, H. (1976): Endurance and Failure Characteristics of Main-Shaft Jet Engine Bearing at 3×10^6 *DN*. *J. Lubr. Technol.*, vol. 98, no. 4, pp. 580–585.

Bamberger, E. N., et al. (1980): Materials for Rolling Element Bearings. *Bearing Design–Historical Aspects, Present Technology and Future Problems*, W. J. Anderson (ed.), *ASME*, New York, pp. 1–46.

Booser, E. R. (1966): Bearing Materials and Properties. *Mach. Des.*, vol. 38, Mar., pp. 22–28.

Clauser, H. R. (1948): Bearing Metals. (Materials and Methods Manual 40), *Mater. Methods*, vol. 28, no. 2, pp. 75–86.

Cundill, R. T., and Giordano, F. (1982): Lightweight Materials for Rolling Elements in Aircraft Bearings. *Problems in Bearings and Lubrication*, *AGARD/Conf. Proc.* 323, pp. 6–1 to 6–11.

Dowson, D. (1979): *History of Tribology*, Longman, London and New York.

Engineering Sciences Data Unit (ESDU) (1984): *Properties of Common Engineering Materials*. Item 84041, London.

Kaufman, H. N. (1980): Bearing Materials. *Tribology–Friction, Lubrication, and Wear*, A. Z. Szeri (ed.). Hemisphere Publishing Corp., Washington, D. C., pp. 477–505.

O'Conner, J. J., Boyd, J., and Avellone, E. A. (eds.) (1968): *Standard Handbook of Lubrication Engineering*. McGraw-Hill, New York.

Sibley, L. B. (1982): *Silicon Nitride Bearing Elements for High-Speed High-Temperature Applications*. *AGARD/Conf. Proc.* 323, pp. 5–1 to 5–15.

Zaretsky, E. V., and Anderson, W. J. (1966): Material Properties and Processing Variables and Their Effect on Rolling-Element Fatigue. *NASA Tech. Memo.* X–52227.

CHAPTER 6

VISCOUS FLOW

This chapter focuses on aspects of fluid mechanics that are important in understanding fluid film lubrication. Four aspects of viscous flow are important:

1. A fluid's viscous resistance increases with the deformation rate. Making a fluid flow fast requires a greater force than making it flow slowly.
2. Molecules do not go back to their original positions when the applied force is removed. The flow involves a nonreversible change, and the work done in producing viscous flow appears as heat in the liquid.
3. A liquid becomes less viscous as its temperature is raised. The greater thermal energy enables the molecules to escape from their neighbors (less external force is required to hurry them on).
4. The viscosity of a liquid in nonconformal conjunctions generally increases as the pressure increases. The viscosity may increase by several orders of magnitude. This is a fortunate state of affairs, for it implies that the harder one tries to squeeze out a lubricant, the higher will be its viscosity and the greater its resistance to extrusion.

6.1 PETROV'S EQUATION

Here Newton's postulate (Sec. 4.3) is applied to a full journal running concentrically with the bearing, as shown in Fig. 6.1. It is shown later that the journal will run concentrically with the bearing only when one of the following conditions prevails: (1) the radial load acting on the bearing is zero, (2) the viscosity of the lubricant is infinite, or (3) the speed of the journal is infinite. None of these

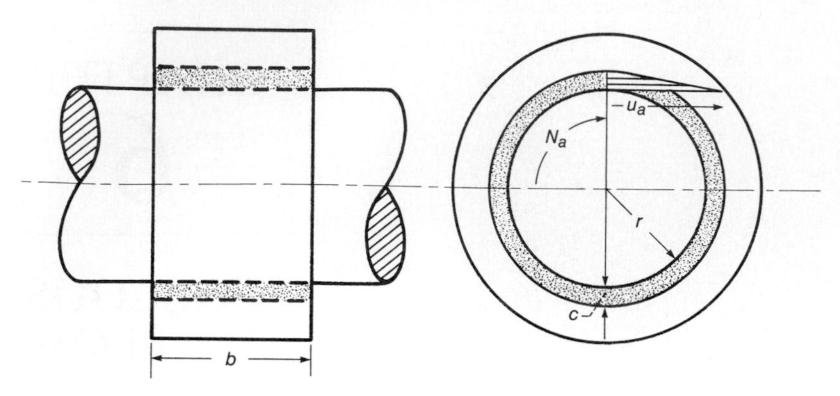

FIGURE 6.1
Concentric journal bearing.

conditions are practically possible. However, if the load is light enough, if the journal has a sufficiently high speed, and if the viscosity is sufficiently high, the eccentricity of the journal relative to the bearing may be so small that the oil film around the journal can be considered practically to be of uniform thickness.

The oil film in a journal bearing is always thin relative to the radius of the bearing. Therefore, the curvature of the bearing surface may be ignored, and the film may be considered as an unwrapped body having a thickness equal to the radial clearance, a length equal to $2\pi r$, and b equal to the width of the bearing. Assume that the viscosity throughout the oil film is constant. In Fig. 6.2 the bottom surface is stationary and the top surface is moving with constant velocity u_a. Petrov (1883) assumed no slip at the interface between the lubricant and the solids.

Making use of Newton's postulate as expressed in Eq. (4.2) gives the friction force in a concentric journal bearing as

$$f = \eta_0 A \frac{u_a}{c} = \eta_0 2\pi rb \frac{2\pi r N_a}{c} = \frac{4\pi^2 \eta_0 r^2 b N_a}{c} \tag{6.1}$$

where N_a is in revolutions per second and η_0 is the viscosity at $p = 0$ and at a constant temperature. The coefficient of friction for a concentric journal bearing

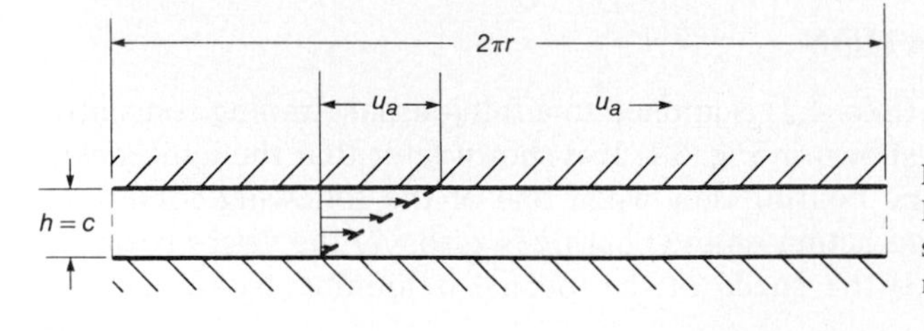

FIGURE 6.2
Developed journal and bearing surfaces for a concentric journal bearing.

can be written as

$$\mu = \frac{f}{w_z} = \frac{4\pi^2 \eta_0 r^2 b N_a}{c w_z} \tag{6.2}$$

where w_z is the normal applied load. The friction torque for a concentric journal bearing can be written as

$$t_q = fr = \frac{4\pi^2 \eta_0 r^3 b N_a}{c} = \frac{2\pi \eta_0 r^3 b \omega}{c} \tag{6.3}$$

where $\omega = 2\pi N_a$, the angular velocity in radians per second. Equation (6.3) is generally called "Petrov's equation" (after N. Petrov, who suggested a similar equation for torque in his work published in 1883).

The power loss is just the velocity multiplied by the frictional force. The power loss for a concentric (lightly loaded) journal bearing can be expressed in horsepower as

$$H_p = \frac{8\pi^3}{(12)(550)} \frac{\eta_0 r^3 b N_a^2}{c} = (0.03758) \frac{\eta_0 r^3 b N_a^2}{c} \tag{6.4}$$

where η_0 = viscosity at $p = 0$ and at a constant temperature, lbf · s/in^2
r = radius of journal, in
b = width of journal, in
N_a = speed, r/s
c = radial clearance, in

Note that Eq. (6.4) is valid for only these units.

6.2 NAVIER-STOKES EQUATIONS

Lubricants in hydrodynamic lubrication analyses and in many elastohydrodynamic lubrication analyses are assumed to behave as Newtonian fluids. As given in Chap. 4 the shear rate is linearly related to the shear stress. Besides assuming that the fluid behavior is Newtonian, assume also that laminar flow exists. Navier (1823) derived the equations of fluid motion for these conditions from molecular considerations and by introducing Newton's postulate for a viscous fluid. Stokes (1845) also derived the governing equations of motion for a viscous fluid in a slightly different form, and the basic equations are thus known as "Navier-Stokes equations of motion."

The Navier-Stokes equations can be derived by considering the dynamic equilibrium of a fluid element. It is necessary to consider surface forces, body forces, and inertia forces.

6.2.1 SURFACE FORCES. Figure 6.3 shows the stresses on the surface of a fluid element in a viscous fluid. Across each of the three mutually perpendicular surfaces there are three stresses, yielding a total of nine stress components. Of

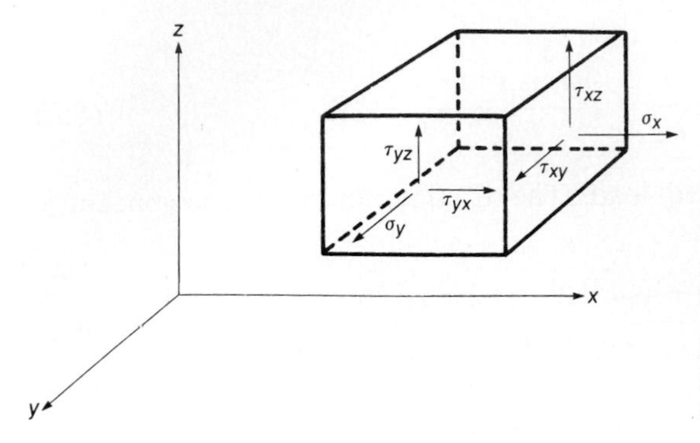

FIGURE 6.3
Stresses on two surfaces of a fluid element.

the three stresses acting on a given surface the normal stress is denoted by σ and the shear stress by τ. The stresses on the surface perpendicular to the z axis have been omitted to avoid overcrowding, the first subscript on the shear stresses refers to the coordinate direction perpendicular to the plane in which the stress acts, and the second designates the coordinate direction in which the stress acts. The following five relationships should be noted in relation to the surface stresses:

1. For equilibrium of the moments of the fluid element the stresses must be symmetric; that is, the subscripts on the shear stresses can be reversed in order.

$$\tau_{xy} = \tau_{yx} \qquad \tau_{xz} = \tau_{zx} \qquad \tau_{yz} = \tau_{zy} \tag{6.5}$$

2. The hydrostatic pressure p in the fluid is considered to be the average of the three normal stress components

$$\sigma_x + \sigma_y + \sigma_z = -3p \tag{6.6}$$

The minus sign is used because hydrostatic pressures are compressive, whereas positive stresses are tensile.

3. The magnitude of the shear stresses depends on the rate at which the fluid is being distorted. For most fluids the dependence is of the form

$$\tau_{ij} = \eta\left(\frac{\partial u_i}{\partial x_j} + \frac{\partial u_j}{\partial x_i}\right) \tag{6.7}$$

where η = absolute viscosity, $\text{N} \cdot \text{s/m}^2$
u_i = components of velocity vector, m/s ($u_x = u$, $u_y = v$, $u_z = w$)
x_i = components of coordinate vector, m ($x_x = x$, $x_y = y$, $x_z = z$)

Note the similarity of Eq. (6.7) to Newton's postulate [Eq. (4.1)]. The terms in parentheses in Eq. (6.7) are a measure of the distortion of the fluid element.

4. The magnitude of the normal stresses can be written as

$$\sigma_i = -p + \lambda_a \xi_a + 2\eta \frac{\partial u_i}{\partial x_i} \tag{6.8}$$

where

$$\xi_a = \frac{\partial u}{\partial x} + \frac{\partial v}{\partial y} + \frac{\partial w}{\partial z} \tag{6.9}$$

and λ_a is a second viscosity coefficient. The divergence of the velocity vector, the dilatation ξ_a, measures the rate at which fluid is flowing out from each point; that is, it measures the expansion of the fluid.

5. From Eq. (6.8)

$$\sigma_x = -p + \lambda_a \xi_a + 2\eta \frac{\partial u}{\partial x}$$

$$\sigma_y = -p + \lambda_a \xi_a + 2\eta \frac{\partial v}{\partial y}$$

$$\sigma_z = -p + \lambda_a \xi_a + 2\eta \frac{\partial w}{\partial z}$$

Substituting these expressions into Eq. (6.6) gives

$$-3p + 3\lambda_a \xi_a + 2\eta\left(\frac{\partial u}{\partial x} + \frac{\partial v}{\partial y} + \frac{\partial w}{\partial z}\right) = -3p$$

or

$$3\lambda_a \xi_a + 2\eta \xi_a = 0$$

$$\therefore \lambda_a = -2\eta/3 \tag{6.10}$$

Therefore, the second viscosity coefficient can simply be expressed in terms of the absolute viscosity.

The conclusions about the surface stresses can be expressed as

$$\tau_{ij} = \tau_{ji} \tag{6.11}$$

$$\tau_{ij} = \eta\left(\frac{\partial u_i}{\partial x_j} + \frac{\partial u_j}{\partial x_i}\right) \tag{6.12}$$

$$\sigma_i = -p - 2\eta \xi_a/3 + 2\eta \frac{\partial u_i}{\partial x_i} \tag{6.13}$$

The normal and shear stresses tend to move the element in the x, y, and z directions. The surface forces resulting from these stresses can be expressed as

$$\frac{\partial \sigma_i}{\partial x_i}\, dx\, dy\, dz \tag{6.14a}$$

and

$$\frac{\partial \tau_{ij}}{\partial x_j}\, dx\, dy\, dz \tag{6.14b}$$

6.2.2 BODY FORCES. The forces needed to accelerate a fluid element may be supplied in part by an external force field, perhaps gravity, associated with the whole body of the element. If the components of the external force field per unit mass are X_a, Y_a, and Z_a, these forces acting on an element are

$$X_a \rho\, dx\, dy\, dz \qquad Y_a \rho\, dx\, dy\, dz \qquad Z_a \rho\, dx\, dy\, dz \tag{6.15}$$

Note that the units of X_a, Y_a, and Z_a are meters per second squared since ρ, the force density, has units of newton-seconds squared per meter to the fourth power ($N \cdot s^2/m^4$).

6.2.3 INERTIA FORCES. The three components of fluid acceleration are the three total derivatives Du/Dt, Dv/Dt, and Dw/Dt. The significance of the total derivative can be seen from the following. Consider only the x component of velocity u.

$$\therefore u = f(x, y, z, t) \tag{6.16}$$

The change in u that occurs in time dt is

$$Du = \frac{\partial u}{\partial t} dt + \frac{\partial u}{\partial x} dx + \frac{\partial u}{\partial y} dy + \frac{\partial u}{\partial z} dz \tag{6.17}$$

In the limit as $dt \to 0$, $dx/dt = u$, $dy/dt = v$, and $dz/dt = w$. Therefore, if Eq. (6.17) is divided throughout by dt while making use of the preceding, the total derivative for the u component can be written as

$$\frac{Du}{Dt} = \frac{\partial u}{\partial t} + u\frac{\partial u}{\partial x} + v\frac{\partial u}{\partial y} + w\frac{\partial u}{\partial z} \tag{6.18}$$

Similarly, for the v and w components of velocity,

$$\frac{Dv}{Dt} = \frac{\partial v}{\partial t} + u\frac{\partial v}{\partial x} + v\frac{\partial v}{\partial y} + w\frac{\partial v}{\partial z} \tag{6.19}$$

$$\frac{Dw}{Dt} = \frac{\partial w}{\partial t} + u\frac{\partial w}{\partial x} + v\frac{\partial w}{\partial y} + w\frac{\partial w}{\partial z} \tag{6.20}$$

The total time derivative measures the change in velocity of one fluid element as it moves about in space. The term $\partial/\partial t$ is known as the "local derivative," since it gives the variation of velocity with time at a fixed point. The last three terms are grouped together under the heading "convective differential."

The resultant forces required to accelerate the elements are

$$\rho\frac{Du}{Dt}\, dx\, dy\, dz \qquad \rho\frac{Dv}{Dt}\, dx\, dy\, dz \qquad \rho\frac{Dw}{Dt}\, dx\, dy\, dz \tag{6.21}$$

6.2.4 EQUILIBRIUM. The surface, body, and inertia forces acting on a fluid element having been defined, the requirement for dynamic equilibrium can now be stated mathematically. When the common factor $dx\,dy\,dz$ is eliminated from each term and the resultant inertia force is set equal to the sum of the body and surface forces,

$$\rho\frac{Du}{Dt} = \rho X_a + \frac{\partial\sigma_x}{\partial x} + \frac{\partial\tau_{xy}}{\partial y} + \frac{\partial\tau_{xz}}{\partial z} \tag{6.22}$$

$$\rho\frac{Dv}{Dt} = \rho Y_a + \frac{\partial\tau_{yx}}{\partial x} + \frac{\partial\sigma_y}{\partial y} + \frac{\partial\tau_{yz}}{\partial z} \tag{6.23}$$

$$\rho\frac{Dw}{Dt} = \rho Z_a + \frac{\partial\tau_{zx}}{\partial x} + \frac{\partial\tau_{zy}}{\partial y} + \frac{\partial\sigma_z}{\partial z} \tag{6.24}$$

Making use of Eqs. (6.11) to (6.13), the Navier-Stokes equations in Cartesian coordinates are

$$\rho\frac{Du}{Dt} = \rho X_a - \frac{\partial p}{\partial x} - \frac{2}{3}\frac{\partial}{\partial x}(\eta\xi_a) + 2\frac{\partial}{\partial x}\left(\eta\frac{\partial u}{\partial x}\right) + \frac{\partial}{\partial y}\left[\eta\left(\frac{\partial u}{\partial y} + \frac{\partial v}{\partial x}\right)\right] + \frac{\partial}{\partial z}\left[\eta\left(\frac{\partial u}{\partial z} + \frac{\partial w}{\partial x}\right)\right] \tag{6.25}$$

$$\rho\frac{Dv}{Dt} = \rho Y_a - \frac{\partial p}{\partial y} - \frac{2}{3}\frac{\partial}{\partial y}(\eta\xi_a) + 2\frac{\partial}{\partial y}\left(\eta\frac{\partial v}{\partial y}\right) + \frac{\partial}{\partial x}\left[\eta\left(\frac{\partial u}{\partial y} + \frac{\partial v}{\partial x}\right)\right] + \frac{\partial}{\partial z}\left[\eta\left(\frac{\partial v}{\partial z} + \frac{\partial w}{\partial y}\right)\right] \tag{6.26}$$

$$\rho\frac{Dw}{Dt} = \rho Z_a - \frac{\partial p}{\partial z} - \frac{2}{3}\frac{\partial}{\partial z}(\eta\xi_a) + 2\frac{\partial}{\partial z}\left(\eta\frac{\partial w}{\partial z}\right) + \frac{\partial}{\partial x}\left[\eta\left(\frac{\partial u}{\partial z} + \frac{\partial w}{\partial x}\right)\right] + \frac{\partial}{\partial y}\left[\eta\left(\frac{\partial v}{\partial z} + \frac{\partial w}{\partial y}\right)\right] \tag{6.27}$$

The terms on the left side of these equations represent inertia effects, and those on the right side are the body force, pressure gradient, and viscous terms in that order. Equations (6.25) to (6.27) are the most general form of the Navier-Stokes equations as expressed in Cartesian coordinates for a Newtonian fluid. These equations play a central role in fluid mechanics, and nearly all analytical work involving a viscous fluid is based on them. These equations so far have not been restricted to either constant density or constant viscosity. They are valid for viscous compressible flow with varying viscosity.

Note that if the inertia terms in Eqs. (6.25) to (6.27) are neglected (left side of these equations set equal to zero), this form of the equations is sometimes referred to as the "Stokes equations."

6.2.5 STANDARD FORMS. For all the forms of the Navier-Stokes equations to be presented in this section, the viscosity is assumed to be constant ($\eta = \eta_0$).

6.2.5.1 Cartesian coordinates. If the viscosity is assumed to be constant ($\eta = \eta_0$), the Navier-Stokes equations in Cartesian coordinates can be simplified and rearranged to give

$$\rho \frac{Du}{Dt} = \rho X_a - \frac{\partial p}{\partial x} + \eta_0 \left(\frac{\partial^2 u}{\partial x^2} + \frac{\partial^2 u}{\partial y^2} + \frac{\partial^2 u}{\partial z^2} \right) + \frac{\eta_0}{3} \frac{\partial \xi_a}{\partial x} \tag{6.28}$$

$$\rho \frac{Dv}{Dt} = \rho Y_a - \frac{\partial p}{\partial y} + \eta_0 \left(\frac{\partial^2 v}{\partial x^2} + \frac{\partial^2 v}{\partial y^2} + \frac{\partial^2 v}{\partial z^2} \right) + \frac{\eta_0}{3} \frac{\partial \xi_a}{\partial y} \tag{6.29}$$

$$\rho \frac{Dw}{Dt} = \rho Z_a - \frac{\partial p}{\partial z} + \eta_0 \left(\frac{\partial^2 w}{\partial x^2} + \frac{\partial^2 w}{\partial y^2} + \frac{\partial^2 w}{\partial z^2} \right) + \frac{\eta_0}{3} \frac{\partial \xi_a}{\partial z} \tag{6.30}$$

If, besides the viscosity being constant, the force density is also constant ($\rho = \rho_0$), the last term on the right side of these equations is zero. Earlier [Eq. (6.9)] $\xi_a = \partial u/\partial x + \partial v/\partial y + \partial w/\partial z$ was defined as the dilatation, or the measure of the rate at which the fluid is flowing out from each point, that is, a measure of the fluid expansion. If the fluid density equals the force density at $p = 0$ and at a fixed temperature ($\rho = \rho_0$), then $\xi_a = 0$.

6.2.5.2 Cylindrical polar coordinates. In cylindrical polar coordinates with r, θ, z such that $x = r\cos\theta$, $y = r\sin\theta$, and $z = z$, the Navier-Stokes equations for constant viscosity and constant density can be expressed as

$$\rho_0 \left(\frac{\partial v_r}{\partial t} + v_r \frac{\partial v_r}{\partial r} + \frac{v_\theta}{r} \frac{\partial v_r}{\partial \theta} + v_z \frac{\partial v_r}{\partial z} - \frac{v_\theta^2}{r} \right) = \rho_0 X_c - \frac{\partial p}{\partial r} + \eta_0 \left(\nabla^2 v_r - \frac{v_r}{r^2} - \frac{2}{r^2} \frac{\partial v_\theta}{\partial \theta} \right) \tag{6.31}$$

$$\rho_0 \left(\frac{\partial v_\theta}{\partial t} + v_r \frac{\partial v_\theta}{\partial r} + \frac{v_\theta}{r} \frac{\partial v_\theta}{\partial \theta} + v_z \frac{\partial v_\theta}{\partial z} + \frac{v_r v_\theta}{r} \right) = \rho_0 Y_c - \frac{1}{r} \frac{\partial p}{\partial \theta} + \eta_0 \left(\nabla^2 v_\theta - \frac{v_\theta}{r^2} + \frac{2}{r^2} \frac{\partial v_r}{\partial \theta} \right) \tag{6.32}$$

$$\rho_0 \left(\frac{\partial v_z}{\partial t} + v_r \frac{\partial v_z}{\partial r} + \frac{v_\theta}{r} \frac{\partial v_z}{\partial \theta} + v_z \frac{\partial v_z}{\partial z} \right) = \rho_0 Z_c - \frac{\partial p}{\partial z} + \eta_0 \nabla^2 v_z \tag{6.33}$$

where

$$\nabla^2 = \frac{\partial^2}{\partial r^2} + \frac{1}{r} \frac{\partial}{\partial r} + \frac{1}{r^2} \frac{\partial^2}{\partial \theta^2} + \frac{\partial^2}{\partial z^2} \tag{6.34}$$

6.2.5.3 Spherical polar coordinates. The comparable equations to (6.31) to (6.34) for the Navier-Stokes equations expressed in spherical polar coordinates r, θ, ϕ where

$$x = r \sin\theta \cos\phi$$
$$y = r \sin\theta \sin\phi \qquad (6.35)$$
$$z = r\cos\theta$$

are

$$\frac{\partial v_r}{\partial t} + v_r \frac{\partial v_r}{\partial r} + \frac{v_\theta}{r}\frac{\partial v_r}{\partial \theta} + \frac{v_\phi}{r\sin\theta}\frac{\partial v_r}{\partial \phi} - \frac{v_\theta^2 + v_\phi^2}{r}$$
$$= X_s + \frac{1}{\rho_0}\frac{\partial p}{\partial r} + \frac{\eta_0}{\rho_0}\left(\nabla^2 v_r - \frac{2v_r}{r^2} - \frac{2}{r^2}\frac{\partial v_\theta}{\partial \theta} - \frac{2v_\theta \cot\theta}{r^2} - \frac{2}{r^2 \sin\theta}\frac{\partial v_\phi}{\partial \phi}\right) \qquad (6.36)$$

$$\frac{\partial v_\theta}{\partial t} + v_r \frac{\partial v_\theta}{\partial r} + \frac{v_\theta}{r}\frac{\partial v_\theta}{\partial \theta} + \frac{v_\phi}{r\sin\theta}\frac{\partial v_\theta}{\partial \phi} + \frac{v_r v_\theta}{r} - \frac{v_\phi^2 \cot\theta}{r}$$
$$= Y_s - \frac{1}{\rho_0}\frac{1}{r}\frac{\partial p}{\partial \theta} + \frac{\eta_0}{\rho_0}\left(\nabla^2 v_\theta + \frac{2}{r^2}\frac{\partial v_r}{\partial \theta} - \frac{v_\theta}{r^2\sin^2\theta} - \frac{2\cos\theta}{r^2\sin^2\theta}\frac{\partial v_\phi}{\partial \phi}\right) \qquad (6.37)$$

$$\frac{\partial v_\phi}{\partial t} + v_r \frac{\partial v_\phi}{\partial r} + \frac{v_\theta}{r}\frac{\partial v_\phi}{\partial \theta} + \frac{v_\phi}{r\sin\theta}\frac{\partial v_\phi}{\partial \phi} + \frac{v_\phi v_r}{r} + \frac{v_\theta v_\phi \cot\theta}{r}$$
$$= Z_s - \frac{1}{\rho_0}\frac{1}{r\sin\theta}\frac{\partial p}{\partial \phi}$$
$$+ \frac{\eta_0}{\rho_0}\left(\nabla^2 v_\phi - \frac{v_\phi}{r^2\sin^2\theta} + \frac{2}{r^2\sin\theta}\frac{\partial v_r}{\partial \phi} + \frac{2\cos\theta}{r^2\sin^2\theta}\frac{\partial v_\theta}{\partial \phi}\right) \qquad (6.38)$$

where

$$\nabla^2 = \frac{1}{r^2}\frac{\partial}{\partial r}\left(r^2\frac{\partial}{\partial r}\right) + \frac{1}{r^2\sin\theta}\frac{\partial}{\partial \theta}\left(\sin\theta\frac{\partial}{\partial \theta}\right) + \frac{1}{r^2\sin^2\theta}\frac{\partial^2}{\partial \phi^2} \qquad (6.39)$$

6.2.5.4 Cartesian coordinates—turbulent flow. Up to this point laminar flow has been assumed. Most of the book will be concerned only with laminar flow. However, it is important to describe the Navier-Stokes equations when confronted with turbulent flow.

For laminar flow the velocity pattern is steady and very predictable, whereas for turbulent flow the velocity pattern changes where eddies and vortices are present and the flow is irregular. Thus, flow properties such as velocities and pressure show random variation with time and position. Because of this the instantaneous value has little practical significance, and it is the mean value that is of interest.

Thus, we may express the pressure and velocities as

$$\begin{aligned} p &= p^* + p' \qquad & v &= v^* + v' \\ u &= u^* + u' \qquad & w &= w^* + w' \end{aligned} \tag{6.40}$$

where asterisked values are mean and primed values are fluctuations.

Substituting Eq. (6.40) into Eqs. (6.28) to (6.30) and following the procedure given, for example, in Szeri (1980) gives

$$\begin{aligned} \rho^* \frac{\partial u^*}{\partial t} &+ \rho^* u^* \frac{\partial u^*}{\partial x} + \rho^* v^* \frac{\partial u^*}{\partial y} + \rho^* w^* \frac{\partial u^*}{\partial z} \\ &= \rho^* X_a - \frac{\partial p^*}{\partial x} + \eta_0 \frac{\partial^2 u^*}{\partial z^2} - \frac{\partial}{\partial x}[\rho^*(u'u')^*] \\ &\quad - \frac{\partial}{\partial y}[\rho^*(u'v')^*] - \frac{\partial}{\partial z}[\bar{\rho}(u'w')^*] \end{aligned} \tag{6.41}$$

$$\begin{aligned} \rho^* \frac{\partial v^*}{\partial t} &+ \rho^* u^* \frac{\partial v^*}{\partial x} + \rho^* v^* \frac{\partial^*}{\partial y} + \rho^* w^* \frac{\partial v^*}{\partial z} \\ &= \rho^* Y_a - \frac{\partial p^*}{\partial y} + \eta_0 \frac{\partial^2 v^*}{\partial z^2} - \frac{\partial v^*}{\partial x}[\rho^*(u'v')^*] \\ &\quad - \frac{\partial}{\partial y}[\rho^*(v'v')^*] - \frac{\partial}{\partial z}[\rho^*(v'w')^*] \end{aligned} \tag{6.42}$$

$$\begin{aligned} \rho^* \frac{\partial w^*}{\partial t} &+ \rho^* u^* \frac{\partial w^*}{\partial x} + \rho^* v^* \frac{\partial w^*}{\partial z} + \rho^* w^* \frac{\partial w^*}{\partial z} \\ &= \rho^* Z_a - \frac{\partial p^*}{\partial z} + \eta_0 \frac{\partial^2 u^*}{\partial z^2} - \frac{\partial}{\partial x}[\rho^*(u'w')^*] \\ &\quad - \frac{\partial}{\partial y}[\rho^*(v'w')^*] - \frac{\partial}{\partial z}[\rho^*(w'w')^*] \end{aligned} \tag{6.43}$$

If we neglect inertia terms and body force terms and retain only dominant turbulent stresses, the above equations become

$$0 = -\frac{\partial p^*}{\partial x} + \eta_0 \frac{\partial^2 u^*}{\partial z^2} - \frac{\partial}{\partial z}[\rho^*(u'w')^*] \tag{6.44}$$

$$0 = -\frac{\partial p^*}{\partial y} + \eta_0 \frac{\partial^2 v^*}{\partial z^2} - \frac{\partial}{\partial z}[\rho^*(v'w')^*] \tag{6.45}$$

$$0 = -\frac{\partial p^*}{\partial z} + \eta_0 \frac{\partial^2 w^*}{\partial z^2} - \frac{\partial}{\partial z}[\rho^*(w'w')^*] \tag{6.46}$$

Equations (6.44) to (6.46) show that in turbulent flow the pressure distribution across the film is no longer hydrostatic, as it was in laminar flow.

6.3 CONTINUITY EQUATION

The Navier-Stokes equations contain three equations and four unknowns: u, v, w, and p. The viscosity and density can be written as functions of pressure and temperature. A fourth equation is supplied by the continuity equation. The principle of mass conservation requires that the net outflow of mass from a volume of fluid must be equal to the decrease of mass within the volume. This is readily calculated with reference to Fig. 6.4. The flow of mass per unit time and area through a surface is a product of the velocity normal to the surface and the density. Thus, the x component of mass flux per unit area at the center of the volume is ρu. This flux, however, changes from point to point as indicated in Fig. 6.4. The net outflow of mass per unit time therefore is

$$\left[\rho u + \frac{1}{2}\frac{\partial(\rho u)}{\partial x}dx\right]dz + \left[\rho w + \frac{1}{2}\frac{\partial(\rho w)}{\partial z}dz\right]dx$$
$$-\left[\rho u - \frac{1}{2}\frac{\partial(\rho u)}{\partial x}dx\right]dz - \left[\rho w - \frac{1}{2}\frac{\partial(\rho w)}{\partial z}dz\right]dx$$

and this must be equal to the rate of mass decrease within the element,

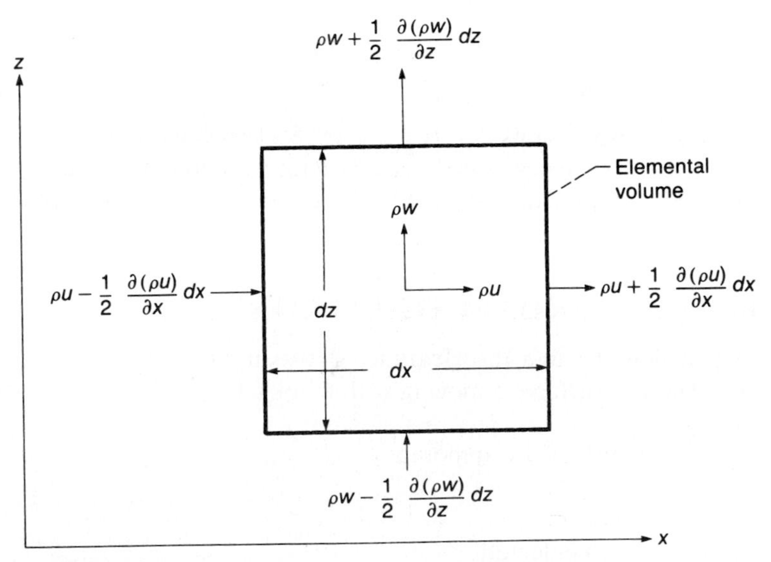

FIGURE 6.4
Velocities and densities for mass flow balance through a flux volume element in two dimensions.

$$-\frac{\partial \rho}{\partial t} dx\, dz$$

Upon simplification this becomes

$$\frac{\partial \rho}{\partial t} + \frac{\partial}{\partial x}(\rho u) + \frac{\partial}{\partial z}(\rho w) = 0$$

When the y direction is included, the continuity equation results.

$$\frac{\partial \rho}{\partial t} + \frac{\partial}{\partial x}(\rho u) + \frac{\partial}{\partial y}(\rho v) + \frac{\partial}{\partial z}(\rho w) = 0 \tag{6.47}$$

If the force density is a constant, the continuity equation becomes

$$\frac{\partial u}{\partial x} + \frac{\partial v}{\partial y} + \frac{\partial w}{\partial z} = 0 \tag{6.48}$$

This equation is valid whether or not the velocity is time dependent.

The continuity equation in cylindrical polar coordinates where $z = z$, $x = r \cos\theta$, and $y = r \sin\theta$ can be expressed as

$$\frac{\partial \rho}{\partial t} + \frac{1}{r}\frac{\partial}{\partial r}(\rho r v_r) + \frac{1}{r}\frac{\partial}{\partial \theta}(\rho v_\theta) + \frac{\partial}{\partial z}(\rho v_z) = 0 \tag{6.49}$$

The continuity equation in spherical polar coordinates (r, θ, ϕ) while making use of Eq. (6.35) is

$$\frac{\partial \rho}{\partial t} + \frac{1}{r^2}\frac{\partial}{\partial r}(\rho r^2 v_r) + \frac{1}{r \sin\theta}\frac{\partial}{\partial \theta}(\rho v_\theta \sin\theta) + \frac{1}{r \sin\theta}\frac{\partial}{\partial \phi}(\rho v_\phi) = 0 \tag{6.50}$$

The continuity equation for turbulent flow is

$$\frac{\partial \rho^*}{\partial t} + \frac{\partial(\rho^* u^*)}{\partial x} + \frac{\partial(\rho^* v^*)}{\partial y} + \frac{\partial(\rho^* w^*)}{\partial z} = 0 \tag{6.51}$$

Now that general expressions for the Navier-Stokes equations and the continuity equation have been developed, the next four sections will illustrate how these equations in simplified form can be used for some specific applications.

6.4 FLOW BETWEEN PARALLEL FLAT PLATES

Consider the rate of flow through the clearance between two parallel surfaces shown in Fig. 6.5. The top surface is moving with a velocity u_a and the bottom surface is at rest.

The following assumptions are imposed:

1. The inertia effect is small.
2. Body force terms can be neglected.
3. Viscosity and density can be considered constant.
4. $dp/dz = dp/dy = 0$.
5. The film thickness is much smaller than other dimensions.

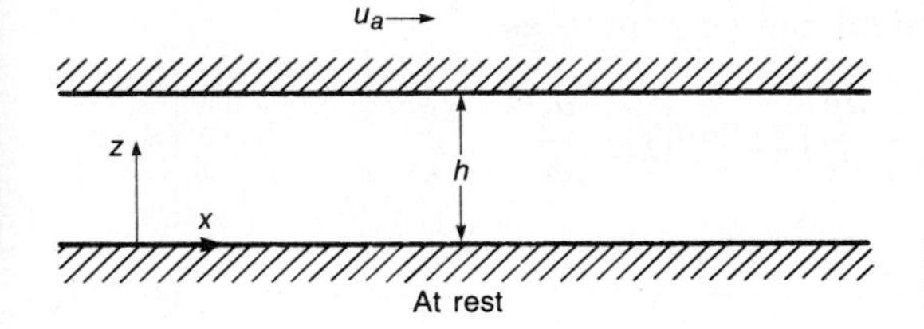

FIGURE 6.5
Flow between parallel flat plates.

For these considerations the reduced Navier-Stokes equation is

$$\frac{\partial^2 u}{\partial z^2} = \frac{1}{\eta_0}\frac{dp}{dx}$$

Integrating twice gives

$$u = \frac{1}{\eta_0}\frac{dp}{dx}\frac{z^2}{2} + \tilde{A}z + \tilde{B} \tag{6.52}$$

where $\tilde{A}$ and $\tilde{B}$ are integration constants. The no-slip boundary conditions are imposed:

1. $z = 0,\ u = 0$
2. $z = h,\ u = u_a$

From boundary condition 1, $\tilde{B} = 0$. Boundary condition 2 gives

$$\tilde{A} = \frac{u_a}{h} - \frac{h}{2\eta_0}\frac{dp}{dx}$$

$$\therefore u = \frac{1}{2\eta_0}\frac{dp}{dx}(z^2 - zh) + \frac{u_a z}{h} \tag{6.53}$$

For $dp/dx = 0$ the *Couette term* is $u = u_a z/h$. For $u_a = 0$ the *Poiseuille term* is $u = -(dp/dx)\,z(h - z)/2\eta_0$. Figure 6.6 shows the Couette and Poiseuille

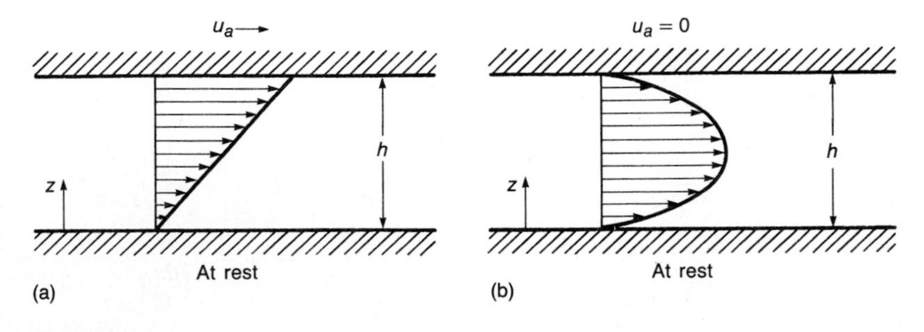

FIGURE 6.6
(a) Couette and (b) Poiseuille velocity profiles.

velocity profiles. The velocity gradient can be written as

$$\frac{du}{dz} = \frac{1}{2\eta_0}\frac{dp}{dx}(2z - h) + \frac{u_a}{h} \tag{6.54}$$

$$\left(\frac{du}{dz}\right)_{z=0} = -\frac{h}{2\eta_0}\frac{dp}{dx} + \frac{u_a}{h} \tag{6.55}$$

$$\left(\frac{du}{dz}\right)_{z=h} = \frac{h}{2\eta_0}\frac{dp}{dx} + \frac{u_a}{h} \tag{6.56}$$

Some other interesting velocity profiles are shown in Fig. 6.7.

The volume flow rate per unit width can be written as

$$q' = \int_0^h u\,dz \tag{6.57}$$

Substituting Eq. (6.53) into this equation gives

$$q' = \frac{1}{2\eta_0}\frac{dp}{dx}\left(\frac{z^3}{3} - \frac{z^2h}{2}\right)_{z=0}^{z=h} + \frac{u_a}{h}\left(\frac{z^2}{2}\right)_{z=0}^{z=h}$$

$$= \underset{\text{Poiseuille}}{-\frac{h^3}{12\eta_0}\frac{dp}{dx}} + \underset{\text{Couette}}{\frac{u_a h}{2}} \tag{6.58}$$

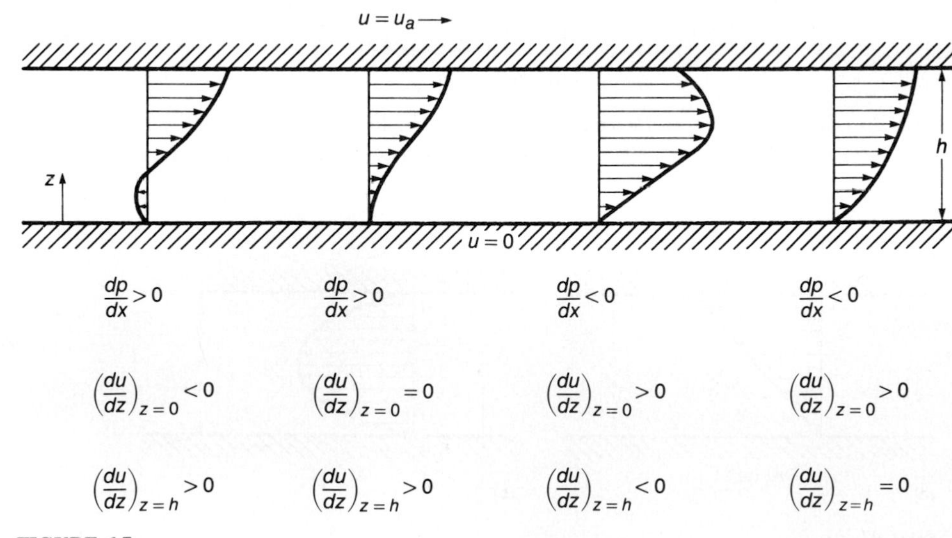

FIGURE 6.7
Some interesting velocity profiles.

6.5 FLOW IN A CIRCULAR PIPE

Consider the flow in a circular pipe as shown in Fig. 6.8. Cylindrical coordinates will be used with their origin at the tube center. The fluid velocity is zero at the pipe walls. The pressure at the left end of the tube is higher than that at the right end and drops gradually along the tube length. This pressure causes the fluid to flow from left to right.

The following assumptions are imposed:

1. Viscosity and density can be considered constant.
2. The inertia effect is small.
3. Body force terms can be neglected.
4. $dp/dr = dp/d\theta = 0$.
5. $v_r = v_\theta = 0$ and $v_z = f(r)$.

With these assumptions the Navier-Stokes equations in cylindrical polar coordinates as expressed in Eqs. (6.31) to (6.33) reduce to the following equations:

$$0 = -\frac{\partial p}{\partial z} + \eta_0\left(\frac{d^2 v_z}{dr^2} + \frac{1}{r}\frac{dv_z}{dr}\right)$$

or

$$\frac{r}{\eta_0}\frac{dp}{dz} = r\frac{d^2 v_z}{dr^2} + \frac{dv_z}{dr} = \frac{d}{dr}\left(r\frac{dv_z}{dr}\right)$$

Integrating once gives

$$\frac{dv_z}{dr} = \frac{r}{2\eta_0}\frac{dp}{dz} + \frac{\tilde{A}}{r} \tag{6.59}$$

Integrating again gives

$$v_z = \frac{r^2}{4\eta_0}\frac{dp}{dz} + \tilde{A}\ln r + \tilde{B} \tag{6.60}$$

The boundary conditions are

1. $v_z = 0$ when $r = a$.
2. From considerations of symmetry $dv_z/dr = 0$ when $r = 0$.

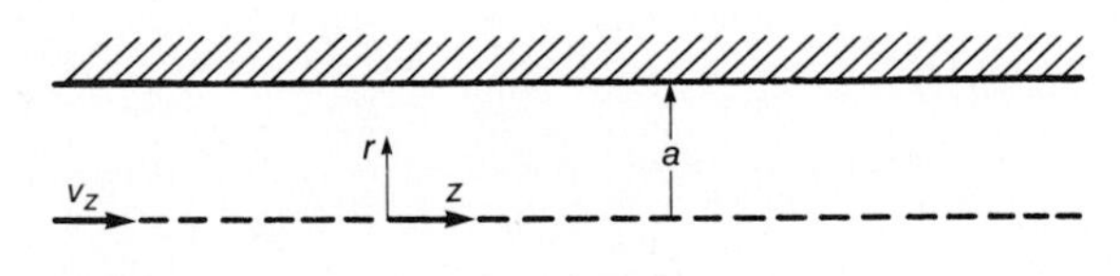

FIGURE 6.8
Flow in a circular pipe.

Making use of boundary condition 2 and Eq. (6.59) results in $\tilde{A} = 0$. Making use of boundary condition 1 gives

$$\tilde{B} = -\frac{a^2}{4\eta_0}\frac{dp}{dz}$$

$$\therefore v_z = -\frac{1}{4\eta_0}\frac{dp}{dz}(a^2 - r^2) \tag{6.61}$$

The volume flow rate can be written as

$$q = 2\pi\int_0^a v_z r\,dr$$

Substituting Eq. (6.61) into this equation gives

$$q = -\frac{\pi}{2\eta_0}\frac{dp}{dz}\int_0^a (a^2 r - r^3)\,dr$$

or

$$q = -\frac{\pi a^4}{8\eta_0}\frac{dp}{dz} \tag{6.62}$$

Note that a negative pressure gradient is required to get a positive flow in the z direction.

6.6 FLOW DOWN A VERTICAL PLANE

Consider a vertical plane and a fluid moving down the plane due to gravity. Figure 6.9 describes the physical situation. The fluid has a uniform thickness of h along the plane.

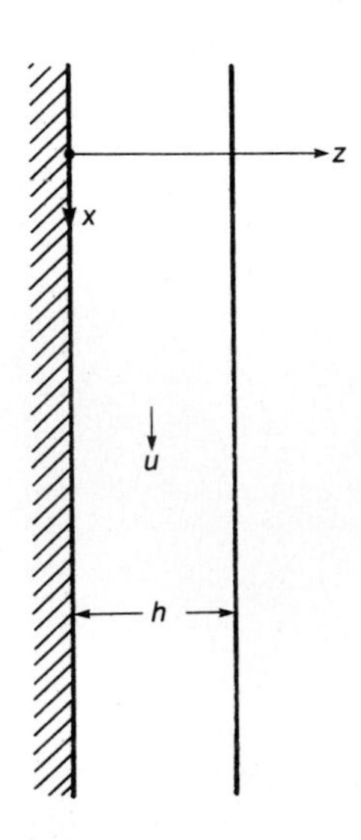

FIGURE 6.9
Flow down a vertical plane.

The following assumptions are imposed:

1. Viscosity and density can be considered constant.
2. The inertia effect is small.
3. The body force term in the x direction contains gravitational acceleration; the body force terms in the y, z directions are zero ($Y_a = Z_a = 0$).
4. There are no pressure gradients.
5. The fluid velocity varies only in the z direction.
6. The film thickness is much smaller than the other dimensions.

With these assumptions the Navier-Stokes equations as expressed in Eqs. (6.28) to (6.30) reduce to

$$\frac{d^2u}{dz^2} = -\frac{\rho_0 g}{\eta_0}$$

Integrating gives

$$\frac{du}{dz} = -\frac{\rho_0 g}{\eta_0}z + \tilde{A} \tag{6.63}$$

Integrating again gives

$$u = -\frac{\rho_0 g}{\eta_0}\frac{z^2}{2} + \tilde{A}z + \tilde{B} \tag{6.64}$$

The boundary conditions are

1. $u = 0$ when $z = 0$.
2. Assuming negligible air resistance, the shear stress at the free surface must be zero.

$$\therefore \frac{du}{dz} = 0 \qquad \text{when } z = h$$

From boundary condition 2 and Eq. (6.63),

$$\tilde{A} = \frac{\rho_0 gh}{\eta_0}$$

$$\therefore u = -\frac{\rho_0 g}{\eta_0}\frac{z^2}{2} + \frac{\rho_0 ghz}{\eta_0} + \tilde{B} \tag{6.65}$$

From boundary condition 1, $\tilde{B} = 0$;

$$\therefore u = \frac{\rho_0 gz}{2\eta_0}(2h - z) \tag{6.66}$$

The volume flow rate per unit width can be expressed as

$$q' = \int_0^h u \, dz$$

Substituting Eq. (6.66) into this equation gives

$$q' = \frac{\rho_0 g h^3}{3\eta_0} \tag{6.67}$$

The main goal of presenting the three simple viscous flow examples just covered is to show the importance of the various terms in the Navier-Stokes equations. These simple solutions are also used in describing the various types of viscometer in the next section.

6.7 VISCOMETERS

The viscosity of fluids can be measured by many methods based on different principles. Only the more typical or most important types of viscometers are discussed here. Also, the emphasis is on the principles by which these viscometers operate. The following classifications of viscometers are considered: capillary, rotational, and falling sphere. Each of these is considered separately.

6.7.1 CAPILLARY VISCOMETERS. This type of viscometer (shown in Fig. 6.10) is based on measuring the rate at which a fluid flows through a small-diameter

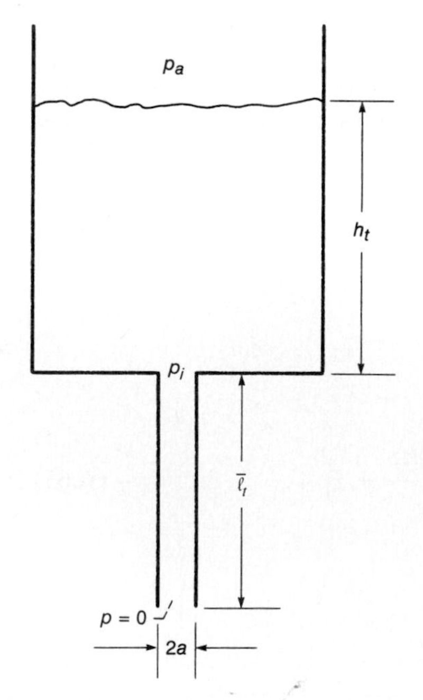

FIGURE 6.10
Important features of a capillary viscometer.

tube. Usually, this takes the form of measuring the time taken to discharge a given quantity of fluid.

From the flow in a circular pipe covered earlier in this chapter [Eq. (6.62)],

$$\frac{dp}{dz} = -\frac{8\eta_0 q}{\pi a^4}$$

If p_i is the pressure at the inlet of the capillary tube and $\bar{\ell}_t$ is the length of the capillary tube,

$$-\frac{dp}{dz} = \frac{p_i}{\bar{\ell}_t}$$

$$\therefore\ p_i = \frac{8\eta_0 q \bar{\ell}_t}{\pi a^4}$$

But the pressure head developed is simply

$$p_i = \rho_0 g h_t$$

where h_t is the height of the capillary tube and ρ_0 is the force density at $p = 0$ and a constant temperature.

$$\therefore\ \rho_0 g h_t = \frac{8\eta_0 q \bar{\ell}_t}{\pi a^4}$$

or

$$h_t = \frac{8\eta_0 q \bar{\ell}_t}{\pi a^4 \rho_0 g} = A^* \eta_{k,0} q \tag{6.68}$$

where $\eta_{k,0} = \eta_0/\rho_0$ is the kinematic viscosity at $p = 0$ and a fixed temperature and

$$A^* = \frac{8\bar{\ell}_t}{\pi g a^4}$$

Recall that q is the volume flow rate per unit time.

$$q \propto \frac{1}{t}$$

$$\therefore\ \eta_{k,0} = \frac{h_t}{A^* q} = B^* t \tag{6.69}$$

where B^* is a constant that is a function of the apparatus.

6.7.2 ROTATIONAL VISCOMETERS. Two different types of rotational viscometer are considered: the rotational cylindrical viscometer and the cone-and-plate viscometer.

6.7.2.1 Rotational cylindrical viscometer. As shown in Fig. 6.11 the rotational cylindrical viscometer consists of two concentric cylinders with a fluid contained

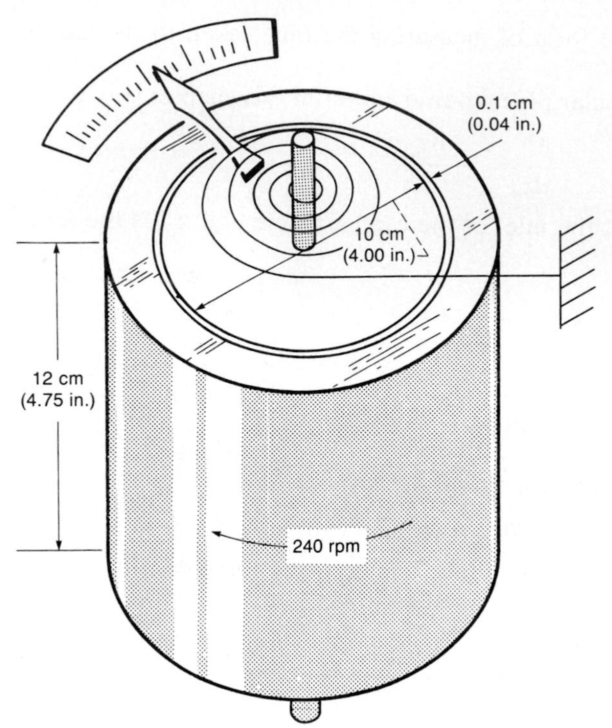

FIGURE 6.11
Rotational cylindrical viscometer.

between them. The outer cylinder rotates and torque is measured at the inner cylinder. Let

r_i	inner cylinder radius
r_o	outer cylinder radius
ℓ_a	length of annulus
c	radial clearance, $r_o - r_i$ ($c \ll r_i$)
ω	angular velocity

From Newton's postulate [Eq. (4.2)]

$$f = \eta_0 A \frac{u}{c} \tag{4.2}$$

where A = area, $2\pi r_o \ell_a$
u = velocity, $r_o \omega$

$$\therefore f = \eta_0 (2\pi r_o \ell_a) \frac{\omega r_o}{c}$$

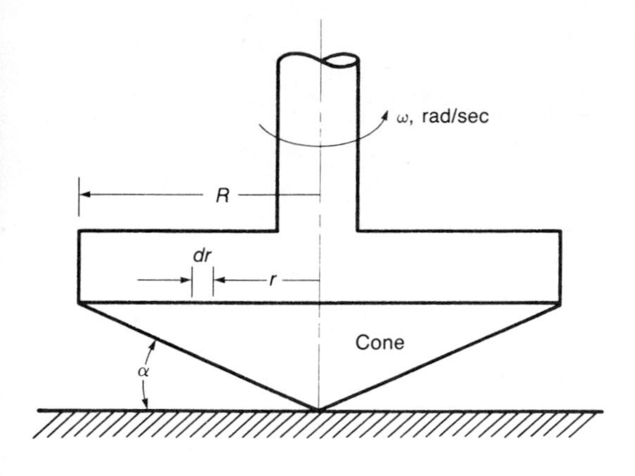

FIGURE 6.12
Cone-and-plane viscometer.

The torque on the inner cylinder is

$$t_q = fr_i = \frac{2\pi\eta_0\omega r_o^2 r_i \ell_a}{c}$$

or

$$\eta_0 = \frac{t_q c}{2\pi\omega r_o^2 r_i \ell_a} \tag{6.70}$$

6.7.2.2 Cone-and-plane viscometer. Figure 6.12 shows the essentials of a cone-and-plane viscometer. The angle α is small. The surface speed of the cone at any radius r is $u = \omega r$. The film thickness is $h = r \tan \alpha \approx r\alpha$. From Newton's postulate

$$f = \eta_0 A \frac{u}{h} = \int_0^R \eta_0 (2\pi r\, dr) \frac{\omega r}{r\alpha} = \int_0^R \frac{2\pi\eta_0\omega}{\alpha} r\, dr$$

The torque is

$$t_q = \frac{2\pi\eta_0\omega}{\alpha} \int_0^R r^2\, dr = \frac{2\pi\eta_0\omega R^3}{3\alpha}$$

$$\therefore \eta_0 = \frac{3t_q\alpha}{2\pi\omega R^3} \tag{6.71}$$

6.7.3 FALLING-SPHERE VISCOMETERS. The absolute viscosity η_0 of a fluid at $p = 0$ and at a constant temperature can be determined by measuring the time it takes for a ball to fall through a tube (preferably of glass so that the ball can be easily observed). If a sphere is falling through a fluid under a constant force, it will assume a constant velocity. Stokes' formula can be applied for a sphere moving through an infinite fluid. A sphere falling freely under gravity in a liquid will be attaining a velocity u_a given by

$$u_a = \frac{2r^2(\rho_s - \rho_f)g}{\eta_0} \tag{6.72}$$

where r = radius of sphere, m
ρ_s = force density of sphere, $N \cdot s^2/m^4$
ρ_f = force density of fluid, $N \cdot s^2/m^4$
g = gravitational acceleration, m/s^2
η_0 = absolute viscosity at $p = 0$ and a fixed temperature, $N \cdot s/m^2$

The Stokes formula given in Eq. (6.72) is for an infinite fluid and not for a fluid in a glass tube as is the case in a falling-sphere viscometer. The correction to account for the tube diameter is

$$u_a = \frac{2r^2(\rho_s - \rho_f)g}{\eta_0} c_f \tag{6.73}$$

where

$$c_f = 1 - 2.104\left(\frac{r}{R}\right) + 2.09\left(\frac{r}{R}\right)^3 - 0.9\left(\frac{r}{R}\right)^5 \tag{6.74}$$

and R is the radius of the tube.

If the sphere falls at a constant velocity $u_a = \bar{h}_t/t$, where $\bar{h}_t$ is the height from which the sphere falls and t is the time it takes to fall this distance, the absolute viscosity at $p = 0$ and a fixed temperature can be expressed as

$$\eta_0 = \frac{2r^2(\rho_s - \rho_f)gtc_f}{\bar{h}_t} \tag{6.75}$$

Since all the other parameters are fixed for this viscometer, for a falling-sphere viscometer the absolute viscosity is directly proportional to the time it takes for the sphere to travel a fixed distance.

6.8 CLOSURE

This chapter focused on viscous fluid flow, which is important in understanding fluid film lubrication. Petrov's equation was derived and gave the friction torque in a concentric journal bearing. Navier-Stokes equations were derived in general form and gave the dynamic equilibrium of a fluid element. It was necessary to consider the surface forces, body forces, and inertia of a fluid element. The continuity equation was also derived by applying the principle of mass conservation to a fluid element. The Navier-Stokes equations and the continuity equations were expressed in Cartesian, cylindrical, and spherical polar coordinates.

Having the general expressions for the Navier-Stokes and continuity equations, the next development in this chapter was to illustrate how these equations can be applied to various physical conditions. Some of the physical conditions considered were flow between parallel flat plates, flow in a circular pipe, flow down a vertical plane, and various types of viscometers. The emphasis in considering these applications was to point out the significance of the various terms in the Navier-Stokes equations. The presence of the Couette and Poiseuille terms in the velocity and flow was emphasized.

6.9 PROBLEMS

6.9.1 A viscous fluid flows parallel to the axis in the annular space between two coaxial cylinders of radii R_1 and R_2 and infinite length. Show that the flow rate between these coaxial cylinders when both cylinders are at rest and there is an axial pressure gradient is

$$q = -\frac{\pi R_1^4}{8\eta_0}\frac{dp}{dz}\left\{\left(\frac{R_2}{R_1}\right)^4 - 1 - \frac{\left[(R_2/R_1)^2 - 1\right]^2}{\ln(R_2/R_1)}\right\}$$

6.9.2 Two immiscible liquids of viscosities η_1 and η_2 flow together down a vertical plane in distinct layers of thicknesses h_1 and h_2. The thinner fluid is assumed to be closest to the plane.

What are the boundary conditions that determine the velocity distribution? Determine the velocity and flow rate for the two regions. For the special case of $\eta_2 = 2\eta_1$, $\rho_1 = \rho_2$, and $h_1 = h_2$, show that $q_2/q_1 = 2$.

6.9.3 Viscous fluid is contained between two parallel flat plates that are a distance h apart. One of the plates moves in its own plane with velocity $3u_a$, and the other in the opposite direction with velocity u_a. If the flow in the direction of motion of the faster plate is $u_a h/2$ per unit width, determine the pressure gradient.

6.9.4 A viscous fluid is contained between two parallel flat plates that are a distance h apart, one of which is fixed while the other moves in its own plane with velocity u_a. There is a pressure gradient dp/dx in the direction of motion.

Show that the velocity distribution can be expressed in the form

$$\frac{u}{u_a} = \lambda(\zeta - \zeta^2) + \zeta$$

where

$$\zeta = \frac{z}{h} \qquad \lambda = 6\frac{u_m}{u_a}$$

and u_m is the mean velocity due to the pressure gradient alone. Find the value of λ corresponding to

(*a*) Couette flow
(*b*) Poiseuille flow
(*c*) A flow for which the volume flow rate is zero
(*d*) Zero shear stress on the fixed plate
(*e*) Zero shear stress on the moving plate

6.9.5 A laminar flow takes place along a pipe of radius a that is included at an angle of α to the horizontal. Define the appropriate Navier-Stokes equation. Determine what the velocity (v_z), mean velocity ($\tilde{v}_z$), and volume flow rate are in the pipe.

Show also that the pressure will remain constant along the pipe when

$$\alpha = \sin^{-1} \frac{8\eta_0 \tilde{v}_z}{\rho_0 g a^2}$$

The viscosity (η_0) and density (ρ_0) are assumed constant throughout the pipe.

6.9.6 A shaft of radius R_1 is held concentric within a sleeve of radius R_2. If the shaft is stationary while the sleeve moves with an axial velocity of v_0, determine an equation for the volume flow rate. Also, what is the force per unit length required to move the sleeve? The pressure gradient in the direction of motion is assumed to be zero. Constant viscosity and density are assumed. Inertia and body force terms are neglected. Assume $(R_2 - R_1)/R_1 \approx 1 \times 10^{-3}$.

6.10 REFERENCES

Navier, C. L. M. H. (1823): Memoire sur les lois du mouvement des fluides. *Mem. Acad. Sci. Inst. Fr.*, vol. 6, pp. 389–416.

Petrov, N. P. (1883): Friction in Machines and the Effect of the Lubricant. *Inzh. Zh. St. Petersburg*, vol. 1, pp. 71–140; vol. 2, pp. 227–279; vol. 3, pp. 377–463; vol. 4, pp. 535–564.

Stokes, G. G. (1845): On the Theories of the Internal Friction of Fluids in Motion, and of the Equilibrium and Motion of Elastic Solids. *Trans. Cambridge Philos. Soc.*, vol. 8, pp. 287–341.

Szeri, A. Z. (ed.) (1980): *Tribology—Friction, Lubrication, and Wear*. Hemisphere Publishing Corp., Washington, D.C.

CHAPTER

7

REYNOLDS EQUATION

The full Navier-Stokes equations presented in the last chapter, in which inertia, body, pressure, and viscous terms are included, are sufficiently complicated to prohibit analytical solutions to most practical problems. There is, however, a class of flow condition known as "slow viscous motion" in which the pressure and viscous terms predominate. Fluid film lubrication problems are in this class. It will first be demonstrated that indeed the pressure and viscous terms of the Navier-Stokes equations are the most important terms.

7.1 DIMENSIONLESS NUMBERS

The following characteristic parameters may be defined:

b_0 characteristic length in y direction, m
h_0 characteristic length in z direction, m
ℓ_0 characteristic length in x direction, m
t_0 characteristic time, s
u_0 characteristic velocity in x direction, m/s
v_0 characteristic velocity in y direction, m/s
w_0 characteristic velocity in z direction, m/s
ρ_0 characteristic force density, $\mathrm{N \cdot s^2/m^4}$
η_0 characteristic absolute viscosity, $\mathrm{N \cdot s/m^2}$

These characteristic parameters are used to define the following dimensionless parameters:

$$X = \frac{x}{\ell_0} \quad Y = \frac{y}{b_0} \quad Z = \frac{z}{h_0} \quad T = \frac{t}{t_0} \quad \bar{u} = \frac{u}{u_0}$$
$$\bar{v} = \frac{v}{v_0} \quad \bar{w} = \frac{w}{w_0} \quad \bar{\rho} = \frac{\rho}{\rho_0} \quad \bar{\eta} = \frac{\eta}{\eta_0} \quad P = \frac{h_0^2 p}{\eta_0 u_0 \ell_0} \tag{7.1}$$

Substituting Eqs. (7.1) into the first Navier-Stokes equation expressed in Eq. (6.25) while making use of Eqs. (6.9) and (6.18) as well as X_a in the body force term being equivalent to gravitational acceleration g gives

$$\frac{\ell_0}{u_0 t_0}\frac{\partial \bar{u}}{\partial T} + \bar{u}\frac{\partial \bar{u}}{\partial X} + \frac{\ell_0}{b_0}\frac{v_0}{u_0}\bar{v}\frac{\partial \bar{u}}{\partial Y} + \frac{\ell_0}{h_0}\frac{w_0}{u_0}\bar{w}\frac{\partial \bar{u}}{\partial Z}$$
$$= \frac{\ell_0 g}{u_0^2} - \frac{\eta_0}{\rho_0 u_0 \ell_0}\left(\frac{\ell_0}{h_0}\right)^2 \frac{1}{\bar{\rho}}\frac{\partial P}{\partial X} - \frac{2}{3}\frac{\eta_0}{\rho_0 u_0 \ell_0}\frac{1}{\bar{\rho}}$$
$$\times \frac{\partial}{\partial X}\left[\bar{\eta}\left(\frac{\partial \bar{u}}{\partial X} + \frac{v_0}{u_0}\frac{\ell_0}{b_0}\frac{\partial \bar{v}}{\partial Y} + \frac{w_0}{u_0}\frac{\ell_0}{h_0}\frac{\partial \bar{w}}{\partial Z}\right)\right]$$
$$+ 2\frac{\eta_0}{\rho_0 u_0 \ell_0}\frac{1}{\bar{\rho}}\frac{\partial}{\partial X}\left(\bar{\eta}\frac{\partial \bar{u}}{\partial X}\right) + \frac{\eta_0}{\rho_0 u_0 \ell_0}\left(\frac{\ell_0}{b_0}\right)^2 \frac{1}{\bar{\rho}}$$
$$\times \frac{\partial}{\partial Y}\left[\bar{\eta}\left(\frac{\partial \bar{u}}{\partial Y} + \frac{v_0}{u_0}\frac{b_0}{\ell_0}\frac{\partial \bar{v}}{\partial X}\right)\right] + \frac{\eta_0}{\rho_0 u_0 \ell_0}\left(\frac{\ell_0}{h_0}\right)^2 \frac{1}{\bar{\rho}}$$
$$\times \frac{\partial}{\partial Z}\left[\bar{\eta}\left(\frac{\partial \bar{u}}{\partial Z} + \frac{w_0}{u_0}\frac{h_0}{\ell_0}\frac{\partial \bar{w}}{\partial X}\right)\right] \tag{7.2}$$

The inertia, pressure, viscous, and gravitational effects from Eq. (7.2) are compared in defining several dimensionless numbers.

7.1.1 REYNOLDS NUMBER. The relative importance of inertia to viscous forces in any flow problem can be judged from the value of the Reynolds number $\mathscr{R}$.

$$\mathscr{R} = \frac{\text{inertia}}{\text{viscous}} = \frac{\rho_0 u_0 \ell_0}{\eta_0} \tag{7.3}$$

Note that the inverse of the Reynolds number occurs throughout Eq. (7.2). The Reynolds number expressed in Eq. (7.3) is the conventional Reynolds number found in fluid mechanics. However, in fluid film lubrication because of the dominance of the viscous term $\partial^2 \bar{u}/\partial Z^2$ the modified Reynolds number $\mathscr{R}_x$ is used. It is defined as

$$\mathscr{R}_x = \frac{\text{inertia}}{\text{viscous}} = \frac{\rho_0 u_0 h_0^2}{\eta_0 \ell_0} \tag{7.4}$$

Note the occurrence of the modified Reynolds number in Eq. (7.2). The modified Reynolds numbers are also defined in the y and z directions as

$$\mathscr{R}_y = \frac{\rho_0 v_0 h_0^2}{\eta_0 b_0} \tag{7.5}$$

$$\mathscr{R}_z = \frac{\rho_0 w_0 h_0}{\eta_0} \tag{7.6}$$

The squeeze number is also defined as

$$\sigma_s = \frac{\rho_0 h_0^2}{\eta_0 t_0} \tag{7.7}$$

Recall that $\mathscr{R}_x$, $\mathscr{R}_y$, $\mathscr{R}_z$, and σ_s are all dimensionless and of order h_0/ℓ_0. Two sample problems illustrate that this is true.

Sample problem 7.1. Typical journal bearing. Typical values of parameters used in defining the modified Reynolds number for a journal bearing are

$$d = 0.05 \text{ m}$$

$$\ell_0 \approx \pi d = 0.157 \text{ m}$$

$$\eta_0 = 0.5 \text{ N} \cdot \text{s/m}^2$$

$$\rho_0 = 850 \text{ N} \cdot \text{s}^2/\text{m}^4$$

$$N_a = 2000 \text{ r/min}$$

$$u_0 = 2000 \text{ r/min } (\pi d/1 \text{ r})(1 \text{ min}/60 \text{ s}) = 5.24 \text{ m/s}$$

$$h_0 = c = d/1000 = 5 \times 10^{-5} \text{ m}$$

From Eq. (7.4)

$$\mathscr{R}_x = \frac{\rho_0 u_0 h_0^2}{\eta_0 \ell_0} = \frac{(850)(5.24)(5 \times 10^{-5})^2}{(0.5)(0.157)} = 0.142 \times 10^{-3}$$

Sample problem 7.2 Typical thrust bearing pad.

$$\ell_0 = 0.03 \text{ m}$$

$$h_0/\ell_0 = 1 \times 10^{-3}$$

$$u_0 = 20 \text{ m/s}$$

$$\eta_0 = 0.5 \text{ N} \cdot \text{s/m}^2$$

$$\rho_0 = 850 \text{ N} \cdot \text{s}^2/\text{m}^4$$

From Eq. (7.4)

$$\mathscr{R}_x = \frac{\rho_0 u_0 h_0^2}{\eta_0 \ell_0} = \frac{(850)(20)(0.03)(10^{-6})}{(0.5)} = 1.02 \times 10^{-3}$$

In both sample problems the modified Reynolds number is considerably less than unity and of the order h_0/ℓ_0. It is clear that in typical hydrodynamically lubricated bearings the viscous forces are much greater than the inertia forces.

Substituting Eqs. (7.4) to (7.7) into Eq. (7.2) gives the first Navier-Stokes equation as

$$\sigma_s \frac{\partial \bar{u}}{\partial T} + \mathscr{R}_x \bar{u} \frac{\partial \bar{u}}{\partial X} + \mathscr{R}_y \bar{v} \frac{\partial \bar{u}}{\partial Y} + \mathscr{R}_z \bar{w} \frac{\partial \bar{u}}{\partial Z}$$

$$= g \frac{\ell_0}{u_0^2} \mathscr{R}_x - \frac{1}{\bar{\rho}} \frac{\partial P}{\partial X} + \frac{1}{\bar{\rho}} \frac{\partial}{\partial Z}\left(\bar{\eta} \frac{\partial \bar{u}}{\partial Z}\right) - \frac{2}{3}\left(\frac{h_0}{\ell_0}\right)^2 \frac{1}{\bar{\rho}}$$

$$\times \frac{\partial}{\partial X}\left[\bar{\eta}\left(\frac{\partial \bar{u}}{\partial X} + \frac{v_0}{u_0}\frac{\ell_0}{b_0}\frac{\partial \bar{v}}{\partial Y} + \frac{w_0}{u_0}\frac{\ell_0}{h_0}\frac{\partial \bar{w}}{\partial Z}\right)\right] + \left(\frac{h_0}{b_0}\right)^2 \frac{1}{\bar{\rho}}$$

$$\times \frac{\partial}{\partial Y}\left[\bar{\eta}\left(\frac{\partial \bar{u}}{\partial Y} + \frac{v_0}{u_0}\frac{b_0}{\ell_0}\frac{\partial \bar{v}}{\partial X}\right)\right] + 2\left(\frac{h_0}{\ell_0}\right)^2 \frac{1}{\bar{\rho}}$$

$$\times \frac{\partial}{\partial X}\left(\bar{\eta}\frac{\partial \bar{u}}{\partial X}\right) + \frac{1}{\bar{\rho}}\frac{\partial}{\partial Z}\left(\bar{\eta}\frac{w_0}{u_0}\frac{h_0}{\ell_0}\frac{\partial \bar{w}}{\partial X}\right) \tag{7.8}$$

In Eq. (7.8) the inertia terms and the gravity term are of order h_0/ℓ_0. The term w_0/u_0 is also of order h_0/ℓ_0. The pressure gradient term and the first viscous term are of order 1. The remaining viscous terms are of order $(h_0/\ell_0)^2$ or $(h_0/b_0)^2$. Therefore, neglecting terms of order $(h_0/\ell_0)^2$ or $(h_0/b_0)^2$ in Eq. (7.8) gives

$$\sigma_s \frac{\partial \bar{u}}{\partial T} + \mathscr{R}_x \bar{u} \frac{\partial \bar{u}}{\partial X} + \mathscr{R}_y \bar{v} \frac{\partial \bar{u}}{\partial Y} + \mathscr{R}_z \bar{w} \frac{\partial \bar{u}}{\partial Z} = \frac{g\ell_0}{u_0^2}\mathscr{R}_x - \frac{1}{\bar{\rho}}\frac{\partial P}{\partial X} + \frac{1}{\bar{\rho}}\frac{\partial}{\partial Z}\left(\bar{\eta}\frac{\partial \bar{u}}{\partial Z}\right) \tag{7.9}$$

Similarly, for the second and third Navier-Stokes equations, neglecting terms of order $(h_0/\ell_0)^2$ or $(h_0/b_0)^2$ gives

$$\sigma_s \frac{\partial \bar{v}}{\partial T} + \mathscr{R}_x \bar{u} \frac{\partial \bar{v}}{\partial X} + \mathscr{R}_y \bar{v} \frac{\partial \bar{v}}{\partial Y} + \mathscr{R}_z \bar{w} \frac{\partial \bar{v}}{\partial Z} = \frac{gb_0}{v_0^2}\mathscr{R}_y - \frac{1}{\bar{\rho}}\frac{\partial P}{\partial Y} + \frac{1}{\bar{\rho}}\frac{\partial}{\partial Z}\left(\bar{\eta}\frac{\partial \bar{v}}{\partial Z}\right) \tag{7.10}$$

$$\frac{\partial P}{\partial Z} = 0 \qquad \rightarrow \qquad P = f(X, Y, T) \tag{7.11}$$

Also, the continuity equation can be expressed as

$$\sigma_s \frac{\partial \bar{\rho}}{\partial T} + \mathscr{R}_x \frac{\partial}{\partial X}(\bar{\rho}\bar{u}) + \mathscr{R}_y \frac{\partial}{\partial Y}(\bar{\rho}\bar{v}) + \mathscr{R}_z \frac{\partial}{\partial Z}(\bar{\rho}\bar{w}) = 0 \tag{7.12}$$

Therefore, Eqs. (7.9) to (7.12) are the Navier-Stokes and continuity equations to be used when higher order effects are considered, as in Chap. 15.

7.1.2 TAYLOR NUMBER. In journal bearings a regular toroidal vortex flow may occur before the laminar flow breaks down into turbulence. This phenomenon was studied by G. I. Taylor in relation to concentric cylinders, and in 1923 he reported that vortices formed at a Reynolds number of

$$\frac{\rho_0 u_0 c}{\eta_0} > 41.3\left(\frac{r_o}{c}\right)^{1/2} \tag{7.13}$$

Thus, the Taylor number (T_a) that describes when vortex first starts to flow occur can be expressed as

$$T_a = \frac{\rho_0 u_0^2 c^3}{r_o \eta_0^2} \geq 1700 \tag{7.14}$$

This is called the "Taylor number" and describes when the vortices appear. Relating this to the modified Reynolds number defined in Eq. (7.4) requires that both sides of Eq. (7.13) be multiplied by $h_0^2/c\ell_0$, which gives

$$\mathscr{R}_x = \frac{\rho_0 u_0 \ell_0}{\eta_0}\left(\frac{h_0}{\ell_0}\right)^2 > 41.3\left(\frac{r_o}{c}\right)^{1/2} \frac{h_0^2}{c\ell_0} \tag{7.15}$$

If $h_0 = c$ and $r_o/c = \ell_0/c = 1 \times 10^3$,

$$\mathscr{R}_x = \frac{\rho_0 u_0 \ell_0}{\eta_0}\left(\frac{h_0}{\ell_0}\right)^2 > 1$$

This indicates that once the inertia terms approach the viscous terms, laminar flow conditions do not hold and vortices are formed.

7.1.3 FROUDE NUMBER. The only body forces normally encountered in lubrication are gravity and magnetic forces. The Froude number shows the relation of inertia to gravity forces.

$$\text{Froude number} = \frac{\text{inertia}}{\text{gravity}} = \frac{u_0^2}{g\ell_0} \tag{7.16}$$

A direct ratio of gravity to viscous forces is found by dividing the Reynolds number by the Froude number.

$$\frac{\text{Reynolds number}}{\text{Froude number}} = \frac{\dfrac{\text{inertia}}{\text{viscous}}}{\dfrac{\text{inertia}}{\text{gravity}}} = \frac{\text{gravity}}{\text{viscous}} = \frac{\rho_0 h_0^2 g}{\eta_0 u_0} \tag{7.17}$$

For example, for a typical journal bearing, besides the information given earlier, $g = 9.8 \text{ m/s}^2$.

$$\therefore \text{Froude number} = \frac{u_0^2}{g \ell_0} = \frac{(5.24)^2}{(9.8)(0.157)} = 17.8$$

This implies that the inertia forces are larger than the gravity forces. Also,

$$\frac{\text{Reynolds number}}{\text{Froude number}} = \frac{\rho_0 h_0^2 g}{\eta_0 u_0} = \frac{(850)(25 \times 10^{-10})(9.8)}{(0.5)(5.24)} = 0.8 \times 10^{-5}$$

Therefore, the gravity forces can be neglected in relation to the viscous forces.

7.1.4 EULER NUMBER. The importance of the pressure term relative to the inertia term can be judged from the value of the Euler number, which is defined as

$$\text{Euler number} = \frac{\text{pressure}}{\text{inertia}} = \frac{p_0}{\rho_0 u_0^2} \tag{7.18}$$

A direct ratio of the pressure to the viscous forces can be obtained by multiplying the Euler number by the Reynolds number, or

$$\begin{aligned}(\text{Euler number})(\text{Reynolds number}) &= \frac{\text{pressure}}{\text{inertia}}\,\frac{\text{inertia}}{\text{viscous}} = \frac{\text{pressure}}{\text{viscous}} \\ &= \frac{p_0}{\rho_0 u_0^2}\,\frac{\rho_0 u_0 \ell_0}{\eta_0}\left(\frac{h_0}{\ell_0}\right)^2 = \frac{p_0 \ell_0}{\eta_0 u_0}\left(\frac{h_0}{\ell_0}\right)^2\end{aligned} \tag{7.19}$$

For example, for a typical journal bearing, besides the information given earlier, $p_0 = 5 \text{ MPa} = 5 \times 10^6 \text{ N/m}^2$.

$$\therefore \text{Euler number} = \frac{p_0}{\rho_0 u_0^2} = \frac{5 \times 10^6}{(850)(5.24)^2} = 214.3$$

Therefore, the pressure term is much larger than the inertia term. Furthermore, from Eq. (7.19)

$$(\text{Euler number})(\text{Reynolds number}) = \frac{(5 \times 10^6)(0.157)}{(3142)^2(0.5)(5.24)} = 0.03$$

Therefore, the viscous term is larger than the pressure term, but both terms need to be considered. Furthermore, in elastohydrodynamic lubrication, where in Chap. 1 we discovered that the pressure is generally three orders of magnitude larger than in hydrodynamic lubrication, we might find the pressure term to be of more importance than the viscous term.

7.2 REYNOLDS EQUATION DERIVED

The differential equation governing the pressure distribution in fluid film lubrication is known as the "Reynolds equation." This equation was first derived in a remarkable paper by Osborne Reynolds in 1886. Reynolds' classical paper contained not only the basic differential equation of fluid film lubrication, but also a direct comparison between his theoretical predictions and the experimental results obtained by Tower (1883). Reynolds, however, restricted his analysis to an incompressible fluid. This is an unnecessary restriction, and Harrison (1913) included the effects of compressibility. In this section the Reynolds equation is derived in two different ways, from the Navier-Stokes and continuity equations and directly from the principle of mass conservation.

7.2.1 FROM NAVIER-STOKES AND CONTINUITY EQUATIONS. From the order-of-magnitude analysis in the preceding section it was discovered that the general Navier-Stokes equations given in Eqs. (6.25) to (6.27) reduce to Eqs. (7.9) to (7.11) when terms of order $(h_0/\ell_0)^2$ and $(h_0/b_0)^2$ and smaller are neglected. This then is the starting point of the derivation. Further neglecting terms of order h_0/ℓ_0 or h_0/b_0 and only keeping terms of order 1 reduces equations (6.25) to (6.27) to

$$\frac{\partial p}{\partial x} = \frac{\partial}{\partial z}\left(\eta \frac{\partial u}{\partial z}\right) \tag{7.20}$$

$$\frac{\partial p}{\partial y} = \frac{\partial}{\partial z}\left(\eta \frac{\partial v}{\partial z}\right) \tag{7.21}$$

From Eq. (7.11) for steady-state conditions the pressure has been shown to be a function of only x and y. Thus, Eqs. (7.20) and (7.21) can be integrated directly to give general expressions for the velocity gradients

$$\frac{\partial u}{\partial z} = \frac{z}{\eta}\frac{\partial p}{\partial x} + \frac{\tilde{A}}{\eta} \tag{7.22}$$

$$\frac{\partial v}{\partial z} = \frac{z}{\eta}\frac{\partial p}{\partial y} + \frac{\tilde{C}}{\eta} \tag{7.23}$$

where $\tilde{A}$ and $\tilde{C}$ are integration constants.

The viscosity of the lubricant may change considerably across the thin film (z direction) as a result of temperature variations that arise in some bearing problems. In this case, progress toward a simple Reynolds equation is considerably complicated.

An approach that is satisfactory in most fluid film applications is to treat η as the average value of the viscosity across the film. Note that this does not restrict the variation of viscosity in the x and y directions. This approach is pursued here.

With η representing an average value of viscosity across the film, integrating Eqs. (7.22) and (7.23) gives the velocity components as

$$u = \frac{z^2}{2\eta}\frac{\partial p}{\partial x} + \tilde{A}\frac{z}{\eta} + \tilde{B} \tag{7.24}$$

$$v = \frac{z^2}{2\eta}\frac{\partial p}{\partial y} + \tilde{C}\frac{z}{\eta} + \tilde{D} \tag{7.25}$$

If zero slip at the fluid-solid interface is assumed, the boundary values for velocity are

1. $z = 0,\ u = u_b,\ v = v_b$
2. $z = h,\ u = u_a,\ v = v_a$

The subscripts a and b refer to conditions on the upper (curved) and lower (plane) surfaces, respectively. Therefore, u_a, v_a, and w_a refer to the velocity components of the upper surface in the x, y, and z directions, respectively, and u_b, v_b, and w_b refer to the velocity components of the lower surface in the same directions.

With the boundary conditions applied to Eqs. (7.24) and (7.25), the velocity gradients and velocity components are

$$\frac{\partial u}{\partial z} = \left(\frac{2z - h}{2\eta}\right)\frac{\partial p}{\partial x} - \frac{u_b - u_a}{h} \tag{7.26}$$

$$\frac{\partial v}{\partial z} = \left(\frac{2z - h}{2\eta}\right)\frac{\partial p}{\partial y} - \frac{v_b - v_a}{h} \tag{7.27}$$

$$u = -z\left(\frac{h - z}{2\eta}\right)\frac{\partial p}{\partial x} + u_b\frac{h - z}{h} + u_a\frac{z}{h} \tag{7.28}$$

$$v = -z\left(\frac{h - z}{2\eta}\right)\frac{\partial p}{\partial y} + v_b\frac{h - z}{h} + v_a\frac{z}{h} \tag{7.29}$$

Note that, if $u_b = 0$, Eq. (7.26) is exactly Eq. (6.54) and Eq. (7.28) is exactly Eq. (6.53). With these expressions for the velocity gradients and the velocity components, expressions can be derived for the surface shear stresses and the volume flow rate.

The viscous shear stresses acting on the solids as defined in Eq. (6.7) can be expressed as

$$\tau_{zx} = \eta\left(\frac{\partial w}{\partial x} + \frac{\partial u}{\partial z}\right)$$

$$\tau_{zy} = \eta\left(\frac{\partial w}{\partial y} + \frac{\partial v}{\partial z}\right)$$

In the order-of-magnitude evaluation, $\partial w/\partial x$ and $\partial w/\partial y$ are much smaller than $\partial u/\partial z$ and $\partial v/\partial z$. Therefore,

$$\tau_{zx} = \eta\frac{\partial u}{\partial z} \tag{7.30}$$

$$\tau_{zy} = \eta\frac{\partial v}{\partial z} \tag{7.31}$$

and the viscous shear stresses acting on the solid surfaces can be expressed, while making use of Eqs. (7.26) and (7.27), as

$$(\tau_{zx})_{z=0} = \left(\eta\frac{\partial u}{\partial z}\right)_{z=0} = -\frac{h}{2}\frac{\partial p}{\partial x} - \frac{\eta(u_b - u_a)}{h} \tag{7.32}$$

$$(-\tau_{zx})_{z=h} = -\left(\eta\frac{\partial u}{\partial z}\right)_{z=h} = -\frac{h}{2}\frac{\partial p}{\partial x} + \frac{\eta(u_b - u_a)}{h} \tag{7.33}$$

$$(\tau_{zy})_{z=0} = \left(\eta\frac{\partial v}{\partial z}\right)_{z=0} = -\frac{h}{2}\frac{\partial p}{\partial y} - \frac{\eta(v_b - v_a)}{h} \tag{7.34}$$

$$(-\tau_{zy})_{z=h} = -\left(\eta\frac{\partial v}{\partial z}\right)_{z=h} = -\frac{h}{2}\frac{\partial p}{\partial y} + \frac{\eta(v_b - v_a)}{h} \tag{7.35}$$

The negative signs on the viscous shear stress indicate that it acts opposite to the direction of motion.

The volume flow rates per unit width in the x and y directions are defined as

$$q'_x = \int_0^h u\,dz \tag{7.36}$$

$$q'_y = \int_0^h v\,dz \tag{7.37}$$

Substituting Eqs. (7.28) and (7.29) in these equations gives

$$q'_x = -\frac{h^3}{12\eta}\frac{\partial p}{\partial x} + \frac{u_a + u_b}{2}h \tag{7.38}$$

$$q'_y = -\frac{h^3}{12\eta}\frac{\partial p}{\partial y} + \frac{v_a + v_b}{2}h \tag{7.39}$$

Note that if $u_b = 0$, Eq. (7.38) is exactly Eq. (6.58), derived for the volume flow rate between parallel flat plates. The first term on the right side of Eqs. (7.38) and (7.39) represents the well-known Poiseuille (or pressure) flow, and the second term represents the Couette (or velocity) flow.

Returning to Eqs. (7.28) and (7.29), the Reynolds equation is formed by introducing these expressions into the continuity equation derived in Eq. (6.47). Before doing so, however, it is convenient to express the continuity equation in integral form.

$$\int_0^h \left[\frac{\partial \rho}{\partial t} + \frac{\partial}{\partial x}(\rho u) + \frac{\partial}{\partial y}(\rho v) + \frac{\partial}{\partial z}(\rho w)\right] dz = 0$$

Now a general rule of integration is that

$$\int_0^h \frac{\partial}{\partial x}[f(x,y,z)]\,dz = -f(x,y,h)\frac{\partial h}{\partial x} + \frac{\partial}{\partial x}\left[\int_0^h f(x,y,z)\,dz\right] \tag{7.40}$$

Hence, if ρ is assumed to be the mean force density across the film (as was done earlier for the viscosity across the film), the u component term in the integrated continuity equation is

$$\int_0^h \frac{\partial}{\partial x}(\rho u)\,dz = -(\rho u)_{z=h}\frac{\partial h}{\partial x} + \frac{\partial}{\partial x}\left(\int_0^h \rho u\,dz\right) = -\rho u_a\frac{\partial h}{\partial x} + \frac{\partial}{\partial x}\left(\rho\int_0^h u\,dz\right)$$

Similarly, for the v component

$$\int_0^h \frac{\partial}{\partial y}(\rho v)\,dz = -\rho v_a\frac{\partial h}{\partial y} + \frac{\partial}{\partial y}\left(\rho\int_0^h v\,dz\right)$$

The w component term can be integrated directly to give

$$\int_0^h \frac{\partial}{\partial z}(\rho w)\,dz = \rho(w_a - w_b)$$

Therefore, the integrated continuity equation becomes

$$h\frac{\partial \rho}{\partial t} - \rho u_a\frac{\partial h}{\partial x} + \frac{\partial}{\partial x}\left(\rho\int_0^h u\,dz\right) - \rho v_a\frac{\partial h}{\partial y} + \frac{\partial}{\partial y}\left(\rho\int_0^h v\,dz\right) + \rho(w_a - w_b) = 0 \tag{7.41}$$

The integrals in this equation represent the volume flow rates per unit width

(q'_x and q'_y) described in Eqs. (7.38) and (7.39). Introducing these flow rate expressions into the integrated continuity equation yields the general Reynolds equation

$$0 = \frac{\partial}{\partial x}\left(-\frac{\rho h^3}{12\eta}\frac{\partial p}{\partial x}\right) + \frac{\partial}{\partial y}\left(-\frac{\rho h^3}{12\eta}\frac{\partial p}{\partial y}\right) + \frac{\partial}{\partial x}\left[\frac{\rho h(u_a + u_b)}{2}\right]$$
$$+ \frac{\partial}{\partial y}\left[\frac{\rho h(v_a + v_b)}{2}\right] + \rho(w_a - w_b) - \rho u_a\frac{\partial h}{\partial x} - \rho v_a\frac{\partial h}{\partial y} + h\frac{\partial \rho}{\partial t} \quad (7.42)$$

7.2.2 FROM LAWS OF VISCOUS FLOW AND PRINCIPLE OF MASS CONSERVATION. The Reynolds equation can be derived directly by considering a control volume fixed in space and extended across the lubricating film. Consider the rate of mass flow through a rectangular section of sides Δx and Δy, thus fixing the coordinate system and extending the lubricating film between the surfaces as shown in Fig. 7.1. Note that one surface is represented by the plane $z = 0$ and the other by a curved surface so that the film thickness at any instant is a function of x and y only. This is exactly the coordinate system used in the previous derivation of the Reynolds equation.

The mass of lubricant in the control volume at any instant is $\rho h\,\Delta x\,\Delta y$. The rate of change within the control volume arises from the change in the difference between the rate of mass flowing into the control volume and the rate leaving the control volume, which is $-(\partial \rho q'_x/\partial x)\,\Delta x\,\Delta y$ in the x direction and $-(\partial \rho q'_y/\partial y)\,\Delta x\,\Delta y$ in the y direction.

The principle of mass conservation demands that the rate at which mass is accumulating in the control volume $\partial(\rho h)/\partial t$ must be equal to the difference between the rates at which mass enters and leaves. Therefore,

$$-\frac{\partial \rho q'_x}{\partial x} - \frac{\partial \rho q'_y}{\partial y} = \frac{\partial}{\partial t}(\rho h) \quad (7.43)$$

But
$$\frac{\partial}{\partial t}(\rho h) = \rho\frac{\partial h}{\partial t} + h\frac{\partial \rho}{\partial t}$$

and
$$\frac{\partial}{\partial t}(\rho h) = \rho\left(w_a - w_b - u_a\frac{\partial h}{\partial x} - v_a\frac{\partial h}{\partial y}\right) + h\frac{\partial \rho}{\partial t} \quad (7.44)$$

By making use of Eqs. (7.38), (7.39), and (7.44), Eq. (7.43) becomes

$$0 = \frac{\partial}{\partial x}\left(-\frac{\rho h^3}{12\eta}\frac{\partial p}{\partial x}\right) + \frac{\partial}{\partial y}\left(-\frac{\rho h^3}{12\eta}\frac{\partial p}{\partial y}\right) + \frac{\partial}{\partial x}\left[\frac{\rho h(u_a + u_b)}{2}\right]$$
$$+ \frac{\partial}{\partial y}\left[\frac{\rho h(v_a + v_b)}{2}\right] + \rho(w_a - w_b) - \rho u_a\frac{\partial h}{\partial x} - \rho v_a\frac{\partial h}{\partial y} + h\frac{\partial \rho}{\partial t} \quad (7.45)$$

This is exactly Eq. (7.42), the general Reynolds equation derived from the Navier-Stokes and continuity equations.

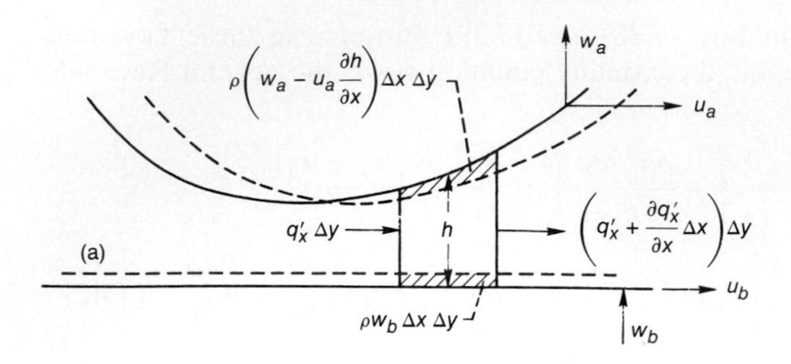

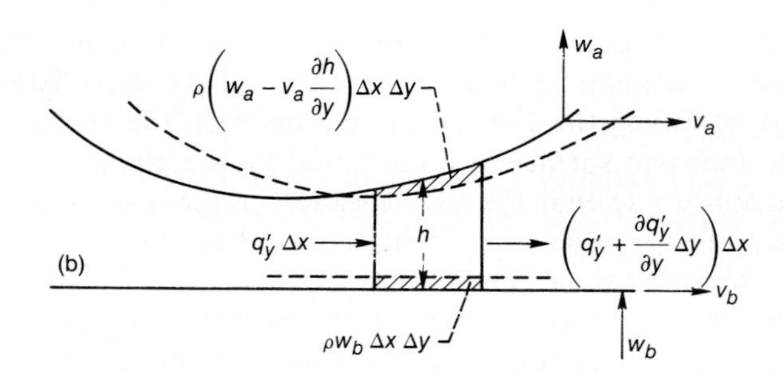

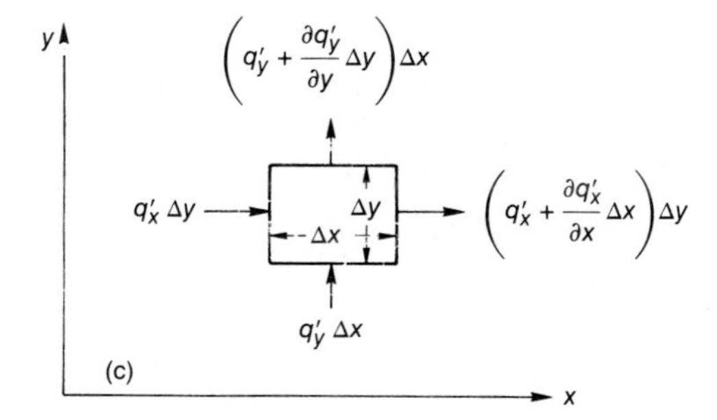

FIGURE 7.1
Mass flow through rectangular-section control volume. (a) x, z plane; (b) y, z plane; (c) x, y plane. [*From Hamrock and Dowson (1981).*]

7.3 PHYSICAL SIGNIFICANCE OF TERMS IN REYNOLDS EQUATION

The first two terms of Eq. (7.45) are the Poiseuille terms and describe the net flow rates due to pressure gradients within the lubricated area; the third and fourth terms are the Couette terms and describe the net entraining flow rates due to surface velocities. The fifth to seventh terms describe the net flow rates due to a squeezing motion, and the last term describes the net flow rate due to local expansion. The flows or "actions" can be considered, without any loss of generality, by eliminating the side-leakage terms $(\partial/\partial y)$ in Eq. (7.45).

$$\underbrace{\frac{\partial}{\partial x}\left(\frac{\rho h^3}{12\eta}\frac{\partial p}{\partial x}\right)}_{\text{Poiseuille}} = \underbrace{\frac{\partial}{\partial x}\left[\frac{\rho h(u_a+u_b)}{2}\right]}_{\text{Couette}} + \underbrace{\rho\left(w_a - w_b - u_a\frac{\partial h}{\partial x}\right)}_{\text{Squeeze}} + \underbrace{h\frac{\partial \rho}{\partial t}}_{\text{Local expansion}} \tag{7.46}$$

Couette:

$$\underbrace{\frac{h(u_a+u_b)}{2}\frac{\partial \rho}{\partial x}}_{\text{Density wedge}} \qquad \underbrace{\frac{\rho h}{2}\frac{\partial}{\partial x}(u_a+u_b)}_{\text{Stretch}} \qquad \underbrace{\frac{\rho(u_a+u_b)}{2}\frac{\partial h}{\partial x}}_{\text{Physical wedge}}$$

It can be seen that the Couette term leads to three distinct actions. The physical significance of each term within the Reynolds equation is now discussed in detail.

7.3.1 DENSITY WEDGE TERM $[(u_a+u_b)h/2](\partial\rho/\partial x)$. The density wedge action is concerned with the rate at which lubricant density changes in the sliding direction as shown in Fig. 7.2. If the lubricant density decreases in the sliding direction, the Couette mass flows for each location of the three velocity profiles of Fig. 7.2 differ. For continuity of mass flow this decrepancy must be eliminated by generating a balancing Poiseuille flow.

Note from Fig. 7.2 that the density must decrease in the direction of sliding if positive pressures are to be generated. This effect could be introduced by raising the temperature of the lubricant as it passes through the bearing. The density wedge (sometimes called the "thermal wedge") mechanism is not important in most bearings. It has been suggested that it may play a significant

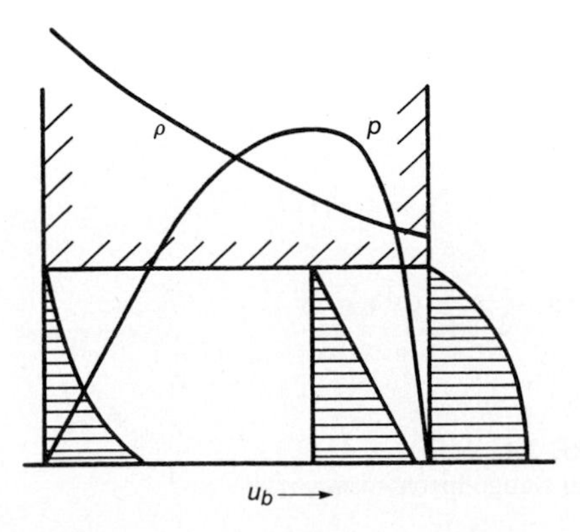

FIGURE 7.2
Density wedge.

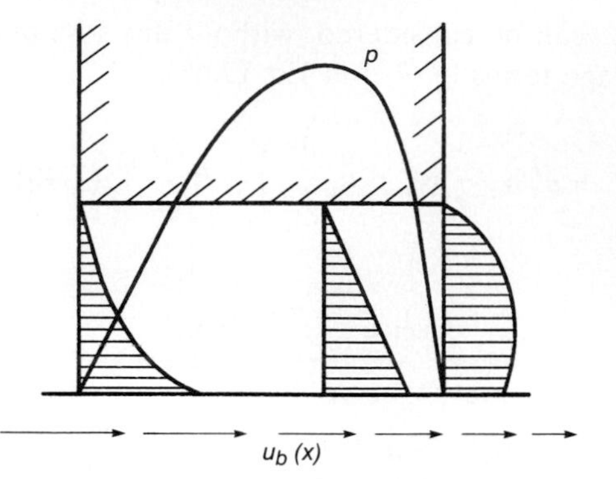

FIGURE 7.3
Stretch mechanism.

role in the performance of parallel-surface thrust bearings, where the major pressure-generating actions are absent.

7.3.2 STRETCH TERM $(\rho h/2)[\partial(u_a + u_b)/\partial x]$. The strength action considers the rate at which surface velocity changes in the sliding direction. This effect is produced if the bounding solids are elastic and the extent to which the surfaces are stretched varies through the bearing. For positive pressures to be developed, the surface velocities have to decrease in the sliding direction, as shown in Fig. 7.3. This action is not encountered in conventional bearings.

7.3.3 PHYSICAL WEDGE TERM $[\rho(u_a + u_b)/2](\partial h/\partial x)$. The physical wedge action is extremely important and is the best known device for pressure generation. This action is illustrated for the case of a plane slider and a stationary bearing pad in Fig. 7.4. At each of the three sections considered, the Couette volume flow rate is proportional to the area of the triangle of height h and base u. Since h varies along the bearing, there is a different Couette flow rate at each section, and flow continuity can be achieved only if a balancing

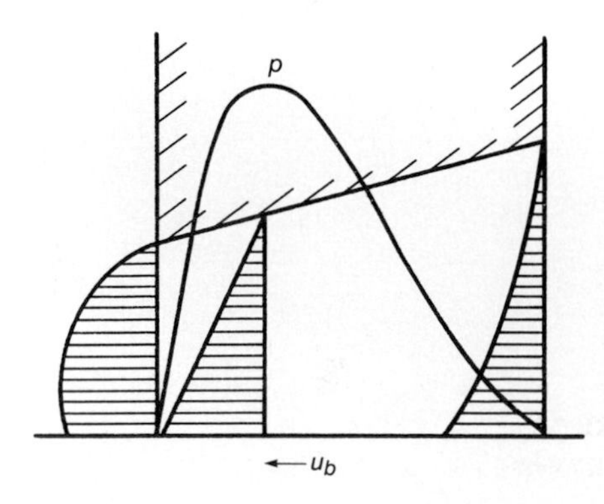

FIGURE 7.4
Physical wedge mechanism.

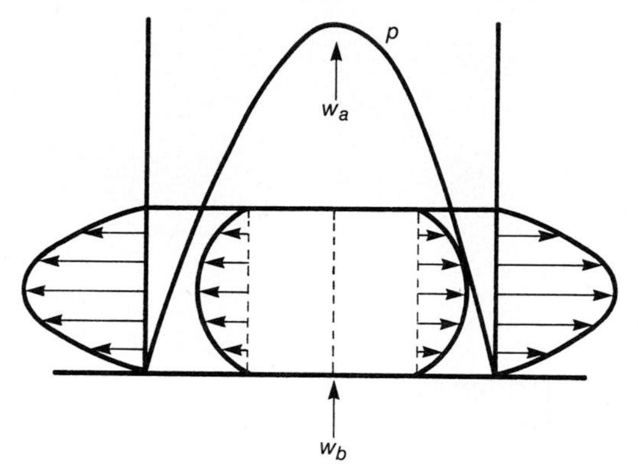

FIGURE 7.5
Normal squeeze mechanism.

Poiseuille flow is superimposed. For a positive load-carrying capacity the thickness of the lubricant film must decrease in the sliding direction.

7.3.4 NORMAL SQUEEZE TERM $\rho(w_a - w_b)$. Normal squeeze action provides a valuable cushioning effect when bearing surfaces tend to be pressed together. Positive pressures will be generated when the film thickness is diminishing. The physical wedge and normal squeeze actions are the two major pressure-generating devices in hydrodynamic or self-acting fluid film bearings. In the absence of sliding, the effect arises directly from the difference in normal velocities $(w_a - w_b)$, as illustrated in Fig. 7.5. Positive pressures will clearly be achieved if the film thickness is decreasing $(w_b > w_a)$.

7.3.5 TRANSLATION SQUEEZE TERM $-\rho u_a(\partial h / \partial x)$. The translation squeeze action results from the translation of inclined surfaces. The local film thickness may be squeezed by the sliding of the inclined bearing surface, as shown in Fig. 7.6. The rate at which the film thickness is decreasing is shown in the figure. Note that in this case the pressure profile is moving over the space covered by

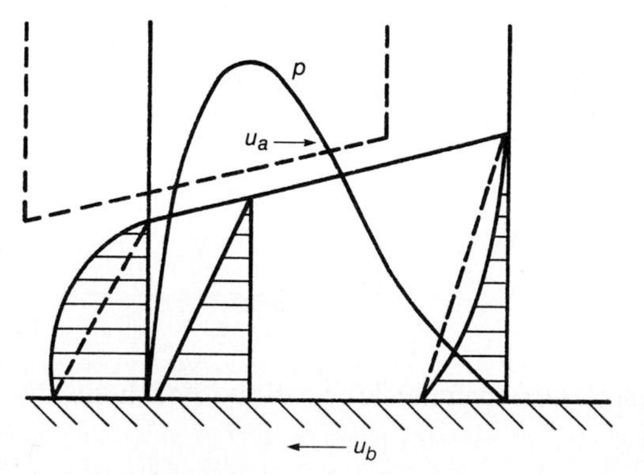

FIGURE 7.6
Translation squeeze mechanism.

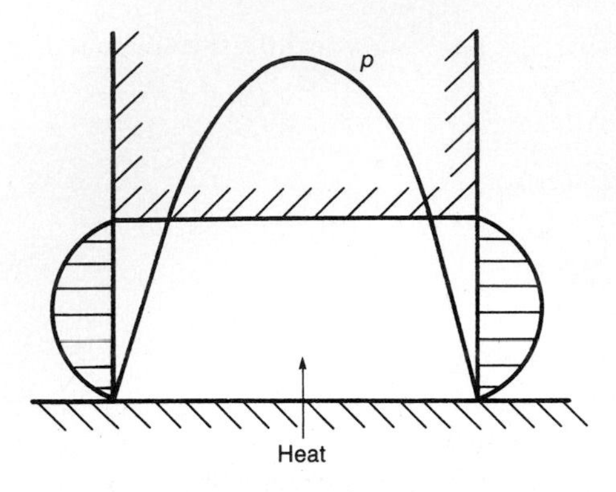

FIGURE 7.7
Local expansion mechanism.

the fixed coordinate system, the pressure at any fixed point being a function of time.

7.3.6 LOCAL EXPANSION TERM $h(\partial\rho/\partial t)$. The local time rate of density change governs the local expansion term. The pressure-generating mechanism can be visualized by considering the thermal expansion of the lubricant contained between stationary bearing surfaces, as shown in Fig. 7.7. If heat is supplied to the lubricant, it will expand and the excess volume will have to be expelled from the space between the bearing surfaces. In the absence of surface velocities the excess lubricant volume must be expelled by a pressure (Poiseuille) flow action. Pressures are thus generated in the lubricant, and for a positive load-carrying capacity, $\partial\rho/\partial t$ must be negative (i.e., the volume of a given mass of lubricant must increase). Local expansion, which is a transient mechanism of pressure generation, is generally of no significance in bearing analysis.

7.4 STANDARD REDUCED FORMS OF REYNOLDS EQUATION

For only tangential motion, where $w_a = u_a\,\partial h/\partial x + v_a\,\partial h/\partial y$ and $w_b = 0$, the Reynolds equation given in Eq. (7.45) becomes

$$\frac{\partial}{\partial x}\left(\frac{\rho h^3}{\eta}\frac{\partial p}{\partial x}\right) + \frac{\partial}{\partial y}\left(\frac{\rho h^3}{\eta}\frac{\partial p}{\partial y}\right) = 12\tilde{u}\frac{\partial(\rho h)}{\partial x} + 12\tilde{v}\frac{\partial(\rho h)}{\partial y} \qquad (7.47)$$

where

$$\tilde{u} = \frac{u_a + u_b}{2} = \text{constant} \qquad \tilde{v} = \frac{v_a + v_b}{2} = \text{constant}$$

This equation is applicable for elastohydrodynamic lubrication. For hydrodynamic lubrication the fluid properties do not vary significantly throughout the bearing and thus may be considered to be a constant. Also, for hydrodynamic lubrication the motion is pure sliding so that $\tilde{v}$ is zero. Thus, the corresponding Reynolds equation is

$$\frac{\partial}{\partial x}\left(h^3 \frac{\partial p}{\partial x}\right) + \frac{\partial}{\partial y}\left(h^3 \frac{\partial p}{\partial y}\right) = 12\tilde{u}\eta_0 \frac{\partial h}{\partial x} \tag{7.48}$$

Equation (7.47) not only allows the fluid properties to vary in the x and y directions but also permits the bearing surfaces to be of finite length in the y direction. Side leakage, or flow in the y direction, is associated with the second term in Eqs. (7.47) and (7.48). If the pressure in the lubricant film has to be considered as a function of x and y, the solution of Eq. (7.47) can rarely be achieved analytically.

In many conventional lubrication problems side leakage can be neglected, and this often leads to analytical solutions. If side leakage is neglected, Eq. (7.47) becomes,

$$\frac{\partial}{\partial x}\left(\frac{\rho h^3}{\eta} \frac{\partial p}{\partial x}\right) = 12\tilde{u} \frac{\partial(\rho h)}{\partial x} \tag{7.49}$$

This equation can be integrated with respect to x to give

$$\frac{1}{\eta} \frac{dp}{dx} = \frac{12\tilde{u}}{h^2} + \frac{\tilde{A}}{\rho h^3} \tag{7.50}$$

Making use of the boundary condition that

$$\frac{dp}{dx} = 0 \qquad \text{when } x = x_m \qquad \rho = \rho_m \qquad h = h_m$$

gives

$$\tilde{A} = -12\tilde{u}\rho_m h_m$$

Substituting this into Eq. (7.50) gives

$$\frac{dp}{dx} = 12\tilde{u}\eta \frac{\rho h - \rho_m h_m}{\rho h^3} \tag{7.51}$$

This is the *integrated form of the Reynolds equation*. Note that the subscript m refers to the condition at all points where $dp/dx = 0$, such as the point of maximum pressure. No assumptions were made about the density or viscosity of the fluid in Eq. (7.51). If the density does not vary much throughout the conjunction, it can be considered constant and Eq. (7.51) reduces to

$$\frac{dp}{dx} = 12\tilde{u}\eta\frac{h - h_m}{h^3} \tag{7.52}$$

The Reynolds equation that is valid for gas-lubricated bearings is discussed next. The equation of state for a perfect gas is

$$p = \rho\bar{R}t_m \tag{7.53}$$

where $\bar{R}$ = gas constant (universal gas constant ÷ molecular weight)
t_m = absolute temperature

Therefore, from Eq. (7.53)

$$\rho = \frac{p}{\bar{R}t_m} \tag{7.54}$$

Substituting this equation into Eq. (7.47) yields the Reynolds equation normally used for gas-lubricated bearings for only tangential motion. Because the viscosity of a gas does not vary much, it can be considered to be constant.

$$\frac{\partial}{\partial x}\left(ph^3\frac{\partial p}{\partial x}\right) + \frac{\partial}{\partial y}\left(ph^3\frac{\partial p}{\partial y}\right) = 12\tilde{u}\eta_0\frac{\partial(ph)}{\partial x} \tag{7.55}$$

The comparable equation to (7.47) as expressed in cylindrical polar coordinates is

$$\frac{\partial}{\partial r}\left(\frac{r\rho h^3}{\eta}\frac{\partial p}{\partial r}\right) + \frac{1}{r}\frac{\partial}{\partial \theta}\left(\frac{\rho h^3}{\eta}\frac{\partial p}{\partial \theta}\right) = 12\left[\tilde{v}_r\frac{\partial}{\partial r}(\rho r h) + \tilde{v}_\theta\frac{\partial}{\partial \theta}(\rho h)\right] \tag{7.56}$$

where $\tilde{v}_r = (v_{ra} + v_{rb})/2$ and $\tilde{v}_\theta = (v_{\theta a} + v_{\theta b})/2$. If the viscosity and the density are assumed to be constant, this equation reduces to

$$\frac{\partial}{\partial r}\left(rh^3\frac{\partial p}{\partial r}\right) + \frac{1}{r}\frac{\partial}{\partial \theta}\left(h^3\frac{\partial p}{\partial \theta}\right) = 12\eta_0\left[\tilde{v}_r\frac{\partial}{\partial r}(rh) + \tilde{v}_\theta\frac{\partial h}{\partial \theta}\right] \tag{7.57}$$

It should be pointed out that Eqs. (7.56) and (7.57) express the Reynolds equation for cylindrical polar coordinates as applied to a thrust bearing where the film direction is in the z direction and the bearing dimensions are in the r, θ directions. If one is interested in expressing the cylindrical polar coordinates as applied to a journal bearing, the Reynolds equation would be different in that θ and z describe the dimensions of the bearing and r describes the film shape.

The Reynolds equation that occurs for the general representation of Eq. (7.45) is

$$\frac{\partial}{\partial x}\left(\frac{\rho h^3}{12\eta}\frac{\partial p}{\partial x}\right) + \frac{\partial}{\partial y}\left(\frac{\rho h^3}{12\eta}\frac{\partial p}{\partial y}\right) = \frac{\partial}{\partial x}\left[\frac{\rho h(u_a + u_b)}{2}\right] + \frac{\partial}{\partial y}\left[\frac{\rho h(v_a + v_b)}{2}\right] + \frac{\partial(\rho h)}{\partial t} \tag{7.58}$$

This equation is exactly the same as (7.45) if

$$\rho(w_a - w_b) - \rho u_a \frac{\partial h}{\partial x} - \rho v_a \frac{\partial h}{\partial y} + h \frac{\partial \rho}{\partial t} = \frac{\partial(\rho h)}{\partial t}$$

This implies that

$$\frac{\partial h}{\partial t} = w_a - w_b - u_a \frac{\partial h}{\partial x} - v_a \frac{\partial h}{\partial y} \tag{7.59}$$

An attempt will be made to prove that Eq. (7.59) is true. First, observe that the film thickness h is a function of x, y, and t.

$$h = f(x, y, t)$$

From the definition of a total derivative,

$$Dh = \frac{\partial h}{\partial t} dt + \frac{\partial h}{\partial x} dx + \frac{\partial h}{\partial y} dy$$

or

$$\frac{Dh}{Dt} = \frac{\partial h}{\partial t} + \frac{\partial h}{\partial x}\frac{dx}{dt} + \frac{\partial h}{\partial y}\frac{dy}{dt}$$

But

$$u_a = \frac{dx}{dt} \qquad v_a = \frac{dy}{dt} \qquad \text{and} \qquad \frac{Dh}{Dt} = w_a - w_b$$

$$\therefore w_a - w_b = \frac{\partial h}{\partial t} + u_a \frac{\partial h}{\partial x} + v_a \frac{\partial h}{\partial y}$$

or

$$\frac{\partial h}{\partial t} = w_a - w_b - u_a \frac{\partial h}{\partial x} - v_a \frac{\partial h}{\partial y} \tag{7.60}$$

thus proving that Eqs. (7.45) and (7.58) are the same.

Making use of Eqs. (6.51) and (6.44) to (6.46) while following the procedure given in Szeri (1980), the Reynolds equation for turbulent flow is

$$\frac{\partial}{\partial x}\left(\frac{h^3}{\eta k_x}\frac{\partial p^*}{\partial x}\right) + \frac{\partial}{\partial y}\left(\frac{h^3}{\eta k_y}\frac{\partial p^*}{\partial y}\right) = \frac{u^*}{2}\frac{\partial h}{\partial x} \tag{7.61}$$

Equation (7.61) is applicable for thrust or journal bearings. Constantinescu (1962) found that

$$k_x = 12 + 0.53(k^2 R_{eh})^{0.725} \tag{7.62}$$

$$k_y = 12 + 0.296(k^2 R_{eh})^{0.65} \tag{7.63}$$

where
$$R_{eh} = \frac{r\omega h\rho}{\eta} \tag{7.64}$$

$$k \approx 0.125R_{eh}^{0.07} \tag{7.65}$$

7.5 DIFFERENT NORMAL SQUEEZE AND SLIDING MOTIONS

The last section described various Reynolds equations when only tangential motion exists. This section describes how the Reynolds equation is altered when considering various tangential and normal squeeze velocity components. The velocity components and the coordinates used are shown in Fig. 7.8. If the density is assumed to be constant and side leakage is neglected for simplicity, the Reynolds equation expressed in Eq. (7.45) can be rewritten as

$$\frac{\partial}{\partial x}\left(\frac{h^3}{12\eta}\frac{\partial p}{\partial x}\right) = \frac{u_a + u_b}{2}\frac{\partial h}{\partial x} + \frac{h}{2}\frac{\partial}{\partial x}(u_a + u_b) + w_a - w_b - u_a\frac{\partial h}{\partial x}$$

Collecting terms gives

$$\frac{\partial}{\partial x}\left(\frac{h^3}{12\eta}\frac{\partial p}{\partial x}\right) = \underbrace{\frac{u_b - u_a}{2}\frac{\partial h}{\partial x}}_{\text{I}} + \underbrace{\frac{h}{2}\frac{\partial}{\partial x}(u_a + u_b)}_{\text{II}} + \underbrace{w_a - w_b}_{\text{III}} \tag{7.66}$$

A number of different tangential and normal squeeze motions may occur and are depicted in Table 7.1. The terms on the right side of Eq. (7.66) are designated as I, II, and III. As the table shows, care must be taken to make sure that the physical motion experienced in a particular application is represented by the appropriate Reynolds equation. Note that some quite different geometries and velocities produce the same equation. However, an understanding of why this is so is important in order to avoid improper interpretation.

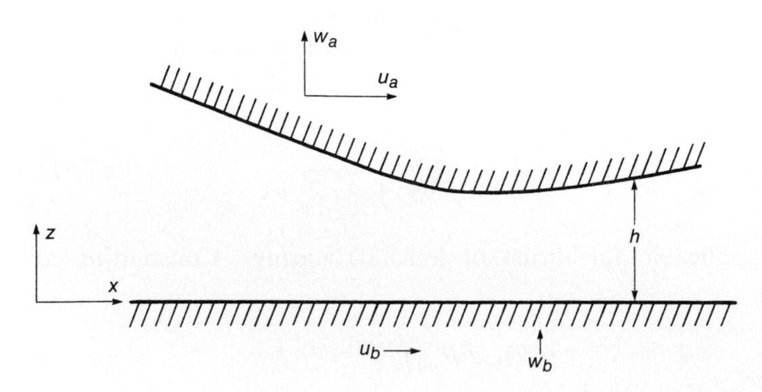

FIGURE 7.8
Normal squeeze and sliding velocities.

TABLE 7.1
Different tangential and normal squeeze motions that may occur in bearings

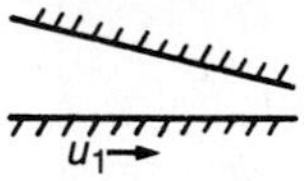 $u_1 \rightarrow$	$u_a = w_a = 0$ $u_b = u_1,\ w_a = 0$	$\text{I} + \text{II} + \text{III} = \frac{u_1}{2}\frac{dh}{dx} + 0 + 0 = \frac{u_1}{2}\frac{dh}{dx}$
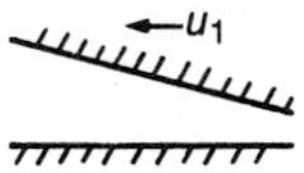 $\leftarrow u_1$	$u_a = -u_1,\ w_a = 0$ $u_b = 0,\ w_b = 0$	$\text{I} + \text{II} + \text{III} = \frac{u_1}{2}\frac{dh}{dx} + 0 + 0 = \frac{u_1}{2}\frac{dh}{dx}$
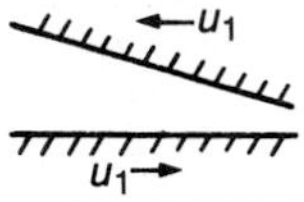 $\leftarrow u_1$ $u_1 \rightarrow$	$u_a = -u_1,\ w_a = 0$ $u_b = u_1,\ w_b = 0$	$\text{I} + \text{II} + \text{III} = \frac{u_1 + u_1}{2}\frac{dh}{dx} + 0 + 0 = u_1\frac{dh}{dx}$
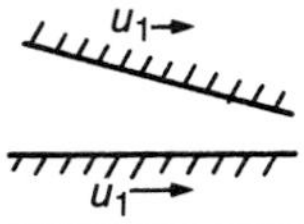 $u_1 \rightarrow$ $u_1 \rightarrow$	$u_a = u_1,\ w_a = 0$ $u_b = u_1,\ w_b = 0$	$\text{I} + \text{II} + \text{III} = \frac{u_1 - u_1}{2}\frac{dh}{dx} + 0 + 0 = 0$
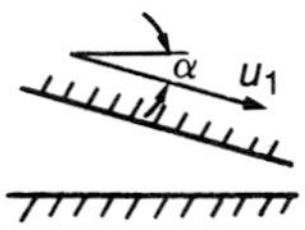 α u_1	$u_a = u_1 \cos\alpha \approx u_1$ $w_a = -u_1 \sin\alpha = u_1\frac{dh}{dx}$ $u_b = 0,\ w_b = 0$	$\text{I} + \text{II} + \text{III} = -\frac{u_1}{2}\frac{dh}{dx} + 0 + u_1\frac{dh}{dx} = \frac{u_1}{2}\frac{dh}{dx}$
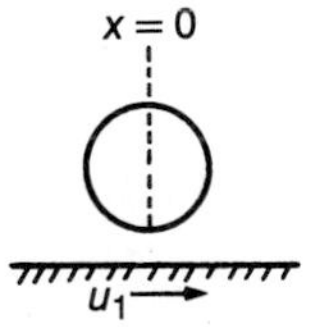 x = 0 $u_1 \rightarrow$	$u_a = 0,\ w_a = 0$ $u_b = u_1,\ w_b = 0$ Note that $x < 0 \qquad \frac{dh}{dx} < 0$ $x > 0 \qquad \frac{dh}{dx} > 0$	$\text{I} + \text{II} + \text{III} = \frac{u_1}{2}\frac{dh}{dx} + 0 + 0 = \frac{u_1}{2}\frac{dh}{dx}$
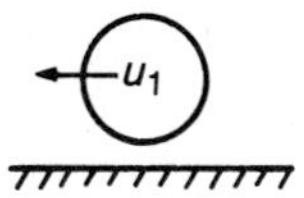 $\leftarrow u_1$	$u_a = -u_1,\ w_a = 0$ $u_b = 0,\ w_b = 0$	$\text{I} + \text{II} + \text{III} = \frac{u_1}{2}\frac{dh}{dx} + 0 + 0 = \frac{u_1}{2}\frac{dh}{dx}$

TABLE 7.1 *Concluded*

u_1	$u_a = u_1,\ w_a = u_1 \dfrac{\partial h}{\partial x}$ $u_b = 0,\ w_b = 0$	$\mathrm{I} + \mathrm{II} + \mathrm{III} = -\dfrac{u_1}{2}\dfrac{dh}{dx} + 0 + u_1 \dfrac{dh}{dx} = \dfrac{u_1}{2}\dfrac{dh}{dx}$
u_1 u_1	$u_a = -u_1 + u_1 = 0,$ $w_a = u_1 \dfrac{\partial h}{\partial x}$ $u_b = 0,\ w_b = 0$	$\mathrm{I} + \mathrm{II} + \mathrm{III} = 0 + 0 + u_1 \dfrac{dh}{dx} = u_1 \dfrac{dh}{dx}$
u_1 u_1	$u_a = u_1,\ w_a = u_1 \dfrac{\partial h}{\partial x}$ $u_b = u_1,\ w_b = 0$	$\mathrm{I} + \mathrm{II} + \mathrm{III} = \dfrac{u_1 - u_1}{2} + 0 + u_1 \dfrac{\partial h}{\partial x} = u_1 \dfrac{\partial h}{\partial x}$
u_1 h_a h_b u_1	$u_a = u_1,\ w_a = u_1 \dfrac{dh_a}{dx}$ $u_b = u_1,\ w_b = u_1 \dfrac{dh_b}{dx}$	$\mathrm{I} + \mathrm{II} + \mathrm{III} = \dfrac{u_1 - u_1}{2}\dfrac{dh}{dx} + 0 + u_1 \dfrac{dh_a}{dx}$ $+ u_1 \dfrac{dh_b}{dx} = u_1 \dfrac{dh}{dx}$
u_1 u_1	$u_a = u_1,\ w_a = u_1 \dfrac{dh_a}{dx}$ $u_b = -u_1,\ w_b = -u_1 \dfrac{dh_b}{dx}$	$\mathrm{I} + \mathrm{II} + \mathrm{III} = \dfrac{-u_1 - u_1}{2}\dfrac{dh}{dx} + 0 + u_1 \dfrac{dh_a}{dx}$ $+ u_1 \dfrac{dh_b}{dx} = 0$
u_1	$u_a = u_1,\ w_a = u_1 \dfrac{dh}{dx}$ $u_b = 0,\ w_b = 0$	$\mathrm{I} + \mathrm{II} + \mathrm{III} = -\dfrac{u_1}{2}\dfrac{dh}{dx} + 0 + u_1 \dfrac{dh}{dx} = \dfrac{u_1}{2}\dfrac{dh}{dx}$

7.6 CLOSURE

The chapter began with the exploration of various dimensionless numbers that describe the significance of the terms within the Reynolds equation. The Reynolds number compares the inertia and viscous terms; the Froude number compares the inertia and gravity terms. The ratio of the Reynolds number to the Froude number compares the gravity and viscous terms.

The Reynolds equation was derived by coupling the Navier-Stokes equations with the continuity equation and by using laws of viscous flow and the principle of mass conservation. The Reynolds equation contains Poiseuille, physical wedge, stretch, local compression, and normal and transverse squeeze terms. Each of these terms describes a specific type of physical motion, and the physical significance of each term was brought out. Standard forms of the Reynolds equations that are used throughout the text were also discussed. The chapter closed with a description of 13 different sliding and/or normal squeeze motions with the corresponding Reynolds equation.

7.7 PROBLEMS

7.7.1 Starting from the Navier-Stokes equations expressed in cylindrical polar coordinates [Eqs. (6.31) and (6.33)] derive the Reynolds equation given in Eq. (7.56). Assume you are applying the cylindrical polar coordinates (r, θ, z) to a thrust bearing where z is in the direction of the lubricating film (h) and $h \ll r$.

7.7.2 From relationships between Cartesian and cylindrical polar coordinates prove that Eq. (7.47) when the viscosity and density are constant ($\eta = \eta_0$ and $\rho = \rho_0$) is equivalent to Eq. (7.57).

7.7.3 Compare the Reynolds equation for laminar flow conditions with that appropriate for turbulent flow conditions. Also list operating conditions and applications where turbulence in fluid film distribution is most likely to occur.

7.7.4 A water-lubricated journal bearing in a boiler feed pump has a shaft of 0.10 m which rotates at 10 r/s. The kinematic viscosity in the full fluid film region may be taken as directly proportional to the film thickness and has a value of 4×10^{-7} m^2/s at a film thickness equal to the radial clearance of 0.10 mm. Determine if the bearing is operating in the laminar or turbulent flow regime. If laminar flow is predicted, what change in these operating conditions would produce the onset of vortex flow.

7.7.5 Write the Reynolds equation for the situations shown below. The circles represent infinitely long cylinders, and all velocities are in relation to a fixed coordinate system. The lubricant can be assumed to be Newtonian, incompressible, and isoviscous.

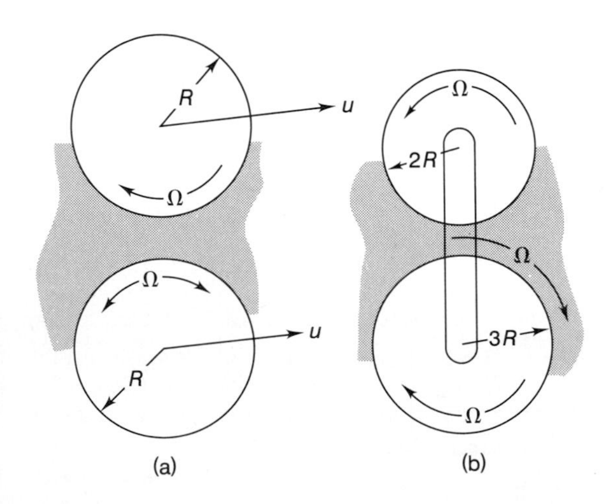

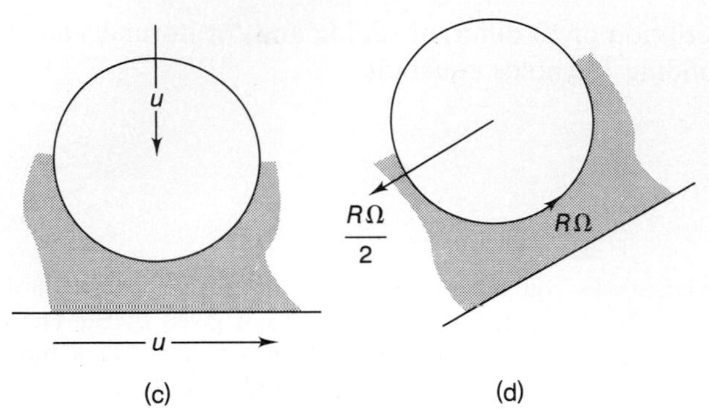

Continued.

7.7.6 For the situation described in each diagram given below, express the appropriate Reynolds equation and sketch the expected pressure distributions. It can be assumed that the bearings are of infinite width and that the lubricant is Newtonian, isoviscous, and incompressible and does not experience cavitation. The shaded members shown in the diagrams are at rest.

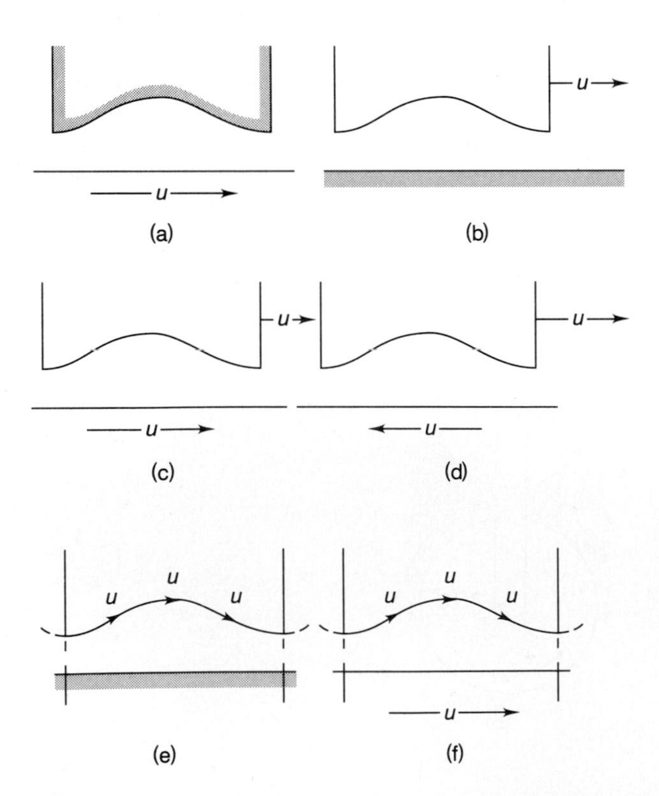

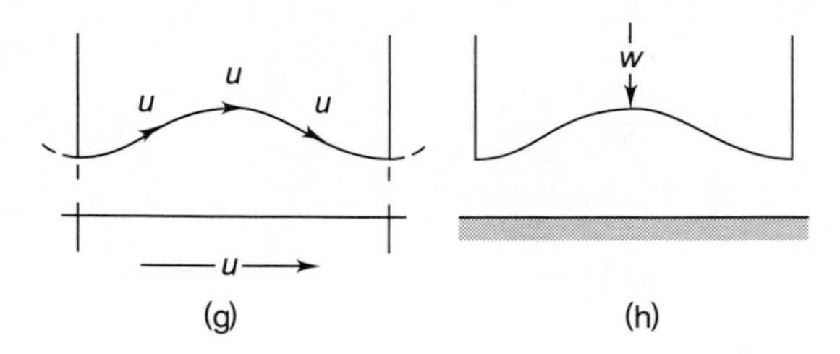

Continued.

7.8 REFERENCES

Constantinescu, V. N. (1962): Analysis of Bearings Operating in Turbulent Regime, Trans of ASME, Series D, *J. of Basic Engr*., vol. 84, no. 1, pp. 139–151.

Hamrock, B. J., and Dowson, D. (1981): *Ball Bearing Lubrication—The Elastohydrodynamics of Elliptical Contacts*. Wiley-Interscience, New York.

Harrison, W. J. (1913): The Hydrodynamical Theory of Lubrication With Special Reference to Air as a Lubricant. *Trans. Cambridge Philos. Soc*., xxii (1912–1925), pp. 6–54.

Reynolds, O. (1886): On the Theory of Lubrication and Its Application to Mr. Beauchamp Tower's Experiments, Including an Experimental Determination of the Viscosity of Olive Oil. *Philos. Trans. R. Soc*., vol. 177, pp. 157–234.

Szeri, A. Z. (ed.) (1980): *Tribology-Friction, Lubrication, and Wear*. Hemisphere Publishing Corp., Washington, D.C.

Taylor, G. I. (1923): Stability of a Viscous Liquid Contained Between Two Rotating Cylinders. *Philos. Trans. R. Soc. London Ser. A*, vol. 223, pp. 289–343.

Tower, B. (1883): First Report on Friction Experiments (Friction of Lubricated Bearings). *Proc. Inst. Mech. Eng. (London)*, pp. 632–659.

CHAPTER 8

HYDRODYNAMIC THRUST BEARINGS—ANALYTICAL SOLUTIONS

A hydrodynamically lubricated bearing is a bearing that develops load-carrying capacity by virtue of the relative motion of two surfaces separated by a fluid film. The processes occurring in a bearing with fluid film lubrication can be better understood by considering qualitatively the development of oil pressure in such a bearing.

8.1 MECHANISM OF PRESSURE DEVELOPMENT

An understanding of the development of load-supporting pressures in hydrodynamic bearings can be gleaned by considering, from a purely physical point of view, the conditions of geometry and motion required to develop pressure. An understanding of the physical situation can make the mathematics of hydrodynamic lubrication much more meaningful. By considering only what must happen to maintain continuity of flow, much of what the mathematical equations tell us later in this chapter can be deduced.

Figure 8.1 shows velocity profiles for two plane surfaces separated by a constant lubricating film thickness. The plates are extremely wide so that side-leakage flow (into and out of the paper) can be neglected. The upper plate is moving with a velocity u_a, and the bottom plate is held stationary. No slip

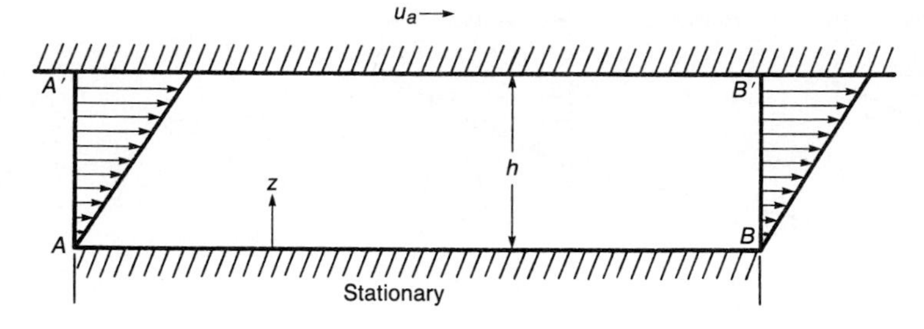

FIGURE 8.1
Velocity profiles in a parallel-surface slider bearing.

occurs at the surfaces. The velocity varies uniformly from zero at surface AB to u_a at surface $A'B'$, thus implying that the rate of shear du/dz throughout the oil film is constant. The volume of fluid flowing across section AA' in unit time is equal to that flowing across section BB'. The flow crossing the two boundaries results only from velocity gradients, and since they are equal, the flow continuity requirement is satisfied without any pressure buildup within the film. Because the ability of a lubricating film to support a load depends on the pressure buildup in the film, a slider bearing with parallel surfaces is not able to support a load by a fluid film. If any load is applied to the surface AB, the lubricant will be squeezed out and the bearing will operate under conditions of boundary lubrication.

Consider now the case of two nonparallel plates as shown in Fig. 8.2(a). Again the width of the plates in the direction perpendicular to the motion is large so that lubricant flow in this direction is negligibly small. The volume of lubricant that the surface $A'B'$ tends to carry into the space between the surfaces AB and $A'B'$ through section AA' during unit time is $AC'A'$. The volume of lubricant that the surface tends to discharge from the space through section BB' during the same time is $BD'B'$. Because the distance AA' is greater than the distance BB', the volume $AC'A'$ is greater than the volume $BD'B'$ by the volume AEC'. From flow continuity the actual volume of oil carried into the space must equal the volume discharged from this space.

It can be surmised that there will be a pressure buildup in the lubricating film until flow continuity is satisfied. The velocity profiles due to Poiseuille flow are shown in Fig. 8.2(b). The flow is outward from both the leading and trailing edges of the bearing because flow is always from a region of higher pressure to a region of lower pressure. Note that the pressure flow at boundary AA' opposes the velocity flow but that the pressure flow at BB' is in the same direction as the velocity flow.

The results of superimposing Couette and Poiseuille flows are shown in Fig. 8.2(c). The form of the velocity distribution curves obtained in this way must satisfy the condition that the flow rate through section AA' is equal to the flow rate through section BB'. Therefore, the area $AHC'A'$ must be equal to the

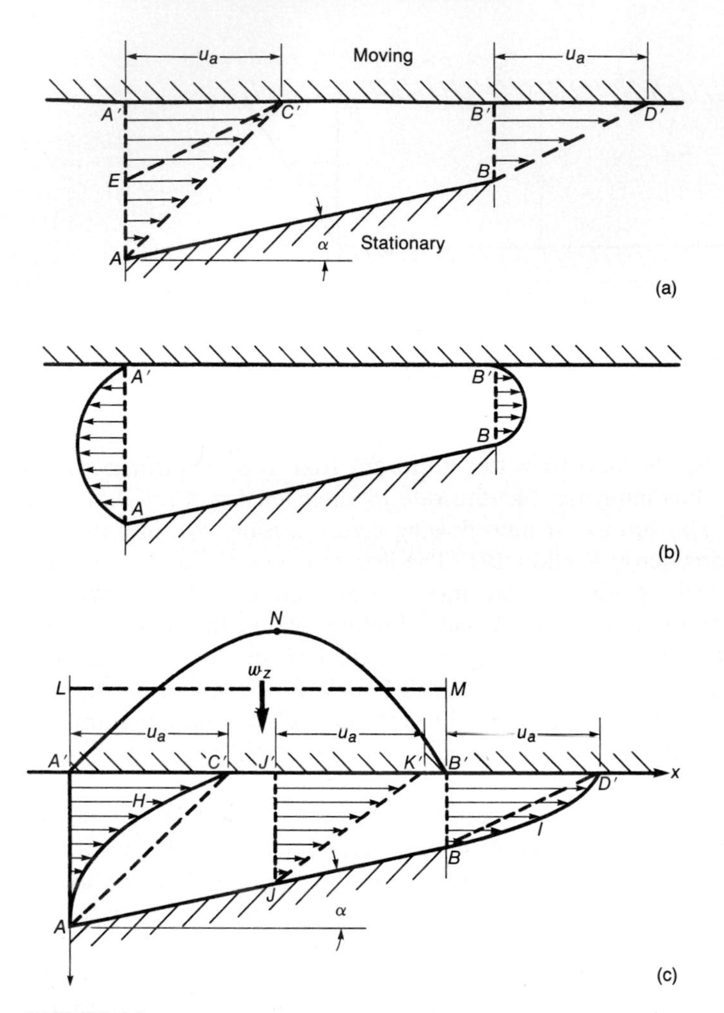

FIGURE 8.2
Flow within a fixed-incline slider bearing. (a) Couette flow; (b) Poiseuille flow; (c) resulting velocity profile.

area $BID'B'$. The area between the straight, dashed line AC' and the curve AHC' in section AA' and the area between the dashed line BD' and the curve BID' represent the pressure-induced flow through these areas.

The pressure is maximum in section JJ', somewhere between sections AA' and BB'. There is no Poiseuille flow contribution in section JJ', since the pressure gradient is zero at this location and all the flow is Couette. Note that continuity of flow is satisfied in that triangle $JK'J'$ is equal to the areas $AHC'A'$ and $BID'B'$.

8.2 GENERAL THRUST BEARING THEORY

Solutions of the Reynolds equation for real bearing configurations are usually obtained in approximate numerical form. Analytical solutions are possible only for the simplest problems. By restricting the flow to two dimensions, say the xz plane, analytical solutions for many common bearing forms become available. The quantitative value of these solutions is limited, since flow in the third dimension y, which is known as "side leakage," plays an important part in fluid film bearing performance. The two-dimensional solutions have a definite value, since they provide a good deal of information about the general characteristics of bearings, information that leads to a clear physical picture of the performance of lubricating films.

Besides neglecting side leakage, another simplification is achieved by neglecting the pressure and temperature effects of the lubricant properties, namely, viscosity and density. The viscosity of common lubricants is particularly sensitive to temperature, and since the heat generated in hydrodynamic bearings is often considerable, the limitation imposed by this assumption is at once apparent.

Introducing variable viscosity and density into the analysis creates considerable complications, even in the case of two-dimensional flow. The temperature rise within the film can be calculated if it is assumed that all the heat produced by the viscous action is carried away by the lubricant (the adiabatic assumption).

From Chap. 7, Eq. (7.52), the two-dimensional Reynolds equation expressed in integrated form for constant density can be written as

$$\frac{dp}{dx} = 12\tilde{u}\eta \frac{h - h_m}{h^3} \tag{7.52}$$

This equation is used in this chapter as the starting point for analyzing various thrust bearing configurations. Before proceeding with the various film configurations in thrust bearings, more needs to be said about thrust bearings in general.

Many loads carried by rotary machinery have components that act in the direction of the shaft's axis of rotation. These thrust loads are frequently carried by self-acting or hydrodynamic bearings of the form shown in Fig. 8.3. A thrust plate attached to, or forming part of, the rotating shaft is separated from the sector-shaped bearing pads by a lubricant film. The load-carrying capacity of the bearing arises entirely from the pressures generated by the geometry of the thrust plate over the bearing pads. What was physically observed earlier in this chapter will be demonstrated analytically, that this lubricating action is achieved only if the clearance space between the stationary and moving components is provided with certain geometrical forms.

It is clear that the lubricant flow between the thrust plate and the bearing pads represents a three-dimensional flow problem. However, for the present

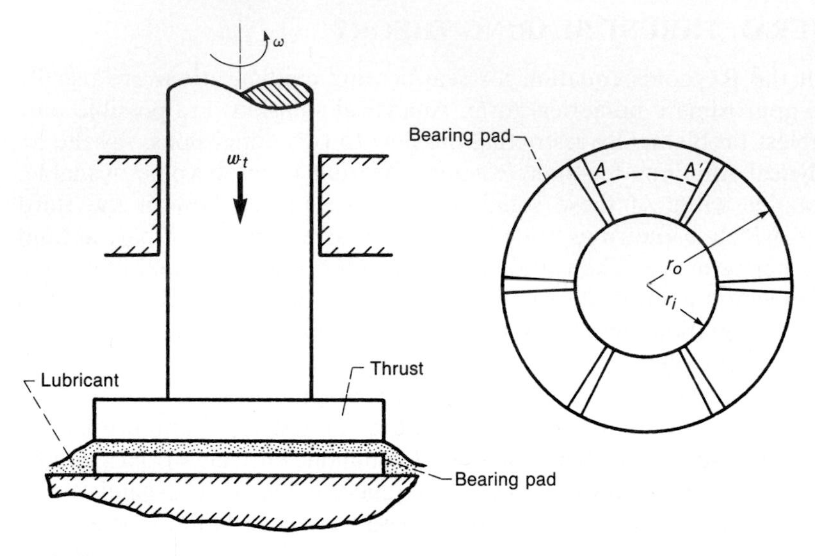

FIGURE 8.3
Thrust bearing geometry.

purpose, flow in the radial direction is neglected. In this chapter each pad is analyzed along a section formed by the mean arc AA' as shown in Fig. 8.3. The effect of curvature on this section is neglected, the oil film geometry being represented by the xz plane. This simple representation allows the total load-carrying capacity of the thrust bearing to be written as

$$w_t = N_0 w_z'(r_o - r_i) \tag{8.1}$$

where N_0 = number of pads in thrust bearing
w_z' = normal load-carrying capacity per unit width of one pad, N/m

This expression will lead to an overestimate of the load-carrying capacity for a given oil film thickness, since lubricant flow in the radial direction (side-leakage direction) will tend to reduce the mean oil film pressure. Side-leakage effects are considered in Chap. 9.

Results from the analysis of various bearing configurations are expressed in dimensionless form for ease in comparing the important bearing characteristics. The following general definitions and relationships are employed. The definitions should be related to the bearing geometry shown in Fig. 8.4.

The forces acting on the solids can be considered in two groups. The loads, which act in the direction normal to the surface, yield normal loads that can be resolved into components w_x' and w_z'. The viscous surface stresses, which act in the direction tangent to the surface, yield shear forces on the solids that have components f' in the x direction. The component of the shear forces in the z direction is negligible.

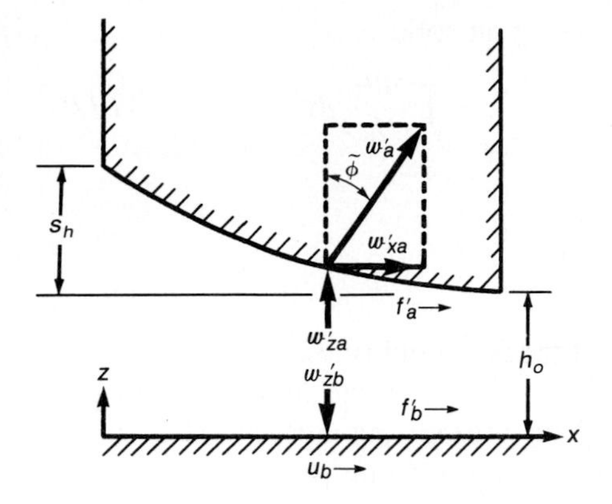

FIGURE 8.4
Force components and oil film geometry in a hydrodynamically lubricated thrust sector.

Once the pressure is obtained for a particular film configuration from the Reynolds equation, the following pressure and force components act on the solids:

$$w'_{za} = w'_{zb} = \int_0^{\ell} p\,dx \tag{8.2}$$

$$w'_{xb} = 0 \tag{8.3}$$

$$w'_{xa} = -\int_{h_o+s_h}^{h_o} p\,dh = -\int_0^{\ell} p\frac{dh}{dx}\,dx$$

$$\therefore w'_{xa} = -(ph)_0^{\ell} + \int_0^{\ell} h\frac{dp}{dx}\,dx = \int_0^{\ell} h\frac{dp}{dx}\,dx \tag{8.4}$$

$$w'_b = \left(w'^2_{zb} + w'^2_{xb}\right)^{1/2} = w'_{zb} \tag{8.5}$$

$$w'_a = \left(w'^2_{za} + w'^2_{xa}\right)^{1/2} \tag{8.6}$$

$$\tilde{\phi} = \tan^{-1}\frac{w'_{xa}}{w'_{za}} \tag{8.7}$$

Shear forces per unit width acting on the solids are

$$f'_b = \int_0^{\ell} (\tau_{zx})_{z=0}\,dx$$

Substituting Eq. (7.32) into this equation gives

$$f'_b = \int_0^{\ell}\left(-\frac{h}{2}\frac{dp}{dx} - \frac{\eta u_b}{h}\right)dx$$

Making use of Eq. (8.4) gives

$$f'_b = -\frac{w'_{xa}}{2} - \int_0^{\ell}\frac{\eta u_b}{h}\,dx \tag{8.8}$$

Similarly, the shear force per unit width acting on solid a is

$$f_a' = -\int_0^\ell (\tau_{zx})_{z=h}\, dx = -\frac{w_{xa}'}{2} + \int_0^\ell \frac{\eta u_b}{h}\, dx \tag{8.9}$$

Note from Fig. 8.4 that

$$f_b' + f_a' + w_{xa}' = 0 \tag{8.10}$$

$$w_{zb}' - w_{za}' = 0 \tag{8.11}$$

These equations represent the condition of static equilibrium.

The viscous stresses generated by the shearing of the lubricant film give rise to a resisting force of magnitude $-f_b$ on the moving surface. The rate of working against the viscous stresses, or power loss, for one pad is

$$\hbar_p = -f_b u_b = -f_b'(r_o - r_i)u_b \tag{8.12}$$

The work done against the viscous stresses appears as heat within the lubricant. Some of this heat may be transferred to the surroundings by radiation or by conduction, or it may be convected from the clearance space by the lubricant flow.

The bulk temperature rise of the lubricant for the case in which all the heat is carried away by convection is known as the "adiabatic temperature rise." This bulk temperature increase can be calculated by equating the rate of heat generated within the lubricant to the rate of heat transferred by convection

$$\hbar_p = J\rho q C_p(\Delta t_m) g$$

or the adiabatic temperature rise in degrees Celsius may be expressed as

$$\Delta t_m = \frac{\hbar_p}{J\rho q C_p g} \tag{8.13}$$

where J = Joule's mechanical equivalent of heat, $\mathrm{N \cdot m/J}$
ρ = force density of lubricant, $\mathrm{N \cdot s^2/m^4}$
q = volume flow rate in direction of motion, $\mathrm{m^3/s}$
C_p = specific heat of material at constant pressure, $\mathrm{J/(N \cdot {}^\circ C)}$
g = gravitational acceleration, $9.8\ \mathrm{m/s^2}$

Some of the relevant equations used in thrust bearing analysis having been defined, attention will be focused on three simple slider bearings: (1) parallel surface, (2) fixed incline, and (3) parallel step. The same nondimensionalization is used to define the resulting performance parameters for all three types of bearing so that they can be directly compared. Throughout this chapter it is assumed that the pressure-generating mechanism is the physical wedge as dicussed in Sec. 7.3.3.

8.3 PARALLEL-SURFACE SLIDER BEARING

Figure 8.5 shows a parallel-surface slider bearing. The film thickness is constant for the entire bearing length. Using the Reynolds equation defined in Eq. (7.48)

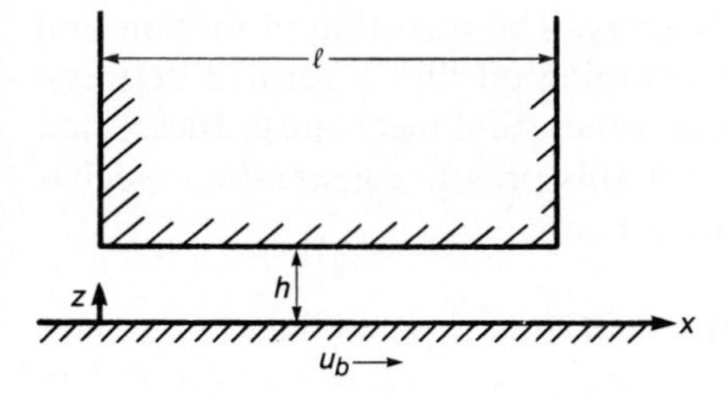

FIGURE 8.5
Parallel-surface slider bearing.

while neglecting the side-leakage term and assuming a constant film thickness gives the reduced Reynolds equation of

$$\frac{d^2p}{dx^2} = 0 \tag{8.14}$$

Integrating twice gives

$$p = \tilde{A}x + \tilde{B}$$

The boundary conditions are:

1. $p = 0$ at $x = 0$.
2. $p = 0$ at $x = \ell$.

Making use of these boundary conditions gives $p = 0$. Hence, a parallel-surface slider bearing does not develop pressure due to the physical wedge mechanism. Having no pressure development implies that this type of bearing will not be able to support a radial load. This is the same conclusion arrived at earlier in the chapter (Sec. 8.1) from physical arguments.

8.4 FIXED-INCLINE SLIDER BEARING

Figure 8.6 shows a fixed-incline slider bearing. A fixed-incline slider consists of two nonparallel plane surfaces separated by an oil film. One surface is station-

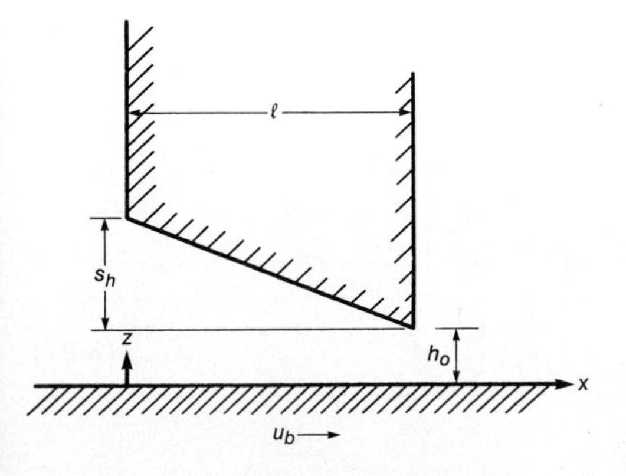

FIGURE 8.6
Fixed-incline slider bearing.

ary while the other moves with a uniform velocity. The direction of motion and the inclination of planes are such that a converging oil film is formed between the surfaces and the physical wedge pressure-generating mechanism (described in Sec. 7.3.3) is developed in the oil film. It is this pressure-generating mechanism that makes the bearing able to support a load.

8.4.1 PRESSURE DISTRIBUTION. The analysis of this type of bearing begins with the integrated form of the Reynolds equation as expressed in Eq. (7.52). The only change to this equation is that $u_a = 0$ and a constant viscosity η_0 is assumed.

$$\frac{dp}{dx} = 6\eta_0 u_b \frac{h - h_m}{h^3} \tag{8.15}$$

where h_m is the film thickness when $dp/dx = 0$.

The oil film thickness can be written as a function of x.

$$h = h_o + s_h\left(1 - \frac{x}{\ell}\right) \tag{8.16}$$

Choosing to write the film thickness and pressure in dimensionless terms where

$$P = \frac{ps_h^2}{\eta_0 u_b \ell} \qquad H = \frac{h}{s_h} \qquad H_m = \frac{h_m}{s_h} \qquad H_o = \frac{h_o}{s_h} \qquad X = \frac{x}{\ell} \tag{8.17}$$

causes Eqs. (8.15) and (8.16) to become

$$\frac{dP}{dX} = 6\left(\frac{H - H_m}{H^3}\right) \tag{8.18}$$

$$H = \frac{h}{s_h} = H_o + 1 - X \tag{8.19}$$

$$\frac{dH}{dX} = -1 \tag{8.20}$$

Integrating Eq. (8.18) gives

$$P = 6\int\left(\frac{1}{H^2} - \frac{H_m}{H^3}\right) dX$$

Making use of Eq. (8.20) in the preceding equation gives

$$P = -6\int\left(\frac{1}{H^2} - \frac{H_m}{H^3}\right) dH$$

$$\therefore P = 6\left(\frac{1}{H} - \frac{H_m}{2H^2}\right) + \tilde{A} \tag{8.21}$$

The boundary conditions are

1. $P = 0$ when $X = 0 \rightarrow H = H_o + 1$.
2. $P = 0$ when $X = 1 \rightarrow H = H_o$.

Making use of boundary conditions 1 and 2 gives

$$H_m = \frac{2H_o(1 + H_o)}{1 + 2H_o} \tag{8.22}$$

and

$$\tilde{A} = -\frac{6}{1 + 2H_o} \tag{8.23}$$

Substituting Eqs. (8.22) and (8.23) into (8.21) gives

$$P = \frac{6X(1 - X)}{(H_o + 1 - X)^2(1 + 2H_o)} \tag{8.24}$$

Note that the dimensionless pressure is a function of X and H_o. The variation of P with X for various values of H_o is shown in Fig. 8.7. It can be seen that the pressure distribution increases with decreasing H_o. Recall that $H_o = h_o/s_h$. Therefore, if the shoulder height s_h remains fixed, Fig. 8.7 indicates that as the outlet film thickness h_o becomes smaller, the pressure profile

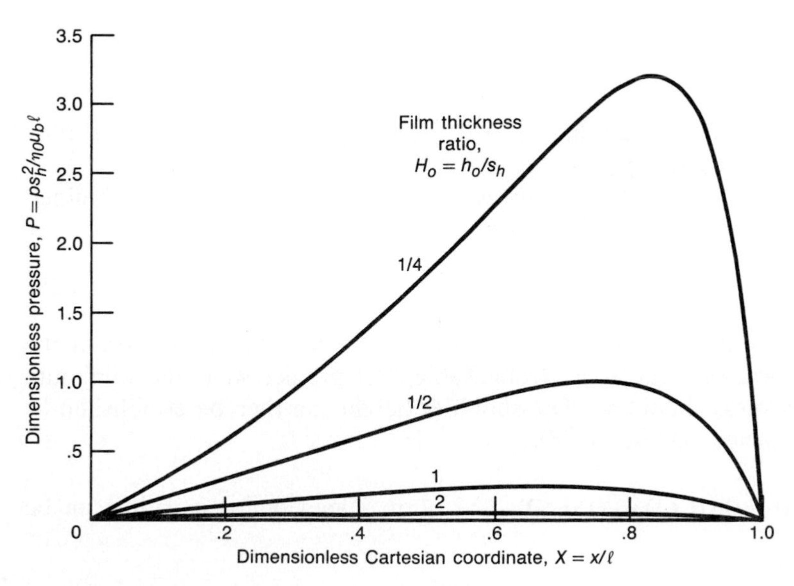

FIGURE 8.7
Pressure distributions of fixed-incline slider bearing.

increases without any limits. This figure also shows that for large H_o there is little pressure buildup in a fixed-incline thrust bearing.

Evaluating the film thickness at the location where the pressure gradient is equal to zero gives

$$H_m = H_o + 1 - X_m$$

Making use of Eq. (8.22) gives

$$X_m = \frac{1 + H_o}{1 + 2H_o} \tag{8.25}$$

As $H_o \to 0$, the location of the maximum pressure $X_m \to 1$, but as $H_o \to \infty$, $X_m \to \frac{1}{2}$. Note that $H_o \to 0$ implies that either $h_o \to 0$ or $s_h \to \infty$, but $H_o \to \infty$ implies that either $h_o \to \infty$ or $s_h \to 0$. The situation of $s_h \to 0$ implies parallel surfaces, but as concluded in Sec. 8.3, parallel surfaces do not develop pressure.

With the location of the maximum pressure defined by Eq. (8.25), the maximum pressure can be found directly from Eq. (8.24), which when $X = X_m$ gives

$$P_m = \frac{3}{2H_o(1 + H_o)(1 + 2H_o)} \tag{8.26}$$

Note that as $H_o \to 0$, $P_m \to \infty$ and that as $H_o \to \infty$, $P_m \to 0$. This corresponds to the conclusion made earlier while discussing Fig. 8.7. Equation (8.26) can be expressed in dimensional terms as

$$p_m = \frac{3\eta_0 u_b \ell s_h}{2h_o(s_h + h_o)(s_h + 2h_o)} \tag{8.27}$$

From Eq. (8.27) it is observed that $s_h \to 0$, which corresponds to a parallel film, and $s_h \to \infty$ both produce $p_m \to 0$.

The shoulder height that produces a maximum pressure can be obtained from $\partial p_m/\partial s_h = 0$. Evaluating this gives

$$(s_h)_{\text{opt}} = \sqrt{2}\, h_o \tag{8.28}$$

Equation (8.28) will be helpful in practical designs, since if a description of the surface is known, one can (e.g., from Table 9.2) predict what the minimum outlet film thickness should be. The shoulder height can then be established by using a safety factor and Eq. (8.28).

8.4.2 NORMAL LOAD COMPONENT. The normal load per unit width can be written as

$$w'_z = \int_0^\ell p\, dx$$

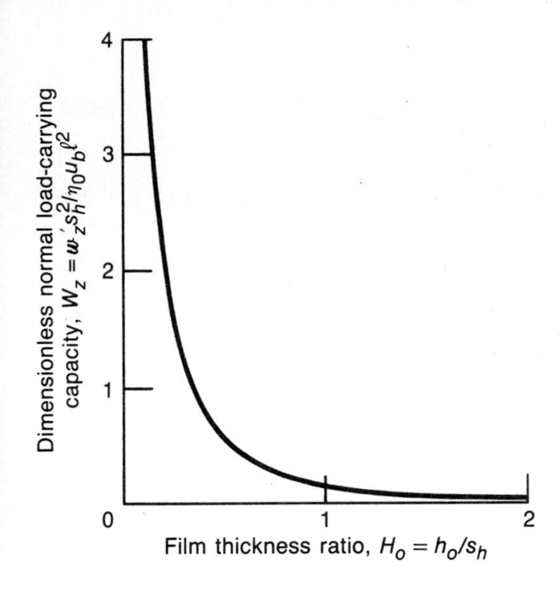

FIGURE 8.8
Effect of film thickness ratio on normal load-carrying capacity.

This equation can be written in dimensionless form by making use of Eq. (8.17).

$$W_z = \frac{w_z' s_h^2}{\eta_0 u_b \ell^2} = \int_0^1 P\,dX$$

Because $dH/dX = -1$, as shown in Eq. (8.20),

$$W_z = -\int_{H_o+1}^{H_o} P\,dH \tag{8.29}$$

Substituting Eqs. (8.21) and (8.23) into this equation gives

$$W_z = 6\ln\left(\frac{H_o+1}{H_o}\right) - \frac{12}{1+2H_o} \tag{8.30}$$

The variation of W_z with H_o, shown in Fig. 8.8, suggests that as $H_o \to 0$, this bearing has a tremendous potential to support a radial load. These results should be tempered by the knowledge of the assumptions imposed in getting this result, namely, that side leakage was neglected and smooth surfaces and isothermal conditions were assumed.

8.4.3 TANGENTIAL FORCE COMPONENTS. The force per unit width in the direction of motion due to pressure being developed is

$$w_{xb}' = 0$$

$$w_{xa}' = -\int_{h_o+s_h}^{h_o} p\,dh$$

By making use of Eq. (8.17) the preceding equation may be expressed in dimensionless form as

$$W_{xa} = \frac{w'_{xa}}{\eta_0 u_b} \frac{s_h}{\ell} = -\int_{H_o+1}^{H_o} P\, dH = W_z \tag{8.31}$$

8.4.4 SHEAR FORCE COMPONENTS. The shear force components per unit width acting on the solids are

$$f'_b = \int_0^\ell (\tau_{zx})_{z=0}\, dx = \int_0^\ell \left(\eta_0 \frac{\partial u}{\partial z}\right)_{z=0} dx$$

$$f'_a = \int_0^\ell (-\tau_{zx})_{z=h}\, dx = -\int_0^\ell \left(\eta_0 \frac{\partial u}{\partial z}\right)_{z=h} dx$$

The viscous shear stresses were defined in Chap. 7 [Eqs. (7.32) and (7.33)]. Making use of these equations while letting $u_a = 0$ and assuming a constant viscosity gives

$$f'_b = -\int_0^\ell \left(\frac{h}{2}\frac{dp}{dx} + \frac{u_b \eta_0}{h}\right) dx$$

$$f'_a = -\int_0^\ell \left(\frac{h}{2}\frac{dp}{dx} - \frac{u_b \eta_0}{h}\right) dx$$

Equation (8.17) can be used to put these equations in dimensionless form:

$$F_b = \frac{f'_b}{\eta_0 u_b}\frac{s_h}{\ell} = -\int_0^1 \left(\frac{H}{2}\frac{dP}{dX} + \frac{1}{H}\right) dX$$

$$F_a = \frac{f'_a}{\eta_0 u_b}\frac{s_h}{\ell} = -\int_0^1 \left(\frac{H}{2}\frac{dP}{dX} - \frac{1}{H}\right) dX$$

By making use of Eqs. (8.18) and (8.19) the preceding equations can be expressed as

$$F_b = 4 \ln\left(\frac{H_o}{H_o + 1}\right) + \frac{6}{1 + 2H_o} \tag{8.32}$$

$$F_a = 2 \ln\left(\frac{H_o}{H_o + 1}\right) + \frac{6}{1 + 2H_o} \tag{8.33}$$

The dimensionless force components W_{xa}, F_b, and F_a are plotted as functions of H_o in Fig. 8.9. Note that the dimensionless force components W_{xa}, F_b, and F_a have the term s_h/ℓ, whereas W_z has the term $(s_h/\ell)^2$.

8.4.5 FRICTION COEFFICIENT. The friction coefficient can be expressed as

$$\mu = -\frac{f'_b}{w'_{zb}} = \frac{f'_a + w'_{xa}}{w'_{za}}$$

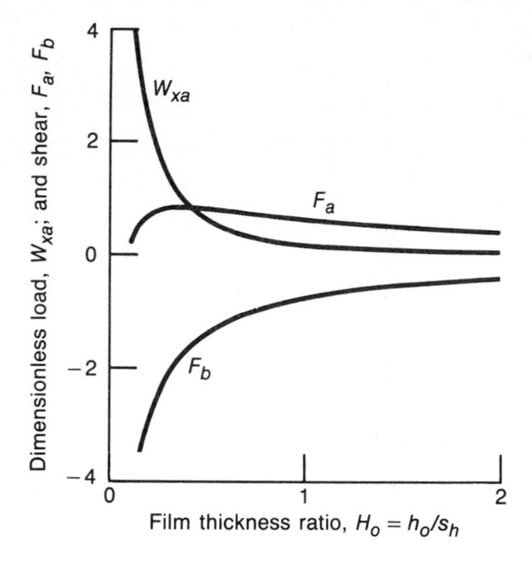

FIGURE 8.9
Effect of film thickness ratio on force components.

Making use of Eqs. (8.30) and (8.32) gives

$$\mu = \frac{2s_h \ln\left(\dfrac{H_o}{H_o + 1}\right) + \dfrac{3s_h}{1 + 2H_o}}{3\ell \ln\left(\dfrac{H_o}{H_o + 1}\right) + \dfrac{6\ell}{1 + 2H_o}} \tag{8.34}$$

The variation of $\mu \ell / s_h$ with H_o is shown in Fig. 8.10. Note that as $h_o \to 0$, the friction coefficient approaches zero. This implies that the normal load component becomes much larger than the tangential load component as h_o decreases.

8.4.6 VOLUME FLOW RATE. The volume flow rate per unit width can be expressed from Chap. 7 as

$$q'_x = -\frac{h^3}{12\eta}\frac{\partial p}{\partial x} + \frac{h(u_b + u_a)}{2} \tag{7.38}$$

Evaluating the flow rate where $dp/dx = 0$ and setting $u_a = 0$ gives the volume

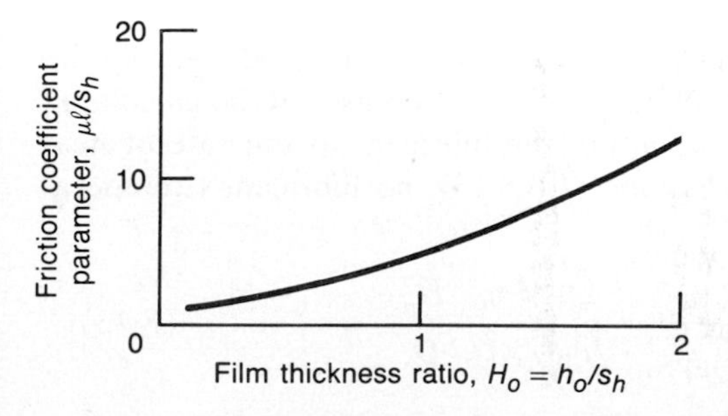

FIGURE 8.10
Effect of film thickness ratio on friction coefficient parameter.

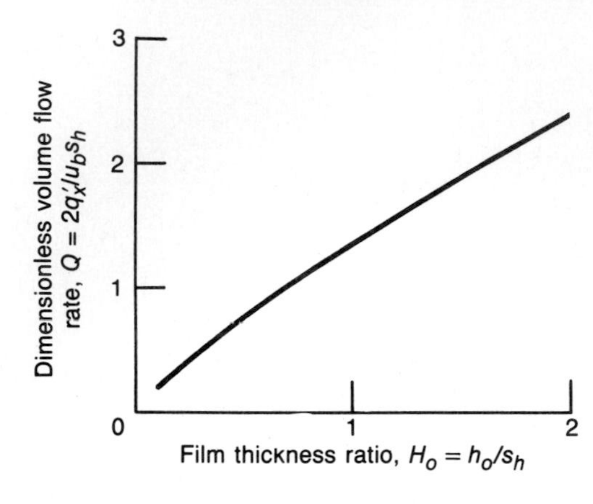

FIGURE 8.11
Effect of film thickness ratio on dimensionless volume flow rate.

flow rate as

$$q'_x = \frac{u_b h_m}{2} \tag{8.35}$$

The dimensionless volume flow rate can be expressed as

$$Q = \frac{2q'_x}{u_b s_h} = H_m = \frac{2H_o(1 + H_o)}{1 + 2H_o} \tag{8.36}$$

The dimensionless volume flow rate Q is plotted as a function of H_o in Fig. 8.11. This figure shows that as H_o increases, the dimensionless volume flow rate increases.

8.4.7 POWER LOSS AND TEMPERATURE RISE. The total rate of working against the viscous stresses, or the power loss, can be expressed from Eq. (8.12) as

$$\hbar_p = -f_b u_b = -f'_b(r_o - r_i)u_b \tag{8.12}$$

Expressed in dimensionless form,

$$H_p = \frac{\hbar_p s_h}{\eta_0 u_b^2 \ell (r_o - r_i)} = -\frac{f'_b s_h}{\eta_0 u_b \ell} = -F_b = -4 \ln\left(\frac{H_o}{H_o + 1}\right) - \frac{6}{1 + 2H_o} \tag{8.37}$$

All the heat produced by viscous shearing is assumed to be carried away by the lubricant (adiabatic condition). The bulk temperature increase can be calculated by equating the rate of heat generated within the lubricant to the rate of heat transferred by convection. Therefore, from Eq. (8.13) the lubricant's temperature rise is

$$\Delta t_m = \frac{\hbar_p}{J\rho_0 q'_x C_p g} = \frac{2u_b \ell \eta_0}{J\rho_0 C_p s_h^2 g} \frac{H_p}{Q} \tag{8.13}$$

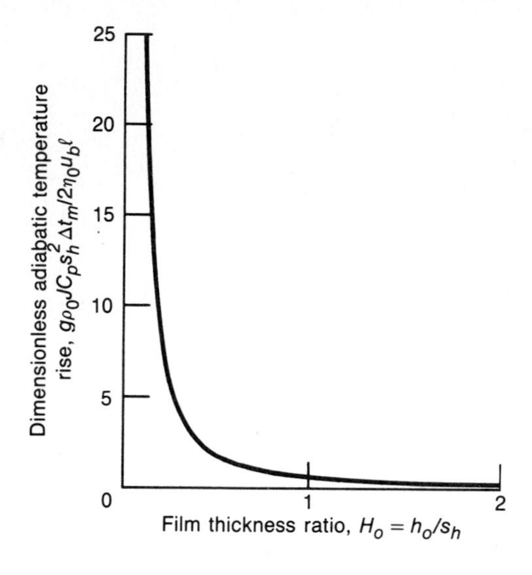

FIGURE 8.12
Effect of film thickness ratio on dimensionless adiabatic temperature rise.

where J = Joule's mechanical equivalent of heat, $\text{N} \cdot \text{m/J}$
ρ_0 = constant lubricant force density, $\text{N} \cdot \text{s}^2/\text{m}^4$
q_x' = volume flow rate per unit width in sliding direction, m^2/s
C_p = specific heat at constant pressure, $\text{J}/(\text{N} \cdot {}^\circ\text{C})$
g = gravitational acceleration, 9.8 m/s^2

The dimensionless temperature rise may be expressed as

$$\frac{gJ\rho_0 C_p s_h^2}{2u_b \ell \eta_0} \Delta t_m = \frac{H_p}{Q} = \frac{2(1 + 2H_o)}{H_o(1 + H_o)} \ln\left(\frac{H_o + 1}{H_o}\right) - \frac{3}{(1 + H_o)H_o} \quad (8.38)$$

Figure 8.12 shows the effect of film thickness ratio on dimensionless adiabatic temperature rise. As $H_o \to 0$, the dimensionless adiabatic temperature rise approaches infinity.

8.4.8 CENTER OF PRESSURE. The location of the center of pressure x_{cp} indicates the position at which the resulting force is acting. The expression for calculating the location is

$$w_z' x_{\text{cp}} = \int_0^\ell px\,dx = \frac{\eta_0 u_b \ell^3}{s_h^2} \int_0^1 PX\,dX$$

Therefore, the dimensionless center of pressure can be written as

$$X_{\text{cp}} = \frac{x_{\text{cp}}}{\ell} = \frac{1}{W_z} \int_0^1 PX\,dX \quad (8.39)$$

Substituting Eqs. (8.19) to (8.21) into (8.39) gives

$$X_{\text{cp}} = \frac{-6}{W_z(1 + 2H_o)} \left[(H_o + 1)(3H_o + 1) \ln\left(\frac{H_o}{H_o + 1}\right) + 3H_o + \frac{5}{2} \right] \quad (8.40)$$

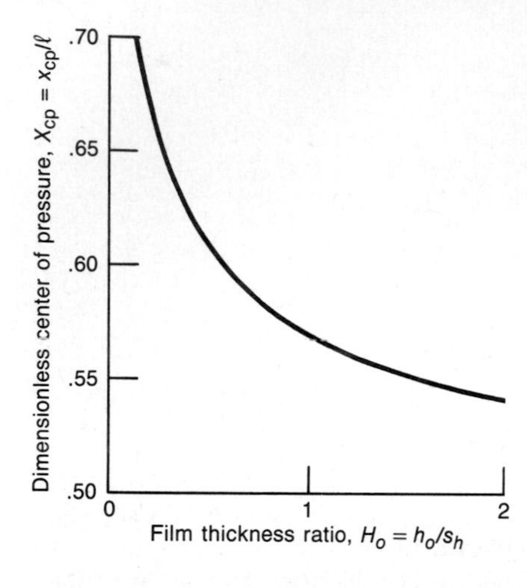

FIGURE 8.13
Effect of film thickness ratio on dimensionless center of pressure.

Figure 8.13 shows the effect of film thickness ratio on center of pressure. The center of pressure is always more toward the outlet than toward the inlet ($X_{cp} > 0.5$).

8.4.9 VELOCITY PROFILE AND STREAM FUNCTION. From Eq. (7.28) for a stationary top surface ($u_a = 0$), the fluid velocity can be written as

$$u = -\frac{z(h-z)}{2\eta_0}\frac{dp}{dx} + \frac{u_b(h-z)}{h} \tag{8.41}$$

By using Eq. (8.17) this equation can be expressed in dimensionless form as

$$\frac{u}{u_b} = \left(1 - \frac{Z}{H}\right)\left(1 - \frac{ZH}{2}\frac{dP}{dX}\right) \tag{8.42}$$

where

$$Z = \frac{z}{s_h} \qquad \text{and} \qquad 0 \le Z \le H \tag{8.43}$$

From Eq. (8.42), $u/u_b = 0$ when

1. $Z_{cr} = H$

or at the top surface

2. $$Z_{cr} = \frac{2}{H(dP/dX)} \tag{8.44}$$

Note that condition 2 can only exist when $dP/dX > 0$ and thus when $X < X_m$.

Substituting Eq. (8.18) into Eq. (8.44) while making sure that the inequality in Eq. (8.43) is satisfied gives

$$0 \leq 2H - 3H_m \tag{8.45}$$

Making use of Eqs. (8.19) and (8.22) gives this inequality as

$$X \leq \frac{1 - H_o^2}{2H_o + 1} \tag{8.46}$$

This inequality is only satisfied if $H_o \leq 1$. This then implies that reverse flow exists when

$$H_o \leq 1 \qquad \text{and} \qquad X \leq \frac{1 - H_o^2}{2H_o + 1} \tag{8.47}$$

If these inequalities are satisfied, then $dP/dX > 0$ and $X < X_m$.

Substituting Eq. (8.18) into Eq. (8.42) gives

$$\frac{u}{u_b} = \left(1 - \frac{Z}{H}\right)\left[1 - \frac{3Z}{H}\left(1 - \frac{H_m}{H}\right)\right] \tag{8.48}$$

where

$$\frac{Z}{H} = \frac{Z}{H_o + 1 - X} \tag{8.49}$$

$$\frac{H_m}{H} = \frac{2H_o(1 + H_o)}{(1 + 2H_o)(H_o + 1 - X)} \tag{8.50}$$

Thus, u/u_b is just a function of X, Z, and H_o.

A "streamline" is a type of curve that has been used to describe the flow of fluid. More specifically, a streamline is a curve everywhere parallel to the direction of the fluid flow. Surface boundaries are streamlines, since the fluid cannot cross the surface boundary. The definition of a streamline may be given mathematically, while neglecting the side-leakage term, as

$$\frac{dx}{u} = \frac{dz}{w} \tag{8.51}$$

The continuity equation given by Eq. (6.48) can be satisfied while neglecting the side-leakage term by introducing a new function defined by

$$u = \frac{\partial \bar{\phi}}{\partial z} \qquad \text{and} \qquad w = -\frac{\partial \bar{\phi}}{\partial x} \tag{8.52}$$

where $\bar{\phi}$ is a function of x and z and is called the "stream function." By using the chain rule of partial differentiation the total derivative of $\bar{\phi}$ can be expressed as

$$d\bar{\phi} = \frac{\partial \bar{\phi}}{\partial x}\,dx + \frac{\partial \bar{\phi}}{\partial z}\,dz \tag{8.53}$$

Making use of Eq. (8.52) gives

$$d\bar{\phi} = -w\,dx + u\,dz \tag{8.54}$$

If $d\bar{\phi}$ is set equal to zero, we obtain the definition of a streamline given in Eq. (8.51). That is, lines of constant $\bar{\phi}$ represent streamlines.

Equation (8.52) can also be expressed in dimensionless terms of a stream function $\bar{\Phi}$ as

$$\frac{u}{u_b} = \frac{1}{u_b}\frac{\partial \bar{\phi}}{\partial z} = \frac{\partial \bar{\Phi}}{\partial Z} \tag{8.55}$$

where $\bar{\Phi} = \bar{\phi}/u_b s_h$. Substituting Eq. (8.42) into this equation and integrating gives

$$\bar{\Phi}(X, Z) = \frac{Z^3}{6}\frac{dP}{dX} - \frac{Z^2}{2H}\left(1 + \frac{H^2}{2}\frac{dP}{dX}\right) + Z = \text{constant} \tag{8.56}$$

This equation can be easily evaluated by using Eqs. (8.18), (8.19), and (8.22).

Also from the continuity equation, while neglecting the side-leakage term, the velocity in the z direction can be written as

$$w = \int\left(-\frac{\partial u}{\partial x}\right)dz = \frac{s_h u_b}{\ell}\int\left[-\frac{\partial(u/u_b)}{\partial X}\right]dZ$$

$$\therefore \frac{w}{u_b}\frac{\ell}{s_h} = -\int_0^Z\left[\frac{\partial(u/u_b)}{\partial H}\right]\frac{\partial H}{\partial X}\,dZ = \int_0^Z\left[\frac{\partial(u/u_b)}{\partial H}\right]dZ$$

Substituting Eq. (8.48) into this equation gives

$$\frac{w\ell}{u_b s_h} = \left(Z^2 - \frac{Z^3}{H}\right)\left(\frac{2}{H^2} - \frac{3H_m}{H^3}\right) = \frac{Z^2}{H^4}(Z - H)(3H_m - 2H) \tag{8.57}$$

Note from this equation that $w/u_b(\ell/s_h) = 0$ when $Z = 0$, when $Z = H$, and at $H = \frac{3}{2}H_m$. Making use of Eqs. (8.19) and (8.22) gives the critical value of X, where $H = \frac{3}{2}H_m$, as

$$X_{\text{cr}} = \frac{1 - H_o^2}{1 + 2H_o} \tag{8.58}$$

Note that $0 \le X_{\text{cr}} \le 1$ when $H_o \le 1$.

By making use of Eq. (8.52) the stream function can be written as

$$\frac{w\ell}{u_b s_h} = -\frac{\partial \bar{\Phi}}{\partial X} = -\frac{\partial \bar{\Phi}}{\partial H}\frac{\partial H}{\partial X} = \frac{\partial \bar{\Phi}}{\partial H} \tag{8.59}$$

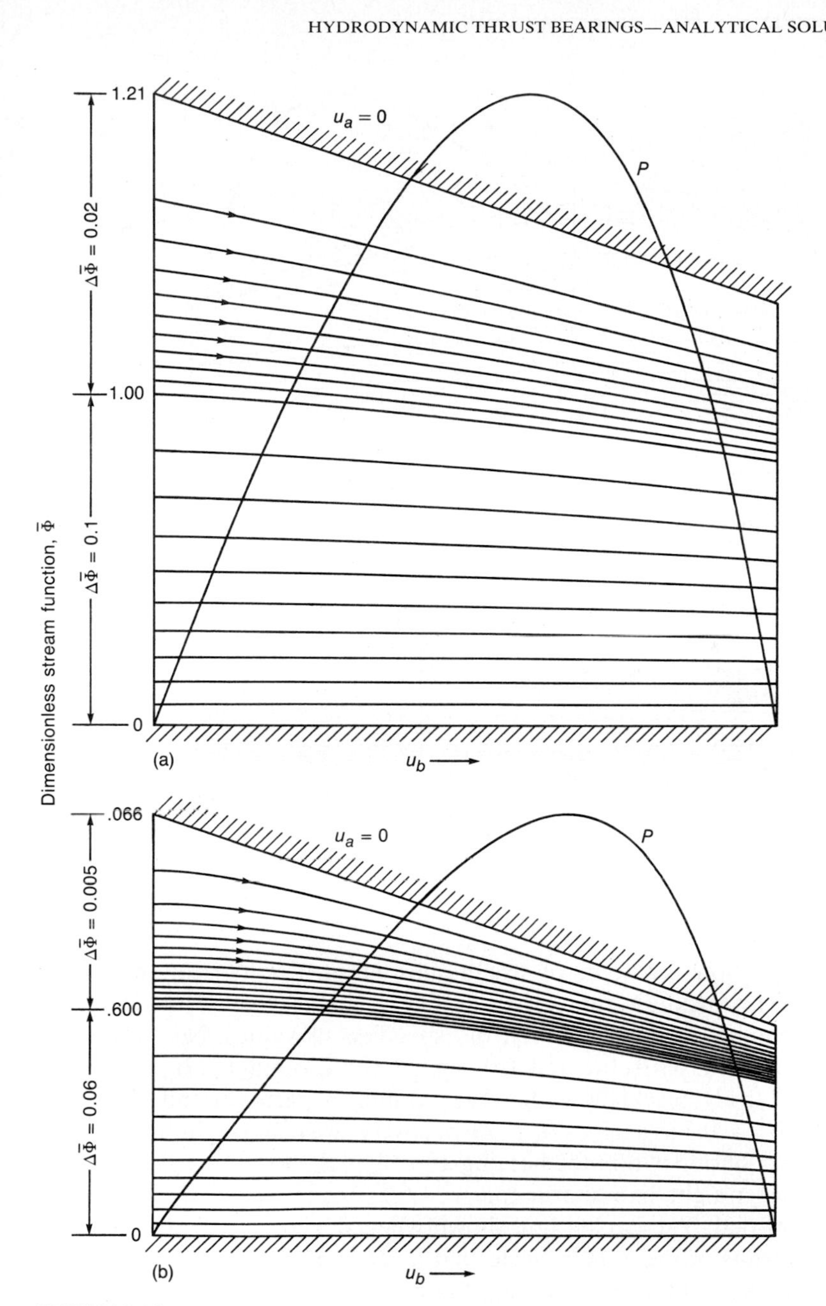

FIGURE 8.14
Streamlines in fixed-incline bearing at four film thickness ratios H_o. (a) $H_o = 2$; (b) $H_o = 1$ (critical value).

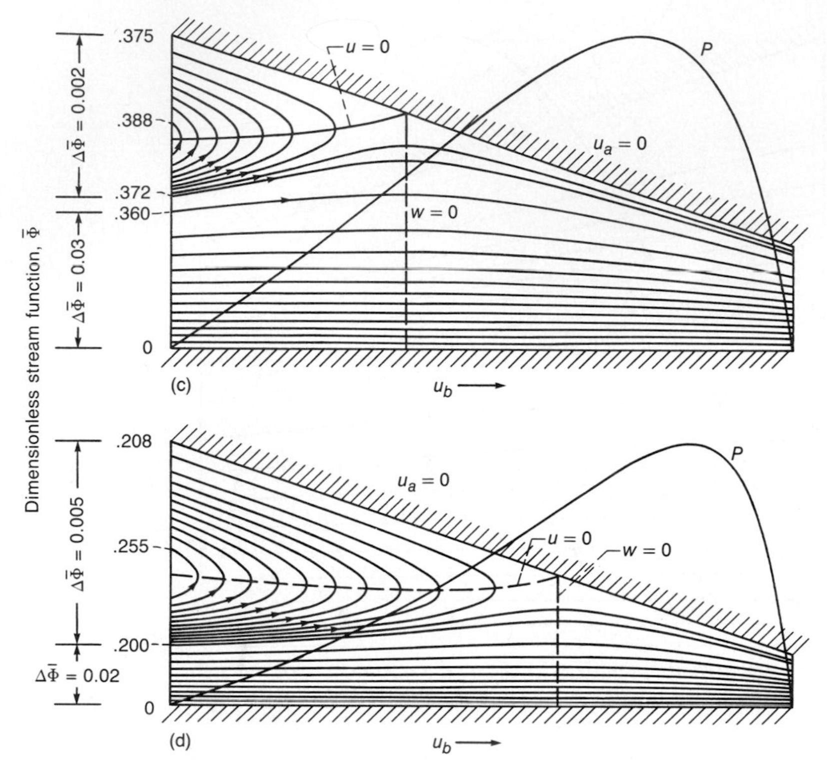

FIGURE 8.14
Concluded. (c) $H_o = 0.5$; (d) $H_o = 0.25$.

Figure 8.14 shows contours of streamlines within the lubrication film for four different film thickness ratios. In this figure the $\Delta\overline{\Phi}$ value indicates the constant incrementation of $\overline{\Phi}$ between the specified $\overline{\Phi}$ values. No backflow is observed to occur in parts (a) and (b), which correspond to $H_o = 2.0$ and $H_o = 1.0$, respectively. Backflow is, however, found in parts (c) and (d), which correspond to $H_o = 0.5$ and $H_o = 0.25$, respectively. Also shown in this figure is the pressure distribution within the bearing as well as the locations where $w = 0$ and $u = 0$. The straight line shown for $w = 0$ is when $H = \frac{3}{2}H_m$ and when Eq. (8.58) is satisfied. The curved line shown when $u = 0$ is when Eq. (8.44) is satisfied. The results of Fig. 8.14 indicate that as H_o becomes smaller, the more difficult it is for the fluid to pass through the bearing.

8.5 PARALLEL-STEP SLIDER BEARING

Lord Rayleigh, as long ago as 1918, demonstrated that a parallel-step geometry produced the optimum load-carrying capacity when side leakage was neglected. This bearing has not, however, enjoyed the same development and applications

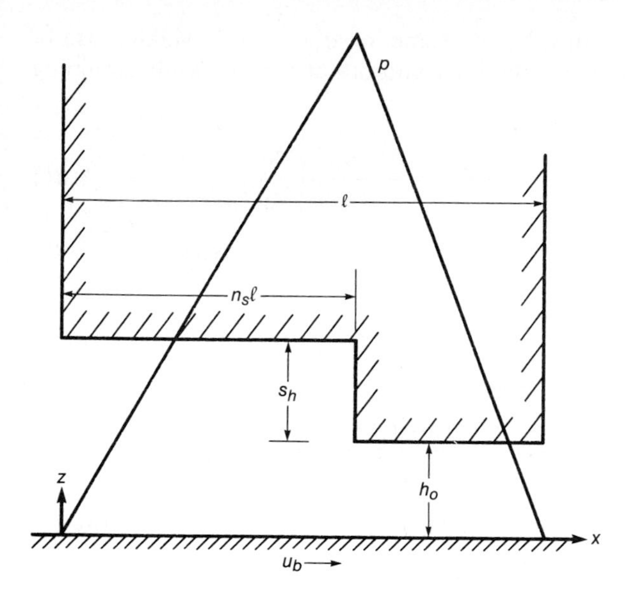

FIGURE 8.15
Parallel-step slider bearing.

as the pivoted-pad slider bearing. Past neglect of this mathematically preferable configuration has been due to doubts about the relative merits of this bearing when side leakage is considered. Figure 8.15 illustrates the geometric shape of the film in this bearing as well as a typical pressure profile resulting from the analysis.

The parallel-step slider bearing is analyzed here by considering two connected parallel-surface bearings. The subscripts i and o will be used to denote conditions in the inlet and outlet films, respectively.

8.5.1 PRESSURE DISTRIBUTION. From Eq. (7.48), with side leakage neglected and a constant film thickness in the inlet and outlet regions, the appropriate Reynolds equation is

$$\frac{d^2p}{dx^2} = 0$$

This equation can be integrated to give

$$\frac{dp}{dx} = \text{constant}$$

This result shows that the pressure gradients in the inlet and outlet regions are constant. Since the film thicknesses in the two regions are different, their pressure gradients are also different. Thus, since there will be no discontinuity in pressure at the step,

$$p_m = n_s \ell \left(\frac{dp}{dx}\right)_i = -(\ell - n_s \ell)\left(\frac{dp}{dx}\right)_o \tag{8.60}$$

Also the flow rate at the step must be the same, or $q'_{x,o} = q'_{x,i}$. Making use of Eq. (7.38) and equating the flow for the inlet and outlet regions while assuming constant viscosity gives

$$-\frac{(h_o + s_h)^3}{12\eta_0}\left(\frac{dp}{dx}\right)_i + \frac{u_b(h_o + s_h)}{2} = -\frac{h_o^3}{12\eta_0}\left(\frac{dp}{dx}\right)_o + \frac{u_b h_o}{2} \quad (8.61)$$

Equations (8.60) and (8.61) represent a pair of simultaneous equations with unknowns $(dp/dx)_i$ and $(dp/dx)_o$. The solutions are

$$\left(\frac{dp}{dx}\right)_i = \frac{6\eta_0 u_b(1 - n_s)s_h}{(1 - n_s)(h_o + s_h)^3 + n_s h_o^3} \quad (8.62)$$

$$\left(\frac{dp}{dx}\right)_o = \frac{-6\eta_0 u_b n_s s_h}{(1 - n_s)(h_o + s_h)^3 + n_s h_o^3} \quad (8.63)$$

The maximum pressure (at the step) can be found directly by substituting Eqs. (8.62) and (8.63) into (8.60) to give

$$p_m = \frac{6\eta_0 u_b \ell n_s(1 - n_s)s_h}{(1 - n_s)(h_o + s_h)^3 + n_s h_o^3} \quad (8.64)$$

By making use of Eq. (8.17), Eq. (8.64) can be written in dimensionless form as

$$P_m = \frac{p_m s_h^2}{\eta_0 u_b \ell} = \frac{6n_s(1 - n_s)}{(1 - n_s)(H_o + 1)^3 + n_s H_o^3} \quad (8.65)$$

The bearing configuration that produces the largest p_m is achieved when

$$\frac{\partial p_m}{\partial n_s} = 0 \qquad \text{and} \qquad \frac{\partial p_m}{\partial s_h} = 0$$

Using these conditions in Eq. (8.64) gives

$$0 = (1 - n_s)^2(h_o + s_h)^3 - n_s^2 h_o^3 \quad (8.66)$$

and

$$0 = (1 - n_s)(h_o + s_h)^2(h_o - 2s_h) + n_s h_o^3 \quad (8.67)$$

Solving for n_s and s_h in these equations gives

$$H_o = \frac{h_o}{s_h} = 1.155 \quad (8.68)$$

$$n_s = 0.7182 \quad (8.69)$$

This is the optimum parallel-step bearing configuration that Lord Rayleigh (1918) described in his classic paper. Knowing the maximum pressure and that the pressure gradients are constants allows the dimensionless pressure in the inlet and outlet regions to be written as

$$P_i = \frac{XP_m}{n_s} = \frac{6X(1-n_s)}{(1-n_s)(H_o+1)^3+n_sH_o^3} \qquad 0 \le X \le n_s \tag{8.70}$$

$$P_o = \frac{(1-X)P_m}{1-n_s} = \frac{6(1-X)n_s}{(1-n_s)(H_o+1)^3+n_sH_o^3} \qquad n_s \le X \le 1 \tag{8.71}$$

The pressure distribution is shown diagrammatically in Fig. 8.15. Note that $P = 0$ everywhere if $H_o \to \infty$ or if $n_s = 0$ or 1. Also note that $H_o \to \infty$ implies that $s_h \to 0$ (a parallel film) or that $h_o \to \infty$.

8.5.2 NORMAL AND TANGENTIAL LOAD COMPONENTS. The normal load per unit width can easily be determined for this simple form of pressure distribution. That is, w'_z is directly proportional to the triangular area formed by the pressure distributions.

$$w'_z = \frac{p_m \ell}{2} = \frac{3\eta_0 u_b \ell^2 n_s(1-n_s)s_h}{(1-n_s)(h_o+s_h)^3+n_sh_o^3} \tag{8.72}$$

The dimensionless normal load can be expressed as

$$W_z = \frac{w'_z}{\eta_0 u_b}\left(\frac{s_h}{\ell}\right)^2 = \frac{3n_s(1-n_s)}{(1-n_s)(H_o+1)^3+n_sH_o^3} = \frac{P_m}{2} \tag{8.73}$$

Note that $W_z = 0$ if $H_o \to \infty$ or if $n_s = 0$ or 1 and is thus a function of n_s and H_o.

The tangential load components per unit width acting on the bearing in the direction of motion are equal to the pressure at the step multiplied by the step height.

$$w'_{xb} = W_{xb} = 0$$

$$w'_{xa} = -\int_{h_o+s_h}^{h_o} p\,dh = p_m s_h \tag{8.74}$$

or

$$w'_{xa} = \frac{6\eta_0 u_b \ell n_s(1-n_s)s_h^2}{(1-n_s)(h_o+s_h)^3+n_sh_o^3}$$

In dimensionless form this equation becomes

$$W_{xa} = \frac{w'_{xa}}{\eta_0 u_b}\frac{s_h}{\ell} = \frac{6n_s(1-n_s)}{(1-n_s)(H_o+1)^3+n_sH_o^3} = P_m \tag{8.75}$$

Note that $W_{xa} = 0$ if $H_o \to \infty$ or if $n_s = 0$ or 1.

The general expression for the shear force per unit width at the moving surface can be written as

$$f_b' = \int_0^{n\ell} (\tau_{zx,i})_{z=0}\, dx + \int_{n\ell}^{\ell} (\tau_{zx,o})_{z=0}\, dx$$

Making use of Eq. (7.34) yields

$$f_b = \left[-\frac{h_o + s_h}{2}\left(\frac{dp}{dx}\right)_i - \frac{\eta_0 u_b}{h_o + s_h} \right] n_s \ell + \left[-\frac{h_o}{2}\left(\frac{dp}{dx}\right)_o - \frac{\eta_0 u_b}{h_o} \right] (1 - n_s)\ell$$

Making use of Eq. (8.60) gives

$$f_b' = -\frac{p_m s_h}{2} - \eta_0 u_b \ell \left(\frac{n_s}{h_o + s_h} - \frac{1 - n_s}{h_o} \right)$$

Making use of Eq. (8.74) gives

$$f_b' = -\frac{w_{xa}'}{2} - \frac{\eta_0 u_b \ell [n_s h_o + (1 - n_s)(h_o + s_h)]}{h_o(h_o + s_h)} \tag{8.76}$$

Similarly,

$$f_a' = -\frac{w_{xa}'}{2} + \frac{\eta_0 u_b \ell [n_s h_o + (1 - n_s)(h_o + s_h)]}{h_o(h_o + s_h)} \tag{8.77}$$

Equations (8.76) and (8.77) can be expressed in dimensionless form as

$$F_b = \frac{f_b'}{\eta_0 u_b} \frac{s_h}{\ell} = -\frac{P_m}{2} - \frac{H_o + 1 - n_s}{H_o(1 + H_o)} \tag{8.78}$$

$$F_a = \frac{f_a'}{\eta_0 u_b} \frac{s_h}{\ell} = -\frac{P_m}{2} + \frac{H_o + 1 - n_s}{H_o(1 + H_o)} \tag{8.79}$$

8.5.3 FRICTION COEFFICIENT AND VOLUME FLOW RATE. The friction coefficient for a parallel-step slider bearing can be directly written from Eqs. (8.78) and (8.73) as

$$\mu = -\frac{f_b'}{w_z'} = -\frac{s_h}{\ell}\frac{F_b}{W_z} = \frac{s_h}{\ell}\left[1 + \frac{2(H_o + 1 - n_s)}{P_m H_o(1 + H_o)}\right] \tag{8.80}$$

The volume flow rate per unit width can be expressed by making use of Eqs. (7.38) and (8.60) as

$$q_x' = -\frac{(h_o + s_h)^3 p_m}{12\eta_0 n_s \ell} + \frac{u_b(h_o + s_h)}{2}$$

The volume flow rate expressed in dimensionless form is given by

$$Q = \frac{2q_x'}{u_b s_h} = -\frac{P_m(H_o + 1)^3}{6n_s} + H_o + 1 \tag{8.81}$$

This expression reduces to the simple parallel-step slider bearing result ($Q = 1$), corresponding to Couette flow between flat plates if $H_o = \infty$ or if $n_s = 0$ or 1.

8.5.4 POWER LOSS, TEMPERATURE RISE, AND CENTER OF PRESSURE. The rate of work produced against viscous shearing, or the power loss, can be expressed from Eq. (8.12) in dimensionless form as

$$H_p = \frac{h_p s_h}{\eta_0 u_b^2 \ell (r_o - r_i)} = -\frac{f_b' s_h}{\eta_0 u_b \ell} = -F_b \tag{8.82}$$

The adiabatic temperature rise can be written from Eq. (8.13) as

$$\Delta t_m = \frac{h_p}{J\rho_0 q_x' C_p g} = \frac{2u_b \ell \eta_0}{J\rho_0 C_p g s_h^2}\frac{H_p}{Q} \tag{8.13}$$

Equations (8.82) and (8.81) were used in this equation. The center of pressure when both the inlet and outlet regions are considered can be expressed as

$$w_z' x_{\text{cp}} = \int_0^{n_s \ell} p_i x\, dx + \int_{n_s \ell}^{\ell} p_o x\, dx$$

This equation can be expressed in dimensionless form as

$$W_z X_{\text{cp}} = \int_0^{n_s} P_i X\, dX + \int_{n_s}^{1} P_o X\, dX$$

Making use of Eqs. (8.70), (8.71), and (8.73) gives

$$X_{\text{cp}} = \frac{x_{\text{cp}}}{\ell} = \frac{2}{n_s}\int_0^{n_s} X^2\, dX + \frac{2}{1 - n_s}\int_{n_s}^{1} X(1 - X)\, dX$$

$$\therefore X_{\text{cp}} = \frac{1 + n_s}{3} \tag{8.83}$$

Therefore, the center of pressure coincides with the step location when $n_s = \frac{1}{2}$. Also, $X_{\text{cp}} \to \frac{1}{3}$ as $n_s \to 0$ and $X_{\text{cp}} \to \frac{2}{3}$ as $n_s = 1$. These results correspond to pressure profiles that approach the form of right triangles in pads having steps near their ends.

8.6 CLOSURE

This chapter has considered film shapes in three different slider bearings: parallel surface, fixed incline, and parallel step. Only flow in the *xz* plane, or in the direction of motion, has been considered. The properties of the lubricant, namely, density and viscosity, were assumed to be constant. A positive pressure profile was obtained from the physical wedge effect in all the film shapes except a parallel film, where the pressure was zero throughout the bearing length. The appropriate Reynolds equation used to evaluate these film shapes was

$$\frac{d}{dx}\left(h^3 \frac{dp}{dx}\right) = 6\eta_0 u_b \frac{\partial h}{\partial x}$$

or

$$\frac{dp}{dx} = \frac{6\eta_0 u_b (h - h_m)}{h^3}$$

The pressure profile was determined from this equation and the particular film shape to be evaluated and then used to evaluate load components, shear force components, power loss, adiabatic temperature rise, friction coefficient, volume flow rate, stream functions, and center of pressure.

Of all these calculations the most important in most design considerations is the normal load-carrying capacity per unit width w'_z. For the fixed-incline slider bearing the dimensionless normal load-carrying capacity was found to be

$$W_z = \frac{w'_z}{\eta_0 u_b}\left(\frac{s_h}{\ell}\right)^2 = f(H_o)$$

This implies that the normal applied load per unit width is directly proportional to the viscosity η_0, the velocity u_b, and the length ℓ squared while also being inversely proportional to the shoulder height s_h squared. The film thickness ratio is defined as $H_o = h_o/s_h$. The stream functions within the lubricated film in an infinitely long, fixed-incline slider bearing is presented. Locations of reverse flow within the conjunction are clearly visible.

For a parallel-step slider bearing the dimensionless normal applied load is a function not only of the film thickness ratio, as was found for the fixed-incline slider bearing, but also of the location of the step. Recall that in this chapter side leakage was neglected.

8.7 PROBLEMS

8.7.1 A slider bearing has an exponential film shape, as shown in the sketch, and a film thickness defined as $h = (h_o + s_h)e^{\alpha x}$. An incompressible fluid is assumed, and no side flow is considered (only flow in the xz plane).

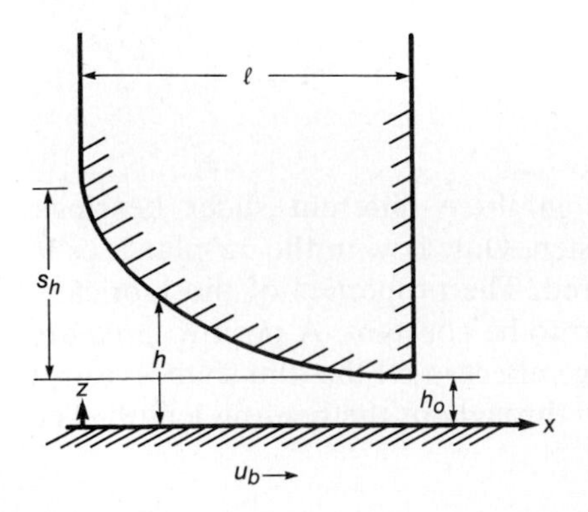

Determine the pressure within the bearing as well as the normal load components, the shear force components, the friction coefficient, the volume flow rate, the power loss, and the adiabatic temperature rise within the bearing.

8.7.2 A flat strip of metal emerges from a liquid bath of viscosity η_0 and pressure p_i above ambient and has velocity u_b on passing through a slot of the form shown in the sketch. In the initial convergent part of the slot the film thickness decreases linearly from $h_o + s_h$ to h_o over a length $n_s \ell$ on each side of the strip. In the final section of the slot the film on each side of the strip has a constant thickness h_o over a length $\ell(1 - n_s)$.

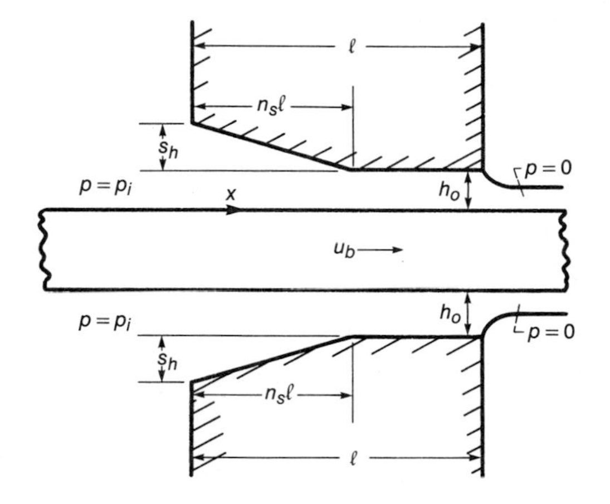

State clearly the boundary conditions required to determine the pressure distribution along the slot length in the sliding direction on the assumption that the liquid is isoviscous and incompressible and the slot is infinitely wide.

Sketch the pressure distribution along the x axis and show that the volume flow rate per unit width q'_x on each side of the strip can be written in dimensionless form as

$$Q = \frac{2q'_x}{u_b s_h} = \frac{(H_o^3 P_i/3)(H_o + 1)^2 + 2H_o(H_o + 1)(1 + H_o - n_s)}{n_s H_o(2H_o + 1) + 2(1 - n_s)(H_o + 1)^2}$$

where

$$P_i = \frac{p_i s_h^2}{\eta_0 u_b \ell} \qquad \text{and} \qquad H_o = \frac{h_o}{s_h}$$

Demonstrate that for P_i equal to zero the expression for Q reduces to forms appropriate to the parallel-surface and fixed-incline-surface bearing situations as n_s approaches zero and unity, respectively.

Determine the minimum value of P_i required to ensure that the peak pressure in the slot is located at the inlet, where $h = h_0 + s_h$ if $n_s = \frac{1}{2}$ and $H_o = 1$.

8.7.3 In designing a fixed-incline self-acting thrust pad when the width of the pad is much larger than the length, it is of interest to know the magnitude and location of the maximum pressure. The viscosity of the lubricant is 0.05 N · s/m^2, the sliding velocity is 10 m/s, the pad length is 0.3 m, the minimum film thickness is 15 μm, and the inlet film thickness is twice the outlet film thickness.

8.7.4 A parallel-step thrust bearing is shown below. Side leakage is neglected. The inlet pressure is greater than zero but less than p_m, the pressure at the step. The outlet pressure is zero. Determine the velocity of the fluid across the film while in the inlet and outlet regions. That is,

$$U_i = \frac{u_i}{u_b} = f(Z, P_i, P_m, H_i, n_s)$$

$$U_0 = \frac{u_0}{u_b} = g(Z, P_i, P_m, H_i, n_s)$$

where

$$Z = \frac{z}{h_o} \qquad P_i = \frac{p_i h_o^2}{\eta_o u_b \ell} \qquad P_m = \frac{p_m h_o^2}{\eta_0 u_b \ell} \qquad H_i = \frac{h_i}{h_o}$$

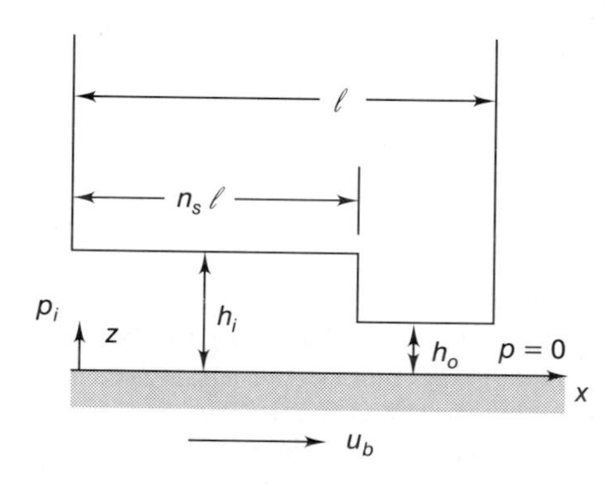

8.7.5 From the results of Prob. 8.7.4 evaluate and plot the velocity profile in the inlet and outlet regions when $n_s = 0.75$, $H_i = 2.0$, $P_m = 1.0$, and $P_i = 0.5$.

8.7.6 A thrust bearing analysis is to be performed on the bearing shown below.

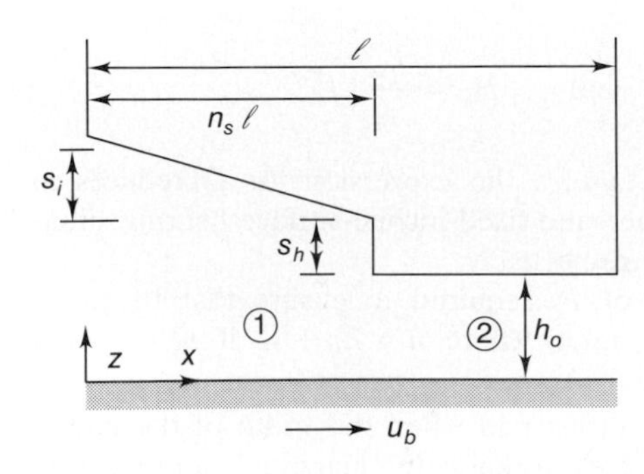

The pressure at $x = 0$ and at $x = \ell$ is zero. The analysis should consist of considering the lubrication in the two separate regions ① $0 \le x \le n_s \ell$ and ② $n_s \ell < x \le \ell$ and equated at the common boundary ($x = n_s \ell$). Side leakage is to be neglected. Starting with the appropriate Reynolds equation, determine the pressure throughout the bearing. Also show that the pressure at the common boundary ($x = n_s \ell$), when $s_i = 0$ is exactly Eq. (8.70) or (8.71) for a parallel-step bearing.

8.8 REFERENCE

Lord Rayleigh (1918): Notes on the Theory of Lubrication, *Philos. Mag.*, vol. 35, no. 1, pp. 1–12.

CHAPTER 9

HYDRODYNAMIC THRUST BEARINGS—NUMERICAL SOLUTIONS

The analysis of thrust slider bearings in Chap. 8 restricted the lubricant flow to two directions (x, z). Flow of lubricant in the third direction (y) is known as "side leakage." The x coordinate is in the direction of sliding motion, y is transverse to the direction of sliding motion, and z is across the fluid film. This chapter is concerned with the side-leakage effect in slider bearings.

For bearings of finite width the pressure along the bearing side edges is ambient, and if hydrodynamic pressures are generated within the lubricant film due to the physcial wedge mechanism, some flow will take place in the third (y) direction. In this case the flow perpendicular to the direction of motion (side-leakage effect) has to be included, and the appropriate Reynolds equation is Eq. (7.48) with $u_a = 0$.

$$\frac{\partial}{\partial x}\left(h^3 \frac{\partial p}{\partial x}\right) + \frac{\partial}{\partial y}\left(h^3 \frac{\partial p}{\partial y}\right) = 6\eta_0 u_b \frac{\partial h}{\partial x} \tag{9.1}$$

Analytical solutions for slider bearings that consider side leakage *are not* normally available. Numerical solutions generally have to be resorted to when considering the side-leakage direction. One of the few analytical solutions that is available is for a parallel-step slider bearing.

For a thrust bearing, as considered in Chap. 8 and also in this chapter, a thrust plate attached to, or forming part of, the rotating shaft is separated from the sector-shaped bearing pads (see Fig. 8.3) by a lubricant film. The load-carrying capacity of the bearing arises entirely from the pressure generated by the motion of the thrust plate over the bearing pads. This action is achieved only if the clearance space between the stationary and moving components is convergent in the direction of motion (physical wedge pressure-generating mechanism).

The pad geometry parameters affecting the pressure generation and therefore the load-carrying capacity of the bearing are (1) pad length-to-width ratio $\lambda = \ell/b$, (2) film thickness ratio $H_o = h_o/s_h$, and (3) step or pivot location parameter n_s. This last parameter is not applicable to a fixed-incline thrust bearing. Recall that the pad length-to-width ratio *did not* appear in Chap. 8 because side leakage was neglected. Side leakage is considered in this chapter, and three different thrust bearings are investigated: two fixed-pad types (namely, a parallel-step and a fixed-incline) and a pivoted-pad bearing.

9.1 FINITE-WIDTH, PARALLEL-STEP-PAD SLIDER BEARING

A sketch of the parallel-step-pad slider bearing is shown in Fig. 9.1. In the solution the inlet and outlet regions are first considered separately and then combined at the common boundary. Therefore, the film thickness is considered to be constant in the two regions.

9.1.1 PRESSURE DISTRIBUTION.

The Reynolds equation given in Eq. (9.1) reduces to

$$\frac{\partial^2 p}{\partial x^2} + \frac{\partial^2 p}{\partial y^2} = 0 \tag{9.2}$$

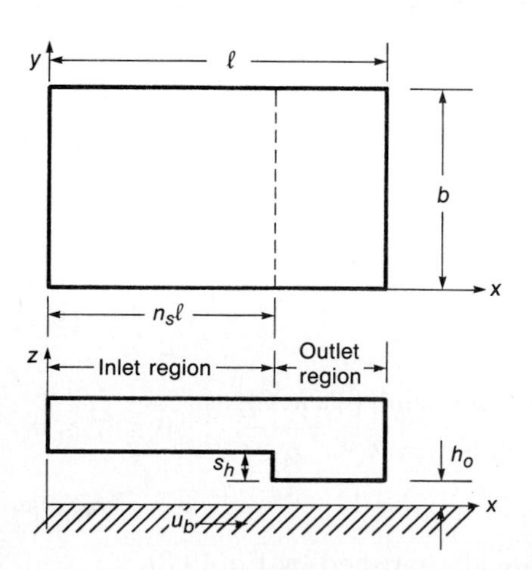

FIGURE 9.1
Finite parallel-step-pad slider bearing.

when the film thickness is constant. Equation (9.2) is a Laplace partial differential equation. Letting

$$x = \ell X \qquad y = bY \qquad \lambda = \frac{\ell}{b} \qquad p = \frac{\eta_0 u_b \ell}{s_h^2} P \tag{9.3}$$

causes Eq. (9.2) to become

$$\frac{\partial^2 P}{\partial X^2} + \lambda^2 \frac{\partial^2 P}{\partial Y^2} = 0 \tag{9.4}$$

The pressure on the exterior boundaries of the inlet and outlet regions is zero, but the pressure on the common boundary is not known. It is assumed that the pressure on the common boundary is given by the following Fourier sine series:

$$(P)_{X=n_s} = \sum_{j=1,3,\ldots}^{\infty} F_j^* \sin(j\pi Y) \tag{9.5}$$

The summation is over the odd values because of symmetry, and F_j^* is the Fourier coefficient to be evaluated later.

The general solution for P can be put in the following form for the inlet region:

$$P_i = \sum_{j=1,3,\ldots}^{\infty} \left[\tilde{A}_j \cos(j\pi Y) + \tilde{B}_j \sin(j\pi Y)\right] \times \left[\tilde{C}_j \cosh(j\pi\lambda X) + \tilde{D}_j \sinh(j\pi\lambda X)\right] \tag{9.6}$$

Note that Eq. (9.6) does satisfy Eq. (9.4).

The boundary conditions for the *inlet region* are:

1. $P_i = 0$ when $X = 0$ for all values of Y.
2. $P_i = 0$ when $Y = 0$ for all values of X.
3. $P_i = 0$ when $Y = 1$ for all values of X.
4. $P_i = \sum_{j=1,3,\ldots}^{\infty} F_j^* \sin(j\pi Y)$ when $X = n_s$.

From boundary conditions 1 and 2, $\tilde{A}_j = \tilde{C}_j = 0$. From boundary condition 4,

$$\sum_{j=1,3,\ldots}^{\infty} F_j^* \sin(j\pi Y) = \sum_{j=1,3,\ldots}^{\infty} \tilde{B}_j \tilde{D}_j \sin(j\pi Y) \sinh(j\pi\lambda n_s)$$

$$\therefore \tilde{B}_j \tilde{D}_j = \frac{F_j^*}{\sinh(j\pi\lambda n_s)} \tag{9.7}$$

$$\therefore P_i = \sum_{j=1,3,\ldots}^{\infty} \frac{F_j^* \sin(j\pi Y) \sinh(j\pi\lambda X)}{\sinh(j\pi\lambda n_s)} \qquad 0 \le X \le n_s \tag{9.8}$$

Note that boundary condition 3 is automatically satisfied by Eq. (9.8).

For the outlet region the solution is taken to be of the same form as Eq. (9.6).

$$P_o = \sum_{j=1,3,\ldots}^{\infty} \left[\tilde{E}_j \cos(j\pi Y) + \tilde{G}_j \sin(j\pi Y)\right] \times \left[\tilde{H}_j \cosh(j\pi\lambda X) + \tilde{I}_j \sinh(j\pi\lambda X)\right] \tag{9.9}$$

The boundary conditions for the *outlet region* are

1. $P_o = 0$ when $X = 1$ for all values of Y.
2. $P_o = 0$ when $Y = 0$ for all values of X.
3. $P_o = 0$ when $Y = 1$ for all values of X.
4. $P_o = \sum_{j=1,3,\ldots}^{\infty} F_j^* \sin(j\pi Y)$ when $X = n_s$.

From boundary condition 2, $\tilde{E}_j = 0$. From boundary condition 1,

$$\tilde{H}_j = -\tilde{I}_j \tanh(j\pi\lambda) \tag{9.10}$$

Therefore, Eq. (9.9) can be rewritten as

$$P_o = \sum_{j=1,3,\ldots}^{\infty} \tilde{G}_j\tilde{I}_j \sin(j\pi Y)\left[\sinh(j\pi\lambda X) - \tanh(j\pi\lambda)\cosh(j\pi\lambda X)\right] \tag{9.11}$$

From boundary condition 4,

$$\sum_{j=1,3,\ldots}^{\infty} F_j^* \sin(j\pi Y) = \sum_{j=1,3,\ldots}^{\infty} \tilde{G}_j\tilde{I}_j \sin(j\pi Y) \times \left[\sinh(j\pi\lambda n_s) - \tanh(j\pi\lambda)\cosh(j\pi\lambda n_s)\right]$$

$$\therefore \tilde{G}_j\tilde{I}_j = \frac{F_j^*}{\sinh(j\pi\lambda n_s) - \tanh(j\pi\lambda)\cosh(j\pi\lambda n_s)} \tag{9.12}$$

Substituting Eq. (9.12) into Eq. (9.11) gives

$$P_o = \sum_{j=1,3\ldots}^{\infty} \frac{F_j^* \sin(j\pi Y)\sinh\left[j\pi\lambda(1-X)\right]}{\sinh\left[j\pi\lambda(1-n_s)\right]} \qquad n_s \le X \le 1 \tag{9.13}$$

Note that boundary condition 3 is satisfied by Eq. (9.13).

For the pressure to be calculated, the Fourier coefficient F_j^* must be evaluated. This is done by making use of the principle of flow continuity. It will be seen that at any point on the common boundary ($X = n_s$) the volume rate of inlet flow is the same as the volume rate of outlet flow.

$$\therefore q_{xi}|_{X=n_s} = q_{xo}|_{X=n_s} \tag{9.14}$$

Making use of Eq. (7.38) while letting $u_a = 0$ and assuming a constant viscosity yields

$$-\frac{(h_o+s_h)^3}{12\eta_0}\left(\frac{dp_i}{dx}\right)_{n_s\ell}+\frac{u_b(h_o+s_h)}{2}=-\frac{h_o^3}{12\eta_0}\left(\frac{dp_o}{dx}\right)_{n_s\ell}+\frac{u_bh_o}{2}$$

Then making use of Eq. (9.3) gives

$$-H_o^3\left(\frac{dP_o}{dX}\right)_{X=n_s}+(1+H_o)^3\left(\frac{dP_i}{dX}\right)_{X=n_s}=6 \tag{9.15}$$

where $H_o = h_o/s_h$. From Eqs. (9.8) and (9.13)

$$\left(\frac{\partial P_i}{\partial X}\right)_{X=n_s}=\sum_{j=1,3,\ldots}^{\infty}\frac{j\pi\lambda F_j^*\sin(j\pi Y)}{\tanh(j\pi\lambda n_s)} \tag{9.16}$$

$$\left(\frac{\partial P_o}{\partial X}\right)_{X=n_s}=\sum_{j=1,3,\ldots}^{\infty}-\frac{j\pi\lambda F_j^*\sin(j\pi Y)}{\tanh[j\pi\lambda(1-n_s)]} \tag{9.17}$$

Substituting Eqs. (9.16) and (9.17) into (9.15) gives

$$H_o^3\sum_{j=1,3,\ldots}^{\infty}\frac{j\pi\lambda F_j^*\sin(j\pi Y)}{\tanh[j\pi\lambda(1-n_s)]}+(1+H_o)^3\sum_{j=1,3,\ldots}^{\infty}\frac{j\pi\lambda F_j^*\sin(j\pi Y)}{\tanh(j\pi\lambda n_s)}=6 \tag{9.18}$$

But

$$6=\sum_{j=1,3,\ldots}^{\infty}6\frac{4}{j\pi}\sin(j\pi Y) \tag{9.19}$$

Then substituting Eq. (9.19) into Eq. (9.18) gives

$$\sum_{j=1,3,\ldots}^{\infty}j\pi\lambda F_j^*\sin(j\pi Y)\{H_o^3\coth[j\pi\lambda(1-n_s)]+(1+H_o)^3\coth(j\pi\lambda n_s)\}=\sum_{j=1,3,\ldots}^{\infty}\frac{24\sin(j\pi Y)}{j\pi}$$

$$\therefore F_j^*=\frac{24}{j^2\pi^2\lambda\{H_o^3\coth[j\pi\lambda(1-n_s)]+(1+H_o)^3\coth(j\pi\lambda n_s)\}} \tag{9.20}$$

Note that as $H_o\to\infty$, which is a parallel-surface bearing, $F_j^*\to 0$ and $P\to 0$. Once an expression for the Fourier coefficient is known, the pressure in the inlet and outlet regions can be calculated by substituting Eq. (9.20) into Eqs. (9.8) and (9.13).

9.1.2 NORMAL LOAD COMPONENT. The normal applied load can be written as

$$w_z = \int_0^b \int_0^{n_s \ell} p_i \, dx \, dy + \int_0^b \int_{n_s \ell}^{\ell} p_0 \, dx \, dy$$

This equation can be written in dimensionless form as

$$W_z = \frac{w_z}{\eta_0 u_b b} \left(\frac{s_h}{\ell} \right)^2 = \int_0^1 \int_0^{n_s} P_i \, dX \, dY + \int_0^1 \int_{n_s}^1 P_o \, dX \, dY \qquad (9.21)$$

Substituting Eqs. (9.8) and (9.13) into this equation gives

$$W_z = \frac{w_z}{\eta_0 u_b b} \left(\frac{s_h}{\ell} \right)^2$$

$$= \sum_{j=1,3,\ldots}^{\infty} \frac{2F_j^*}{j^2 \pi^2 \lambda} \left\{ \frac{\cosh(j\pi\lambda n_s) - 1}{\sinh(j\pi\lambda n_s)} + \frac{\cosh[j\pi\lambda(1-n_s)] - 1}{\sinh[j\pi\lambda(1-n_s)]} \right\} \qquad (9.22)$$

Equations (9.20) and (9.22) reveal that when side leakage is considered, the dimensionless normal applied load W_z is a function of λ as well as H_o and n_s.

9.1.3 RESULTS. The inclusion of side-leakage effects has introduced an additional parameter, the length-to-width ratio $\lambda = \ell/b$. Figure 9.2 shows the effect of the film thickness ratio H_o on the dimensionless load-carrying capacity W_z for five step locations n_s and four λ values ($\frac{1}{4}$, $\frac{1}{2}$, 1, and 2). The following can be concluded from Fig. 9.2:

1. For any combination of H_o and λ the maximum load condition never occurs when the step location is less than 0.5. The location of the step should be closer to the outlet than to the inlet.
2. For small λ, where side leakage becomes less important, the step location that produces a maximum dimensionless load is larger than when side leakage is considered.

Table 9.1 tabulates the results shown in Fig. 9.2 as $H_o \to 0$ as well as results for three additional λ. From this table it is observed that as λ becomes smaller, the maximum load increases. This concurs with the results of Chap. 8 [Eq. (8.73)], which indicated that when side leakage is neglected and $H_o \to \infty$, the dimensionless normal load-carrying capacity approaches zero.

The parallel-step bearing shown in Fig. 9.1 has considerable side leakage, especially for large λ, which reduces its normal load-carrying capacity. Shrouds are sometimes introduced to restrict this side leakage and thereby increase the bearing's normal load-carrying capacity. Two types of shrouded-step bearings

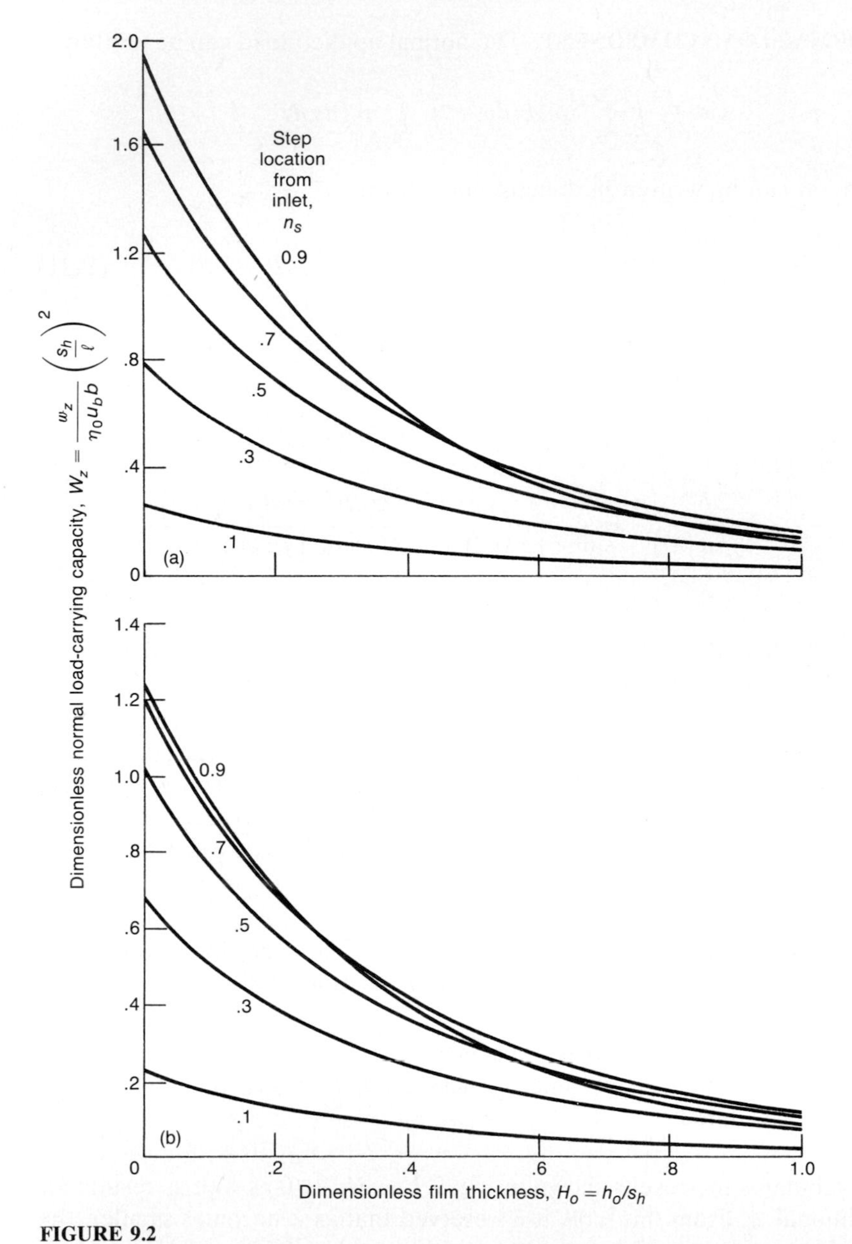

FIGURE 9.2
Effect of film thickness on dimensionless normal load-carrying capacity at five step locations n_s and four length-to-width ratios λ. (a) $\lambda = \frac{1}{4}$; (b) $\lambda = \frac{1}{2}$.

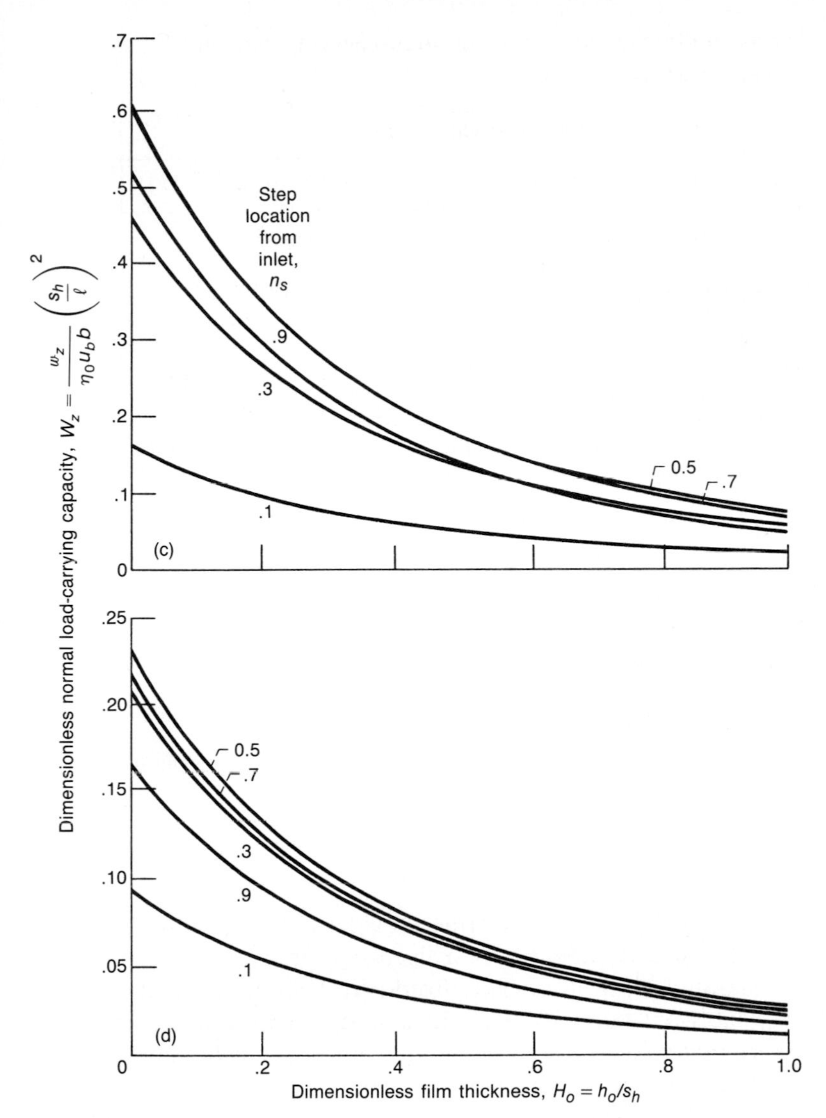

FIGURE 9.2
Concluded. (c) $\lambda = 1$; (d) $\lambda = 2$.

are shown in Fig. 9.3. The semicircular-step shrouded bearing [Fig. 9.3(a)] has a step radius of $\ell/2$, a film thickness ratio of 1.43, and a length-to-width ratio of 1. The resulting normal load-carrying capacity for this bearing is a 45 percent increase over that for an optimum unshrouded parallel-step thrust bearing. For a truncated triangular-step shrouded bearing [Fig. 9.3(b)] with $H_o = 0.89$, the resulting W_z is a 67 percent increase over that for an unshrouded bearing.

TABLE 9.1 Maximum dimensionless normal load-carrying capacity
[Dimensionless outlet film thickness $H_o \to 0$]

Step location, n_s	Length-to-width ratio, λ						
	1/100	1/4	1/2	3/4	1	3/2	2
	Maximum dimensionless normal load-carrying capacity, W_z						
0.1	0.298570	0.264269	0.228659	0.194657	0.164811	0.120508	0.092064
.2	.597193	.529825	.459748	.391479	.329074	.230908	.165224
.3	.895500	.787513	.675112	.564927	.463443	.305444	.205006
.4	1.193245	1.031140	.862610	.700004	.555898	.346178	.223081
.5	1.490253	1.256341	1.014517	.791103	.607007	.362142	.228742
.6	1.786400	1.460020	1.127112	.839600	.622694	.359732	.225744
.7	2.081597	1.640043	1.200150	.850630	.609714	.342147	.214448
.8	2.375785	1.795099	1.236016	.830670	.574190	.311116	.193714
.9	2.668928	1.924660	1.238860	.786357	.521621	.268646	.162993

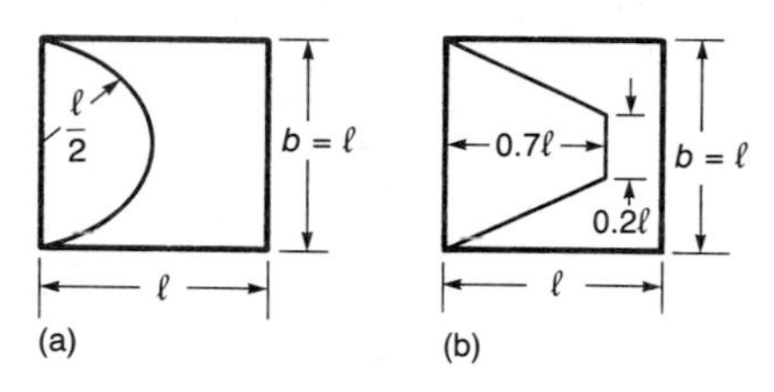

FIGURE 9.3
Shrouded-step slider bearings. (a) Semicircular step; (b) truncated triangular step.

9.2 FIXED-INCLINE-PAD SLIDER BEARING

The simplest form of fixed-pad thrust bearing provides only straight-line motion and consists of a flat surface sliding over a fixed pad or land having a profile similar to that shown in Fig. 9.4. The fixed-pad bearing depends for its operation on the lubricant being drawn into a wedge-shaped space and thus producing pressure that counteracts the load and prevents contact between the sliding parts. Since the wedge action takes place only when the sliding surface moves in the direction in which the lubricant film converges, the fixed-incline bearing (Fig. 9.4) can carry load only in this direction. If reversibility is desired, a combination of two or more pads with their surfaces sloped in opposite directions is required. Fixed-incline pads are used in multiples as in the thrust bearing shown in Fig. 9.5.

The following procedure assists in designing a fixed-incline-pad thrust bearing:

1. Choose a pad length-to-width ratio. A square pad ($\lambda = 1$) is generally felt to give good performance. If it is known whether maximum load or minimum power loss is more important in the particular application, a value of the minimum film thickness ratio can be determined from Fig. 9.6.

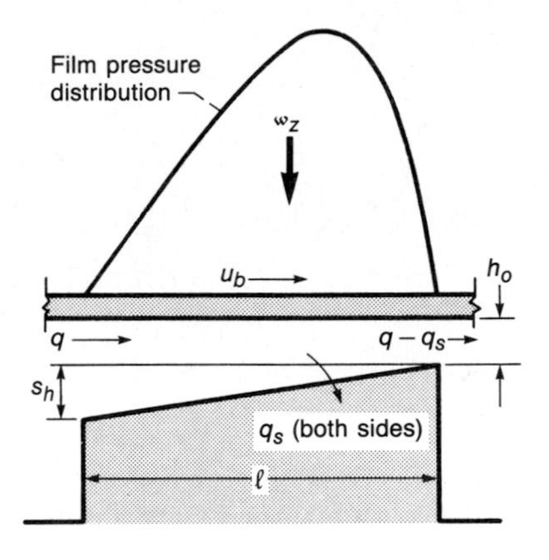

FIGURE 9.4
Side view of fixed-incline-pad bearing. [*From Raimondi and Boyd (1955).*]

2. Determine lubricant temperature. Lubricant temperature can be expressed as

$$t_m = t_i + \frac{\Delta t_m}{2} \tag{9.23}$$

where t_i = inlet temperature
Δt_m = change in temperature due to viscous shear heating

The inlet temperature is usually known beforehand, but the change in temperature due to viscous shear heating must be initially guessed. Once the temperature t_m is known, it can be used in Fig. 4.6 to determine the viscosity of SAE oils.

3. Determine outlet film thickness. Once the viscosity is known, the next least likely parameter to be preassigned is the outlet film thickness. That is, once a length-to-width ratio λ and a dimensionless minimum film thickness (from Fig. 9.7) are known, the dimensionless load can be determined. The outlet film thickness can be determined from

$$h_o = H_o \ell \left(\frac{W_z \eta_0 u_b b}{w_z} \right)^{1/2} \tag{9.24}$$

4. Check Table 9.2 to see if the minimum (outlet) film thickness is sufficient for the preassigned surface finish. If $(h_o)_{\text{Eq. (9.24)}} \geq (h_o)_{\text{Table 9.2}}$ go to step 5. If

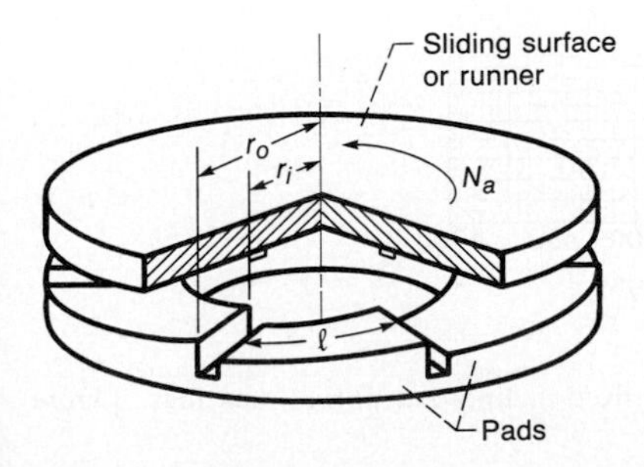

FIGURE 9.5
Configuration of multiple fixed-incline-pad thrust bearing. [*From Raimondi and Boyd (1955).*]

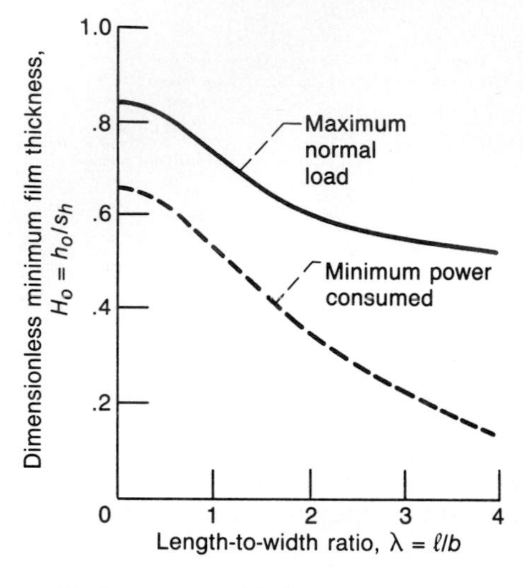

FIGURE 9.6
Chart for determining minimum film thickness corresponding to maximum load or minimum power loss for various pad proportions—fixed-incline-pad bearings. [*From Raimondi and Boyd (1955).*]

$(h_o)_{(\text{Eq. 9.24})} < (h_o)_{\text{Table 9.2}}$ consider one or more of the following steps:

a. Increase the speed of the bearing.

b. Decrease the load, the surface finish, or the inlet temperature.

Upon making this change, return to step 2.

5. From Fig. 9.8 determine the temperature rise due to shear heating for a given length-to-width ratio and bearing characteristic number. Recall from Sec. 4.8 that the volumetric specific heat $C_s = \rho^* C_p$, which is in the dimensionless temperature rise parameter, is relatively constant for mineral oils and is equivalent to 1.36×10^6 N/(m$^2 \cdot$°C). If the temperature rise as obtained from Fig. 9.8 is within 5 percent of the previous value, go to step 6; otherwise return to step 2 with a new value of temperature rise.

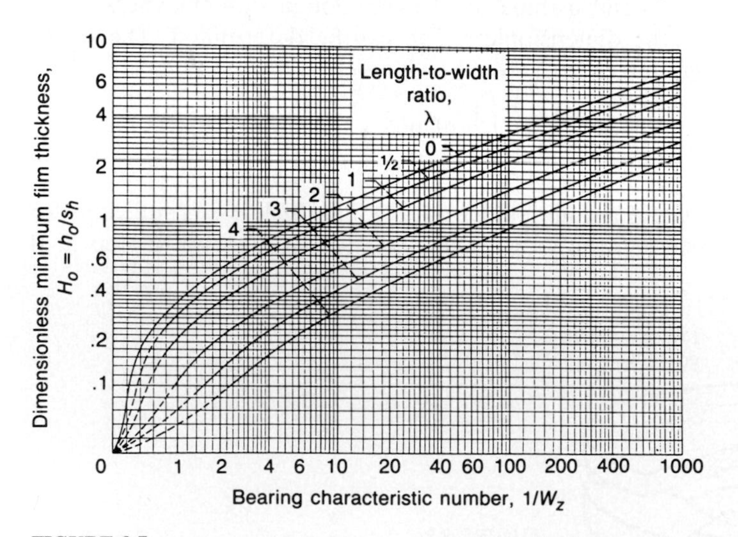

FIGURE 9.7
Chart for determining minimum film thickness for fixed-incline-pad thrust bearings. [*From Raimondi and Boyd (1955).*]

TABLE 9.2

Allowable minimum outlet film thickness for a given surface finish

[From Engineering Sciences Data Unit (1967)]

Surface finish (centerline average, R_a)		Description of surface	Examples of manufacturing methods	Approximate relative costs	Allowable minimum outlet film thickness[a], h_o	
μm	μin.				μm	μin.
0.1-0.2	4-8	Mirror-like surface without toolmarks; close tolerances	Grind, lap, and super-finish	17-20	2.5	100
.2-.4	8-16	Smooth surface without scratches; close tolerances	Grind and lap	17-20	6.2	250
.4-.8	16-32	Smooth surface; close tolerances	Grind, file, and lap	10	12.5	500
.8-1.6	32-63	Accurate bearing surface without toolmarks	Grind, precision mill, and file	7	25	1000
1.6-3.2	63-125	Smooth surface without objectionable toolmarks; moderate tolerances	Shape, mill, grind, and turn	5	50	2000

[a]The values of film thickness are given only for guidance. They indicate the film thickness required to avoid metal-to-metal contact under clean oil conditions with no misalignment. It may be necessary to take a larger film thickness than that indicated (e.g., to obtain an acceptable temperature rise). It has been assumed that the average surface finish of the pads is the same as that of the runner.

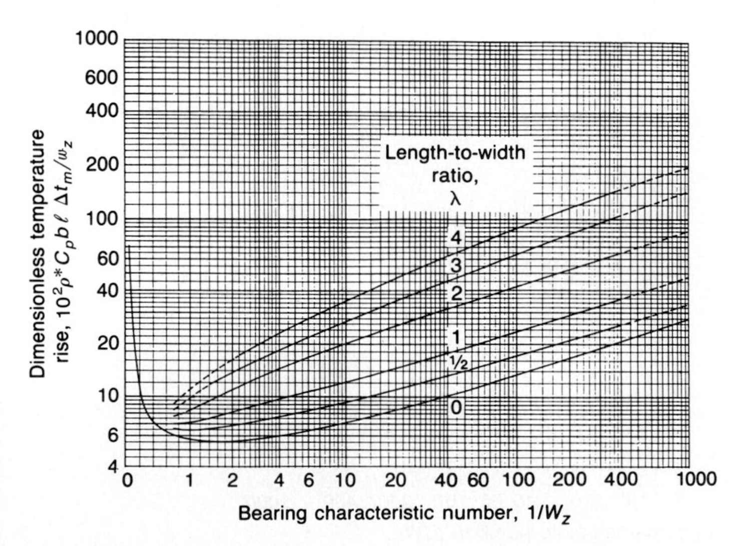

FIGURE 9.8

Chart for determining dimensionless temperature rise due to viscous shear heating of lubricant in fixed-incline-pad thrust bearings. [*From Raimondi and Boyd (1955).*]

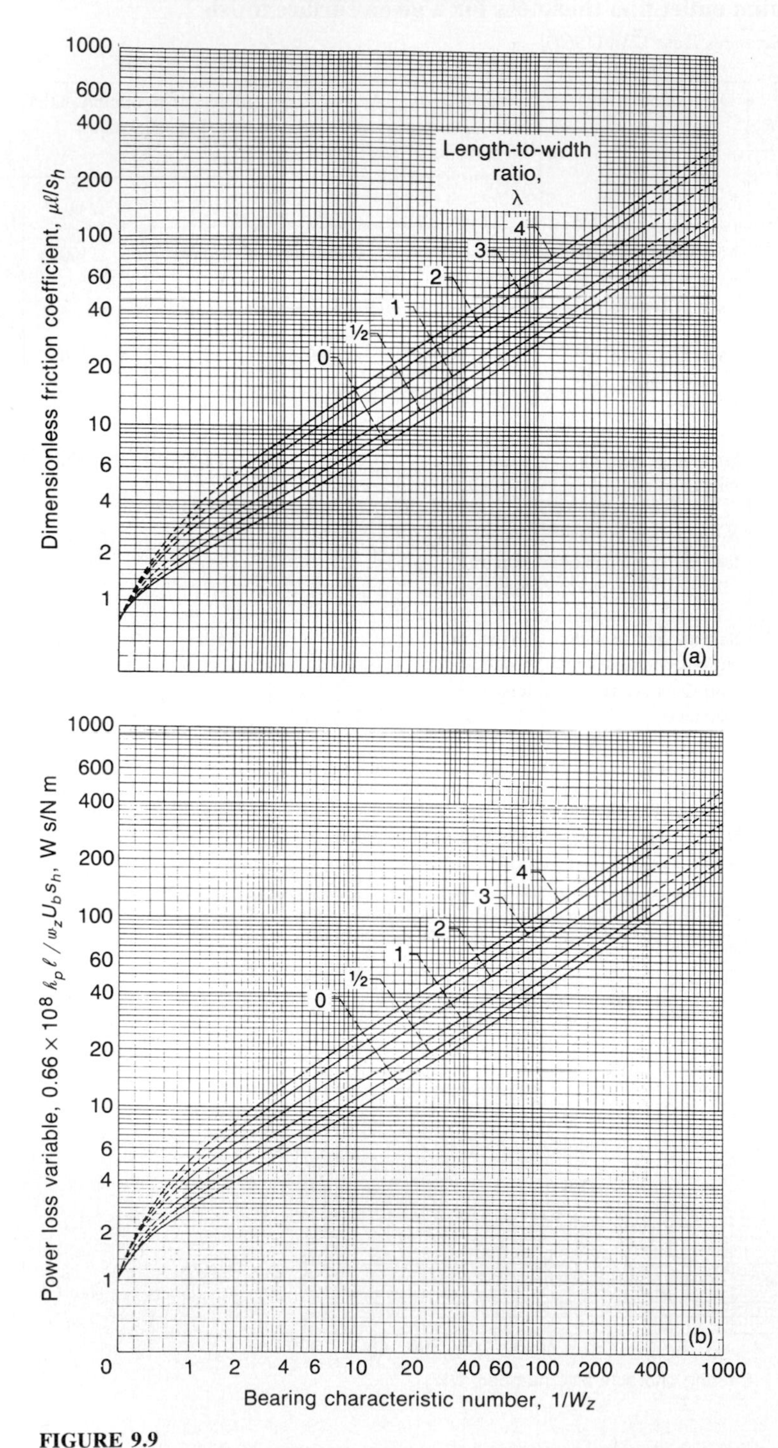

FIGURE 9.9
Chart for determining performance parameters of fixed-incline-pad thrust bearings. (a) Friction coefficient; (b) power loss. [*From Raimondi and Boyd (1955).*]

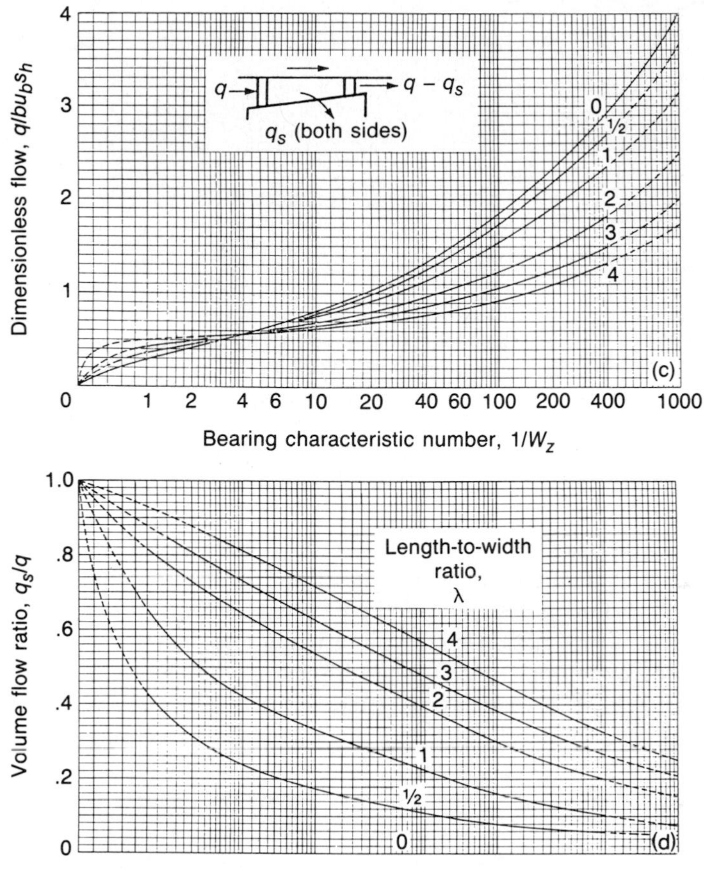

FIGURE 9.9
Concluded. (c) Lubricant flow; (d) lubricant side flow.

6. Evaluate the performance parameters. Once an adequate minimum film thickness and proper lubricant temperature have been determined, the performance parameters can be evaluated. Specifically, from Fig. 9.9 the power loss, the friction coefficient, and the total and side flow can be determined.

9.3 PIVOTED-PAD SLIDER BEARING

The simplest form of pivoted-pad bearing provides only for straight-line motion and consists of a flat surface sliding over a pivoted pad as shown in Fig. 9.10. If the pad is assumed to be in equilibrium under a given set of operating conditions, any change in these conditions, such as a change in load, speed, or viscosity, will alter the pressure distribution and thus momentarily shift the center of pressure, creating a moment that causes the pad to change its inclination and shoulder height s_h. A pivoted-pad slider bearing is thus sup-

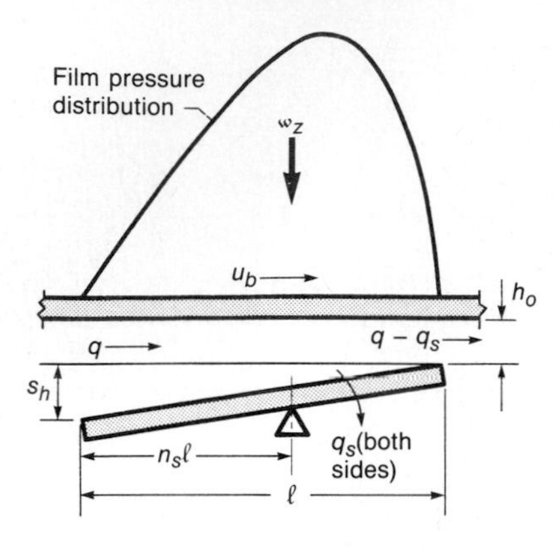

FIGURE 9.10
Side view of pivoted-pad thrust bearing. [*From Raimondi and Boyd (1955).*]

ported at a single point so that the angle of inclination becomes a variable and has much better stability than a fixed-incline slider under varying conditions of operation. The location of the shoe's pivot point can be found from the equilibrium of moments acting on the shoe about the point. For all practical purposes, only two significant forces may be considered in the moment equation: the resultant due to film pressure and the reaction force normal to the shoe surface. The force due to friction in the pivot is ignored.

Pivoted pads are sometimes used in multiples as pivoted-pad thrust bearings, shown in Fig. 9.11. Calculations are carried through for a single-pad, and the properties for the complete bearing are found by combining these calculations in the proper manner, as discussed in Sec. 9.4.

Normally, a pivoted pad will only carry load if the pivot is placed somewhere between the center of the pad and the outlet edge ($0.5 \leq n_s \leq 1.0$). For bidirectional operation the pivot is located at the center of the pad or at $n_s = 0.5$.

The following procedure helps in designing pivoted-pad thrust bearings:

1. Having established whether minimum power or maximum load is more critical in the particular application and having chosen a pad length-to-width ratio, establish the pivot location from Fig. 9.12.

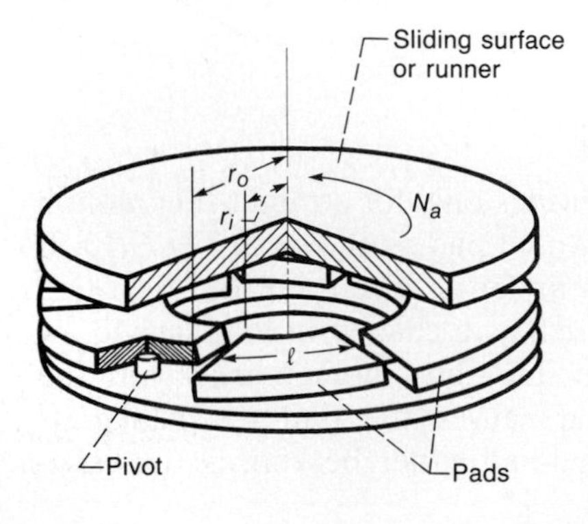

FIGURE 9.11
Configuration of multiple pivoted-pad thrust bearing. [*From Raimondi and Boyd (1955).*]

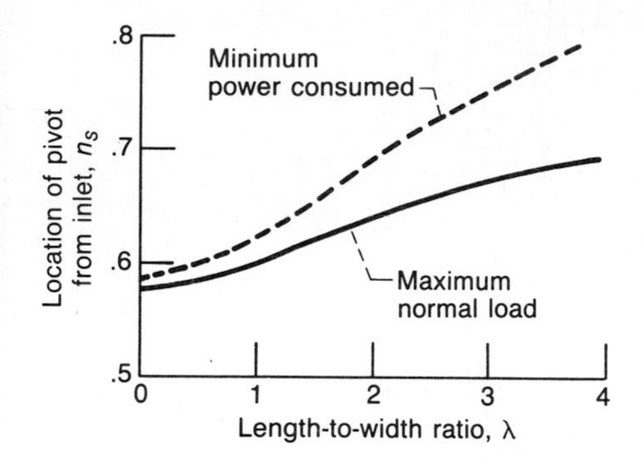

FIGURE 9.12
Chart for determining pivot location corresponding to maximum load or minimum power loss for various pad proportions—pivoted-pad bearings. [*From Raimondi and Boyd (1955).*]

2. Just as was done for a fixed-incline-pad thrust bearing, establish the lubricant temperature by using Eq. (9.23). Once the temperature is known, the viscosity can be obtained from Fig. 4.6.
3. Determine the dimensionless load from Fig. 9.13 and the outlet or minimum film thickness from Eq. (9.24).
4. Check Table 9.2 to see if the outlet film thickness is sufficient for the preassigned surface finish. If $(h_o)_{(\text{Eq. }9.24)} \geq (h_o)_{\text{Table }9.2}$ go to step 5. If $(h_o)_{\text{Eq. }(9.24)} < (h_o)_{\text{Table }9.2}$ consider
   a. Increasing the speed of the bearing
   b. Decreasing the inlet temperature of the bearing
   c. Decreasing the load on the bearing
   d. Decreasing the finish on the bearing lubricating surfaces, thereby making them smoother.

   Upon making this change return to step 2.

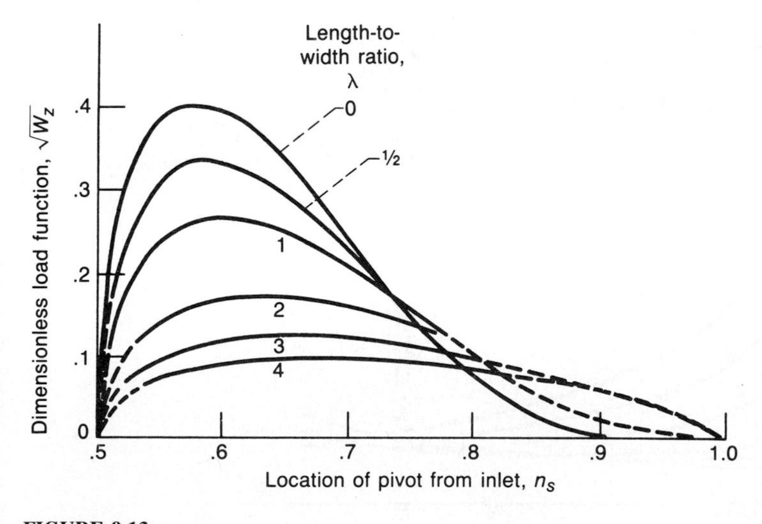

FIGURE 9.13
Chart for determining outlet film thickness for pivoted-pad thrust bearings. [*From Raimondi and Boyd (1955).*]

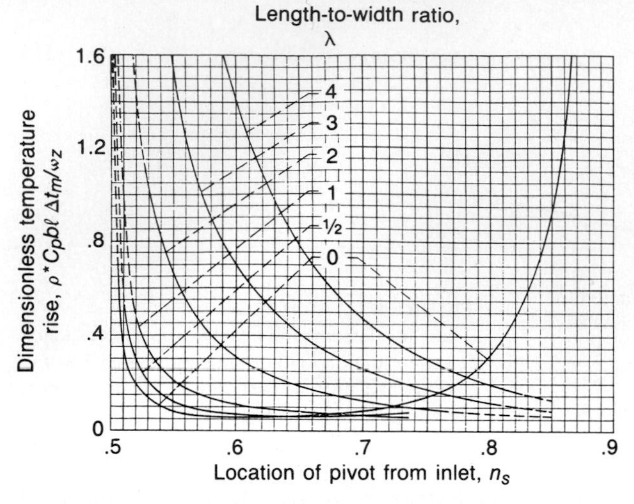

FIGURE 9.14
Chart for determining dimensionless temperature rise due to viscous shear heating of lubricant for pivoted-pad thrust bearings. [*From Raimondi and Boyd (1955).*]

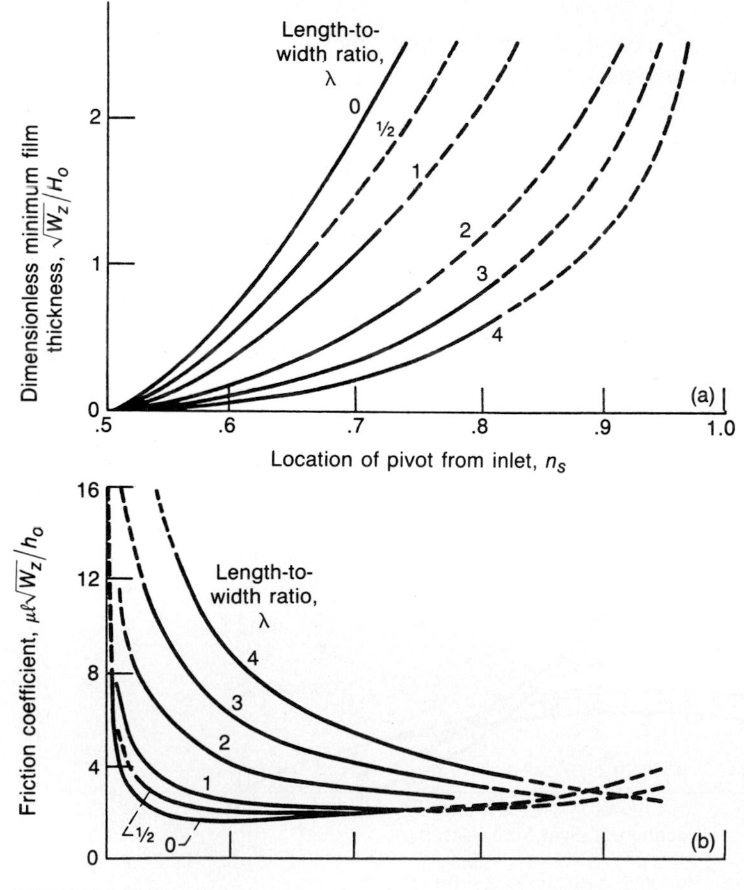

FIGURE 9.15
Chart for determining performance parameters of pivoted-pad thrust bearings. (a) Film thickness ratio; (b) friction coefficient. [*From Raimondi and Boyd (1955).*]

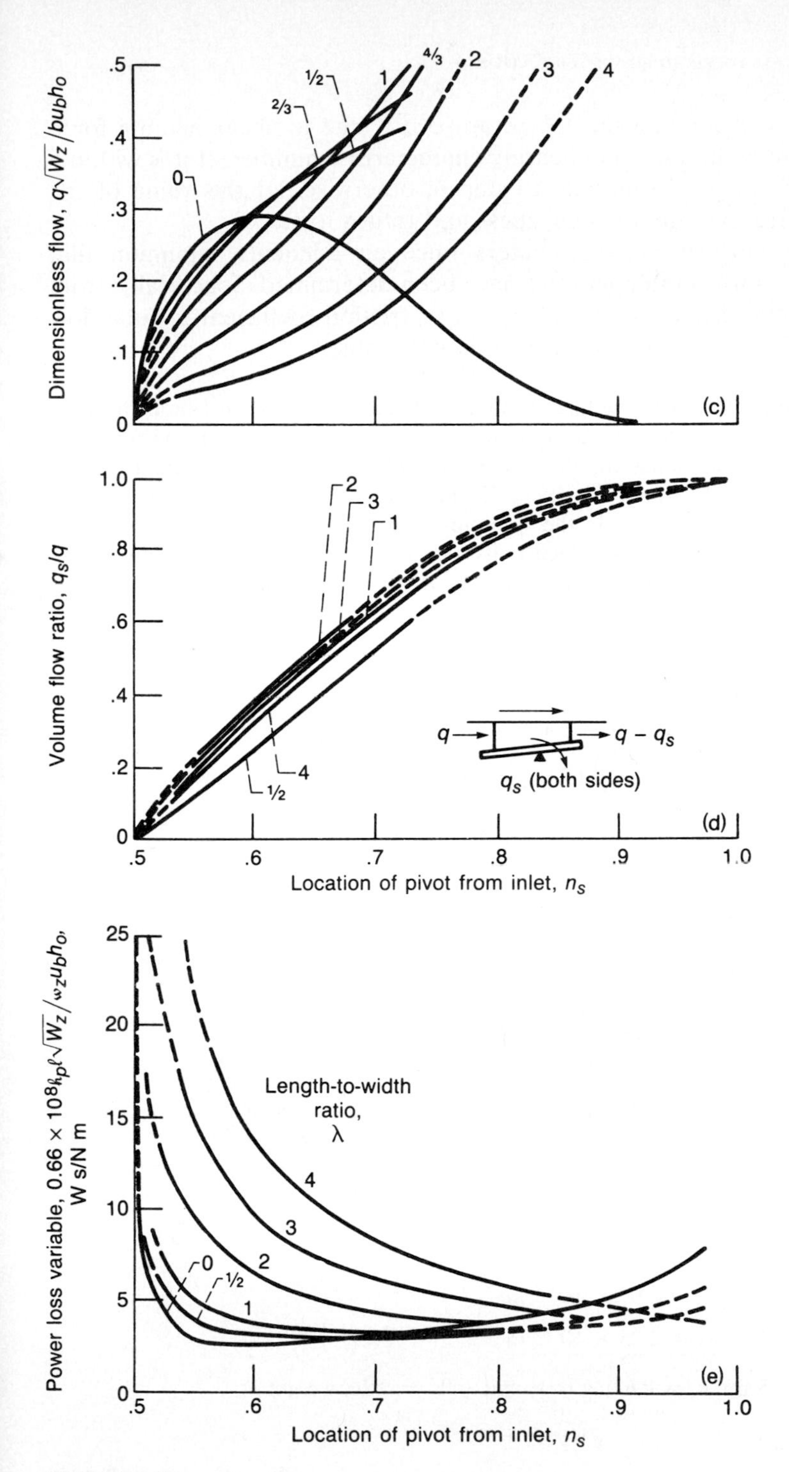

FIGURE 9.15
Concluded. (c) Lubricant flow; (d) lubricant side flow; (e) power loss.

5. From Fig. 9.14 determine the temperature rise due to shear heating for a given length-to-width ratio and bearing characteristic number. If it is within 5 percent of the guessed value, go to step 6; otherwise, let this value of the temperature rise become the new guess and return to step 2.
6. Evaluate the performance parameters once an adequate minimum film thickness and lubricant temperature have been determined. Specifically, from Fig. 9.15 the film thickness ratio $H_o = h_o/s_h$, friction coefficient μ, total flow q, side flow q_s, and power loss $\hbar_p$ can be determined.

Sample problem 9.1. A tilting thrust pad has a length and width of 1 in. Determine the pivot location if the bearing operates with a minimum power consumption. Determine what the minimum film thickness is when supporting a normal applied load of 2000 lbf at a surface velocity of 300 in/s. The lubricant viscosity is assumed constant at 5×10^{-6} reyn. Is the minimum film thickness obtained from your calculations adequate if the surface finish is 20 μin? If not, could you give some suggestions to meet this requirement?

Solution. From Fig. 9.12 for $\lambda = 1$ and minimum power consumption assumed,

$$n_s = 0.62$$

From Fig. 9.13 for $n_s = 0.62$ and $\lambda = 1$

$$\sqrt{W_z} = 0.26$$

or

$$W_z = (0.26)^2 = 0.0676$$

But

$$W_z = \frac{w_z}{\eta_0 u_b b}\left(\frac{s_h}{\ell}\right)^2$$

It is given that

$$\eta_0 = 5 \times 10^{-6} \text{ reyn} = 5 \times 10^{-6} \text{ lbf} \cdot \text{s/in}^2$$

$$w_z = 2000 \text{ lbf}$$

$$u_b = 300 \text{ in/s}$$

$$\ell = b = 1 \text{ in}$$

$$\therefore (s_h)^2 = \frac{W_z \eta_0 u_b b \ell^2}{w_z} = \frac{(0.0676)(5)(10^{-6})(300)(1)}{2000}$$

$$= 5.07 \times 10^{-8} \text{ in}^2$$

$$\underline{s_h = 2.25 \times 10^{-4} \text{ in}}$$

From Fig. 9.15 for $n_s = 0.62$, $\lambda = 1$, and $\sqrt{W_z} = 0.26$, we get

$$\sqrt{W_z}\,H_o = 0.45$$

or

$$H_o = \frac{0.45}{0.26} = 1.73$$

But
$$H_o = \frac{h_o}{s_h}$$

$$\therefore h_o = H_o s_h = (1.73)(2.25)(10^{-4}) \text{ in}$$
$$= 3.89 \times 10^{-4} \text{ in}$$

From Table 9.2 for a surface finish of 20 μin, the allowable minimum film thickness is 500 μin. Thus

$$(h_o)_{\min} = 5 \times 10^{-4} \text{ in}$$

Since the minimum film thickness that we calculated from Eq. (9.24) is less than the allowable minimum film thickness $(h_o)_{\min}$ obtained from Table 9.2, the pad must be redesigned.

Some possible suggestions for the redesign are:

1. Change the surface finish from 20 to 12 μin.
2. Increase the fluid viscosity from 5×10^{-6} to 2×10^{-5} reyn.
3. Increase the speed from 300 to 1200 in/s.
4. Decrease the load from 2000 to 500 lbf.

Only one of these changes is necessary to meet the requirements, but combinations of these changes would make the overall redesign less demanding.

9.4 THRUST BEARING GEOMETRY

This chapter, as well as Chap. 8, has dealt with the performance of an individual pad of a thrust bearing. Normally, a number of identical pads are assembled in a thrust bearing as shown, for example, in Figs. 9.5 and 9.11. The length, width, speed, and load of an individual pad can be related to the geometry of a thrust bearing by the following formulas:

$$b = r_o - r_i \tag{9.25}$$

$$\ell = \frac{r_o + r_i}{2}\left(\frac{2\pi}{N_0} - \frac{\pi}{36}\right) \tag{9.26}$$

$$u_b = \frac{(r_o + r_i)\omega}{2} \tag{9.27}$$

$$w_t = N_0 w_z \tag{9.28}$$

where N_0 is the number of identical pads placed in the thrust bearing (usually N_0 is between 3 and 20). The $\pi/36$ portion of Eq. (9.26) accounts for feed grooves between the pads. These are deep grooves which ensure that ambient pressure is maintained between the pads. Also in Eq. (9.28) w_t is the total thrust load on the bearing.

In the design of thrust bearings the total normal load-carrying capacity w_t and the angular velocity ω are usually specified along with the overall bearing dimensions r_o and r_i. The lubricant is usually specified to suit the design requirements of the associated journal bearing of other components. It therefore remains only to determine the number of pads N_0 and the pad geometry (H_o, n_s, and λ). The bearing performance characteristics may then be evaluated to check if the design meets the required specifications.

9.5 CLOSURE

The appropriate form of the Reynolds equation for a hydrodynamically lubricated thrust bearing when side leakage is considered is

$$\frac{\partial}{\partial x}\left(h^3\frac{\partial p}{\partial x}\right) + \frac{\partial}{\partial y}\left(h^3\frac{\partial p}{\partial y}\right) = 6\eta u_b\frac{\partial h}{\partial x}$$

The only film shape that can be solved analytically when considering side leakage is that for a parallel-step slider bearing. Since the film shape is constant within the inlet and outlet regions, the preceding equation reduces to a Laplacian equation that can be readily solved for the pressure. Integrating the pressure over the bearing area enables the normal applied load to be obtained. Side leakage, which reduces the normal load-carrying capacity, is considerable for a parallel-step bearing. It therefore was suggested that shrouds be placed at the sides to restrict the side leakage and thereby increase the bearing's load-carrying capacity.

Results were presented for fixed-incline-pad and pivoted-pad thrust bearings from numerical evaluation of the Reynolds equation. A procedure was outlined to assist in designing these bearings. These procedures provide an optimum pad configuration as well as describe such performance parameters as normal applied load, friction coefficient, power loss, and lubricant flow through the bearing.

9.6 PROBLEMS

9.6.1 Compare the dimensionless normal applied load W_z for the following three types of thrust pad: (*a*) parallel step, (*b*) fixed incline, and (*c*) tilting pad for $\lambda = \ell/b$ of $\frac{1}{4}$, $\frac{1}{2}$, 1, and 2. Consider side-leakage effects. Use optimal geometry for each type of pad at each λ. Also compare these results with the W_z when side leakage is neglected.

9.6.2 Draw the pressure profile of an unshrouded parallel-step pad when $\lambda = 1$, $H_o = 1$, $n_s = 0.5$, and $Y = 0.5$ for $X = 0.1, 0.2, \ldots, 0.9$. Use Eqs. (9.8) and (9.13) in your evaluation. Give numerical values of pressure at the various values of X. Also for the film shape given, determine the dimensionless normal applied load W_z.

9.6.3 A thrust bearing carries a load of 50,000 N at a rotational speed of 2000 r/min. The bearing has an outside radius of 0.25 m and an inside radius of 0.15 m. A surface finish of 1 μm CLA is recommended. The oil is SAE 10 and the operating inlet temperature is 50°C. Design a suitable tilting-pad bearing for maximum thrust load. The selected pad geometry, pivot location, number of pads, and minimum film thickness should be stated together with an estimate of Δt_m, s_h, $\hbar_p$, μ, q, and q_s.

9.6.4 For a fixed-incline-pad slider thrust bearing with $w_z = 3600$ lbf, $u_b = 1200$ in/s, $\ell = 3$ in, $b = 3$ in, and SAE 10 oil with inlet temperature of 40°C, determine the following for a maximum normal load: s_h, h_o, Δt_m, μ, $\hbar_p$, q, and q_s.

9.6.5 For a fixed-incline-pad thrust bearing with a total normal load of 12,000 lbf, $r_o = 4$ in, $r_i = 2$ in, $\omega = 30$ r/s $\ell/b = 1$, SAE 10 oil, and $t_i = 100$°F, determine the following: N_o, s_h, h_o, Δt_m, μ, $\hbar_p$, q, and q_s.

9.6.6 For the input parameters given in Prob. 9.6.5, determine the same output parameters given in Prob. 9.6.5 but for a tilted-pad thrust bearing.

9.6.7 By starting with Eq. (9.22) show that as $\lambda \to 0$ or as the situation is approached where side leakage can be neglected, the normal load is the same as discovered in Chap. 8, namely, Eq. (8.73).

9.7 REFERENCES

Engineering Sciences Data Unit (ESDU) (1967): *General Guide to the Choice of Thrust Bearing Type.* Item 67033. Institution of Mechanical Engineers, London.

Raimondi, A. A., and Boyd, J. (1955): Applying Bearing Theory to the Analysis and Design of Pad-Type Bearings. *ASME Trans.*, vol. 77, no. 3, pp. 287–309.

CHAPTER 10

HYDRODYNAMIC JOURNAL BEARINGS—ANALYTICAL SOLUTIONS

The past two chapters have dealt with slider bearing pads as used in thrust bearings. The surfaces of thrust bearings are perpendicular to the axis of rotation as shown in Fig. 8.3. This chapter and the next deal with journal bearings, where the bearing surfaces are parallel to the axis of rotation. Journal bearings are used to support shafts and to carry radial loads with minimum power loss and minimum wear. The journal bearing can be represented by a plain cylindrical sleeve (bushing) wrapped around the journal (shaft), but the bearings can adopt a variety of forms. The lubricant is supplied at some convenient point in the bearing through a hole or a groove. If the bearing extends around the full 360° of the journal, it is described as a "full journal bearing." If the angle of wrap is less than 360°, the term "partial journal bearing" is used.

Journal bearings rely on shaft motion to generate the load-supporting pressures in the lubricant film. The geometry of the journal bearing is shown in Fig. 10.1. The shaft does not normally run concentric with the bearing. The displacement of the shaft center relative to the bearing center is known as the "eccentricity." The shaft's eccentric position within the bearing clearance is

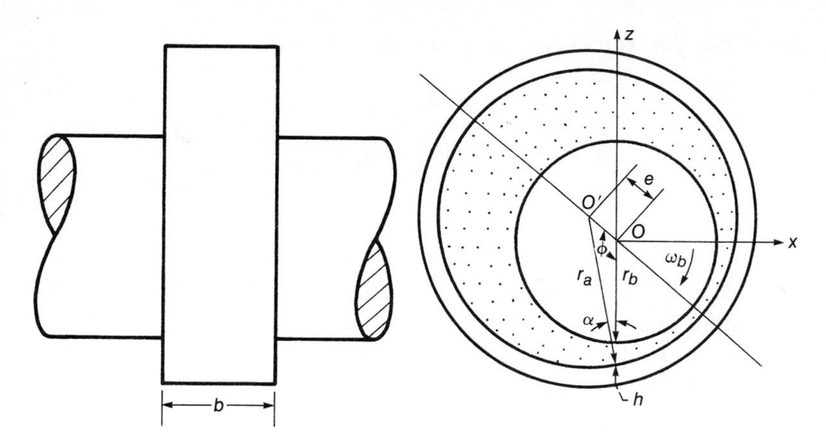

FIGURE 10.1
Hydrodynamic journal bearing geometry.

influenced by the load that it carries. The amount of eccentricity adjusts itself until the load is balanced by the pressure generated in the converging lubricating film. The line drawn through the shaft center and the bearing center is called the "line of centers."

The pressure generated and therefore the load-carrying capacity of the bearing depend on the shaft eccentricity, the angular velocity, the effective viscosity of the lubricant, and the bearing dimensions and clearance.

$$w_z = f(e, \omega, \eta_0, r, b, c)$$

The load and the angular velocity are usually specified and the minimum shaft diameter is often predetermined. To complete the design, it will be necessary to calculate the bearing dimensions and clearance and to choose a suitable lubricant if this is not already specified.

The approach used in this chapter is to present two approximate journal bearing solutions: (1) an infinite-width solution (side leakage neglected) and (2) a short-width-journal-bearing theory. These two approximate solutions will illustrate the many important characteristics of journal bearings.

10.1 INFINITELY WIDE-JOURNAL-BEARING SOLUTION

For an infinitely wide-journal-bearing solution the pressure in the axial direction is assumed to be constant. This approach is valid for diameter-to-width ratios ($\lambda_k = 2r_b/b$) less than 0.5. The integrated form of the Reynolds equation can be written from Eq. (7.52), while assuming a constant viscosity, as

$$\frac{dp}{dx} = \frac{6\eta_0 r_b \omega_b (h - h_m)}{h^3} \tag{7.52}$$

where h_m denotes the film thickness when $dp/dx = 0$. Now

$$dx = r_b\, d\phi$$

$$\therefore \frac{dp}{d\phi} = \frac{6\eta_0 r_b^2 \omega_b (h - h_m)}{h^3} \tag{10.1}$$

The transition from Eq. (7.52) to (10.1) is acceptable, since the film thickness is small relative to the shaft radius and the curvature of the lubricant film can be neglected. This implies that the film shape can be unwrapped from around the shaft and viewed as a periodic stationary profile with wavelength $2\pi r_b$ and that the plane surface of the shaft is moving with velocity $r\omega$ as shown in Fig. 10.2. From Fig. 10.1

$$\cos\alpha = \frac{1}{r_a}\left[h + r_b + e\cos(\pi - \phi)\right]$$

$$\therefore h = r_a \cos\alpha - r_b + e\cos\phi \tag{10.2}$$

From the law of sines

$$\frac{e}{\sin\alpha} = \frac{r_a}{\sin\phi}$$

$$\therefore \sin\alpha = \frac{e\sin\phi}{r_a}$$

and

$$\cos\alpha = (1 - \sin^2\alpha)^{1/2} = \left[1 - \left(\frac{e}{r_a}\right)^2 \sin^2\phi\right]^{1/2} \tag{10.3}$$

Substituting Eq. (10.3) into Eq. (10.2) gives

$$h = r_a\left[1 - \left(\frac{e}{r_a}\right)^2 \sin^2\phi\right]^{1/2} - r_b + e\cos\phi$$

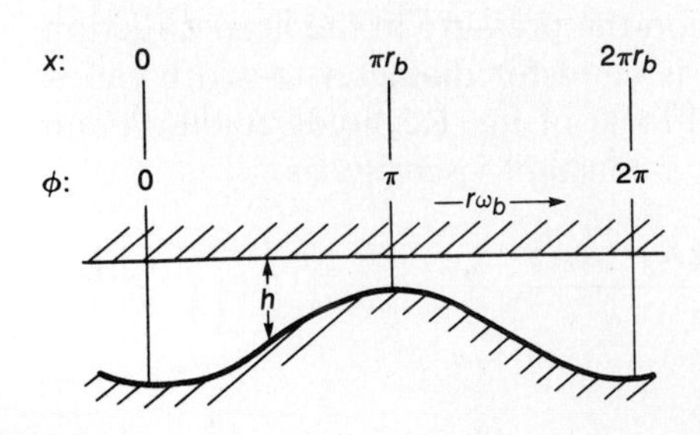

FIGURE 10.2
Unwrapped film shape in a journal bearing.

But

$$\left[1 - \left(\frac{e}{r_a}\right)^2 \sin^2\phi\right]^{1/2} = 1 - \frac{1}{2}\left(\frac{e}{r_a}\right)^2 \sin^2\phi - \frac{1}{8}\left(\frac{e}{r_a}\right)^4 \sin^4\phi - \cdots$$

$$\therefore h = r_a\left[1 - \frac{1}{2}\left(\frac{e}{r_a}\right)^2 \sin^2\phi - \frac{1}{8}\left(\frac{e}{r_a}\right)^4 \sin^4\phi - \cdots\right] - r_b + e\cos\phi$$

or, since $r_a - r_b = c$,

$$h = c + e\left[\cos\phi - \frac{1}{2}\left(\frac{e}{r_a}\right)\sin^2\phi - \frac{1}{8}\left(\frac{e}{r_a}\right)^3 \sin^4\phi - \cdots\right] \tag{10.4}$$

Since the ratio of e/r_a is of the order of magnitude 10^{-3}, Eq. (10.4) can be safely reduced to

$$h = c(1 + \varepsilon\cos\phi) \tag{10.5}$$

where

$$\varepsilon = \frac{e}{c} \tag{10.6}$$

is the eccentricity ratio. Note that $0 \le \varepsilon \le 1$.

Substituting the film thickness equation (10.5) into Eq. (10.1) gives

$$\frac{dp}{d\phi} = 6\eta_0\omega_b\left(\frac{r_b}{c}\right)^2\left[\frac{1}{(1 + \varepsilon\cos\phi)^2} - \frac{h_m}{c(1 + \varepsilon\cos\phi)^3}\right] \tag{10.7}$$

An expression for the pressure distribution can be obtained by direct integration of Eq. (10.7)

$$p = 6\eta_0\omega_b\left(\frac{r_b}{c}\right)^2 \int\left[\frac{1}{(1 + \varepsilon\cos\phi)^2} - \frac{h_m}{c(1 + \varepsilon\cos\phi)^3}\right] d\phi + \tilde{A} \tag{10.8}$$

10.1.1 FULL SOMMERFELD SOLUTION. The procedure for evaluating integrals of the type

$$\int \frac{d\phi}{(1 + \varepsilon\cos\phi)^n}$$

is to introduce a new variable $\gamma = \tan(\phi/2)$. With this procedure the pressure can be evaluated, but the expression is not particularly useful, since it is difficult to obtain the load components from a further integration.

Sommerfeld in 1904 neatly overcame these difficulties by using the substitution

$$1 + \varepsilon\cos\phi = \frac{1 - \varepsilon^2}{1 - \varepsilon\cos\gamma} \tag{10.9}$$

This relationship is known as the “Sommerfeld substitution,” and γ is known as the "Sommerfeld variable." Table 10.1 shows the relationship between the circumferential coordinate angle ϕ and γ for a number of eccentricity ratios. From Eq. (10.9) the following can be written:

$$\sin\phi = \frac{(1-\varepsilon^2)^{1/2}\sin\gamma}{1-\varepsilon\cos\gamma} \tag{10.10}$$

$$\cos\phi = \frac{\cos\gamma-\varepsilon}{1-\varepsilon\cos\gamma} \tag{10.11}$$

$$\sin\gamma = \frac{(1-\varepsilon^2)^{1/2}\sin\phi}{1+\varepsilon\cos\phi} \tag{10.12}$$

$$\cos\gamma = \frac{\varepsilon+\cos\phi}{1+\varepsilon\cos\phi} \tag{10.13}$$

$$d\phi = \frac{(1-\varepsilon^2)^{1/2}\,d\gamma}{1-\varepsilon\cos\gamma} \tag{10.14}$$

Making use of the Sommerfeld substitution and the periodic boundary condition yields

$$p - p_0 = \frac{6\eta_0\omega_b(r_b/c)^2\varepsilon\sin\phi\,(2+\varepsilon\cos\phi)}{(2+\varepsilon^2)(1+\varepsilon\cos\phi)^2} \tag{10.15}$$

where p_0 is the pressure at the point of maximum film thickness. This equation represents the Sommerfeld solution for pressure distribution in a full journal bearing. In dimensionless form this equation becomes

$$P = \frac{p-p_0}{\eta_0\omega_b}\left(\frac{c}{r_b}\right)^2 = \frac{6\varepsilon\sin\phi\,(2+\varepsilon\cos\phi)}{(2+\varepsilon^2)(1+\varepsilon\cos\phi)^2} \tag{10.16}$$

Booker (1965) has developed a useful tabulation of integrals normally encountered in journal bearing analysis.

The pressure distribution is shown in Fig. 10.3 for a full Sommerfeld solution. Note that positive pressures are generated in the convergent film $(0 \le \phi \le \pi)$ and negative pressures in the divergent film $(\pi \le \phi \le 2\pi)$. Figure 10.3 shows that the pressure distribution is skewed symmetrically, with the numerical values of the maximum and minimum pressures and their locations relative to the point of minimum film thickness being equal. In the derivation of Eq. (10.15), which is a problem at the end of this chapter,

$$h_m = \frac{2c(1-\varepsilon^2)}{2+\varepsilon^2} \tag{10.17}$$

Making use of Eqs. (10.17) and (10.5) gives the value of ϕ where $dp/dx = 0$ as

$$\phi_m = \cos^{-1}\frac{-3\varepsilon}{2+\varepsilon^2} \tag{10.18}$$

TABLE 10.1 Relationship between angle ϕ and Sommerfeld variable γ for various eccentricity ratios ε

[When $\varepsilon = 0$, $\gamma = 0$. When $\varepsilon = 1$, $\gamma = 0$. For $180° \le \phi \le 360°$ the relationship $\gamma(-\phi) = -\gamma(\phi)$]

Circumferential coordinate angle, ϕ, deg	Eccentricity ratio, ϵ							
	0.2	0.4	0.5	0.6	0.7	0.8	0.9	0.95
	Sommerfeld variable, γ, deg							
0	0	0	0	0	0	0	0	0
10	8.172	6.556	5.783	5.010	4.210	3.341	2.300	1.605
20	16.385	13.169	11.626	10.077	8.473	6.727	4.633	3.235
30	24.681	19.899	17.588	15.261	12.844	10.208	7.035	4.914
40	33.102	26.804	23.735	20.628	17.368	13.835	9.546	6.671
50	41.688	33.952	30.136	26.249	22.166	17.670	12.212	8.541
60	50.479	41.410	36.870	32.204	27.266	21.787	15.090	10.564
70	59.515	49.253	44.024	38.591	32.782	26.276	18.252	12.795
80	68.832	57.562	51.696	45.521	38.834	31.253	21.793	15.305
90	78.463	66.422	60.000	53.130	45.573	36.870	28.842	18.195
100	88.436	75.922	69.061	61.579	53.188	43.331	30.583	21.608
110	98.769	86.149	79.014	71.059	61.923	50.914	36.282	25.763
120	109.471	97.181	90.000	81.787	72.080	60.000	43.342	31.003
130	120.538	109.076	102.147	93.994	84.030	71.117	52.393	37.905
140	131.948	121.854	115.544	107.895	98.187	84.969	64.448	47.494
150	143.663	135.482	130.208	123.626	114.937	102.412	81.140	61.726
160	155.628	149.851	146.034	141.149	134.460	124.244	104.909	84.487
170	167.768	164.776	162.767	160.150	156.471	150.587	138.251	122.699
180	180.000	180.000	180.000	180.000	180.000	180.000	180.000	180.000

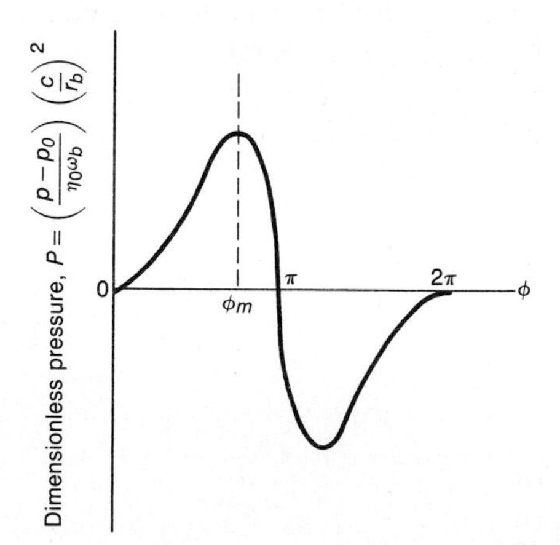

FIGURE 10.3
Pressure distribution for full Sommerfeld solution.

Note that $\phi_m \to \pm\pi/2$ as $\varepsilon \to 0$ and $\phi_m \to \pm\pi$ as $\varepsilon \to 1$. From Fig. 10.3 the maximum pressure occurs in the second quadrant, and the minimum pressure in the third quadrant. The maximum and minimum pressures are equidistant from the line of centers.

The maximum dimensionless pressure can be written from Eq. (10.16) as

$$P_m = \frac{p_m - p_0}{\eta_0 \omega_b}\left(\frac{c}{r_b}\right)^2 = \frac{6\varepsilon \sin\phi_m (2 + \varepsilon\cos\phi_m)}{(2+\varepsilon^2)(1+\varepsilon\cos\phi_m)^2}$$

Making use of Eq. (10.18) and

$$\sin\phi_m = \left(1 - \cos^2\phi_m\right)^{1/2} = \frac{\left(4 - 5\varepsilon^2 + \varepsilon^4\right)^{1/2}}{2+\varepsilon^2}$$

gives

$$P_m = \frac{3\varepsilon\left(4 - 5\varepsilon^2 + \varepsilon^4\right)^{1/2}(4-\varepsilon^2)}{2(2+\varepsilon^2)(1-\varepsilon^2)^2} \tag{10.19}$$

Note that $P_m \to 0$ as $\varepsilon \to 0$ and $P_m \to \infty$ as $\varepsilon \to 1$.

Once the pressure is known, the load components can be evaluated. It is convenient to determine the components of the resultant load along and perpendicular to the line of centers. The coordinate system and the load components are shown in Fig. 10.4, where

w'_x = load component per unit width perpendicular to line of centers, N/m
w'_z = load component per unit width along line of centers, N/m
w'_r = resultant load per unit width (equal but acting in opposite direction to applied load), N/m
Φ = attitude angle (angle through which load vector has to be rotated in direction of journal rotation to bring it into line of centers)

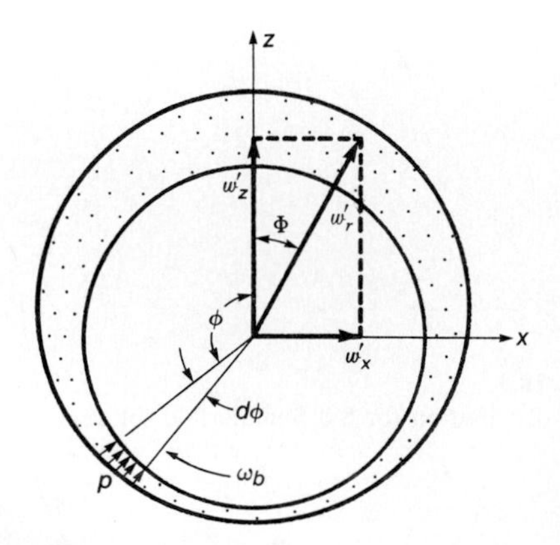

FIGURE 10.4
Coordinate system and force components in a journal bearing.

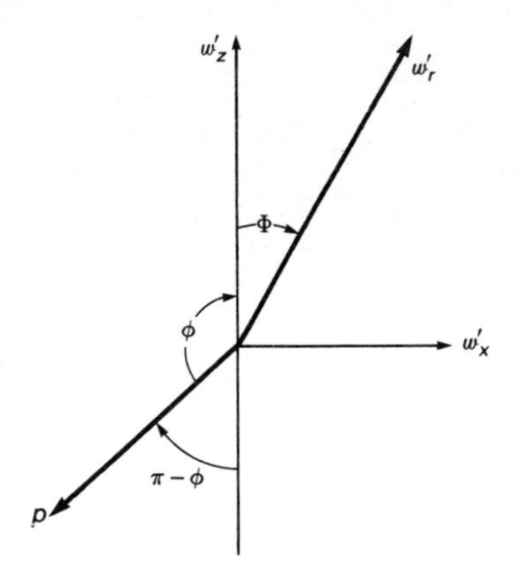

FIGURE 10.5
Vector forces acting on a journal.

From Fig. 10.5, which shows the vector forces acting on the journal,

$$w'_x = \int_0^{2\pi} p r_b \sin(\pi - \phi)\, d\phi$$

$$w'_z = \int_0^{2\pi} p r_b \cos(\pi - \phi)\, d\phi$$

These equations reduce to

$$w'_x = \int_0^{2\pi} p r_b \sin\phi\, d\phi \tag{10.20}$$

$$w'_z = -\int_0^{2\pi} p r_b \cos\phi\, d\phi \tag{10.21}$$

Equations (10.20) and (10.21) can be integrated by parts to give

$$w'_x = r_b \int_0^{2\pi} \cos\phi \frac{dp}{d\phi}\, d\phi \tag{10.22}$$

$$w'_z = r_b \int_0^{2\pi} \sin\phi \frac{dp}{d\phi}\, d\phi \tag{10.23}$$

Substituting Eq. (10.7) into these equations gives

$$w'_x = 6\eta_0 \omega_b r_b \left(\frac{r_b}{c}\right)^2 \int_0^{2\pi} \left[\frac{\cos\phi}{(1+\varepsilon\cos\phi)^2} - \frac{h_m \cos\phi}{c(1+\varepsilon\cos\phi)^3}\right] d\phi \tag{10.24}$$

$$w'_z = 6\eta_0 \omega_b r_b \left(\frac{r_b}{c}\right)^2 \int_0^{2\pi} \left[\frac{\sin\phi}{(1+\varepsilon\cos\phi)^2} - \frac{h_m \sin\phi}{c(1+\varepsilon\cos\phi)^3}\right] d\phi \tag{10.25}$$

Now

$$\int_0^{2\pi} \frac{\cos\phi}{(1+\varepsilon\cos\phi)^2}\,d\phi = \frac{1}{\varepsilon}\left[\frac{\gamma}{(1-\varepsilon^2)^{1/2}} - \frac{\gamma - \varepsilon\sin\gamma}{(1-\varepsilon^2)^{3/2}}\right]_{\gamma=0}^{\gamma=2\pi}$$

$$\int_0^{2\pi} \frac{\cos\phi\,d\phi}{(1+\varepsilon\cos\phi)^3}$$

$$= \frac{1}{\varepsilon}\left[\frac{\gamma - \varepsilon\sin\gamma}{(1-\varepsilon^2)^{3/2}} + \frac{-2\gamma(2+\varepsilon^2) + 8\varepsilon\sin\gamma - \varepsilon^2\sin 2\gamma}{4(1-\varepsilon^2)^{5/2}}\right]_{\gamma=0}^{\gamma=2\pi}$$

$$\int_0^{2\pi} \frac{\sin\phi\,d\phi}{(1+\varepsilon\cos\phi)^2} = \left[\frac{1}{\varepsilon(1+\varepsilon\cos\phi)}\right]_{\phi=0}^{\phi=2\pi}$$

$$\int_0^{2\pi} \frac{\sin\phi\,d\phi}{(1+\varepsilon\cos\phi)^3} = \left[\frac{1}{2\varepsilon(1+\varepsilon\cos\phi)^2}\right]_{\phi=0}^{\phi=2\pi}$$

Substituting these integrals into Eqs. (10.24) and (10.25) gives

$$w_x' = 6\eta_0\omega_b r_b\left(\frac{r_b}{c}\right)^2 \frac{1}{\varepsilon}\left[\frac{\gamma}{(1-\varepsilon^2)^{1/2}} - \frac{\gamma - \varepsilon\sin\gamma}{(1-\varepsilon^2)^{3/2}}\right] - \frac{h_m}{c\varepsilon}$$

$$\times\left[\frac{\gamma + \varepsilon\sin\gamma}{(1-\varepsilon^2)^{3/2}} + \frac{-2\gamma(2+\varepsilon^2) + 8\varepsilon\sin\gamma - \varepsilon^2\sin 2\gamma}{4(1-\varepsilon^2)^{5/2}}\right]_{\gamma=0}^{\gamma=2\pi}$$

$$w_z' = 6\eta_0\omega_b r_b\left(\frac{r_b}{c}\right)^2\left[-\frac{2\pi\varepsilon}{(1-\varepsilon^2)^{3/2}} + \frac{3\pi\varepsilon h_m}{c(1-\varepsilon^2)^{5/2}}\right]$$

Making use of Eq. (10.17) yields

$$w_x' = 12\pi\eta_0\omega_b r_b\left(\frac{r_b}{c}\right)^2 \frac{\varepsilon}{(2+\varepsilon^2)(1-\varepsilon^2)^{1/2}} \tag{10.26}$$

and

$$w_z' = 6\eta_0\omega_b r_b\left(\frac{r_b}{c}\right)^2\left[\frac{1}{\varepsilon(1+\varepsilon\cos\phi)} - \frac{h_m}{2c\varepsilon(1+\varepsilon\cos\phi)^2}\right]_{\phi=0}^{\phi=2\pi}$$

$$\therefore\ w_z' = 0 \tag{10.27}$$

This result demonstrates that for a full Sommerfeld solution the resultant normal load acts at right angles to the line of centers; that is, the attitude angle is 90°. As load is applied to the bearing, the journal center moves away from the bearing center at right angles to the load vector.

$$\therefore \Phi = 90^\circ \tag{10.28}$$

$$w'_r = w'_x \tag{10.29}$$

Making use of Eq. (10.17) yields

$$W_r = \frac{w'_r}{r_b \omega_b \eta_0}\left(\frac{c}{r_b}\right)^2 = \frac{12\pi\varepsilon}{(2+\varepsilon^2)(1-\varepsilon^2)^{1/2}} \tag{10.30}$$

where the dimensionless resultant load-carrying capacity is a function of the eccentricity ratio alone.

Attention is drawn to two important cases for full Sommerfeld solutions:

1. $\varepsilon \to 0$ as $W_r \to 0$.
2. $\varepsilon \to 1$ as $W_r \to \infty$.

The first case demonstrates that when the shaft is concentric relative to the bearing (constant clearance around the bearing), the bearing does not have any load-carrying capacity. The second case demonstrates the tremendous potential of a fluid film journal bearing for supporting radial loads. The result suggests that load increases can be accommodated by operating at higher eccentricity ratios, but this suggestion must be tempered by the knowledge that side leakage has been neglected, that bearing surfaces are not perfectly smooth, and that restrictively high temperatures will occur in the thin oil films that exist at high eccentricities. The operating eccentricity must be selected with several design points in mind, but as a design guide it is worth noting that most journal bearings operate with eccentricity ratios between 0.5 and 0.8.

10.1.2 HALF SOMMERFELD SOLUTION. It has been noted that the Sommerfeld solution for a full 360° journal bearing leads to the skew-symmetrical pressure distribution shown in Fig. 10.3. The pressures in the divergent film are all lower than ambient pressure. Such pressures are rarely encountered in real bearings. Mineral oils contain between 8 and 12 percent dissolved air. This air will start to come out of solution whenever the pressure falls below the saturation pressure. In many situations the saturation pressure is similar to the ambient pressure surrounding the bearing, and in these cases gas liberation will maintain the pressure in the divergent clearance space at close to the ambient level.

This hindrance to predicting subambient pressures by the normal Sommerfeld analysis has led to the suggestion that the subambient pressures predicted by the analysis should be ignored. This approach, which limits the analysis to the convergent film, is known as the "half Sommerfeld solution." The approximation does, in fact, lead to more realistic predictions of some bearing characteristics, but the simple approach leads to a violation of the continuity of mass flow condition at the end of the pressure curve. The boundary condition to be applied at the outlet end of the pressure curve, where the full lubricant film gives way to a ruptured or cavitated film composed of a mixture of gas and liquid, is discussed later.

The pressure distribution assumed for the half Sommerfeld solution is exactly that shown in Fig. 10.3 for the region $\phi = 0$ to π. However, from $\phi = \pi$ to 2π, instead of the negative pressures shown in the figure, the pressures are zero.

$$\therefore P = \frac{p - p_0}{\eta_0 \omega_b}\left(\frac{c}{r_b}\right)^2 = \frac{6\varepsilon \sin\phi\,(2 + \varepsilon\cos\phi)}{(2 + \varepsilon^2)(1 + \varepsilon\cos\phi)^2} \qquad \text{for } 0 \le \phi \le \pi \tag{10.31}$$

and

$$P = 0 \qquad \text{for } \pi \le \phi \le 2\pi \tag{10.32}$$

The equations for the film thickness at the location of maximum pressure h_m, the angle of maximum pressure ϕ_m, and the maximum dimensionless pressure P_m are exactly the same as those developed for the full Sommerfeld solution. That is, Eqs. (10.17) to (10.19) provide values of h_m, ϕ_m, and P_m for the half Sommerfeld solution.

The load components per unit width w'_x and w'_z, given in Eqs. (10.20) and (10.21), are exactly the same for the half Sommerfeld solution except that the integration limits must be written as 0 to π. When this is done, the load components turn out to be

$$w'_x = 6\eta_0 \omega_b r_b \left(\frac{r_b}{c}\right)^2 \left[\frac{-\pi\varepsilon}{(1 - \varepsilon^2)^{3/2}} + \frac{3\pi\varepsilon h_m}{2c(1 - \varepsilon^2)^{5/2}}\right] \tag{10.33}$$

$$w'_z = 6\eta_0 \omega_b r_b \left(\frac{r_b}{c}\right)^2 \left[\frac{2}{1 - \varepsilon^2} - \frac{2h_m}{c(1 - \varepsilon^2)^2}\right] \tag{10.34}$$

Introducing Eq. (10.17) for h_m into Eqs. (10.33) and (10.34),

$$w'_x = 6\eta_0 \omega_b r_b \left(\frac{r_b}{c}\right)^2 \frac{\pi\varepsilon}{(2 + \varepsilon^2)(1 - \varepsilon^2)^{1/2}} \tag{10.35}$$

$$w'_z = 12\eta_0 \omega_b r_b \left(\frac{r_b}{c}\right)^2 \frac{\varepsilon^2}{(2 + \varepsilon^2)(1 - \varepsilon^2)} \tag{10.36}$$

Note that w'_x in Eq. (10.35) is one-half the corresponding full Sommerfeld solution, since the resultant contribution of subambient pressures perpendicular to the line of centers is neglected in the present case. In addition, note that for the half Sommerfeld solution w'_z is not zero as it was for the full Sommerfeld solution, since the contributions from the convergent and divergent films do not now cancel.

In dimensionless form these components become

$$W_x = \frac{w'_x}{\eta_0 \omega_b r_b}\left(\frac{c}{r_b}\right)^2 = \frac{6\pi\varepsilon}{(2+\varepsilon^2)(1-\varepsilon^2)^{1/2}} \tag{10.37}$$

$$W_z = \frac{w'_z}{\eta_0 \omega_b r_b}\left(\frac{c}{r_b}\right)^2 = \frac{12\varepsilon^2}{(2+\varepsilon^2)(1-\varepsilon^2)} \tag{10.38}$$

The resultant load is

$$w'_r = \left(w'^2_x + w'^2_z\right)^{1/2} = \eta_0 \omega_b r_b \left(\frac{r_b}{c}\right)^2 \frac{6\varepsilon\left[\pi^2 - \varepsilon^2(\pi^2 - 4)\right]^{1/2}}{(2+\varepsilon^2)(1-\varepsilon^2)} \tag{10.39}$$

In dimensionless form the resultant load is

$$W_r = \frac{w'_r}{\eta_0 \omega_b r_b}\left(\frac{c}{r_b}\right)^2 = \frac{6\varepsilon\left[\pi^2 - \varepsilon^2(\pi^2 - 4)\right]^{1/2}}{(2+\varepsilon^2)(1-\varepsilon^2)} \tag{10.40}$$

The attitude angle can be written as

$$\Phi = \tan^{-1}\frac{w'_x}{w'_z} = \tan^{-1}\left[\frac{\pi}{2\varepsilon}(1-\varepsilon^2)^{1/2}\right] \tag{10.41}$$

Note that $\Phi = 90°$ when $\varepsilon = 0$ and $\Phi = 0°$ when $\varepsilon = 1$. The shaft starts to move at right angles to the applied load at extremely light loads, but ultimately the shaft meets the bearing along the load line. The locus of the shaft center for full and half Sommerfeld conditions is shown in Fig. 10.6.

10.1.3 REYNOLDS BOUNDARY CONDITIONS. As described earlier, the half Sommerfeld solution results in more realistic predictions of load components, but this simple approach leads to a violation of the continuity of mass flow at

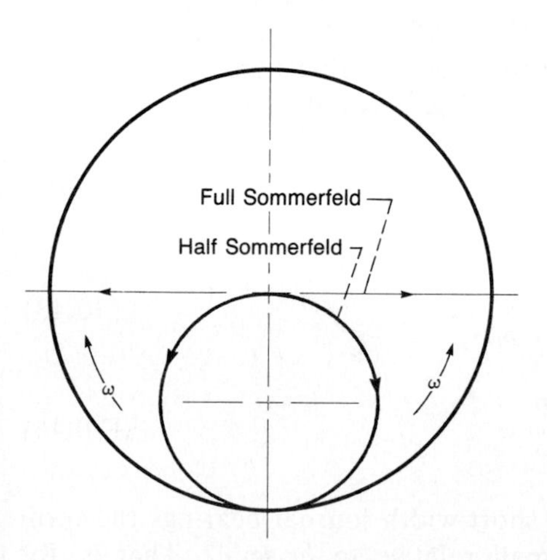

FIGURE 10.6
Locus of shaft center for full and half Sommerfeld journal bearing solutions.

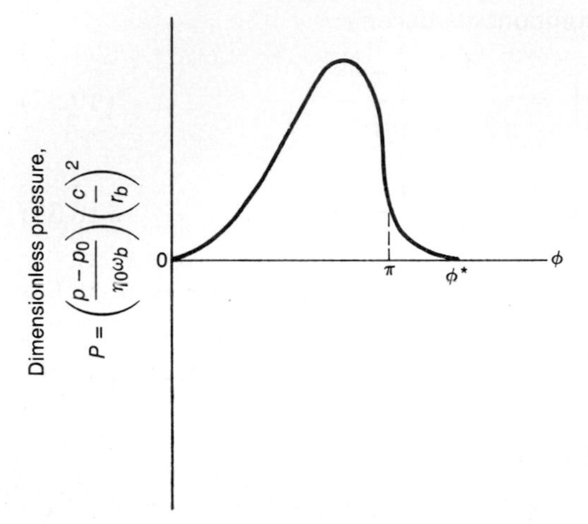

FIGURE 10.7
Pressure profile for a journal bearing using Reynolds boundary condition.

the outlet end of the pressure curve. That is, in Fig. 10.3 the pressure suddenly becomes zero at $\Phi = \pi$ and then stays at zero from π to 2π. The pressure distribution approaching π violates the continuity of mass flow condition. A better boundary condition is the Reynolds boundary condition, where

$$p = 0 \qquad \text{and} \qquad \frac{dp}{dx} = 0 \qquad \text{at } \phi = \phi^* \tag{10.42}$$

Figure 10.7 shows a pressure profile for a bearing using the Reynolds boundary condition.

10.2 SHORT-WIDTH-JOURNAL-BEARING THEORY

It is shown in Chap. 9, for thrust bearings, that side leakage can account for a substantial reduction in the theoretical prediction of fluid film bearing load-carrying capacity. The complete solution of the Reynolds equation [Eq. (7.48)] for three-dimensional flow generally requires considerable numerical effort. A useful approximate analytical solution that takes account of side leakage was presented in 1953 by Dubois and Ocvirk.

When $u_a = v_a = v_b = 0$, $\partial x = r_b \partial\phi$, $u_b = r_b\omega_b$, and the viscosity is constant, Eqs. (7.38) and (7.39) become

$$q'_\phi = -\frac{h^3}{12\eta_0 r_b}\frac{\partial p}{\partial \phi} + \frac{h r_b \omega_b}{2} \tag{10.43}$$

$$q'_y = -\frac{h^3}{12\eta_0}\frac{\partial p}{\partial y} \tag{10.44}$$

Dubois and Ocvirk (1953) state that for short-width journal bearings the term $(h^3/12\eta_0 r_b)(\partial p/\partial\phi)$ in Eq. (10.43) is small relative to $hr_b\omega_b/2$. That is, for short-width journal bearings the pressure-induced flow in the circumferential direction is small relative to the Couette flow term. Therefore, they assume that

$$q'_\phi = \frac{hr_b\omega_b}{2} \tag{10.45}$$

It should be emphasized that this assumption implies not that $\partial p/\partial x = 0$ but that $(h^3/12\eta_0 r_b)(\partial p/\partial \phi)$ is small in terms of $hr_b\omega_b/2$ for short-width journal bearings. This assumption further implies that the Poiseuille (pressure) flow is more significant in the y direction than in the circumferential (ϕ) direction. The result of this assumption is that the Reynolds equation given in Eq. (7.48) reduces to

$$\frac{\partial}{\partial y}\left(h^3\frac{\partial p}{\partial y}\right) = 6\eta_0\omega_b\frac{\partial h}{\partial \phi} \tag{10.46}$$

The short-width-journal-bearing theory is valid as long as the diameter-to-width ratio is greater than 2 ($\lambda_k > 2$). Of course, the larger λ_k is, the better the agreement with the exact theory.

Assuming no misalignment, the film thickness in a function of ϕ only, and hence the right side of Eq. (10.46) is independent of y. Integrating twice gives

$$p = \frac{6\eta_0\omega_b}{h^3}\frac{\partial h}{\partial \phi}\frac{y^2}{2} + \frac{\tilde{A}y}{h^3} + \tilde{B} \tag{10.47}$$

Now the y axis is chosen to be in the center of the bearing so that the boundary conditions, as they relate to the y coordinate, can be written as

$$p = 0 \qquad \text{when } y = \pm\frac{b}{2}$$

Making use of these boundary conditions results in

$$\tilde{A} = 0$$

$$\tilde{B} = -\frac{6\eta_0\omega_b}{h^3}\frac{\partial h}{\partial \phi}\frac{b^2}{8}$$

$$p = \frac{3\eta_0\omega_b}{h^3}\frac{\partial h}{\partial \phi}\left(y^2 - \frac{b^2}{4}\right) \tag{10.48}$$

The film thickness in a journal bearing was defined in Eq. (10.5). The film thickness gradient can be written as

$$\frac{dh}{d\phi} = -e\sin\phi \tag{10.49}$$

Substituting Eqs. (10.5) and (10.49) into (10.48) gives

$$p = \frac{3\eta_0\omega_b\varepsilon}{c^2}\left(\frac{b^2}{4} - y^2\right)\frac{\sin\phi}{(1+\varepsilon\cos\phi)^3} \qquad \text{for } 0 \le \phi \le \pi \tag{10.50}$$

This equation shows that the parabolic function governs the axial variation of pressure, whereas the trigonometric function dictates the circumferential variation of pressure. Subambient pressures predicted by Eq. (10.50) are ignored, and it is assumed that the positive pressure region from $\phi = 0$ to $\phi = \pi$ carries the total load of the bearing (half Sommerfeld assumption).

The location of the maximum pressure is obtained when $\partial p/\partial \phi = 0$. From Eq. (10.48)

$$\frac{\partial p}{\partial \phi} = 3\eta_0 \omega_b \left(y^2 - \frac{b^2}{4} \right) \frac{\partial}{\partial \phi} \left(\frac{1}{h^3} \frac{\partial h}{\partial \phi} \right) = 0$$

or

$$-3h^{-4} \left(\frac{\partial h}{\partial \phi} \right)^2 + h^{-3} \frac{\partial^2 h}{\partial \phi^2} = 0$$

Substituting Eqs. (10.5) and (10.49) into this equation gives

$$3\varepsilon \sin^2 \phi_m + \cos \phi_m (1 + \varepsilon \cos \phi_m) = 0$$

But $\sin^2 \phi_m = 1 - \cos^2 \phi_m$

$$\therefore \phi_m = \cos^{-1} \left[\frac{1 - (1 + 24\varepsilon^2)^{1/2}}{4\varepsilon} \right] \tag{10.51}$$

Note that $\phi_m \to \pm\pi/2$ as $\varepsilon \to 0$ and $\phi_m \to \pm\pi$ as $\varepsilon \to 1$. The maximum pressure occurs when $\phi = \phi_m$ and $y = 0$. Therefore, from Eq. (10.50) the equation for the maximum pressure is

$$p_m = \frac{3\eta_0 \omega_b \varepsilon b^2 \sin \phi_m}{4c^2 (1 + \varepsilon \cos \phi_m)^3} \tag{10.52}$$

The load components resulting from the pressure development parallel and perpendicular to the line of centers under the half Sommerfeld assumption are

$$w_x = 2 \int_0^{\pi} \int_0^{b/2} p r_b \sin \phi \, dy \, d\phi \tag{10.53}$$

$$w_z = -2 \int_0^{\pi} \int_0^{b/2} p r_b \cos \phi \, dy \, d\phi \tag{10.54}$$

Substituting Eq. (10.48) into these equations gives

$$w_x = \frac{\eta_0 \omega_b \varepsilon r_b b^3}{2c^2} \int_0^{\pi} \frac{\sin^2 \phi \, d\phi}{(1 + \varepsilon \cos \phi)^3}$$

$$w_z = -\frac{\eta_0 \omega_b \varepsilon r_b b^3}{2c^2} \int_0^{\pi} \frac{\sin \phi \cos \phi}{(1 + \varepsilon \cos \phi)^3} \, d\phi$$

When the Sommerfeld substitution given in Eq. (10.9) is used,

$$w_x = \frac{\eta_0 \omega_b \varepsilon r_b b^3}{2c^2(1-\varepsilon^2)^{3/2}} \int_0^\pi \sin^2 \gamma \, d\gamma$$

$$w_z = -\frac{\eta_0 \omega_b \varepsilon r_b b^3}{2c^2(1-\varepsilon^2)^2} \int_0^\pi (\sin\gamma \cos\gamma - \varepsilon \sin\gamma) \, d\gamma$$

Evaluating these definite integrals yields

$$w_x = \frac{\eta_0 \omega_b r_b b^3}{4c^2} \frac{\pi\varepsilon}{(1-\varepsilon^2)^{3/2}} \tag{10.55}$$

$$w_z = \frac{\eta_0 \omega_b r_b b^3}{c^2} \frac{\varepsilon^2}{(1-\varepsilon^2)^2} \tag{10.56}$$

The resultant load vector is

$$w_r = \left(w_x^2 + w_z^2\right)^{1/2} = \frac{\eta_0 \omega_b r_b b^3}{4c^2} \frac{\varepsilon}{(1-\varepsilon^2)^2} \left[16\varepsilon^2 + \pi^2(1-\varepsilon^2)\right]^{1/2} \tag{10.57}$$

The attitude angle is

$$\Phi = \tan^{-1}\frac{w_x}{w_z} = \tan^{-1}\frac{\pi(1-\varepsilon^2)^{1/2}}{4\varepsilon} \tag{10.58}$$

Note that the attitude angle depends directly on the eccentricity rate ε so that a single polar curve of ε against Φ applies for all diameter-to-width ratios ($\lambda_k = 2r_b/b$). Equations (10.55) to (10.57) can be written in dimensionless form as

$$W_x = \frac{w_x}{\eta_0 \omega_b r_b b}\left(\frac{c}{r_b}\right)^2 = \left(\frac{b}{r_b}\right)^2 \frac{\pi\varepsilon}{4(1-\varepsilon^2)^{3/2}} \tag{10.59}$$

$$W_z = \frac{w_z}{\eta_0 \omega_b r_b b}\left(\frac{c}{r_b}\right)^2 = \left(\frac{b}{r_b}\right)^2 \frac{\varepsilon^2}{(1-\varepsilon^2)^2} \tag{10.60}$$

$$W_r = \frac{w_r}{\eta_0 \omega_b r_b b}\left(\frac{c}{r_b}\right)^2 = \left(\frac{b}{r_b}\right)^2 \frac{\varepsilon}{4(1-\varepsilon^2)^2}\left[16\varepsilon^2 + \pi^2(1-\varepsilon^2)\right]^{1/2} \tag{10.61}$$

The volume flow of lubricant supplied to the bearing through a central hole or groove must be equal to the net rate of outflow along the bearing axis (y direction). The total leakage from the sides of the bearing in the convergent film region (half Sommerfeld assumption) can be expressed as

$$q_y = -2r_b \int_0^{\pi} \left(\frac{h^3}{12\eta_0} \frac{\partial p}{\partial y} \right)_{y=b/2} d\phi$$

Introducing Eq. (10.48) gives

$$q_y = -\frac{\omega_b r_b b}{2} \int_0^{\pi} \frac{\partial h}{\partial \phi} \, d\phi$$

Making use of Eq. (10.49) gives

$$q_y = \frac{\omega_b r_b b e}{2} \int_0^{\pi} \sin \phi \, d\phi = -\frac{\omega_b r_b b e}{2} (\cos \phi)_{\phi=0}^{\phi=\pi} = \omega_b r_b b e \tag{10.62}$$

In dimensionless form this equation becomes

$$Q_y = \frac{2 q_y \pi}{\omega_b r_b b c} = 2\varepsilon\pi \tag{10.63}$$

It can be seen from this equation that $q_y \to 0$ as $\varepsilon \to 0$ (no side leakage) and $q_y \to \omega_b r_b b c$ as $\varepsilon \to 1$ (complete side leakage).

The load-carrying capacity as obtained from the long-width-bearing solution can be compared with the short-width bearing theory results by comparing Eqs. (10.40) and (10.61) to give

$$\frac{W \text{ (short-width bearing)}}{W \text{ (long-width bearing)}} = \left(\frac{b}{r_b} \right)^2 \frac{(2+\varepsilon^2)\left[16\varepsilon^2 + \pi^2(1-\varepsilon^2)\right]^{1/2}}{24(1-\varepsilon^2)\left[\pi^2 - \varepsilon^2(\pi^2 - 4)\right]^{1/2}} \tag{10.64}$$

Note from this equation that the load ratios are functions of b/r_b and ε. The long-width-bearing analysis overestimates the load-carrying capacity for all ε and should be used for $\lambda_k < 0.5$. The short-width-bearing theory provides a much better estimate for finite bearings of diameter-to-width ratios greater than 2 ($\lambda_k = 2r_b/b > 2$). The useful range of b/r_b for the short-width-bearing theory depends on the eccentricity ratio. Short-width-bearing theory leads to excessive load-carrying capacities (sometimes exceeding the long-width-bearing values) at small diameter-to-width ratios ($\lambda_k = 2r_b/b$).

10.3 CLOSURE

Just as was true in Chap. 8 for the thrust bearings, in this chapter the Reynolds equations for real journal bearings were obtained only in approximate form. Analytical solutions were possible only for the simplest problems. One of these solutions (infinitely wide solution) was obtained by restricting the flow to two dimensions, the circumferential and cross-film directions, neglecting the axial flow. The two-dimensional solutions have a definite value, since they provide a good deal of information about the general characteristics of journal bearings. Three types of boundary condition were imposed on the two-dimensional solutions. A full Sommerfeld solution produced a skew-symmetrical pressure distribution. The pressures in the divergent film were all lower than the ambient pressure, and such pressures are rarely encountered in real bearings. This led to the half Sommerfeld solution, which simply equates the negative pressures to zero. The half Sommerfeld solution is more realistic in predicting journal bearing characteristics than the full Sommerfeld solution, but this simple approach leads to a violation of the continuity of mass flow at the exit. This violation of continuity of mass flow leads to a third type of boundary condition used in analyzing journal bearings, namely, the Reynolds boundary conditions, $p = 0$ and $dp/dx = 0$ at the outlet end. This type of boundary condition gives excellent agreement with experimental results.

A useful approximate analytical approach also covered in this chapter was the short-width-journal-bearing theory. It is asserted that the circumferential Poiseuille flow term is less important for short-width journal bearings than either the axial Poiseuille flow term or the Couette flow term and therefore can be neglected. For diameter-to-width ratios greater than 2 ($\lambda_k = 2r_b/b > 2$) the short-width journal bearings give a good estimate of load-carrying capacity for finite-width journal bearings. However, the useful range of short-width-journal-bearing theory depends not only on $2r_b/b$ but also on the eccentricity ratio ε. Short-width-bearing theory predicts excessive load-carrying capacities, sometimes exceeding the long-width-bearing solution at small $2r_b/b$. The half Sommerfeld boundary conditions used for the infinitely wide bearing overestimate the normal load-carrying capacity for all eccentricity ratios.

10.4 PROBLEMS

10.4.1 Starting with Eq. (10.7) show all the steps in arriving at Eq. (10.15) by using the Sommerfeld substitution.

10.4.2 Show how the resulting load-carrying capacity differs when using the full Sommerfeld solution from that when using the half Sommerfeld solution when considering an infinitely wide bearing. Show results in tabular and graphical form for the complete range of eccentricity ratios $0 \leq \varepsilon \leq 1$.

10.4.3 Show how the resulting load-capacity differs when using infinitely wide- and short-width-bearing analysis while considering the complete range of eccentricity ratios $0 \leq \varepsilon \leq 1$. Assume the half-Sommerfeld boundary condition for both analysis. Show results in tabular and graphical form for diameter to width ratios (λ_k) of 1, 2, and 4.

10.5 REFERENCES

Booker, J. F. (1965): A Table of the Journal-Bearing Integrals. *J. Basic Eng.*, vol. 87, no. 2, pp. 533–535.

DuBois, G. B., and Ocvirk, F. W. (1953): Analytical Derivation and Experimental Evaluation of Short-Bearing Approximation for Full Journal Bearings. *NACA Rep.* 1157.

Sommerfeld, A. (1904): Zur Hydrodynamischen Theorie der Schmiermittelreibung. *Z. Angew. Math. Phys.*, vol. 50, pp. 97–155.

CHAPTER 11

DYNAMICALLY LOADED JOURNAL BEARINGS

The design procedures for steadily loaded journal bearings given in Chap. 10 enable the designer to estimate the performance parameters in terms of the operating parameters. For example, the attitude angle and the eccentricity ratio can be calculated for any steady-state operating condition. From these values the minimum film thickness, a most important quantity affecting the performance of the bearing, can be calculated.

In many important bearing operating situations the load varies in both magnitude and direction, often cyclically. Examples include reciprocating machinery such as steam, diesel, and gasoline engines, reciprocating gas compressors, and out-of-balance rotating machinery such as turbine rotors. Bearings are generally dynamically loaded. Furthermore, it must be stressed that journal bearings are not inherently stable. For certain combinations of steady-state operating parameters, self-excited whirl of the journal can be sustained. If this occurs in a case with varying load, the whirl orbit will increase rapidly until the journal and sleeve come into contact. Journal bearing stability is an important consideration in high-speed rotating machinery, and unstable operation should always be avoided.

With dynamically loaded journal bearings the eccentricity and the attitude angle will vary throughout the loading cycle, and care must be taken to ensure that the combination of load and speed does not yield a dangerously small minimum film thickness. It is not easy to state a unique value of minimum film thickness that can be assumed to be safe, since a great deal depends on the manufacturing process, the alignment of the machine elements associated with the bearings, and the general operating conditions, including the environment of the machine.

It is also important to recognize the difference between dynamic effects in hydrodynamically lubricated bearings and in rolling-element bearings, which are dealt with in Chap. 24. Although the supporting structure formed by the rolling elements is discontinuous and moving, the bearing as a whole may still be treated as though it were a solid, elastic, springlike element. Spring constants for rolling-element bearings usually fall in the range 1×10^8 to 4×10^8 N/m in the direction of the load application. The rolling elements act in series with the shaft and support stiffnesses and combine according to the reciprocal summation equation. Thus, the dynamic effects as they relate to the fluid film effects in rolling-element bearings are not important and are generally not considered.

Hydrodynamic fluid film bearings are quite another matter, and thus the need for the present chapter. Unfortunately, they cannot be treated as a simple, direct spring. Although the hydrodynamic fluid film bearing does exhibit a springlike resistance that is dependent on journal displacement relative to the sleeve, this force is not linearly related to the displacement nor is it collinear with it. A hydrodynamic fluid film bearing exhibits damping effects that play a very important role in the stability of this type of bearing.

Much of this chapter uses the approach given in Lund (1966, 1979, 1987).

11.1 RELEVANT REYNOLDS EQUATION

From Eq. (7.45) the general Reynolds equation can be written as

$$\frac{\partial}{\partial x}\left(\frac{\rho h^3}{12\eta}\frac{\partial p}{\partial x}\right)+\frac{\partial}{\partial y}\left(\frac{\rho h^3}{12\eta}\frac{\partial p}{\partial y}\right)=\frac{\partial}{\partial x}\left[\frac{\rho h(u_a+u_b)}{2}\right]+\frac{\partial}{\partial y}\left[\frac{\rho h(v_a+v_b)}{2}\right]$$
$$+\rho(w_a-w_b)-\rho u_a\frac{\partial h}{\partial x}-\rho v_a\frac{\partial h}{\partial y}+h\frac{\partial \rho}{\partial t} \tag{7.45}$$

The film geometry and velocity components in a dynamically loaded journal bearing are shown in Fig. 11.1. This figure introduces a new parameter Φ_ℓ, the time-dependent direction of the load relative to the fixed coordinate system. Note that the attitude angle is measured from the fixed x axis, so the x axis should be oriented in the direction of the steady-state part of the load in order

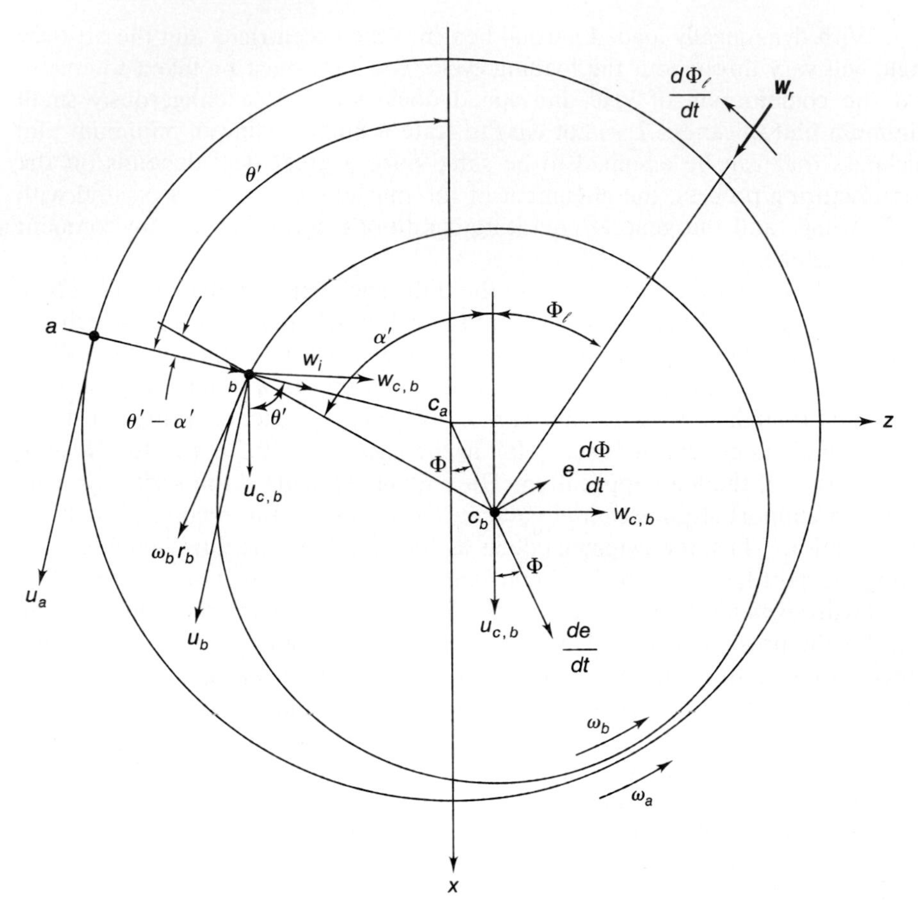

FIGURE 11.1
Film geometry and velocity components in a dynamically loaded journal bearing.

to agree with the concept of attitude angles introduced in Chap. 10. Note also the difference between the x coordinate used in Eq. (7.45), which is in the direction of the unwrapped film thickness (see Fig. 10.2), and the one used in Fig. 11.1. The film thickness will be described in this chapter from $\theta' = 0$ to $\theta' = 2\pi$, where θ' relates to the coordinate ϕ used in Chap. 10 by

$$\theta' = \phi + \Phi \tag{11.1}$$

This equation indicates the position in the film thickness relative to the fixed coordinate system and independent of the immediate attitude angle. Note also from Fig. 11.1 that the radius of the sleeve r_a is equal to the radius of the journal r_b plus the radial clearance c.

The general Reynolds equation given in Eq. (7.45) can be applied to any section of the oil film. The surface velocities at points a and b in Fig. 11.1 can be written as

$$u_a = \omega_a r_a \tag{11.2}$$

$$u_b = \omega_b r_b \cos(\theta' - \alpha') + u_{c,b} \sin\theta' - w_{c,b} \cos\theta' \tag{11.3}$$

$$v_a = v_b = 0 \qquad \text{(no movement in } y \text{ direction)} \tag{11.4}$$

$$w_a = 0 \qquad \text{(sleeve rotates about its own center)} \tag{11.5}$$

$$w_b = -\omega_b r_b \sin(\theta' - \alpha') + u_{c,b} \cos\theta' + w_{c,b} \sin\theta' \tag{11.6}$$

where $u_{c,b}$ = velocity at journal center parallel to x axis
$w_{c,b}$ = velocity at journal center perpendicular to x axis

Note from Fig. 11.1 that the eccentricity denoted by e is the distance between the journal center c_b and the sleeve center c_a. From Fig. 11.1

$$u_{c,b} = \frac{de}{dt}\cos\Phi - e\frac{d\Phi}{dt}\sin\Phi \tag{11.7}$$

$$w_{c,b} = \frac{de}{dt}\sin\Phi + e\frac{d\Phi}{dt}\cos\Phi \tag{11.8}$$

Figure 11.1 also shows the following relationship between θ' and α':

$$r_b \cos\alpha' = e\cos\Phi + |c_a a|\cos\theta' \tag{11.9}$$

$$r_b \sin\alpha' = e\sin\Phi + |c_a a|\sin\theta' \tag{11.10}$$

Multiplying Eq. (11.9) by $\sin\theta'$, multiplying Eq. (11.10) by $\cos\theta'$, and substracting the two gives

$$r_b \sin(\theta' - \alpha') = e\sin(\theta' - \Phi) \tag{11.11}$$

Substituting Eqs. (11.11), (11.7), and (11.8) into Eqs. (11.3) and (11.6) gives

$$u_b = \omega_b r_b \left[1 - \frac{e^2}{r_b^2}\sin^2(\theta' - \Phi)\right]^{1/2} + \frac{de}{dt}\sin(\theta' - \Phi) - e\frac{d\Phi}{dt}\cos(\theta' - \Phi) \tag{11.12}$$

$$w_b = -\omega_b e\sin(\theta' - \Phi) + \frac{de}{dt}\cos(\theta' + \Phi) + e\frac{d\Phi}{dt}\sin(\theta' - \Phi) \tag{11.13}$$

The film thickness can be described from Eqs. (10.5) and (11.1) as

$$h = c + e\cos(\theta' - \Phi) \tag{11.14}$$

Note that in Eq. (11.14) all terms of order e^2/r_a^2 have been neglected. Likewise, in Eq. (11.12) neglecting terms of order e^2/r_a^2 gives

$$u_b = \omega_b r_b + \frac{de}{dt}\sin(\theta' - \Phi) - e\frac{d\Phi}{dt}\cos(\theta' - \Phi) \tag{11.15}$$

Furthermore, neglecting the curvature of the film gives the x coordinate in the unwrapped film as

$$x = r_a \theta'$$

and thus

$$dx = r_a \, d\theta' \tag{11.16}$$

Substituting Eqs. (11.2), (11.4), (11.5), (11.13), (11.15), and (11.16) into Eq. (7.45) while assuming an incompressible fluid gives

$$\begin{aligned}
&\frac{1}{r_a^2} \frac{\partial}{\partial \theta'} \left(\frac{h^3}{12\eta} \frac{\partial p}{\partial \theta'} \right) + \frac{\partial}{\partial y} \left(\frac{h^3}{12\eta} \frac{\partial p}{\partial y} \right) \\
&\quad = \frac{1}{2r_a} \frac{\partial}{\partial \theta'} \left\{ h \left[\omega_b (r_a - c) + \frac{de}{dt} \sin(\theta' - \Phi) - e \frac{d\Phi}{dt} \cos(\theta' - \Phi) + \omega_a r_a \right] \right\} \\
&\quad - \omega_b e \sin(\theta' - \Phi) + \frac{de}{dt} \cos(\theta' - \Phi) + e \frac{d\Phi}{dt} \sin(\theta' - \Phi) \\
&\quad - \frac{1}{r_a} \frac{\partial h}{\partial \theta'} \left[\omega_b (r_a - c) + \frac{de}{dt} \sin(\theta' - \Phi) - e \frac{d\Phi}{dt} \cos(\theta' - \Phi) \right]
\end{aligned} \tag{11.17}$$

Retaining only first-order terms on the right side of this equation gives

$$\begin{aligned}
&\frac{1}{r_a^2} \frac{\partial}{\partial \theta'} \left(\frac{h^3}{12\eta} \frac{\partial p}{\partial \theta'} \right) + \frac{\partial}{\partial y} \left(\frac{h^3}{12\eta} \frac{\partial p}{\partial y} \right) \\
&\quad = \frac{1}{2} (\omega_a + \omega_b) \frac{\partial h}{\partial \theta'} + \frac{de}{dt} \cos(\theta' - \Phi) + e \frac{d\Phi}{dt} \sin(\theta' - \Phi)
\end{aligned} \tag{11.18a}$$

or

$$\begin{aligned}
&\frac{1}{r_a^2} \frac{\partial}{\partial \theta'} \left(\frac{h^3}{12\eta} \frac{\partial p}{\partial \theta'} \right) + \frac{\partial}{\partial y} \left(\frac{h^3}{12\eta} \frac{\partial p}{\partial y} \right) \\
&\quad = \left[\frac{1}{2} (\omega_a + \omega_b) - \omega \right] \frac{\partial h}{\partial \theta'} + \frac{de}{dt} \cos(\theta' - \Phi)
\end{aligned} \tag{11.18b}$$

where $\omega = d\Phi/dt$ = rotational velocity of journal about sleeve center when eccentricity ratio is constant.

Equation (11.18) is the governing equation for the pressure distribution in journal bearings when considering the time-dependent journal position. This equation reduces to the following important forms for particular situations.

11.1.1 STEADY-STATE CONDITIONS. For $de/dt = d\Phi/dt = 0$, Eq. (11.18) reduces to

$$\frac{1}{r_a^2}\frac{\partial}{\partial \theta'}\left(\frac{h^3}{12\eta}\frac{\partial p}{\partial \theta'}\right) + \frac{\partial}{\partial y}\left(\frac{h^3}{12\eta}\frac{\partial p}{\partial y}\right) = \frac{1}{2}(\omega_a + \omega_b)\frac{\partial h}{\partial \theta'} \tag{11.19}$$

This is the form of the Reynolds equation employed in the analysis of steadily loaded journal bearings that is covered in Chap. 10 [Eq. (10.1)] but with the side-leakage term neglected. By using the short-width-journal-bearing assumptions, Eq. (10.46) reduces from Eq. (11.19) when only one surface is moving. Note that Eq. (11.19) describes three different physical actions of the journal relative to the sleeve:

1. If the sleeve and journal rotate in the same direction, the Couette term increases and results in greater load-carrying capacity.
2. If ω_a and ω_b are in opposite directions, the load-carrying capacity is reduced.
3. If $\omega_b = -\omega_a$, implying that the speeds are equal but opposite in sign, the bearing will have no load-carrying capacity.

11.1.2 ABSENCE OF ROTATION. For $\omega_a = \omega_b = 0$ and

$$\omega = \frac{\partial \Phi}{\partial t} = 0 \tag{11.20}$$

Eq. (11.18) reduces to

$$\frac{\partial}{\partial \phi}\left(\frac{h^3}{\eta}\frac{\partial p}{\partial \phi}\right) + r_a^2\frac{\partial}{\partial y}\left(\frac{h^3}{\eta}\frac{\partial p}{\partial y}\right) = 12 r_a^2 \frac{\partial e}{\partial t}\cos(\theta' - \Phi) \tag{11.21}$$

This equation corresponds to the normal squeeze film for a journal bearing.

11.1.3 HALF-FREQUENCY WHIRL. If the journal center rotates about the sleeve center at one-half the shaft rotational speed while the eccentricity remains constant and the sleeve is stationary, half-frequency whirl occurs.

$$\therefore \frac{\partial e}{\partial t} = \omega_a = 0 \qquad \text{and} \qquad \omega = \frac{\partial \Phi}{\partial t} = \frac{\omega_b}{2} \tag{11.22}$$

When the preceding occurs, the right side of Eq. (11.18) reduces to zero, giving

$$\frac{\partial}{\partial \phi}\left(\frac{h^3}{\eta}\frac{\partial p}{\partial \phi}\right) + r_a^2\frac{\partial}{\partial y}\left(\frac{h^3}{\eta}\frac{\partial p}{\partial y}\right) = 0 \tag{11.23}$$

This implies a constant (zero) pressure throughout the bearing. If the shaft precesses about the bearing center at a rotational speed equal to one-half the shaft speed, the theoretical load-carrying capacity is zero and the phenomenon is known as "half-speed whirl."

11.2 FULL SOMMERFELD INFINITELY WIDE-JOURNAL-BEARING SOLUTION

Analytic solutions to Eq. (11.18) are possible for both the infinitely wide- and short-width-bearing assumptions discussed in Chap. 10. In this chapter only the dynamically loaded infinitely wide-journal-bearing solution will be presented. If the side-leakage term is neglected, Eq. (11.18) can be rewritten as

$$\frac{\partial}{\partial \phi}\left(\frac{h^3}{\eta}\frac{\partial p}{\partial \phi}\right) = 12 r_a^2\left[\left(\frac{\omega_a + \omega_b}{2} - \omega\right)\frac{\partial h}{\partial \theta'} + \frac{de}{dt}\cos(\theta' - \Phi)\right] \tag{11.24}$$

Integrating Eq. (11.24) while making use of Eqs. (11.14) and (11.1) gives

$$\frac{\partial p}{\partial \phi} = \frac{12\eta\left(\frac{r_a}{c}\right)^2\left[\frac{\partial \varepsilon}{\partial t}\sin\phi - \varepsilon\cos\phi\left(\omega - \frac{\omega_a + \omega_b}{2}\right) - \tilde{A}\right]}{(1 + \varepsilon\cos\phi)^3} \tag{11.25}$$

Integrating again while assuming that the viscosity does not vary in the circumferential direction results in

$$p = 12\eta_0\left(\frac{r_a}{c}\right)^2 \int \frac{\frac{\partial \varepsilon}{\partial t}\sin\phi - \varepsilon\cos\phi\left(\omega - \frac{\omega_a + \omega_b}{2}\right) - \tilde{A}}{(1 + \varepsilon\cos\phi)^3}\, d\phi + \text{constant} \tag{11.26}$$

From the Sommerfeld substitution presented in Chap. 10 the pressure can be written as

$$p = 12\eta_0\left(\frac{r_a}{c}\right)^2 \left\{ \frac{\partial\varepsilon/\partial t}{2\varepsilon(1 + \varepsilon\cos\phi)^2} - \left(\omega - \frac{\omega_a + \omega_b}{2}\right) \times \left[\frac{\gamma - \varepsilon\sin\gamma}{(1 - \varepsilon^2)^{3/2}} - \frac{\left(\frac{2 + \varepsilon^2}{2}\right)\gamma - 2\varepsilon\sin\gamma + \frac{\varepsilon^2}{4}\sin 2\gamma}{(1 - \varepsilon^2)^{5/2}}\right] - \frac{\tilde{A}\left[\frac{(2 + \varepsilon^2)}{2} - 2\varepsilon\sin\gamma + \frac{\varepsilon^2}{4}\sin 2\gamma\right]}{(1 - \varepsilon^2)^{5/2}} + \tilde{B} \right\} \tag{11.27}$$

The boundary condition for a full Sommerfeld solution assumes that a continuous oil film exists all around the bearing (i.e., no cavitation). Thus,

$p_\phi = p_{\phi+2\pi}$ at all points including $\phi = 0$. From Eq. (11.27)

$$(p)_{\substack{\phi=0\\ \gamma=0}} = 12\eta_0\left(\frac{r_a}{c}\right)^2\left[\frac{\partial\varepsilon/\partial t}{2\varepsilon(1+\varepsilon)^2} + \bar{B}\right] \tag{11.28}$$

$$(p)_{\substack{\phi=2\pi\\ \gamma=2\pi}} = 12\eta_0\left(\frac{r_a}{c}\right)^2\left[\frac{\partial\varepsilon/\partial t}{2\varepsilon(1+\varepsilon)^2} + \frac{3\pi\varepsilon^2\left(\omega - \dfrac{\omega_a+\omega_b}{2}\right)}{(1-\varepsilon^2)^{5/2}} - \frac{\pi(2+\varepsilon^2)\tilde{A}}{(1-\varepsilon^2)^{5/2}} + \tilde{B}\right] \tag{11.29}$$

Therefore, if $(p)_{\phi=0} = (p)_{\phi=2\pi} = p_0$,

$$\tilde{A} = \frac{3\varepsilon^2}{2+\varepsilon^2}\left(\omega - \frac{\omega_a+\omega_b}{2}\right) \tag{11.30}$$

$$\tilde{B} = \frac{p_0}{12\eta_0(r_a/c)^2} - \frac{\partial\varepsilon/\partial t}{2\varepsilon(1+\varepsilon)^2} \tag{11.31}$$

Thus, substituting equations (11.30) and (11.31) into (11.27) gives

$$p - p_0 = 6\eta_0\left(\frac{r_a}{c}\right)^2\left\{\frac{\partial\varepsilon/\partial t}{\varepsilon}\left[\frac{1}{(1+\varepsilon\cos\phi)^2} - \frac{1}{(1+\varepsilon)^2}\right] + [2\omega - (\omega_a+\omega_b)]\frac{\varepsilon\sin\gamma(\varepsilon\cos\gamma - 2 + \varepsilon^2)}{(2+\varepsilon^2)(1-\varepsilon^2)^{3/2}}\right\} \tag{11.32}$$

Converting from the Sommerfeld variable γ back to ϕ gives the pressure as

$$p - p_0 = 6\eta_0\left(\frac{r_a}{c}\right)^2\left\{\frac{\partial\varepsilon/\partial t}{\varepsilon}\left[\frac{1}{(1+\varepsilon\cos\phi)^2} - \frac{1}{(1+\varepsilon)^2}\right] + (\omega_a+\omega_b-2\omega)\frac{\varepsilon\sin\phi(2+\varepsilon\cos\phi)}{(2+\varepsilon^2)(1+\varepsilon\cos\phi)^2}\right\} \tag{11.33}$$

Note that if instead of the full Sommerfeld solution the half Sommerfeld solution is of interest, Eq. (11.33) is valid as long as $p > 0$. If $p < 0$, the pressure is set equal to zero ($p = 0$). Equation (11.33) enables the pressure in a dynamically loaded journal bearing to be determined when $d\varepsilon/dt$, ω_a, ω_b, and $\omega = d\Phi/dt$ are known at any eccentricity ratio. Equation (11.33) reduces to Eq. (10.15), the steady-state equation for a full Sommerfeld solution, if $\omega_a = \omega = 0$ and $\partial\varepsilon/\partial t = 0$.

The force components perpendicular and along the line of centers can be written from Eqs. (10.22) and (10.23). Substituting Eqs. (11.25) and (11.30) into these equations while making use of the Sommerfeld substitution covered in

Chap. 10 gives

$$w'_x = 12\pi\eta_0 r_a\left(\frac{r_a}{c}\right)^2 \frac{\varepsilon(\omega_a + \omega_b - 2\omega)}{(2+\varepsilon^2)(1-\varepsilon^2)^{1/2}} \tag{11.34}$$

$$w'_z = 12\pi\eta_0 r_a\left(\frac{r_a}{c}\right)^2 \frac{\partial\varepsilon/\partial t}{(1-\varepsilon^2)^{3/2}} \tag{11.35}$$

These equations reduce to the steady-state form for the full Sommerfeld solution given in Eqs. (10.26) and (10.27) when $\omega = \omega_a = \partial\varepsilon/\partial t = 0$.

From Fig. 11.1 the following can be expressed:

$$w'_z = \frac{w_r}{b}\sin\Phi_\ell$$

$$w'_x = \frac{w_r}{b}\cos\Phi_\ell$$

Consider the case of a rotor of mass $2m_a$ supported in two identical and well-alligned journal bearings. Using the infinitely wide-bearing approximation, the equation of motion for the journal may be written as

$$\begin{Bmatrix} m_a & 0 \\ 0 & m_a \end{Bmatrix} \frac{\partial^2}{\partial t^2}\begin{Bmatrix} e\cos\Phi \\ e\sin\Phi \end{Bmatrix} = \begin{Bmatrix} w_r\cos\Phi_\ell \\ w_r\sin\Phi_\ell \end{Bmatrix} - \begin{Bmatrix} w_x \\ w_z \end{Bmatrix} \tag{11.36}$$

where w_x and w_z are obtained from Eqs. (11.34) and (11.35). These equations are highly nonlinear even when w_r and Φ_ℓ are known functions of time. Explicit expressions for ε and Φ cannot be derived. The same problems arise when one considers the short-width-journal-bearing theory.

11.3 LINEARIZATION OF BEARING REACTION

A method used to deal with the highly nonlinear equations of motion described in the preceding section is to linearize the bearing reaction about a quasi-steady-state journal position. Figure 11.2 depicts the effect of changing load on the bearing's shaft position. The relevant parameters associated with these changes are also shown in this figure, where subscript 0 refers to the quasi-steady-state position and Δx and Δz are the displacement of the shaft away from this position.

The resultant reaction load presented in Fig. 11.2 has components w_x and w_z. Performing a first-order Taylor expansion of these components gives

$$w_x = (w_x)_0 + \left(\frac{\partial w_x}{\partial x}\right)_0 \Delta x + \left(\frac{\partial w_x}{\partial z}\right)_0 \Delta z + \left(\frac{\partial w_x}{\partial \dot{x}}\right)_0 \Delta\dot{x} + \left(\frac{\partial w_x}{\partial \dot{z}}\right)_0 \Delta\dot{z} \tag{11.37}$$

$$w_z = (w_z)_0 + \left(\frac{\partial w_z}{\partial x}\right)_0 \Delta x + \left(\frac{\partial w_z}{\partial z}\right)_0 \Delta z + \left(\frac{\partial w_z}{\partial \dot{x}}\right)_0 \Delta\dot{x} + \left(\frac{\partial w_z}{\partial \dot{z}}\right)_0 \Delta\dot{z} \tag{11.38}$$

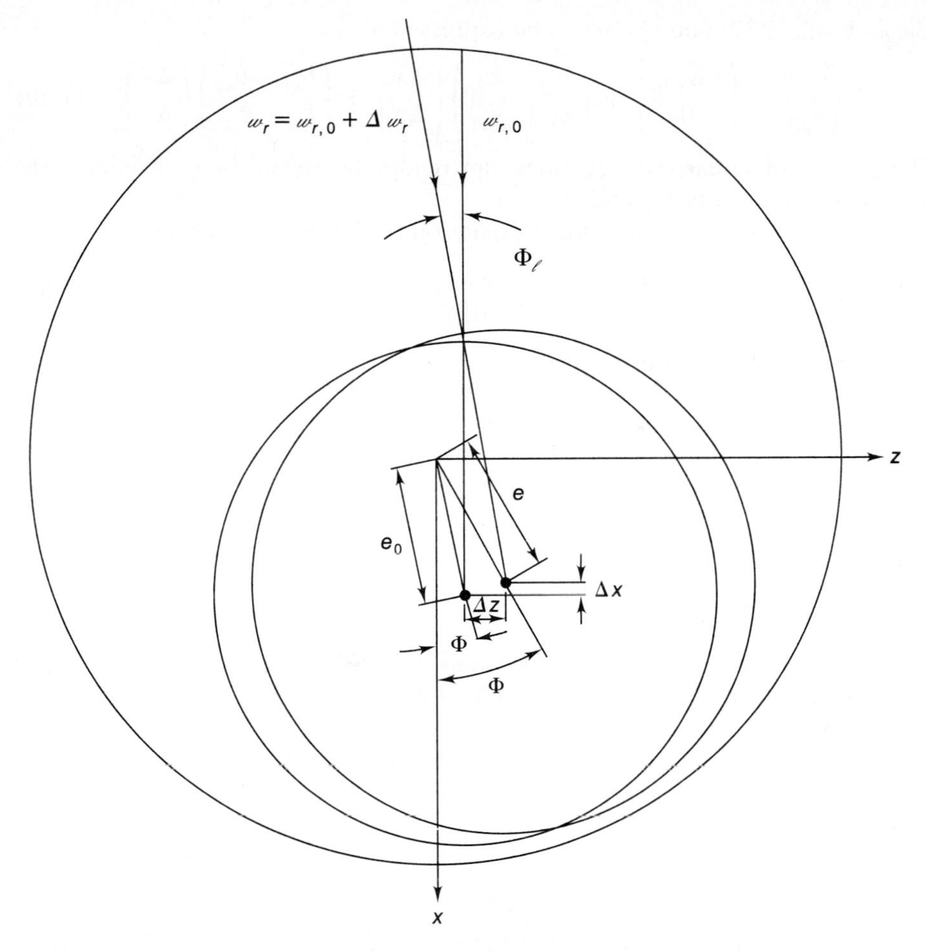

FIGURE 11.2
Effect of changing load on bearing's shaft position and relevant parameters associated with these changes.

where the dots indicate time derivatives. Recall that the direction of the x axis was chosen so that $(w_z)_0 = 0$. Letting

$$k_{xx} = \left(\frac{\partial w_x}{\partial x}\right)_0 \qquad k_{xz} = \left(\frac{\partial w_x}{\partial z}\right)_0 \qquad k_{zx} = \left(\frac{\partial w_z}{\partial x}\right)_0 \qquad k_{zz} = \left(\frac{\partial w_z}{\partial z}\right)_0$$

$$b_{xx} = \left(\frac{\partial w_x}{\partial \dot{x}}\right)_0 \qquad b_{xz} = \left(\frac{\partial w_x}{\partial \dot{z}}\right)_0 \qquad b_{zx} = \left(\frac{\partial w_z}{\partial \dot{x}}\right)_0 \qquad b_{zz} = \left(\frac{\partial w_z}{\partial \dot{z}}\right)_0 \tag{11.39}$$

allows Eqs. (11.37) and (11.38) to be expressed as

$$\begin{Bmatrix} w_x \\ w_z \end{Bmatrix} = \begin{Bmatrix} (w_x)_0 \\ 0 \end{Bmatrix} + \begin{pmatrix} k_{xx} & k_{xz} \\ k_{zx} & k_{zz} \end{pmatrix} \begin{Bmatrix} \Delta x \\ \Delta z \end{Bmatrix} + \begin{pmatrix} b_{xx} & b_{xz} \\ b_{zx} & b_{zz} \end{pmatrix} \begin{Bmatrix} \Delta \dot{x} \\ \Delta \dot{z} \end{Bmatrix} \tag{11.40}$$

These types of linearized equations can readily be solved by determining the linearized coefficients k_{ij} and b_{ij}.

Consider a similar first-order expansion of the pressure profile:

$$p = (p)_0 + \left(\frac{\partial p}{\partial x}\right)_0 \Delta x + \left(\frac{\partial p}{\partial z}\right)_0 \Delta z + \left(\frac{\partial p}{\partial \dot{x}}\right)_0 \Delta \dot{x} + \left(\frac{\partial p}{\partial \dot{z}}\right) \Delta \dot{z} \tag{11.41}$$

For simplicity of notation let

$$(p)_0 = p_0 \qquad \left(\frac{\partial p}{\partial x}\right)_0 = p_x \qquad \left(\frac{\partial p}{\partial z}\right)_0 = p_z \qquad \left(\frac{\partial p}{\partial \dot{x}}\right)_0 = p_{\dot{x}} \qquad \left(\frac{\partial p}{\partial \dot{z}}\right)_0 = p_{\dot{z}}$$

The bearing reaction components are found by integrating Eq. (11.41) over the bearing area or

$$\begin{Bmatrix} w_x \\ w_z \end{Bmatrix} = \int_y \int_{\theta'} \left(p_0 + p_x \,\Delta x + p_z \,\Delta z + p_{\dot{x}} \,\Delta \dot{x} + p_{\dot{z}} \Delta \dot{z} \right) \begin{Bmatrix} \cos\theta' \\ \sin\theta' \end{Bmatrix} r_a \, d\theta' \, dy \tag{11.42}$$

The perturbation terms Δx, Δz, $\Delta \dot{x}$, and $\Delta \dot{z}$ are independent of the integration variables, and thus from Eqs. (11.40) and (11.42)

$$\begin{Bmatrix} (w_x)_0 \\ 0 \end{Bmatrix} = \begin{Bmatrix} \displaystyle\int_y \int_{\theta'} p_0 \cos\theta' \, r_a \, dy \, d\theta' \\ \displaystyle\int_y \int_{\theta'} p_0 \sin\theta' \, r_a \, dy \, d\theta' \end{Bmatrix} \tag{11.43}$$

$$\begin{Bmatrix} k_{xx} & k_{xz} \\ k_{zx} & k_{zz} \end{Bmatrix} = \begin{Bmatrix} \displaystyle\int_y \int_{\theta'} p_x \cos\theta' \, r_a \, dy \, d\theta' & \displaystyle\int_y \int_{\theta'} p_z \cos\theta' \, r_a \, dy \, d\theta' \\ \displaystyle\int_y \int_{\theta'} p_x \sin\theta' \, r_a \, dy \, d\theta' & \displaystyle\int_y \int_{\theta'} p_z \sin\theta' \, r_a \, dy \, d\theta' \end{Bmatrix} \tag{11.44}$$

$$\begin{Bmatrix} b_{xx} & b_{xz} \\ b_{zx} & b_{zz} \end{Bmatrix} = \begin{Bmatrix} \displaystyle\int_y \int_{\theta'} p_{\dot{x}} \cos\theta' \, r_a \, dy \, d\theta' & \displaystyle\int_y \int_{\theta'} p_{\dot{z}} \cos\theta' \, r_a \, dy \, d\theta' \\ \displaystyle\int_y \int_{\theta'} p_{\dot{x}} \sin\theta' \, r_a \, dy \, d\theta' & \displaystyle\int_y \int_{\theta'} p_{\dot{z}} \sin\theta' \, r_a \, dy \, d\theta' \end{Bmatrix} \tag{11.45}$$

Equation (11.43) is the steady-state solution. From these equations we need to determine p_x, p_z, $p_{\dot{x}}$, and $p_{\dot{z}}$.

From Fig. 11.2 we get

$$e_0 \cos\Phi_0 + \Delta x = e \cos\Phi \tag{11.46}$$

$$e_0 \sin\Phi_0 + \Delta z = e \sin\Phi \tag{11.47}$$

Substituting Eqs. (11.46) and (11.47) into Eq. (11.14) gives

$$h = h_0 + \Delta x \cos\theta' + \Delta z \sin\theta' \tag{11.48}$$

where

$$h_0 = c + e_0 \cos(\theta' - \Phi_0) \tag{11.49}$$

Also from Eq. (11.14)

$$\frac{dh}{dt} = \frac{de}{dt}\cos(\theta' - \Phi) + e\frac{d\Phi}{dt}\sin(\theta' - \Phi) \tag{11.50}$$

Or directly from Eq. (11.48)

$$\frac{dh}{dt} = \Delta\dot{x}\cos\theta' + \Delta\dot{z}\sin\theta' \tag{11.51}$$

When no misalignment, $h \neq f(y)$, and an incompressible fluid are assumed and $\omega_a = 0$, Eq. (7.58) can be expressed as

$$\frac{1}{r_a^2}\frac{\partial}{\partial\theta'}\left(\frac{h^3}{12\eta}\frac{\partial p}{\partial\theta'}\right) + h^3\frac{\partial}{\partial y}\left(\frac{1}{12\eta}\frac{\partial p}{\partial y}\right) = \frac{\omega_b}{2}\frac{\partial h}{\partial\theta'} + \frac{\partial h}{\partial t} \tag{11.52}$$

This equation is Eq. (11.18a) when $\omega_a = 0$ and making use of Eq. (11.50). Making use of Eqs. (11.41) and (11.48) gives Eq. (11.52) as

$$\begin{aligned}
&\frac{1}{r_a^2}\frac{\partial}{\partial\theta'}\left[\frac{(h_0 + \Delta x\cos\theta' + \Delta z\sin\theta')^3}{12\eta}\right.\\
&\qquad\left.\times\frac{\partial}{\partial\theta'}\left(p_0 + p_x\,\Delta x + p_z\Delta z + p_{\dot{x}}\Delta\dot{x} + p_{\dot{z}}\Delta\dot{z}\right)\right]\\
&\qquad + (h_0 + \Delta x\cos\theta' + \Delta z\sin\theta')^3\\
&\qquad\times\frac{\partial}{\partial y}\left[\frac{1}{12\eta}\frac{\partial}{\partial y}\left(p_0 + p_x\Delta x + p_z\Delta z + p_{\dot{x}}\Delta\dot{x} + p_{\dot{z}}\Delta\dot{z}\right)\right]\\
&= \frac{\omega_b}{2}\frac{\partial}{\partial\theta'}(h_0 + \Delta x\cos\theta' + \Delta z\sin\theta') + \Delta\dot{x}\cos\theta' + \Delta\dot{z}\sin\theta'
\end{aligned} \tag{11.53}$$

Collecting terms of like order gives

$O(1)$:

$$\frac{1}{r_a^2}\frac{\partial}{\partial\theta'}\left(\frac{h_0^3}{12\eta}\frac{\partial p_0}{\partial\theta'}\right) + h_0^3\frac{\partial}{\partial y}\left(\frac{1}{12\eta}\frac{\partial p_0}{\partial y}\right) = \frac{\omega_b}{2}\frac{\partial h_0}{\partial\theta'} \tag{11.54}$$

$O(\Delta x)$:

$$\frac{1}{r_a^2}\frac{\partial}{\partial\theta'}\left(\frac{h_0^3}{12\eta}\frac{\partial p_x}{\partial\theta'}\right)+h_0^3\frac{\partial}{\partial y}\left(\frac{1}{12\eta}\frac{\partial p_x}{\partial y}\right)$$

$$= -\tfrac{1}{2}\omega_b\sin\theta' - \left[\frac{1}{r_a^2}\frac{\partial}{\partial\theta'}\left(\frac{3h_0^2}{12\eta}\cos\theta'\frac{\partial p_0}{\partial\theta'}\right)+3h_0^2\cos\theta'\frac{\partial}{\partial y}\left(\frac{1}{12\eta}\frac{\partial p_0}{\partial y}\right)\right] \tag{11.55}$$

$O(\Delta z)$:

$$\frac{1}{r_a^2}\frac{\partial}{\partial\theta'}\left(\frac{h_0^3}{12\eta}\frac{\partial p_z}{\partial\theta'}\right)+h_0^3\frac{\partial}{\partial y}\left(\frac{1}{12\eta}\frac{\partial p_z}{\partial y}\right)$$

$$= \tfrac{1}{2}\omega_b\cos\theta' - \left[\frac{1}{r_a^2}\frac{\partial}{\partial\theta'}\left(\frac{3h_0^2}{12\eta}\sin\theta'\frac{\partial p_0}{\partial\theta'}\right)+\frac{h_0^2}{4}\frac{\partial}{\partial y}\left(\frac{\sin\theta'}{\eta}\frac{\partial p_0}{\partial y}\right)\right] \tag{11.56}$$

$O(\Delta\dot{x})$:

$$\frac{1}{r_a^2}\frac{\partial}{\partial\theta'}\left(\frac{h_0^3}{12\eta}\frac{\partial p_{\dot{x}}}{\partial\theta'}\right)+h_0^3\frac{\partial}{\partial y}\left(\frac{1}{12\eta}\frac{\partial p_{\dot{x}}}{\partial y}\right)=\cos\theta' \tag{11.57}$$

$O(\Delta\dot{z})$:

$$\frac{1}{r_a^2}\frac{\partial}{\partial\theta'}\left(\frac{h_0^3}{12\eta}\frac{\partial p_{\dot{z}}}{\partial\theta'}\right)+h_0^3\frac{\partial}{\partial y}\left(\frac{1}{12\eta}\frac{\partial p_{\dot{z}}}{\partial y}\right)=\sin\theta' \tag{11.58}$$

Expanding the second and third terms on the right side of Eq. (11.56) gives

$$\frac{1}{r_a^2}\frac{\partial}{\partial\theta'}\left(\frac{3h_0^2}{12\eta}\sin\theta'\frac{\partial p_0}{\partial\theta}\right)+\frac{h_0^2}{4}\frac{\partial}{\partial y}\left(\frac{\sin\theta'}{\eta}\frac{\partial p_0}{\partial y}\right)$$

$$=\frac{3\cos\theta'}{h_0}\left[\frac{1}{r_a^2}\frac{\partial}{\partial\theta'}\left(\frac{h_0^3}{12\eta}\frac{\partial p_0}{\partial\theta'}\right)+h_0^3\frac{\partial}{\partial y}\left(\frac{1}{12\eta}\frac{\partial p_0}{\partial y}\right)\right]$$

$$+\frac{h_0^3}{12\eta}\left[\frac{3}{r_a^2}\frac{\partial p_0}{\partial\theta'}\frac{\partial}{\partial\theta'}\left(\frac{\cos\theta'}{h_0}\right)\right] \tag{11.59}$$

The first two terms on the right side of Eq. (11.59) are exactly the two terms of the left side of Eq. (11.54). Therefore, Eq. (11.59) becomes

$$\frac{1}{r_a^2}\frac{\partial}{\partial\theta'}\left(\frac{h_0^3}{12\eta}\frac{\partial p_x}{\partial\theta'}\right)+h_0^3\frac{\partial}{\partial y}\left(\frac{1}{12\eta}\frac{\partial p_x}{\partial y}\right)$$

$$=-\frac{\omega_b}{2}\left(\sin\theta'+\frac{3\cos\theta'}{h_0}\frac{\partial h_0}{\partial\theta'}\right)-\frac{h_0^3}{4\eta r_a^2}\frac{\partial p_0}{\partial\theta'}\frac{\partial}{\partial\theta'}\left(\frac{\cos\theta'}{h_0}\right) \tag{11.60}$$

Continuing with this approach, the following is true:

$$\left[\frac{1}{r_a^2}\frac{\partial}{\partial\theta'}\left(\frac{h_0^3}{12\eta_0}\frac{\partial}{\partial\theta'}\right)+h_0^3\frac{\partial}{\partial y}\left(\frac{1}{12\eta_0}\frac{\partial}{\partial y}\right)\right]\begin{Bmatrix} p_0 \\ p_x \\ p_z \\ p_{\dot{x}} \\ p_{\dot{z}} \end{Bmatrix}$$

$$= \begin{Bmatrix} \dfrac{\omega_b}{2}\dfrac{\partial h_0}{\partial\theta'} \\ -\dfrac{\omega_b}{2}\left(\sin\theta' + \dfrac{3\cos\theta'}{h_0}\dfrac{\partial h_0}{\partial\theta'} - \dfrac{h_0^3}{4\eta r_b^2}\dfrac{\partial p_0}{\partial\theta'}\dfrac{\partial}{\partial\theta'}\left(\dfrac{\cos\theta'}{h_0}\right)\right) \\ \dfrac{\omega_b}{2}\left(\cos\theta' - \dfrac{3\sin\theta'}{h_0}\dfrac{\partial h_0}{\partial\theta'} - \dfrac{h_0^3}{4\eta r_b^2}\dfrac{\partial p_0}{\partial\theta'}\dfrac{\partial}{\partial\theta'}\left(\dfrac{\sin\theta'}{h_0}\right)\right) \\ \cos\theta' \\ \sin\theta' \end{Bmatrix} \tag{11.61}$$

Therefore, once the steady-state pressure is obtained, it can be used in obtaining the perturbation pressure. Furthermore, once the perturbation pressures are known, the dynamic coefficients given in Eqs. (11.44) and (11.45) can be evaluated.

Taking the double derivative with respect to time in Eqs. (11.46) and (11.47) gives

$$\frac{\partial}{\partial t^2}\begin{Bmatrix} e\cos\Phi \\ e\sin\Phi \end{Bmatrix} = \begin{Bmatrix} \Delta\ddot{x} \\ \Delta\ddot{z} \end{Bmatrix}$$

Substituting this equation, as well as the linearized bearing reaction Eq. (11.40), into the equation of motion given in Eq. (11.36) while grouping all terms involving dynamic positioning on the left side of the equation gives

$$\begin{Bmatrix} m_a & 0 \\ 0 & m_a \end{Bmatrix}\begin{Bmatrix} \Delta\ddot{x} \\ \Delta\ddot{z} \end{Bmatrix} + \begin{Bmatrix} b_{xx} & b_{xz} \\ b_{zx} & b_{zz} \end{Bmatrix}\begin{Bmatrix} \Delta\dot{x} \\ \Delta\dot{z} \end{Bmatrix} + \begin{Bmatrix} k_{xx} & k_{xz} \\ k_{zx} & k_{zz} \end{Bmatrix}\begin{Bmatrix} \Delta x \\ \Delta z \end{Bmatrix}$$

$$= \begin{Bmatrix} w_r\cos\Phi_\ell \\ w_r\sin\Phi_\ell \end{Bmatrix} - \begin{Bmatrix} (w_x)_0 \\ 0 \end{Bmatrix} \tag{11.62}$$

This is a linear differential equation for which a variety of methods can be used to obtain the dynamic positioning for known values of w_r and Φ_ℓ.

The equation of motion given in (11.62) and resulting from the linearization of the bearing reaction is a good approximation for small displacements from the initial position given by ε_0 and Φ_0. For very large displacements the nonlinear effects become more dominant, but for many practical purposes solving Eq. (11.62) will be a good approximation. The eight dynamic coefficients are only functions of the bearing's operating parameters, which are characterized by the static equilibrium eccentricity ratio and the attitude angle. Thus, linearizing the bearing reaction forces has the obvious advantage of decoupling the rotor and the bearing. In an exact analysis where the bearing forces are nonlinear, the rotor equations must be integrated simultaneously with the lubrication equations, but making the forces linear allows the bearing to be solved without any regard for the particular rotor.

11.4 JOURNAL BEARING STABILITY

An important application of the linearization technique presented in Sec. 11.3 is the stability analysis for journal bearings operating at steady-state conditions.

Consider again a shaft of mass $2m_a$ supported by two identical well-aligned journal bearings. Let the load be stationary, $w_r = w_{x,0}$ and $\Phi_\ell = 0$; then the linearized equation of journal motion (11.62) reduces to

$$\begin{Bmatrix} m_a & 0 \\ 0 & m_a \end{Bmatrix} \begin{Bmatrix} \Delta\ddot{x} \\ \Delta\ddot{z} \end{Bmatrix} + \begin{Bmatrix} b_{xx} & b_{xz} \\ b_{zx} & b_{zz} \end{Bmatrix} \begin{Bmatrix} \Delta\dot{x} \\ \Delta\dot{z} \end{Bmatrix} + \begin{Bmatrix} k_{xx} & k_{xz} \\ k_{zx} & k_{zz} \end{Bmatrix} \begin{Bmatrix} \Delta x \\ \Delta z \end{Bmatrix} = \begin{Bmatrix} 0 \\ 0 \end{Bmatrix} \tag{11.63}$$

The stiffness and damping coefficients can be made dimensionless by the following equations:

$$\begin{pmatrix} K_{xx} & K_{xz} \\ K_{zx} & K_{zz} \end{pmatrix} = \frac{c}{w_r} \begin{pmatrix} k_{xx} & k_{xz} \\ k_{zx} & k_{zz} \end{pmatrix} \tag{11.64}$$

$$\begin{pmatrix} B_{xx} & B_{xz} \\ B_{zx} & B_{zz} \end{pmatrix} = \frac{c\omega_b}{w_r} \begin{pmatrix} b_{xx} & b_{xz} \\ b_{zx} & b_{zz} \end{pmatrix} \tag{11.65}$$

Substituting Eqs. (11.64) and (11.65) into equation (11.63) gives

$$\begin{Bmatrix} m_a & 0 \\ 0 & m_a \end{Bmatrix} \begin{Bmatrix} \Delta\ddot{x} \\ \Delta\ddot{z} \end{Bmatrix} + \frac{w_r}{c\omega_b} \begin{Bmatrix} B_{xx} & B_{xz} \\ B_{zx} & B_{zz} \end{Bmatrix} \begin{Bmatrix} \Delta\dot{x} \\ \Delta\dot{z} \end{Bmatrix}$$

$$+ \frac{w_r}{c} \begin{Bmatrix} K_{xx} & K_{xz} \\ K_{zx} & K_{zz} \end{Bmatrix} \begin{Bmatrix} \Delta x \\ \Delta z \end{Bmatrix} = \begin{Bmatrix} 0 \\ 0 \end{Bmatrix} \tag{11.66}$$

Equation (11.66) has the obvious solution $\Delta x = \Delta z = \Delta\dot{x} = \Delta\dot{z} = \Delta\ddot{x} = \Delta\ddot{z} = 0$ that agrees with the steady-state conditions. We will solve Eq. (11.66) and show that under certain circumstances a nontrivial solution exists.

The solution to Eq. (11.66) is of the type

$$\begin{Bmatrix} \Delta x \\ \Delta z \end{Bmatrix} = \begin{Bmatrix} x_h \\ z_h \end{Bmatrix} \exp(\overline{\Omega} t \omega_b) \tag{11.67}$$

Substituting Eq. (11.67) into equation (11.66) gives

$$\begin{Bmatrix} M_a + \overline{\Omega} B_{xx} + K_{xx} & \overline{\Omega} B_{xz} + K_{xz} \\ \overline{\Omega} B_{zx} + K_{zx} & M_a + \overline{\Omega} B_{zz} + K_{zz} \end{Bmatrix} \begin{Bmatrix} x_h \\ z_h \end{Bmatrix} \exp(\overline{\Omega} t \omega_b) = \begin{Bmatrix} 0 \\ 0 \end{Bmatrix} \quad (11.68)$$

where

$$M_a = \frac{c m_a \Omega^2}{w_r} \quad \text{and} \quad \overline{\Omega} = \frac{\Omega}{\omega_b} \quad (11.69)$$

Thus either $x_h = z_h = 0$, which is the trivial steady-state solution, or

$$\left(M_a + \overline{\Omega} B_{xx} + K_{xx}\right)\left(M_a + \overline{\Omega} B_{zz} + K_{zz}\right) - \left(\overline{\Omega} B_{zx} + K_{zx}\right)\left(\overline{\Omega} B_{xz} + K_{xz}\right) = 0 \quad (11.70)$$

The solution given in Eq. (11.70) is an eigenvalue problem stating that if the system should dislodge itself from the steady-state position, a transient vibration will result although the external load is constant.

The eigenvalue $\overline{\Omega}$ will generally be complex such that

$$\overline{\Omega} = -\overline{\Omega}_d + i\overline{\Omega}_v \quad (11.71)$$

and the transient shaft motion will be

$$\begin{Bmatrix} \Delta x \\ \Delta z \end{Bmatrix} = \begin{Bmatrix} x_h \\ z_h \end{Bmatrix} \exp(-\overline{\Omega}_d t \omega_b)\left[\cos(\overline{\Omega}_v t \omega_b) + i \sin(\overline{\Omega}_v t \omega_b)\right] \quad (11.72)$$

where x_h and z_h are also complex. The left side of Eq. (11.72) is real so that only the real part of the right side will be necessary to describe the physical motion. Note that when $\overline{\Omega}_d > 0$, the shaft displacement away from the equilibrium position will decrease continuously, since x_h and z_h are constants, and finally reach the steady-state position. Such a behavior is obviously stable in that deviations from the steady-state conditions disappear due to the damping action of the hydrodynamic film.

When $\overline{\Omega}_d < 0$, the shaft vibration will increase and will be restricted only by the bearing sleeve. Bearing instability occurs under these conditions, and the transition from stable ($\overline{\Omega}_d > 0$) to unstable ($\overline{\Omega}_d < 0$) condition is called the "threshold of instability," which occurs at $\overline{\Omega}_d = 0$. Steady-state condition is not possible when $\overline{\Omega}_d < 0$.

Substituting Eq. (11.71) into Eq. (11.70) results in the two equations given below, the first being for the real part and the second for the imaginary part:

$$\begin{aligned} M_a^2 &- M_a\overline{\Omega}_d(B_{zz} + B_{xx}) + M_a(K_{zz} + K_{xx}) + \overline{\Omega}_d^2(B_{xx}B_{zz} - B_{zx}B_{xz}) \\ &+ \overline{\Omega}_d(-B_{xx}K_{zz} - B_{zz}K_{xx} + B_{zx}K_{xz} + B_{xz}K_{zx}) \\ &+ \overline{\Omega}_v^2(B_{zx}B_{xz} - B_{xx}B_{zz}) + K_{xx}K_{zz} - K_{zx}K_{xz} = 0 \end{aligned} \quad (11.73)$$

and

$$M_a(B_{xx} + B_{zz}) + 2\overline{\Omega}_d(B_{xz}B_{zx} - B_{xx}B_{zz}) \\ + B_{xx}K_{zz} + K_{xx}B_{zz} - B_{zx}K_{xz} - B_{xz}K_{zx} = 0 \tag{11.74}$$

At the threshold of instability, $\overline{\Omega}_d = 0$ and Eq. (11.74) produces

$$(M_a)_{cr} = \left(\frac{cm_a\omega_b^2}{w_r}\right)(\overline{\Omega}_v)_{cr}^2 = \frac{B_{xx}K_{zz} + K_{xx}B_{zz} - B_{zx}K_{xz} - B_{xz}K_{zx}}{B_{xx} + B_{zz}} \tag{11.75}$$

Making use of Eq. (11.73) at the threshold of instability ($\overline{\Omega}_d = 0$) gives

$$(\overline{\Omega}_v)_{cr}^2 = \frac{[K_{xx} - (M_a)_{cr}][K_{zz} - (M_a)_{cr}] - K_{xz}K_{zx}}{B_{xx}B_{zz} - B_{xz}B_{zx}} \tag{11.76}$$

If M_a is smaller than $(M_a)_{cr}$, the system will be stable ($\overline{\Omega}_d > 0$), but it will be unstable for M_a larger than $(M_a)_{cr}$. Thus, whether the bearing is susceptible to instability obviously depends on the values of the bearing coefficients, which in turn depend on the bearing type and the various performance parameters of these bearings.

Throughout this chapter only plain journal bearings are discussed, but the linearization technique is applicable for any bearing configuration consisting of circular segments. This restriction must be imposed because Eq. (11.14) is used to describe the film thickness.

11.5 LINEARIZATION OF BEARING REACTION APPLIED TO SHORT-WIDTH JOURNAL BEARINGS

The short-width-journal-bearing theory developed in Sec. 10.2 will be used in defining the bearing reaction. Recall from Chap. 10 that short-width-journal-bearing theory is valid for diameter-to-width ratios greater than 2 ($\lambda_k = 2r_b/b > 2$). For this theory, Eq. (11.61) reduces to

$$h_0^3 \frac{\partial}{\partial y}\left(\frac{1}{12\eta}\frac{\partial}{\partial y}\right)\begin{pmatrix} p_0 \\ p_x \\ p_z \\ p_{\dot{x}} \\ p_{\dot{z}} \end{pmatrix} = \begin{Bmatrix} \dfrac{\omega_b}{2}\dfrac{\partial h_0}{\partial \theta'} \\ -\dfrac{\omega_b}{2}\left(\sin\theta' + \dfrac{3\cos\theta'}{h_0}\dfrac{\partial h_0}{\partial \theta'}\right) \\ \dfrac{\omega_b}{2}\left(\cos\theta' - \dfrac{3\sin\theta'}{h_0}\dfrac{\partial h_0}{\partial \theta'}\right) \\ \cos\theta' \\ \sin\theta' \end{Bmatrix} \tag{11.77}$$

Note that the right side of Eq. (11.77) is independent of the perturbed pressure p_0, which was not the situation in Eq. (11.61).

No misalignment is assumed to occur in the bearing, so the pressure profile is symmetric about the center plane or

$$\left(\frac{\partial p}{\partial y}\right)_{y=0} = 0 \tag{11.78}$$

Also

$$(p)_{y=b/2} = 0 \tag{11.79}$$

Reexpressing the preceding boundary condition in terms of Eq. (11.41) gives

$$\frac{\partial p_0}{\partial y} = \frac{\partial p_x}{\partial y} = \frac{\partial p_z}{\partial y} = \frac{\partial p_{\dot{x}}}{\partial y} + \frac{\partial p_{\dot{z}}}{\partial y} = 0 \qquad \text{at } y = 0 \tag{11.80}$$

$$p_0 = p_x = p_z = p_{\dot{x}} = p_{\dot{z}} = 0 \qquad \text{at } y = \frac{b}{2} \tag{11.81}$$

Substituting these expressions into Eq. (11.77) gives

$$\begin{Bmatrix} p_0 \\ \\ p_x \\ \\ p_z \\ \\ p_{\dot{x}} \\ p_{\dot{z}} \end{Bmatrix} = \frac{6\eta_0}{h_0^3}\left(y^2 - \frac{b^2}{4}\right) \begin{Bmatrix} \dfrac{\omega_b}{2}\dfrac{\partial h_0}{\partial \theta'} \\ -\dfrac{\omega_b}{2}\left(\sin\theta' + \dfrac{3\cos\theta'}{h_0}\dfrac{\partial h_0}{\partial \theta'}\right) \\ \dfrac{\omega_b}{2}\left(\cos\theta' - \dfrac{3\sin\theta'}{h_0}\dfrac{\partial h_0}{\partial \theta'}\right) \\ \cos\theta' \\ \sin\theta' \end{Bmatrix} \tag{11.82}$$

Recall from Chap. 10 for a short-width-journal bearing that using the half Sommerfeld boundary condition gives

$$\frac{1}{W_r \lambda_k^2} = \frac{\left(1 - \varepsilon_0^2\right)^2}{\varepsilon_0\left[16\varepsilon_0^2 + \pi^2\left(1 - \varepsilon_0^2\right)\right]^{1/2}} \tag{10.61}$$

$$\tan\Phi_0 = \frac{\pi\left(1 - \varepsilon_0^2\right)^{1/2}}{4\varepsilon_0} \tag{10.58}$$

where $\lambda_k = 2r_b/b$. From Eq. (11.49)

$$\frac{\partial h_0}{\partial \theta'} = -e_0 \sin(\theta' - \Phi_0) \tag{11.83}$$

From Eqs. (11.82), (11.49), and (11.83)

$$p_x = \left(\frac{\partial p}{\partial x}\right)_0$$

$$= -\frac{3\eta_0\omega_b}{c^3}\frac{y^2 - b^2/4}{\left[1 + \varepsilon_0\cos(\theta' - \Phi_0)\right]^3}\left[\sin\theta' - \frac{3\varepsilon_0\cos\theta'\sin(\theta' - \Phi_0)}{1 + \varepsilon_0\cos(\theta' - \Phi_0)}\right] \tag{11.84}$$

In a similar manner p_z, $p_{\dot{x}}$, and $p_{\dot{z}}$ can be obtained.

The half Sommerfeld boundary condition covered in Chap. 10 assumes that the pressure is zero when the film thickness of the bearing is divergent. The dynamic stiffness coefficient k_{xx} can be given by using Eq. (11.44) as

$$k_{xx} = 2\int_0^{b/2}\int_{\Phi_0}^{\pi+\Phi_0}\frac{3\eta_0\omega_b}{c^3}\left(y^2 - \frac{b^2}{4}\right)\frac{\left[\dfrac{3\varepsilon_0\cos\theta'\sin(\theta' - \Phi_0)}{1 + \varepsilon_0\cos(\theta' - \Phi_0)} - \sin\theta'\right]}{\left[1 + \varepsilon_0\cos(\theta' - \Phi_0)\right]^3}\cos\theta'\, r_b\, d\theta'\, dy$$

Going back to the ϕ coordinate, using Eq. (11.1), and integrating with respect to dy while using Eq. (11.64) to express in dimensionless form gives

$$K_{xx} = \frac{1}{W_r}\left(\frac{b}{r_b}\right)^2\int_0^{\pi}\left[\frac{3\varepsilon_0\left(\cos\phi\cos\Phi_0 - \sin\phi\sin\Phi_0\right)^2\sin\phi}{\left(1 + \varepsilon_0\cos\phi\right)^4}\right.$$

$$\left. - \frac{\left(\sin\phi\cos\Phi_0 + \cos\phi\sin\Phi_0\right)\left(\cos\phi\cos\Phi_0 - \sin\phi\sin\Phi_0\right)}{\left(1 + \varepsilon_0\cos\phi\right)^3}\right]d\phi$$

or

$$K_{xx} = \frac{4}{W_r\lambda_k^2}\left[\frac{\varepsilon_0\sin^2\Phi_0}{\left(1 - \varepsilon_0^2\right)^2} + \frac{3\pi\varepsilon_0^2\sin\Phi_0\cos\Phi_0}{4\left(1 - \varepsilon_0^2\right)^{5/2}} + \frac{2\varepsilon_0\left(1 + \varepsilon_0^2\right)\cos^2\Phi_0}{\left(1 - \varepsilon_0^2\right)^3}\right] \tag{11.85}$$

In a similar manner making use of Eqs. (11.44), (11.45), and (11.82) gives

$$K_{xz} = \frac{4}{W_r\lambda_k^2}\left[\frac{\pi\left(1 + 2\varepsilon_0^2\right)}{4\left(1 - \varepsilon_0^2\right)^{5/2}}\sin^2\Phi_0 + \frac{\varepsilon_0\left(1 + 3\varepsilon_0^2\right)}{\left(1 - \varepsilon_0^2\right)^3}\sin\Phi_0\cos\Phi_0 + \frac{\pi\cos^2\Phi_0}{4\left(1 - \varepsilon_0^2\right)^{3/2}}\right] \tag{11.86}$$

$$K_{zx} = \frac{4}{W_r\lambda_k^2}\left[-\frac{\pi\sin^2\Phi_0}{4\left(1 - \varepsilon_0^2\right)^{3/2}} + \frac{\varepsilon_0\left(1 + 3\varepsilon_0^2\right)}{\left(1 - \varepsilon_0^2\right)}\sin\Phi_0\cos\Phi_0 - \frac{\pi\left(1 + 2\varepsilon_0^2\right)\cos^2\Phi_0}{4\left(1 - \varepsilon_0^2\right)^{5/2}}\right] \tag{11.87}$$

$$K_{zz} = \frac{4}{W_r \lambda_k^2}\left[\frac{2\varepsilon_0\left(1+\varepsilon_0^2\right)}{\left(1-\varepsilon_0^2\right)^3}\sin^2\Phi_0 - \frac{3\pi\varepsilon_0^2 \sin\Phi_0 \cos\Phi_0}{4\left(1-\varepsilon_0^2\right)^{5/2}} + \frac{\varepsilon_0 \cos^2\Phi_0}{\left(1-\varepsilon_0^2\right)^2}\right] \tag{11.88}$$

$$B_{xx} = \frac{4}{W_r \lambda_k^2}\left[\frac{\pi \sin^2\Phi_0}{2\left(1-\varepsilon_0^2\right)^{3/2}} + \frac{4\varepsilon_0 \sin\Phi_0 \cos\Phi_0}{\left(1-\varepsilon_0^2\right)^2} + \frac{\pi\left(1+2\varepsilon_0^2\right)\cos^2\Phi_0}{2\left(1-\varepsilon_0^2\right)^{5/2}}\right] \tag{11.89}$$

$$B_{zx} = B_{xz} = \frac{4}{W_r \lambda_k^2}\left[\frac{-4\varepsilon_0 \sin^2\Phi_0}{\left(1-\varepsilon_0^2\right)^2} + \frac{3\pi\varepsilon_0^2 \sin\Phi_0 \cos\Phi_0}{2\left(1-\varepsilon_0^2\right)^{5/2}} - \frac{\varepsilon_0 \cos^2\Phi_0}{2\left(1-\varepsilon_0^2\right)^2}\right] \tag{11.90}$$

$$B_{zz} = \frac{4}{W_r \lambda_k^2}\left[\frac{\pi\left(1+2\varepsilon_0^2\right)\sin^2\Phi_0}{2\left(1-\varepsilon_0^2\right)^{5/2}} - \frac{4\varepsilon_0 \sin\Phi_0 \cos\Phi_0}{\left(1-\varepsilon_0^2\right)^2} + \frac{\pi \cos^2\Phi_0}{2\left(1-\varepsilon_0^2\right)^{3/2}}\right] \tag{11.91}$$

The procedure one might follow in establishing if a bearing is stable is the following:

1. Make sure your journal bearing is a short-width journal bearing, implying that $2r_b/b = \lambda_k > 2$.
2. For a fixed steady-state eccentricity ratio, calculate the steady-state attitude angle and the dimensionless resultant load from Eqs. (10.58) and (10.61), respectively.
3. Calculate the dimensionless damping coefficients B_{xx}, B_{xz}, B_{zx}, and B_{zz} and the dimensionless stiffness coefficients K_{xx}, K_{xz}, K_{zx}, and K_{zz} by using Eqs. (11.85) to (11.91).
4. Calculate $(M_a)_{cr}$ and $(\overline{\Omega}_v)_{cr}$ from Eqs. (11.75) and (11.76), respectively. For stable operation make sure that M_a does not exceed $(M_a)_{cr}$.

These calculations are given in Table 11.1 for a range of steady-state eccentricity ratios.

Figure 11.3 gives a stability map for a short-width journal bearing. From this figure it is observed that short-width-journal-bearing theory predicts that plain journal bearings are inherently stable when operating above a static eccentricity ratio of about 0.756. This is quite a large eccentricity to be operating under, and stability is therefore an important consideration.

TABLE 11.1
Steady-state and dynamic parameters for plain short-width journal bearing
[From Lund (1966)]

Steady-state parameters			...ers										
Eccentricity ratio	Dimensionless load parameter	Attitude angle, deg	Dimensionless critical mass parameter	Dimensionless speed parameter	Dimensionless stiffness coefficients				Dimensionless damping coefficients			
ε_0	$\frac{1}{W_r}\left(\frac{2\pi}{\lambda_k}\right)^2$	Φ_0	$(M_a)_{cr}$	$(\bar{\Omega}_r)_{cr}$	K_{xx}	K_{xz}	K_{zx}	K_{zz}	B_{xx}	B_{xz}	B_{zx}	B_{zz}
.02	15.90	88.54	7.634	.5001	1.275	50.08	−49.95	2.546	100.1	2.546	2.546	99.96
.1	3.110	82.71	7.515	.5029	1.328	10.38	−9.758	2.531	20.49	2.532	2.532	19.79
.2	1.449	75.43	7.190	.5106	1.498	5.763	−4.520	2.485	10.98	2.492	2.492	9.587
.3	.8551	68.18	6.790	.5194	1.795	4.482	−2.625	2.413	8.153	2.427	2.427	6.061
.4	.5355	60.94	6.475	.5238	2.249	4.048	−1.571	2.319	7.022	2.343	2.343	4.216
.5	.3332	53.68	6.460	.5146	2.923	3.977	−.8577	2.210	6.615	2.245	2.245	3.054
.6	.1964	46.32	7.295	.4740	3.951	4.138	−.3071	2.092	6.651	2.138	2.138	2.239
.7	.1036	38.70	13.16	.3446	5.659	4.535	.1734	1.970	7.099	2.027	2.027	1.624
.8	.04362	30.50			9.042	5.326	.6739	1.848	8.177	1.915	1.915	1.128
.9	.01041	20.83			19.10	7.264	1.421	1.729	11.00	1.806	1.806	.6869

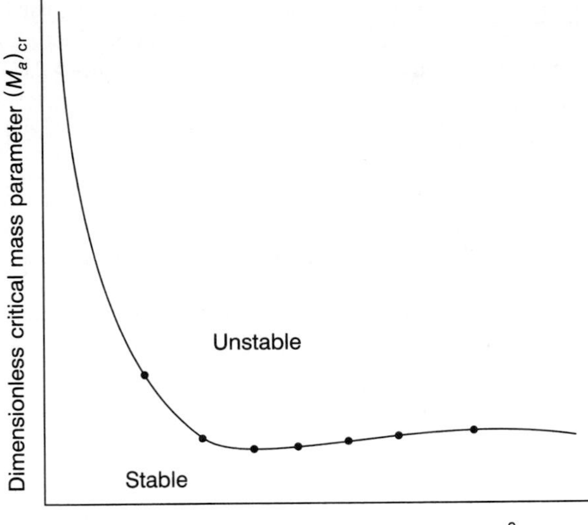

FIGURE 11.3
Stability map for short-width plain journal bearings.

11.6 CLOSURE

Equations describing the journal bearing shaft motion are highly nonlinear. Linearizing the bearing reaction enables the shaft motion to be approximated, but only for small displacements. The linear model is adequate for establishing a stability criterion, since unstable whirl starts from steady-state situations where $\Delta x = \Delta z = 0$. The stability analysis only allows a certain amount of mass to be supported in the journal bearing at given operating conditions. In practice, W_r^{-1} is the parameter to be monitored but is elusive, since $(M_a)_{\mathrm{cr}}$ depends on the same parameters as W_r^{-1}. A fast method of integrating shaft motion and still acknowledging nonlinearity effects consists of updating the dimensionless dynamic coefficients (K's and B's) at each time step.

11.7 PROBLEMS

11.7.1 Determine the exact equation of motion for short-width-journal-bearing theory. Determine the load components w_x and w_z during one cycle of

$$x = x_0 + \Delta x \cos \omega t$$

$$z = z_0 + \Delta z \sin \omega t$$

given ω, Δx, and Δz. Compare results with the linearized bearing reaction derived in the text.

11.7.2 Derive the dynamic coefficients (K and B) for an infinitely wide journal bearing. Assume that $p_0 = 0$ and accept the half Sommerfeld assumption. Compare $(M_a)_{cr}$ with short-width-journal-bearing theory when $b/2r_b = 0.5$, 1.0, and 2.0.

11.8 REFERENCES

Lund, J. W. (1966): "Self-Excited, Stationary Whirl Orbits of a Journal in a Sleeve Bearing." Ph.D. thesis. Rensselaer Polytechnic Institute, Troy, N.Y.

Lund, J. W. (1979): "Rotor-Bearing Dynamics." Lecture notes. Technical University of Denmark, ISBN 83-04-00267-1.

Lund, J. W. (1987): Review of the Concept of Dynamic Coefficients for Fluid Film Journal Bearings. *ASME Trans., J. Tribology*, vol. 109, no. 1, pp. 37–41.

CHAPTER 12

HYDRODYNAMIC JOURNAL BEARINGS— NUMERICAL SOLUTIONS

The preceding two chapters and this chapter focus on hydrodynamically lubricated journal bearings. Chapter 10 dealt with solutions that could be obtained analytically. These included an infinitely wide journal bearing [applicable for diameter-to-width ratios less than $\frac{1}{2}$ ($\lambda_k = 2r_b/b < \frac{1}{2}$)] and short-width journal bearings [applicable for diameter-to-width ratios greater than 2 ($\lambda_k = 2r_b/b > 2$)]. In Chap. 11 the same analytical solutions were used, but dynamic loading effects were considered, whereas in Chap. 10 steady loading was assumed. With dynamically loaded journal bearings the eccentricity and the attitude angle vary throughout the loading cycle. The present chapter utilizes numerical solutions in obtaining results for the complete range of diameter-to-width ratios. Steady loading conditions are considered throughout most of the chapter, but in the latter part of the chapter dynamic loading effects are considered.

12.1 OPERATING AND PERFORMANCE PARAMETERS

From Eq. (7.48) the Reynolds equation appropriate when considering the finite journal bearing can be expressed as

$$\frac{\partial}{\partial x}\left(h^3\frac{\partial p}{\partial x}\right) + \frac{\partial}{\partial y}\left(h^3\frac{\partial p}{\partial y}\right) = 12\tilde{u}\eta_0\frac{\partial h}{\partial x} \tag{7.48}$$

Now for a journal bearing $x = r_b\phi$ and $\tilde{u} = u_b/2 = r_b\omega_b/2$.

$$\therefore \frac{\partial}{\partial \phi}\left(h^3 \frac{\partial p}{\partial \phi}\right) + r_b^2 \frac{\partial}{\partial y}\left(h^3 \frac{\partial p}{\partial y}\right) = 6\eta_0\omega_b r_b^2 \frac{\partial h}{\partial \phi} \tag{12.1}$$

In Chap. 10 the film thickness around the journal is expressed as

$$h = c(1 + \varepsilon \cos \phi) \tag{10.5}$$

Therefore, Eq. (12.1) can be expressed as

$$\frac{\partial}{\partial \phi}\left(h^3 \frac{\partial p}{\partial \phi}\right) + r_b^2 h^3 \frac{\partial^2 p}{\partial y^2} = -6\eta_0\omega_b r_b^2 e \sin \phi \tag{12.2}$$

Analytical solutions to Eq. (12.2) are not normally available, and numerical methods are needed. Equation (12.2) is often solved by using a relaxation method. In the relaxation process the first step is to replace the derivatives in Eq. (12.2) by finite difference approximations. The lubrication area is covered by a mesh, and the numerical method relies on the fact that a function can be represented with sufficient accuracy over a small range by a quadratic expression. The Reynolds boundary condition covered in Sec. 10.1.3 is used. Only the results from using this numerical method are presented in this chapter.

The three dimensionless groupings normally used to define the operating parameters in journal bearings are

1. The eccentricity ratio $\varepsilon = e/c$
2. The angular extent of the journal (full or partial)
3. The diameter-to-width ratio $\lambda_k = 2r_b/b$

Recall from Chap. 10 that when the side-leakage term was neglected in Eq. (12.2), λ_k did not exist in the formulation, whereas for the short-width-journal-bearing theory all three parameters occurred although the region of applicability was somewhat limited. The results presented in this chapter are valid for the complete range of operating parameters.

This chapter focuses on the following performance parameters:

1. Dimensionless load $W_r = w_r/[\eta_0\omega_b b r_b (r_b/c)^2]$
2. Location of minimum film thickness, sometimes referred to as "attitude angle," Φ
3. Friction coefficient μ
4. Total and side flow q and q_s
5. Angle of maximum pressure ϕ_m

6. Location of terminating pressure ϕ_0
7. Temperature rise due to lubricant shearing, Δt_m

The parameters Φ, ϕ_m, and ϕ_0 are described in Fig. 12.1, which gives the pressure distribution around a journal bearing. Note from this figure that if the bearing is concentric ($e = 0$), the film shape around the journal is constant and equal to c and no fluid film pressure is developed. At the other extreme, at heavy loads the journal is forced downward and the limiting position is reached when $h_{\min} = 0$ and $e = c$; that is, the journal is touching the bearing.

Temperature rise due to lubricant shearing will be considered in this chapter as was done in Chap. 8 for thrust bearings. In Eq. (12.2) the viscosity of the lubricant corresponds to the viscosity when $p = 0$ but can vary as a function of temperature. Since work is done on the lubricant as the fluid is being sheared, the temperature of the lubricant is higher when it leaves the conjunction than on entry. In Chap. 4 (Figs. 4.5 and 4.6) it was shown that the viscosity of oils drops off significantly with rising temperature. This is compensated for by using a mean of the inlet and outlet temperatures.

$$t_m = t_i + \frac{\Delta t_m}{2} \tag{12.3}$$

where t_i = inlet temperature
Δt_m = temperature rise of lubricant from inlet to outlet

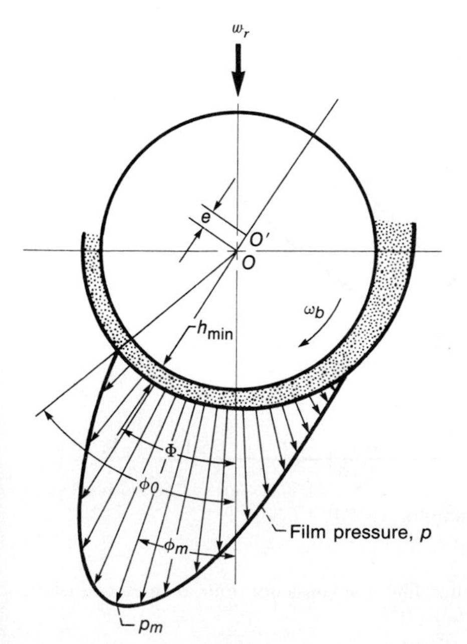

FIGURE 12.1
Pressure distribution around a journal bearing.

The viscosity used in the dimensionless load parameter W_r and other performance parameters is the mean temperature t_m. The temperature rise of the lubricant from inlet to outlet, Δt_m, can be determined from the performance charts provided in this chapter.

12.2 PERFORMANCE PARAMETER RESULTS

The performance parameters having been defined, the results will be presented as a function of the operating parameters discussed earlier. The results presented are for a full journal bearing. Results for a partial journal bearing can be obtained from Raimondi and Boyd (1958).

Figure 12.2 shows the effect of the dimensionless load parameter $(\pi W_r)^{-1}$ on the minimum film thickness h_{min}, where $h_{min} = c - e$, for λ_k of 0, 1, 2, and 4. Recall that the load is made dimensionless in the following manner, $W_r = w_r/[\eta_0 \omega_b b r_b (r_b/c)^2]$. In this figure a recommended operating eccentricity ratio, or minimum film thickness, is indicated as well as a preferred operating area. The left boundary of the shaded zone defines the optimum eccentricity ratio for a minimum friction coefficient; and the right boundary, the optimum eccentricity ratio for maximum load. The recommended operating eccentricity is midway between these two boundaries.

Figure 12.3 shows the effect of the dimensionless load parameter on the attitude angle Φ (angle between the load direction and a line drawn through the centers of the bearing and journal (see Fig. 12.1)) for various λ_k. This angle

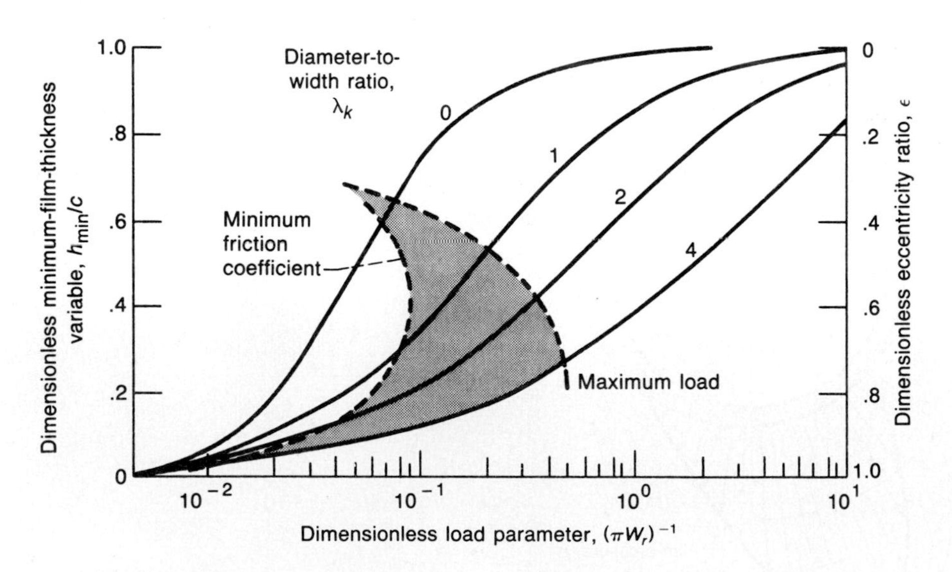

FIGURE 12.2
Effect of dimensionless load parameter on minimum film thickness for four diameter-to-width ratios. [*From Raimondi and Boyd (1958).*]

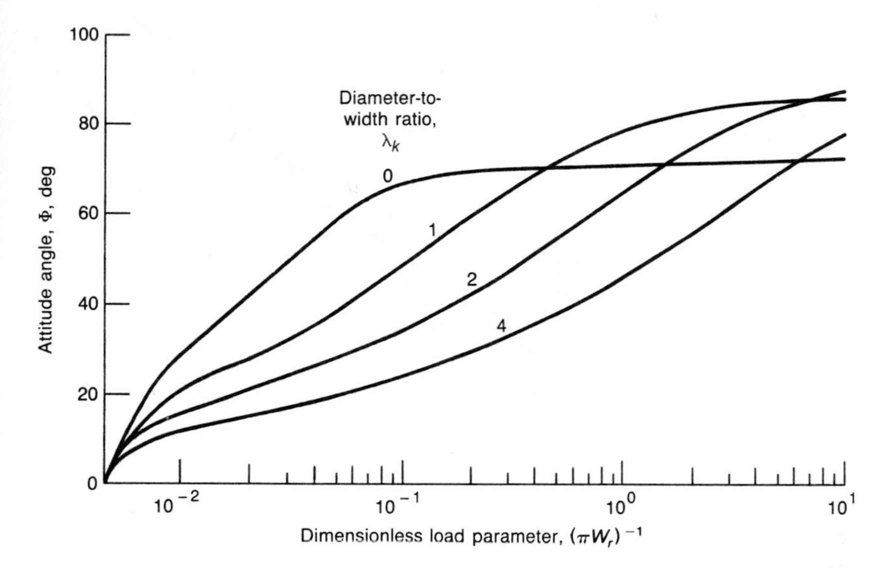

FIGURE 12.3
Effect of dimensionless load parameter on attitude angle for four diameter-to-width ratios. [*From Raimondi and Boyd (1958).*]

establishes where the minimum and maximum film thicknesses are located within the bearing.

Figure 12.4 shows the effect of the dimensionless load parameter on the friction coefficient for four values of λ_k. The effect is small for a complete range of dimensionless load parameters.

Figure 12.5 shows the effect of the dimensionless load parameter on the dimensionless volume flow rate $Q = 2\pi q/r_b c b \omega_b$ for four values of λ_k. The dimensionless volume flow rate Q that is pumped into the converging space by the rotating journal can be obtained from this figure. Of the volume of oil q pumped by the rotating journal, an amount q_s flows out the ends and hence is called "side-leakage flow." This side leakage can be computed from the volume flow ratio q_s/q of Fig. 12.6.

Figure 12.7 illustrates the maximum pressure developed in the journal bearing. In this figure the maximum pressure is made dimensionless with the load per unit area. The maximum pressure as well as its location are shown in Fig. 12.1. Figure 12.8 shows the effect of the dimensionless load parameter on the location of the terminating and maximum pressures for four values of λ_k.

The temperature rise in degrees Celsius of the lubricant from the inlet to the outlet can be obtained from Shigley and Mitchell (1983) as

$$\Delta t_m = \frac{8.3 W_r^* (r_b/c)\mu}{Q(1 - 0.5 q_s/q)} \tag{12.4}$$

where $W_r^* = w_r/2r_b b$ is in megapascals. Therefore, the temperature rise can be

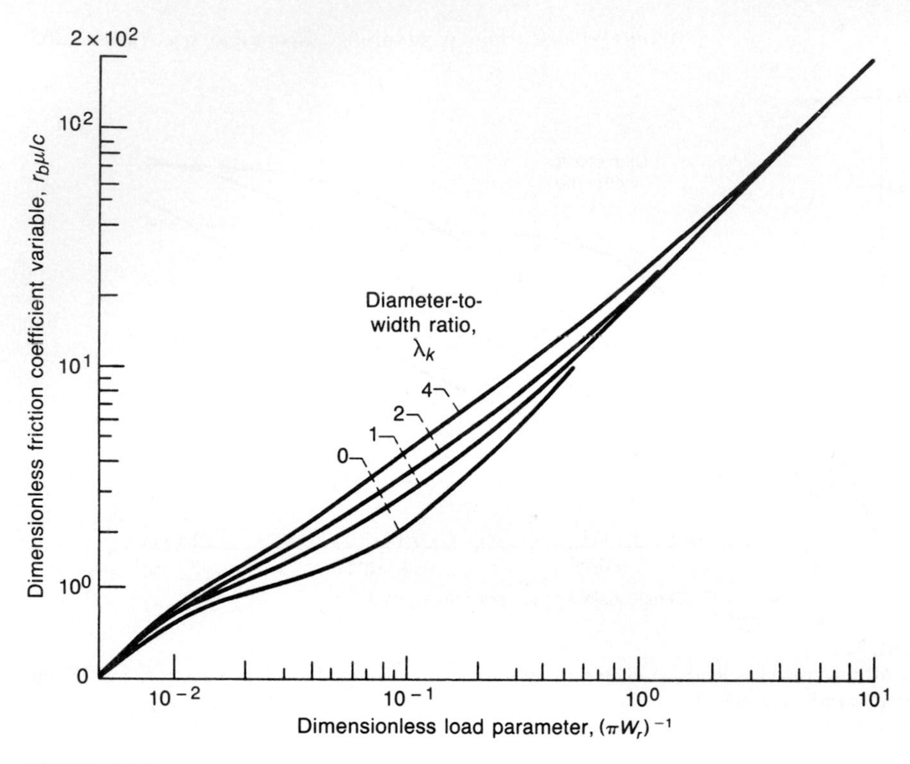

FIGURE 12.4
Effect of dimensionless load parameter on friction coefficient for four diameter-to-width ratios. [*From Raimondi and Boyd (1958).*]

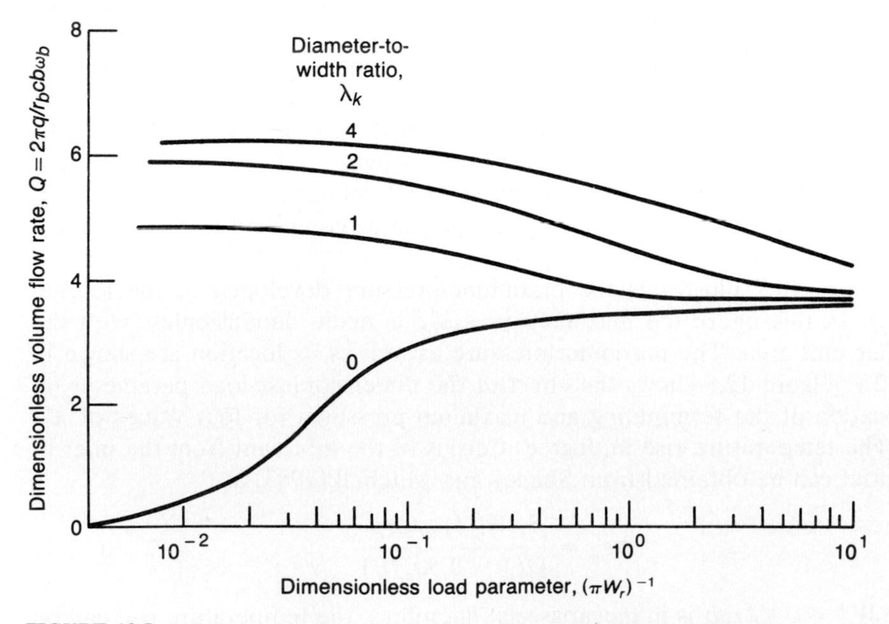

FIGURE 12.5
Effect of dimensionless load parameter on dimensionless volume flow rate for four diameter-to-width ratios. [*From Raimondi and Boyd (1958).*]

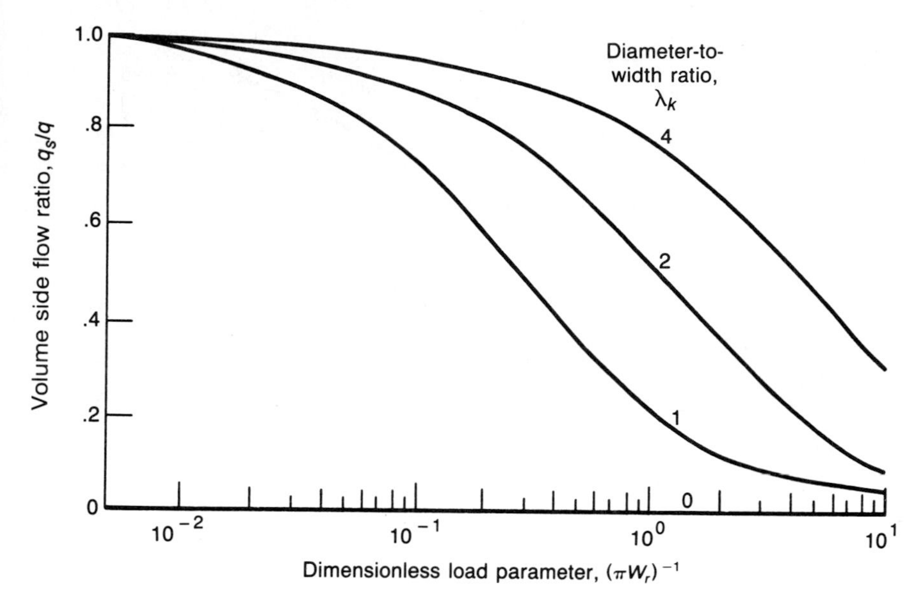

FIGURE 12.6
Effect of dimensionless load parameter on volume side flow ratio for four diameter-to-width ratios. [*From Raimondi and Boyd (1958).*]

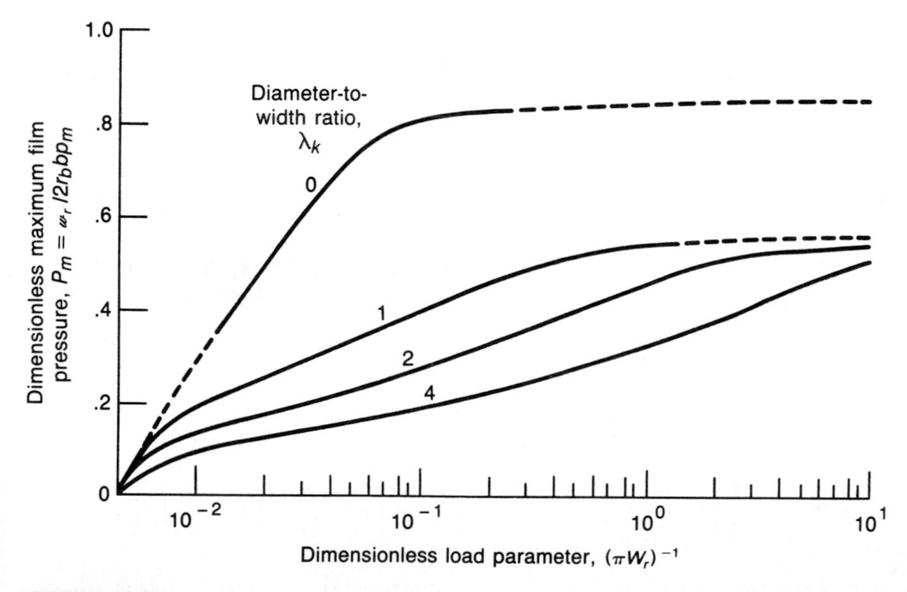

FIGURE 12.7
Effect of dimensionless load parameter on dimensionless maximum film pressure for four diameter-to-width ratios. [*From Raimondi and Boyd (1958).*]

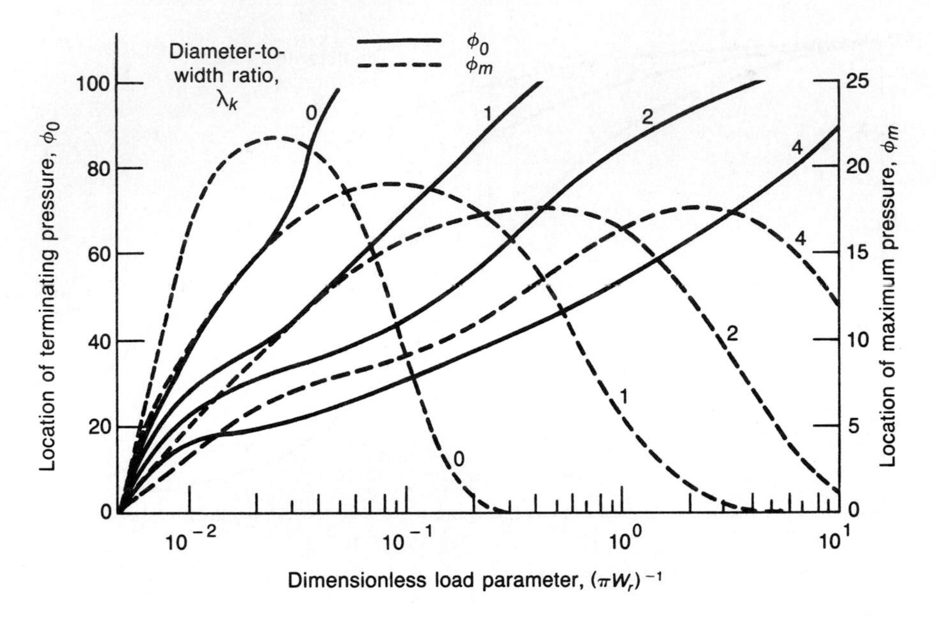

FIGURE 12.8
Effect of dimensionless load parameter on location of terminating and maximum pressures for four diameter-to-width ratios. [*From Raimondi and Boyd (1958).*]

directly obtained by substituting the values of $(r_b/c)\mu$ obtained from Fig. 12.4, Q from Fig. 12.5, and q_s/q from Fig. 12.6 into Eq. (12.4). If the value of Δt_m differs from that initially assumed, recalculate the mean temperature in the conjunction from Eq. (12.3) and obtain the new viscosity from Fig. 4.5 or 4.6. Once the viscosity is known, the dimensionless load can be calculated and then the performance parameters can be obtained from Figs. 12.2 to 12.8 and Eq. (12.4). Continue this process until there is little change in Δt_m from one iteration to the next.

The results presented thus far have been for λ_k of 0, 1, 2, and 4. If λ_k is some other value, use the following formula for establishing the performance parameter:

$$y = \frac{1}{(b/2r_b)^3}\left[-\frac{1}{8}\left(1-\frac{b}{2r_b}\right)\left(1-\frac{b}{r_b}\right)\left(1-2\frac{b}{r_b}\right)y_0 + \frac{1}{3}\left(1-\frac{b}{r_b}\right)\left(1-\frac{2b}{r_b}\right)y_1 - \frac{1}{4}\left(1-\frac{b}{2r_b}\right)\left(1-\frac{2b}{r_b}\right)y_2 + \frac{1}{24}\left(1-\frac{b}{2r_b}\right)\left(1-\frac{b}{r_b}\right)y_4\right] \tag{12.5}$$

where y is any one of the performance parameters (W_r, Φ, $r_b\mu/c$, Q, q_s/q, p_m, ϕ_0, or ϕ_m) and where the subscript on y is the λ_k value; for example, y_1 is equivalent to y evaluated at $\lambda_k = 1$. All the results presented are valid for a full journal bearing. If a partial journal bearing (180° or 120° bearing) is desired, use

Raimondi and Boyd (1958). The same procedure developed for the full journal bearing should be used for the partial journal bearing.

Sample problem 12.1. Given a full journal bearing with the specifications of $\eta_0 = 4\mu\text{reyn}$, $N_a = 30$ r/s, $w_r = 2220$ N, $r_b = 2$ cm, $c = 40$ μm, and $b = 4$ cm, use the figures given within this chapter to establish the operating and performance parameters for this bearing.

Solution $\eta_0 = 4\mu\text{reyn} = (4 \times 10^{-6})(6.9)(10^3)\ \text{Ns/m}^2 = 0.0276\ \text{Ns/m}^2$ (see Table 4.6)

$$\lambda_k = \frac{2r_b}{b} = \frac{2(2)}{4} = 1$$

The dimensionless load parameter is

$$W_r = \frac{w_r}{\eta_0 \omega_b b r_b (r_b/c)^2} = \frac{(2.2)(10^3)}{(0.0276)(60\pi)(0.04)(0.02)(2 \times 10^{-2}/4 \times 10^{-5})^2}$$

$$= 2.358$$

$$\therefore (\pi W_r)^{-1} = 0.151$$

For $\lambda_k = 1$ and $(\pi W_r)^{-1} = 0.151$, we get the following:

(*a*) From Fig. 12.2 we get that

$$\varepsilon = e/c = 0.55 \qquad \text{or} \qquad \frac{h_{\min}}{c} = 0.45$$

(*b*) From Fig. 12.3 we get that the attitude angle is 54°.

(*c*) From Fig. 12.4 we get

$$\frac{r_b \mu}{c} = 3.8$$

or

$$\mu = 7.6 \times 10^{-3}$$

(*d*) From Fig. 12.5 the dimensionless volume flow rate

$$Q = \frac{2\pi q}{b r_b c \omega_b} = 4.5$$

or

$$q = \frac{4.3 b r_b c \omega_b}{2\pi} = \frac{(4.5)(0.04)(0.02)(4 \times 10^{-5})(60\pi)}{2\pi}$$

$$= 4.32 \times 10^{-6}\ \text{m}^3/\text{s}$$

(*e*) From Fig. 12.6

$$\frac{q_s}{q} = 0.67$$

$$q_s = 2.89 \times 10^{-6}\ \text{m}^3/\text{s}$$

(f) From Fig. 12.7

$$P_m = \frac{w_r}{2r_b b p_{\mathrm{m}}} = 0.43$$

$$p_{\mathrm{m}} = \frac{w_r}{2(0.43)r_b b} = \frac{2{,}200}{2(0.43)(0.02)(0.04)}$$
$$= 3.2 \text{ MPa}$$

(g) From Fig. 12.8 we get

$$\phi_0 = 77.5^\circ \qquad \phi_m = 18.7^\circ$$

(h) The temperature rise due to the shearing of the lubricant is

$$\Delta t_m = 0.83 \frac{w_r}{2r_b b} \frac{(r_b/c)\mu}{Q}$$

But
$$\frac{w_r}{2r_b b} = \frac{(2.2)(10^3)}{2(2)(10^{-2})(4)(10^{-2})} = 1.375 \text{ MPa}$$

Recall that in calculating the temperature rise $w_r/2r_b b$ must be expressed in megapascals.

$$\Delta t_m = 8.3(1.375)\left(\frac{3.8}{4.5}\right) = 9.6\ ^\circ\mathrm{C}$$

12.3 OPTIMIZATION TECHNIQUES

The most difficult of the parameters in the operating conditions to control is the radial clearance c. The radial clearance is difficult to control accurately during manufacturing, and it may increase because of wear. Figure 12.9 shows the performance of a particular bearing calculated for a range of radial clearances and plotted with radial clearance as the independent variable. If the clearance is too tight, the temperature will be too high and the minimum film thickness too low. High temperature may cause the bearing to fail by fatigue. If the oil film is too thin, dirt particles may not pass without scoring or may embed themselves in the bearing. In either event there will be excessive wear and friction, resulting in high temperatures and possible seizing. A large clearance will permit dirt particles to pass through and also permit a large flow of oil. This lowers the temperature and lengthens bearing life. However, if the clearance becomes too large, the bearing becomes noisy and the minimum film thickness begins to decrease again.

Figure 12.9 shows the best compromise, when both the production tolerance and the future wear on the bearing are considered, to be a clearance range slightly to the left of the top of the minimum-film-thickness curve. In this way, future wear will move the operating point to the right, increasing the film thickness and decreasing the operating temperature.

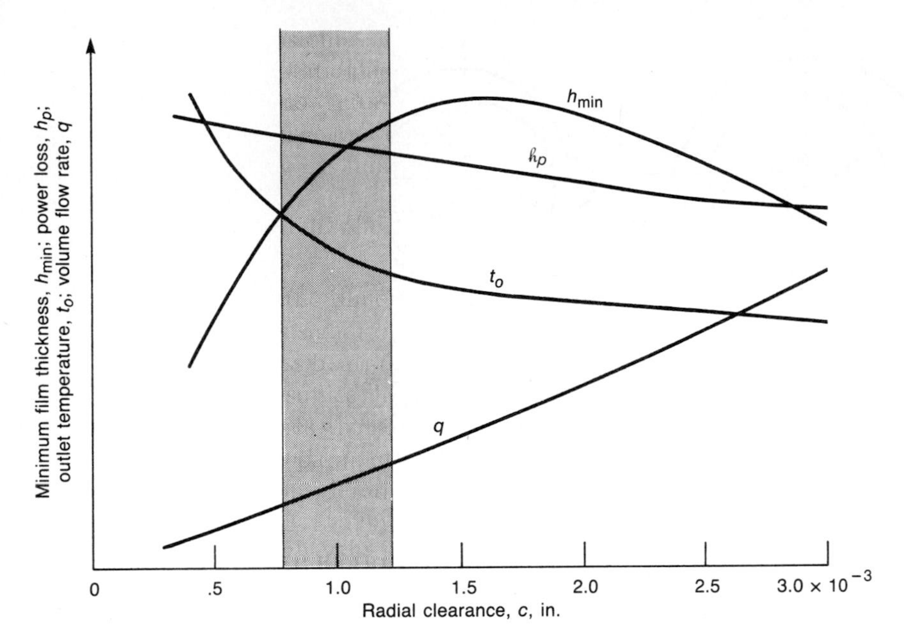

FIGURE 12.9
Effect of radial clearance on some performance parameters for a particular case.

12.4 NONPLAIN CONFIGURATIONS

Thus far this chapter has focused on plain full journal bearings. As applications have demanded higher speeds, vibration problems due to critical speeds, imbalance, and instability have created a need for journal bearing geometries other than plain journal bearings. These geometries have various patterns of variable clearance so as to create pad film thicknesses that have more strongly converging and diverging regions. Figure 12.10 shows elliptical, offset-half, three-lobe, and four-lobe bearings—bearings different from the plain journal bearing. An excellent discussion of the performance of these bearings is provided in Allaire and Flack (1980), and some of their conclusions are presented here. In Fig. 12.10 each pad is moved toward the bearing center some fraction of the pad clearance in order to make the fluid film thickness more converging and diverging than that occurring in a plain journal bearing. The pad center of curvature is indicated by a cross. Generally, these bearings suppress instabilities in the system well but can be subject to subsynchronous vibration at high speeds. They are not always manufactured accurately.

A key parameter used in describing these bearings is the fraction of length in which the film thickness is converging to the full pad length, called the "offset factor" and defined as

$$\alpha_a = \frac{\text{length of pad with converging film thickness}}{\text{full pad length}}$$

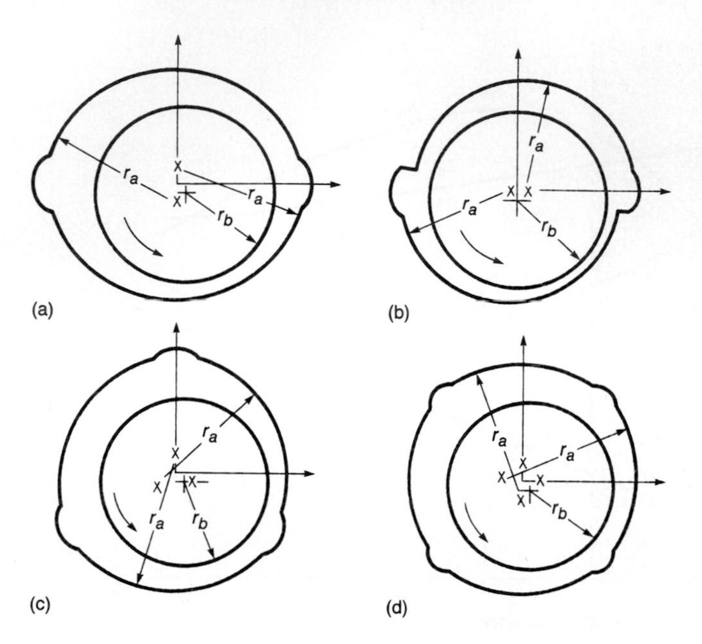

FIGURE 12.10
Types of fixed-incline-pad preloaded journal bearing and their offset factors α_a. Preload factor m_p, 0.4. (a) Elliptical bore bearing ($\alpha_a = 0.5$); (b) offset-half bearing ($\alpha_a = 1.125$); (c) three-lobe bearing ($\alpha_a = 0.5$); (d) four-lobe bearing ($\alpha_a = 0.5$). [*From Allaire and Flack (1980).*]

In an elliptical bearing [Fig. 12.10(a)] the two pad centers of curvature are moved along the vertical axis. This creates a pad with one-half the film shape converging and the other half diverging (if the shaft were centered), corresponding to an offset factor α_a of 0.5. The offset-half bearing [Fig. 12.10(b)] is a two-axial-groove bearing that is split by moving the top half horizontally. This results in low vertical stiffness. Generally, the vibration characteristics of this bearing are such as to avoid oil whirl, which can drive a machine unstable. The offset-half bearing has a purcly converging film thickness with a converged pad arc length of 160° and the point opposite the center of curvature at 180°. Both the three-lobe and four-lobe bearings (Figs. 12.10(c) and (d)] have an α_a of 0.5.

The fractional reduction of the film clearance when the pads are brought in is called the "preload factor" m_p. Let the bearing clearance at the pad minimum film thickness (with the shaft center) be denoted by c_b. Figure 12.11(a) shows that the largest shaft that can be placed in the bearing has a radius $r_b + c_b$, thereby establishing the definition of c_b. The preload factor m_p is given by

$$m_p = \frac{c - c_b}{c}$$

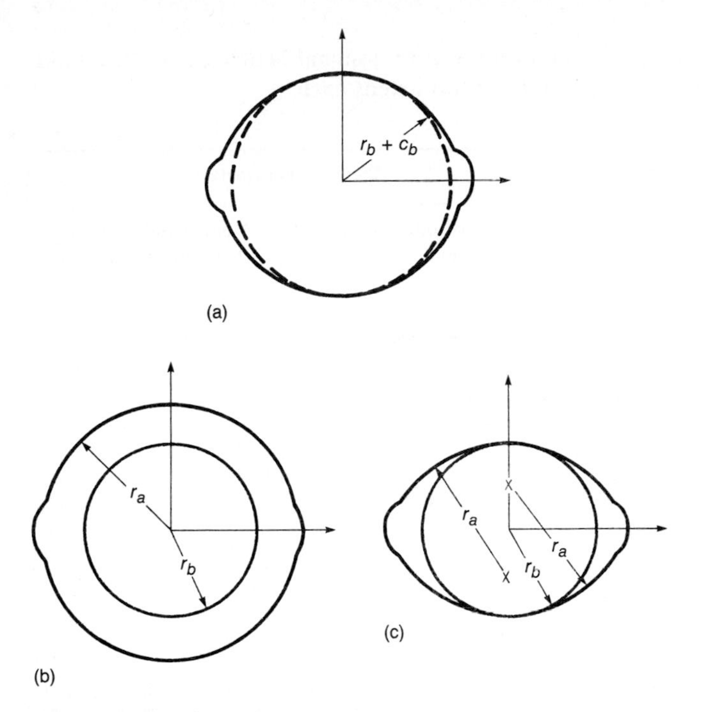

FIGURE 12.11
Effect of preload factor m_p on two-lobe bearings. (a) Largest shaft that fits in bearing. (b) $m_p = 0$; largest shaft, r_a; bearing clearance $c_b = c$. (c) $m_p = 1.0$; largest shaft, r_b; bearing clearance $c_b = 0$. [*From Allaire and Flack (1980).*]

A preload factor of zero corresponds to all the pad centers of curvature coinciding at the bearing center; a preload factor of 1.0 corresponds to all the pads touching the shaft. Figures 12.11(b) and (c) illustrate these extreme situations. For the various types of fixed journal bearing shown in Fig. 12.10 the preload factor is 0.4.

The nonplain configurations covered thus far in this chapter can be evaluated when considering dynamic loading, much like the procedure given in Chap. 11. Because of the complex film geometry and lubrication film grooves, the pressure distributions must be determined numerically before the dynamic coefficients and the load-carrying capacity, flow, friction power loss, and temperature can be calculated. Lund (1979) considered the plain, elliptical, cylindrical, offset, and three-lobe bearings discussed earlier. The results from the numerical evaluation of the dynamic load effects for these four bearing types are given in Tables 12.1 to 12.4. Each table has two parts for diameter-to-width ratios λ_k of 2 and 1. The preload factor was set equal to 0.5 ($m_p = 0.5$) for these tables. The dynamic coefficients, bearing number, eccentricity ratio, and attitude angle are also given in each of these tables.

TABLE 12.1

Steady-state and dynamic parameters for a plain journal bearing with two axial grooves 180° apart and of 20° circumferential extent each

[From Lund (1979)]

Steady-state parameters			Dynamic parameters						
Eccentricity ratio	Dimensionless load parameter	Attitude angle, deg	Dimensionless stiffness coefficients				Dimensionless damping coefficients		
ε_0	$\frac{1}{W_r}\left(\frac{2\pi}{\lambda_k}\right)^2$	Φ_0	K_{xx}	K_{xz}	K_{zx}	K_{zz}	B_{xx}	$B_{xz} = B_{zx}$	B_{zz}
(a) Diameter-to-width ratio is 2 (λ_k = 2.0)									
0.071	6.430	81.89	1.55	14.41	−6.60	1.88	28.75	1.89	13.31
0.114	3.037	77.32	1.57	9.27	−4.20	1.89	18.44	1.93	8.58
0.165	2.634	72.36	1.61	6.74	−3.01	1.91	13.36	2.00	6.28
0.207	2.030	68.75	1.65	5.67	−2.50	1.93	11.18	2.07	5.33
0.244	1.656	65.85	1.69	5.06	−2.20	1.95	9.93	2.15	4.80
0.372	0.917	57.45	2.12	4.01	−1.30	1.85	7.70	2.06	3.23
0.477	0.580	51.01	2.67	3.70	−0.78	1.75	6.96	1.94	2.40
0.570	0.375	45.43	3.33	3.64	−0.43	1.68	6.76	1.87	1.89
0.655	0.244	40.25	4.21	3.74	−0.13	1.64	6.87	1.82	1.54
0.695	0.194	37.72	4.78	3.84	0.01	1.62	7.03	1.80	1.40
0.734	0.151	35.29	5.48	3.98	0.15	1.61	7.26	1.79	1.27
0.753	0.133	33.93	5.89	4.07	0.22	1.60	7.41	1.79	1.20
0.751	0.126	33.42	6.07	4.11	0.25	1.60	7.48	1.79	1.18
0.772	0.116	32.65	6.36	4.17	0.30	1.60	7.59	1.79	1.15
0.809	0.086	30.04	7.51	4.42	0.47	1.59	8.03	1.79	1.03
0.879	0.042	24.41	11.45	5.23	0.92	1.60	9.48	1.80	0.82
(b) Diameter-to-width ratio is 1 (λ_k = 1.0)									
0.103	1.470	75.99	1.53	10.14	−3.01	1.50	20.34	1.53	6.15
0.150	0.991	70.58	1.56	7.29	−2.16	1.52	14.66	1.58	4.49
0.224	0.635	63.54	1.62	5.33	−1.57	1.56	10.80	1.70	3.41
0.352	0.358	55.41	1.95	3.94	−0.97	1.48	8.02	1.63	2.37
0.460	0.235	49.27	2.19	3.57	−0.80	1.55	7.36	1.89	2.19
0.559	0.159	44.33	2.73	3.36	−0.48	1.48	6.94	1.78	1.74
0.650	0.108	39.72	3.45	3.34	−0.23	1.44	6.89	1.72	1.43
0.734	0.071	35.16	4.49	3.50	0.03	1.44	7.15	1.70	1.20
0.773	0.056	32.82	5.23	3.65	0.18	1.45	7.42	1.71	1.10
0.793	0.050	31.62	5.69	3.75	0.26	1.45	7.60	1.71	1.06
0.811	0.044	30.39	6.22	3.88	0.35	1.46	7.81	1.72	1.01
0.883	0.024	25.02	9.77	4.69	0.83	1.53	9.17	1.78	0.83

TABLE 12.2
Steady-state and dynamic parameters for an elliptical bore bearing [see Fig. 12.10(*a*)] with a preload of 0.5 (m_p = 0.5). Angular extent of the two grooves is 20° each
[From Lund (1979)]

Steady-state parameters			Dynamic parameters						
Eccentricity ratio	Dimensionless load parameter	Attitude angle, deg	Dimensionless stiffness coefficients				Dimensionless damping coefficients		
ε_0	$\frac{1}{W_r}\left(\frac{2\pi}{\lambda_k}\right)^2$	Φ_0	K_{xx}	K_{xz}	K_{zx}	K_{zz}	B_{xx}	$B_{xz} = B_{zx}$	B_{zz}
(a) Diameter-to-width ratio is 2 (λ_k = 2.0).									
0.024	7.079	88.79	91.58	40.32	−57.12	1.29	159.20	−63.29	45.50
0.061	2.723	88.58	35.54	15.77	−22.03	0.74	61.63	−23.96	17.80
0.086	1.889	88.33	24.93	11.18	−15.33	0.71	43.14	−16.31	12.59
0.127	1.229	87.75	16.68	7.66	−10.03	0.78	28.65	−10.11	8.57
0.155	0.976	87.22	13.59	6.39	−7.99	0.84	23.20	−7.66	7.08
0.176	0.832	86.75	11.88	5.69	−6.82	0.90	20.14	−6.23	6.23
0.254	0.494	84.36	8.11	4.28	−3.99	1.09	13.26	−2.76	4.27
0.323	0.318	81.08	6.52	3.82	−2.34	1.23	10.03	−0.81	3.15
0.364	0.236	78.09	6.07	3.76	−1.49	1.31	8.80	0.11	2.54
0.391	0.187	75.18	6.03	3.82	−0.92	1.37	8.23	0.66	2.13
0.410	0.153	72.26	6.21	3.92	−0.52	1.41	7.98	1.02	1.82
0.424	0.127	69.31	6.53	4.04	−0.21	1.45	7.91	1.26	1.58
0.444	0.090	63.24	7.55	4.33	0.23	1.50	8.11	1.54	1.23
(b) Diameter-to-width ratio is 1 (λ_k = 1.0).									
0.050	1.442	93.91	38.58	22.65	−22.14	−1.29	79.05	−28.14	18.60
0.100	0.698	93.12	18.93	11.25	−10.79	−0.24	38.73	−12.97	9.40
0.150	0.442	91.97	12.28	7.45	−6.87	0.26	25.00	−7.50	6.36
0.200	0.308	90.37	8.93	5.58	−4.79	0.58	17.99	−4.50	4.82
0.213	0.282	89.87	8.30	5.24	−4.38	0.66	16.66	−3.91	4.53
0.220	0.271	89.61	8.03	5.09	−4.20	0.69	16.08	−3.64	4.40
0.226	0.261	89.37	7.79	4.96	−4.03	0.72	15.57	−3.41	4.28
0.239	0.240	88.80	7.31	4.70	−3.70	0.77	14.54	−2.93	4.04
0.250	0.224	88.28	6.95	4.51	−3.43	0.82	13.74	−2.55	3.86
0.260	0.211	87.79	6.65	4.36	−3.21	0.86	13.09	−2.23	3.70
0.304	0.161	83.29	5.63	3.84	−2.32	1.01	10.75	−1.02	3.07
0.350	0.120	81.80	4.99	3.54	−1.52	1.14	9.04	−0.01	2.49
0.381	0.097	78.65	4.82	3.46	−1.01	1.21	8.26	0.56	2.10
0.403	0.081	75.63	4.87	3.47	−0.65	1.26	7.87	0.92	1.82
0.419	0.069	72.65	5.06	3.52	−0.38	1.31	7.71	1.17	1.60
0.432	0.060	69.69	5.36	3.60	−0.16	1.34	7.67	1.34	1.42
0.451	0.045	63.70	6.25	3.83	0.19	1.40	7.88	1.56	1.16

TABLE 12.3

Steady-state and dynamic parameters for a three-lobe bearing [see Fig. 12.10(c)] with a preload of 0.5 ($m_p = 0.5$). Angular extent of each feed groove is 20°

[From Lund (1979)]

Steady-state parameters			Dynamic parameters						
Eccentricity ratio	Dimensionless load parameter	Attitude angle, deg	Dimensionless stiffness coefficients				Dimensionless damping coefficients		
ε_0	$\frac{1}{W_r}\left(\frac{2\pi}{\lambda_k}\right)^2$	Φ_0	K_{xx}	K_{xz}	K_{zx}	K_{zz}	B_{xx}	$B_{xz} = B_{zx}$	B_{zz}
(a) Diameter-to-width ratio is 2 ($\lambda_k = 2.0$).									
0.018	6.574	55.45	34.58	45.43	−46.78	31.32	97.87	−1.46	93.55
0.031	3.682	56.03	20.35	25.35	−26.57	17.08	56.10	−1.35	51.73
0.045	2.523	56.57	14.75	17.41	−18.48	11.48	39.52	−1.22	35.06
0.070	1.621	57.35	10.53	11.38	−12.20	7.25	26.81	−1.01	22.25
0.094	1.169	57.95	8.56	8.49	−9.06	5.26	20.62	−0.79	15.96
0.144	0.717	58.62	6.85	5.85	−5.92	3.49	14.74	−0.37	9.93
0.192	0.491	58.63	6.27	4.75	−4.34	2.77	12.07	0.02	7.12
0.237	0.356	58.14	6.15	4.26	−3.35	2.41	10.67	0.36	5.51
0.278	0.267	57.30	6.29	4.05	−2.63	2.19	9.87	0.66	4.46
0.314	0.203	56.18	6.62	4.00	−2.05	2.04	9.43	0.91	3.68
0.347	0.156	54.85	7.11	4.05	−1.55	1.90	9.23	1.12	3.06
0.360	0.141	54.26	7.35	4.10	−1.36	1.85	9.20	1.20	2.84
0.377	0.121	53.31	7.77	4.19	−1.09	1.78	9.20	1.30	2.54
0.402	0.093	51.55	8.63	4.39	−0.67	1.67	9.30	1.44	2.10
0.441	0.055	47.10	11.07	4.94	0.14	1.49	9.91	1.61	1.29
(b) Diameter-to-width ratio is 1 ($\lambda_k = 1.0$)									
0.020	3.256	59.21	28.31	43.30	−43.40	25.25	94.58	−1.11	88.33
0.035	1.818	59.68	16.74	24.39	−24.34	13.70	54.59	−0.98	48.27
0.050	1.243	60.09	12.21	16.93	−16.72	9.18	38.75	−0.84	32.37
0.076	0.796	60.62	8.82	11.26	−10.82	5.80	26.62	−0.61	20.18
0.103	0.574	60.95	7.24	8.55	−7.90	4.24	20.73	−0.37	14.27
0.155	0.383	61.00	5.91	6.07	−5.02	2.89	15.15	0.06	8.70
0.203	0.245	60.44	5.48	5.01	−3.60	2.36	12.59	0.43	6.16
0.246	0.181	59.46	5.41	4.49	−2.74	2.09	11.20	0.73	4.73
0.285	0.138	58.22	5.54	4.22	−2.12	1.92	10.39	0.98	3.81
0.320	0.108	56.80	5.83	4.10	−1.65	1.80	9.91	1.18	3.16
0.351	0.085	55.23	6.25	4.08	−1.26	1.71	9.64	1.35	2.67
0.379	0.068	53.54	6.82	4.13	−0.92	1.62	9.54	1.48	2.29
0.389	0.062	52.82	7.09	4.17	−0.79	1.59	9.54	1.52	2.16
0.403	0.054	51.68	7.56	4.25	−0.57	1.54	9.57	1.57	1.92
0.441	0.034	47.19	9.70	4.65	0.11	1.42	10.03	1.67	1.23

TABLE 12.4

Steady-state and dynamic parameters for an offset-half bearing [see Fig. 12.10(b)] with a preload of 0.5 (m_p = 0.5). Angular extent of each feed groove is 30°

[From Lund (1979)]

Steady-state parameters			Dynamic parameters						
Eccentricity ratio	Dimensionless load parameter	Attitude angle, deg	Dimensionless stiffness coefficients				Dimensionless damping coefficients		
ε_0	$\frac{1}{W_r}\left(\frac{2\pi}{\lambda_k}\right)^2$	Φ_0	K_{xx}	K_{xz}	K_{zx}	K_{zz}	B_{xx}	$B_{xz} = B_{zx}$	B_{zz}
(a) Diameter-to-width ratio of 2 (λ_k = 2.0)									
0.025	8.519	−4.87	47.06	82.04	5.48	64.74	97.59	45.00	59.71
0.050	4.240	−4.82	23.60	41.06	2.64	32.32	49.04	22.62	29.94
0.075	2.805	−4.72	15.81	27.42	1.65	21.49	32.97	15.22	20.06
0.100	2.081	−4.59	11.93	20.61	1.12	16.05	25.01	11.56	15.15
0.150	1.339	−4.14	8.08	13.79	0.54	10.56	17.15	7.98	10.25
0.200	0.953	−3.47	6.18	10.39	0.20	7.78	13.34	6.31	7.83
0.250	0.717	−2.76	5.14	8.45	−0.05	6.15	11.29	5.43	6.51
0.300	0.585	−2.02	4.63	7.20	−0.09	5.00	10.00	4.76	5.38
0.325	0.493	−1.78	4.56	6.72	0.01	4.53	9.49	4.38	4.74
0.400	0.383	−1.70	4.63	5.78	0.22	3.53	8.51	3.56	3.40
0.450	0.284	−2.00	4.85	5.40	0.33	3.08	8.17	3.18	2.79
0.500	0.228	−2.51	5.18	5.15	0.42	2.74	7.99	2.88	2.34
0.551	0.182	−3.19	5.65	5.01	0.51	2.48	7.95	2.65	1.98
0.576	0.162	−3.58	5.93	4.97	0.55	2.37	7.97	2.53	1.82
0.601	0.143	−4.02	6.26	4.95	0.60	2.27	8.02	2.46	1.69
0.627	0.126	−4.49	6.64	4.95	0.65	2.19	8.10	2.38	1.56
(b) Diameter-to-width ratio of 1 (λ_k = 1.0).									
0.025	3.780	−3.21	52.13	83.73	8.14	56.69	113.96	42.08	47.10
0.051	1.883	−3.16	26.11	41.89	3.99	28.31	57.20	21.13	23.61
0.076	1.247	−8.08	17.45	27.95	2.57	18.83	38.38	14.19	15.81
0.101	0.927	−7.96	13.13	20.99	1.83	14.08	29.04	10.75	11.93
0.151	0.596	−7.46	8.74	13.89	1.05	9.22	19.61	7.33	8.00
0.201	0.418	−6.58	6.44	10.17	0.62	6.68	14.73	5.64	5.96
0.251	0.316	−5.85	5.22	8.13	0.33	5.26	12.18	4.78	4.90
0.301	0.248	−5.10	4.49	6.87	0.11	4.35	10.71	4.30	4.28
0.351	0.198	−4.29	4.08	6.02	−0.04	3.70	9.80	3.99	3.83
0.401	0.160	−3.59	4.00	5.40	0.01	3.17	9.07	3.57	3.22
0.451	0.130	−3.27	4.13	4.96	0.12	2.76	8.55	3.15	2.65
0.501	0.107	−3.28	4.37	4.68	0.22	2.46	8.23	2.84	2.22
0.551	0.087	−3.54	4.74	4.50	0.31	2.23	8.08	2.60	1.89
0.576	0.078	−3.76	4.98	4.45	0.36	2.14	8.06	2.50	1.75
0.601	0.070	−4.03	5.25	4.42	0.41	2.06	8.07	2.42	1.63

12.5 CLOSURE

The side-leakage term in the Reynolds equation was considered in this chapter for a journal bearing. Analytical solutions to this form of the Reynolds equation are not normally available, and numerical methods are used. When side leakage is considered, an additional operating parameter exists, the diameter-to-width ratio λ_k. Results from numerical solution of the Reynolds equation were presented. These results focused on a full journal bearing, four values of λ_k, and a complete range of eccentricity ratios or minimum film thicknesses. The performance parameters presented for these ranges of operating parameters were

1. Dimensionless load
2. Attitude angle
3. Friction coefficient
4. Total and side flow
5. Maximum pressure and its location
6. Location of terminating pressure
7. Temperature rise due to lubricant shearing

These performance parameters were presented in the form of figures that can easily be used for designing plain journal bearings. An interpolation formulation was provided so that if λ_k is something other than the four specified values, the complete range of λ_k can be considered. Nonplain journal configurations were also considered. It was found that bearing designs with more converging and less diverging film thickness suppressed instabilities of the system. Steady-state and dynamic parameters are given for a plain journal bearing and three nonplain journal bearings.

12.6 PROBLEMS

12.6.1 For the same bearing considered in Sample Problem 12.1 determine what the operating and performance parameters are when (*a*) the half Sommerfeld infinitely long-journal-bearing theory of Chap. 10 is used and (*b*) the short-width-journal-bearing theory of Chap. 10 is used. Compare the results.

12.6.2 For the four types of bearings considered in Tables 12.1 to 12.4 determine the dimensionless critical mass $(M_a)_{cr}$ and dimensionless critical speed $(\overline{\Omega}_v)_{cr}$ for each of the eccentricity ratios ε_0 given in Tables 12.1 to 12.4. Also plot these results of $(M_a)_{cr}$ versus $(\overline{\Omega}_v)_{cr}$ for the four types of bearing. What conclusions can you make from these results as they relate to the stability of these bearings?

12.6.3 Describe the process of transition to turbulence in the flow between concentric cylinders when the outer cylinder is at rest and the inner cylinder rotates. How is the process influenced by (*a*) eccentricity and (*b*) a superimposed axial flow?

12.6.4 A plain journal bearing has a diameter of 2 in and a length of 1 in. The full journal bearing is to operate at a speed of 2000 r/min and carries a load of

750 lbf. If SAE 10 oil at an inlet temperature of 110°F is to be used, what should the radial clearance be for optimum load-carrying capacity? Describe the surface finish that would be sufficient and yet less costly. Also indicate what the temperature rise, coefficient of friction, flow rate, side flow rate, and attitude angle are.

12.6.5 Discuss the stability of flow between eccentric, rotating cylinders with reference to Rayleigh's criterion. Describe the steps involved in the process of transition to turbulence via the Taylor vortex regime in the flow, and compare the experimentally determined critical Taylor numbers with the results of the Rayleigh criterion analysis.

12.7 REFERENCES

Allaire, P. E., and Flack, R. D. (1980): Journal Bearing Design for High Speed Turbomachinery. *Bearing Design—Historical Aspects, Present Technology and Future Problems*. W. J. Anderson (ed.). American Society of Mechanical Engineers, New York, pp. 111–160.

Lund, J. W. (1979): "*Rotor-Bearing Dynamics*," Lecture notes. Technical University of Denmark, ISBN 83–04–00267–1.

Raimondi, A. A., and Boyd, J. (1958): A Solution for the Finite Journal Bearing and Its Application to Analysis and Design–I, –II, and –III. *ASLE Trans.*, vol. 1, no. 1, I- pp. 159–174; II- pp. 175–193; III- pp. 194–209.

Shigley, J. E., and Mitchell, L. D. (1983): *Mechanical Engineering Design*, 4th ed. McGraw-Hill, New York.

CHAPTER 13

HYDRODYNAMIC SQUEEZE FILM BEARINGS

As mentioned in Chap. 1 (Fig. 1.4), a positive pressure can be generated in a fluid contained between two surfaces when the surfaces are moving toward each other. A finite time is required to squeeze the fluid out of the gap, and this action provides a useful cushioning effect in bearings. The reverse effect, which occurs when the surfaces are moving apart, can lead to cavitation in liquid films. For squeeze film bearings a relationship needs to be developed between load and normal velocity at any instant. The time required for the separation of the surfaces to change by a specified amount can then be determined by a single integration with respect to time.

The starting point of the analysis, as for hydrodynamic lubrication of journal and thrust bearings, is the Reynolds equation. If the motion is restricted to normal approach such that the sliding velocities are zero ($u_a = u_b = v_a = v_b = 0$), the general Reynolds equation given in Equation (7.58) reduces to

$$\frac{\partial}{\partial x}\left(\frac{\rho h^3}{12\eta}\frac{\partial p}{\partial x}\right) + \frac{\partial}{\partial y}\left(\frac{\rho h^3}{12\eta}\frac{\partial p}{\partial y}\right) = \frac{\partial(\rho h)}{\partial t} \tag{13.1}$$

If the density and viscosity are assumed to be constant, Eq. (13.1) reduces to

$$\frac{\partial}{\partial x}\left(h^3\frac{\partial p}{\partial x}\right) + \frac{\partial}{\partial y}\left(h^3\frac{\partial p}{\partial y}\right) = 12\eta_0\frac{\partial h}{\partial t} = -12\eta_0 w \tag{13.2}$$

where $w = -\partial h/\partial t$ is the squeeze velocity.

In cylindrical polar coordinates Eq. (13.1) can be expressed as

$$\frac{\partial}{\partial r}\left(\frac{\rho h^3}{12\eta} r \frac{\partial p}{\partial r}\right) + \frac{1}{r}\frac{\partial}{\partial \theta}\left(\frac{\rho h^3}{12\eta}\frac{\partial p}{\partial \theta}\right) = r\frac{\partial(\rho h)}{\partial t} \tag{13.3}$$

For constant density and viscosity this equation reduces to

$$\frac{\partial}{\partial r}\left(rh^3\frac{\partial p}{\partial r}\right) + \frac{1}{r}\frac{\partial}{\partial \theta}\left(h^3\frac{\partial p}{\partial \theta}\right) = 12\eta_0 r\frac{\partial h}{\partial t} = -12\eta_0 rw \tag{13.4}$$

These are the Reynolds equations that are normally used in analyzing squeeze film bearings. This chapter is restricted to situations where the density and the viscosity are considered to be constant [Eqs. (13.2) and (13.4)].

13.1 PARALLEL-SURFACE BEARING OF INFINITE WIDTH

When side leakage is neglected and a parallel film is assumed, Eq. (13.2) reduces to

$$\frac{\partial^2 p}{\partial x^2} = -\frac{12\eta_0 w}{h_o^3} \tag{13.5}$$

Figure 13.1 shows a parallel-surface squeeze film bearing and the coordinate system that will be used. Symmetry of the oil film geometry makes it convenient to place the origin of the coordinates at the midpoint of the bearing.

Integrating Eq. (13.5) gives

$$\frac{dp}{dx} = -\frac{12\eta_0 w}{h_o^3}x + \tilde{A} \tag{13.6}$$

Integrating again gives

$$p = -\frac{6\eta_0 w}{h_o^3}x^2 + \tilde{A}x + \tilde{B} \tag{13.7}$$

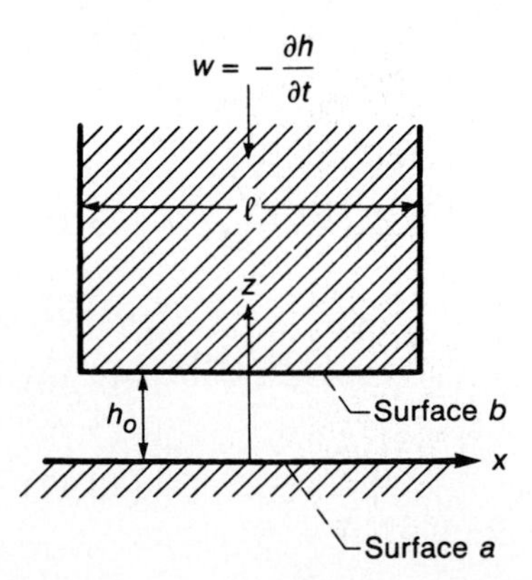

FIGURE 13.1
Parallel-surface squeeze film bearing.

The boundary conditions are

$$p = 0 \qquad \text{when } x = \pm\frac{\ell}{2}$$

Making use of these boundary conditions gives $\tilde{A} = 0$ and $\tilde{B} = 6\eta_0 w \ell^2/4h_o^3$. Substituting these into Eqs. (13.6) and (13.7) gives

$$\frac{dp}{dx} = -\frac{12\eta_0 wx}{h_o^3} \tag{13.8}$$

and

$$p = \frac{3\eta_0 w}{2h_o^3}(\ell^2 - 4x^2) \tag{13.9}$$

Letting

$$P = \frac{ph_o^3}{\eta_0 \ell^2 w} \qquad \text{and} \qquad X = \frac{x}{\ell} \tag{13.10}$$

Eq. (13.9) becomes

$$P = \frac{ph_o^3}{\eta_0 \ell^2 w} = \frac{3}{2}(1 - 4X^2) \tag{13.11}$$

where $-\frac{1}{2} \le X \le \frac{1}{2}$.

The pressure distribution is parabolic and symmetrical about the bearing center, and the maximum pressure is given by

$$P_m = \frac{p_m h_o^3}{\eta_0 \ell^2 w} = \frac{3}{2} \tag{13.12}$$

or

$$p_m = \frac{3\eta_0 \ell^2 w}{2h_o^3} \tag{13.13}$$

The tangential load-carrying capacity as defined at the beginning of Chap. 8 can be expressed as

$$w_{xa} = w_{xb} = 0 \tag{13.14}$$

The normal load-carrying capacity can be written as

$$w_z' = w_{za}' = w_{zb}' = \int_{-\ell/2}^{\ell/2} p\,dx = \frac{3\eta_0 w}{2h_o^3}\int_{-\ell/2}^{\ell/2}(\ell^2 - 4x^2)\,dx = \frac{\eta_0 \ell^3 w}{h_o^3} \tag{13.15}$$

In dimensionless terms

$$W_{xa} = W_{xb} = 0 \tag{13.16}$$

$$W_z = \frac{w_z' h_o^3}{\eta_0 \ell^3 w} = 1 \tag{13.17}$$

The shear stress acting on the solid surfaces can be written as

$$(\tau_{zx})_{z=0} = \left(\eta \frac{\partial u}{\partial z}\right)_{z=0} = \frac{6\eta_0 wx}{h_o^2} \tag{13.18}$$

$$(\tau_{zx})_{z=h} = -\left(\eta \frac{\partial u}{\partial z}\right)_{z=h} = -\frac{6\eta_0 wx}{h_o^2} \tag{13.19}$$

The shear forces at the solid surfaces can be written as

$$f_b = b\int_{-\ell/2}^{\ell/2} (\tau_{zx})_{z=0}\, dx = 0 \tag{13.20}$$

$$f_a = b\int_{-\ell/2}^{\ell/2} (\tau_{zx})_{z=h}\, dx = 0 \tag{13.21}$$

The volume flow rate can be written from Eqs. (7.38) and (13.8) as

$$q_x = \frac{h_o^3 b}{12\eta_0}\frac{12\eta_0 wx}{h_o^3} = bwx \tag{13.22}$$

The volume flow rate increases from zero at the bearing center to a maximum of $wb\ell/2$ at the bearing edge. The dimensionless volume flow rate, making use of Eq. (13.10), is

$$Q = \frac{2q_x}{wb\ell} = 2X \tag{13.23}$$

where $-\frac{1}{2} \le X \le \frac{1}{2}$.

For time-independent loads (w_z' not a function of t), Eq. (13.15) can be used to determine the time taken for the gap between the parallel surfaces to be reduced by a given amount. Since $w = -\partial h/\partial t$ and $h = h_o$, from Eq. (13.15)

$$w_z' = -\frac{\eta_0 \ell^3}{h_o^3}\frac{\partial h_o}{\partial t} \tag{13.24}$$

Rearranging terms and integrating gives

$$-\frac{w_z'}{\eta_0 \ell^3}\int_{t_1}^{t_2} dt = \int_{h_{o,1}}^{h_{o,2}} \frac{dh_o}{h_o^3}$$

$$\therefore \Delta t = t_2 - t_1 = \frac{\eta_0 \ell^3}{2w_z'}\left(\frac{1}{h_{o,2}^2} - \frac{1}{h_{o,1}^2}\right) \tag{13.25}$$

The final outlet film thickness $h_{o,2}$ can be expressed in terms of the initial outlet film thickness $h_{o,1}$ and the time interval Δt as

$$h_{o,2} = \frac{h_{o,1}}{\left[1 + \left(2w_z' \Delta t\, h_{o,1}^2/\eta_0 \ell^3\right)\right]^{1/2}} \tag{13.26}$$

Sample problem 13.1 Calculate the theoretical separation velocity required to reduce the oil film pressure between two parallel plates 0.025 m long and infinitely wide to a pressure of absolute zero if the oil film separating the plates is 25 μm thick and has a viscosity of 0.5 $N \cdot s/m^2$.

Solution. Reducing the pressure to absolute zero implies that $p_m = -0.1$ MPa in Eq. (13.13) and

$$\frac{\partial h_o}{\partial t} = -\frac{2h_o^3}{3\eta_0 \ell^2} p_m = \frac{-2(25)^3(10^{-6})^3(-0.1)(10^6)}{3(0.5)(0.025)^2} - 0.33 \times 10^{-4} \text{ m/s}$$

It is clear from the solution that an extremely small separation velocity will lead to cavitation and rupture of the oil film.

Sample problem 13.2 If a load per width w_z' of 20,000 N/m is applied to the conditions prevailing in Sample Problem 13.1, calculate the time required to reduce the film thickness to (*a*) 2.5 μm, (*b*) 0.25 μm, and (*c*) zero.

Solution

(*a*) Making use of Eq. (13.25) gives

$$\Delta t = \frac{\eta_0 \ell^3}{2 w_z'}\left(\frac{1}{h_{o,2}^2} - \frac{1}{h_{o,1}^2}\right) = \frac{0.5(0.025)^3}{2(20{,}000)}\left[\frac{1}{(2.5 \times 10^{-6})^2} - \frac{1}{(25 \times 10^{-6})^2}\right]$$

$$= 1.95 \times 10^{-10}(0.16 - 0.0016)10^{12} = 30.9 \text{ s}$$

(*b*)

$$\Delta t = 1.95 \times 10^{-10}\left[\frac{1}{(0.25 \times 10^{-6})^2} - \frac{1}{(25 \times 10^{-6})^2}\right]$$

$$= 3120 \text{ s}$$

(*c*)

$$\Delta t = 1.95 \times 10^{-10}\left[\frac{1}{0} - \frac{1}{(25 \times 10^{-6})^2}\right] = \infty$$

This implies that theoretically the oil would never be squeezed out of the space between the parallel plates.

The differences in the way the load is made dimensionless for normal squeeze action in contrast to the way it was made dimensionless for a slider bearing should be observed. For normal squeeze action, as the bearing surfaces move toward each other, the viscous fluid shows great reluctance to be squeezed out the sides of the bearing. The tenacity of a squeeze film is remarkable, and the survival of many modern bearings depends on this phenomenon. The expression for the dimensionless normal load-carrying capacity of a bearing subjected to squeeze action can be written as

$$(W_z)_{\text{squeeze}} = \frac{w_z'}{\eta_0 w}\left(\frac{h_o}{\ell}\right)^3 = O|1| \tag{13.27}$$

In Chap. 8 the normal load-carrying capacity for a slider bearing was made dimensionless in the following way:

$$(W_z)_{\text{sliding}} = \frac{w'_z}{\eta_0 u_b}\left(\frac{h_o}{\ell}\right)^2 = O|1| \tag{13.28}$$

Note the difference in exponent on the h_o/ℓ term in Eqs. (13.27) and (13.28).

Another point to be made about the results presented thus far is associated with the separation of the surfaces. Extremely small separation velocities can yield extremely low pressures and give rise to cavitation in the form of gas release or boiling at reduced pressures. One of the most common possibilities of severe cavitation damage in bearings is associated with changes in operating conditions that might tend to pull the bearing surfaces apart in the presence of the lubricant.

13.2 JOURNAL BEARING

In the absence of rotation, from Chap. 11 [Eqs. (11.20) and (11.21)],

$$\omega = \frac{\partial \Phi}{\partial t} = 0 \tag{11.20}$$

and
$$\frac{\partial}{\partial \phi}\left(\frac{h^3}{\eta}\frac{\partial p}{\partial \phi}\right) + r_a^2\frac{\partial}{\partial y}\left(\frac{h^3}{\eta}\frac{\partial p}{\partial y}\right) = 12 r_a^2 \frac{\partial e}{\partial t}\cos(\theta' - \Phi) \tag{11.21}$$

If side leakage is neglected and the viscosity is assumed to be constant, this equation reduces to

$$\frac{\partial}{\partial \phi}\left(h^3\frac{\partial p}{\partial \phi}\right) = 12 r_a^2 \eta_0 \frac{\partial e}{\partial t}\cos(\theta' - \Phi) \tag{13.29}$$

But the film thickness in a journal bearing was defined in Chap. 10 as

$$h = c(1 + \varepsilon \cos\phi) \tag{10.5}$$

Note that this film thickness is equivalent to that given in Chap. 11 [Eq. (11.14)], since $\phi = \theta' - \Phi$. Making use of Eq. (10.5) results in the following form of Eq. (13.1):

$$\frac{\partial}{\partial \phi}\left(h^3\frac{\partial p}{\partial \phi}\right) = 12\eta_0 r_a^2\frac{\partial h}{\partial t} = -12\eta_0 r_a^2 w \tag{13.30}$$

where $w = -\partial h/\partial t$ is the normal squeeze velocity.

The film shape expressed in Eq. (10.5) and shown in Fig. 11.1 shows the film thickness to be the largest at $\phi = 0$. For the condition of normal squeeze motion in a journal bearing, Fig. 13.2 can be used. Here the ϕ coordinate starts at the minimum film thickness. Therefore, the film shape is defined as

$$h = c(1 - \varepsilon \cos\phi) \tag{13.31}$$

This change does not affect the Reynolds equation expressed in Eq. (13.30).

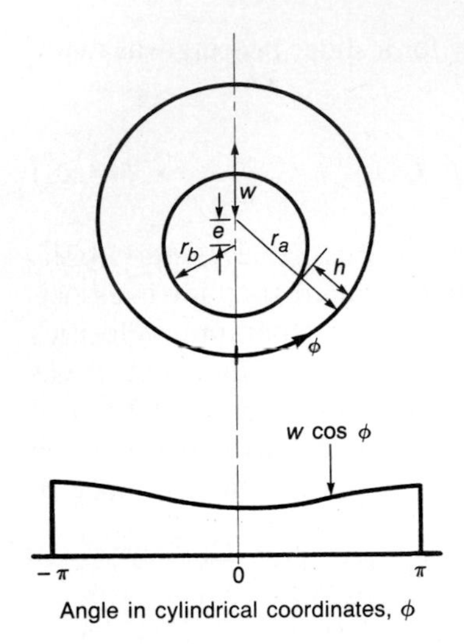

FIGURE 13.2
Journal bearing with normal squeeze film action. Rotational velocities are all zero.

When the journal is unwrapped as shown in Fig. 13.2, it can be seen that the rate at which the film thickness given by Eq. (13.31) is being reduced owing to the normal squeeze motion is $w \cos \phi$. Thus, for the unwrapped journal the Reynolds equation is

$$\frac{d}{d\phi}\left(h^3 \frac{dp}{d\phi}\right) = -12\eta_0 r_a^2 w \cos \phi \tag{13.32}$$

Integrating with respect to ϕ gives

$$\frac{dp}{d\phi} = -\frac{12\eta_0 r_a^2 w \sin \phi}{h^3} + \frac{\tilde{A}}{h^3}$$

Symmetry about the line of centers requires that $dp/d\phi = 0$ when $\phi = 0$. Therefore, $\tilde{A} = 0$ and

$$\frac{dp}{d\phi} = -\frac{12\eta_0 r_a^2 w \sin \phi}{h^3} \tag{13.33}$$

With the film thickness relationship given in Eq. (13.31)

$$\frac{dp}{d\phi} = -\frac{12\eta_0 r_a^2 w \sin \phi}{c^3(1 - \varepsilon \cos \phi)^3}$$

Integrating gives

$$p = \frac{6\eta_0 r_a^2 w}{\varepsilon c^3(1 - \varepsilon \cos \phi)^2} + \tilde{B} \tag{13.34}$$

If the pressure p is to have a finite value when $\varepsilon \to 0$, it is necessary that

$$\tilde{B} = -\frac{6\eta_0 r_a^2 w}{\varepsilon c^3} \tag{13.35}$$

$$\therefore p = \frac{6\eta_0 r_a^2 w}{c^3} \frac{\cos\phi(2 - \varepsilon\cos\phi)}{(1 - \varepsilon\cos\phi)^2} \tag{13.36}$$

The normal load-carrying capacity can be expressed as

$$w'_z = 2\int_0^{\pi} p r_a \cos\phi \, d\phi$$

Substituting Eq. (13.36) into this equation gives

$$w'_z = \frac{12\eta_0 r_a^3 w}{c^3} \int_0^{\pi} \frac{\cos^2\phi(2 - \varepsilon\cos\phi)\, d\phi}{(1 - \varepsilon\cos\phi)^2}$$

This reduces to

$$w'_z = \frac{12\pi\eta_0 r_a^3 w}{c^3(1 - \varepsilon^2)^{3/2}} \tag{13.37}$$

Figure 13.2 shows that the film thickness is a minimum when $\phi = 0$. Therefore, from Eq. (13.31)

$$h_{\text{min}} = c(1 - \varepsilon) \tag{13.38}$$

$$\therefore \frac{dh_{\text{min}}}{dt} = -c\frac{d\varepsilon}{dt} \tag{13.39}$$

The time of approach can be written as

$$w = -\frac{dh_{\text{min}}}{dt} = c\frac{d\varepsilon}{dt} \tag{13.40}$$

Substituting Eq. (13.40) into Eq. (13.37) gives

$$\frac{w'_z c^2}{12\pi\eta_0 r_a^3}\int_{t_1}^{t_2} dt = \int_{\varepsilon_1}^{\varepsilon_2} \frac{d\varepsilon}{(1 - \varepsilon^2)^{3/2}} = \left[\frac{\varepsilon}{(1 - \varepsilon^2)^{1/2}}\right]_{\varepsilon_1}^{\varepsilon_2}$$

$$\therefore t_2 - t_1 = \Delta t = \frac{12\pi\eta_0 r_a^3}{w'_z c^2}\left[\frac{\varepsilon_2}{(1 - \varepsilon_2^2)^{1/2}} - \frac{\varepsilon_1}{(1 - \varepsilon_1^2)^{1/2}}\right] \tag{13.41}$$

Sample problem 13.3. Calculate the time it takes for the eccentricity ratio to increase from 0.6 to 0.9 in a 5-cm-diameter journal bearing having a radial clearance of 25 μm, a lubricant viscosity of 1 P, and a load per unit width of 20,000 N/m.

Solution. From Eq. (13.41)

$$\Delta t = \frac{12\pi(0.1)(0.025)^3}{(25)^2(10^{-12})(20{,}000)}\left[\frac{0.9}{(0.19)^{1/2}} - \frac{0.6}{(0.64)^{1/2}}\right] \text{s}$$

$$= 4.71(2.064 - 0.750) \text{ s}$$

$$= 6.2 \text{ s}$$

13.3 PARALLEL CIRCULAR PLATE

The geometry of a circular plate approaching a plane in a parallel position is shown in Fig. 13.3. The appropriate Reynolds equation for cylindrical coordinates is expressed in Eq. (13.4). If the surfaces are parallel, axial symmetry exists and the pressure is a function only of the radius. Thus,

$$\frac{d}{dr}\left(rh_o^3\frac{dp}{dr}\right) = -12\eta_0 rw \tag{13.42}$$

Integrating gives

$$\frac{dp}{dr} = -\frac{6\eta_0 rw}{h_o^3} + \frac{\tilde{A}}{rh_o^3}$$

Since dp/dr will not acquire an infinite value when $r = 0$, then $\tilde{A} = 0$. Hence,

$$\frac{dp}{dr} = -\frac{6\eta_0 rw}{h_o^3} \tag{13.43}$$

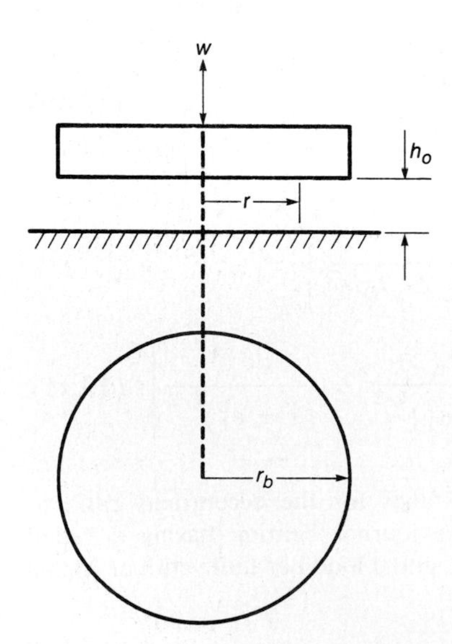

FIGURE 13.3
Parallel circular plate approaching a plane surface.

A further integration gives

$$p = -\frac{3\eta_0 r^2 w}{h_o^3} + \tilde{B}$$

The boundary condition is that $p = 0$ when $r = r_b$.

$$\therefore \tilde{B} = \frac{3\eta_0 r_b^2 w}{h_o^3}$$

and

$$p = \frac{3\eta_0 w}{h_o^3}\left(r_b^2 - r^2\right) \tag{13.44}$$

$$p_m = \frac{3\eta_0 w r_b^2}{h_o^3} \tag{13.45}$$

The normal load component can be written as

$$w_z = \int_0^{r_b} 2\pi r p\, dr = \frac{6\pi\eta_0 w}{h_o^3}\left(\frac{r_b^2 r^2}{2} - \frac{r^4}{4}\right)_{r=0}^{r=r_b}$$

$$\therefore w_z = \frac{3\pi\eta_0 r_b^4 w}{2h_o^3} \tag{13.46}$$

The time of approach can be obtained from this equation while letting $w = -dh/dt$.

$$\int dt = -\frac{3\pi\eta_0 r_b^4}{2\,w_z}\int_{h_{o,1}}^{h_{o,2}} \frac{dh}{h^3}$$

or

$$\Delta t = \frac{3\pi\eta_0 r_b^4}{4\,w_z}\left(\frac{1}{h_{o,2}^2} - \frac{1}{h_{o,1}^2}\right) \tag{13.47}$$

13.4 INFINITELY LONG CYLINDER NEAR A PLANE

The geometry of the solid surfaces being lubricated by normal squeeze motion considered thus far in this chapter has been conformal surfaces. The present section deals with the nonconformal surfaces of an infinitely long cylinder near a plane. The geometry and the coordinate system used to describe these surfaces are shown in Fig. 13.4. The cylinder's length is assumed to be large

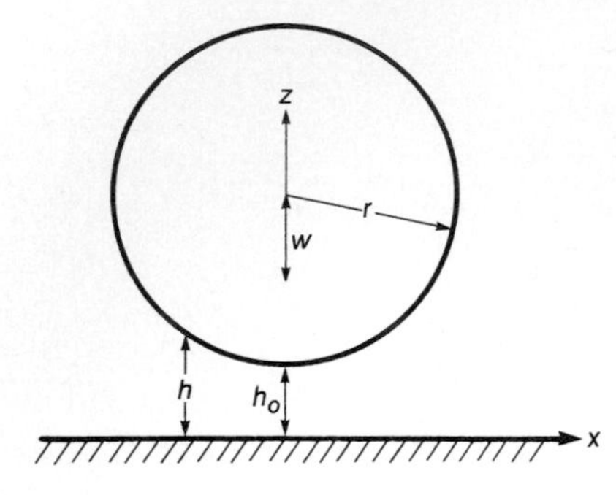

FIGURE 13.4
Rigid cylinder approaching a plane surface.

relative to its radius so that side leakage can be neglected. As with all the other considerations given in this chapter, the viscosity and the density are assumed to be constant and the solid surfaces, rigid. The appropriate Reynolds equation is

$$\frac{d}{dx}\left(h^3\frac{dp}{dx}\right) = -12\eta_0 w$$

Integrating once gives

$$\frac{dp}{dx} = -12\eta_0 w\frac{x}{h^3} + \tilde{A}$$

Making use of the boundary condition that $dp/dx = 0$ when $x = 0$ results in $\tilde{A} = 0$. Integrating again gives

$$p = -12\eta_0 w\int\frac{x}{h^3}\,dx + \tilde{B} \tag{13.48}$$

The film thickness in Fig. 13.4 can be expressed as

$$h = h_o + r - (r^2 - x^2)^{1/2} = h_o + r - r\left[1 - \left(\frac{x}{r}\right)^2\right]^{1/2}$$

The last term on the right side of this equation can be expressed in terms of a series as

$$h = h_o + r - r\left[1 - \frac{1}{2}\left(\frac{x}{r}\right)^2 - \frac{1}{8}\left(\frac{x}{r}\right)^4 - \frac{1}{16}\left(\frac{x}{r}\right)^6 - \cdots\right]$$

or

$$h = h_o + \frac{1}{2}\left(\frac{x^2}{r}\right)\left[1 + \frac{1}{4}\left(\frac{x}{r}\right)^2 + \frac{1}{8}\left(\frac{x}{r}\right)^4 + \cdots\right] \tag{13.49}$$

Because in the lubrication region $x \ll r$ so that $x^2/r^2 \ll 1$, this equation reduces to

$$h = h_o + \frac{x^2}{2r} \tag{13.50}$$

or

$$dh = \frac{x\,dx}{r} \tag{13.51}$$

Thus, Eq. (13.48) becomes

$$p = -12\eta_0 wr \int \frac{dh}{h^3} + \tilde{B}$$

$$= \frac{6\eta_0 wr}{h^2} + \tilde{B}$$

Making use of the boundary condition that h is large when the pressure is zero results in $\tilde{B} = 0$ and

$$\therefore p = \frac{6\eta_0 wr}{h^2} = \frac{6\eta_0 wr}{h_o^2} \frac{1}{\left(1 + x^2/2rh_o\right)^2} \tag{13.52}$$

The normal load component can be written as

$$w_z' = \int_{-r}^{r} p\,dx = \frac{6\eta_0 wr}{h_o^2} \int_{-r}^{r} \frac{dx}{\left(1 + x^2/2rh_o\right)^2} \tag{13.53}$$

Let $x^2/2rh_o = \tan^2\psi$. Then

$$dx = \left(2rh_o\right)^{1/2} \sec^2\psi\,d\psi \tag{13.54}$$

and

$$\left(1 + \frac{x^2}{2rh_o}\right)^2 = \left(1 + \tan^2\psi\right)^2 = \sec^4\psi$$

$$\therefore \int_{-r}^{r} \frac{dx}{\left(1 + x^2/2rh_o\right)^2} = \left(2rh_o\right)^{1/2} \int_{-\pi/2}^{\pi/2} \cos^2\psi\,d\psi = \frac{\pi}{2}\left(2rh_o\right)^{1/2} \tag{13.55}$$

and

$$w_z' = \frac{3\pi\eta_0 wr}{h_o} \left(\frac{2r}{h_o}\right)^{1/2} \tag{13.56}$$

The time of approach can be obtained directly from this equation while making use of the fact that $w = -dh_o/dt$.

$$\frac{w_z'}{3\pi\eta_0 r\sqrt{2r}} \int dt = -\int_{h_{o,1}}^{h_{o,2}} \frac{dh_o}{h_o^{3/2}} = \left(\frac{2}{h_o^{1/2}}\right)_{h_o=h_{o,1}}^{h_o=h_{o,2}}$$

$$\therefore \Delta t = \frac{6\sqrt{2}\,\pi\eta_0 r}{w_z'} \left[\left(\frac{r}{h_{o,2}}\right)^{1/2} - \left(\frac{r}{h_{o,1}}\right)^{1/2}\right] \tag{13.57}$$

If $h_{o,2} \ll h_{o,1}$

$$\Delta t = \frac{6\sqrt{2}\,\pi\eta_0 r}{w_z'}\left(\frac{r}{h_{o,2}}\right)^{1/2} \tag{13.58}$$

13.5 CLOSURE

A positive pressure was found to occur in a fluid contained between two surfaces when the surfaces are moving toward each other. A finite time is required to squeeze the fluid out of the gap, and this action provides a useful cushioning effect in bearings. The reverse effect, which occurs when the surfaces are moving apart, can lead to cavitation in liquid films.

It was found that a parallel film shape produced the largest normal load-carrying capacity of all possible film shapes. This is in contrast with the slider bearing results given in Chap. 8, where the parallel film was shown to produce no positive pressure and therefore no load-carrying capacity. It was also found that in the normal squeeze action, as the bearing surfaces move toward each other, the viscous fluid shows great reluctance to be squeezed out the sides of the bearing. The tenacity of a squeeze film is remarkable, and the survival of many modern bearings depends on this phenomenon. It was found that a relatively small approach velocity will provide an extremely large load-carrying capacity.

The various geometrical shapes considered were a parallel-surface bearing, a journal bearing, a parallel circular plate, and an infinitely long cylinder near a plane. For each of these shapes not only is the pressure given but also the load-carrying capacity and the time of approach.

13.6 PROBLEMS

13.6.1 Given the parallel stepped bearing shown below, neglect side leakage and assume only a normal squeeze motion. Determine the pressure distribution for the inlet and outlet regions. Also determine the location of the peak pressure, the normal load-carrying capacity, and the volume flow rate.

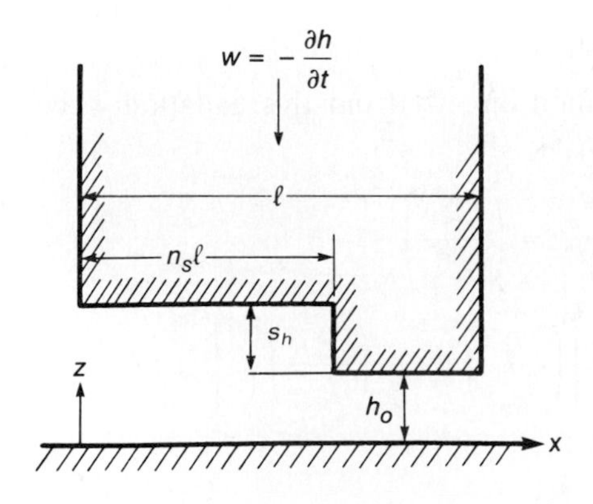

13.6.2 Derive expressions for the pressure distribution, the load-carrying capacity, and the time of approach for a squeeze film situation and the following geometry:

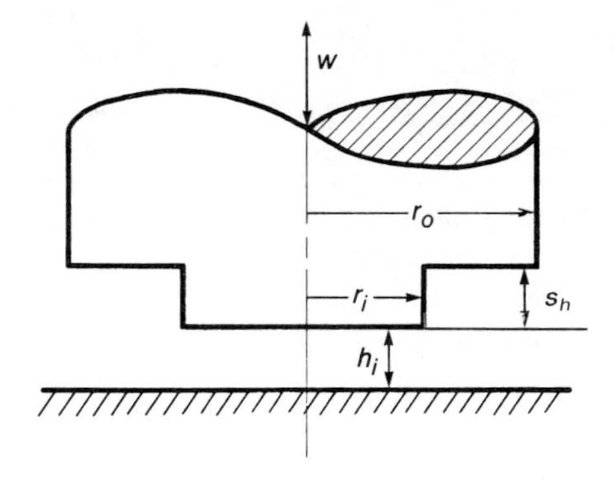

In cylindrical polar coordinates the Reynolds equation for incompressible, isoviscous conditions is expressed in Eq. (13.4). In the present problem axial symmetry is assumed that results in the pressure being a function of radius only. Thus, the preceding Reynolds equation reduces to

$$\frac{\partial}{\partial r}\left(rh^3\frac{\partial p}{\partial r}\right) = -12\eta_0 rw$$

13.6.3 Derive the expression for the pressure distribution, normal applied load capacity, and time of approach for the normal squeeze film bearing shown below.

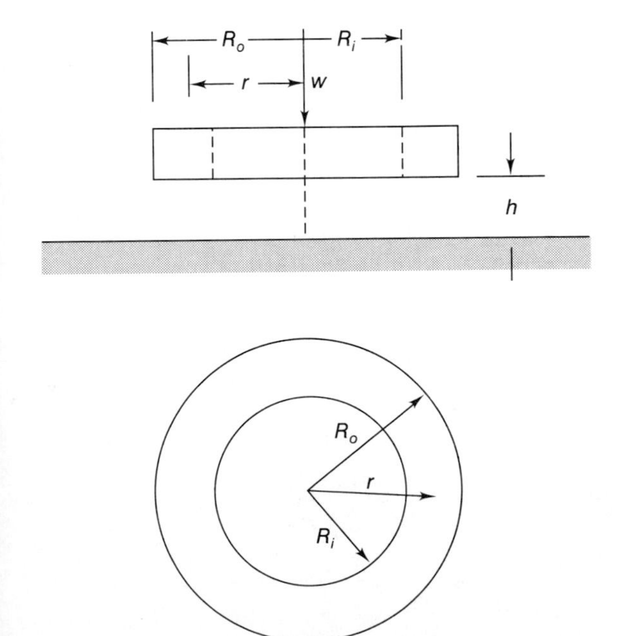

Assume axial symmetry and the the pressure at $r = R_i$ and $r = R_o$ is zero.

13.6.4 The Reynolds equation for squeeze film action in an incompressible lubricant contained between two nonparallel bearing surfaces of infinite width approaching each other with a velocity w_0 is

$$\frac{d}{dx}\left(h^3 \frac{dp}{dx}\right) = 12\eta_0 w_0$$

Side leakage is neglected. Determine the expression for the time it takes for the minimum film thickness in an inclined-pad bearing to fall from an initial value $h_{o,1}$ to a final value $h_{o,2}$ under the action of a constant load per unit width w_z'. The assumptions in the analysis should be clearly stated.

If the shoulder height is initially equal to the outlet film thickness on a 1.5-in-square pad, calculate the time taken for a load of 500 lbf on the pad to reduce the minimum film thickness from 0.001 in to 0.0001 in if the lubricant viscosity is 0.5 P.

CHAPTER 14

HYDROSTATIC LUBRICATION

Hydrodynamic lubrication, where the bearing surfaces are completely separated by a fluid film, was considered in Chaps. 8 to 12. Film separation is achieved by using sliding action, with a physical wedge pressure-generating mechanism, to develop a pressure within the bearing. Such bearings, in addition to having low frictional drag and hence low power loss, have the great advantage that they are basically simple and therefore are reliable and cheap and require little attention. Self-acting hydrodynamically lubricated slider bearings have, however, certain important disadvantages:

1. If the design speed is low, it may not be possible to generate sufficient hydrodynamic pressure.
2. Fluid film lubrication may break down during starting, direction changing, and stopping.
3. In a journal bearing as was considered in Chaps. 10 to 12, the shaft runs eccentrically and the bearing location varies with load, thus implying low stiffness.

In hydrostatic (also called "externally pressurized") lubricated bearings the bearing surfaces are separated by a fluid film maintained by a pressure source outside the bearing. Hydrostatic bearings avoid disadvantages 1 and 2 and reduce the variation of bearing location with load mentioned in disadvantage 3. The characteristics of hydrostatically lubricated bearings are

1. Extremely low friction
2. Extremely high load-carrying capacity at low speeds
3. High positional accuracy in high-speed, light-load applications
4. A lubrication system more complicated than that for self-acting bearings (considered in Chaps. 8 to 12)

Therefore, hydrostatically lubricated bearings are used when the requirements are extreme as in large telescopes and radar tracking units, where extremely heavy loads and extremely low speeds are used, or in machine tools and gyroscopes, where extremely high speeds, light loads, and gas lubricants are used.

14.1 FORMATION OF FLUID FILM

Figure 14.1 shows how a fluid film forms in a hydrostatically lubricated bearing system. In a simple bearing system under no pressure [Fig. 14.1(a)] the runner, under the influence of a load w_z, is seated on the bearing pad. As the source pressure builds up [Fig. 14.1(b)], the pressure in the pad recess also increases. The recess pressure is built up to a point [Fig. 14.1(c)] where the pressure on the runner over an area equal to the pad recess area is just sufficient to lift the load. This is commonly called the "lift pressure" p_ℓ. Just after the runner separates from the bearing pad [Fig. 14.1(d)], the recess pressure is less than that required to lift the bearing runner ($p_r < p_\ell$). After lift, flow commences through the system. Therefore, a pressure drop exists between the pressure source and the bearing (across the restrictor) and from the recess to the bearing outlet. If more load is added to the bearing [Fig. 14.1(e)], the film thickness will decrease and the recess pressure will rise until the integrated pressure across the land equals the load. If the load is then reduced to less than the original [Fig. 14.1(f)], the film thickness will increase to some higher value and the recess pressure will decrease accordingly. The maximum load that can be supported by the pad will be reached theoretically when the recess pressure is equal to the source pressure. If a load greater than this is applied, the bearing will seat and remain seated until the load is reduced and can again be supported by the supply pressure.

14.2 PRESSURE DISTRIBUTION AND FLOW

Consider a bearing of the form shown in Fig. 14.2, where a load w_z is supported by a fluid supplied to the recess at p_r and the lubricant flow is radial. The recesses, or pockets, are provided on one of the surfaces to a depth Δ in order to increase the resultant pressure. The pressure in the recess is constant over

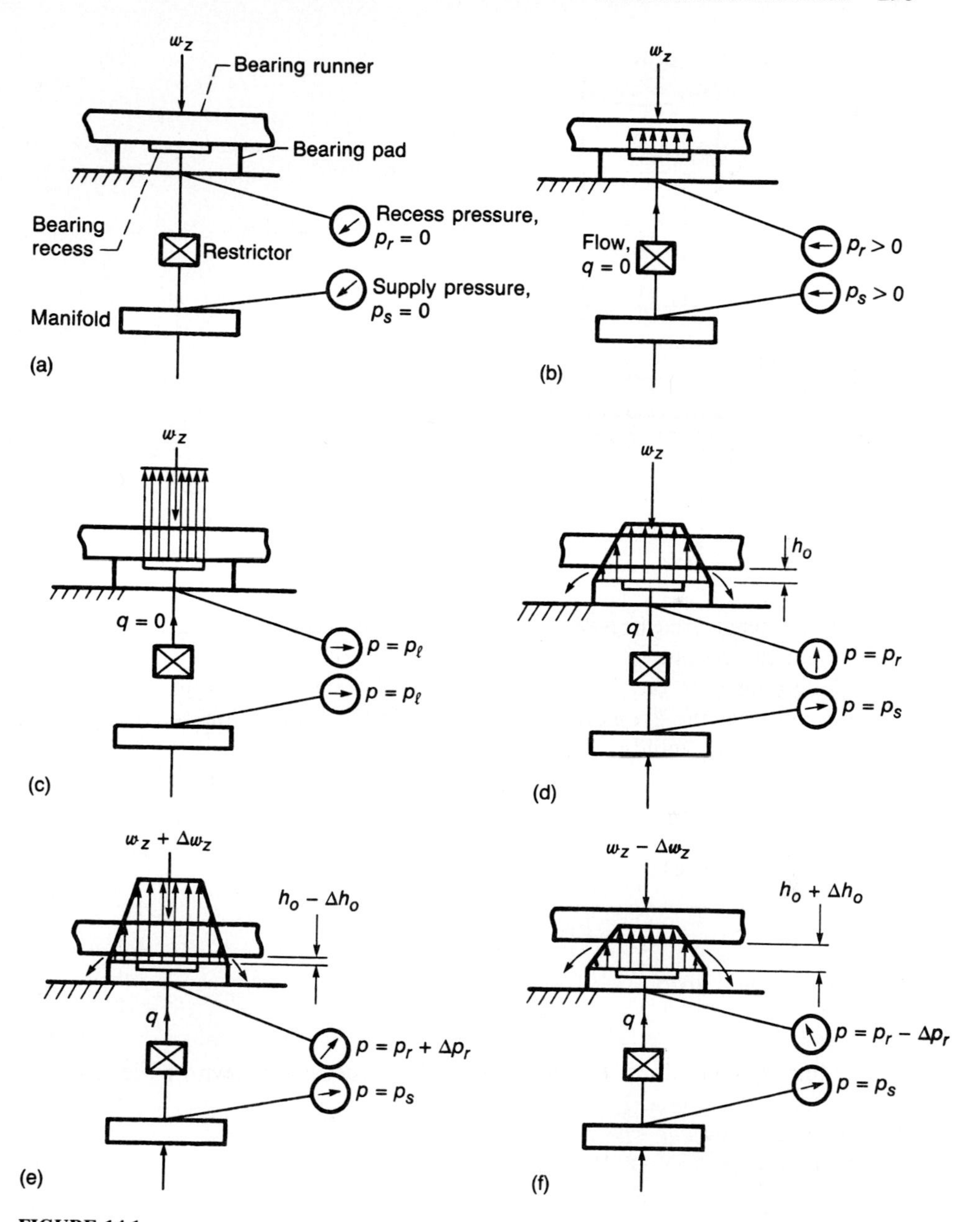

FIGURE 14.1
Formation of fluid in hydrostatic bearing system. (a) Pump off; (b) pressure building up; (c) pressure times recess area equals normal applied load; (d) bearing operating; (e) increased load; (f) decreased load; [*From Rippel (1963).*]

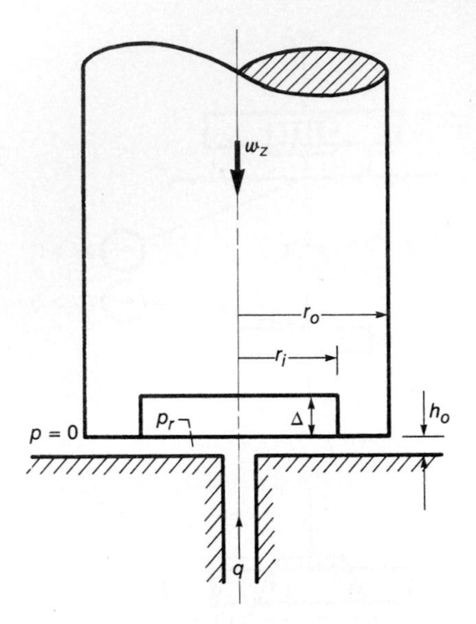

FIGURE 14.2
Radial-flow hydrostatic thrust bearing with circular step pad.

the whole pocket area if $\Delta/h_o > 10$ for liquid lubricants. Assume that $p = p_r$ throughout the recess.

Assuming that the film thickness is the same in any radial or angular position and that the pressure does not vary in the θ direction, from Eq. (7.57) the appropriate Reynolds equation is

$$\frac{\partial}{\partial r}\left(r\frac{\partial p}{\partial r}\right) = 0$$

Integrating once gives

$$\frac{dp}{dr} = \tilde{A}$$

Integrating again gives

$$p = \tilde{A} \ln r + \tilde{B} \tag{14.1}$$

The boundary conditions for the circular thrust bearing shown in Fig. 14.2 are

1. $p = p_r$ at $r = r_i$.
2. $p = 0$ at $r = r_o$.

Making use of these boundary conditions gives

$$p = p_r \frac{\ln(r/r_o)}{\ln(r_i/r_o)} \tag{14.2}$$

and

$$\frac{dp}{dr} = \frac{p_r}{r \ln(r_i/r_o)} \tag{14.3}$$

The radial volumetric flow rate per circumference is given as

$$q_r' = -\frac{h_o^3}{12\eta_0}\frac{dp}{dr} = -\frac{h_o^3 p_r}{12\eta_0 r \ln(r_i/r_o)} \tag{14.4}$$

The total volumetric flow rate is

$$q = 2\pi r q_r' = -\frac{\pi h_o^3 p_r}{6\eta_0 \ln(r_i/r_o)} = \frac{\pi h_o^3 p_r}{6\eta_0 \ln(r_o/r_i)} \tag{14.5}$$

14.3 NORMAL LOAD COMPONENT

Making use of Eq. (14.2) gives the pressure distribution in a radial-flow hydrostatic thrust bearing (shown in Fig. 14.3). The normal load component is balanced by the total pressure force or

$$w_z = \pi r_i^2 p_r + \int_{r_i}^{r_o} \frac{p_r \ln(r/r_o)}{\ln(r_i/r_o)} 2\pi r\, dr$$

This reduces to

$$w_z = \frac{\pi p_r (r_o^2 - r_i^2)}{2\ln(r_o/r_i)} \tag{14.6}$$

Note that the load is not a function of the viscosity; therefore, any fluid that does not attack the bearing materials can be used. Where a fluid is already present in a device, it is an ideal lubricant. Some examples are

1. Kerosene in aircraft engines
2. Water in hydraulic machinery
3. Liquid oxygen in a rocket engine

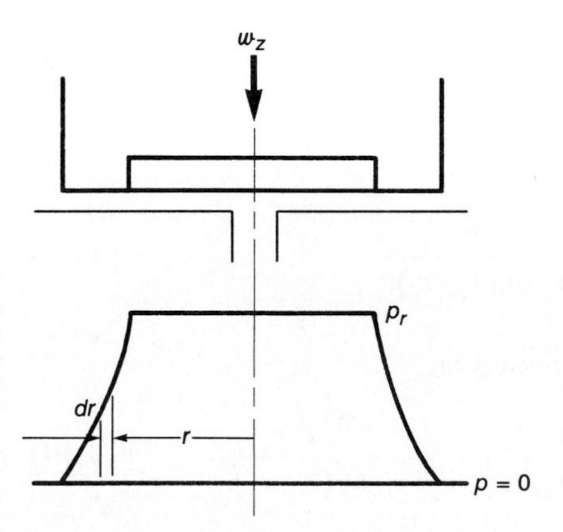

FIGURE 14.3
Pressure distribution in radial-flow hydrostatic thrust bearing.

Some other observations about the equations developed:

1. "Lift-off" will occur if (lift pressure)(πr_i^2) $= w_z$. It is important to check for this in the design of hydrostatic bearings.
2. From Eqs. (14.5) and (14.6) it can be seen that for a given bearing geometry and fluid

$$w_z \propto p_r \qquad \text{and} \qquad p_r \propto \frac{q}{h_o^3}$$

$$\therefore q \propto w_z h_o^3$$

Therefore, for a constant fluid flow rate the normal load-carrying capacity w_z increases as the film thickness decreases. This means that a bearing with a constant flow rate is self-compensating. The limiting practical load is reached when h_o reduces to the size of the surface roughness.

14.4 FRICTIONAL TORQUE AND POWER LOSS

Assuming (1) that the circumferential component of the fluid velocity varies linearly across the film and (2) that viscous friction within the recess is neglected, then from Eq. (6.1) the shear force on a fluid element can be written as

$$f = \eta_0 A \frac{u}{h_o} = \eta_0 (r\, d\phi\, dr) \frac{\omega r}{h_o} = \frac{\eta_0 \omega r^2\, dr\, d\phi}{h_o}$$

The frictional torque is

$$t_q = \frac{\eta_0 \omega}{h_o} \int_0^{2\pi} \int_{r_i}^{r_o} r^3\, dr\, d\phi = \frac{\pi \eta_0 \omega}{2 h_o} \left(r_o^4 - r_i^4 \right) \tag{14.7}$$

The total power loss required will consist of two parts:

1. Viscous dissipation

$$H_v = \omega t_q = \frac{\pi \eta_0 \omega^2}{2 h_o} \left(r_o^4 - r_i^4 \right) \tag{14.8}$$

2. Pumping loss

$$\overline{H}_p = p_r q = \frac{\pi h_o^3 p_r^2}{6 \eta_0 \ln (r_o / r_i)} \tag{14.9}$$

Therefore, the total power loss can be expressed as

$$H_t = H_v + \overline{H}_p = \frac{\pi \eta_0 \omega^2}{2 h_o} \left(r_o^4 - r_i^4 \right) + \frac{\pi h_o^3 p_r^2}{6 \eta_0 \ln (r_o / r_i)} \tag{14.10}$$

Note from this equation that H_v is inversely proportional to h_o and $\overline{H}_p$ is proportional to h_o^3.

Often the bearing velocities are low and only the pumping power is significant. Assuming that the bearing outside diameter $2r_o$ is fixed by physical limitations, many combinations of bearing recess size and pressure will support the load. For large hydrostatic bearings it can be important to minimize power. That is, we would like to know the optimum recess size for minimum pumping power.

Expressing the recess pressure from Eq. (14.6) as

$$p_r = \frac{2w_z \ln(r_o/r_i)}{\pi(r_o^2 - r_i^2)}$$

and using Eq. (14.5) gives the power loss due to pumping power as

$$\overline{H}_p = p_r q = \frac{\pi h_o^3}{6\eta_0 \ln(r_o/r_i)} \frac{4w_z^2[\ln(r_o/r_i)]^2}{\pi^2(r_o^2 - r_i^2)^2} = \frac{2h_o^3 w_z^2 \ln(r_o/r_i)}{3\pi\eta_0(r_o^2 - r_i^2)^2}$$

Differentiating with respect to r_i and equating to zero while assuming a constant load gives

$$\frac{\partial \overline{H}_p}{\partial r_i} = 0 \qquad \text{and} \qquad \ln\frac{r_o}{r_i} = \frac{1}{4}\left(\frac{r_o^2}{r_i^2} - 1\right)$$

Therefore, $r_i/r_o = 0.53$. This produces the minimum pumping power loss.

Sample problem 14.1 Given a circular step thrust bearing that is hydrostatically lubricated as shown in Fig. 14.2 and the following values:

$$r_o = 7.5 \text{ cm}$$

$$r_i = 5.0 \text{ cm}$$

$$w_z = 50 \text{ kN}$$

$$N_a = 15 \text{ r/s}$$

$$\eta_0 = 2.4 \times 10^{-2} \text{ N} \cdot \text{s/m}^2$$

If the desired film thickness is 100 μm, determine the recess pressure, the flow rate, and the power loss for this bearing.

Solution From Eq. (14.6) the normal load is

$$w_z = \frac{\pi p_r(r_o^2 - r_i^2)}{2\ln(r_o/r_i)}$$

From this equation the recess pressure can be expressed as

$$p_r = \frac{2w_z \ln(r_o/r_i)}{\pi(r_o^2 - r_i^2)}$$

$$= \frac{2(50)(10^3)\ln(7.5/5)}{\pi \times 10^{-4}\left[(7.5)^2 - (5)^2\right]}$$

$$= 0.413 \times 10^7 \text{ N/m}^2 = 4.13 \text{ MPa}$$

From Eq. (14.5) the total volumetric flow rate is

$$q = \frac{\pi h_o^3 p_r}{6\eta_0 \ln(r_o/r_i)} = \frac{\pi(100 \times 10^{-6})^3(4.13 \times 10^6)}{6(2.4)(10^{-2})\ln(7.5/5)}$$

$$= 0.0133 \text{ m}^3/\text{min} = 0.8 \text{ m}^3/\text{h}$$

From Eq. (14.10) the total power loss can be expressed as

$$H_t = H_r + H_p = \frac{\pi\eta_0\omega^2}{2h_o}\left(r_o^4 - r_i^4\right) + \frac{\pi h_o^3 p_r^2}{6\eta_0 \ln(r_o/r_i)}$$

Thus, the power loss due to viscous dissipation is

$$H_r = \frac{\pi(2.4 \times 10^{-2})}{2(100 \times 10^{-6})}\left[(15)(2\pi)\right]^2(10^{-4})(10^{-4})\left[(7.5)^4 - 5^4\right]$$

$$= 85 \text{ W} = \underline{0.085 \text{ kW}}$$

The power loss due to pumping loss is

$$H_p = \frac{\pi(100 \times 10^{-6})^3(0.413 \times 10^7)^2}{6(2.4 \times 10^{-2})\ln(7.5/5)}$$

$$= 0.092 \times 10^4 \text{ W} = \underline{0.92 \text{ kW}}$$

$$H_t = 0.085 + 0.92 = \underline{1.005 \text{ kW}}$$

Note that $H_t \approx H_p$ and $H_p \gg H_r$.

14.5 PAD COEFFICIENTS

Equations (14.5), (14.6), and (14.9) express the flow, load, and power loss for a circular step bearing pad. The load-carrying capacity of a bearing pad, regardless of its shape or size can be expressed in a more general form as

$$w_z = a_b A_p p_r \tag{14.11}$$

where a_b = dimensionless bearing pad load coefficient
A_p = total projected pad area, m^2

The amount of lubricant flow across a pad and through the bearing clearance is

$$q = q_b \frac{w_z}{A_p}\frac{h_o^3}{\eta_0} \tag{14.12}$$

where q_b is the dimensionless bearing pad flow coefficient. The pumping power required by the hydrostatic pad can be evaluated by determining the product of recess pressure and flow. Assume the angular speed to be zero so that the viscous dissipation power loss is zero.

$$\overline{H}_p = p_r q = H_b \left(\frac{w_z}{A_p} \right)^2 \frac{h_o^3}{\eta_0} \tag{14.13}$$

where $H_b = q_b / a_b$ is the dimensionless bearing pad power coefficient.

Therefore, the designer of hydrostatic bearings is primarily concerned with the three dimensionless bearing coefficients (a_b, q_b, and H_b). Values of any two of these coefficients suffice to determine the third. Bearing coefficients are dimensionless quantities that relate performance characteristics of load, flow, and power to physical parameters. The bearing coefficients for several types of bearing pad will be considered.

14.5.1 CIRCULAR STEP BEARING PAD. Rewriting Eqs. (14.5), (14.6), and (14.9) in terms of Eqs. (14.11) to (14.13), respectively, gives

$$a_b = \frac{1 - (r_i/r_o)^2}{2 \ln (r_o/r_i)} \tag{14.14}$$

$$q_b = \frac{\pi}{3\left[1 - (r_i/r_o)^2\right]} \tag{14.15}$$

$$H_b = \frac{2\pi \ln (r_o/r_i)}{3\left[1 - (r_i/r_o)^2\right]^2} \tag{14.16}$$

The total projected pad area is

$$A_p = \pi r_o^2 \tag{14.17}$$

Figure 14.4 shows the three bearing pad coefficients for various ratios of recess radius to bearing radius for a circular step thrust bearing. The bearing pad load coefficient a_b varies from zero for extremely small recesses to unity for bearings having large recesses with respect to pad dimensions. In a sense, a_b is a measure of how efficiently the bearing uses the recess pressure to support the applied load. The bearing pad flow coefficient q_b varies from unity for relatively small recesses to a value approaching infinity for bearings with extremely large recesses. Physically, as the recess becomes larger with respect to the bearing, the hydraulic resistance to fluid flow decreases and thus flow increases. Also from Fig. 14.4. the power coefficient H_b approaches infinity for extremely small recesses, decreases to a minimum as the recess size increases, and then approaches infinity again for extremely large recesses. For a circular step thrust bearing the minimum value of H_b occurs at $r_i/r_o = 0.53$.

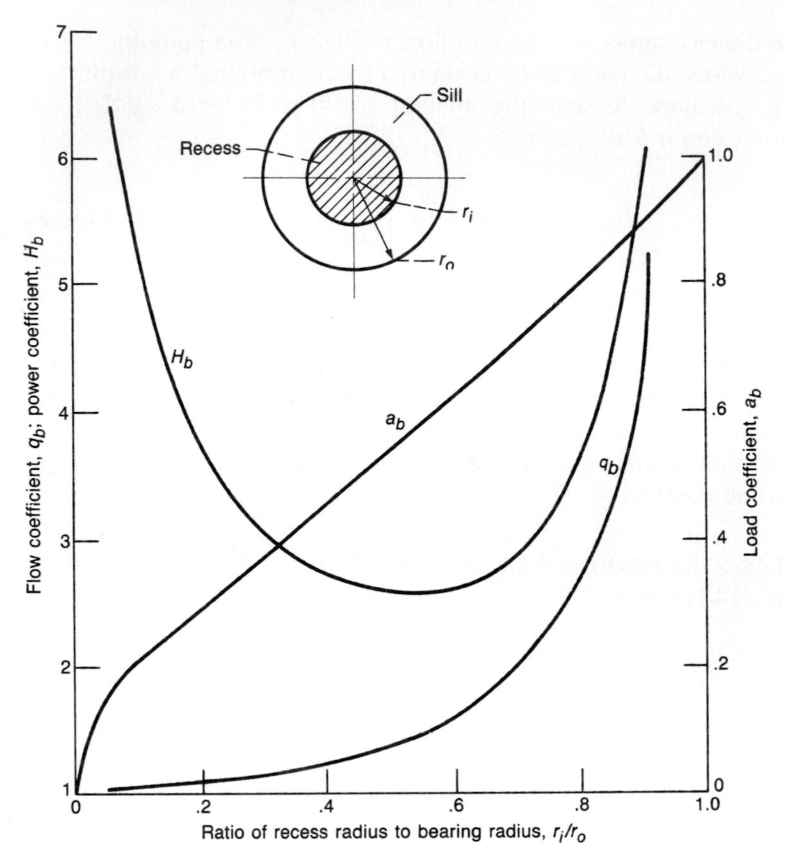

FIGURE 14.4
Chart for determining bearing pad coefficients for circular step thrust bearing. [*From Rippel (1963).*]

14.5.2 ANNULAR THRUST BEARING. Figure 14.5 shows an annular thrust bearing with four different radii used to define the recess and the sills. In this bearing the lubricant flows from the annular recess over the inner and outer sills. An analysis similar to that done in Secs. 14.2 and 14.4 for a circular step thrust bearing yields the following expressions for the pad coefficients:

$$a_b = \frac{1}{2(r_4^2 - r_1^2)}\left[\frac{r_4^2 - r_3^2}{\ln(r_4/r_3)} - \frac{r_2^2 - r_1^2}{\ln(r_2/r_1)}\right] \tag{14.18}$$

$$q_b = \frac{\pi}{6a_b}\left[\frac{1}{\ln(r_4/r_3)} + \frac{1}{\ln(r_2/r_1)}\right] \tag{14.19}$$

$$H_b = \frac{q_b}{a_b} \tag{14.20}$$

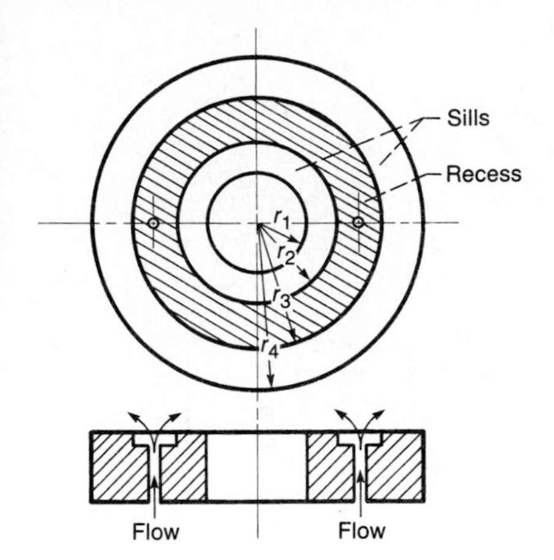

FIGURE 14.5
Configuration of annular thrust pad bearing. [*From Rippel (1963).*]

For this type of bearing the projected pad area is

$$A_p = \pi\left(r_4^2 - r_1^2\right) \tag{14.21}$$

Figure 14.6 shows the pad coefficients for an annular thrust pad bearing for $r_1/r_4 = \frac{1}{4}, \frac{1}{2}$, and $\frac{3}{4}$. These results were obtained directly by evaluating Eqs. (14.18) to (14.20). For this figure it is assumed that the annular recess is centrally located within the bearing width. This therefore implies that $r_1 + r_4 = r_2 + r_3$. Note that the curve for the load coefficient a_b applies for all r_1/r_4 ratios.

14.5.3 RECTANGULAR SECTORS. If the pressure drop across the sill of a rectangular sector is linear, the pad coefficients can be calculated. Figure 14.7 shows a rectangular sector along with the linear pressure distribution. The pad coefficients for the rectangular sector are

$$a_b = \frac{1}{2}\left(1 + \frac{A_r}{A_s}\right) = 1 - \frac{b}{B} - \frac{\ell}{L} + \frac{2b\ell}{BL} \tag{14.22}$$

$$q_b = \frac{1}{6a_b}\left(\frac{B-b}{\ell} + \frac{L-\ell}{b}\right) \tag{14.23}$$

$$H_b = \frac{q_b}{a_b} \tag{14.24}$$

The areas of the bearing, the recess, and the sill are given as

$$\begin{aligned} A_r &= (L - 2\ell)(B - 2b) \\ A_s &= A_b - A_r \\ A_b &= LB \end{aligned} \tag{14.25}$$

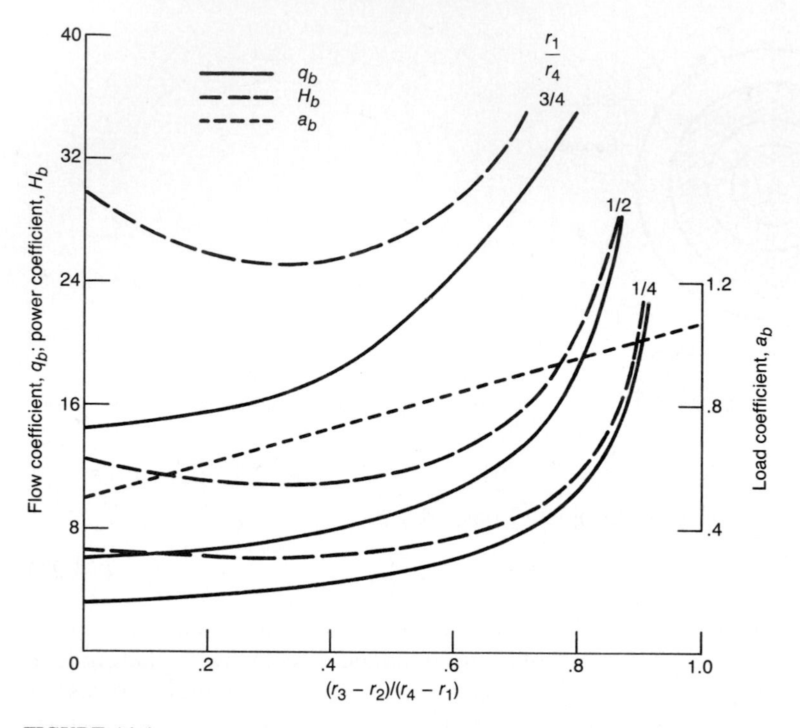

FIGURE 14.6
Chart for determining bearing pad coefficients for annular thrust pad bearings. [*From Rippel (1963).*]

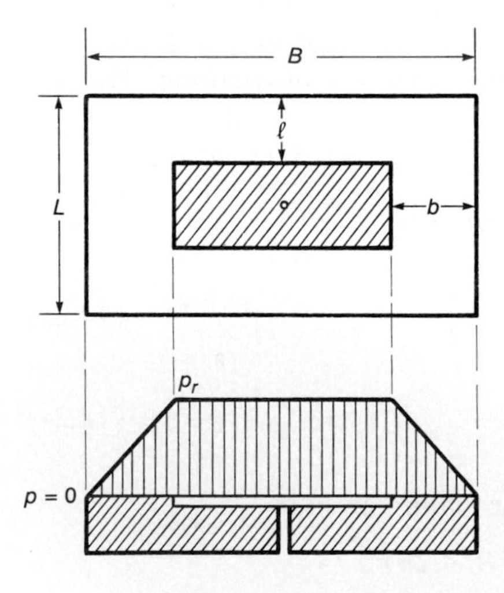

FIGURE 14.7
Rectangular hydrostatic pad.

Equation (14.23) produces a conservative flow rate and larger loads than actually experienced; Eq. (14.22) produces slightly smaller loads.

Assuming that the four corners are not contributing,

$$a_b = 1 - \frac{\ell}{L} - \frac{b}{B} \tag{14.26}$$

$$q_b = \frac{1}{6a_b}\left(\frac{B-2b}{\ell} + \frac{L-2\ell}{b}\right) \tag{14.27}$$

Equations (14.26) and (14.27) are preferred over Eqs. (14.22) and (14.23), since they produce results closer to those actually experienced. For example, consider a square sector having $B = L$ and $b = \ell = B/4$; then

1. $a_b = 0.625$ from Eq. (14.22).
2. $a_b = 0.500$ from Eq. (14.26).
3. $a_b = 0.54$ (exact value).

Consequently, formula (14.26) is preferred in calculating the load coefficient. Similar conclusions can be reached about Eq. (14.27).

Figure 14.8 gives the pad coefficients for a square pad and for a rectangular pad with $B = 2L$ and $b = \ell$. The optimum power loss coefficients are clearly indicated in the figure.

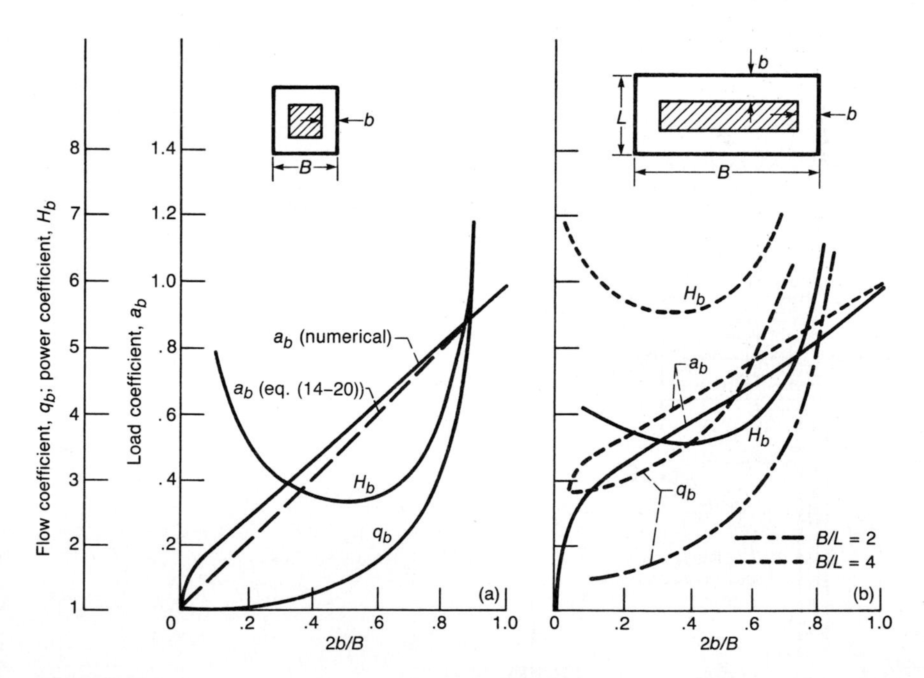

FIGURE 14.8
Pad coefficients. (a) Square pad; (b) rectangular pad with $B = 2L$ and $b = \ell$.

The hydrostatic bearings considered in this book have been limited to flat thrust-loaded bearings. Design information about conical, spherical, and cylindrical pads can be obtained from Rippel (1963). The same approach used for the flat thrust-loaded bearings is used in obtaining the pad coefficients for the more complex geometries.

14.6 COMPENSATING ELEMENTS

Hydrostatic bearings are relatively more complex systems than the hydrodynamically lubricated bearings considered in the preceding chapters. In addition to the bearing pad the system includes a pump and a compensating element, or restrictor. Three common types of compensating elements for hydrostatic bearings are the capillary tube, the sharp-edge orifice, and constant-flow-valve compensation. The purpose of the compensating element is to bring a pressurized fluid from the supply manifold to the recess. The type of compensation will affect the quantity of the fluid in the recess.

14.6.1 CAPILLARY COMPENSATION. Figure 14.9 shows a capillary-compensated hydrostatic bearing as obtained from Rippel (1963). The small diameter of the capillary tube provides restriction and resultant pressure drop in the bearing pad. The characteristic feature of capillary compensation is a long tube of relatively small diameter ($\ell_c > 20 d_c$). The laminar flow of fluid through such a tube when inlet and outlet effects and viscosity changes due to temperature and pressure are neglected can be expressed as

$$q_c = \frac{k_c(p_s - p_r)}{\eta_0} \tag{14.28}$$

where

$$k_c = \frac{\pi d_c^4}{128\,\ell_c} \tag{14.29}$$

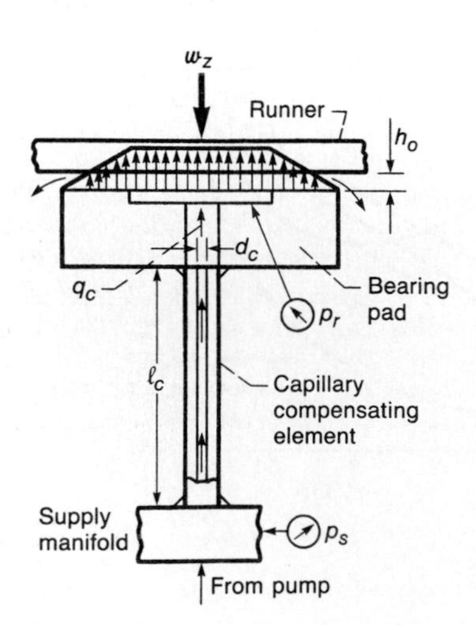

FIGURE 14.9
Capillary-compensated hydrostatic bearing. [*From Rippel (1963).*]

For a given capillary tube, k_c is a constant expressed in cubic meters. Thus, from Eq. (14.28) the flow through a capillary tube is related linearly to the pressure drop across it. In a hydrostatic bearing with capillary compensation and a fixed supply pressure the flow through the bearing will decrease with increasing load, since the recess pressure p_r is proportional to the load. For the assumption of laminar flow to be satisfied, the Reynolds number must be less than 2000 when expressed as

$$\mathscr{R} = \frac{4\rho q_c}{\pi d_c \eta_0} < 2000 \tag{14.30}$$

where ρ is the force density of the lubricant in $\mathrm{N \cdot s^2/m^4}$. Hypodermic needle tubing serves quite well as capillary tubing for hydrostatic bearings. Very small diameter tubing is available, but diameters less than 6×10^{-4} m should not be used because of the tendency to clog.

14.6.2 ORIFICE COMPENSATION. Orifice compensation is illustrated in Fig. 14.10. The flow of an incompressible fluid through a sharp-edge orifice can be expressed as

$$q_o = k_o(p_s - p_r)^{1/2} \tag{14.31}$$

where

$$k_o = \frac{\pi c_d (\tilde{d}_o)^2}{\sqrt{8\rho}} \tag{14.32}$$

and c_d is the orifice discharge coefficient. For a given orifice size and a given lubricant, k_o is a constant expressed in $\mathrm{m^4/(s \cdot N^{1/2})}$. Thus, from Eq. (14.31) flow through an orifice is proportional to the square root of the pressure difference across the orifice.

The orifice discharge coefficient c_d is a function of Reynolds number. For an orifice the Reynolds number is

$$\mathscr{R} = \frac{\tilde{d}_o}{\eta_0}[2\rho(p_s - p_r)]^{1/2} \tag{14.33}$$

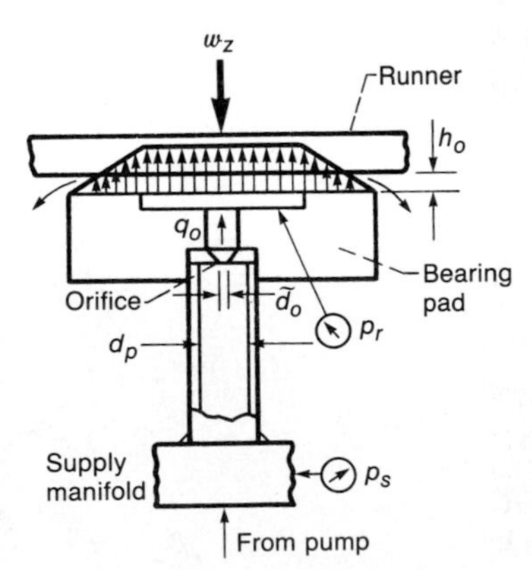

FIGURE 14.10
Orifice-compensated hydrostatic bearing. [*From Rippel (1963).*]

For a Reynolds number greater than approximately 15, which is the usual case in orifice-compensated hydrostatic bearings, c_d is about 0.6 for $\tilde{d}_o/d_p < 0.1$. For a Reynolds number less than 15 the discharge coefficient is approximately

$$c_d = 0.2\sqrt{\mathscr{R}} \tag{14.34}$$

The pipe diameter d_p at the orifice should be at least 10 times the orifice diameter $\tilde{d}_o$. Sharp-edge orifices, depending on their diameters, have a tendency to clog. Therefore, $\tilde{d}_o$ less than 5×10^{-4} m should be avoided.

14.6.3 CONSTANT-FLOW-VALVE COMPENSATION. Constant-flow-valve compensation is illustrated in Fig. 14.11. This type of restrictor has a constant flow regardless of the pressure difference across the valve. Hence, the flow is independent of the recess pressure.

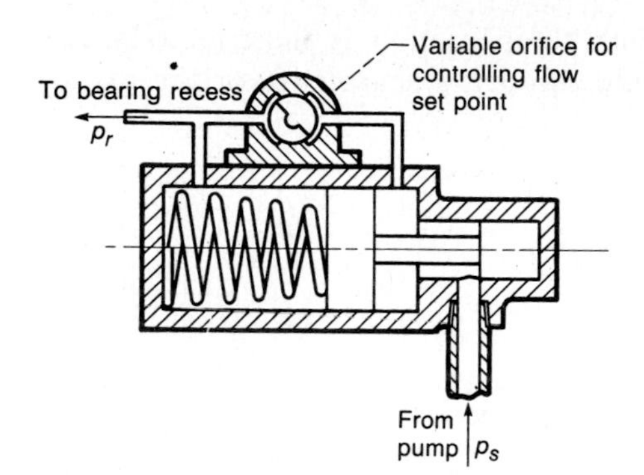

FIGURE 14.11
Constant-flow-valve compensation in hydrostatic bearing. [*From Rippel (1963).*]

TABLE 14.1 Ranking of compensating elements[a]
[From Rippel (1963)]

Consideration	Compensating element		
	Capillary	Orifice	Constant-flow valve
Initial cost	2	1	3
Cost to fabricate and install	2	3	1
Space required	2	1	3
Reliability	1	2	3
Useful life	1	2	3
Commercial availability	2	3	1
Tendency to clog	1	2	3
Serviceability	2	1	3
Adjustability	3	2	1

[a]Ranking of 1 denotes best or most desirable element for that particular consideration.

The relative ranking of the three types of compensating elements with regard to a number of considerations is given in Table 14.1. A ranking of 1 in this table indicates the best or most desirable type of compensation for a particular application. Basically, any type of compensating element can be designed into a hydrostatic bearing system if loads on the bearing never change. But if stiffness, load, or flow varies, the choice of the proper compensating element becomes more difficult and the reader is again referred to Rippel (1963).

14.7 CLOSURE

Hydrostatic bearings offer certain operating advantages over other types of bearing. Probably, the most useful characteristics of hydrostatic bearings are high load-carrying capacity and inherently low friction at any speed, even zero. The principles and basic concepts of the film formation were discussed, and a circular step thrust pad configuration was analyzed. The expressions for flow, normal load-carrying capacity, and pumping power loss for a circular step thrust pad configuration are

$$q = \frac{\pi h_o^3 p_r}{6\eta_0 \ln(r_o/r_i)} \tag{14.5}$$

$$w_z = \frac{\pi p_r (r_o^2 - r_i^2)}{2\ln(r_o/r_i)} \tag{14.6}$$

$$\overline{H}_p = \frac{\pi h_o^3 p_r^2}{6\eta_0 \ln(r_o/r_i)} \tag{14.9}$$

Equation (14.6) shows the load-carrying capacity to be independent of bearing motion and lubricant viscosity. There is no problem of surface contact wear at starting and stopping as was true for slider bearings.

A more general way to express Eqs. (14.5), (14.6), and (14.9) is

$$q = q_b \frac{w_z}{A_p} \frac{h_o^3}{\eta_0} \tag{14.12}$$

$$w_z = a_b A_p p_r \tag{14.11}$$

$$\overline{H}_p = H_b \left(\frac{w_z}{A_p}\right)^2 \frac{h_o^3}{\eta_0} \tag{14.13}$$

where q_b, a_b, and H_b are only functions of the pad geometry being considered and A_p is the total projected pad area. Results were presented for a circular step bearing pad, an annular thrust bearing, and rectangular sectors.

Three common types of compensating element for hydrostatic bearings were considered, namely, the capillary tube, the sharp-edge orifice, and constant-flow-valve compensation:

1. Capillary compensation implies that the length of the tube ℓ_c is much larger than the capillary diameter ($\ell_c > 20d_c$). The flow was found to be proportional to $p_s - p_r$.
2. In orifice compensation the flow was found to be proportional to $(p_s - p_r)^{1/2}$.

14.8 PROBLEMS

14.8.1 For an annular thrust bearing as shown in Fig. 14.5, start with the appropriate Reynolds equation and determine the flow, the load, and the pumping power loss for this bearing. Prove that Eqs. (14.18) to (14.20) are valid.

14.8.2 A 2-in-diameter shaft that acts as a pivot in a research machine carries a radial load of 500 lb maximum and an axial load of 1000 lb. It supports a pair of hydrostatic journal bearings that are centered at A and B as shown and a thrust bearing at C.

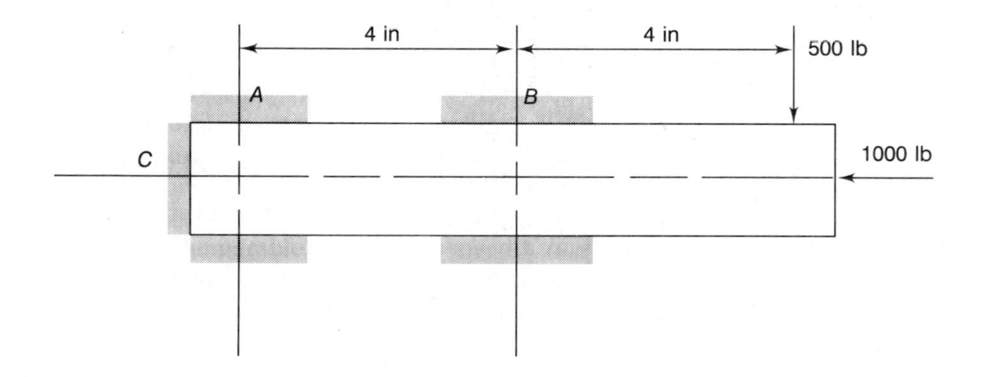

The oil supply is at a pressure of 1500 lb/in^2, and the oil viscosity can be assumed to be 0.8 P. Nominal radial clearance c is 0.001 in.

Select the appropriate pocket sizes, land widths, and capillary dimensions for the journal bearing so that the radial movement at the point of load application does not exceed 0.001 in (exclusive of any deflection due to shaft bending), and select a suitable pocket size for the thrust bearing.

Also determine the full-load and zero-load flow required. Comment on the results that you have obtained, on any assumptions that you have made, and on any additional information that would be desirable when designing this bearing layout.

14.8.3 A shaft sits in two hydrostatic bearings and is centrally loaded by a force of 2400 N as shown in the sketch below.

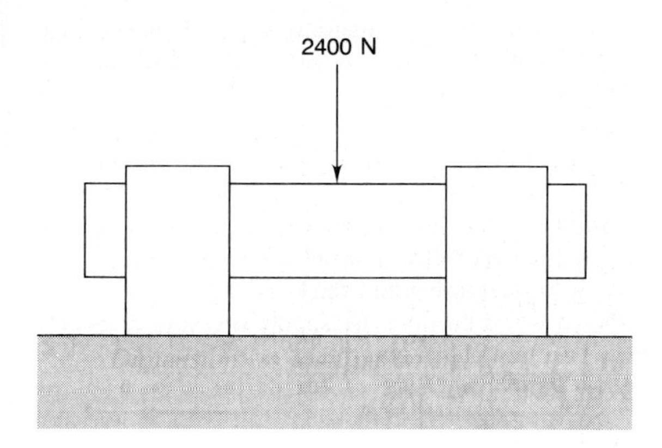

The shaft does not rotate but is simply used as a low-friction pivot for very small oscillatory motion. The diameter of the shaft is 50 mm, and each of the bearings is 50 mm long and has pockets. The bearing clearance is 40 μm, and the supply pressure has to be below 2.4 MN/m^2.

What dimensions would you make the lands of the bearing, what would be the minimum film thickness under load (ignore shaft deflections), and what would be the fluid flow rate for the two bearings if the oil viscosity is 0.05 $N \cdot s/m^2$?

Use capillary restrictors as compensators and specify the size of these. (Specific gravity of the oil is 0.9.)

14.8.4 The stepped-parallel bearing shown below is to be used as a pump with a small volume flow rate of q per unit bearing width of an incompressible lubricant of density ρ and viscosity η being drawn off at the step location. Derive the relationship between the flow rate and supply pressure while neglecting side

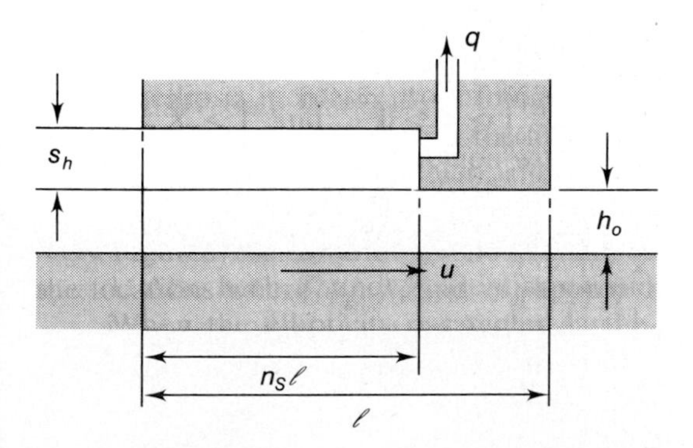

leakage and using the following dimensionalization.

$$Q = \frac{2q'}{us_h} \qquad P_s = \frac{s_h^2 p_s}{\eta ul} \qquad H = \frac{h_o}{s_h}$$

Discuss the practicality of the arrangement, outlining any suggestions you may have for making it more useful from this point of view. Also describe any possible applications for the device.

14.8.5 (*a*) In a circular externally pressurized bearing with capillary restrictor, what radius ratio r_i/r_o would you select as a starting point for the design and why?

(*b*) What ratio would you normally select for p_r/p_s, where p_r is the recess pressure and p_s is the supply pressure? Give reasons for your choice.

(*c*) How does the stiffness vary with decreasing film thickness?

(*d*) If the applied load is reduced while keeping the supply pressure constant, what happens to the recess pressure and the film thickness?

(*e*) If the capillary restrictor has its diameter reduced by part of the system to half its former value, what factor of the original film thickness would the pad's final film thickness be?

14.9 REFERENCE

Rippel, H. C. (1963): *Cast Bronze Hydrostatic Bearing Design Manual*, 2d ed. Cast Bronze Bearing Institute, Inc., Cleveland.

CHAPTER 15

HYDRODYNAMIC BEARINGS—CONSIDERING HIGHER-ORDER EFFECTS

In Chap. 7 the Reynolds number $\mathscr{R}$ was defined as the ratio of inertia effects to viscous effects. It was found that in typical hydrodynamically lubricated thrust and journal bearings the inertia effects are small relative to the viscous effects. However, inertia effects can sometimes be important; for example:

1. Any change in pressure in the entrance region of a foil bearing can greatly affect the behavior of the entire bearing. Such a change in pressure can be caused by inertia effects if the fluid velocity and the inertia are high enough to create a pressure component ($\frac{1}{2}\rho u^2$) comparable to the bearing pressure (T/r_o, where T is the foil tension).
2. Inertia effects can be significant if a separation bubble exists when flow enters the film around a sharp-edge corner as is the case in hydrostatic (also referred to as "externally pressurized") bearings when flow goes from the recessed region to the bearing region. This effect is the same as the vena contracta effect that occurs in the flow through sharp-edge orifices. The Reynolds number is generally larger for hydrostatic bearings than for any other type of hydrodynamic bearing owing to the fact that the film thickness is larger, especially in the recessed region.

These two examples typify the sort of applications where inertia effects might need to be considered. Furthermore, as the operational speed of bearings increases, inertia effects become more important.

Besides the importance of the inertia terms of the Navier-Stokes equations, the neglected viscous terms in deriving the Reynolds equation can also prove to be important in some situations. Two examples are given in Myllerup and Hamrock (1992). The first example is for any surface feature resulting in a film geometry where $dh/dx = O(1)$ or $d^2h/dx^2 = O(1)$. The singularity arising from the curvature is less severe than the one arising from the film gradient, but in both cases using the Reynolds equation will not allow a proper matching between the fluid regions. The second example where the Reynolds equation will not describe what occurs in the lubricant conjunction is when a solid particle is within the conjunction.

This chapter will focus on how to consider the neglected inertia terms of the Navier-Stokes equations, since this is well understood whereas the consideration of the additional viscous terms in the Navier-Stokes equations is a topic of current research. The analysis is as general as possible, with an application at the end of the chapter to a fixed-incline slider bearing. To illustrate the results of considering inertia effects, a simple case is considered where

1. Viscosity is assumed to be constant.
2. Steady-state conditions prevail.
3. Side-leakage effects are neglected.
4. No body forces exist.
5. The fluid is considered to be incompressible.

The Navier-Stokes equations applicable under these conditions can be written from Eqs. (6.28) to (6.30) as

$$u\frac{\partial u}{\partial x} + w\frac{\partial u}{\partial z} = -\frac{1}{\rho_0}\frac{\partial p}{\partial x} + \frac{\eta_0}{\rho_0}\left(\frac{\partial^2 u}{\partial x^2} + \frac{\partial^2 u}{\partial z^2}\right) \tag{15.1}$$

$$u\frac{\partial w}{\partial x} + w\frac{\partial w}{\partial z} = -\frac{1}{\rho_0}\frac{\partial p}{\partial z} + \frac{\eta_0}{\rho_0}\left(\frac{\partial^2 w}{\partial x^2} + \frac{\partial^2 w}{\partial z^2}\right) \tag{15.2}$$

The corresponding continuity equation under these conditions is

$$\frac{\partial u}{\partial x} + \frac{\partial w}{\partial z} = 0 \tag{15.3}$$

The velocity components are defined in terms of stream functions, which are introduced in Sec. 8.4.9. Therefore, the Navier-Stokes and continuity equations are made dimensionless by letting

$$x = \ell X \qquad z = h_o Z \qquad u = u_o \overline{\Phi}_Z(X, Z) \qquad \overline{\Phi}_Z = \frac{\partial \overline{\Phi}}{\partial Z} \tag{15.4}$$

$$p = \frac{\eta_0 \ell u_o}{h_o^2} P \qquad w = -\frac{u_o h_o}{\ell} \overline{\Phi}_X(X, Z) \qquad \overline{\Phi}_X = \frac{\partial \overline{\Phi}}{\partial X}$$

The scale of the velocities is done this way to ensure that $\partial w / \partial z$ does not vanish, since it is required to exist in order to get physically meaningful solutions.

Substituting Eqs. (15.4) into Eqs. (15.1) to (15.3) while making use of Eq. (7.4) gives

$$\mathscr{R}_x \left(\overline{\Phi}_Z \overline{\Phi}_{ZX} - \overline{\Phi}_X \overline{\Phi}_{ZZ} \right) = -\frac{\partial P}{\partial X} + \left(\frac{h_o}{\ell} \right)^2 \overline{\Phi}_{ZXX} + \overline{\Phi}_{ZZZ} \tag{15.5}$$

$$\mathscr{R}_x \left(\frac{h_o}{\ell} \right)^2 \left(\overline{\Phi}_X \overline{\Phi}_{XZ} - \overline{\Phi}_Z \overline{\Phi}_{XX} \right) = -\frac{\partial P}{\partial Z} - \left(\frac{h_o}{\ell} \right)^2 \left[\overline{\Phi}_{XZZ} + \left(\frac{h_o}{\ell} \right)^2 \overline{\Phi}_{XXX} \right] \tag{15.6}$$

$$\overline{\Phi}_{ZX} = \overline{\Phi}_{XZ} \tag{15.7}$$

These then are the first and third Navier-Stokes equations and the continuity equation, respectively, for the earlier defined situation.

Diprima and Stuart (1972) recommend the following method of solution to the nonlinear problem described in these equations:

$$\overline{\Phi} = \overline{\Phi}_{00}(X, Z) + \left(\frac{h_o}{\ell} \right)^2 \overline{\Phi}_{01}(X, Z) + \mathscr{R}_x \overline{\Phi}_{10}(X, Z) + \left(\frac{h_o}{\ell} \right)^4 \overline{\Phi}_{02}(X, Z)$$
$$+ \left(\frac{h_o}{\ell} \right)^2 \mathscr{R}_x \overline{\Phi}_{11}(X, Z) + \mathscr{R}_x^2 \overline{\Phi}_{20}(X, Z) + \cdots \tag{15.8}$$

and

$$P = P_{00}(X, Z) + \left(\frac{h_o}{\ell} \right)^2 P_{01}(X, Z) + \mathscr{R}_x P_{10}(X, Z) + \left(\frac{h_o}{\ell} \right)^4 P_{02}(X, Z)$$
$$+ \left(\frac{h_o}{\ell} \right)^2 \mathscr{R}_x P_{11}(X, Z) + \mathscr{R}_x^2 P_{20}(X, Z) + \cdots \tag{15.9}$$

Substituting Eqs. (15.8) and (15.9) into Eq. (15.5) gives

$$\begin{aligned}
\mathscr{R}_x \Bigg\{ & \left[\frac{\partial \overline{\Phi}_{00}}{\partial Z} + \left(\frac{h_o}{\ell} \right)^2 \frac{\partial \overline{\Phi}_{01}}{\partial Z} + \mathscr{R}_x \frac{\partial \overline{\Phi}_{10}}{\partial Z} + \cdots \right] \\
& \times \left[\frac{\partial^2 \overline{\Phi}_{00}}{\partial Z \, \partial X} + \left(\frac{h_o}{\ell} \right)^2 \frac{\partial^2 \overline{\Phi}_{01}}{\partial Z \, \partial X} + \mathscr{R}_x \frac{\partial^2 \overline{\Phi}_{10}}{\partial Z \, \partial X} + \cdots \right] \\
& - \left[\frac{\partial \overline{\Phi}_{00}}{\partial X} + \left(\frac{h_o}{\ell} \right)^2 \frac{\partial \overline{\Phi}_{01}}{\partial X} + \mathscr{R}_x \frac{\partial \overline{\Phi}_{10}}{\partial X} + \cdots \right] \\
& \times \left[\frac{\partial^2 \overline{\Phi}_{00}}{\partial Z^2} + \left(\frac{h_o}{\ell} \right)^2 \frac{\partial^2 \overline{\Phi}_{01}}{\partial Z^2} + \mathscr{R}_x \frac{\partial^2 \overline{\Phi}_{10}}{\partial Z^2} + \cdots \right] \Bigg\} \\
= & - \frac{\partial P_{00}}{\partial X} - \left(\frac{h_o}{\ell} \right)^2 \frac{\partial P_{01}}{\partial X} - \mathscr{R}_x \frac{\partial P_{10}}{\partial X} + \left(\frac{h_o}{\ell} \right)^2 \\
& \times \left[\frac{\partial^3 \overline{\Phi}_{00}}{\partial Z \, \partial X^2} + \left(\frac{h_o}{\ell} \right)^2 \frac{\partial^3 \overline{\Phi}_{01}}{\partial Z \, \partial X^2} + \mathscr{R}_x \frac{\partial^3 \overline{\Phi}_{10}}{\partial Z \, \partial X^2} + \cdots \right] \\
& + \frac{\partial^3 \overline{\Phi}_{00}}{\partial Z^3} + \left(\frac{h_o}{\ell} \right)^2 \frac{\partial^3 \overline{\Phi}_{01}}{\partial Z^3} + \mathscr{R}_x \frac{\partial^3 \overline{\Phi}_{10}}{\partial Z^3} + \cdots
\end{aligned} \tag{15.10}$$

Collecting terms of O(1), $O[(h_o/\ell)^2]$, and $O(\mathscr{R}_x)$, respectively, gives

$$0 = -\frac{\partial P_{00}}{\partial X} + \frac{\partial^3 \overline{\Phi}_{00}}{\partial Z^3} \tag{15.11}$$

$$0 = -\frac{\partial P_{01}}{\partial X} + \frac{\partial^3 \overline{\Phi}_{00}}{\partial Z \, \partial X^2} + \frac{\partial^3 \overline{\Phi}_{01}}{\partial Z^3} \tag{15.12}$$

$$\frac{\partial \overline{\Phi}_{00}}{\partial Z} \frac{\partial^2 \overline{\Phi}_{00}}{\partial Z \, \partial X} - \frac{\partial \overline{\Phi}_{00}}{\partial X} \frac{\partial^2 \overline{\Phi}_{00}}{\partial Z^2} = -\frac{\partial P_{10}}{\partial X} + \frac{\partial^3 \overline{\Phi}_{10}}{\partial Z^3} \tag{15.13}$$

These are the terms from the first Navier-Stokes equation. From the third Navier-Stokes equation [Eq. (15.6)] in a similar manner, we get terms of O(1), $O[(h_o/\ell)^2]$, $O(\mathscr{R}_x)$, respectively, as

$$0 = -\frac{\partial P_{00}}{\partial Z} \rightarrow P_{00} = P_{00}(X) \tag{15.14}$$

$$0 = -\frac{\partial P_{01}}{\partial Z} - \frac{\partial^3 \overline{\Phi}_{00}}{\partial X \, \partial Z^2} \tag{15.15}$$

$$0 = -\frac{\partial P_{10}}{\partial Z} \rightarrow P_{10} = P_{10}(X) \tag{15.16}$$

The boundary conditions as they pertain to Eqs. (15.11) to (15.16) are

1. When $z = 0$ or $Z = z/h_o = 0$, then $u = u_o$.
2. When $z = h$ or $Z = h/h_o = H$, then $u = 0$.

Making use of the nondimensionalization given in Eq. (15.4) and applying boundary condition 1 gives

$$u = u_o \frac{\partial \overline{\Phi}}{\partial Z} = u_o \qquad \text{when } Z = 0 \tag{15.17}$$

Recalling Eq. (15.8) gives

$$\frac{\partial \overline{\Phi}}{\partial Z} = \frac{\partial \overline{\Phi}_{00}}{\partial Z} + \left(\frac{h_o}{\ell}\right)^2 \frac{\partial \overline{\Phi}_{01}}{\partial Z} + \mathscr{R}_x \frac{\partial \overline{\Phi}_{10}}{\partial Z} + \cdots = 1 \qquad \text{when } Z = 0$$

This implies that

$$\frac{\partial \overline{\Phi}_{00}}{\partial Z} = 1 \qquad \frac{\partial \overline{\Phi}_{01}}{\partial Z} = \frac{\partial \overline{\Phi}_{10}}{\partial Z} = 0 \qquad \text{when } Z = 0 \tag{15.18}$$

The result of Eq. (15.17) implies that

$$\overline{\Phi} = \overline{\Phi}_{00} = \overline{\Phi}_{01} = \overline{\Phi}_{10} = 0 \qquad \text{when } Z = 0 \tag{15.19}$$

Making use of Eq. (15.4) and boundary condition 2 gives

$$u = u_o \frac{\partial \overline{\Phi}}{\partial Z} = 0 \qquad \text{when } Z = H$$

Making use of Eq. (15.8) gives

$$u = u_o \left[\frac{\partial \overline{\Phi}_{00}}{\partial Z} + \left(\frac{h_o}{\ell}\right)^2 \frac{\partial \overline{\Phi}_{01}}{\partial Z} + \mathscr{R}_x \frac{\partial \overline{\Phi}_{10}}{\partial Z} + \cdots \right] = 0 \qquad \text{when } Z = H$$

$$\therefore \frac{\partial \overline{\Phi}_{00}}{\partial Z} = \frac{\partial \overline{\Phi}_{01}}{\partial Z} = \frac{\partial \overline{\Phi}_{10}}{\partial Z} = 0 \qquad \text{when } Z = H \tag{15.20}$$

$$\left.\begin{aligned} \overline{\Phi}_{00} &= \text{constant} = d_{00} \qquad \text{when } Z = H \\ \overline{\Phi}_{01} &= \text{constant} = d_{01} \qquad \text{when } Z = H \\ \overline{\Phi}_{10} &= \text{constant} = d_{10} \qquad \text{when } Z = H \end{aligned}\right\} \tag{15.21}$$

15.1 ORDER 1 SOLUTIONS (P_{00} AND $\overline{\Phi}_{00}$)

Integrating Eq. (15.11) three times with respect to dZ gives

$$\overline{\Phi}_{00}(X, Z) = \frac{1}{6} Z^3 \frac{dP_{00}}{dX} + \overline{A}_{00}(X) Z^2 + \overline{B}_{00}(X) Z + \overline{C}_{00}(X) \tag{15.22}$$

Making use of the boundary conditions in Eqs. (15.19) and (15.18) implies that

$\overline{C}_{00}(X) = 0$ and $\overline{B}_{00}(X) = 1$, respectively. Therefore, Eq. (15.22) becomes

$$\overline{\Phi}_{00}(X,Z) = \frac{Z^3}{6}\frac{dP_{00}}{dX} + \overline{A}_{00}(X)Z^2 + Z \tag{15.23}$$

Making use of the boundary condition expressed in Eq. (15.20) gives

$$0 = \frac{H^2}{2}\frac{dP_{00}}{dX} + 2\overline{A}_{00}(X)H + 1$$

or

$$\overline{A}_{00}(X) = -\frac{1}{2H}\left(1 + \frac{H^2}{2}\frac{dP_{00}}{dX}\right) \tag{15.24}$$

Substituting Eq. (15.24) into Eq. (15.23) gives

$$\overline{\Phi}_{00}(X,Z) = \frac{Z^3}{6}\frac{dP_{00}}{dX} - \frac{Z^2}{2H}\left(1 + \frac{H^2}{2}\frac{dP_{00}}{dX}\right) + Z \tag{15.25}$$

Making use of the boundary condition expressed in Eq. (15.21) gives

$$\overline{\Phi}_{00}(X,Z) = d_{00} = \frac{H^3}{6}\frac{dP_{00}}{dX} - \frac{H}{2}\left(1 + \frac{H^2}{2}\frac{dP_{00}}{dX}\right) + H$$

or

$$d_{00} = -\frac{H^3}{12}\frac{dP_{00}}{dX} + \frac{H}{2} = \text{constant} \tag{15.26}$$

Taking the derivative with respect to dX gives

$$\frac{d}{dX}\left(H^3\frac{dP_{00}}{dX}\right) = 6\frac{dH}{dX} \tag{15.27}$$

This is the Reynolds equation describing the variation of pressure P_{00} with respect to X. Note also that Eq. (15.27) is exactly the Reynolds equation used in Chap. 8 [Eq. (8.15)] when considering thrust bearings while neglecting inertia and side-leakage terms.

From the definition given in Eq. (15.4)

$$\frac{u}{u_o} = \overline{\Phi}_Z = \frac{\partial\overline{\Phi}}{\partial Z} = \frac{\partial\overline{\Phi}_{00}}{\partial Z} + \text{HOT}$$

where HOT denotes higher-order terms. Ignoring the higher-order terms while making use of Eq. (15.25) gives

$$\frac{u}{u_o} = \frac{Z^2}{2}\frac{dP_{00}}{dX} - \frac{Z}{H}\left(1 + \frac{H^2}{2}\frac{dP_{00}}{dX}\right) + 1$$

or

$$\frac{u}{u_o} = \underbrace{\left(1 - \frac{Z}{H}\right)}_{\text{Couette term}} - \underbrace{\frac{ZH}{2}\frac{dP_{00}}{dX}\left(1 - \frac{Z}{H}\right)}_{\text{Poiseuille term}} \tag{15.28}$$

This velocity component is the planar velocity, or the velocity in the direction of sliding. The velocity in the direction of the film can be written from Eq. (15.4) as

$$\frac{w}{u_o} = -\frac{h_o}{\ell}\overline{\Phi} = -\frac{h_o}{\ell}\frac{\partial \overline{\Phi}}{\partial X} \tag{15.29}$$

15.2 ORDER $(h_o/\ell)^2$ SOLUTIONS (P_{01} AND $\overline{\Phi}_{01}$)

Integrating Eq. (15.15) with respect to dZ gives

$$P_{01}(X,Z) = -\frac{\partial^2 \overline{\Phi}_{00}}{\partial Z\,\partial X} + \overline{A}_{01}(X) \tag{15.30}$$

Substituting Eq. (15.30) into Eq. (15.12) gives

$$\frac{\partial^3 \overline{\Phi}_{01}}{\partial Z^3} = \frac{d\overline{A}_{01}}{dX} - 2\frac{\partial^3 \overline{\Phi}_{00}}{\partial Z\,\partial X^2}$$

Substituting Eq. (15.25) into this equation gives

$$\frac{\partial^3 \overline{\Phi}_{01}}{\partial Z^3} = \frac{d\overline{A}_{01}}{dX} - Z^2\frac{d^3 P_{00}}{dX^3} + 2Z\frac{d^2}{dX^2}\left(\frac{H}{2}\frac{dP_{00}}{dX} + \frac{1}{H}\right)$$

Integrating with respect to dZ three times gives

$$\overline{\Phi}_{01}(X,Z) = \frac{Z^3}{6}\frac{d\overline{A}_{01}}{dX} - \frac{Z^5}{60}\frac{d^3 P_{00}}{dX^3} + \frac{Z^4}{12}\frac{d^2}{dX^2}\left(\frac{H}{2}\frac{dP_{00}}{dX} + \frac{1}{H}\right)$$
$$+ \overline{B}_{01}(X)Z^2 + \overline{C}_{01}(X)Z + \overline{E}_{01}(X) \tag{15.31}$$

Making use of the boundary conditions expressed in Eqs. (15.18) and (15.19) results in $\overline{C}_{01} = 0$ and $\overline{E}_{01} = 0$, respectively. Making use of the boundary condition expressed in Eq. (15.20) gives

$$0 = \frac{H^2}{2}\frac{d\overline{A}_{01}}{dX} - \frac{H^4}{12}\frac{d^3 P_{00}}{dX^3} + \frac{H^3}{3}\frac{d^2}{dX^2}\left(\frac{H}{2}\frac{dP_{00}}{dX} + \frac{1}{H}\right) + 2H\overline{B}_{01} \tag{15.32}$$

Substituting the boundary condition expressed in Eq. (15.21) gives

$$d_{01} = \frac{H^3}{6}\frac{d\overline{A}_{01}}{dX} - \frac{H^5}{60}\frac{d^3 P_{00}}{dX^3} + \frac{H^4}{12}\frac{d^2}{dX^2}\left(\frac{H}{2}\frac{dP_{00}}{dX} + \frac{1}{H}\right) + H^2\overline{B}_{01} \tag{15.33}$$

Equations (15.32) and (15.33) have two equations and two unknowns, $\overline{B}_{01}$ and $d\overline{A}_{01}/dX$. Multiplying Eq. (15.32) by $H/2$ while subtracting from Eq. (15.33) gives

$$d_{01} = -\frac{H^3}{12}\frac{d\overline{A}_{01}}{dX} + \frac{H^5}{40}\frac{d^3 P_{00}}{dX^3} - \frac{H^4}{12}\frac{d^2}{dX^2}\left(\frac{H}{2}\frac{dP_{00}}{dX} + \frac{1}{H}\right) \tag{15.34}$$

Differentiating Eq. (15.34) with respect to X turns it into a second-order linear nonhomogeneous equation for $\bar{A}_{01}(X)$.

Solving for $d\bar{A}_{01}/dX$ in Eq. (15.34) while expanding this equation gives

$$\frac{d\bar{A}_{01}}{dX} = -\frac{H^2}{5}\frac{d^3P_{00}}{dX^3} - H\frac{dH}{dX}\frac{d^2P_{00}}{dX^2} - \frac{H}{2}\frac{d^2H}{dX^2}\frac{dP_{00}}{dX} - H\frac{d^2}{dX^2}\left(\frac{1}{H}\right) - \frac{12d_{01}}{H^3} \tag{15.35}$$

From Eq. (15.27)

$$H^3\frac{d^2P_{00}}{dX^2} + 3H^2\frac{dH}{dX}\frac{dP_{00}}{dX} = 6\frac{dH}{dX} \tag{15.36}$$

$$H^3\frac{d^3P_{00}}{dX^3} + 6H^2\frac{dH}{dX}\frac{d^2P_{00}}{dX^2} + \frac{dP_{00}}{dX}\frac{d}{dX}\left(3H^2\frac{dH}{dX}\right) = 6\frac{d^2H}{dX^2} \tag{15.37}$$

From Eq. (15.36)

$$\frac{d^2P_{00}}{dX^2} = \frac{6}{H^3}\frac{dH}{dX} - \frac{3}{H}\frac{dH}{dX}\frac{dP_{00}}{dX} \tag{15.38}$$

Substituting Eq. (15.38) into Eq. (15.37) gives

$$\frac{d^3P_{00}}{dX^3} = \frac{dP_{00}}{dX}\left[\frac{18}{H^2}\left(\frac{dH}{dX}\right)^2 - \frac{1}{H^3}\frac{d}{dX}\left(3H^2\frac{dH}{dX}\right)\right] + \frac{6}{H^3}\frac{d^2H}{dX^2} - \frac{36}{H^4}\left(\frac{dH}{dX}\right)^2 \tag{15.39}$$

The second and third derivatives in Eq. (15.35) can be eliminated by making use of Eqs. (15.38) and (15.39) to give the following:

$$\frac{d\bar{A}_{01}}{dX} = \frac{dP_{00}}{dX}\left[\frac{3}{5}\left(\frac{dH}{dX}\right)^2 + \frac{H}{10}\frac{d^2H}{dX^2}\right] - \frac{4}{5H^2}\left(\frac{dH}{dX}\right)^2 - \frac{1}{5H}\frac{d^2H}{dX^2} - \frac{12d_{01}}{H^3} \tag{15.40}$$

Equation (15.32) can be rewritten as

$$0 = \frac{H^2}{2}\frac{d\bar{A}_{01}}{dX} + \frac{H^4}{12}\frac{d^3P_{00}}{dX^3} + \frac{H^3}{3}\frac{dH}{dX}\frac{d^2P_{00}}{dX^2} + \frac{H^3}{6}\frac{d^2H}{dX^2}\frac{dP_{00}}{dX} + \frac{2}{3}\left(\frac{dH}{dX}\right)^2 - \frac{H}{3}\frac{d^2H}{dX^2} + 2H\bar{B}_{01}$$

Expressions for $d\bar{A}_{01}/dX$, d^3P_{00}/dX^3, and d^2P_{00}/dX^2 having been developed in Eqs. (15.40), (15.39), and (15.38), respectively, the preceding equation can be rewritten as

$$0 = \frac{H^2}{2}\frac{dP_{00}}{dX}\left[\frac{3}{5}\left(\frac{dH}{dX}\right)^2 + \frac{H}{10}\frac{d^2H}{dX^2}\right] - \frac{2}{5}\left(\frac{dH}{dX}\right)^2 - \frac{H}{10}\frac{d^2H}{dX^2} - \frac{6d_{01}}{H}$$

$$+ \frac{H^4}{12}\frac{dP_{00}}{dX}\left[\frac{18}{H^2}\left(\frac{dH}{dX}\right)^2 - \frac{3}{H}\frac{d^2H}{dX^2} - \frac{6}{H^2}\left(\frac{dH}{dX}\right)^2\right] + \frac{H}{2}\frac{d^2H}{dX^2} - 3\left(\frac{dH}{dX}\right)^2$$

$$+ \frac{H^3}{3}\frac{dH}{dX}\left(\frac{6}{H^3}\frac{dH}{dX} - \frac{3}{H}\frac{dH}{dX}\frac{dP_{00}}{dX}\right) + \frac{H^3}{6}\frac{d^2H}{dX^2}\frac{dP_{00}}{dX} + \frac{2}{3}\left(\frac{dH}{dX}\right)^2$$

$$- \frac{H}{3}\frac{d^2H}{dX^2} + 2H\overline{B}_{01}$$

This equation can be rewritten as

$$\overline{B}_{01} = \frac{dP_{00}}{dX}\left[-\frac{3H}{20}\left(\frac{dH}{dX}\right)^2 + \frac{H}{60}\frac{d^2H}{dX^2}\right] + \frac{11}{30H}\left(\frac{dH}{dX}\right)^2 - \frac{1}{30}\frac{d^2H}{dX^2} + \frac{3d_{01}}{H^2} \tag{15.41}$$

Having an expression for $\overline{B}_{01}$ as well as Eqs. (15.35), (15.36), and (15.39) while recalling that $\overline{C}_{01} = \overline{E}_{01} = 0$, we can evaluate the stream function $\overline{\Phi}_{01}(X, Z)$ given in Eq. (15.31).

15.3 INERTIA CORRECTION (P_{10} and $\overline{\Phi}_{10}$)

Substituting Eq. (15.25) into (15.13) while recalling Eq. (15.16) gives

$$\left[\frac{Z^2}{2}\frac{dP_{00}}{dX} - \frac{Z}{H}\left(1 + \frac{H^2}{2}\frac{dP_{00}}{dX}\right) + 1\right]\left[\frac{Z^2}{2}\frac{d^2P_{00}}{dX^2} - Z\frac{d}{dX}\left(\frac{1}{H} + \frac{H}{2}\frac{dP_{00}}{dX}\right)\right]$$

$$-\left[\frac{Z^3}{6}\frac{d^2P_{00}}{dX^2} - \frac{Z^2}{2}\frac{d}{dX}\left(\frac{1}{H} + \frac{H}{2}\frac{dP_{00}}{dX}\right)\right]\left[Z\frac{dP_{00}}{dX} - \frac{1}{H}\left(1 + \frac{H^2}{2}\frac{dP_{00}}{dX}\right)\right]$$

$$= -\frac{\partial P_{10}}{\partial X} + \frac{\partial^3\overline{\Phi}_{10}}{\partial Z^3}$$

Expanding this equation gives

$$\frac{Z^4}{12}\left(\frac{dP_{00}}{dX}\right)\left(\frac{d^2P_{00}}{dX^2}\right) - \frac{Z^3}{3}\left(\frac{1}{H} + \frac{H}{2}\frac{dP_{00}}{dX}\right)\frac{d^2P_{00}}{dX^2} + \frac{Z^2}{2}\left(\frac{1}{H} + \frac{H}{2}\frac{dP_{00}}{dX}\right)$$

$$\times\frac{d}{dX}\left(\frac{1}{H} + \frac{H}{2}\frac{dP_{00}}{dX}\right) + \frac{Z^2}{2}\frac{d^2P_{00}}{dX^2} - Z\frac{d}{dX}\left(\frac{1}{H} + \frac{H}{2}\frac{dP_{00}}{dX}\right)$$

$$= -\frac{dP_{10}}{dX} + \frac{\partial^3\overline{\Phi}_{10}}{\partial Z^3} \tag{15.42}$$

Integrating three times gives

$$\frac{2Z^7}{7!}\left(\frac{dP_{00}}{dX}\right)\left(\frac{d^2P_{00}}{dX^2}\right) - \frac{2Z^6}{6!}\left(\frac{1}{H} + \frac{H}{2}\frac{dP_{00}}{dX}\right)\frac{d^2P_{00}}{dX^2} + \frac{Z^5}{5!}\left(\frac{1}{H} + \frac{H}{2}\frac{dP_{00}}{dX}\right)$$

$$\times\frac{d}{dX}\left(\frac{1}{H} + \frac{H}{2}\frac{dP_{00}}{dX}\right) + \frac{Z^5}{5!}\frac{d^2P_{00}}{dX^2} - \frac{Z^4}{4!}\frac{d}{dX}\left(\frac{1}{H} + \frac{H}{2}\frac{dP_{00}}{dX}\right)$$

$$+ \frac{Z^3}{3!}\frac{dP_{10}}{dX} + \frac{Z^2}{2!}\bar{A}_{10}(X) + Z\bar{B}_{10}(X) + \bar{C}_{10}(X) = \bar{\Phi}_{10}(X, Z) \quad (15.43)$$

The boundary conditions are

1. $Z = 0$ and $\bar{\Phi}_{10} = 0$.
2. $Z = 0$ and $\partial\bar{\Phi}_{10}/\partial Z = 0$.
3. $Z = H$ and $\bar{\Phi}_{10} = \text{constant} = d_{10}$.
4. $Z = H$ and $\partial\bar{\Phi}_{10}/\partial Z = 0$.

Making use of boundary conditions 1 and 2 gives $\bar{C}_{10} = \bar{B}_{10} = 0$. From boundary condition 3

$$\frac{\bar{A}_{10}}{2} = \frac{d_{10}}{H^2} - \frac{2H^5}{7!}\left(\frac{dP_{00}}{dX}\right)\left(\frac{d^2P_{00}}{dX^2}\right) + \frac{2H^4}{6!}\left(\frac{1}{H} + \frac{H}{2}\frac{dP_{00}}{dX}\right)\frac{d^2P_{00}}{dX^2}$$

$$- \frac{H^3}{5!}\left(\frac{1}{H} + \frac{H}{2}\frac{dP_{00}}{dX}\right)\frac{d}{dX}\left(\frac{1}{H} + \frac{H}{2}\frac{dP_{00}}{dX}\right) - \frac{H^3}{5!}\frac{d^2P_{00}}{dX^2}$$

$$+ \frac{H^2}{4!}\frac{d}{dX}\left(\frac{1}{H} + \frac{H}{2}\frac{dP_{00}}{dX}\right) - \frac{H}{3!}\frac{dP_{10}}{dX} \quad (15.44)$$

Boundary condition 4 gives

$$\frac{d_{10}}{H^2} = -\frac{5H^5}{7!}\left(\frac{dP_{00}}{dX}\right)\left(\frac{d^2P_{00}}{dX^2}\right) + \frac{4H^4}{6!}\left(\frac{1}{H} + \frac{H}{2}\frac{dP_{00}}{dX}\right)\frac{d^2P_{00}}{dX^2}$$

$$- \frac{3H^3}{4(5!)}\frac{d}{dX}\left[\left(\frac{1}{H} + \frac{H}{2}\frac{dP_{00}}{dX}\right)^2\right] - \frac{3H^3}{2(5!)}\frac{d^2P_{00}}{dX^2}$$

$$+ \frac{H^2}{4!}\frac{d}{dX}\left(\frac{1}{H} + \frac{H}{2}\frac{dP_{00}}{dX}\right) - \frac{H}{12}\frac{dP_{10}}{dX} \quad (15.45)$$

Making use of the preceding equation gives Eq. (15.43) as

$$\overline{\Phi}_{10} = \left(\frac{dP_{00}}{dX}\right)\left(\frac{d^2P_{00}}{dX^2}\right)\left[\frac{2Z^7}{7!} - \frac{HZ^6}{6!} + \frac{H^2Z^5}{4(5!)} - \frac{7H^5Z^2}{(4!)(5!)}\right]$$
$$+ \frac{d^2P_{00}}{dX^2}\left[-\frac{2Z^6}{6!H} + \frac{9H^3Z^2}{4(5!)} + \frac{3Z^5}{2(5!)} - \frac{HZ^4}{2(4!)}\right]$$
$$+ \frac{dH}{dX}\left(\frac{dP_{00}}{dX}\right)^2 \frac{(2HZ^5 - 5H^4Z^2)}{8(5!)} - \frac{\left(Z^5 - \frac{5}{2}H^3Z^2\right)}{5!H^3}\frac{dH}{dX}$$
$$- \frac{dP_{00}}{dX}\frac{dH}{dX}\frac{(Z^4 - 2H^2Z^2)}{2(4!)} + \frac{dH}{dX}\frac{(Z^4 - 2H^2Z^2)}{4!H^2}$$
$$+ \frac{dP_{10}}{dX}\left(\frac{Z^3 - 3HZ^2/2}{3!}\right) \tag{15.46}$$

Thus far in this chapter the formulation is in a general form. The remainder of the chapter (since it will be related to bearings) is more specific. For example, the pressure is assumed to be known at two end points *a* and *b*. This implies that

1. At $X = 0, P = P_a \rightarrow P = P_{00} + (h_o/\ell)^2 P_{01} + \mathscr{R}_x P_{10} + \cdots = P_a$.
2. At $X = 1, P = P_b \rightarrow P = P_{00} + (h_o/\ell)^2 P_{01} + \mathscr{R}_x P_{10} + \cdots = P_b$.

Therefore, these conditions can be rewritten as

1. At $X = 0$, $P_{00} = P_a$ and $P_{01} = P_{10} = 0$.
2. At $X = 1$, $P_{00} = P_b$ and $P_{01} = P_{10} = 0$.

These conditions will be satisfied in establishing the pressure.

15.4 FORCE COMPONENTS

This section uses the already developed expressions for the stream function and the pressure in determining the force components as they relate to bearing lubrication. From Eq. (6.8) the normal stresses acting on a fluid element when the dilatation term is neglected (since this chapter is concerned with an incompressible fluid) can be written as

$$\sigma_z = -p + 2\eta_0 \frac{\partial w}{\partial z} \tag{15.47}$$

$$\sigma_x = -p + 2\eta_0 \frac{\partial u}{\partial x} \tag{15.48}$$

Also, from Eq. (6.12) the shear stress can be written as

$$\tau_{xz} = \eta_0\left(\frac{\partial u}{\partial z} + \frac{\partial w}{\partial x}\right) \tag{15.49}$$

Note that only the x-dimensional stress field is considered, since at the beginning of the chapter it was assumed that side leakage can be neglected.

Making use of the nondimensionalization given in Eq. (15.4) while making use of Eqs. (15.8) and (15.9) changes Eqs. (15.47) to (15.49) as follows:

$$\sigma_z = -\frac{\eta_0 \ell u_o}{h_o^2}\left[P_{00} + \left(\frac{h_o}{\ell}\right)^2 P_{01} + \mathscr{R}_x P_{10} + 2\left(\frac{h_o}{\ell}\right)^2 \frac{\partial^2 \overline{\Phi}_{00}}{\partial X\,\partial Z} + \mathscr{R}_x \frac{\partial^2 \overline{\Phi}_{10}}{\partial X\,\partial Z} + \cdots\right] \tag{15.50}$$

$$\sigma_x = -\frac{\eta_0 \ell u_o}{h_o^2}\left[P_{00} + \left(\frac{h_o}{\ell}\right)^2 P_{01} + \mathscr{R}_x P_{10} - 2\left(\frac{h_o}{\ell}\right)^2 \frac{\partial^2 \overline{\Phi}_{00}}{\partial X\,\partial Z} - \mathscr{R}_x \frac{\partial^2 \overline{\Phi}_{10}}{\partial X\,\partial Z} + \cdots\right] \tag{15.51}$$

$$\tau_{xz} = \frac{\eta_0 u_o}{h_o}\left[\frac{\partial^2 \overline{\Phi}_{00}}{\partial Z^2} + \left(\frac{h_o}{\ell}\right)^2 \frac{\partial^2 \overline{\Phi}_{01}}{\partial Z^2} + \mathscr{R}_x \frac{\partial^2 \overline{\Phi}_{10}}{\partial Z^2} - \left(\frac{h_o}{\ell}\right) \frac{\partial^2 \overline{\Phi}_{00}}{\partial X^2} - \left(\frac{h_o}{\ell}\right)^3 \frac{\partial^2 \overline{\Phi}_{01}}{\partial X^2} - \left(\frac{h_o}{\ell}\right) \mathscr{R}_x \frac{\partial^2 \overline{\Phi}_{10}}{\partial X^2} + \cdots\right] \tag{15.52}$$

The normal applied load per unit length is

$$w_z' = \int_0^\ell (\sigma_z - p_a)\, dx$$

Substituting Eq. (15.50) gives

$$W_z = \frac{w_z' h_o^2}{\eta_0 u_o \ell^2} = -\int_0^1 \left[P_{00} - P_a + \left(\frac{h_o}{\ell}\right)^2 P_{01} + \mathscr{R}_x P_{10} + 2\left(\frac{h_o}{\ell}\right)^2 \frac{\partial^2 \overline{\Phi}_{00}}{\partial X\,\partial Z} + \mathscr{R}_x \frac{\partial^2 \overline{\Phi}_{10}}{\partial X\,\partial Z} + \cdots\right] dX$$

$$\therefore W_z = W_{z,00} + \left(\frac{h_o}{\ell}\right)^2 W_{z,01} + \mathscr{R}_x W_{z,10} + \cdots \tag{15.53}$$

where
$$W_{z,00} = -\int_0^1 (P_{00} - P_a)\, dX \tag{15.54}$$

$$W_{z,01} = -\int_0^1 \left(P_{01} + 2\frac{\partial^2 \overline{\Phi}_{00}}{\partial X\, \partial Z} \right) dX \tag{15.55}$$

$$W_{z,10} = -\int_0^1 \left(P_{10} + \frac{\partial^2 \overline{\Phi}_{10}}{\partial X\, \partial Z} \right) dX \tag{15.56}$$

Integrating Eq. (15.26) gives

$$\int_0^X \frac{dP_{00}}{dX}\, dX = P_{00}(X) - P_a = 6\int_0^X \frac{dX}{H^2} - 12 d_{00} \int_0^X \frac{dX}{H^3} \tag{15.57}$$

Making use of the boundary conditions gives

$$d_{00} = \frac{1}{2} \frac{\int_0^1 dX/H^2}{\int_0^1 dX/H^3} \tag{15.58}$$

Substituting Eq. (15.58) into (15.57) gives

$$P_{00}(X) - P_a = 6\int_0^X \frac{dX}{H^2} - 6\frac{\int_0^1 dX/H^2}{\int_0^1 dX/H^3} \int_0^X \frac{dX}{H^3} \tag{15.59}$$

Substituting Eq. (15.59) into (15.54) gives

$$W_{z,00} = -6\int_0^1 \left(\int_0^X \frac{dX}{H^2} \right) dX + 6\frac{\int_0^1 dX/H^2}{\int_0^1 dX/H^3} \int_0^1 \left(\int_0^X \frac{dX}{H^3} \right) dX \tag{15.60}$$

Integrating by parts gives

$$\int_0^1 \left(\int_0^X \frac{dX}{H^2} \right) dX = \int_0^1 \frac{dX}{H^2} - \int_0^1 \frac{X}{H^2}\, dX$$

$$\int_0^1 \left(\int_0^X \frac{dX}{H^3} \right) dX = \int_0^1 \frac{dX}{H^3} - \int_0^1 \frac{X}{H^3}\, dX$$

Substituting these two equations into Eq. (15.60) gives

$$W_{z,00} = 6\left(\int_0^1 \frac{X}{H^2}\, dX - \frac{\int_0^1 dX/H^2}{\int_0^1 dX/H^3} \int_0^1 \frac{X}{H^3}\, dX \right) \tag{15.61}$$

The evaluation of $W_{z,01}$ and $W_{z,10}$ would have to be done numerically.

The shearing force per unit length at the moving surface can be expressed as

$$f_b' = \int_0^{\ell} (\tau_{xz})_{z=0}\, dx$$

Making use of Eq. (15.52) while nondimensionalizing the preceding equation gives

$$F_b = \frac{f_b' h_o}{\eta_0 u_o \ell} = F_{b,00} + \left(\frac{h_o}{\ell}\right)^2 F_{b,01} + \mathscr{R}_x F_{b,10} + \cdots \tag{15.62}$$

where

$$F_{b,00} = \int_0^1 \left(\frac{\partial^2 \overline{\Phi}_{00}}{\partial Z^2} - \frac{h_o}{\ell} \frac{\partial^2 \overline{\Phi}_{00}}{\partial X^2} \right)_{Z=0} dX \tag{15.63}$$

$$F_{b,01} = \int_0^1 \left(\frac{\partial^2 \overline{\Phi}_{01}}{\partial Z^2} - \frac{h_o}{\ell} \frac{\partial^2 \overline{\Phi}_{01}}{\partial X^2} \right)_{Z=0} dX \tag{15.64}$$

$$F_{b,10} = \int_0^1 \left(\frac{\partial^2 \overline{\Phi}_{10}}{\partial Z^2} - \frac{h_o}{\ell} \frac{\partial^2 \overline{\Phi}_{10}}{\partial X^2} \right)_{Z=0} dX \tag{15.65}$$

From Eq. (15.25)

$$\left(\frac{\partial^2 \overline{\Phi}_{00}}{\partial X^2} \right)_{Z=0} = 0 \tag{15.66}$$

$$\frac{\partial \overline{\Phi}_{00}}{\partial Z} = \frac{Z^2}{2} \frac{dP_{00}}{dX} - \frac{Z}{H}\left(1 + \frac{H^2}{2} \frac{dP_{00}}{dX}\right) + 1$$

$$\frac{\partial^2 \overline{\Phi}_{00}}{\partial Z^2} = Z \frac{dP_{00}}{dX} - \frac{1}{H}\left(1 + \frac{H^2}{2} \frac{dP_{00}}{dX}\right)$$

$$\left(\frac{\partial^2 \overline{\Phi}_{00}}{\partial Z^2} \right)_{Z=0} = -\frac{1}{H}\left(1 + \frac{H^2}{2} \frac{dP_{00}}{dX}\right) \tag{15.67}$$

Substituting Eqs. (15.66) and (15.67) into (15.63) gives

$$F_{b,00} = -\int_0^1 \left(\frac{1}{H} + \frac{H}{2} \frac{dP_{00}}{dX} \right) dX \tag{15.68}$$

From Eqs. (15.26) and (15.58)

$$\frac{dP_{00}}{dX} = \frac{6}{H^2} - \frac{6}{H^3} \frac{\int_0^1 dX/H^2}{\int_0^1 dX/H^3} \tag{15.69}$$

Substituting Eq. (15.69) into (15.68) gives

$$F_{b,00} = -4\int_0^1 \frac{dX}{H} + \frac{3\left(\int_0^1 dX/H^2\right)^2}{\int_0^1 dX/H^3} \tag{15.70}$$

15.5 FIXED-INCLINE SLIDER BEARING

The equations developed thus far in this chapter were written in a general form so that the results are applicable for any film shape. This section applies the solutions to a fixed-incline slider bearing as was considered in Sec. 8.4. Note that the first-order solution of this chapter is exactly the same as the results obtained in Chap. 8. In particular, Eq. (8.30) gives the same results for W_z as those obtained for $W_{z,00}$ from Eq. (15.61) when the film shape definition given in Eq. (8.19) is used. Furthermore, the dimensionless shear stress at the moving surface F_b for a fixed-incline slider bearing given in Eq. (8.32) is exactly the same as that obtained from Eq. (15.70) for $F_{b,00}$ when the film shape given in Eq. (8.19) is used.

Table 15.1 shows how the inertia correction term compares with the first-order term when both the normal applied load and the shear stress are at the moving surface and the ratio of inlet to outlet film thickness is varied. The inertia correction becomes more significant with decreasing film thickness ratio. For $H_o = 2.0$, the inertia correction is only 3.3 percent of the first-order solution of the normal applied load; for $H_o = 0.2$, the inertia correction is 67.6 percent of the first-order solution. For the shear stress calculations when $H_o = 2.0$, the inertia correction is only 5.6 percent of the first-order solution; for $H_o = 0.2$, the inertia correction is 29.8 percent of the first-order solution. The significance of these corrections may be properly viewed when they are substituted into Eqs. (15.53) and (15.62). Table 15.1 also shows that for an incompressible fluid, as was the restriction of the chapter, and as a parallel surface is

TABLE 15.1 Forces of a fixed-incline slider bearing as obtained from first-order solution and inertia corrections

Film thickness ratio, $H_o = h_o/s_h$	Normal applied load		Shear stress	
	First-order solution, $W_{z,00}$	Inertia correction, $W_{z,10}$	First-order solution, $F_{b,00}$	Inertia correction, $F_{b,10}$
2.0	0.0335	0.0011	−0.424	−0.0237
1.0	.158	.0123	−.772	−.0856
.5	.592	.113	−1.396	−.271
.2	2.175	1.470	−2.885	−.861

being approached ($H_o \to \infty$), the inertia correction as well as the first-order solution approaches zero. Of course, the inertia correction approaches zero at a much faster rate than the first-order solution. Therefore, the flow accelerates in order to maintain a constant mass flow.

15.6 CLOSURE

The Reynolds number was defined as the ratio of inertia to viscous effects on a fluid element. Previous chapters indicated that for hydrodynamically lubricated thrust and journal bearings the inertia effects are small relative to the viscous effects. In this chapter analytical methods were presented showing how solutions of pressure, stream functions, and force components can be obtained while considering inertia effects. The analysis was formulated in the most general manner. The pressure and stream functions were expressed in first-order terms, inertia corrections, and $(h_o/\ell)^2$ terms. Analytical solutions were obtained for each of these terms while neglecting terms of higher order. A number of assumptions were imposed, including neglecting side-leakage and body force terms and assuming incompressible, steady-state, and constant viscosity. General expressions were also obtained for the normal load-carrying capacity and the shear stress at the moving surface. At the end of the chapter the general solutions were applied to a fixed-incline slider thrust bearing. The first-order terms for the normal load-carrying capacity and the shear stress were found to be the same as those developed in Chap. 8. Furthermore, it was found that the inertia corrections approached zero as a parallel film shape was approached ($H_o \to \infty$) and became more significant as the film thickness ratio was decreased. It was, however, concluded that it is in general accurate to neglect inertia effects in self-acting thrust and journal bearings, since their contribution is small relative to the viscous effects. Inertia effects may be important when considering the entrance effects in foil bearings or if a separation bubble exists when flow enters the film around a sharp-edge corner, as is the case in an externally pressurized bearing.

15.7 PROBLEM

15.7.1 Consider a fixed-incline bearing as shown in Fig. 8.6. Let $\alpha = s_h/\ell$ and $h = h_o + s_h - \alpha_x$. For this bearing determine expressions for P_{00}, $\overline{\Phi}_{01}$, $\overline{\Phi}_{10}$, P_{10}, $W_{z,00}$, $W_{z,10}$, $F_{b,00}$, and $F_{b,10}$.

15.8 REFERENCES

Diprima, R. C., and Stuart, J. T. (1972): Flow Between Eccentric Rotating Cylinders. *J. Lubr. Technol.*, vol. 94, no. 3, pp. 266–274.

Myllerup, C. M. and Hamrock, B. J. (1992): "Local Effects in Thin Film Lubrication," Presented and published in the 19th Leeds–Lyon Symposium on Tribology.

CHAPTER

16

GAS-LUBRICATED THRUST BEARINGS

Gas-lubricated film bearings have commanded considerable attention in recent years because they possess characteristics that make their use advantageous in many bearing applications. They are analogous to the hydrodynamic oil-lubricated bearings considered thus far except that the fluid is compressible. Furthermore, since air is 1000 times less viscous than even the thinnest mineral oils, the viscous resistance is very much less. However, the distance of nearest approach between the bearing surfaces is also correspondingly smaller so that special precautions must be taken in manufacturing the bearing.

Some strengths of gas-lubricated bearings are as follows:

1. Their friction or viscous resistance is extremely low.
2. The lubricant is ample and clean.
3. The lubricant does not contaminate surfaces.
4. The lubricant operates well from extremely low to extremely high temperatures.
5. The film does not break down from cavitation or ventilation.

Some weaknesses of gas-lubricated bearings are as follows:

1. For the same size bearing the load-carrying capacity of a gas-lubricated bearing is many times less than that of an oil-lubricated bearing.

2. The surfaces must have an extremely fine finish.
3. The alignment must be extremely good.
4. Dimensions and clearances must be extremely accurate.
5. Speed must be high.
6. Loading must be low.
7. Their stability characteristics are poor.

Gas bearings are finding increasing usage in gas-cycle machinery, where the cycle gas is used in the bearings, thus eliminating the need for a conventional lubrication system; in gyros, where precision and constancy of torque are critical; in food and textile processing machinery, where cleanliness and absence of contaminants are critical; and in high-speed dental drills. It is obvious that gas bearings will find applications only where their strengths outweigh their weaknesses. Although their use will always be limited in scope because of inherent weaknesses, it is broad enough to merit an examination of their behavior as a machine element.

16.1 REYNOLDS EQUATION

From Eq. (7.58), the general form of the Reynolds equations can be expressed as

$$\frac{\partial}{\partial x}\left(\frac{\rho h^3}{12\eta}\frac{\partial p}{\partial x}\right)+\frac{\partial}{\partial y}\left(\frac{\rho h^3}{12\eta}\frac{\partial p}{\partial y}\right)=\frac{\partial}{\partial x}\left[\frac{\rho h(u_a+u_b)}{2}\right]+\frac{\partial}{\partial y}\left[\frac{\rho h(v_a+v_b)}{2}\right]+\frac{\partial(\rho h)}{\partial t} \tag{7.58}$$

Now, if $u_a = v_a = v_b = 0$, this equation reduces to

$$\frac{\partial}{\partial x}\left(\frac{\rho h^3}{\eta}\frac{\partial p}{\partial x}\right)+\frac{\partial}{\partial y}\left(\frac{\rho h^3}{\eta}\frac{\partial p}{\partial y}\right)=6u_b\frac{\partial(\rho h)}{\partial x}+12\frac{\partial(\rho h)}{\partial t} \tag{16.1}$$

The viscosity of gases varies little with pressure. The viscosity thus can be considered a function of temperature only and can be accounted for in η_0. Therefore, the viscosity can be assumed to be constant throughout the bearing.

$$\therefore \eta = \eta_0 = \text{constant} \tag{16.2}$$

The equation of state for a perfect gas gives the density as

$$\frac{p}{\rho} = \overline{R}t_m \tag{16.3}$$

where $\overline{R}$ = gas constant = universal gas constant ÷ molecular weight
t_m = gas temperature

Gas bearings normally operate isothermally. In such a case Eq. (16.3) reduces to

$$\frac{p}{\rho} = \text{constant} \tag{16.4}$$

Note that a more general, polytropic law can be used where

$$\frac{p}{\rho^{\bar{n}}} = \text{constant} \tag{16.5}$$

and where

$$1 \le \bar{n} \le \frac{C_p}{C_v} \tag{16.6}$$

and C_p = specific heat at constant pressure, J/(kg · K)
C_v = specific heat at constant volume, J/(kg · K)
$\bar{n}$ = polytropic gas-expansion exponent

When flow is "adiabatic" (i.e., there is no transferred heat and the change in internal energy equals the compression work), $\bar{n} = C_p/C_v$. When flow is "isothermal," $\bar{n} = 1$ and we have the equation of state, Eq. (16.4). Experience shows that Eq. (16.5) applies to many thermodynamic processes.

Inserting Eqs. (16.2) and (16.5) into (16.1) gives

$$\frac{\partial}{\partial x}\left(p^{1/\bar{n}}h^3\frac{\partial p}{\partial x}\right) + \frac{\partial}{\partial y}\left(p^{1/\bar{n}}h^3\frac{\partial p}{\partial y}\right) = 6\eta_0 u_b \frac{\partial}{\partial x}\left(p^{1/\bar{n}}h\right) + 12\eta_0\frac{\partial}{\partial t}\left(p^{1/\bar{n}}h\right) \tag{16.7}$$

Letting $p = p_a P$, $h = h_{\text{min}}H$, $x = \ell X$, $y = bY$, and $t = T/\omega$ gives

$$\frac{\partial}{\partial X}\left(P^{1/\bar{n}}H^3\frac{\partial P}{\partial X}\right) + \lambda^2\frac{\partial}{\partial Y}\left(P^{1/\bar{n}}H^3\frac{\partial P}{\partial Y}\right) = \Lambda_g\frac{\partial}{\partial X}\left(P^{1/\bar{n}}H\right) + \sigma_g\frac{\partial}{\partial T}\left(P^{1/\bar{n}}H\right) \tag{16.8}$$

where

$$\lambda = \frac{\ell}{b} \tag{16.9}$$

is the length-to-width ratio,

$$\Lambda_g = \frac{6\eta_0 u_b \ell}{p_a h_{\text{min}}^2} \tag{16.10}$$

is the dimensionless bearing number, and

$$\sigma_g = \frac{12\eta_0\omega\ell^2}{p_a h_{\text{min}}^2} \tag{16.11}$$

is the dimensionless squeeze number. For isothermal conditions, $\bar{n} = 1$ and Eq.

(16.8) becomes

$$\frac{\partial}{\partial X}\left(PH^3\frac{\partial P}{\partial X}\right)+\lambda^2\frac{\partial}{\partial Y}\left(PH^3\frac{\partial P}{\partial Y}\right)=\Lambda_g\frac{\partial(PH)}{\partial X}+\sigma_g\frac{\partial(PH)}{\partial T} \tag{16.12}$$

Either Eq. (16.8) or (16.12) is normally used in gas lubrication analysis when laminar flow conditions prevail. As can be seen from these equations, the compressibility of gas complicates the analysis by making the Reynolds equation nonlinear. That is, the left sides of Eqs. (16.8) and (16.12) are nonlinear in the variable P.

The first simplification to Eq. (16.12) is to consider only time-invariant lubrication and thereby drop the $\partial(PH)/\partial T$ term. The remaining equation is often reduced further by ignoring either the side-leakage term $\partial(PH^3\,\partial P/\partial Y)/\partial Y$ or the so-called parabolic portion $\partial(PH^3\,\partial P/\partial X)/\partial X$ of the main bearing flow. The first of these approximations is the infinitely wide-bearing approximation; the second is the short-width-bearing approximation. Both were covered while considering journal bearings in Chap. 10. If there is no relative motion, as in hydrostatic bearings, $u_b = 0$ and the Λ_g term in Eq. (16.12) vanishes.

16.1.1 LIMITING SOLUTIONS. For steady-state operation, $\omega = \sigma_g = 0$ and Eq. (16.12) reduces to

$$\frac{\partial}{\partial X}\left(PH^3\frac{\partial P}{\partial X}\right)+\lambda^2\frac{\partial}{\partial Y}\left(PH^3\frac{\partial P}{\partial Y}\right)=\Lambda_g\frac{\partial(PH)}{\partial X} \tag{16.13}$$

There are two important limiting cases in the solution of the compressible Reynolds equation given in (16.13).

16.1.1.1 Extremely low velocities ($u_b \to 0$, $\Lambda_g \to 0$). Equation (16.13) can be differentiated to yield

$$\frac{\partial}{\partial X}\left(H^3\frac{\partial P}{\partial X}\right)+\lambda^2\frac{\partial}{\partial Y}\left(H^3\frac{\partial P}{\partial Y}\right)+\frac{H^3}{P}\left[\left(\frac{\partial P}{\partial X}\right)^2+\lambda^2\left(\frac{\partial P}{\partial Y}\right)^2\right]$$
$$=\Lambda_g\left(\frac{\partial H}{\partial X}+\frac{H}{P}\frac{\partial P}{\partial X}\right) \tag{16.14}$$

Now as $u_b \to 0$ and $\Lambda_g \to 0$, $P \to 1.0$ and the pressure rise $\Delta P \to 0$. Therefore, terms such as $\partial P/\partial X$ are small. In particular, $(\partial P/\partial X)^2 \ll \partial P/\partial X$ and is therefore negligible. Similarly, $(\partial P/\partial Y)^2 \ll \partial P/\partial Y$ and is therefore negligible. Furthermore, it can also be shown that $H/P(\partial P/\partial X) \ll \partial H/\partial X$. Therefore, Eq. (16.14), when $u_b \to 0$ and $\Lambda_g \to 0$, reduces to

$$\frac{\partial}{\partial X}\left(H^3\frac{\partial P}{\partial X}\right)+\lambda^2\frac{\partial}{\partial Y}\left(H^3\frac{\partial P}{\partial Y}\right)=\Lambda_g\frac{\partial H}{\partial X} \tag{16.15}$$

Equation (16.15) is the incompressible form of the Reynolds equation as described in Eq. (7.48). This brings us to a *first* important principle: at extremely low velocities self-acting gas lubrication can be considered incompressible, and the results derived for complete film lubrication for liquids apply directly to low-velocity gas lubrication.

16.1.1.2 Extremely high velocities ($u_b \to \infty$, $\Lambda_g \to \infty$). From Eq. (16.13) we can observe that as $u_b \to \infty$ or as $\Lambda_g \to \infty$ the only way for the pressure to remain finite is for $\partial PH/\partial X \to 0$. This implies that

$$PH = \text{constant} \tag{16.16}$$

and that

$$ph = \text{constant} = p_a h_{\min} \tag{16.17}$$

A fundamental difference exists in Eq. (16.17) between liquid lubrication and gas lubrication. In liquid lubrication the pressure and load (integrated pressure) are directly proportional to velocity and viscosity and independent of ambient pressure. In self-acting, gas-lubricated bearings a *second* important principle is that an extremely high velocity the pressure and load (integrated pressure) become independent of velocity and viscosity but are directly proportional to ambient pressure. Note that the high-velocity asymptote is determined solely from the right side of Eq. (16.13) independent of the terms appearing on the left side of the equation. This means that the high-velocity asymptote for finite-width bearings is the same as that for an infinitely wide bearing. This leads to a *third* important principle as related to gas lubrication, namely, that side leakage can be neglected even on short-width bearings if the velocity is sufficiently high.

These three fundamental principles of self-acting gas lubrication are presented graphically in Fig. 16.1. Repeating these important points:

1. At extremely low velocities gas bearing behavior is closely approximated by incompressible or liquid lubrication solutions for the same bearing geometry.
2. At extremely high velocities ph = constant and load becomes independent of speed.
3. At extremely high velocities side-leakage effects are negligible, since finite- and infinite-width-bearing solutions both approach ph = constant.

16.1.2 SLIP FLOW. The material covered thus far assumes laminar flow. In slip flow the velocity profile through the film is different from that seen in laminar flow. In slip flow a difference exists between surface velocity and average fluid velocity at the surface. Slip flow only becomes important in gas-lubricating films.

The Knudsen number $\mathscr{K}$ is an inverse measure of the average number of molecular collisions in a given film thickness.

$$\mathscr{K} = \frac{\lambda_m}{h} \tag{16.18}$$

where λ_m = mean free molecular path of gas, m
h = film thickness, m

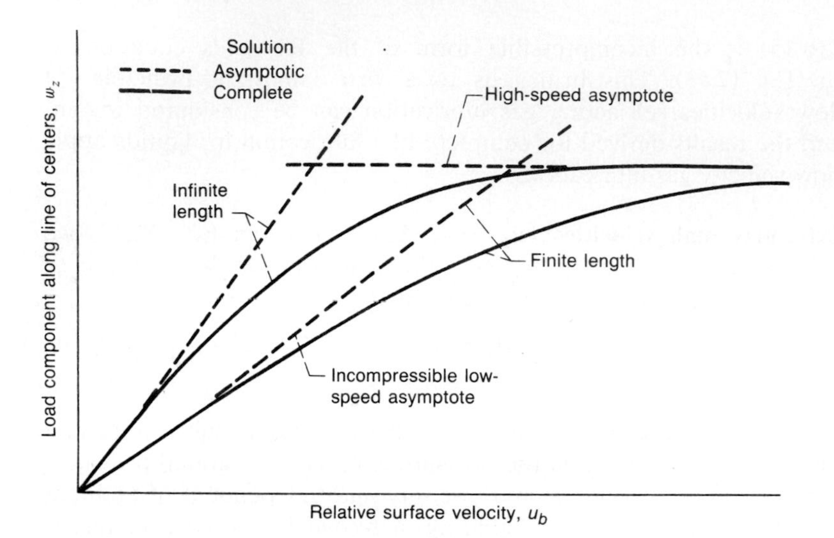

FIGURE 16.1
Effect of speed on load for self-acting, gas-lubricated bearings. [*From Ausman (1961).*]

When $\mathscr{K} < 0.01$, the flow may be treated as a continuum or as laminar flow; when $0.01 < \mathscr{K} < 15$, slip flow becomes significant; and when $\mathscr{K} > 15$, fully developed molecular flow results. The mean free path of air molecules at room temperature and atmospheric pressure is about 0.064 μm. Thus, a film thickness of 2.54 μm yields $\mathscr{K} = 0.025$, and slip is not quite negligible.

Burgdorfer (1959) showed that incorporating slip flow in the boundary conditions results in the following Reynolds equation:

$$\frac{\partial}{\partial X}\left[(1 + 6\mathscr{K})PH^3\frac{\partial P}{\partial X}\right] + \lambda^2\frac{\partial}{\partial Y}\left[(1 + 6\mathscr{K})PH^3\frac{\partial P}{\partial Y}\right] = \Lambda_g\frac{\partial(PH)}{\partial X} + \sigma_g\frac{\partial(PH)}{\partial T} \tag{16.19}$$

For gases $\lambda_m p = \tilde{C}$, a constant,

$$\therefore \mathscr{K} = \frac{\tilde{C}}{ph} \tag{16.20}$$

If $\hat{m}$ is defined as a representative Knudsen number such that

$$\hat{m} = \frac{\tilde{C}}{p_a h_{\min}} \tag{16.21}$$

then

$$\mathscr{K} = \frac{\hat{m}}{PH} \tag{16.22}$$

Substituting Eq. (16.22) into Eq. (16.19) gives

$$\frac{\partial}{\partial X}\left[\left(1+\frac{6\hat{m}}{PH}\right)PH^3\frac{\partial P}{\partial X}\right]+\lambda^2\frac{\partial}{\partial Y}\left[\left(1+\frac{6\hat{m}}{PH}\right)PH^3\frac{\partial P}{\partial Y}\right]$$

$$=\Lambda_g\frac{\partial(PH)}{\partial X}+\sigma_g\frac{\partial(PH)}{\partial T} \tag{16.23}$$

16.2 PARALLEL-SURFACE BEARING

Assuming no slip at the gas-surface interface, isothermal conditions, and time-invariant lubrication and neglecting side leakage, the appropriate Reynolds equation for gas-lubricated bearings is

$$\frac{d}{dX}\left(PH^3\frac{dP}{dX}\right)=\Lambda_g\frac{d(PH)}{dX} \tag{16.24}$$

For a parallel-surface bearing, $H = 1$; therefore,

$$\frac{d}{dX}\left(P\frac{dP}{dX}\right)=\Lambda_g\frac{dP}{dX}$$

or

$$\frac{d^2(P^2)}{dX^2}=2\Lambda_g\frac{dP}{dX} \tag{16.25}$$

Integrating gives

$$\frac{d(P^2)}{dX}=2\Lambda_g P+\tilde{A} \tag{16.26}$$

16.2.1 LOW-BEARING-NUMBER RESULTS.

As $\Lambda_g \to 0$, Eq. (16.26) reduces to

$$\frac{d(P^2)}{dX}=\tilde{A}$$

Integrating gives

$$P^2=\tilde{A}X+\tilde{B}$$

The boundary conditions are

1. $P = P_i$ at $X = 0$. (16.27)
2. $P = P_o$ at $X = 1$. (16.28)

Making use of these boundary conditions gives

$$\tilde{B}=P_i^2$$

$$\tilde{A}=P_o^2-P_i^2$$

$$\therefore P^2=P_i^2+\left(P_o^2-P_i^2\right)X \tag{16.29}$$

Note that if $P_i = P_o$, then $P = P_i = P_o$ everywhere. The results for an incompressible parallel-surface bearing, in Sec. 8.3, are similar. The normal load-carrying capacity is just the integration of the pressure, or

$$w_z = b\int_0^\ell (p - p_a)\, dx$$

or

$$W_z = \frac{w_z}{b\ell p_a} = \int_0^1 (P - 1)\, dX \tag{16.30}$$

Substituting Eq. (16.29) into Eq. (16.30) gives

$$W_z = \int_0^1 \left\{\left[P_i^2 + (P_o^2 - P_i^2)X\right]^{1/2} - 1\right\} dX$$

$$= \frac{2P_o^2 + 2P_oP_i + 2P_i^2 - 3P_o - 3P_i}{3(P_o + P_i)} \tag{16.31}$$

Note that if $P_i = P_o = 1$, $W_z = 0$ for a parallel-surface bearing.

16.2.2 HIGH-BEARING-NUMBER RESULTS. Equation (16.24) reveals that as $\Lambda_g \to \infty$

$$\frac{d(PH)}{dX} \to 0$$

But $H = 1$ for a parallel-surface bearing and

$$\therefore \frac{dP}{dX} \to 0 \tag{16.32}$$

or

$$P = \text{constant} = P_i$$

The normal load-carrying capacity can be obtained by substituting Eq. (16.32) into Eq. (16.30) to give

$$W_z = \int_0^1 (P_i - 1)\, dX = P_i - 1 \tag{16.33}$$

16.2.3 INTERMEDIATE-BEARING-NUMBER RESULTS. Integrating Eq. (16.24) while letting $H = 1$ for a parallel-surface bearing gives

$$P\frac{dP}{dX} = \Lambda_g P + \tilde{A} = \Lambda_g(P + \tilde{B})$$

or

$$\left(1 - \frac{\tilde{B}}{P + \tilde{B}}\right) dP = \Lambda_g\, dX$$

Integrating gives

$$P - \tilde{B}\ln(P + \tilde{B}) = \Lambda_g X + \tilde{C} \tag{16.34}$$

Making use of the boundary conditions expressed in Eqs. (16.27) and (16.28) gives

$$P_i - \tilde{B} \ln(P_i + \tilde{B}) = \tilde{C} \tag{16.35}$$

$$P_o - \tilde{B} \ln(P_o + \tilde{B}) = \Lambda_g + \tilde{C} \tag{16.36}$$

or

$$P_i - P_o + \tilde{B} \ln \frac{P_o + \tilde{B}}{P_i + \tilde{B}} = -\Lambda_g \tag{16.37}$$

This equation shows that $\tilde{B}$ can be determined numerically for given values of P_i, P_o, and Λ_g.

Substituting Eq. (16.35) into Eq. (16.34) gives

$$P - \tilde{B} \ln \frac{P + \tilde{B}}{P_i + \tilde{B}} = P_i + \Lambda_g X \tag{16.38}$$

The normal applied load must be obtained numerically by using Eq. (16.37) to solve for $\tilde{B}$, Eq. (16.38) to solve for P, and Eq. (16.30) to solve for the dimensionless normal load-carrying capacity.

Note that even for the simplest of film shapes a complete analytical solution is not possible for gas-lubricated thrust bearings. Therefore, a linearized analysis needs to be employed if analytical solutions are to be obtained.

16.3 PARALLEL-STEP BEARING

Just as in Sec. 9.1, where a finite-width, parallel-step slider bearing was analyzed for incompressible lubrication, this section analyzes the same bearing (Fig. 16.2) but with gas as a lubricant. In the figure the ridge is where the film thickness is

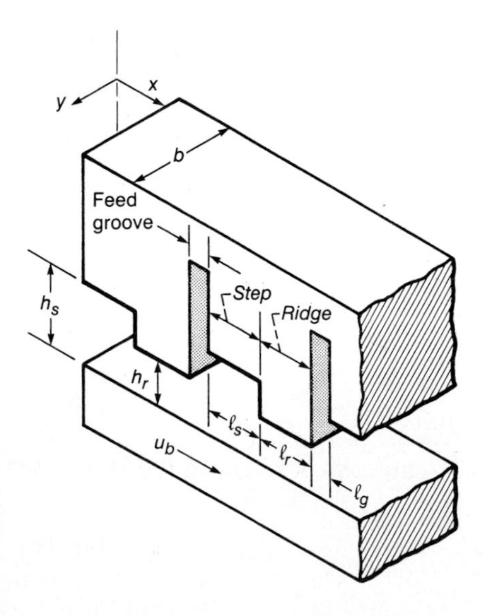

FIGURE 16.2
Rectangular-step thrust bearing. [*From Hamrock (1972)*.]

h_r, and the step is where the film thickness is h_s, or $h_r + \Delta$. The feed groove is the deep groove separating the end of a ridge and the beginning of the next step. Although not shown in this figure, the feed groove is orders of magnitude deeper than h_r. A "pad" is defined as including a ridge, a step, and a feed groove. The feed groove is short relative to the length of the pad. Note that each pad acts independently, since the pressure profile is broken at the lubrication feed groove.

16.3.1 PRESSURE DISTRIBUTION. With no slip at the gas-surface interface, isothermal conditions, time-invariant lubrication, and constant film thickness, Eq. (16.7) becomes

$$\frac{\partial}{\partial x}\left(p\frac{\partial p}{\partial x}\right) + \frac{\partial}{\partial y}\left(p\frac{\partial p}{\partial y}\right) = \frac{6\eta_0 u_b}{h^2}\frac{\partial p}{\partial x} \tag{16.39}$$

Expanding and rearranging terms give

$$\frac{\partial^2 p}{\partial x^2} + \frac{\partial^2 p}{\partial y^2} - \frac{6\eta_0 u_b}{ph^2}\frac{\partial p}{\partial x} = -\frac{1}{p}\left[\left(\frac{\partial p}{\partial x}\right)^2 + \left(\frac{\partial p}{\partial y}\right)^2\right] \tag{16.40}$$

To solve this equation analytically, linearization assumptions need to be imposed. Ausman (1961) first introduced this linearization, which states that

1. The terms on the right side of Eq. (16.40) are small relative to the terms on the left side and therefore can be neglected.
2. p, which appears as a coefficient in the third term of Eq. (16.40), is replaced by the ambient pressure p_a.

Making use of these linearizations reduces Eq. (16.40) to

$$\frac{\partial^2 p}{\partial x^2} + \frac{\partial^2 p}{\partial y^2} = \frac{6\eta_0 u_b}{p_a h^2}\frac{\partial p}{\partial x} \tag{16.41}$$

Recall from Sec. 9.1.1 that for the same type of bearing but with an incompressible lubricant the appropriate Reynolds equation is Eq. (16.41) but with the right side of the equation equal to zero [see Eq. (9.2)].

From Eq. (16.41) the separate linearized Reynolds equations for the ridge and step regions of the finite-width, parallel-step slider are

$$\frac{\partial^2 p_r}{\partial x^2} + \frac{\partial^2 p_r}{\partial y^2} = \frac{6\eta_0 u_b}{p_a h_r^2}\frac{\partial p_r}{\partial x} \tag{16.42}$$

$$\frac{\partial^2 p_s}{\partial x^2} + \frac{\partial^2 p_s}{\partial y^2} = \frac{6\eta_0 u_b}{p_a h_s^2}\frac{\partial p_s}{\partial x} \tag{16.43}$$

Subscript r refers to the ridge (see Fig. 16.2), subscript s refers to the step, and subscript g refers to the feed groove. Letting

$$x = bX \qquad y = bY \qquad p_r = p_a(P_r + 1) \qquad p_s = p_a(P_s + 1) \tag{16.44}$$

changes Eqs. (16.42) and (16.43) to

$$\frac{\partial^2 P_r}{\partial X^2} + \frac{\partial^2 P_r}{\partial Y^2} = \Lambda_a \frac{\partial P_r}{\partial X} \tag{16.45}$$

$$\frac{\partial^2 P_s}{\partial X^2} + \frac{\partial^2 P_s}{\partial Y^2} = \frac{\Lambda_a}{H_a^2} \frac{\partial P_s}{\partial X} \tag{16.46}$$

where
$$\Lambda_a = \frac{6\eta_0 u_b b}{p_a h_r^2} \tag{16.47}$$

$$H_a = \frac{h_s}{h_r} \tag{16.48}$$

Using separation of variables while solving for the pressure gives

$$P_r = \exp\left(\frac{\Lambda_a X}{2}\right)\left\{\tilde{A}_r \exp\left[X\left(\frac{\Lambda_a^2}{4} + J_r^2\right)^{1/2}\right] + \tilde{B}_r \exp\left[-X\left(\frac{\Lambda_a^2}{4} + J_r^2\right)^{1/2}\right]\right\}$$
$$\times\left[\tilde{D}_r \sin(J_r Y) + \tilde{E}_r \cos(J_r Y)\right] \tag{16.49}$$

$$P_s = \exp\left(\frac{\Lambda_a X}{2H_a^2}\right)\left\{\tilde{A}_s \exp\left[X\left(\frac{\Lambda_a^2}{4H_a^4} + J_s^2\right)^{1/2}\right] + \tilde{B}_s \exp\left[-X\left(\frac{\Lambda_a^2}{4H_a^4} + J_s^2\right)^{1/2}\right]\right\}$$
$$\times\left[\tilde{D}_s \sin(J_s Y) + \tilde{E}_s \cos(J_s Y)\right] \tag{16.50}$$

where J_r and J_s are separation constants. The boundary conditions are

1. $P_s = 0$ when $X = 0$.
2. $P_r = 0$ when

$$X = \frac{\ell_s + \ell_r}{b} = \frac{\ell_s + \ell_r}{\ell_s + \ell_r + \ell_g} \frac{\ell_s + \ell_r + \ell_g}{b} = \beta_g \lambda$$

3. $P_r = P_s = \sum_{m=1,3,\ldots}^{\infty} F_m^* \cos(m\pi Y)$ when

$$X = \frac{\ell_s}{b} = \frac{\ell_s}{\ell_s + \ell_r + \ell_g} \frac{\ell_s + \ell_r + \ell_g}{b}$$

   or $X = \psi_g \lambda$, where F_m^* is the Fourier coefficient.
4. $\dfrac{\partial P_r}{\partial Y} = \dfrac{\partial P_s}{\partial Y} = 0$ when $Y = 0$.
5. $P_r = P_s = 0$ when $Y = \frac{1}{2}$.
6. $q_r = q_s$ when $X = \psi_g \lambda$.

For boundary conditions 1 to 5, Eqs. (16.49) and (16.50) become

$$P_r = \sum_{m=1,3,\ldots}^{\infty} \frac{F_m^* \cos(m\pi Y) e^{(\Lambda_a/2)(X-\psi_g\lambda)}}{e^{-\psi_g\lambda\xi_r} - e^{-\lambda\xi_r(2\beta_g-\psi_g)}} \left[e^{-X\xi_r} - e^{\xi_r(2\lambda\beta_g - X)}\right] \quad (16.51)$$

$$P_s = \sum_{m=1,3,\ldots}^{\infty} \frac{F_m^* \cos(m\pi Y) e^{(\Lambda_a/2H_a^2)(X-\psi_g\lambda)}}{e^{-\psi_g\lambda\xi_s} - e^{\psi_g\lambda\xi_s}} \left[e^{-X\xi_s} - e^{X\xi_s}\right] \quad (16.52)$$

where

$$\xi_r = \left[\left(\frac{\Lambda_a}{2}\right)^2 + m^2\pi^2\right]^{1/2}$$

$$\xi_s = \left[\left(\frac{\Lambda_a}{2H_a^2}\right)^2 + m^2\pi^2\right]^{1/2}$$

The linearized equations describing the mass flow across the ridge and the step can be written as

$$q_{m,r} = \frac{\rho_a}{p_a}\left(\frac{p_r u_b h_r}{2} - \frac{p_a h_r^3}{12\eta_0}\frac{\partial p_r}{\partial x}\right)$$

$$q_{m,s} = \frac{\rho_a}{p_a}\left(\frac{p_s u_b h_s}{2} - \frac{p_a h_s^3}{12\eta_0}\frac{\partial p_s}{\partial x}\right).$$

Note that $q_{m,r}$ and $q_{m,s}$ are mass flow rates in the ridge and step regions having units of newton-seconds per square meter. This is quite different from volume flow rates in Chap. 9 having units of cubic meters per second. By making use of Eq. (16.44) these equations can be made dimensionless, where

$$q_{m,r} = \frac{\rho_a p_a h_r^3}{12\eta_0 b}\left[\Lambda_a(P_r + 1) - \frac{\partial P_r}{\partial X}\right]$$

$$q_{m,s} = \frac{\rho_a p_a h_s^3}{12\eta_0 b}\left[\frac{\Lambda_a}{H_a^2}(P_s + 1) - \frac{\partial P_s}{\partial X}\right]$$

Making use of boundary conditions 3 and 6 gives

$$H_a^3\left(\frac{\partial P_s}{\partial X}\right)_{X=\lambda\psi_g} - \left(\frac{\partial P_r}{\partial X}\right)_{X=\lambda\psi_g} = \Lambda_a(H_a - 1)$$

$$\times\left[1 + \sum_{m=1,3,\ldots}^{\infty} F_m^* \cos(m\pi Y)\right] \quad (16.53)$$

By making use of Eqs. (16.51) to (16.53) the Fourier coefficient F_m^* can be solved.

$$F_m^* = \frac{4(H_a - 1)\sin(m\pi/2)}{m\pi\left[\dfrac{1-H_a}{2} + \dfrac{\xi_s H_a^3}{\Lambda_a}\left(\dfrac{1+e^{-2\xi_s\lambda\psi_g}}{1-e^{-2\xi_s\lambda\psi_g}}\right) + \dfrac{\xi_r}{\Lambda_a}\left(\dfrac{1+e^{-2\lambda\xi_r(\beta_g-\psi_g)}}{1-e^{-2\lambda\xi_r(\beta_g-\psi_g)}}\right)\right]} \tag{16.54}$$

16.3.2 NORMAL LOAD COMPONENT AND STIFFNESS.

The dimensionless load components for the ridge and the step can be written as

$$W_r = \frac{w_r}{p_a b \ell} = \frac{2}{\lambda}\int_0^{1/2}\int_{\psi_g\lambda}^{\beta_g\lambda} P_r\, dX\, dY$$

$$W_s = \frac{w_s}{p_a b \ell} = \frac{2}{\lambda}\int_0^{1/2}\int_0^{\psi_g\lambda} P_s\, dX\, dY$$

Substituting Eqs. (16.51) and (16.52) into these equations gives

$$W_r = \sum_{m=1,3,\ldots}^{\infty} \frac{2F_m^* \sin(m\pi/2)}{m^3\pi^3\lambda} \times \left[\frac{\Lambda_a}{2} + \frac{\xi_r\left(1 - 2e^{-\lambda(\beta_g-\psi_g)(\xi_r-\Lambda_a/2)} + e^{-2\lambda\xi_r(\beta_g-\psi_g)}\right)}{1-e^{-2\lambda\xi_r(\beta_g-\psi_g)}}\right] \tag{16.55}$$

$$W_s = \sum_{m=1,3,\ldots}^{\infty} \frac{2F_m^* \sin(m\pi/2)}{m^3\pi^3\lambda} \times \left[-\frac{\Lambda_a}{2H_a^2} + \frac{\xi_s\left(1 - 2e^{-\lambda\psi_g(\xi_s+\Lambda_a/2H_a^2)} + e^{-2\psi_g\lambda\xi_s}\right)}{1-e^{-2\psi_g\lambda\xi_s}}\right] \tag{16.56}$$

The total dimensionless load supported by a rectangular step pad can be written as

$$W = \frac{w_r + w_s}{p_a \ell b} = W_r + W_s \tag{16.57}$$

The equation for the dimensionless stiffness is

$$K_g = -h_r \frac{\partial W}{\partial h_r} \tag{16.58}$$

Therefore, with Eqs. (16.54) to (16.58) the dimensionless load component and stiffness for a self-acting, gas-lubricated, finite-width-step thrust pad is completely defined. From these equations it is evident that the dimensionless load component and stiffness are functions of the following five parameters:

1. The dimensionless bearing number $\Lambda_a = \dfrac{6\eta_0 u_b b}{p_a h_r^2}$
2. The length-to-width ratio of the pad $\lambda = \dfrac{\ell_s + \ell_r + \ell_g}{b}$
3. The film thickness ratio $H_a = \dfrac{h_s}{h_r}$
4. The step location parameter $\psi_g = \dfrac{\ell_s}{\ell_s + \ell_r + \ell_g}$
5. The groove width ratio $\beta_g = \dfrac{\ell_s + \ell_r}{\ell_s + \ell_r + \ell_g}$

16.3.3 OPTIMIZING PROCEDURE. The problem is to find the optimal step bearing for maximum load-carrying capacity or stiffness at various bearing numbers. This means, given the dimensionless bearing number Λ_a, find the optimal λ, the optimal H_a, and the optimal ψ_g. The groove width ratio β_g is much less significant than the other parameters. Therefore, for all evaluations β_g will be set equal to 0.97.

The basic problem in optimizing λ, H_a, and ψ_g for maximum load-carrying capacity and stiffness is essentially that of finding values of λ, H_a, and ψ_g that satisfy the following equations:

$$\frac{\partial W}{\partial \lambda} = \frac{\partial W}{\partial H_a} = \frac{\partial W}{\partial \psi_g} = 0 \tag{16.59}$$

$$\frac{\partial K_g}{\partial \lambda} = \frac{\partial K_g}{\partial H_a} = \frac{\partial K_g}{\partial \psi_g} = 0 \tag{16.60}$$

A Newton-Raphson method is used for solving these simultaneous equations.

16.3.4 STEP SECTOR THRUST BEARING. Figure 16.3 shows the transformation of a rectangular slider bearing into a circular sector bearing. For optimization of a step sector thrust bearing, parameters for the sector must be found that are analogous to those for the rectangular step bearing. The following substitution accomplishes this transformation:

$$b \rightarrow r_o - r_i$$

$$N_0(\ell_s + \ell_r + \ell_g) \rightarrow \pi(r_o + r_i)$$

$$u_b \rightarrow \frac{\omega}{2}(r_o + r_i)$$

where N_0 is the number of pads placed in the step sector. By making use of this

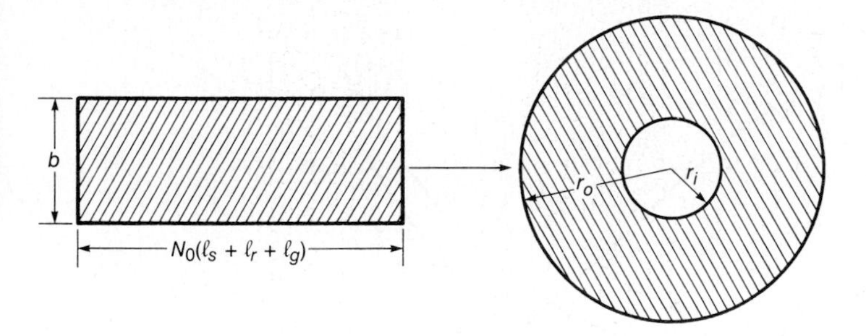

FIGURE 16.3
Transformation of rectangular slider bearing into circular sector bearing.

equation, the dimensionless bearing number can be rewritten as

$$\Lambda_a = \frac{3\eta_0\omega\left(r_o^2 - r_i^2\right)}{p_a h_r^2} \tag{16.61}$$

The optimal number of pads to be placed in the sector is obtained from the following formula:

$$N_0 = \frac{\pi(r_o + r_i)}{\lambda_{\text{opt}}(r_o - r_i)} \tag{16.62}$$

In this equation λ_{opt} is the optimum value for the length-to-width ratio. Since N_0 will normally not be an integer, rounding it to the nearest integer is required.

16.3.5 RESULTS. Figure 16.4(a) shows the effect of Λ_a on λ, H_a, and ψ_g for the *maximum load-carrying capacity* and a range of Λ_a from 0 to 410. The optimum step parameters (λ, H_a, and ψ_g) are seen to approach asymptotes as the dimensionless bearing number Λ_a becomes small. This asymptotic condition corresponds to the incompressible solution of $\lambda = 0.918$, $\psi_g = 0.555$, and $H_a = 1.693$. Recall that for the incompressible solution of a step bearing the right sides of Eqs. (16.42) and (16.43) are zero.

Figure 16.4(b) shows the effect of Λ_a on λ, H_a, and ψ_g for the *maximum stiffness*. As in Fig. 16.4(a) the optimum step parameters approach asymptotes as the incompressible solution is reached. The asymptotes for the maximum stiffness are $\lambda = 0.915$, $\psi_g = 0.557$, and $H_a = 1.470$. Note that there is a difference in the asymptote for the film thickness ratio but virtually no change in λ and ψ_g when compared with the results obtained for the maximum load-carrying capacity.

Figures 16.4(a) and (b) shows that for bearing numbers Λ_a greater than the locations where ψ_g, λ, and H_a approach asymptotic values

1. The length-to-width ratio λ increases (i.e., the length of the pad increases relative to its width).

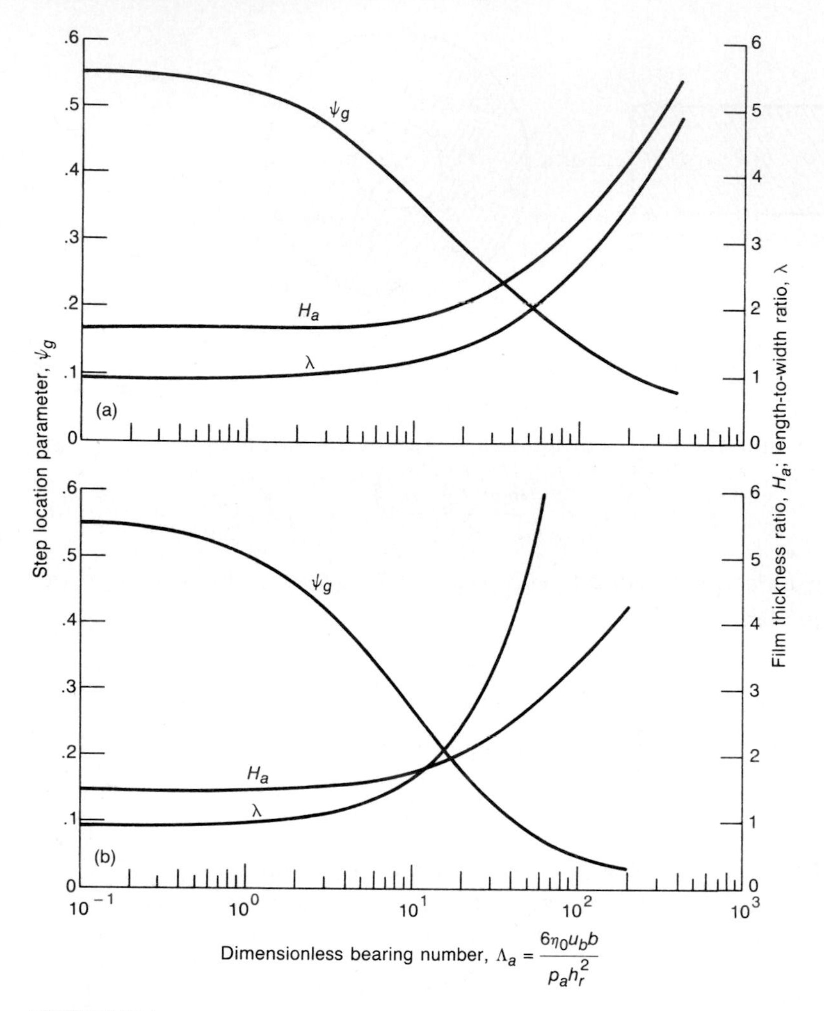

FIGURE 16.4
Effect of dimensionless bearing number on optimum step parameters. (a) For maximum dimensionless load-carrying capacity; (b) for maximum dimensionless stiffness. [*From Hamrock (1972).*]

2. The step location parameter ψ_g decreases (i.e., the length of the step decreases relative to the length of the pad).
3. The film thickness ratio H_a increases (i.e., the depth of the step increases relative to the clearance).

Figure 16.5(a) and (b) shows the effect of dimensionless bearing number Λ_a on dimensionless load-carrying capacity and stiffness. The difference in these figures is that the optimal step parameters are used in Fig. 16.5(a) as obtained for the maximum load-carrying capacity and in Fig. 16.5(b) as obtained

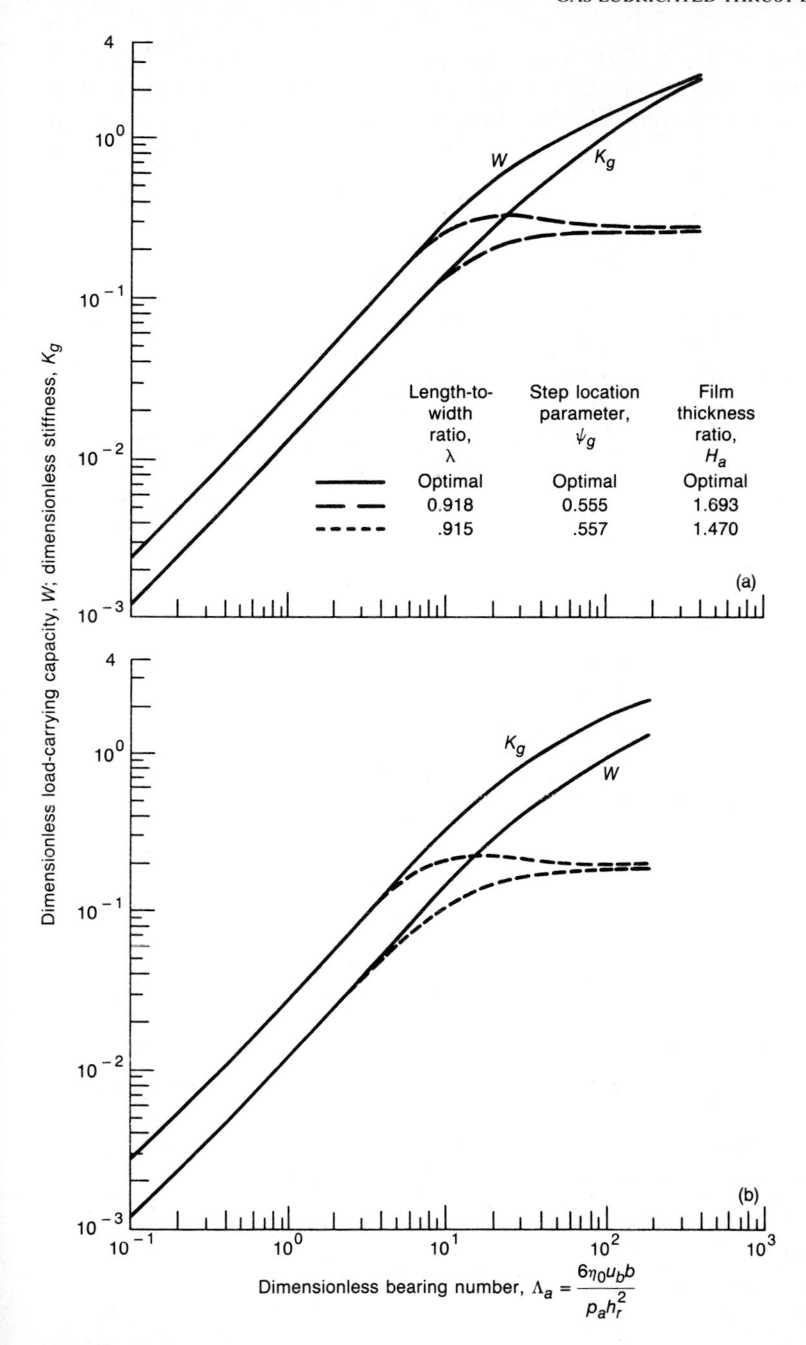

FIGURE 16.5
Effect of dimensionless bearing number on dimensionless load-carrying capacity and dimensionless stiffness. (a) For maximum dimensionless load-carrying capacity; (b) for maximum dimensionless stiffness. [*From Hamrock (1972).*]

for the maximum stiffness. Also shown in these figures are values of K_g and W when the step parameters are held fixed at the optimal values as obtained for the incompressible solution. The significant difference between these results does not occur until $\Lambda_a > 8$.

16.4 SPIRAL-GROOVE BEARING

An inward-pumping, spiral-groove thrust bearing is shown in Fig. 16.6. The dimensionless parameters normally associated with a spiral-groove thrust bearing are (1) groove angle β_a, (2) groove width ratio $\psi_a = \theta_r/\theta_g$, (3) film thickness ratio $H_a = h_s/h_r$, (4) radius ratio $\bar{\alpha}_r = r_i/r_o$, (5) groove length fraction $R_g = (r_o - r_m)/(r_o - r_i)$, (6) number of grooves N_0, and (7) dimensionless bearing number $\Lambda_s = 3\eta_0\omega(r_o^2 - r_i^2)/p_a h_r^2$. The first six parameters are geometrical parameters, and the last parameter is an operating parameter.

The performance of spiral-groove thrust bearings is represented by the following dimensionless parameters:

Dimensionless load:

$$W_\infty = \frac{1.5 G_f w_z}{\pi p_a\left(r_o^2 - r_i^2\right)} \tag{16.63}$$

where G_f is the groove factor

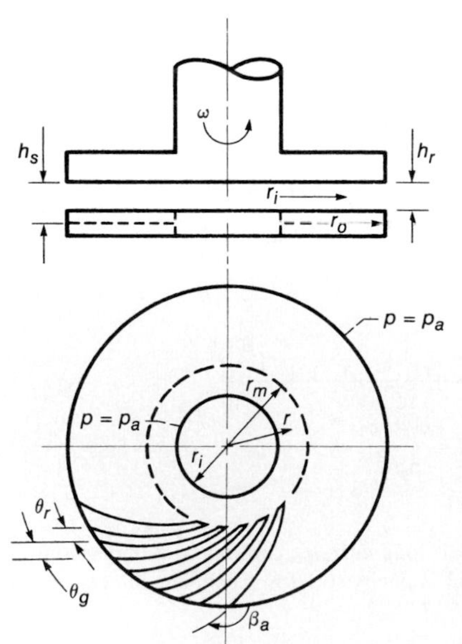

FIGURE 16.6
Spiral-groove thrust bearing. [*From Malanoski and Pan (1965).*]

Dimensionless stiffness:

$$K_\infty = \frac{1.5 h_r G_f k_\infty}{\pi p_a \left(r_o^2 - r_i^2\right)} \tag{16.64}$$

Dimensionless mass flow rate:

$$Q_m = \frac{3\eta_0 q_m}{\pi p_a h_r^3} \tag{16.65}$$

Dimensionless torque:

$$T_q = \frac{6 t_q}{\pi p_a \left(r_o^2 + r_i^2\right) h_r \Lambda_s} \tag{16.66}$$

The design charts of Reiger (1967) are reproduced as Fig. 16.7. These were obtained by solving for the pressure in the Reynolds equation [Eq. (16.13)] for the spiral-groove thrust bearing given in Fig. 16.6 and then solving for the various performance parameters.

In a typical design problem the given factors are load, speed, bearing envelope, gas viscosity, ambient pressure, and an allowable radius-to-clearance ratio. The maximum value of the radius-to-clearance ratio is usually dictated by the distortion likely to occur to the bearing surfaces. Typical values are 5000 to 10,000. The procedure normally followed in designing a spiral-groove thrust bearing by using the design curves given in Fig. 16.7 is as follows:

1. Select the number of grooves N_0.
2. From Fig. 16.7(a) determine the groove factor G_f for a given radius ratio $\bar{\alpha}_r = r_i/r_o$ and N_0.
3. Calculate the dimensionless load W_∞ [Eq. (16.63)].
4. If $W_\infty > 0.8$, then r_o must be increased. Return to step 2.
5. From Fig. 16.7(b), given W_∞ and $\bar{\alpha}_r$, establish Λ_s.
6. Calculate

$$\frac{r_o}{h_r} = \left\{ \frac{\Lambda_s p_a}{3\eta_0 \omega \left[1 - (r_i/r_o)^2\right]} \right\}^{1/2}$$

   If $r_o/h_r > 10{,}000$ (or whatever preassigned radius-to-clearance ratio), a larger bearing or higher speed is required. Return to step 2. If these changes cannot be made, an externally pressurized bearing must be used.
7. Having established what $\bar{\alpha}_r$ and Λ_s should be, obtain values of K_∞, T_q, and Q_m from Fig. 16.7(c), (d), and (e), respectively. From Eqs. (16.64) to (16.66) calculate k_∞, q_m, and t_q.

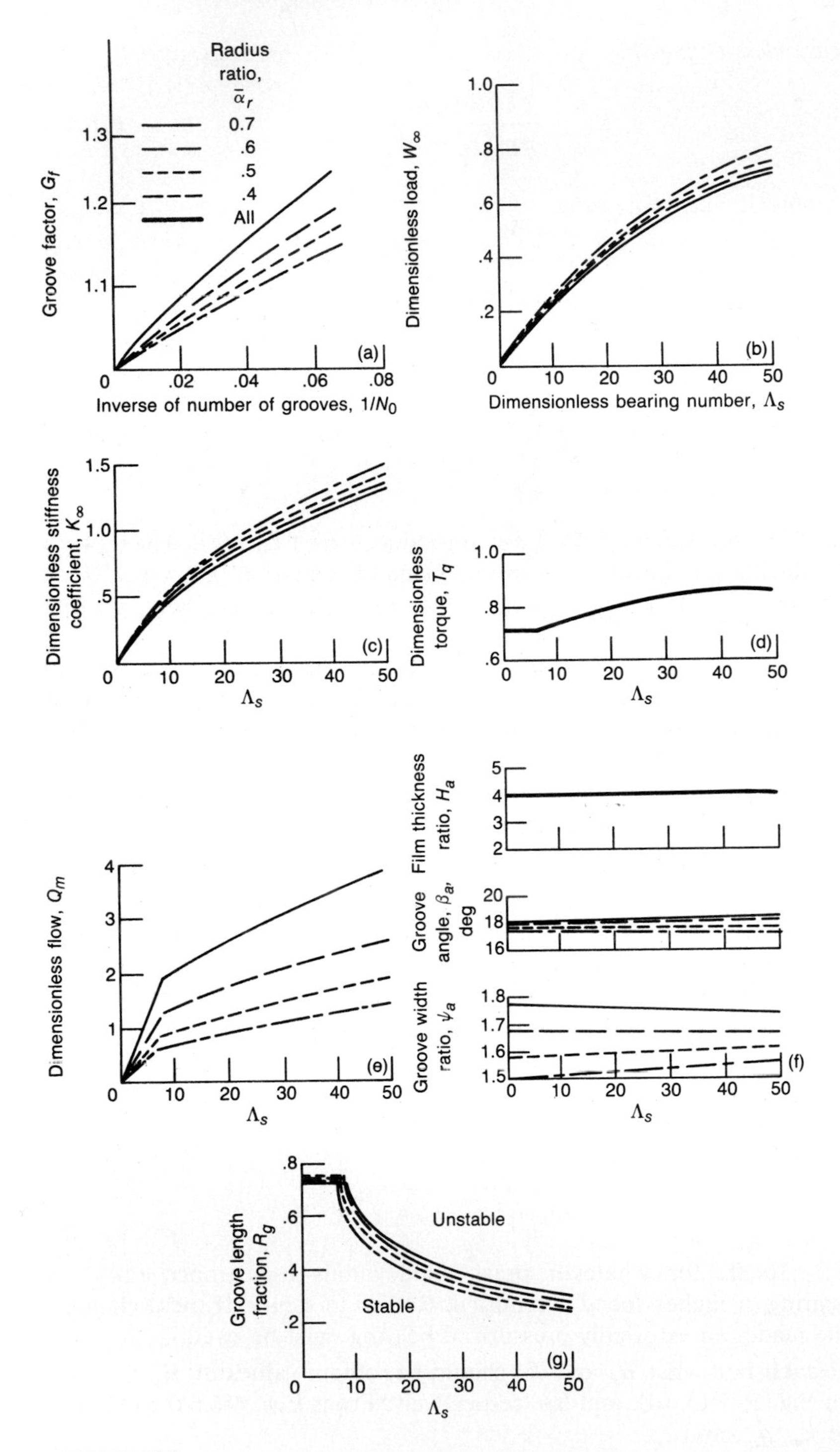

FIGURE 16.7
Charts for determining characteristics of spiral-groove thrust bearings. (a) Groove factor; (b) load; (c) stiffness; (d) torque; (e) flow; (f) optimal groove geometry; (g) groove length factor. [*From Reiger (1967)*.]

8. From Fig. 16.7(f) obtain the film thickness ratio H_a, the groove angle β_a, and the groove width ratio ψ_a. From Fig. 16.7(g) obtain the groove length fraction R_g.

16.5 CLOSURE

Gas bearings are analogous to hydrodynamic oil-lubricated bearings except that the fluid is compressible. Furthermore, since air is 1000 times less viscous than even the thinnest mineral oils, the viscous resistance is much less.

The Reynolds equation in dimensionless terms for a gas-lubricated bearing where isothermal conditions and no slip flow prevail can be written as

$$\frac{\partial}{\partial X}\left(PH^3\frac{\partial P}{\partial X}\right) + \lambda^2\frac{\partial}{\partial Y}\left(PH^3\frac{\partial P}{\partial Y}\right) = \Lambda_g\frac{\partial(PH)}{\partial X} + \sigma_g\frac{\partial(PH)}{\partial T}$$

where λ = length-to-width ratio, ℓ/b
Λ_g = dimensionless bearing number for thrust bearings, $6\eta_0 u_b \ell/p_a h_{\min}^2$
σ_g = squeeze number, $12\eta_0 \omega \ell^2/p_a h_{\min}^2$

In the preceding equation the compressibility of the gas complicates the analysis by making the Reynolds equation nonlinear in the pressure P.

When only time-invariant lubrication is assumed, two limiting solutions of extremely small velocities ($\Lambda_g \to 0$) and extremely large velocities ($\Lambda_g \to \infty$) produce three important principles:

1. At extremely low speeds gas bearing behavior is closely approximated by incompressible or liquid lubrication.
2. At extremely high speeds ph = constant and load becomes independent of speed.
3. At extremely high speeds side-leakage effects are negligible, since finite- and infinite-width-bearing solutions both approach ph = constant.

In slip flow a difference exists between surface velocity and average fluid velocity at the surface. Slip flow becomes important only in gas-lubricated films. The Knudsen number $\mathcal{K}$ is an inverse measure of the average number of molecular collisions in a given film thickness.

$$\mathcal{K} = \frac{\lambda_m}{h}$$

where λ_m is the mean free molecular path of the gas. For $0.01 < \mathcal{K} < 15$, slip flow becomes important.

Complete analytical results could not be obtained for a gas-lubricated, parallel-surface bearing with no slip at the fluid-surface interface, time-invariant lubrication, isothermal conditions, and no side leakage. This led to a linearization approach first proposed by Ausman (1961), which was used to analyze a gas-lubricated, rectangular-step thrust bearing.

16.6 PROBLEMS

16.6.1 For gas-lubricated, parallel-surface thrust bearings, covered in this chapter, evaluate the pressure and normal load-carrying capacity by using Eqs. (16.37), (16.38), and (16.30). Calculations are to be done on a digital computer for
(*a*) $P_i = 1$, $P_o = 4$, and $\Lambda_g = 1, 3, 6, 10, 30$, and 100.
(*b*) $P_i = 4$, $P_o = 1$, and $\Lambda_g = 1, 3, 6, 10, 30$, and 100.
Plot pressure profiles and give tabular results of pressure and load-carrying capacity.

16.6.2 Discuss the effect of side leakage on normal load-carrying capacity for gas bearings found in this chapter and that found for liquid-lubricated bearings covered in Chaps. 8 and 10.

16.7 REFERENCES

Ausman, J. S. (1961): An Approximate Analytical Solution for Self-Acting Gas Lubrication of Stepped Sector Thrust Bearings. *ASLE Trans.*, vol. 4, no. 2, pp. 304–313.

Burgdorfer, A. (1959): The Influence of the Molecular Mean Free Path on the Performance of Hydrodynamic Gas Lubricated Bearings. *J. Basic. Eng.*, vol. 81, no. 1, pp. 94–100.

Hamrock, B. J. (1972): Optimization of Self-Acting Step Thrust Bearings for Load Capacity and Stiffness. *ASLE Trans.*, vol. 15, no. 23, pp. 159–170.

Malanoski, S. B., and Pan, C. H. T. (1965): The Static and Dynamic Characteristics of the Spiral-Groove Thrust Bearing. *J. Basic Eng.*, vol. 87, no. 3, pp. 547–558.

Reiger, N. F. (1967): *Design of Gas Bearings*. Mechanical Technology Inc., Latham, New York.

CHAPTER 17

GAS-LUBRICATED JOURNAL BEARINGS

In Chap. 16 the focus was on the fundamentals of gas lubrication in general and on applying these fundamentals to thrust bearings. This chapter continues with gas-lubricated bearings, but the concern here is with journal bearings. Recall that journal bearing surfaces are parallel to the axis of rotation, whereas thrust bearing surfaces are perpendicular to that axis. Self-acting journal bearings, considered in this chapter, rely on shaft motion to generate the load-supporting pressures in the lubricant film. General information about journal bearing operation is given at the beginning of Chap. 10 and therefore will not be repeated here.

17.1 REYNOLDS EQUATION

Assuming no slip at the fluid-surface interface, isothermal conditions, and time-invariant lubrication, the compressible Reynolds equation in Cartesian coordinates can be written from Eq. (7.55) as

$$\frac{\partial}{\partial x}\left(ph^3\frac{\partial p}{\partial x}\right)+\frac{\partial}{\partial y}\left(ph^3\frac{\partial p}{\partial y}\right)=6\eta_0 u_b\frac{\partial(ph)}{\partial x} \tag{7.55}$$

For journal bearings, as considered in Chap. 10 (see Fig. 10.2), it is convenient to change the variables in the preceding equation to

$$x=r\phi \qquad y=r\zeta \qquad u_b=r\omega \tag{17.1}$$

Substituting Eq. (17.1) into Eq. (7.55) gives

$$\frac{\partial}{\partial\phi}\left(ph^3\frac{\partial p}{\partial\phi}\right)+\frac{\partial}{\partial\zeta}\left(ph^3\frac{\partial p}{\partial\zeta}\right)=6\eta_0\omega r^2\frac{\partial(ph)}{\partial\phi} \tag{17.2}$$

If

$$p=p_aP \quad \text{and} \quad h=cH,$$

this equation becomes

$$\frac{\partial}{\partial\phi}\left(PH^3\frac{\partial P}{\partial\phi}\right)+\frac{\partial}{\partial\zeta}\left(PH^3\frac{\partial P}{\partial\zeta}\right)=\Lambda_j\frac{\partial(PH)}{\partial\phi} \tag{17.3}$$

where

$$\Lambda_j=\frac{6\eta_0\omega r^2}{p_ac^2} \tag{17.4}$$

is the dimensionless bearing number for journal bearings.

17.2 LIMITING SOLUTIONS

Just as for gas-lubricated thrust bearings in Chap. 16, two limiting cases are considered here.

17.2.1 LOW BEARING NUMBERS. As in Chap. 16 as the speed $\omega\to 0$ and $\Lambda_j\to 0$, $P\to 1$ and the pressure rise $\Delta P\to 0$.

$$\therefore\left(\frac{\partial P}{\partial\phi}\right)^2\ll\frac{\partial^2P}{\partial\phi^2}\qquad\left(\frac{\partial P}{\partial\zeta}\right)^2\ll\frac{\partial^2P}{\partial\zeta^2}\qquad\text{and}\qquad H\frac{\partial P}{\partial\phi}\ll P\frac{\partial H}{\partial\phi}$$

Therefore, expanding Eq. (17.3) and neglecting these terms results in

$$\frac{\partial}{\partial\phi}\left(H^3\frac{\partial P}{\partial\phi}\right)+\frac{\partial}{\partial\zeta}\left(H^3\frac{\partial P}{\partial\zeta}\right)=\Lambda_j\frac{\partial H}{\partial\phi} \tag{17.5}$$

This is the same equation dealt with in Chap. 10 for incompressibly lubricated journal bearings. Neglecting the side-leakage term in Eq. (17.5) and assuming a full Sommerfeld solution from Eq. (10.30) gives

$$(W_r)_{\Lambda_j\to 0}=\frac{w_r'}{2rp_a}=\frac{12\pi\varepsilon}{(2+\varepsilon^2)(1-\varepsilon^2)^{1/2}} \tag{17.6}$$

From the friction force the friction coefficient is

$$(\mu)_{\Lambda_j\to 0}=\frac{c(1+2\varepsilon^2)}{3\varepsilon r} \tag{17.7}$$

These results are applicable for $\omega \to 0$ or $\Lambda_j \to 0$ for a self-acting, gas-lubricated journal bearing.

17.2.2 HIGH BEARING NUMBERS. As in Chap. 16, as $\omega \to \infty$ or $\Lambda_j \to \infty$, the only way for the pressure to remain finite in Eq. (17.3) is for

$$\frac{\partial(PH)}{\partial\phi} \to 0 \tag{17.8}$$

implying that PH = constant. A possible solution is $P \propto 1/H$, and P is symmetrical about the line of centers. In Chap. 10 the film thickness was derived as

$$h = c(1 + \varepsilon \cos\phi) \tag{10.5}$$

or

$$H = \frac{h}{c} = 1 + \varepsilon \cos\phi \tag{17.9}$$

$$\therefore P = \frac{\tilde{A}}{1 + \varepsilon \cos\phi} \tag{17.10}$$

and therefore the constant $\tilde{A}$ must be evaluated.

One way to evaluate $\tilde{A}$ is to evaluate the mass of gas trapped in an infinitely wide bearing.

$$m'_a = 2\int_0^{\pi} \rho h r \, d\phi \tag{17.11}$$

where m'_a is the mass of gas per unit bearing width. The density in this equation can be expressed as shown in Eq. (16.3).

$$\therefore m'_a = \frac{2r}{\bar{R}t_m}\int_0^{\pi} ph\, d\phi = \frac{2rcp_a}{\bar{R}t_m}\int_0^{\pi} PH\, d\phi$$

But when $\omega \to \infty$ or $\Lambda_j \to \infty$, PH = constant = $\tilde{A}$.

$$\therefore m'_a = \frac{2\pi rcp_a\tilde{A}}{\bar{R}t_m}$$

or

$$m'_a \propto \tilde{A} \tag{17.12}$$

From Eqs. (17.9) and (17.10) for a concentric journal ($\varepsilon = 0$)

$$P = H = \tilde{A} = 1 \tag{17.13}$$

Therefore, Eq. (17.10) gives

$$P = \frac{1}{1 + \varepsilon \cos\phi} \tag{17.14}$$

The normal load-carrying capacity for an infinitely wide journal bearing when $\Lambda_j \to \infty$ is

$$w_z' = -2p_a r \int_0^{\pi} P \cos\phi \, d\phi = -2p_a r \int_0^{\pi} \frac{\cos\phi \, d\phi}{1 + \varepsilon \cos\phi}$$

Using the Sommerfeld substitution covered in Chap. 10 [Eqs. (10.9), (10.11), and (10.14)] gives

$$W_z = \frac{w_z'}{2p_a r} = \frac{\pi\left[1 - (1 - \varepsilon^2)^{1/2}\right]}{\varepsilon(1 - \varepsilon^2)^{1/2}} \tag{17.15}$$

The friction force for an infinitely wide journal bearing when $\omega \to \infty$ or $\Lambda_j \to \infty$ is

$$f' = \int_0^{2\pi} \tau r \, d\phi = -r\int_0^{2\pi} \left(\frac{\eta_0 r \omega}{h} + \frac{h}{2r}\frac{dp}{d\phi} \right) d\phi$$

In dimensionless form this equation becomes

$$f' = -\frac{p_a c}{2} \int_0^{2\pi} \left(\frac{\Lambda_j}{3H} + H\frac{dP}{d\phi} \right) d\phi$$

But

$$\int_0^{2\pi} \frac{d\phi}{H} = \int_0^{2\pi} \frac{d\phi}{1 + \varepsilon\cos\phi} = \frac{2\pi}{(1 - \varepsilon^2)^{1/2}}$$

$$\int_0^{2\pi} H\frac{dP}{d\phi} \, d\phi = \int_0^{2\pi} H\left(-\frac{1}{H^2}\frac{dH}{d\phi} \right) d\phi = -\int_0^{2\pi} \frac{dH}{H} = (-\ln H)_{\phi=0}^{\phi=2\pi} = 0$$

$$F = \frac{f'}{p_a c} = \frac{\pi \Lambda_j}{3(1 - \varepsilon^2)^{1/2}} \tag{17.16}$$

The friction coefficient when $\omega \to \infty$ and $\Lambda_j \to \infty$ is

$$(\mu)_{\Lambda_j \to \infty} = \frac{(F)_{\Lambda_j \to \infty}}{(W_z)_{\Lambda_j \to \infty}} = \frac{\pi \Lambda_j}{3(1 - \varepsilon^2)^{1/2}} \frac{\varepsilon(1 - \varepsilon^2)^{1/2}}{\pi\left[1 - (1 - \varepsilon^2)^{1/2}\right]}$$

$$= \frac{\varepsilon \Lambda_j}{3\left[1 - (1 - \varepsilon^2)^{1/2}\right]} \tag{17.17}$$

These limiting solutions are important information when verifying exact numerical solutions.

17.3 PRESSURE PERTURBATION SOLUTION

Ausman (1959) used the perturbation method to linearize Eq. (17.2) to obtain an approximate solution. The general concept of the perturbation method is to substitute Eq. (10.5) and

$$p = p_a + \varepsilon p_1 + \varepsilon^2 p_2 + \cdots \tag{17.18}$$

into Eq. (17.2) and neglect all terms of order ε^2 or higher. This gives

$$\frac{\partial^2 p_1}{\partial \phi^2} + \frac{\partial^2 p_1}{\partial \zeta^2} = \Lambda_j \left(\frac{\partial p_1}{\partial \phi} - p_a \sin \phi \right) \tag{17.19}$$

Equation (17.19) can be solved (see Ausman, 1959) for the first-order perturbation pressure, which in turn can be integrated in the usual fashion to obtain parallel and perpendicular components of the load $\bar{w}_x$ and $\bar{w}_z$. The results are

$$\bar{W}_x = \frac{\bar{w}_x}{p_a b(2r)} = \frac{\pi \varepsilon \Lambda_j}{2(1 + \Lambda_j^2)} [\Lambda + f_x(\Lambda_j, \lambda_j)] \tag{17.20}$$

$$\bar{W}_z = \frac{\bar{w}_z}{p_a b(2r)} = \frac{\pi \varepsilon \Lambda_j}{2(1 + \Lambda_j^2)} [1 - f_z(\Lambda_j, \lambda_j)] \tag{17.21}$$

where

$$\lambda_j = \frac{b}{2r} \tag{17.22}$$

$$f_x(\Lambda_j, \lambda_j) = \frac{(\Lambda_b - \Lambda_c \Lambda_j) \sin(2\Lambda_c \lambda_j) - (\Lambda_b \Lambda_j + \Lambda_c) \sinh(2\Lambda_b \lambda_j)}{\lambda_j (1 + \Lambda_j^2)^{1/2} [\cosh(2\Lambda_b \lambda_j) + \cos(2\Lambda_c \lambda_j)]} \tag{17.23}$$

$$f_z(\Lambda_j, \lambda_j) = \frac{(\Lambda_b - \Lambda_c \Lambda_j) \sinh(2\Lambda_b \lambda_j) + (\Lambda_b \Lambda_j + \Lambda_c) \sin(2\Lambda_c \lambda_j)}{\lambda_j (1 + \Lambda_j^2)^{1/2} [\cosh(2\Lambda_b \lambda_j) + \cos(2\Lambda_c \lambda_j)]} \tag{17.24}$$

$$\Lambda_b = \left[\frac{(1 + \Lambda_j^2)^{1/2} + 1}{2} \right]^{1/2} \tag{17.25}$$

$$\Lambda_c = \left[\frac{(1 + \Lambda_j^2)^{1/2} - 1}{2} \right]^{1/2} \tag{17.26}$$

The equations for the total load and the attitude angle can be written as

$$\bar{W}_r = \frac{\bar{w}_r}{\pi p_a r b \varepsilon} = (\bar{W}_x^2 + \bar{W}_z^2)^{1/2} \tag{17.27}$$

$$\bar{\Phi} = \tan^{-1} \frac{\bar{W}_z}{\bar{W}_x} \tag{17.28}$$

Note from Eqs. (17.20) and (17.21) that the first-order perturbation solution

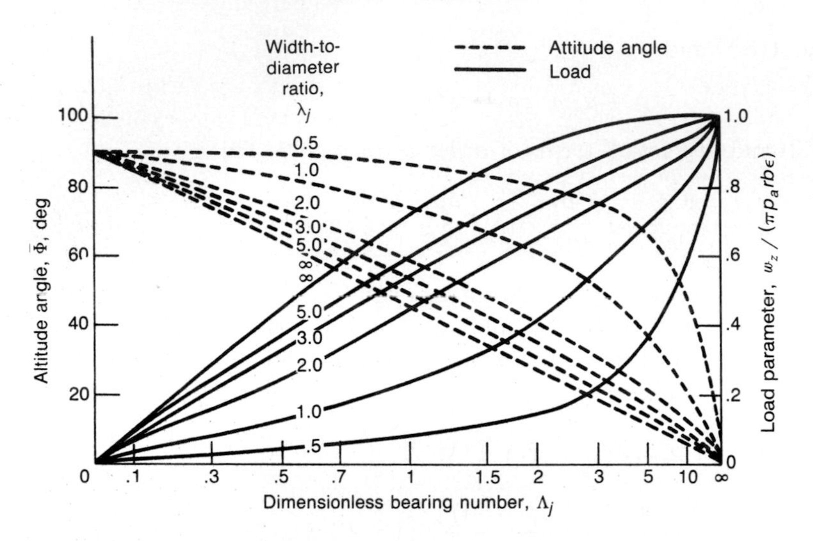

FIGURE 17.1
Design chart for radially loaded, self-acting, gas-lubricated journal bearings (isothermal first-order perturbation solution.) [*From Ausman (1959).*]

yields a load linearly related to the eccentricity ratio ε. This is a consequence of the linearization and is valid only for small ε, say $\varepsilon < 0.3$, although as a conservative engineering approximation it may be used for higher values.

Figure 17.1 plots the load parameter and the attitude angle for the first-order perturbation analysis at several λ_j values.

17.4 LINEARIZED *ph* SOLUTION

Ausman (1961) introduced a linearized *ph* solution to the self-acting, gas-lubricated journal bearing problem that corrected for the deficiency of the first-order perturbation solution at high-eccentricity ratios given in the last section. The general method of linearization is essentially the same as the perturbation method except that the product *ph* is considered to be the dependent variable. To do this, Eq. (17.2) is arranged so that p always appears multiplied by h. Equation (10.5) and

$$ph = p_a c + \Delta(ph) \tag{17.29}$$

are then substituted into this equation, and only first-order terms in eccentricity ratio ε are retained. It is assumed that $\Delta(ph)$ is of order ε. The resulting "linearized *ph*" equation is

$$\frac{\partial^2(ph)}{\partial \phi^2} + \frac{\partial^2(ph)}{\partial \zeta^2} = \Lambda_j \frac{\partial(ph)}{\partial \phi} - \varepsilon p_a c \cos \phi \tag{17.30}$$

Equation (17.30) is essentially the same form as the first-order pressure perturbation equation [Eq. (17.19)] and can be solved in the same manner for ph. Once ph is found, the pressure can be obtained by dividing by h. The resulting expression for p is then put into the load component integrals to obtain w_x and w_z. The results are

$$W_x = \frac{w_x}{p_a b(2r)} = \frac{2\left[1 - (1 - \varepsilon^2)^{1/2}\right]}{\varepsilon^2(1 - \varepsilon^2)^{1/2}} \overline{W}_x \tag{17.31}$$

$$W_z = \frac{w_z}{p_a b(2r)} = \frac{2}{\varepsilon^2}\left[1 - (1 - \varepsilon^2)^{1/2}\right] \overline{W}_z \tag{17.32}$$

The total load and the attitude angle for the linearized ph solution are

$$W_r = \frac{w_r}{p_a b(2r)} = \left(W_x^2 + W_z^2\right)^{1/2} = \frac{2\overline{W}_r\left[1 - (1 - \varepsilon^2)^{1/2}\right]}{\varepsilon^2(1 - \varepsilon^2)^{1/2}} \left(1 - \varepsilon^2 \sin^2 \overline{\Phi}\right)^{1/2} \tag{17.33}$$

$$\tan \Phi = \tan \frac{W_z}{W_x} = (1 - \varepsilon^2)^{1/2} \tan \overline{\Phi} \tag{17.34}$$

where $\overline{W}_r$ and $\overline{\Phi}$ are the values obtained from the first-order perturbation solution given in Eqs. (17.20), (17.21), (17.27), and (17.28), respectively.

Sample problem 17.1 Estimate the maximum load that can be carried by a 2.5-cm-diameter by 2.5-cm-wide self-acting, gas-lubricated journal bearing operating at 1500 rad/s (14,400 r/min) with an average radial clearance of 7.5 μm and a minimum clearance of 1.5 μm. Assume ambient pressure of 1×10^5 Pa and $\eta_0 = 2 \times 10^{-5}$ Pa · s.

Solution

(a) Compute

$$\Lambda_j = \frac{6\eta_0 \omega r^2}{p_a c^2} = \frac{6(2 \times 10^{-5})(1.5 \times 10^3)(1.25^2 \times 10^{-4})}{(1 \times 10^5)(7.5^2 \times 10^{-12})} = 5$$

$$\lambda_j = \frac{b}{2r} = \frac{2.5}{2.5} = 1$$

(b) From Fig. 17.1 for $\Lambda_j = 5$ and $\lambda_j = 1$

$$\overline{W}_r = 0.7 \qquad \text{or} \qquad \overline{w}_r = 0.7 \frac{\pi}{2} p_a b(2r)\varepsilon = 68.7\varepsilon \text{ N}$$

$$\overline{\Phi} = 35°$$

(c) Compute ε and w_r.

$$\varepsilon = \frac{e}{c} = \frac{7.5 - 1.5}{7.5} = 0.80$$

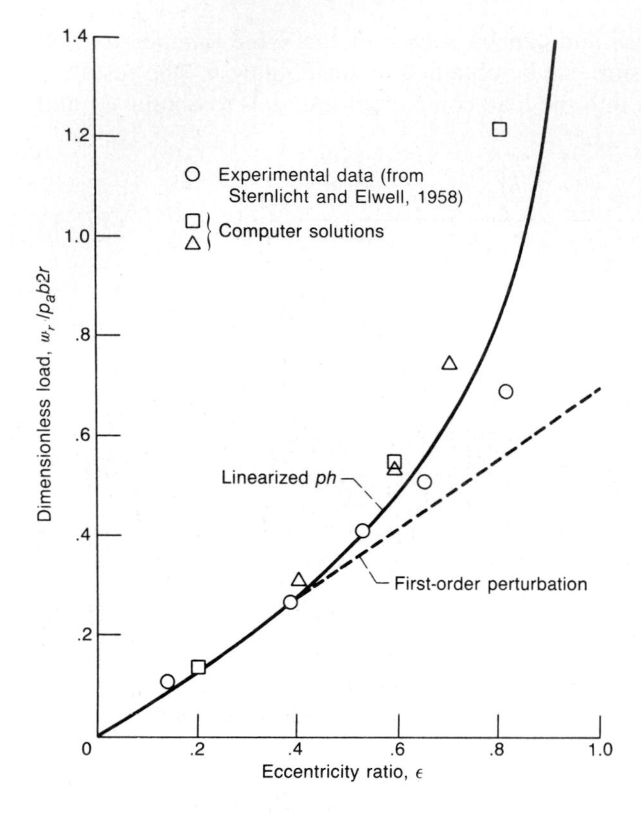

FIGURE 17.2
Effect of dimensionless load on eccentricity ratio for finite-length, self-acting, gas-lubricated journal bearing. Dimensionless bearing number Λ_j, 1.3; width-to-diameter ratio λ_j, 1.5. [*From Ausman (1961).*]

From Eq. (17.33)

$$w_r = 137.4\frac{\left[1-(1-0.64)^{1/2}\right]}{0.80(1-0.64)^{1/2}}\left[1-(0.64)(0.33)\right]^{1/2} = 92\text{ N}$$

Figure 17.2 shows the effectiveness of this method in estimating the load versus eccentricity ratio for particular Λ_j and λ_j. The linearized *ph* and the first-order perturbation are the same for $\varepsilon < 0.4$ but differ for larger values of ε. For $\varepsilon < 0.4$ the linearized *ph* solution and the numerical exact solutions are in good agreement.

The bearing frictional torque and the power dissipated in a journal bearing are easily estimated by the simple but fairly accurate formulas given here.

$$t_q = \frac{\pi}{4}\frac{\eta_0 \omega b(2r)^3}{c(1-\varepsilon^2)^{1/2}} \tag{17.35}$$

$$h_p = \omega t_q \tag{17.36}$$

17.5 NONPLAIN JOURNAL BEARINGS

Thus far this chapter has been concerned with plain gas-lubricated journal bearings. A problem with using these bearings is their poor stability characteristics. Lightly loaded bearings that operate at low eccentricity ratios are subjected to fractional frequency whirl, which can result in bearing destruction. Two types of nonplain gas-lubricated journal bearings that find widespread use are the pivoted pad and the herringbone groove.

17.5.1 PIVOTED-PAD JOURNAL BEARINGS. Pivoted-pad bearings were first introduced in Sec. 9.3 as a type of incompressibly lubricated thrust bearing. Pivoted-pad journal bearings are most frequently used as shaft supports in gas-lubricated machinery because of their excellent stability characteristics. An individual pivoted-pad assembly is shown in Fig. 17.3. A three-pad pivoted-pad bearing assembly is shown in Fig. 17.4. Generally, each pad provides pad rotational degrees of freedom about three orthogonal axes (pitch, roll, and yaw). Pivoted-pad bearings are complex because of the many geometric variables

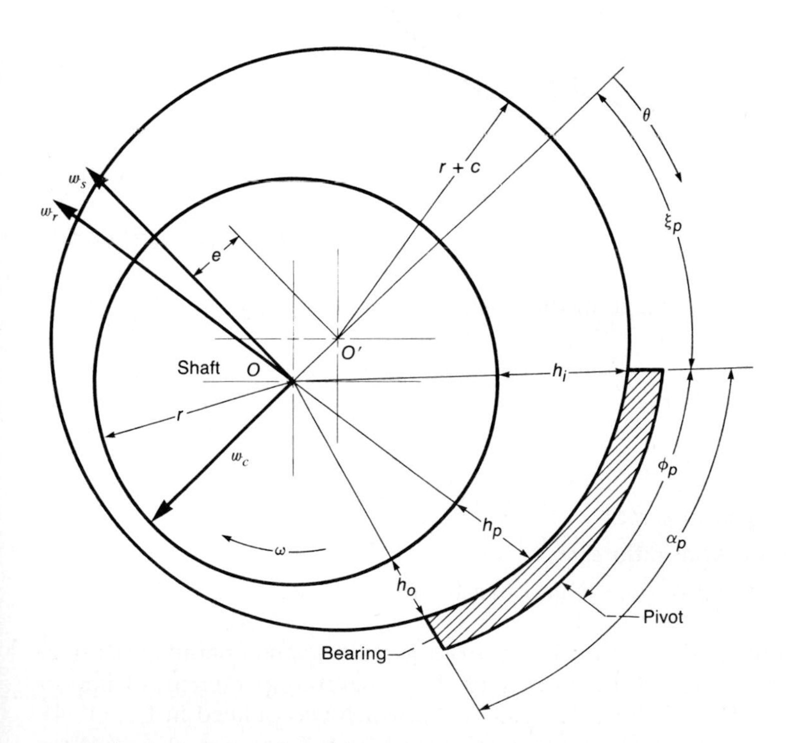

FIGURE 17.3
Geometry of individual pivoted-pad bearing. [*From Gunter et al. (1964).*]

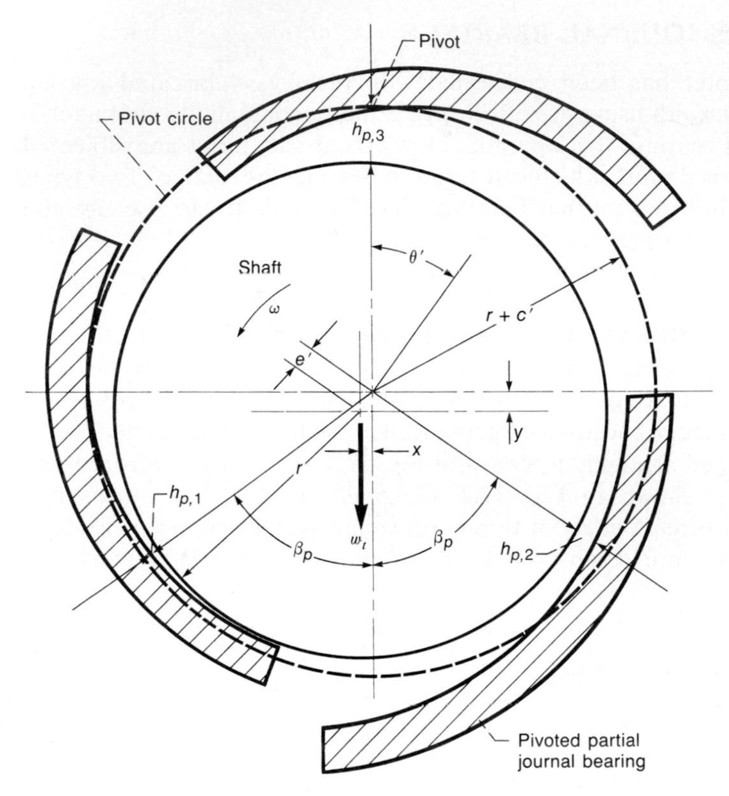

FIGURE 17.4
Geometry of pivoted-pad journal bearing with three pads. [*From Gunter et al. (1964).*]

involved in their design. Some of these variables are

1. Number of pads N_0
2. Angular extent of pad α_p
3. Aspect ratio of pad r/b
4. Pivot location ϕ_p/α_p
5. Machined-in clearance ratio c/r
6. Pivot circle clearance ratio c'/r
7. Angle between line of centers and pad leading edge ξ_p

An individual pad is analyzed first. Both geometric and operating parameters influence the design of a pivoted pad. The operating parameter of importance is the dimensionless bearing number Λ_j, which was defined in Eq. (17.4).

The results of computer solutions obtained from Gunter et al. (1964) for the performance of a single pad are shown in Fig. 17.5. This figure shows dimensionless load, dimensionless pivot film thickness, and dimensionless outlet film thickness as functions of pivot location and eccentricity ratio. These field maps apply for a pad with the following held fixed:

1. Aspect ratio of pad $r/b = 0.606$
2. Angular extent of pad $\alpha_p = 94.5°$
3. Dimensionless bearing number $\Lambda_j = 3.5$

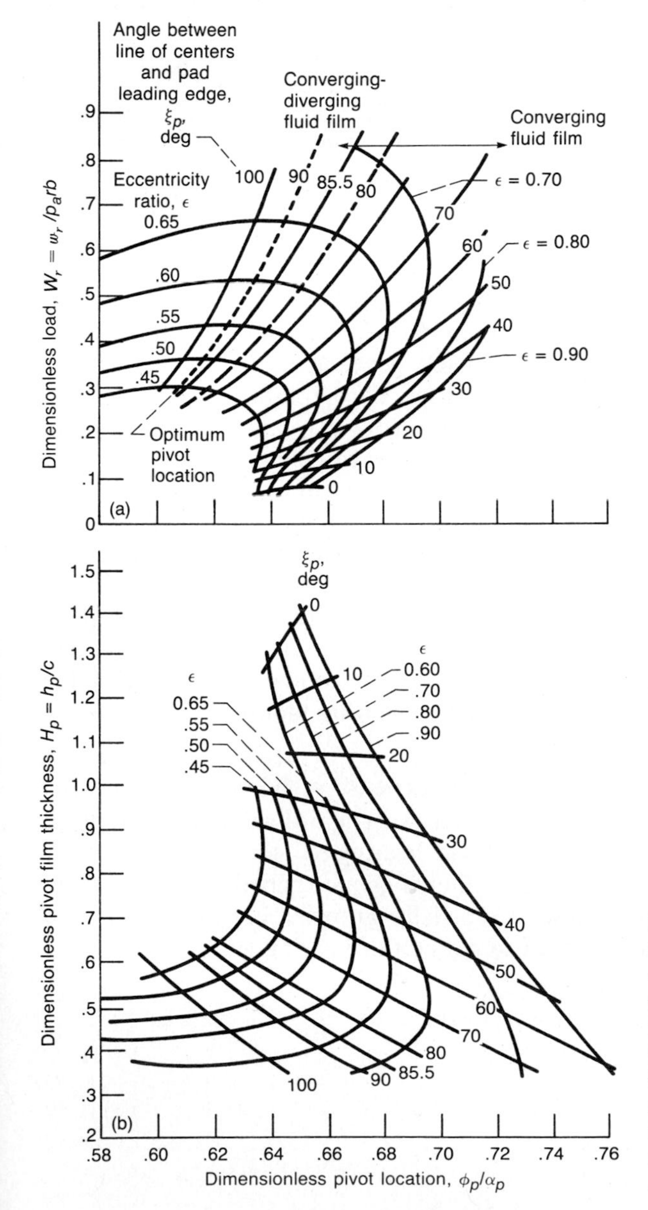

FIGURE 17.5
Charts for determining load coefficient, pivot film thickness, and trailing-edge film thickness. Bearing radius-to-length ratio r/b, 0.6061; angular extent of pad α_p, 94.5°; dimensionless bearing number Λ_j, 3.5. (a) Dimensionless load; (b) dimensionless pivot film thickness. [*From Gunter et al. (1964).*]

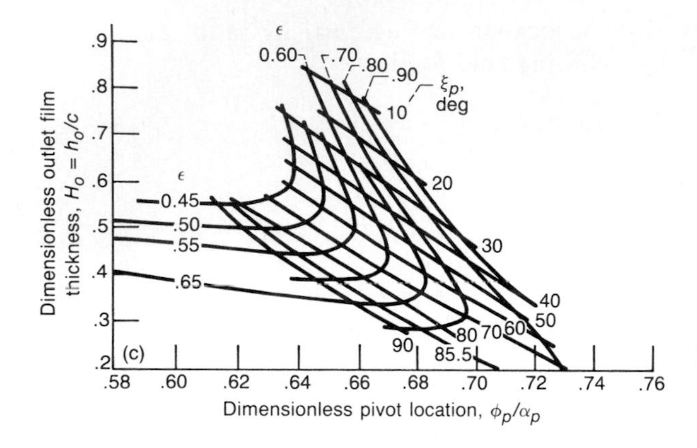

FIGURE 17.5 Concluded. (c) Dimensionless outlet film thickness.

Field maps must be generated for other geometries and bearing numbers. Additional field maps are given in Gunter et al. (1964).

When the individual pad characteristics are known, the characteristics of the multipad bearing shown in Fig. 17.4 can be determined. With the arrangement shown in Fig. 17.4 the load is directed between the two lower pivoted pads. For this case the load carried by each of the lower pads is initially assumed to be $w_r \cos \beta_p$. The pivot film thicknesses $h_{p,1}$ and $h_{p,2}$ are obtained from Fig. 17.5(b). Furthermore, the upper-pad pivot film thickness $h_{p,3}$, the eccentricity ratio ε, and the dimensionless load $W_{r,3}$ can be determined.

Pivoted-pad journal bearings are usually assembled with a pivot circle clearance c' somewhat less than the machined-in clearance c. When $c'/c < 1$, the bearing is said to be "preloaded." Preload is usually given in terms of a preload coefficient, which is equal to $(c - c')/c$. Preloading is used to increase bearing stiffness and to prevent complete unloading of one or more pads. The latter condition can lead to pad flutter and possible contact of the pad leading edge and the shaft, which in turn can result in bearing failure.

17.5.2 HERRINGBONE-GROOVE JOURNAL BEARINGS. A fixcd-geometry bearing that has demonstrated good stability characteristics and thus promise for use in high-speed gas bearings is the herringbone bearing. It consists of a circular journal and bearing sleeve with shallow, herringbone-shaped grooves cut into either member. Figure 17.6 shows a partially grooved herringbone journal bearing. The groove parameters used to define this bearing are

1. Groove angle β_a
2. Groove width ratio $\alpha_b = \ell_s/(\ell_r + \ell_s)$
3. Film thickness ratio $H_a = h_s/h_r$
4. Number of grooves N_0
5. Groove width ratio $\gamma_g = b_1/b$

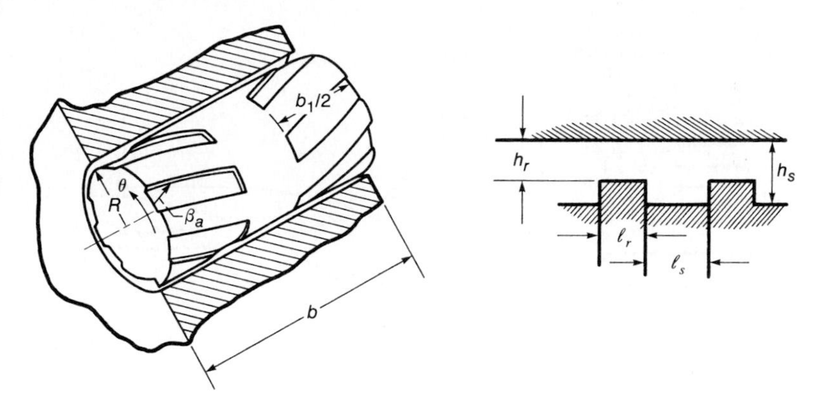

FIGURE 17.6
Configuration of concentric herringbone-groove journal bearing.

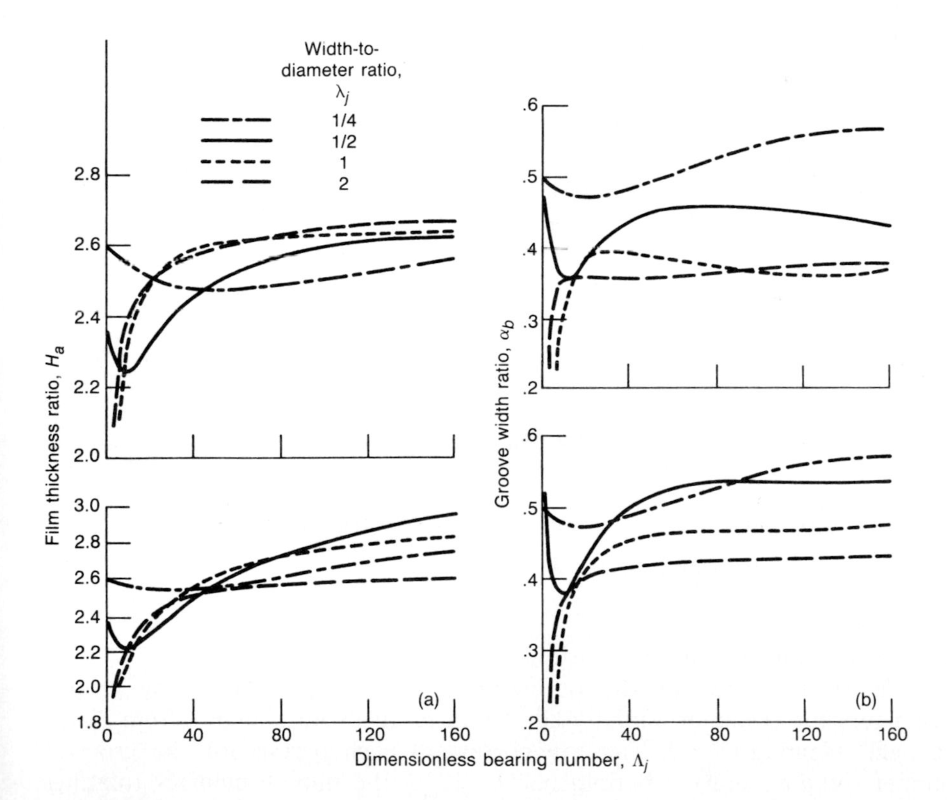

FIGURE 17.7
Charts for determining optimal herringbone-journal-bearing groove parameters for maximum radial load. Top plots are for grooved member rotating; bottom plots are for smooth member rotating. (a) Optimal film thickness ratio; (b) optimal groove width ratio. [*From Hamrock and Fleming (1971).*]

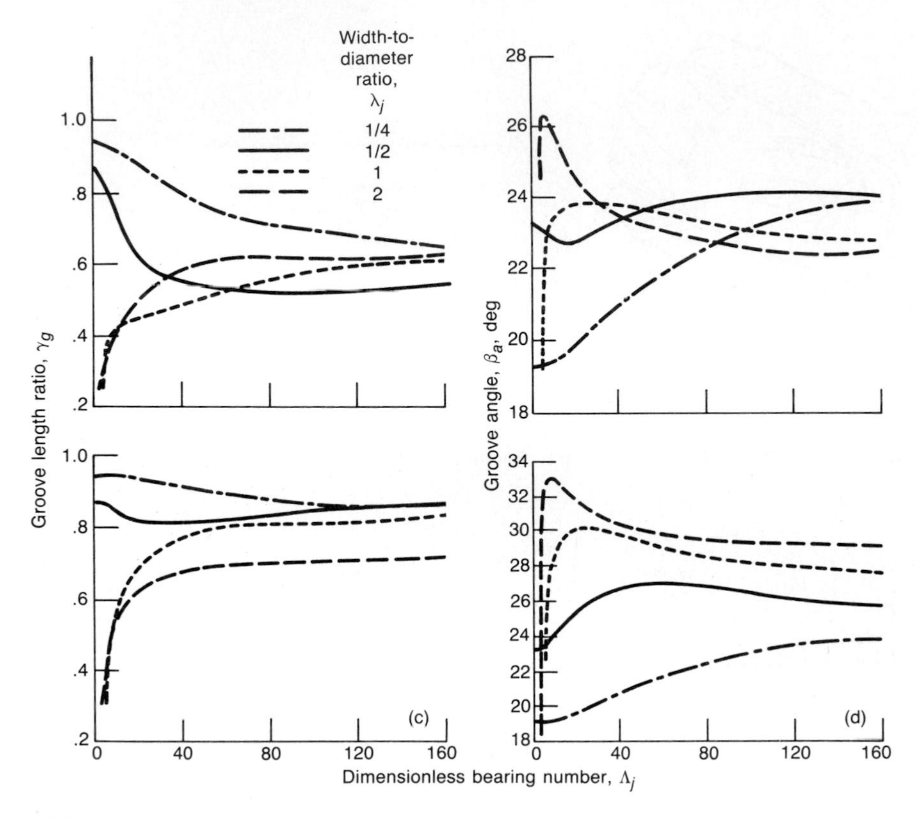

FIGURE 17.7
Concluded. (c) Optimal groove length ratio; (d) optimal groove angle.

The operating parameters used for herringbone journal bearings are

1. Width-to-diameter ratio $\lambda_j = b/2r$
2. Dimensionless bearing number $\Lambda_j = 6\eta_0\omega r^2/p_a h_r^2$.

Both these parameters are dimensionless.

Figure 17.7 presents the optimum herringbone-journal-bearing groove parameters for maximum radial load. These results were obtained from Hamrock and Fleming (1971). The top portion of each part is for the grooved member rotating, and the bottom portion is for the smooth member rotating. The only groove parameter not represented in this figure is the number of grooves to be used. Hamrock and Fleming (1971) found that the *minimum* number of grooves to be placed around the journal can be represented by $N_0 \geq \Lambda_j/15$.

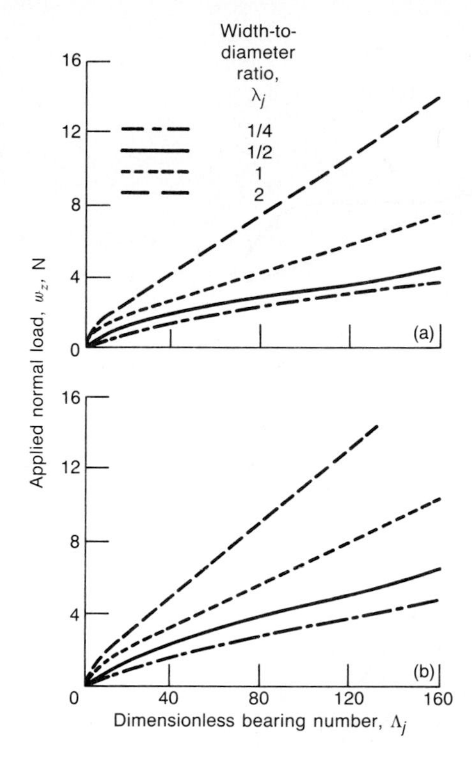

FIGURE 17.8
Chart for determining maximum normal load-carrying capacity. (a) Grooved member rotating; (b) smooth member rotating. [*From Hamrock and Fleming (1971).*]

Figure 17.8 establishes the maximum normal load-carrying capacity for the operating parameters of a gas-lubricated herringbone journal bearing. The optimum groove parameters obtained from Fig. 17.7 are assumed.

More than any other factors, self-excited whirl instability and low load-carrying capacity limit the usefulness of gas-lubricated journal bearings. The whirl problem is the tendency of the journal center to orbit the bearing center at an angular speed less than or equal to one-half that of the journal about its own center. In many cases the whirl amplitude is large enough to cause destructive contact of the bearing surfaces.

Figure 17.9, obtained from Fleming and Hamrock (1974), shows the stability attained by the optimized herringbone journal bearing. In this figure the dimensionless stability parameter is introduced, where

$$\overline{M} = \frac{m_a p_a h_r^5}{2r^5 b \eta_0^2} \tag{17.37}$$

where m_a is the mass supported by the bearing. In Fig. 17.9 the bearings with the grooved member rotating are substantially more stable than those with the smooth member rotating, especially at high bearing numbers.

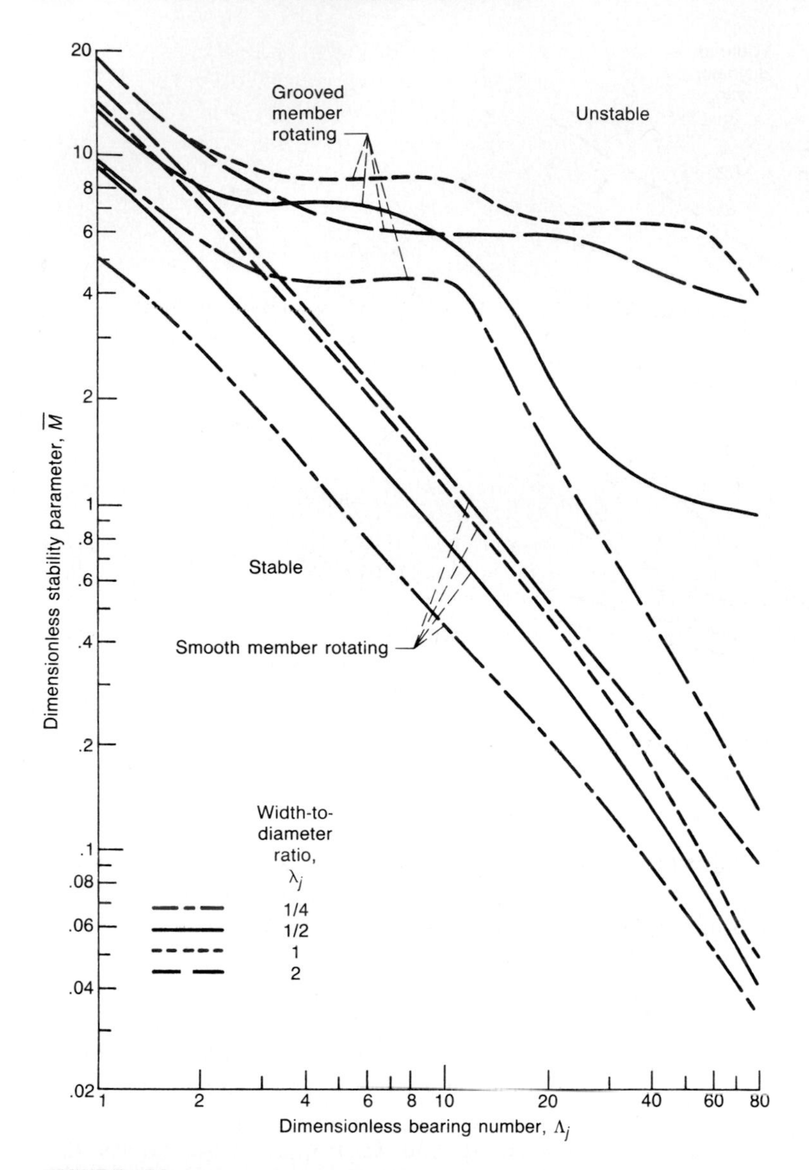

FIGURE 17.9
Chart for determining maximum stability of herringbone-groove bearings. [*From Fleming and Hamrock (1974).*]

17.6 CLOSURE

The discussion of gas-lubricated bearings was continued from Chap. 16 with the focus of this chapter being journal bearings. The appropriate Reynolds equation was presented, and limiting solutions of extremely low- and high-speed results were discussed. Because of the nonlinearity of the Reynolds equation, two approximate methods were introduced. The first of these approaches is a perturbation solution where the pressure is perturbed with the eccentricity ratio

ε. When terms of order ε^2 and higher are neglected, the nonlinear Reynolds equation becomes linear, and thus analytical solutions can be obtained. It was found that the first-order perturbation solution yields a load relationship that is linearly related to the eccentricity ratio. This result was found to be valid only for small eccentricity ratios, say $\varepsilon \leq 0.3$, although as a conservative engineering approximation it may be used for higher values.

The second approximate method covered in this chapter was the linearized *ph* approach of Ausman. The linearization is essentially the same as the perturbation method except that the product *ph* is considered as the dependent variable. This method predicts the load-carrying capacity more accurately for the complete range of eccentricity ratios. Furthermore, the results agree well with those found experimentally.

The latter part of the chapter covered nonplain journal bearings. Two types of nonplain gas-lubricated journal bearing were considered, namely, pivoted pad and herringbone groove. Charts were presented to help in the design these bearings.

17.7 PROBLEMS

17.7.1 Discuss the effect of side leakage on normal load-carrying capacity of gas- and liquid-lubricated journal bearings.

17.7.2 Starting with Eq. (17.2) show all the steps in arriving at the "linearized *ph*" solution given in Eq. (17.30).

17.7.3 Starting with Eq. (17.3) derive the Reynolds equation for a perturbed $PH^{3/2}$ solution.

17.7.4 A self-acting, air-lubricated journal bearing with a 400-mm diameter and 100-mm width is used in gas turbine aircraft engines. It may be assumed that an eccentricity ratio of 0.5 is its operating position. Estimate the theoretical maximum load capacity of the bearing: (*a*) Under full-speed conditions at sea level (engine speed, 20,000 r/min; ambient pressure, 101.3 kN/m^2). (*b*) Under engine idling conditions at 13-km altitude (speed, 10,000 r/min; pressure, 16.58 kN/m^2).

17.8 REFERENCES

Ausman, J. S. (1959): Theory and Design of Self-Acting, Gas-Lubricated Journal Bearings Including Misalignment Effects. International Symposium on Gas-Lubricated Bearings, D. D. Fuller (ed.). Office of Naval Research, Dept. of the Navy, Washington, pp. 161–192.

Ausman, J. S. (1961): An Improved Analytical Solution for Self-Acting, Gas-Lubricated Journal Bearings of Finite Length. *J. Basic Eng*. vol. 83, no. 2, pp. 188–194.

Fleming, D. P., and Hamrock, B. J. (1974): Optimization of Self-Acting Herringbone Journal Bearings for Maximum Stability. Proceedings of the Sixth International Gas Bearing Symposium, Southampton, N. G. Coles (ed.). BHRA Fluid Engineering, pp. C1–C11.

Gunter, E. J., Jr., Hinkle, J. G., and Fuller, D. D. (1964): The Effects of Speed, Load, and Film Thickness on the Performance of Gas-Lubricated, Tilting-Pad Journal Bearings. *ASLE Trans*., vol. 7, no. 4, pp. 353–365.

Hamrock, B. J., and Fleming D. P. (1971): Optimization of Self-Acting Herringbone Journal Bearings for Minimum Radial Load Capacity. Fifth International Gas Bearing Symposium, University of Southampton, paper #13.

Sternlicht, B., and Elwell, R. C. (1958): Theoretical and Experimental Analysis of Hydrodynamic Gas-Lubricated Journal Bearings. *Trans. ASME*, vol. 80, no. 4, pp. 865–878.

CHAPTER

18

HYDRODYNAMIC LUBRICATION OF NONCONFORMAL SURFACES

The focus of the last 10 chapters has been on conformal lubricated conjunctions such as those in journal and thrust bearings. The remainder of the book focuses on nonconformal conjunctions such as those in rolling-element bearings, gears, and cams and followers. In these conjunctions the common feature is that the solid surfaces do not fit well geometrically, as was found to be true for conformal surfaces such as journal bearings. When these nonconformal conjunctions are very lightly loaded, the hydrodynamic pressure is not high enough to cause appreciable elastic deformation of the solid surfaces or to significantly change the lubricant viscosity so that the lubrication mode is hydrodynamic. Later chapters deal with elastic deformation of surfaces and how pressure affects viscosity; this chapter serves as a link between the two parts of the book. The particular nonconformal surfaces investigated here are cylindrical surfaces of finite width and elliptical surfaces in general.

18.1 INFINITELY WIDE-RIGID-CYLINDER SOLUTION

From Eq. (7.58) the Reynolds equation can be written as

$$\frac{\partial}{\partial x}\left(\frac{\rho h^3}{12\eta}\frac{\partial p}{\partial x}\right)+\frac{\partial}{\partial y}\left(\frac{\rho h^3}{12\eta}\frac{\partial p}{\partial y}\right)=\frac{\partial}{\partial x}\left[\frac{\rho h(u_a+u_b)}{2}\right]+\frac{\partial}{\partial y}\left[\frac{\rho h(v_a+v_b)}{2}\right]+\frac{\partial(\rho h)}{\partial t} \qquad (7.58)$$

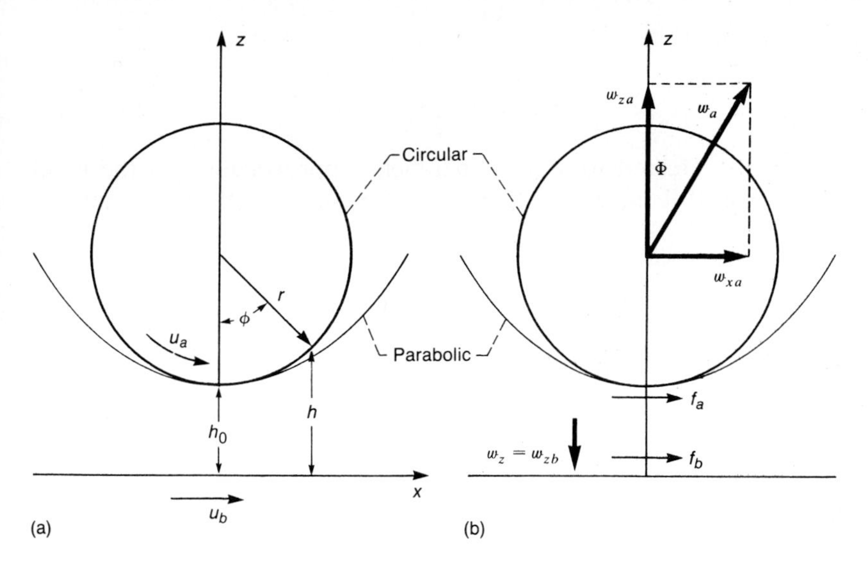

FIGURE 18.1
Lubrication of a rigid cylinder near a plane. (a) Coordinates and surface velocities; (b) forces.

Neglecting side leakage and assuming isoviscous, pure rolling, and time-invariant conditions reduces this equation to

$$\frac{d}{dx}\left(\rho h^3 \frac{dp}{dx}\right) = 6\eta_0(u_a + u_b)\frac{d(\rho h)}{dx} \tag{18.1}$$

As was found in Sec. 10.1, the solutions from making the above approximation are felt to be valid as long as the diameter-to-width ratio is less than 0.5 ($\lambda_b = 2r_b/b < 0.5$). Integrating gives

$$\frac{dp}{dx} = 6\eta_0(u_a + u_b)\frac{\rho h - \rho_m h_m}{\rho h^3} \tag{18.2}$$

where

$$\begin{matrix} \rho = \rho_m \\ h = h_m \end{matrix} \qquad \text{when } \frac{dp}{dx} = 0$$

Figure 18.1 shows the coordinates, surface velocities, and forces associated with the lubrication of a rigid cylinder near a plane. Recall that in Sec. 13.4 the film shape between a rigid cylinder and a plane was found to be

$$h = h_0 + \frac{x^2}{2r}\left[1 + \frac{1}{4}\left(\frac{x}{r}\right)^2 + \frac{1}{8}\left(\frac{x}{r}\right)^4 + \cdots\right] \tag{13.49}$$

If $x \ll r$ so that $(x/r)^2 \ll 1$,

$$h = h_0 + \frac{x^2}{2r} \tag{13.50}$$

Equation (13.50) is referred to as the "parabolic approximation to the exact solution." From Fig. 18.1 the x coordinate and the exact film thickness can also be written as

$$x = r \sin \phi \tag{18.3}$$

and

$$h = h_0 + r - r \cos \phi \tag{18.4}$$

This equation can be rewritten as

$$h = (h_0 + r)\left(1 - \frac{r}{h_0 + r} \cos \phi\right)$$

or

$$h = \frac{-r}{a_1}(1 + a_1 \cos \phi) \tag{18.5}$$

where

$$a_1 = \frac{-r/h_0}{1 + r/h_0} = \frac{-s_1}{s_1 + 1} \tag{18.6}$$

$$s_1 = \frac{r}{h_0} \tag{18.7}$$

The advantage that Eq. (18.5) has over Eq. (13.49) or (13.50) is that the Sommerfeld substitution introduced in Chap. 10 for journal bearings can be employed directly.

18.1.1 PRESSURE DISTRIBUTION. Assuming incompressible lubrication reduces Eq. (18.2) to

$$\frac{dp}{dx} = 6\eta_0(u_a + u_b)\frac{h - h_m}{h^3} \tag{18.8}$$

Substituting Eqs. (18.3) and (18.5) into the preceding equation gives

$$\frac{dp}{d\phi} = \frac{6\eta_0(u_a + u_b)a_1^2}{r}\left[\frac{\cos \phi}{(1 + a_1 \cos \phi)^2} + \frac{h_m a_1 \cos \phi}{r(1 + a_1 \cos \phi)^3}\right] \tag{18.9}$$

Making use of the Sommerfeld substitution covered in Chap. 10, in particular Eqs. (10.9), (10.11), and (10.14), while integrating Eq. (18.9) gives

$$p = \frac{6\eta_0(u_a + u_b)a_1^2}{r}\left[B_2(a_1, \gamma) + \frac{a_1 h_m}{r}B_3(a_1, \gamma) + \tilde{A}\right] \tag{18.10}$$

where γ is the Sommerfeld variable

$$B_2(a_1, \gamma) = \frac{\sin \gamma - a_1 \gamma}{\left(1 - a_1^2\right)^{3/2}} \tag{18.11}$$

and

$$B_3(a_1, \gamma) = \frac{1}{\left(1 - a_1^2\right)^{5/2}} \left[-\frac{3 a_1 \gamma}{2} + \left(1 + a_1^2\right) \sin \gamma - \frac{a_1}{2} \sin \gamma \cos \gamma \right] \tag{18.12}$$

The boundary conditions assumed by the Martin (1916) analysis are

1. $p = 0$ when $\phi = \phi_i$.
2. $p = \partial p / \partial x = 0$ when $\phi = \phi_o$.

The choice of ϕ_i has little effect on the solution as long as ϕ_i is not too small, and hence it is often assumed that $\phi_i = -90°$. The second boundary condition given above is the Reynolds boundary condition. Unfortunately, an analytical expression for the pressure distribution cannot be obtained, and numerical methods must be used.

18.1.2 LOAD COMPONENTS. Figure 18.1(b) shows the loads acting on the solid surfaces. Note that these load components are essentially the same as those presented in Fig. 8.4 for thrust bearings. The only difference is that Fig. 8.4 is for conformal surfaces but Fig. 18.1(b) is for nonconformal surfaces. The loads acting on a unit width (in the y direction) of the cylinder and plane are defined by the following relationships:

$$w'_{xb} = 0 \tag{18.13}$$

$$w'_{xa} = -\int_{h_i}^{h_o} p \, dh = -r \int_{\phi_i}^{\phi_o} p \sin \phi \, d\phi \tag{18.14}$$

$$w'_{zb} = w'_{za} = \int_{x_i}^{x_o} p \, dx = r \int_{\phi_i}^{\phi_o} p \cos \phi \, d\phi \tag{18.15}$$

$$w'_b = \left[\left(w'_{xb} \right)^2 + \left(w'_{zb} \right)^2 \right]^{1/2} = w'_{zb} \tag{18.16}$$

$$\Phi = \tan^{-1} \frac{w'_{xa}}{w'_{za}} \tag{18.17}$$

A different stress equation must be used in the load-carrying and cavitated parts of the conjunction. Owing to the presence of air bubbles, the shear stresses act over a small portion of the cavitated region. If Y_ℓ is the fraction of the clearance space width in the y direction occupied by lubricant, flow continuity yields

$$Y_\ell = \frac{h_m}{h} \tag{18.18}$$

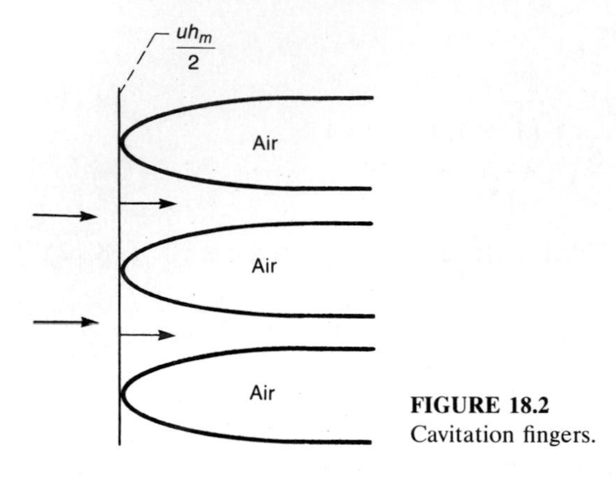

FIGURE 18.2
Cavitation fingers.

Figure 18.2 shows the cavitation fingers that normally occur. In general, the cavitation zone will terminate at some point between x_o and R (ϕ_2 and $\pi/2$). The general expressions for the viscous forces acting on the solids are

$$f_b' = \int_{x_i}^{x_o} (\tau)_{z=0}\, dx + \int_{x_o}^{R} Y_\ell (\tau)_{z=0}\, dx$$

$$= r\int_{\phi_i}^{\phi_o} (\tau)_{z=0} \cos\phi\, d\phi + r\int_{\phi_o}^{\pi/2} Y_\ell (\tau)_{z=0} \cos\phi\, d\phi \tag{18.19}$$

$$f_a' = -\int_{x_i}^{x_o} (\tau)_{z=h}\, dx - \int_{x_o}^{R} Y_\ell (\tau)_{z=h}\, dx$$

$$= -r\int_{\phi_i}^{\phi_o} (\tau)_{z=h} \cos\phi\, d\phi - r\int_{\phi_o}^{\pi/2} Y_\ell (\tau)_{z=h} \cos\phi\, d\phi \tag{18.20}$$

It can be shown that

$$f_b' = -\frac{w_{xa}'}{2} - \eta(u_b - u_a)A_1 \tag{18.21}$$

$$f_a' = -\frac{w_{xa}'}{2} + \eta(u_b - u_a)A_1 \tag{18.22}$$

where

$$A_1 = (\phi_i - \phi_o) + \frac{\gamma_o - \gamma_i}{\left(1 - a_1^2\right)^{1/2}} + \frac{a_1^2 h_m}{r}\left[\frac{\sin\gamma - a_1\gamma}{\left(1 - a_1^2\right)^{3/2}}\right]_{\gamma_o}^{\gamma=\cos^{-1}(a_1)} \tag{18.23}$$

Solutions have been presented for these force components for a wide range of r/h_0 by Dowson and Higginson (1966). For r/h_0 greater than 10^4 they found that a good approximation to the force components is given by

$$w'_{xb} = 0 \tag{18.24}$$

$$w'_{xa} = 4.58\eta_0(u_b + u_a)\left(\frac{r}{h_0}\right)^{1/2} \tag{18.25}$$

$$w'_{za} = w'_{zb} = 2.44\eta_0(u_b + u_a)\frac{r}{h_0} \tag{18.26}$$

$$f'_b = -\frac{w'_{xa}}{2} - 2.84\eta_0(u_b - u_a)\left(\frac{r}{h_0}\right)^{1/2} \tag{18.27}$$

$$f'_a = -\frac{w'_{xa}}{2} + 2.84\eta_0(u_b - u_a)\left(\frac{r}{h_0}\right)^{1/2} \tag{18.28}$$

The results given in Eqs. (18.24) to (18.28) are referred to as the "Martin" (1916) solutions. Note from Eq. (18.26) that the minimum film thickness is directly proportional to the speed and inversely proportional to the load.

18.2 SHORT-WIDTH-RIGID-CYLINDER SOLUTION

This analysis follows the approach developed by Dubois and Ocvirk (1953) and used in the short-width-journal-bearing solution covered in Sec. 10.2. It is assumed that the width of the cylinder b is small relative to the radius r and that the first term on the left side of Eq. (7.48) can be neglected. As found in Chap. 10 this condition can only be applied if $\lambda_k = 2r_b/b > 2$.

18.2.1 PRESSURE DISTRIBUTION. Neglecting the first term on the left side of Eq. (7.48) givcs

$$\frac{\partial}{\partial y}\left(h^3\frac{\partial p}{\partial y}\right) = 6\eta_0(u_a + u_b)\frac{\partial h}{\partial x} \tag{18.29}$$

Since h is only a function of x, the preceding equation can be integrated to give

$$\frac{dp}{dy} = \frac{6\eta_0(u_a + u_b)}{h^3}\frac{\partial h}{\partial x}y + \frac{\tilde{A}}{h^3} \tag{18.30}$$

and

$$p = \frac{6\eta_0(u_a + u_b)}{h^3}\frac{\partial h}{\partial x}\frac{y^2}{2} + \frac{\tilde{A}y}{h^3} + \tilde{B} \tag{18.31}$$

The y coordinate is in the axial center of the bearing so that the boundary conditions are $p = 0$ when $y = \pm b/2$. Making use of these boundary conditions gives

$$\tilde{A} = 0 \qquad \text{and} \qquad \tilde{B} = -\frac{3\eta_0}{4}\frac{(u_a + u_b)b^2}{h^3}\frac{\partial h}{\partial x}$$

$$\therefore p = \frac{3\eta_0(u_a + u_b)}{h^3}\frac{dh}{dx}\left(y^2 - \frac{b^2}{4}\right) \tag{18.32}$$

Difficulties are encountered if a circular profile for the film thickness is used as shown in Eq. (18.5), since dh/dx and hence p achieve an infinite value when $\phi = -90°$. However, if the parabolic profile given in Eq. (13.50) is used, the difficulty is removed.

$$\therefore \frac{dh}{dx} = \frac{x}{r}$$

$$\therefore p = \frac{3\eta_0(u_a + u_b)x(y^2 - b^2/4)}{r(h_0 + x^2/2r)^3} \tag{18.33}$$

Note that the pressure is zero (or ambient) along the line of closest approach, or $x = 0$.

18.2.2 LOAD COMPONENTS. If the subambient pressures predicted for the divergent clearance space are neglected, a normal load component can be defined as

$$w_{za} = w_{zb} = \int_{-b/2}^{b/2}\int_{-\infty}^{0} p\,dx\,dy \tag{18.34}$$

Substituting Eq. (18.33) into this equation gives

$$w_{za} = w_{zb} = \frac{3\eta_0(u_a + u_b)}{h_0^3 r}\int_{-b/2}^{b/2}\int_{-\infty}^{0}\frac{x}{(1 + x^2/2rh_0)^3}\left(y^2 - \frac{b^2}{4}\right)dx\,dy$$

$$= \frac{\eta_0(u_a + u_b)b^3}{4h_0^2} \tag{18.35}$$

The comparable solution for a width b of the infinitely wide cylinder given in Eq. (18.26) is

$$w'_{za} = w'_{zb} = 2.44\eta_0(u_b + u_a)\frac{r}{h_0} \tag{18.36}$$

An important difference between the two solutions given in Eqs. (18.35) and (18.36) is that the short-width cylinder's load-carrying capacity is independent of r in Eq. (18.35). Also note in Eq. (18.35) that the central (also minimum) film thickness is proportional to the normal applied load raised to the -0.5 power. This is contrasted with the results given in Eq. (18.36) for an infinitely wide cylinder where $h_0 \propto 1/w'_z$.

18.3 EXACT RIGID-CYLINDER SOLUTION

The infinitely wide-cylinder and short-width-cylinder solutions represent extremes that are clearly different from each other and that are likely to lead to erroneous predictions for cylinders of realistic size. Therefore, for most applications both Poiseuille terms are important and need to be considered.

18.3.1 PRESSURE DISTRIBUTION. The full solution of the Reynolds equation for rigid cylinders of finite length has been obtained numerically by Dowson and Whomes (1967). The appropriate Reynolds equation is

$$\frac{\partial}{\partial x}\left(h^3\frac{\partial p}{\partial x}\right)+\frac{\partial}{\partial y}\left(h^3\frac{\partial p}{\partial y}\right)=6\eta_0(u_a+u_b)\frac{\partial h}{\partial x} \tag{18.37}$$

Let

$$x=Xr-r \qquad y=\frac{bY}{2} \qquad h=h_0H \qquad p=\frac{6\eta_0(u_a+u_b)rP}{h_0^2} \tag{18.38}$$

Substituting Eqs. (18.38) into (18.37) gives

$$\frac{\partial}{\partial X}\left(H^3\frac{\partial P}{\partial X}\right)+\frac{1}{\lambda_j^2}\frac{\partial}{\partial Y}\left(H^3\frac{\partial P}{\partial Y}\right)=\frac{\partial H}{\partial X} \tag{18.39}$$

where

$$\lambda_j=\frac{b}{2r} \tag{18.40}$$

Assuming a parabolic film shape such as given in Eq. (13.50) while making use of Eq. (18.38) gives

$$H=1+\frac{r}{2h_0}(X-1)^2 \tag{18.41}$$

Note that using Eqs. (18.38) and (18.41) in (18.33) gives the dimensionless pressure for the short-width-cylinder analysis as

$$P=\frac{(\alpha_h\lambda_j^2)(1-X)(1-Y^2)}{2\left[1+(\alpha_h/2)(X-1)^2\right]^3} \tag{18.42}$$

where

$$\alpha_h=\frac{r}{h_0} \tag{18.43}$$

and $0\le X\le 1$ and $-1\le Y\le 1$. The pressure distribution exhibits an extremely localized pressure region with high values of dP/dX. Such a condition is not favorable for using a numerical method in solving Eq. (18.37). To produce a much gentler curve, a parameter Γ is introduced such that

$$\Gamma=PH^{3/2} \tag{18.44}$$

Expressing the Reynolds equation in terms of Γ produces a much gentler curve than just using P. If P is large, then H is small, and vice versa, so that the new parameter Γ is much gentler.

Substituting Eq. (18.44) into Eq. (18.39) gives

$$\frac{\partial^2 \Gamma}{\partial X^2} + \frac{1}{\lambda_j^2}\frac{\partial^2 \Gamma}{\partial Y^2} - \frac{3}{2}\frac{\Gamma}{H^{3/2}}\frac{\partial}{\partial X}\left(H^{1/2}\frac{\partial H}{\partial X}\right) = \frac{1}{H^{3/2}}\frac{\partial H}{\partial X} \tag{18.45}$$

Note that substituting Eq. (18.44) into (18.39) eliminated all terms containing derivatives of products of H and Γ. This turns out to be an advantage for the numerical calculations that need to be performed.

Relaxation methods are used to solve Eq. (18.45) numerically. In the relaxation process the first step is to replace the differential equation (18.45) by finite difference approximations. The relaxation method relies on the fact that a function can be represented with sufficient accuracy over a small range by a quadratic expression. A full account of the relaxation procedure employed in solving Eq. (18.45) can be found in Dowson and Whomes (1967). A finite difference representation is introduced that is based on having equal increments in the width or axial direction and variable increments in the x direction.

The boundary conditions imposed are

1. $x = x_i$ and $P = 0$.
2. $y = \pm b/2$, $Y = \pm 1$, and $P = 0$.
3. $x = x_o$ and $P = \partial P/\partial X = \partial P/\partial Y = 0$.

By making use of these boundary conditions the pressure distribution can be obtained.

18.3.2 LOAD COMPONENTS. Once an acceptable distribution has been obtained, the load components w_x and w_z can be obtained from the integrals given in the short-width-bearing analysis [Eq. (18.34)]. From Eqs. (18.39) and (18.41) the operating parameters are (1) the width-to-diameter ratio $\lambda_j = b/2r$ and (2) the radius-to-central-film-thickness ratio $\alpha_h = r/h_0$. The range of α_h is from 10^2 to 10^6.

Figure 18.3 shows the effect of side leakage on the normal load component for five values of α_h. For increasing values of α_h the asymptotic value of the normal load ratio is reached for smaller values of λ_j. Similar results were found for the tangential load ratio (Fig. 18.4). From the load ratio $\hat{w}$ and the infinite-width solutions from Eqs. (18.25) and (18.26), the load component for a finite-width bearing can be obtained. That is,

$$w_{\text{finite}} = \hat{w} w_{\text{infinite}} \tag{18.46}$$

The normal and tangential load components of a finite-width bearing can be obtained by making use of Figs. 18.3 and 18.4 and Eqs. (18.46), (18.25), and (18.26).

Sample problem 18.1 Given a cylinder with $b = 2$ in, $r = 10$ in, $h_0 = 10^{-2}$ in, $\eta_0 = 1$ P, $u_b = 500$ in/s, and $u_a = 0$, find w_z.

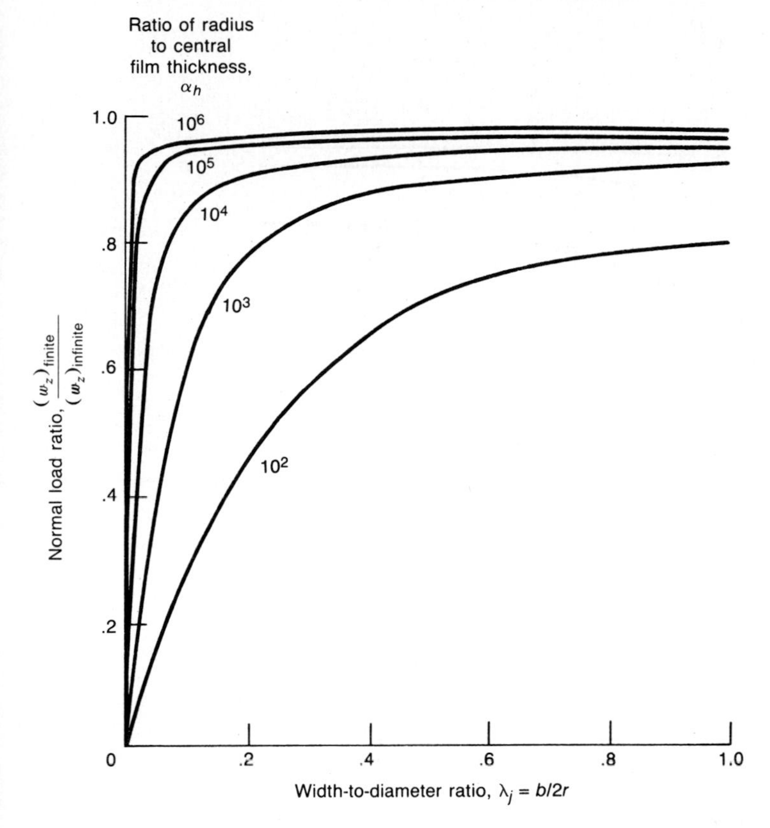

FIGURE 18.3
Side-leakage effect on normal load component.

Solution

$$\lambda_j = \frac{b}{2r} = \frac{2}{2(10)} = 0.1$$

and

$$\alpha_h = \frac{r}{h_0} = \frac{10}{10^{-2}} = 10^3$$

From Fig. 18.3 for $\lambda_j = 0.1$ and $\alpha_h = 10^3$

$$\hat{w}_z = 0.5$$

$$\therefore (w_z)_{\text{finite}} = 0.5(w_z)_{\text{infinite}}$$

From Eq. (18.26)

$$(w_z)_{\text{infinite}} = b w'_z = 2.45\eta_0(u_a + u_b)br/h_0$$

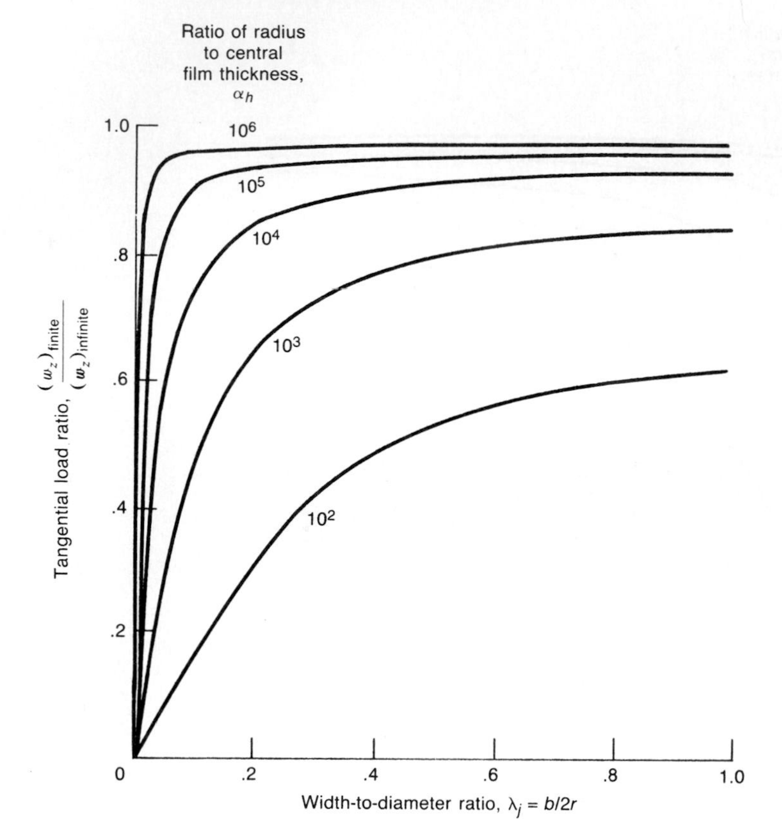

FIGURE 18.4
Effect of leakage on tangential load component.

From Table 4.1

$$1 \text{ P} = 100 \text{ cP}$$

$$= 100(1.45 \times 10^{-7}) \text{ lb} \cdot \text{s/in}^2 = 1.45 \times 10^{-5} \text{ lb} \cdot \text{s/in}^2$$

$$(w_z)_{\text{infinite}} = 2.45(1.45 \times 10^{-5} \text{ lb} \cdot \text{s/in}^2)(500 \text{ in/s})(2 \text{ in})(10^3) = 35.5 \text{ lb}$$

$$(w_z)_{\text{finite}} = 8.88 \text{ lb}$$

18.4 GENERAL RIGID-BODY SOLUTION

The three previous sections having been concerned with the lubrication of a rigid cylinder near a plane, the present section describes the lubrication between two rigid bodies that have any nonconformal shape. The effect of the normal load component on the ratio of transverse radius to rolling radius (radius ratio) for different film thicknesses is investigated. The conjunction is assumed to be fully immersed in lubricant, thereby implying that a fully flooded condition exists.

18.4.1 FILM SHAPE. The thickness of a hydrodynamic film between two rigid bodies in rolling can be written as

$$h = h_0 + S(x, y) \tag{18.47}$$

where h_0 = central (also minimum) film thickness due to hydrodynamic lubrication action

$S(x, y)$ = separation due to geometry of solids

The separation of two rigid solids is shown in Fig. 18.5(a), in which the principal axes of rotation of the two bodies are parallel.

$$S(x, y) = S_{ax} + S_{bx} + S_{ay} + S_{by} \tag{18.48}$$

where

$$S_{ax} = r_{ax} - \left(r_{ax}^2 - x^2\right)^{1/2} \tag{18.49}$$

$$S_{bx} = r_{bx} - \left(r_{bx}^2 - x^2\right)^{1/2} \tag{18.50}$$

$$S_{ay} = r_{ay} - \left(r_{ay}^2 - y^2\right)^{1/2} \tag{18.51}$$

$$S_{by} = r_{by} - \left(r_{by}^2 - y^2\right)^{1/2} \tag{18.52}$$

A simplifying transformation can be made by summing the curvatures in the $x = 0$ and $y = 0$ planes. In terms of the effective radius of curvature

$$\frac{1}{R_x} = \frac{1}{r_{ax}} + \frac{1}{r_{bx}} \tag{18.53}$$

$$\frac{1}{R_y} = \frac{1}{r_{ay}} + \frac{1}{r_{by}} \tag{18.54}$$

The resulting equivalent system is shown in Fig. 18.5(b). The separation in terms of the coordinates and the effective radius of curvature is

$$S(x, y) = R_x - \left(R_x^2 - x^2\right)^{1/2} + R_y - \left(R_y^2 - y^2\right)^{1/2} \tag{18.55}$$

Letting $x = XR_x$, $y = YR_x$, and $h = HR_x$ while substituting Eq. (18.55) into Eq. (18.47) gives

$$H = H_0 + 1 - (1 - X^2)^{1/2} + \alpha_r\left[1 - \left(1 - \frac{Y^2}{\alpha_r^2}\right)^{1/2}\right] \tag{18.56}$$

where

$$\alpha_r = R_y/R_x \tag{18.57}$$

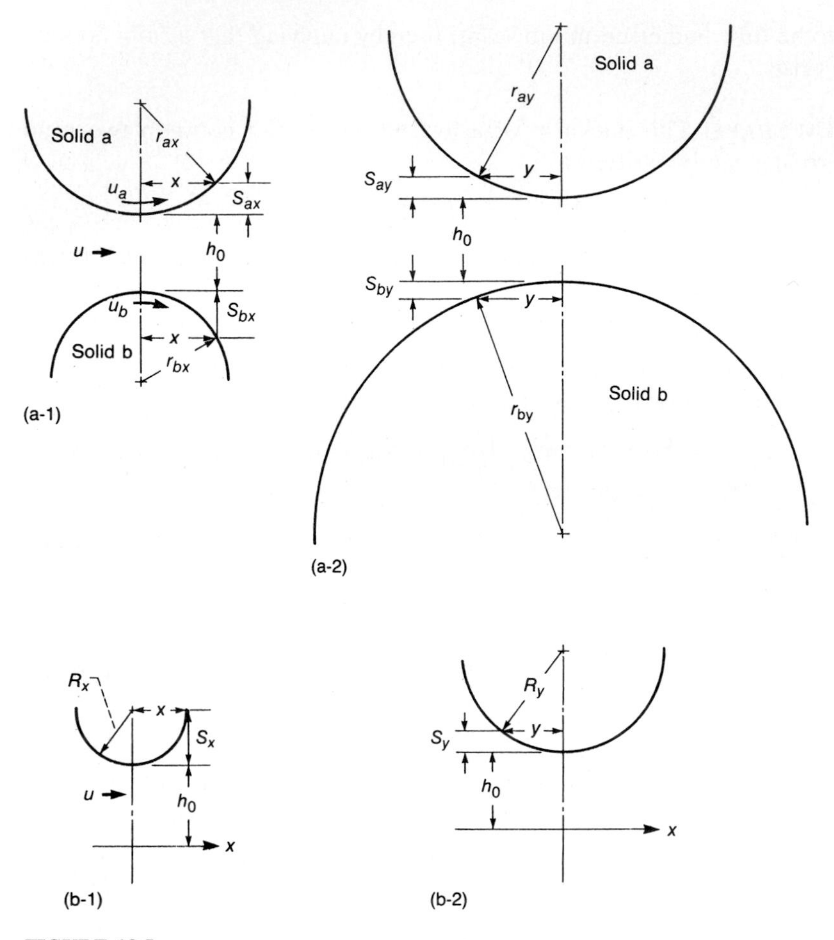

FIGURE 18.5
Contact geometry. (a) Two rigid solids separated by a lubricant film: (a-1) $y = 0$ plane; (a-2) $x = 0$ plane. (b) Equivalent system of a rigid solid near a plane separated by a lubricant film: (b-1) $y = 0$ plane; (b-2) $x = 0$ plane. [*From Brewe et al. (1979)*.]

Thus, Eq. (18.56) completely determines the film shape when the hydrodynamic effects on the central film thickness are known. For situations in which $X^2 \ll 1$ and $(Y/\alpha_r)^2 \ll 1$, it is convenient to expand H in a two-dimensional Taylor series to give

$$H \approx H_0 + \frac{X^2}{2} + \frac{Y^2}{2\alpha_r} \tag{18.58}$$

Equation (18.58) is called the "parabolic approximation." In the analysis that follows, Eq. (18.58) is used to describe the film shape. Figure 18.1 illustrates the difference between the parabolic approximation and the actual circular shape.

18.4.2 PRESSURE DISTRIBUTION. The hydrodynamic lubrication effect developed in establishing the central film thickness in Eq. (18.58) can be obtained from the Reynolds equation (7.48) while assuming an incompressible Newtonian fluid operating under laminar, isothermal, isoviscous, steady-state conditions. The resultant Reynolds equation is

$$\frac{\partial}{\partial x}\left(h^3\frac{\partial p}{\partial x}\right)+\frac{\partial}{\partial y}\left(h^3\frac{\partial p}{\partial y}\right)=6\eta_0(u_a+u_b)\frac{\partial h}{\partial x} \tag{7.48}$$

Making use of

$$x=XR_x \qquad y=YR_x \qquad h=HR_x \qquad \alpha_r=\frac{R_y}{R_x} \qquad p=\frac{\eta_0(u_a+u_b)P}{2R_x} \tag{18.59}$$

changes Eq. (7.48) to

$$\frac{\partial}{\partial X}\left(H^3\frac{\partial P}{\partial X}\right)+\frac{\partial}{\partial Y}\left(H^3\frac{\partial P}{\partial Y}\right)=12\frac{\partial H}{\partial X} \tag{18.60}$$

This equation has both a homogeneous solution and a particular solution; that is,

$$P=P_h+P_p \tag{18.61}$$

for which P_h is a solution to the homogeneous equation and satisfies the condition that $P_h=-P_p$ at the boundaries

$$\frac{\partial}{\partial X}\left(H^3\frac{\partial P_h}{\partial X}\right)+\frac{\partial}{\partial Y}\left(H^3\frac{\partial P_h}{\partial Y}\right)=0 \tag{18.62}$$

For the parabolic film approximation the particular solution for the pressure is simply proportional to X/H^2; that is,

$$P_p=-\frac{4\varphi X}{H^2} \tag{18.63}$$

where

$$\varphi=\frac{1}{1+2/3\alpha_r} \tag{18.64}$$

In the preceding equation φ is the side-leakage factor established by Archard

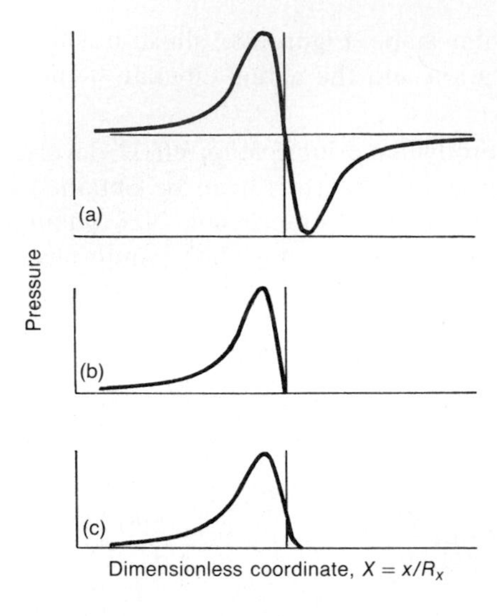

FIGURE 18.6
Effect of boundary conditions. (a) Solution using full Sommerfeld boundary conditions; (b) solution using half Sommerfeld boundary condition; (c) solution using Reynolds boundary conditions. [*From Brewe et al. (1979).*]

and Cowking (1965–66) and can be verified by inserting P_p back into Eq. (18.60). If we define P_h as $4\varphi\hat{P}_h$, using Eq. (18.61) results in the full solution

$$P = 4\varphi\left(-\frac{X}{H^2} + \hat{P}_h\right) \tag{18.65}$$

In general, the homogeneous solution of the dimensionless pressure $\hat{P}_h$ is an unknown function of X and Y. Consequently, the pressure distribution must be determined numerically.

The effect of three different types of boundary condition that can be used in association with this problem are shown in Fig. 18.6. Sommerfeld (1904) assumed the pressure to be ambient or zero at the point of closest approach. The resulting antisymmetric solution with respect to X is shown in Fig. 18.6(a). As discussed in Chap. 10 for journal bearings, the lubricant is unable to sustain the negative pressures predicted by the full Sommerfeld solution.

A simple approach taken by Kapitza (1955) was to ignore the negative pressures, that is, to employ the half Sommerfeld boundary condition. Figure 18.6(b) shows that using this boundary condition gives a reasonable estimate of load-carrying capacity. However, Kapitza's solution does not satisfy continuity conditions at the outlet (cavitation) boundary; that is, the pressure gradient normal to the cavitation boundary must be zero.

To insist on $P = \partial P/\partial X = 0$ at $X = 0$ would be overspecifying the problem mathematically. However, the Reynolds boundary condition shown in Fig. 18.6(c) insists on $P = \partial P/\partial X = 0$ at the outlet boundary, which is undetermined.

The variable-mesh nodal structure shown in Fig. 18.7 was used to provide close spacing in and around the pressure peak. This helped to minimize the

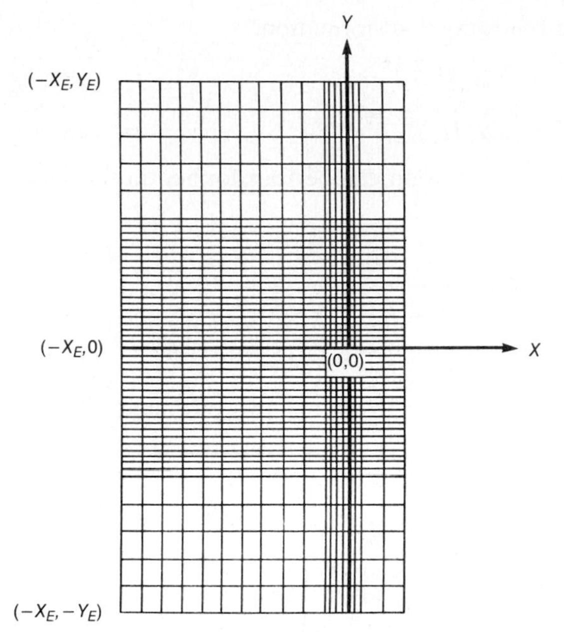

FIGURE 18.7
Variable nodal structure used for numerical calculations. [*From Brewe et al. (1979).*]

errors that can occur because of large gradients in the high-pressure region. The grid spacing used in terms of coordinates X and Y was varied depending on the anticipated pressure distribution. That is, for a highly peaked and localized pressure distribution, the fine mesh spacing was about 0.002 and the coarse mesh spacing about 0.1. For a relatively flat pressure distribution the fine mesh spacing was about 0.005 and the coarse mesh spacing about 0.13.

18.4.3 NORMAL LOAD COMPONENT. Once the pressure is known, the normal load component can be calculated from

$$w_z = \int\int p\,dx\,dy$$

In dimensionless form and making use of Eqs. (18.59) this equation becomes

$$w_z = \frac{\eta_0(u_a + u_b)R_x}{2}\int\int P\,dX\,dY$$

Substituting Eq. (18.65) into this equation gives

$$w_z = 2\eta_0(u_a + u_b)R_x\varphi\int\int\left(-\frac{X}{H^2} + \hat{P}_h\right)dX\,dY \qquad (18.66)$$

For the parabolic film assumption the central film thickness can be isolated from the integral by defining the following transformation:

$$X = (2H_0)^{1/2}\bar{X} \tag{18.67}$$

$$Y = (2\alpha_r H_0)^{1/2}\bar{Y} \tag{18.68}$$

If it is assumed that the homogeneous solution can be transformed in the same manner as the particular solution,

$$w_z = 4\varphi\eta_0(u_a + u_b)R_x\left(\frac{2\alpha_r}{H_0}\right)^{1/2}\iint\left[\frac{-\bar{X}}{\left(1+\bar{X}^2+\bar{Y}^2\right)^2} + \hat{P}_h(\bar{X},\bar{Y})\right]d\bar{X}\,d\bar{Y} \tag{18.69}$$

Kapitza (1955) refers to this integral as the "reduced hydrodynamic lift L." Thus,

$$L = \iint\left[\frac{-\bar{X}}{\left(1+\bar{X}^2+\bar{Y}^2\right)^2} + \hat{P}_h(\bar{X},\bar{Y})\right]d\bar{X}\,d\bar{Y} \tag{18.70}$$

Therefore, the dimensionless minimum film thickness is

$$H_{\min} = H_0 = 128\alpha_r\left[\frac{\varphi\eta_0(u_a+u_b)R_x L}{2w_z}\right]^2 \tag{18.71}$$

The ratio of the dimensionless speed to the dimensionless load can be defined as

$$\frac{U}{W} = \frac{\eta_0(u_a+u_b)R_x}{2w_z} \tag{18.72}$$

where

$$U = \frac{\eta_0\tilde{u}}{E'R_x} \qquad W = \frac{w_z}{E'R_x^2} \qquad \tilde{u} = \frac{u_a+u_b}{2}$$

Therefore, Eq. (18.71) can be rewritten as

$$H_{\min} = H_0 = 128\alpha_r\left(\frac{\varphi UL}{W}\right)^2 \tag{18.73}$$

For the parabolic film approximation, L needs to be determined only as a function of the geometry; that is,

$$L = L(\alpha_r) \qquad \text{if}\begin{cases}\bar{X}^2 \ll \dfrac{1}{2H_0}\\[2ex] \bar{Y}^2 \ll \dfrac{1}{2H_0}\end{cases} \tag{18.74}$$

This will be determined numerically.

18.4.4 FILM THICKNESS FORMULAS. The film thickness formula can be determined, while assuming the parabolic film thickness equation given in Eq. (18.58), by numerically determining the pressure distribution from Eqs. (18.60) and (18.65), integrating the pressure to establish the load-carrying capacity from Eq. (18.69), and then inserting the load w_z into Eq. (18.71) and solving for L for a specific α_r. If the Reynolds boundary conditions are assumed, the resulting formula obtained from curve fitting the data as obtained from Brewe et al. (1979) is

$$H_0 = H_{\min} = 128\alpha_r\left(\frac{\varphi UL}{W}\right)^2 \tag{18.75}$$

where

$$L = 0.131\tan^{-1}\left(\frac{\alpha_r}{2}\right) + 1.683 \tag{18.76}$$

The minimum-film-thickness equation derived by Kapitza using half Sommerfeld boundary conditions and assuming a parabolic film approximation is

$$H_0 = H_{\min} = 128\alpha_r\left(\frac{\varphi U}{W}\frac{\pi}{2}\right)^2 \tag{18.77}$$

By equating Eqs. (18.75) and (18.77) for a given speed parameter, the load-carrying capacities for the two theories can be compared; that is,

$$\frac{W}{W_K} = \frac{L}{\pi/2} \tag{18.78}$$

The effect of the geometry on the reduced hydrodynamic lift is shown in Fig. 18.8. This figure shows that L, and hence the load-carrying capacity [Eq. (18.78)], is 11 to 20 percent greater than that predicted by Kapitza. The least difference occurs for a ball-on-plane contact. As α_r is increased, the difference in load-carrying capacities approaches a constant 20 percent. The alteration of

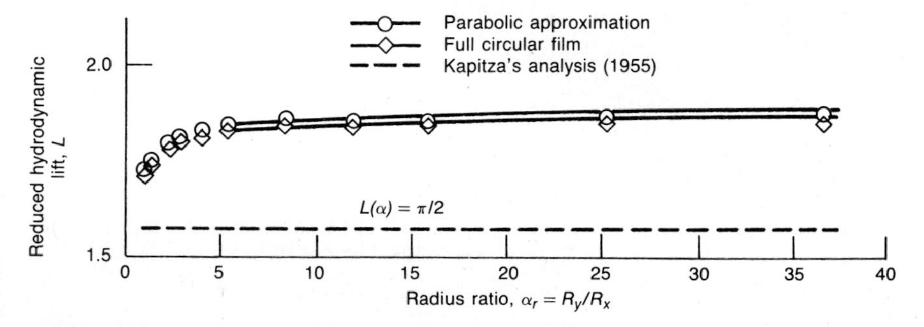

FIGURE 18.8
Effect of radius ratio on reduced hydrodynamic lift. [*From Brewe et al. (1979).*]

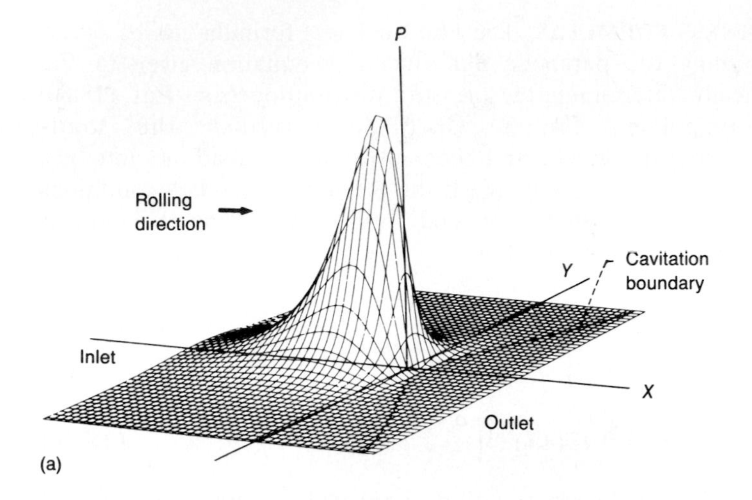

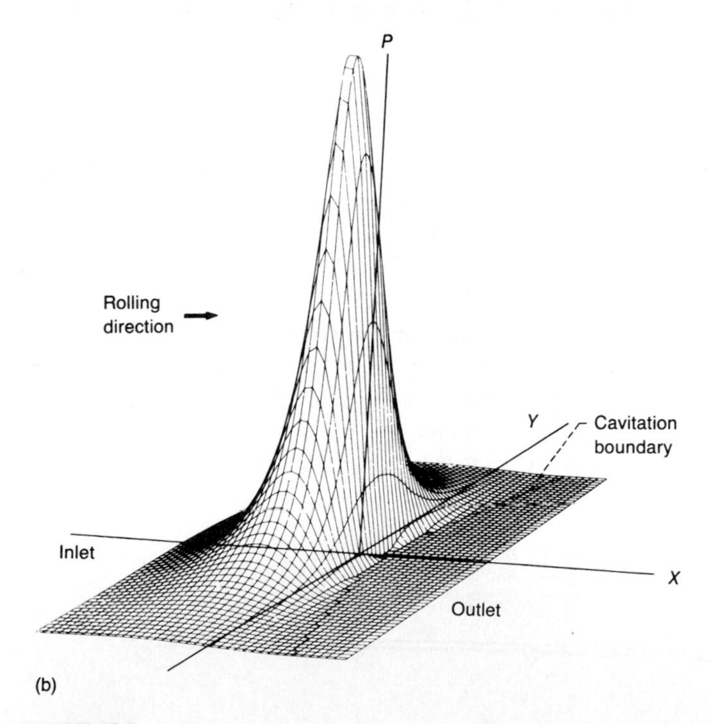

FIGURE 18.9
Three-dimensional representations of pressure distributions as viewed from outlet region for two radius ratios α_r. (a) $\alpha_r = 1.00$; (b) $\alpha_r = 36.54$. [*From Brewe et al. (1979).*]

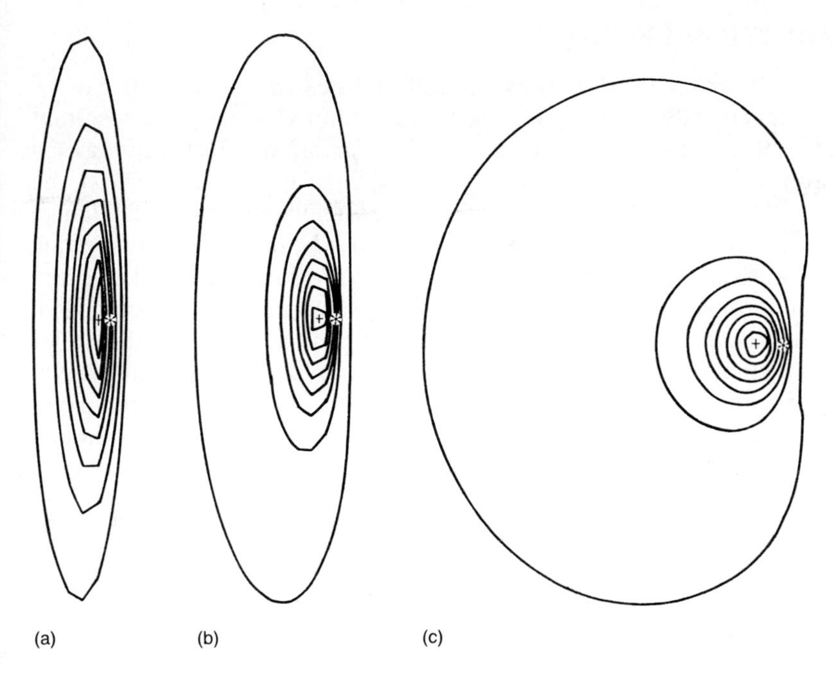

FIGURE 18.10
Pressure contours for three radius ratios α_r. (a) $\alpha_r = 25.29$; (b) $\alpha_r = 8.30$; (c) $\alpha_r = 1.00$. [*From Brewe et al. (1979).*]

the pressure distribution due to the Reynolds boundary conditions at the cavitation boundary is responsible for this geometry effect.

Figure 18.9 is a three-dimensional representation of a pressure distribution for α_r of 1.00 and 36.54. This figure also illustrates the shape of the cavitation boundary. As α_r becomes large, the cavitation boundary tends to straighten out, accompanied by decreasing changes in L. The scale along the Y axis in Fig. 18.9(a) has been magnified three times to improve the resolution. Consequently, the differences in the shapes of the cavitation boundary are actually subdued as they are presented.

Isobar plots for three radius ratios (that is, α_r of 25.29, 8.30, and 1.00) are shown in Fig. 18.10. The contours belong to the family of curves defined by Eq. (18.65). The center of the contact is represented by the asterisk. The pressure peak builds up in the inlet region, which is located to the left of the contact center and is indicated by the plus sign. Since the isobars in each case are evenly spaced, the pressure gradients can be easily depicted. Note that as the radius ratio increases, the steeper pressure gradients are predominantly along the rolling direction. This implies that the amount of side leakage decreases as α_r increases. A decrease in side leakage is reflected in an increase in φ.

18.5 STARVATION EFFECTS

The effect of starvation in a hydrodynamically lubricated conjunction can be studied by systematically reducing the inlet supply and observing the resultant pressure distribution and film thickness. This starvation effect can have a significant role in the operation of machine elements. It is desirable that the hydrodynamic film generated between the roller end and the guide flange provide stiffness and damping to limit the amplitude of the roller skewing motion. However, at high rotational speeds the roller end and the flange are often subjected to a depletion in the lubricant supply owing to centrifugal effects. In such cases the minute amount of lubricant available at the roller end–flange conjunction might well represent an example of steady-state starvation. Starvation effects in hydrodynamically lubricated contacts are important for calculating the rolling and sliding resistance and traction encountered in ball and roller bearings (see Chiu, 1974).

In another example the effect of restricting the lubricant to a roller bearing is seen experimentally and theoretically to reduce the amount of cage and roller slip (see Boness, 1970). The theoretical analysis was accomplished by changing the location of the boundary where the pressure begins to build up and noting the effect on the hydrodynamic forces.

18.5.1 FILM THICKNESS FORMULAS. The difficulty with the equation developed by Brewe et al. (1979), Eq. (18.75), is that its range of applicability is limited. In a later publication by Brewe and Hamrock (1982) a minimum film thickness valid for the full range of dimensionless film thickness normally encountered in a fully flooded undeformed contact is given as

$$H_0 = H_{\min} = \left[\frac{W/U}{\varphi L (128 \alpha_r)^{1/2}} + 3.02 \right]^{-2} \tag{18.79}$$

Brewe and Hamrock (1982) found that the reduction in minimum film thickness from the fully flooded value, if the fluid inlet level is known, is

$$H_{\min,s} = \left\{ \frac{1}{H_{\min}^{1/2}} + 3.02 \left[\left(\frac{2 - H_{\text{in}}}{H_{\text{in}}} \right)^{1/2} e^{H_{\text{in}} - 1} - 1 \right] \right\}^{-2} \tag{18.80}$$

where H_{in}, the dimensionless fluid inlet level, is equal to h_{in}/R_x. Dividing both sides of Eq. (18.80) by $H_{\min}$ gives

$$\beta_s = \frac{H_{\min,s}}{H_{\min}} = \left\{ 1 + 3.02 H_{\min}^{1/2} \left[\left(\frac{2 - H_{\text{in}}}{H_{\text{in}}} \right)^{1/2} e^{H_{\text{in}} - 1} - 1 \right] \right\}^{-2} \tag{18.81}$$

where β_s is the reduction in minimum film thickness due to starvation.

18.5.2 PRESSURE DISTRIBUTION. The discussion of lubricant starvation can be facilitated by focusing on one of the simplest geometric arrangements (i.e., a

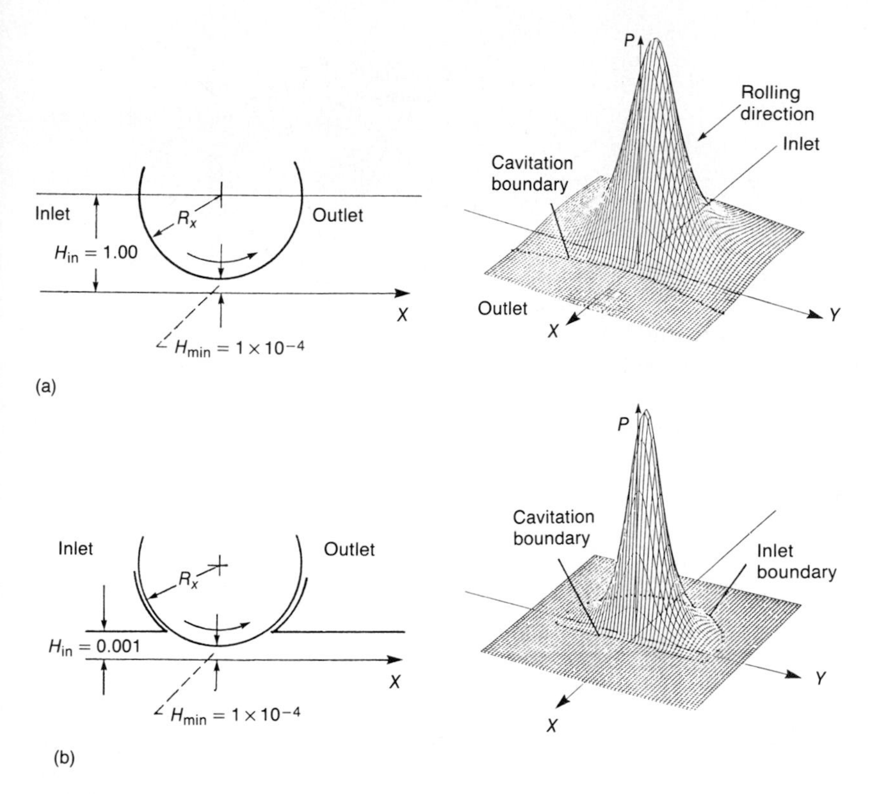

FIGURE 18.11
Three-dimensional representation of pressure distributions for dimensionless minimum film thickness H_{min} of 1.0×10^{-4}. (a) Fully flooded condition; (b) starved condition. [*From Brewe and Hamrock (1982).*]

ball rolling or sliding against a flat plane) as shown in Fig. 18.11. The figure compares the pressure distribution determined numerically for the fully flooded inlet ($H_{in} = 1.00$) with that for the most severely starved inlet ($H_{in} = 0.001$). The comparison is made for a constant minimum film thickness of $H_{min} = 1.0 \times 10^{-4}$. Note that the pressure peak built up in the starved inlet is only slightly smaller than that of the fully flooded inlet. However, the area of pressure buildup is considerably smaller so that the starved inlet is unable to support as much load for a given film thickness as the fully flooded inlet.

Figure 18.12 provides the same sort of comparison but for a thicker minimum film, $H_{min} = 1.0 \times 10^{-3}$. The significant difference between the two figures is that the starved inlet for the thicker film has a more pronounced effect on the pressure peak. The fluid inlet level ($H_{in} = 0.002$) for the thicker film represents a more highly starved inlet since H_{in} is of the order of H_{min} in this case. The other feature to be noticed in comparing Figs. 18.11 and 18.12 is that the pressure distribution is more evenly spread out for the thicker film. Thus,

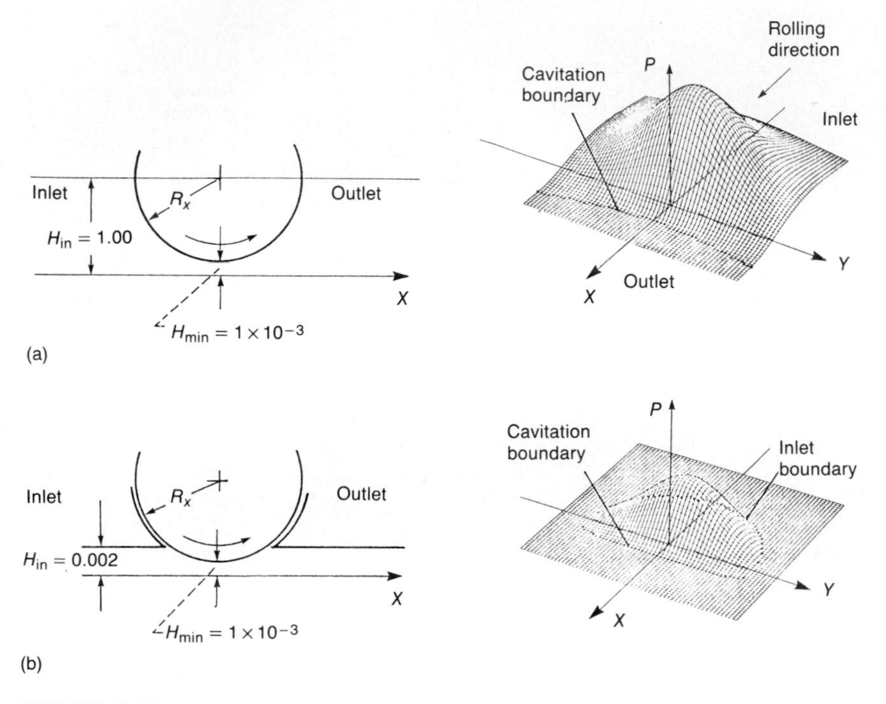

FIGURE 18.12
Three-dimensional representation of pressure distribution for dimensionless minimum film thickness H_{min} of 1.0×10^{-3}. (a) Fully flooded condition; (b) starved condition. [*From Brewe and Hamrock (1982).*]

changes in the meniscus (or integration domain) are going to have a more noticeable effect on the load-carrying capacity. Note also that because of the boundary conditions the integration domain takes on a kidney-shaped appearance. This is more clearly shown in the isobaric contour plot in Fig. 18.13.

18.5.3 FULLY FLOODED–STARVED BOUNDARY. Of practical importance to lubricant starvation is the decrease in the film thickness reduction factor β_s from the fully flooded value. Equation (18.81) is a derived expression for β_s in terms of the fluid inlet level and the fully flooded film thickness. Figure 18.14 is a plot of β_s as a function of H_{in} for four values of H_{min}. It is of interest to determine a fully flooded–starved boundary (i.e., the fluid inlet level after which any further decrease causes a significant reduction in the film thickness). The definition of this boundary is established by

$$(1 - \beta_s)_{H_{in}=H_{in}^*} = 0.03 \qquad (18.82)$$

The value of 0.03 is used in Eq. (18.82) because it was ascertained that the numerical data obtained were accurate to only ± 3 percent.

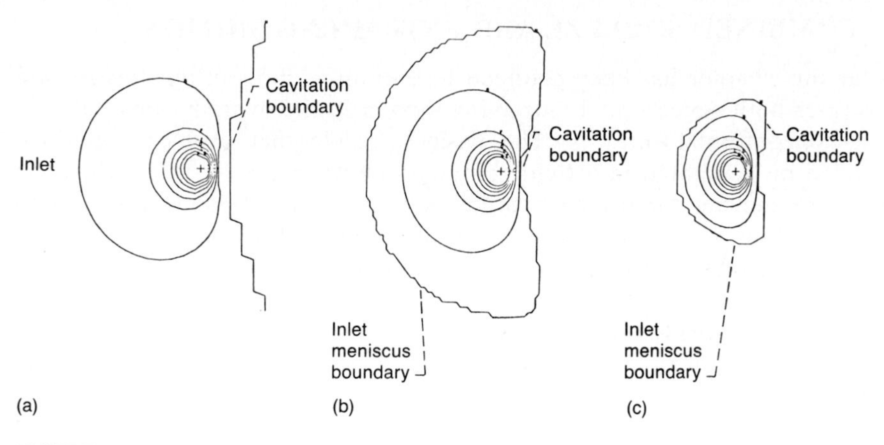

FIGURE 18.13
Isobaric contour plots for three fluid inlet levels for dimensionless minimum film thickness H_{min} of 1.0×10^{-4}. (a) Fully flooded condition: dimensionless fluid inlet level H_{in}, 1.00; dimensionless pressure, where $dP/dX = 0$, P_m, 1.20×10^6; dimensionless load-speed ratio W/U, 1153.6. (b) Starved condition: H_{in}, 0.004; $P_m = 1.19 \times 10^6$; $W/U = 862.6$. (c) Starved condition: $H_{in} = 0.001$; $P_m = 1.13 \times 10^6$; $W/U = 567.8$. [*From Brewe and Hamrock (1982).*]

Thus, for a given value of H_{min} a value of H_{in}^* can be determined that satisfies Eq. (18.82). A suitable relationship between H_{in}^* and H_{min} was obtained by Brewe and Hamrock (1982) and is given as

$$H_{in}^* = 4.11(H_{min})^{0.36} \tag{18.83}$$

This equation determines the onset of starvation. That is, for $H_{in} \geq H_{in}^*$ a fully flooded condition exists, whereas for $H_{in} < H_{in}^*$ a starved condition exists.

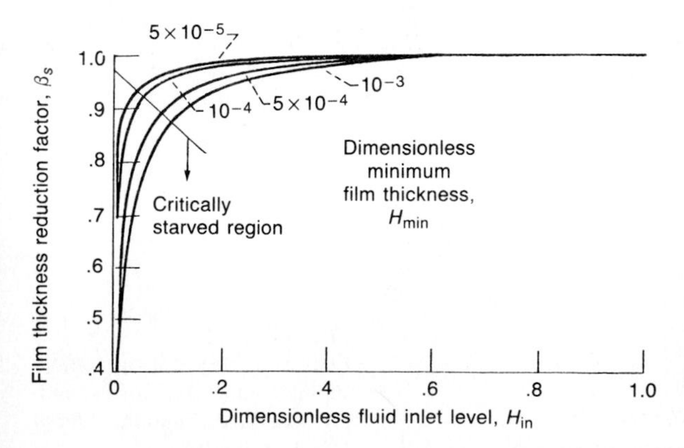

FIGURE 18.14
Effect of fluid inlet level on film thickness reduction factor in flooded conjunctions. [*From Brewe and Hamrock (1982).*]

18.6 COMBINED SQUEEZE AND ENTRAINING MOTION

Thus far this chapter has been confined to entraining motion, but this section incorporates both squeeze and entraining motion in the hydrodynamic lubrication of nonconformal surfaces. The basic principle that enables a positive pressure to be generated in a fluid contained between two surfaces when the surfaces are moving toward each other is that a finite time is required to squeeze the fluid out of the gap, as was discovered in Chap. 13. In this section the discussion is limited to fully flooded conjunctions.

18.6.1 PRESSURE DISTRIBUTION AND LOAD. Figure 18.15 shows the flow of lubricant in a conjunction with combined squeeze and entraining motion. the Reynolds equation for the hydrodynamic lubrication of two rigid solids separated by an incompressible, isoviscous lubricant film with entraining and normal squeeze motion can be expressed from Eq. (7.45) as

$$\frac{\partial}{\partial x}\left(h^3\frac{\partial p}{\partial x}\right)+\frac{\partial}{\partial y}\left(h^3\frac{\partial p}{\partial y}\right)=12\eta\left(\frac{u_a+u_b}{2}\right)\frac{\partial h}{\partial x}+12\eta\frac{\partial h}{\partial t} \qquad (18.84)$$

Making use of the dimensionalization given in Eq. (18.59) and

$$w_a=\frac{\partial h}{\partial t} \qquad \alpha_\omega=\frac{w_a}{\tilde{u}} \qquad \tilde{u}=\frac{u_a+u_b}{2} \qquad (18.85)$$

changes Eq. (18.84) to

$$\frac{\partial}{\partial X}\left(H^3\frac{\partial P}{\partial X}\right)+\frac{\partial}{\partial Y}\left(H^3\frac{\partial P}{\partial Y}\right)=12\left(\frac{\partial H}{\partial X}+\alpha_\omega\right) \qquad (18.86)$$

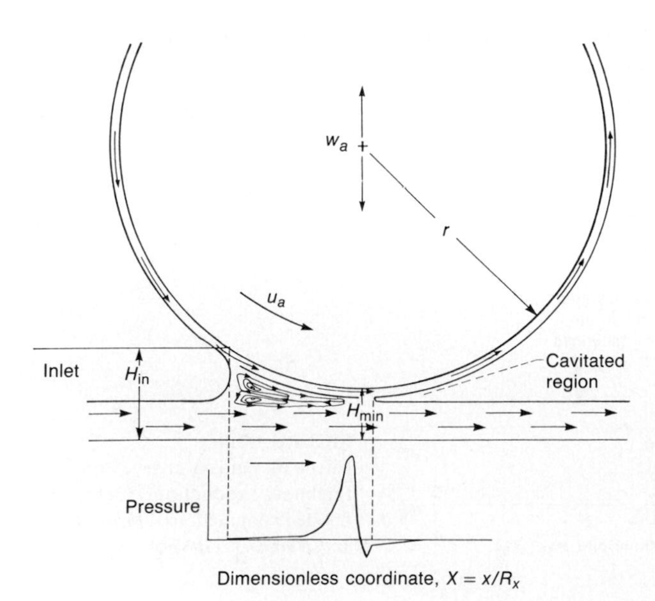

FIGURE 18.15
Lubricant flow for a rolling-sliding contact and corresponding pressure buildup. [*From Ghosh et al. (1985).*]

If

$$\beta_\omega = \frac{\alpha_\omega}{(2H_{\min})^{1/2}} \tag{18.87}$$

this equation becomes

$$\frac{\partial}{\partial X}\left(H^3 \frac{\partial P}{\partial X}\right) + \frac{\partial}{\partial Y}\left(H^3 \frac{\partial P}{\partial Y}\right) = 12\left[\frac{\partial H}{\partial X} + (2H_{\min})^{1/2}\beta_\omega\right] \tag{18.88}$$

The parabolic approximation [Eq. (18.58)], along with the Reynolds boundary conditions, can be used to evaluate Eq. (18.88) numerically.

The load-carrying capacity can be calculated by integrating the pressure in the contact region or

$$w_z = \int_A p\, dx\, dy$$

In dimensionless form it is expressed as

$$W = \frac{w_z}{\eta_0 \tilde{u} R_x} = \int_A P\, dX\, dY \tag{18.89}$$

where A is the dimensionless domain of integration that is dependent on both the fluid level and the cavitation boundary. The instantaneous load-carrying capacity is expressed as a ratio

$$\beta_t = \frac{W}{(W)_{\beta_\omega = 0}} \tag{18.90}$$

18.6.2 RESULTS AND DISCUSSION. The dynamic performance of hydrodynamically lubricated nonconformal contacts in combined entraining and normal squeeze motion is governed by the dimensionless normal velocity parameter β_ω. It incorporates three major parameters, namely, the normal squeeze velocity w_a, the mean entraining velocity $\tilde{u}$, and the central or minimum film thickness $h_{\min}$. The performance parameters affected by the normal velocity parameter are

1. The dynamic load ratio β_t
2. The dynamic peak pressure ratio $\xi_s = P_m/(P_m)_{\beta_\omega = 0}$
3. The dimensionless location of peak pressure $X_{e,s}$
4. The dimensionless location of film rupture $X_{e,r}$

This section attempts to show the effect of the operating parameters on these performance parameters.

Figure 18.16 shows the effect of the dimensionless normal velocity parameter on the dynamic load ratio. The dimensionless normal velocity parameter β_ω clearly has a significant pressure-generating effect during the normal approach, where the dynamic load ratio β_t increases with increasing β_ω. On the contrary, β_t is significantly reduced during normal separation for $\beta_\omega > 0$.

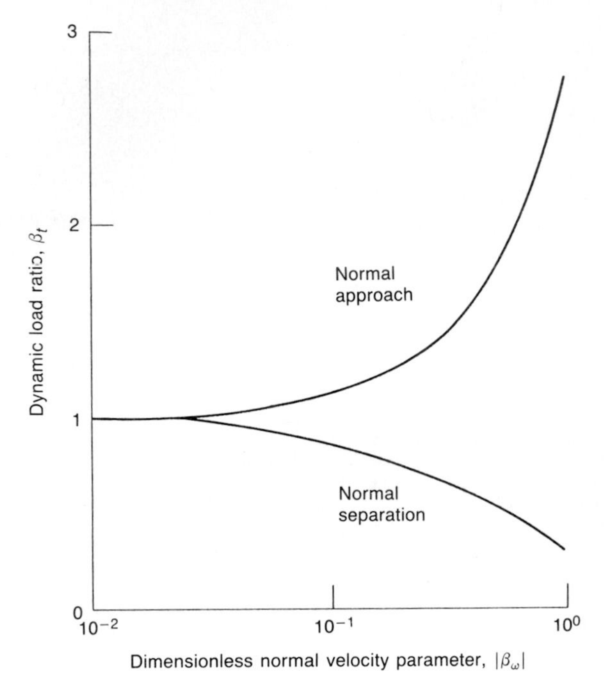

FIGURE 18.16
Effect of dimensionless normal velocity parameter on dynamic load ratio. Dimensionless central film thickness H_{min}, 1.0×10^{-4}; radius ratio α_r, 1.0; dimensionless fluid inlet level H_{in}, 0.035. [*From Ghosh et al. (1985).*]

The magnitude of the peak pressure generated in a conjunction, its location, and the location of the film rupture boundary are also affected significantly by the dimensionless normal velocity parameter. Pressure distributions for five values of β_ω are shown in Fig. 18.17. The inlet meniscus boundary is not shown in this figure. During normal approach the film rupture boundary moves downstream into the outlet region (or the divergent portion of the film) with reference to the minimum-film-thickness position. During separation it moves upstream, reaching the convergent portion of the film for higher values of β_ω. The location of the pressure peak and the entire pressure distribution in the contact also shift accordingly. Thus, superposition of normal motion on the entraining velocity alters both the magnitude and distribution of the pressure in the contact.

The variation of the dynamic peak pressure ratio ξ_s with normal velocity parameter is shown in Fig. 18.18. Pressures approximately three or four times the corresponding peak pressure in the steady situation are generated in the conjunction during normal approach. Relative reductions in the peak pressures of similar magnitude occur during separation.

The influence of the radius ratio α_r on the dynamic load ratio β_t and the dynamic peak pressure ratio ξ_s is shown in Figs. 18.19 and 18.20, respectively. Recall that for $\alpha_r > 1$ the major axis of the film contours is transverse to the entrainment velocity, so for $\alpha_r \gg 1$ the geometry resembles a lubricated cylinder, as covered in Secs. 18.1 to 18.3. During normal approach β_t increases with increasing α_r up to a certain value and then tends to approach a limiting value

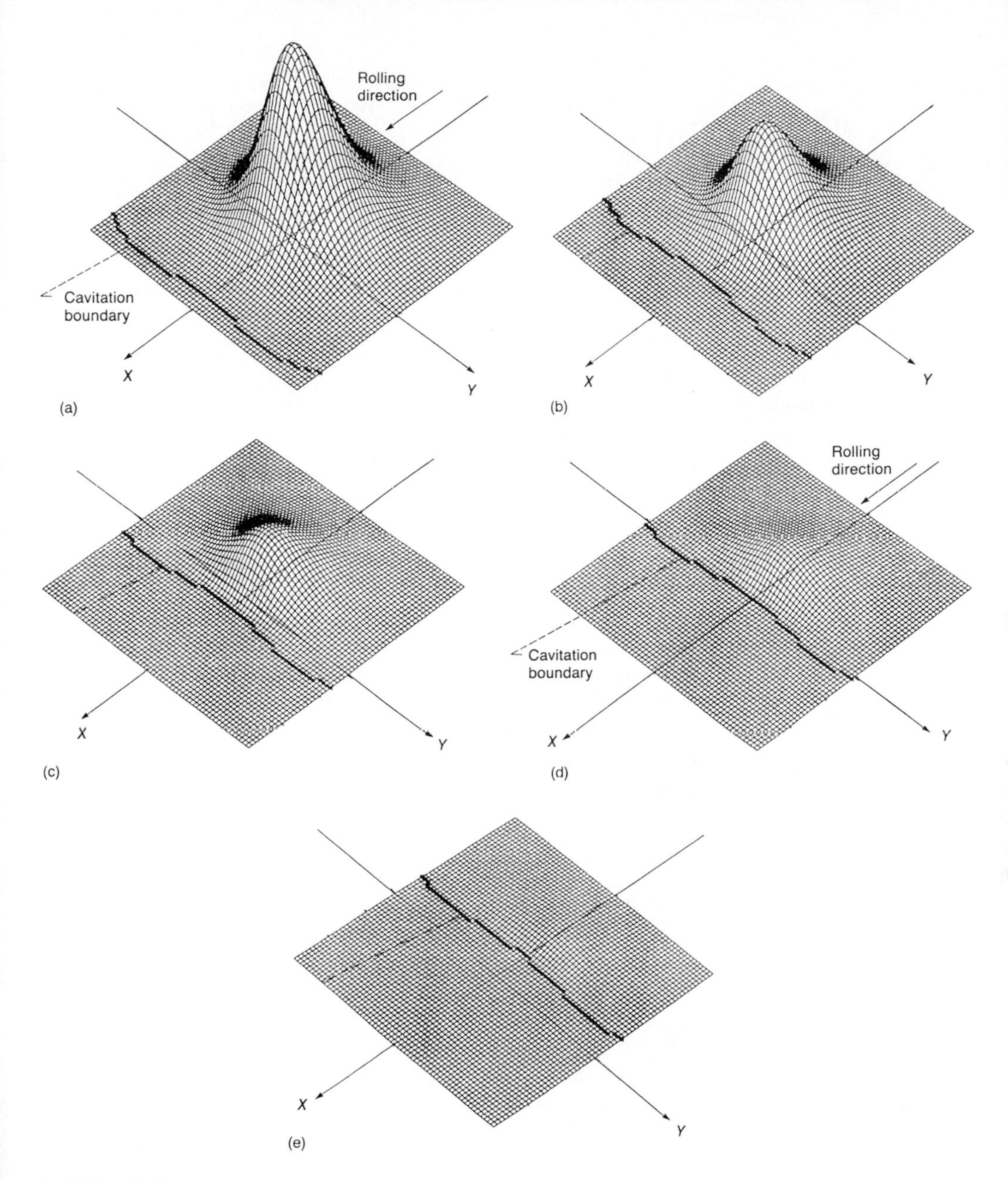

FIGURE 18.17
Pressure distribution in contact for various values of dimensionless normal velocity parameter β_ω. (a) $\beta_\omega = -1.0$; (b) $\beta_\omega = -0.5$; (c) $\beta_\omega = 0$; (d) $\beta_\omega = 0.25$; (e) $\beta_\omega = 0.75$. Dimensionless central film thickness $H_{\min}$, 1.0×10^{-4}; radius ratio α_r, 1.0; dimensionless fluid inlet level H_{in}, 0.0006. [*From Ghosh et al. (1985)*.]

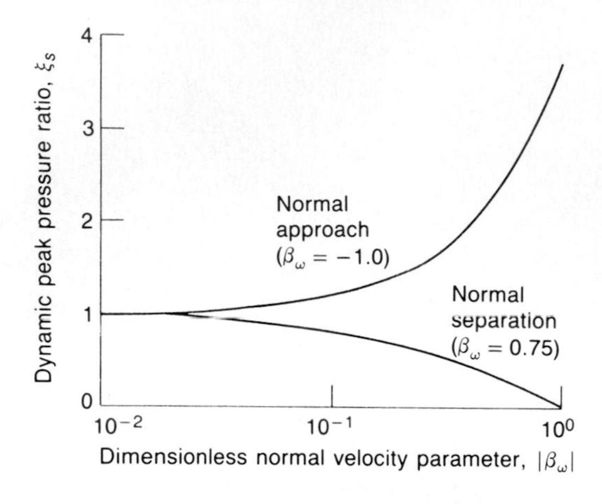

FIGURE 18.18
Effect of dimensionless normal velocity parameter on dynamic peak pressure ratio. Dimensionless central film thickness H_{min}, 1.0×10^{-4}; radius ratio α_r, 1.0; dimensionless fluid inlet level H_{in}, 0.035. [*From Ghosh et al. (1985).*]

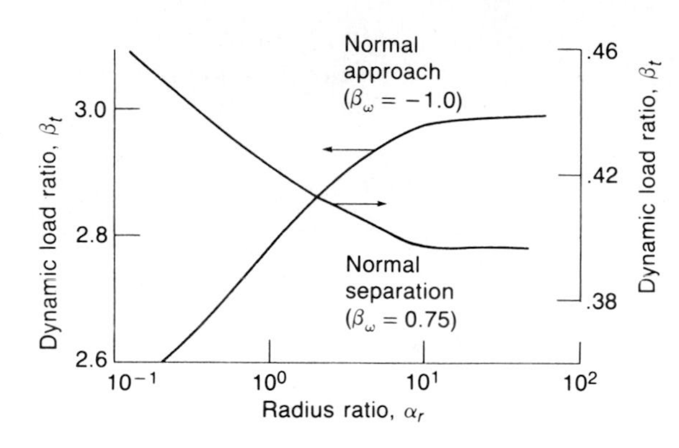

FIGURE 18.19
Effect of radius ratio on dynamic load ratio. Dimensionless central film thickness H_{min}, 1.0×10^{-4}; dimensionless fluid inlet level H_{in}, 0.035. [*From Ghosh et al. (1985).*]

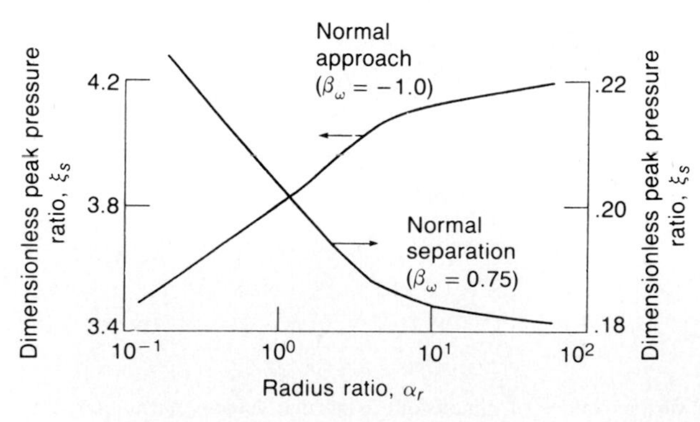

FIGURE 18.20
Effect of radius ratio on dynamic peak pressure ratio. Dimensionless central film thickness H_{min}, 1.0×10^{-4}; dimensionless fluid inlet level H_{in}, 0.035. [*From Ghosh et al. (1985).*]

for further increases in α_r. Similarly, during normal separation β_t initially decreases with increasing α_r and then approaches a limiting value for higher values of α_r. Similar variations with α_r can be observed for ξ_s in Fig. 18.20. Therefore it can be said that the radius ratio significantly affects the dynamic performance of nonconformal conjunctions for α_r between 0.2 and 10. However, for α_r between 10 and 35 the effects are small. By curve fitting 63 cases Ghosh et al. (1985) came up with the following formula for the dynamic load ratio:

$$\beta_t = \left[\alpha_r^{-0.028}\,\mathrm{sech}(1.68\beta_\omega)\right]^{1/\beta_\omega} \qquad \text{for } \beta_\omega \neq 0 \tag{18.91}$$

It has been pointed out that significantly higher peak pressures can be generated in the conjunction during normal approach than during normal separation. Therefore, the estimation of peak pressures in the conjunction is of practical importance for estimating the maximum stress in the contact and the fatigue life of the contact. Dynamic peak pressure ratio ξ_s is expressed by a simple formula as a function of the normal velocity parameter β_ω and the radius ratio α_r or

$$\xi_s = \left[\alpha_r^{-0.032}\,\mathrm{sech}(2\beta_\omega)\right]^{1/\beta_\omega} \qquad \text{for } \beta_\omega \neq 0 \tag{18.92}$$

18.7 CLOSURE

This chapter is the first of a number that will focus on the lubrication of nonconformal surfaces. The first three sections dealt with the lubrication of a cylinder near a plane; the first section with an infinitely wide-cylinder, the second with a short-width cylinder, and the third with the exact solution. The infinitely wide-cylinder theory found the minimum film thickness to be inversely proportional to the normal applied load. The short-width-cylinder theory found the minimum film thickness to be proportional to the normal load-carrying capacity raised to the -0.5 power, and this is much closer to the exact solution results.

For the exact solution it was found that two important operating parameters exist, namely, (1) width-to-diameter ratio $\lambda_j = b/2r$ and (2) radius-to-central-film-thickness ratio $\alpha_h = r/h_0$. Figures were provided that enable easy calculation of the normal and tangential load components when side leakage is considered and the operating parameters and the load component for infinite width are known.

The influence of geometry on the isothermal hydrodynamic film separating two general rigid solids was investigated. The investigation was conducted for a conjunction fully immersed in lubricant (i.e., fully flooded). The effect of the geometry on the film thickness was determined by varying the radius ratio from 1 (a ball-on-plane configuration) to 36 (a ball in a conforming groove). It was found that the minimum film thickness had the same speed, viscosity, and load dependence as Kapitza's classical solution. The central or minimum film thickness for the complete range of operation for a fully flooded conjunction is

$$H_0 = H_{\min} = \left[\frac{W/U}{\varphi L(128\alpha_r)^{1/2}} + 3.02 \right]^{-2}$$

where

$$L = 0.131 \tan^{-1}\left(\frac{\alpha_r}{2} \right) + 1.683$$

$$\varphi = \frac{1}{1 + 2/3\alpha_r}$$

Numerical methods were used to determine the effect of lubricant starvation on the minimum film thickness under conditions of hydrodynamic lubrication and side leakage. Starvation was effected by varying the inlet level. The reduction in minimum film thickness due to starvation can be expressed as

$$\beta_s = \frac{H_{\min,s}}{H_{\min}} = \left\{ 1 + 3.02 H_{\min}^{1/2} \left[\left(\frac{2 - H_{\text{in}}}{H_{\text{in}}} \right)^{1/2} e^{H_{\text{in}} - 1} - 1 \right] \right\}^{-2}$$

A numerical solution to the hydrodynamic lubrication of nonconformal rigid contacts with an incompressible, isoviscous lubricant in combined entraining and normal squeeze motion has been obtained. The following conclusions were reached through a parametric study:

1. Normal squeeze motion combined with entraining motion significantly increases the load-carrying capacity and peak pressure in the conjunction during normal approach. A correspondingly significant decrease in load-carrying capacity and peak pressure occurs during normal separation.
2. The film rupture (or cavitation) boundary moves farther into the outlet region away from the minimum-film-thickness position during normal approach but moves toward the inlet and into the convergent portion of the film for higher normal velocities during separation.
3. Increasing the radius ratio α_r increases the dynamic load ratio β_t and the dynamic peak pressure ratio ξ_s during normal approach. Reverse effects are observed during separation.
4. The ratios β_t and ξ_s are weakly dependent on the minimum film thickness and can be expressed in terms of the dimensionless normal velocity parameter β_ω and the radius ratio α_r by

$$\beta_t = \left[\alpha_r^{-0.028} \operatorname{sech}(1.68\beta_\omega) \right]^{1/\beta_\omega}$$

$$\xi_s = \left[\alpha_r^{-0.032} \operatorname{sech}(2\beta_\omega) \right]^{1/\beta_\omega}$$

18.8 PROBLEMS

18.8.1 Determine the normal load capacity expression for a cylinder near a plane when normal squeeze motion exists and subambient pressures are neglected, since we are only interested in diminishing film thickness. The length of the cylinder is one-half its radius. The cylinder and plane may be considered rigid. An adequate amount of viscous fluid is between the cylinder and the plane.

18.8.2 Discuss the effect of side leakage on the normal load capacity of nonconformal surfaces. Explain why side leakage has such a large effect on the normal load capacity of journal and thrust bearings and a small effect on that of nonconformal surfaces.

18.8.3 Theories for the isoviscous hydrodynamic lubrication of rigid cylinders have been completed and are now regarded as classical examples of the application of Reynolds equations to a practical problem. Discuss these analyses under the following headings:

(*a*) The approximation made
(*b*) The transformation used
(*c*) The boundary conditions applied
(*d*) The extension of the theories to cover cases where the viscosity is regarded as a function of pressure theories
(*e*) The practical significance, limitations, and application of the theories

A steel roller 1.20 in long and weighing 0.3 lb rides on an inclined, well-lubricated, metal foil f, as shown in the diagram. The foil is stretched between rollers that drive it a speed of 60 ft/min. The roller rotates about its axis without measurable slip while otherwise remaining fixed in space. From simple theory estimate the viscosity of the lubricant if the angle of inclination α of the foil is 0.573°.

Discuss any areas of doubt that might arise from the estimation of viscosity in this manner.

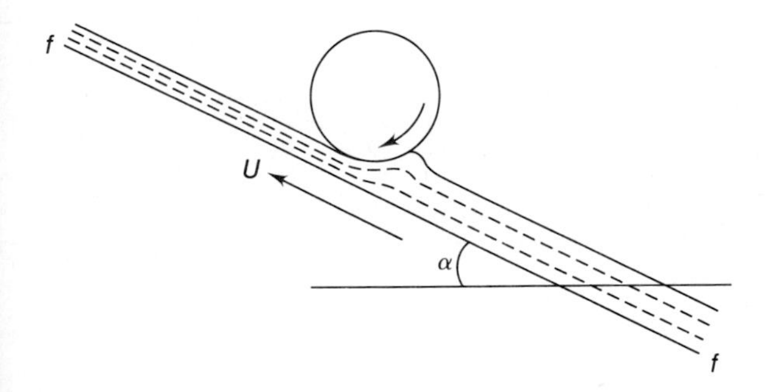

18.8.4 A 2-in-diameter steel roller, 1 in long, is between two parallel steel plates that can be considered infinitely wide and long enough to cover its ends (see figure below). If the load on the top plate is 1 lbf and the oil at the contact surfaces has a constant viscosity of 0.345 P, estimate the minimum oil film thickness between the roller and the top plate if this moves at 10 in/s in the positive x direction, the bottom plate being stationary.

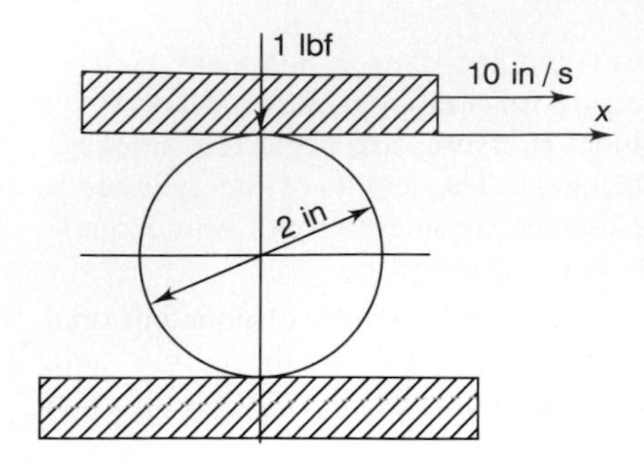

Assume the pressure distribution is $p = -2u\eta x/h^2$, where the symbols have their usual meaning.

$$u = \text{mean velocity}, \frac{u_a + u_b}{2}, m/s$$

$$\eta = \text{viscosity of fluid}, N \cdot s/m^2$$

$$x = \text{coordinate in the direction of motion}, m$$

$$h = \text{film thickness}, m$$

Neglect surface distortion and pressure and temperature effects on the oil.

18.8.5 Derive the hydrodynamic normal and shear force components (w_z, w_x, and F) acting on a rigid cylinder rotating near a moving rigid plane in the presence of an isoviscous lubricant as shown in figure (a). If the horizontal axis of the cylinder of

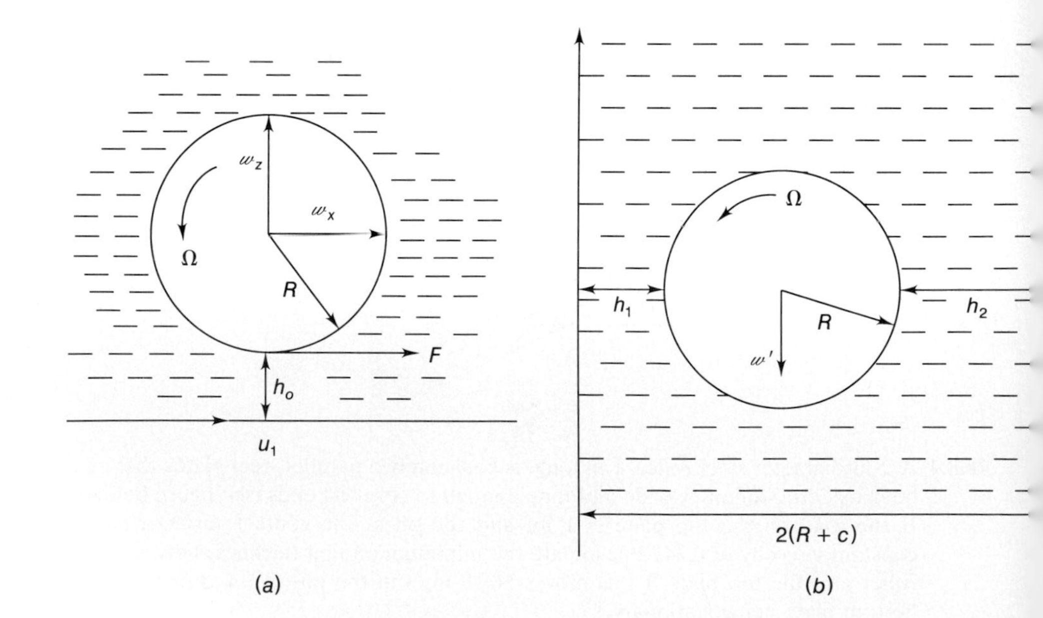

radius R and weight per unit length w' contained between a stationary vertical plate and parallel moving plate [see figure (b)] at distance $2(R + c)$ remains stationary when the velocity of the moving surface is u, show that the angular velocity Ω of the cylinder and the surface speed u of the plate are given by the expressions

$$\Omega = 0.0322 \frac{u}{R}$$

$$u = 0.235 w'/\eta (R/c)^{1/2}$$

where η = viscosity of fluid which surrounds the cylinder. It can be assumed that $c \ll R$. Show also that $h_1 = 0.06c$ and $h_2 = 1.94c$.

18.9 REFERENCES

Archard, J. F., and Cowking, E. W. (1965–66): Elastohydrodynamic Lubrication at Point Contacts. *Elastohydrodynamic Lubrication. Proc. Inst. Mech. Eng. (London)*, vol. 180, pt. 3B, pp. 17–26.

Boness, R. J. (1970): The Effect of Oil Supply on Cage and Roller Motion in a Lubricated Roller Bearing. *J. Lubr. Technol.*, vol. 92, no. 1, pp. 39–53.

Brewe, D. E., and Hamrock, B. J. (1982): Analysis of Starvation Effects on Hydrodynamic Lubrication in Nonconforming Contacts. *J. Lubr. Technol.*, vol. 104, no. 3, pp. 410–417.

Brewe, D. E., Hamrock, B. J., and Taylor, C. M. (1979): Effect of Geometry on Hydrodynamic Film Thickness. *J. Lubr. Technol.*, vol. 101, no. 2, pp. 231–239.

Chiu, Y. P. (1974): A Theory of Hydrodynamic Friction Forces in Starved Point Contacts Considering Cavitation. *J. Lubr. Technol.*, vol. 96, no. 2, pp. 237–246.

Dowson, D., and Higginson, G. R. (1966): *Elastohydrodynamic Lubrication, The Fundamentals of Roller and Gear Lubrication*. Pergamon, Oxford.

Dowson, D., and Whomes, T. L. (1967): Side-Leakage Factors for a Rigid Cylinder Lubricated by an Isoviscous Fluid. *Lubrication and Wear, Fifth Convention, Proc. Inst. Mech. Eng.*, vol. 181, pt. 30, pp. 165–176.

Dubois, G. B., and Ocvirk, F. W. (1953): Analytical Derivation and Experimental Evaluation of Short-Bearing Approximation for Full Journal Bearings. *NACA Rep.* 1157.

Ghosh, M. K., Hamrock, B. J., and Brewe, D. E. (1985): Hydrodynamic Lubrication of Rigid Nonconformal Contacts in Combined Rolling and Normal Motion. *J. Tribol.*, vol. 107, no. 1, pp. 97–103.

Kapitza, P. L. (1955): Hydrodynamic Theory of Lubrication During Rolling. *Zh. Tekh. Fiz.*, vol. 25, no. 4, pp. 747–762.

Martin, H. M. (1961): Lubrication of Gear Teeth. *Engineering (London)*, vol. 102, pp. 119–121.

Sommerfeld, A. (1904): Zur Hydrodynamischen Theorie der Schmiermittelreibung. *Z. Angew. Math. Phys.*, vol. 50, 97–155.

CHAPTER 19

SIMPLIFIED SOLUTIONS FOR STRESSES AND DEFORMATIONS

In the preceding chapters the solid surfaces in a lubricated conjunction have been considered to be rigid. This then is the demarcation point where elastic deformation of the solid surfaces will begin to be considered and remain of concern until the end of the book.

The classical Hertzian solution for deformation requires calculating the ellipticity parameter k and the complete elliptic integrals of the first and second kinds $\mathscr{F}$ and $\mathscr{E}$, respectively. Simplifying expressions for k, $\mathscr{F}$, and $\mathscr{E}$ as a function of the radius ratio α_r were presented by Brewe and Hamrock (1977) in a curve-fit analysis. With these expressions researchers could determine the deformation at the contact center, with a slight sacrifice in accuracy, without using involved mathematical methods or design charts. The simplified expressions were useful for radius ratios ranging from circular point contact to near line contact normal to the rolling direction (that is, $1.0 \le \alpha_r \le 100$).

However, in a number of applications the semimajor axis of the elliptical contact is parallel to the rolling direction, resulting in $\alpha_r < 1$. Some of these applications are (1) Novikov gear contacts, (2) locomotive wheel–rail contacts, and (3) roller-flange contacts in an axial loaded roller bearing. The elliptical contact deformation and stresses presented by Hamrock and Brewe (1983) are applicable for any contact ranging from something similar to a disk rolling on a plane ($\alpha_r = 0.03$) to a ball-on-plane contact ($\alpha_r = 1$) to a contact approaching a nominal line contact ($\alpha_r \rightarrow 100$) such as a barrel-shaped roller bearing against a plane. This chapter focuses on the results obtained by Hamrock and Brewe in their 1983 paper.

19.1 CURVATURE SUM AND DIFFERENCE

The undeformed geometry of nonconformal contacting solids can be represented in general terms by two ellipsoids, as shown in Fig. 19.1. The two solids with different radii of curvature in a pair of principal planes (x and y) passing through the contact between the solids make contact at a single point under the condition of zero applied load. Such a condition is called "point contact" and is shown in Fig. 19.1, where the radii of curvature are denoted by r's. It is assumed throughout this book that convex surfaces, as shown in Fig. 19.1, exhibit positive curvature and concave surfaces, negative curvature. Therefore, if the center of curvature lies within the solid, the radius of curvature is positive; if the center of curvature lies outside the solid, the radius of curvature is negative. Figure 19.2 shows the sign designations for the radii of curvature for various machine elements such as rolling elements and bearing races. The importance of the sign of the radius of curvature presents itself later in the chapter.

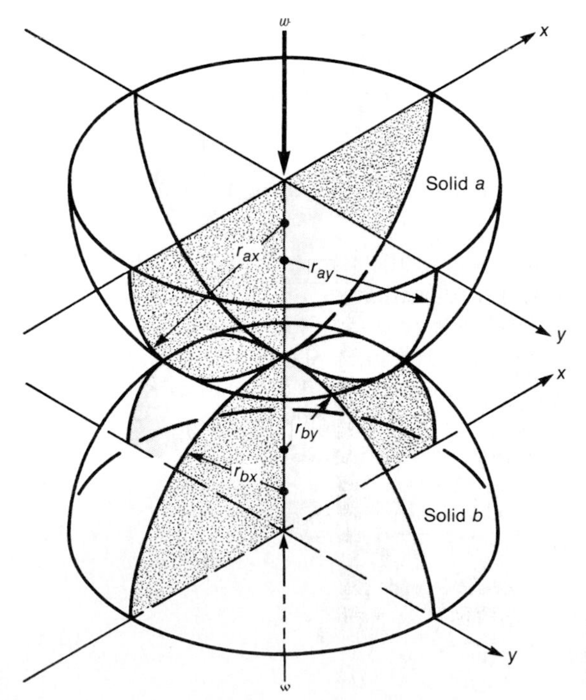

FIGURE 19.1
Geometry of contacting elastic solids.
[*From Hamrock and Dowson (1981).*]

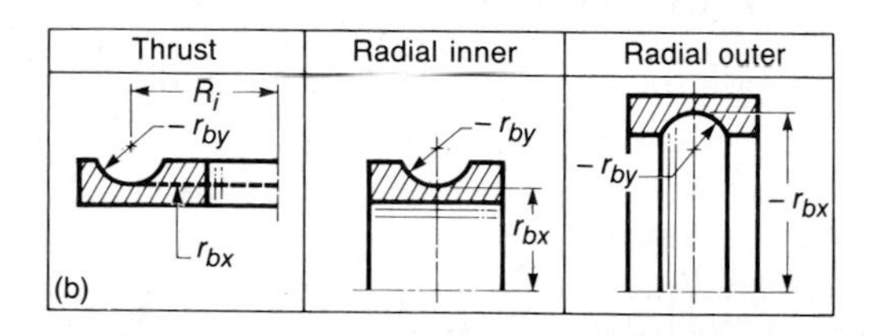

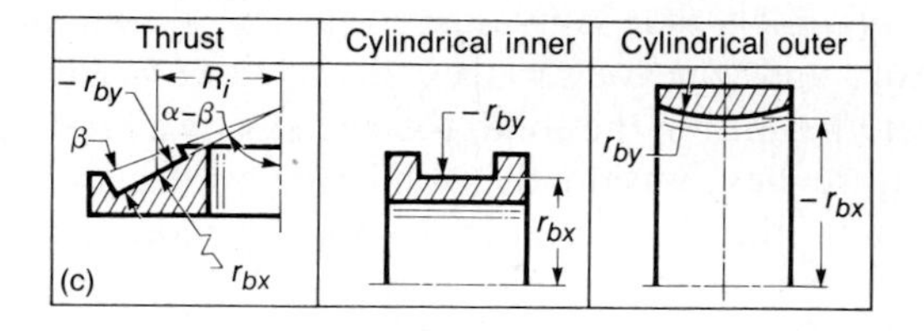

FIGURE 19.2
Sign designations for radii of curvature of various machine elements. (a) Rolling elements; (b) ball bearing races; (c) rolling bearing races.

Note that if coordinates x and y are chosen such that

$$\frac{1}{r_{ax}} + \frac{1}{r_{bx}} \geq \frac{1}{r_{ay}} + \frac{1}{r_{by}} \tag{19.1}$$

coordinate x then determines the direction of the semiminor axis of the contact area when a load is applied and y, the direction of the semimajor axis. The direction of the entraining motion is always considered to be along the x axis. For those situations in which the principal curvature planes of the two contacting bodies are not coincident, refer to Timoshenko and Goodier (1970).

The curvature sum and difference, which are quantities of some importance in analyzing contact stresses and deformation, are

$$\frac{1}{R} = \frac{1}{R_x} + \frac{1}{R_y} \tag{19.2}$$

$$R_d = R\left(\frac{1}{R_x} - \frac{1}{R_y}\right) \tag{19.3}$$

where

$$\frac{1}{R_x} = \frac{1}{r_{ax}} + \frac{1}{r_{bx}} \tag{19.4}$$

$$\frac{1}{R_y} = \frac{1}{r_{ay}} + \frac{1}{r_{by}} \tag{19.5}$$

Equations (19.4) and (19.5) effectively redefine the problem of two ellipsoidal solids approaching one another in terms of an equivalent solid of radii R_x and R_y approaching a plane. Note that the curvature difference expressed in Eq. (19.3) is dimensionless.

The radius ratio α_r defined in Eq. (18.57) is the same for this chapter.

$$\therefore \alpha_r = \frac{R_y}{R_x} \tag{18.57}$$

Thus, if Eq. (19.1) is satisfied, $\alpha_r > 1$; and if it is not satisfied, $\alpha_r < 1$.

19.2 SURFACE STRESSES AND DEFORMATIONS

When an elastic solid is subjected to a load, stresses are produced that increase as the load is increased. These stresses are associated with deformations, which are defined by strains. Unique relationships exist between stresses and their corresponding strains. For elastic solids the stresses are linearly related to the strains, with the constant of proportionality being an elastic constant that adopts different values for different materials as covered in Sec. 5.6.2. The modulus of elasticity E and Poisson's ratio ν are two important parameters described in Chap. 5 that are used in this chapter to describe contacting solids.

As the stresses increase within the material, elastic behavior is replaced by plastic flow in which the material is permanently deformed. The stress state at which the transition from elastic to plastic behavior occurs, known as the "yield stress," has a definite value for a given material at a given temperature. In this book only elastic behavior is considered.

When two elastic solids are brought together under a load, a contact area develops whose shape and size depend on the applied load, the elastic properties of the materials, and the curvatures of the surfaces. When the two solids shown in Fig. 19.1 have a normal load applied to them, the contact area is elliptical. It has been common to refer to elliptical contacts as point contacts, but since under load these contacts become elliptical, they are referred to herein as such. For the special case where $r_{ax} = r_{ay}$ and $r_{bx} = r_{by}$ the resulting contact is a circle rather than an ellipse. Where r_{ay} and r_{by} are both infinite, the initial line contact develops into a rectangle when load is applied.

Hertz (1881) considered the stresses and deformations in two perfectly smooth, ellipsoidal, contacting solids much like those shown in Fig. 19.1. His application of the classical elasticity theory to this problem forms the basis of

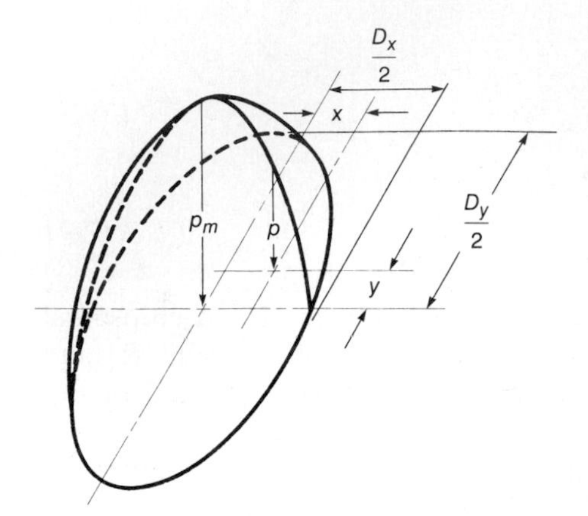

FIGURE 19.3
Pressure distribution in ellipsoidal contact.

stress calculations for machine elements such as ball and roller bearings, gears, and cams and followers. Hertz made the following assumptions:

1. The materials are homogeneous and the yield stress is not exceeded.
2. No tangential forces are induced between the solids.
3. Contact is limited to a small portion of the surface such that the dimensions of the contact region are small in comparison with thc radii of the ellipsoids.
4. The solids are at rest and in equilibrium.

Making use of these assumptions, Hertz (1881) was able to obtain the following expression for the pressure within the ellipsoidal contact (shown in Fig. 19.3):

$$p = p_m \left[1 - \left(\frac{2x}{D_x} \right)^2 - \left(\frac{2y}{D_y} \right)^2 \right]^{1/2} \tag{19.6}$$

where D_x = diameter of contact ellipse in x direction, m
D_y = diameter of contact ellipse in y direction, m

If the pressure is integrated over the contact area, it is found that

$$p_m = \frac{6 w_z}{\pi D_x D_y} \tag{19.7}$$

where w_z is the normal applied load. Equation (19.6) determines the distribution of pressure or compressive stress on the common interface; it is clearly a maximum at the contact center and decreases to zero at the periphery.

The ellipticity parameter k is defined as the elliptical contact diameter in the y direction (transverse direction) divided by the elliptical contact diameter in the x direction (direction of entraining motion), or

$$k = \frac{D_y}{D_x} \tag{19.8}$$

If Eq. (19.1) is satisfied and $\alpha_r \geq 1$, the contact ellipse will be oriented with its major diameter transverse to the direction of motion, and consequently $k \geq 1$. Otherwise, the major diameter would lie along the direction of motion with both $\alpha_r < 1$ and $k < 1$. To avoid confusion, the commonly used solutions to the surface deformation and stresses are presented only for $\alpha_r > 1$. The simplified solutions are presented, and then their application for $\alpha_r < 1$ is discussed.

Harris (1966) has shown that the ellipticity parameter can be written as a transcendental equation relating the curvature difference [Eq. (19.3)] and the elliptic integrals of the first $\mathscr{F}$ and second $\mathscr{E}$ kinds as

$$k = \left[\frac{2\mathscr{F} - \mathscr{E}(1 + R_d)}{\mathscr{E}(1 - R_d)}\right]^{1/2} \tag{19.9}$$

where

$$\mathscr{F} = \int_0^{\pi/2}\left[1 - \left(1 - \frac{1}{k^2}\right)\sin^2\phi\right]^{-1/2} d\phi \tag{19.10}$$

$$\mathscr{E} = \int_0^{\pi/2}\left[1 - \left(1 - \frac{1}{k^2}\right)\sin^2\phi\right]^{1/2} d\phi \tag{19.11}$$

A one-point iteration method that was adopted by Hamrock and Anderson (1973) can be used to obtain the ellipticity parameter, where

$$k_{n+1} \cong k_n \tag{19.12}$$

The iteration process is normally continued until k_{n+1} differs from k_n by less than 1.0×10^{-7}. Note that the ellipticity parameter is a function only of the solids' radii of curvature:

$$k = f(r_{ax}, r_{bx}, r_{ay}, r_{by}) \tag{19.13}$$

That is, as the load increases, the semiaxes in the x and y directions of the contact ellipse increase proportionately to each other so that the ellipticity parameter remains constant.

Figure 19.4 shows the ellipticity parameter and the elliptic integrals of the first and second kinds for a radius ratio ($\alpha_r = R_y/R_x$) range usually encountered in nonconformal conjunctions. Note from Fig. 19.4 that $\mathscr{E} = \mathscr{F}$ when $\alpha_r = 1$. Also both $\mathscr{E}$ and $\mathscr{F}$ are discontinuous at $\alpha_r = 1$.

When the ellipticity parameter k, the normal applied load w_z, Poisson's ratio ν, and the modulus of elasticity E of the contacting solids are known, the major and minor axes of the contact ellipse and the maximum deformation at

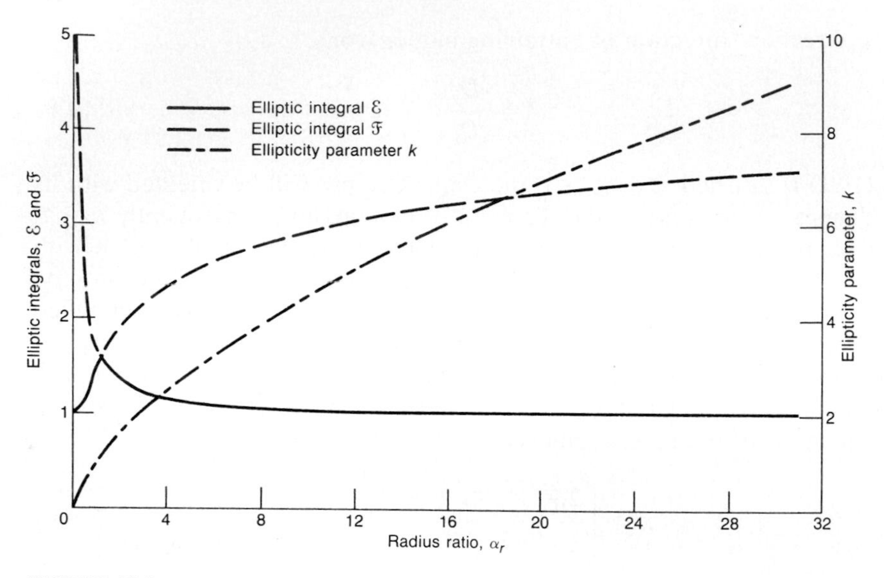

FIGURE 19.4
Variation of ellipticity parameter and elliptic integrals of first and second kinds as function of radius ratio. [*From Hamrock and Brewe (1983).*]

the contact center can be written from the analysis of Hertz (1881) as

$$D_y = 2\left(\frac{6k^2\mathscr{E}w_z R}{\pi E'}\right)^{1/3} \tag{19.14}$$

$$D_x = 2\left(\frac{6\mathscr{E}w_z R}{\pi k E'}\right)^{1/3} \tag{19.15}$$

$$\delta_m = \mathscr{F}\left[\frac{9}{2\mathscr{E}R}\left(\frac{w_z}{\pi k E'}\right)^2\right]^{1/3} \tag{19.16}$$

where

$$E' = \frac{2}{\left(1-\nu_a^2\right)/E_a + \left(1-\nu_b^2\right)/E_b} \tag{19.17}$$

In these equations, D_x and D_y are proportional to $w_z^{1/3}$ and δ_m is proportional to $w_z^{2/3}$.

19.3 SUBSURFACE STRESSES

Fatigue cracks usually start at a certain depth below the surface in planes parallel to the rolling direction. Because of this, special attention must be given to the shear stress amplitude in this plane. Furthermore, a maximum shear stress is reached at a certain depth below the surface. The analysis used by Lundberg and Palmgren (1947) is used here to define this stress.

The stresses are referred to in a rectangular coordinate system with its origin at the contact center, its z axis coinciding with the interior normal of the body considered, its x axis in the rolling direction, and its y axis perpendicular to the rolling direction. In the analysis that follows, it is assumed that $y = 0$.

From Lundberg and Palmgren (1947) the following equations can be written:

$$\tau_{zx} = \frac{6 w_z \cos^2 \phi_a \sin \phi_a \sin \gamma_a}{\pi \left(D_y^2 \tan^2 \gamma_a + D_x^2 \cos^2 \phi_a \right)} \tag{19.18}$$

$$x = \tfrac{1}{2} \left(D_x^2 + D_y^2 \tan^2 \gamma_a \right)^{1/2} \sin \phi_a \tag{19.19}$$

$$z = \frac{D_y}{2} \tan \gamma_a \cos \phi_a \tag{19.20}$$

where ϕ_a and γ_a are auxiliary angles used in place of the coordinate set (x, z). They are defined so as to satisfy the relationship of a conformal ellipsoid to the pressure ellipse (for further details see Hertz, 1881, and Lundberg and Palmgren, 1947).

The maximum shear stress is defined as

$$\tau_0 = (\tau_{zx})_{\max}$$

The amplitude of τ_0 is obtained from

$$\frac{\partial \tau_{zx}}{\partial \phi_a} = 0 \qquad \frac{\partial \tau_{zx}}{\partial \gamma_a} = 0$$

For the point of maximum shear stress

$$\tan^2 \phi_a = t^* \tag{19.21}$$

$$\tan^2 \gamma_a = t^* - 1 \tag{19.22}$$

$$\frac{D_x}{D_y} = \left\{ \left[(t^*)^2 - 1 \right] (2t^* - 1) \right\}^{1/2} \tag{19.23}$$

where t^* is the auxiliary parameter. The position of the maximum point is determined by

$$z_0 = \zeta^* \frac{D_x}{2} \tag{19.24}$$

$$x_0 = \pm \xi^* \frac{D_x}{2} \tag{19.25}$$

where

$$\zeta^* = \frac{1}{(t^* + 1)(2t^* - 1)^{1/2}} \tag{19.26}$$

$$\xi^* = \frac{t^*}{t^* + 1} \left(\frac{2t^* + 1}{2t^* - 1} \right)^{1/2} \tag{19.27}$$

Furthermore, the magnitude of the maximum shear stress expressed in terms of t^* is given by

$$\tau_0 = p_m \frac{(2t^* - 1)^{1/2}}{2t^*(t^* + 1)} \tag{19.28}$$

It should be emphasized that τ_0 represents the maximum half-amplitude of the subsurface orthogonal shear stress and is not to be confused with the maximum subsurface shear stress that occurs below the center of the contact on the plane oriented 45° to the surface. The Lundberg-Palmgren prediction of fatigue life is based on the calculation of τ_0 and was limited to cross sections lying in the plane of symmetry of the roller path ($y = 0$).

19.4 SIMPLIFIED SOLUTIONS

The classical Hertzian solution presented in the previous sections requires the calculation of the ellipticity parameter k and the complete elliptic integrals of the first and second kinds $\mathscr{F}$ and $\mathscr{E}$. This entails finding a solution to a transcendental equation relating k, $\mathscr{F}$, and $\mathscr{E}$ to the geometry of the contacting solids, as expressed in Eq. (19.9). This is usually accomplished by some iterative numerical procedure, as described by Hamrock and Anderson (1973), or with the aid of charts, as shown by Jones (1946).

Table 19.1 shows various radius ratios α_r and corresponding values of k, $\mathscr{F}$, and $\mathscr{E}$ obtained from the numerical procedure given in Hamrock and Anderson (1973). Hamrock and Brewe (1983) used a linear regression by the method of least squares to power fit the set of pairs of data $[(k_i, \alpha_{r,i}),\ i = 1, 2, \ldots, 26]$ shown in Table 19.1. They obtained the following equation:

$$\bar{k} = \alpha_r^{2/\pi} \tag{19.29}$$

The asymptotic behavior of $\mathscr{E}$ and $\mathscr{F}$ ($\alpha_r \to 1$ implies $\mathscr{E} \to \mathscr{F} \to \pi/2$, and $\alpha_r \to \infty$ implies $\mathscr{F} \to \infty$ and $\mathscr{E} \to 1$) was suggestive of the type of functional dependence that $\bar{\mathscr{E}}$ and $\bar{\mathscr{F}}$ might follow. As a result, both inverse and logarithmic curve fits were tried for $\bar{\mathscr{E}}$ and $\bar{\mathscr{F}}$, respectively. Hamrock and Brewe (1983) obtained the following:

$$\bar{\mathscr{E}} = 1 + \frac{q_a}{\alpha_r} \qquad \text{for } \alpha_r \ge 1 \tag{19.30}$$

where

$$q_a = \frac{\pi}{2} - 1 \tag{19.31}$$

and

$$\bar{\mathscr{F}} = \frac{\pi}{2} + q_a \ln \alpha_r \qquad \text{for } \alpha_r \ge 1 \tag{19.32}$$

Values of $\bar{k}$, $\bar{\mathscr{E}}$, and $\bar{\mathscr{F}}$ are also presented in Table 19.1 and compared with the numerically determined values of k, $\mathscr{E}$, and $\mathscr{F}$. Table 19.1 also gives the

TABLE 19.1

Comparison of numerically determined values with curve-fit values for geometrically dependent variables

[From Hamrock and Brewe (1983); R_x = 1.0 cm]

Radius ratio, α_r	Ellipticity			Complete elliptic integral of first kind			Complete elliptic integral of second kind		
	k	$\bar{k}$	Error, e_r, percent	$\mathcal{F}$	$\bar{\mathcal{F}}$	Error, e_r, percent	$\mathcal{E}$	$\bar{\mathcal{E}}$	Error, e_r, percent
1.00	1.00	1.00	0	1.57	1.57	0	1.57	1.57	0
1.25	1.16	1.15	.66	1.68	1.69	−.50	1.46	1.45	.52
1.50	1.31	1.29	1.19	1.78	1.80	−.70	1.39	1.38	.76
1.75	1.45	1.42	1.61	1.87	1.89	−.75	1.33	1.32	.87
2.00	1.58	1.55	1.96	1.95	1.96	−.73	1.29	1.28	.91
3.00	2.07	2.01	2.87	2.18	2.19	−.44	1.20	1.19	.83
4.00	2.50	2.41	3.35	2.35	2.36	−.11	1.15	1.14	.69
5.00	2.89	2.78	3.61	2.49	2.48	.17	1.12	1.11	.57
6.00	3.25	3.12	3.74	2.60	2.59	.40	1.10	1.09	.48
7.00	3.58	3.45	3.80	2.69	2.68	.59	1.08	1.08	.40
8.00	3.90	3.75	3.81	2.77	2.75	.75	1.07	1.07	.35
9.00	4.20	4.05	3.78	2.85	2.82	.88	1.06	1.06	.30
10.00	4.49	4.33	3.74	2.91	2.88	1.00	1.05	1.05	.26
15.00	5.79	5.60	3.32	3.16	3.11	1.38	1.03	1.03	.15
20.00	6.92	6.73	2.81	3.33	3.28	1.60	1.02	1.02	.10
25.00	7.94	7.76	2.29	3.46	3.40	1.74	1.02	1.02	.07
30.00	8.87	8.71	1.79	3.57	3.51	1.84	1.01	1.01	.05
35.00	9.74	9.61	1.32	3.67	3.60	1.90	↓	↓	.04
40.00	10.56	10.46	.87	3.74	3.67	1.95			.03
45.00	11.33	11.28	.44	3.81	3.74	1.99			.02
50.00	12.07	12.06	.03	3.88	3.80	2.02			.02
60.00	13.45	13.52	−.72	3.98	3.90	2.06	1.00	1.00	.01
70.00	14.74	14.94	−1.40	4.08	3.99	2.08	↓	↓	.01
80.00	15.95	16.27	−2.03	4.15	4.07	2.09			.01
90.00	17.09	17.54	−2.61	4.22	4.13	2.10			0
100.00	18.18	18.76	−3.15	4.28	4.19	2.10			0

percentage of error determined as

$$e_r = \frac{(\bar{i} - i)100}{i} \tag{19.33}$$

where

$$i = \{k, \mathscr{E}, or\ \mathscr{F}\} \tag{19.34}$$

$$\bar{i} = \{\bar{k}, \bar{\mathscr{E}}, \text{or } \bar{\mathscr{F}}\} \tag{19.35}$$

The agreement between the exact solution and the approximate formulas is quite good. The best agreement is with $\mathscr{E}$, which is between 0 and 1 percent; the worst agreement is with k, which is ± 4 percent.

TABLE 19.2
Effect of radius ratio on auxiliary parameter used in subsurface stress calculations
[From Hamrock and Brewe (1983)]

Radius ratio, α_r	Auxiliary parameter			Radius ratio, α_r	Auxiliary parameter		
	t^*	$\bar{t}^*$	Error, e_r, percent		t^*	$\bar{t}^*$	Error, e_r, percent
0.01	5.71	7.00	−22.47	2.00	1.14	1.18	−3.71
.02	4.42	4.86	−9.87	3.00	1.09	1.13	−3.78
.03	3.81	3.98	−4.49	4.00	1.06	1.10	−3.45
.04	3.43	3.48	−1.51	5.00	1.05	1.08	−3.03
.05	3.16	3.15	.35	6.00	1.04	1.07	−2.62
.06	2.96	2.91	1.58	7.00	1.03	1.05	−2.24
.07	2.80	2.73	2.42	8.00	1.03	1.05	−1.91
.08	2.67	2.59	3.01	9.00	1.02	1.04	−1.61
.09	2.56	2.47	3.43	10.00	1.02	1.03	−1.35
.10	2.47	2.38	3.72	15.00	1.01	1.01	−.50
.20	1.96	1.88	3.90	20.00	1.00	1.01	−.09
.30	1.73	1.68	2.86	25.00	↓	1.00	.12
.40	1.59	1.56	1.79	30.00		↓	.21
.50	1.50	1.48	.85	35.00			.26
.60	1.43	1.43	.05	40.00			.27
.70	1.38	1.39	−.62	45.00			.27
.80	1.34	1.35	−1.19	50.00			.26
.90	1.30	1.33	−1.66	60.00			.24
1.00	1.28	1.30	−2.05	70.00			.21
1.25	1.22	1.26	−2.78	80.00			.19
1.50	1.19	1.23	−3.25	90.00			.17
1.75	1.16	1.20	−3.54	100.00	↓	↓	.15

Table 19.2 shows various radius ratios α_r and corresponding values of the auxiliary parameter t^* used in calculating the position and value of the maximum subsurface orthogonal shear stress. The exact solution for t^* was obtained from the numerical procedures given in Hamrock and Anderson (1973). For the set of data $[(t_i^*, \alpha_{r,i}), i = 1, 2, \ldots, 44]$ shown in Table 19.2 the following simplified formula was obtained from Hamrock and Brewe (1983):

$$\bar{t}^* = 1 + 0.16 \operatorname{csch} \frac{\bar{k}}{2} \tag{19.36}$$

The percentage of error e_r is given for the auxiliary parameter in Table 19.2. The agreement between the exact and the approximate values of t^* is quite good except at extremely small radius ratios ($\alpha_r \le 0.03$). Once the value of t^* is determined, the position and value of the maximum subsurface orthogonal shear stress can readily be calculated from Eqs. (19.24) to (19.28).

TABLE 19.3
Simplified equations
[From Hamrock and Brewe (1983)]

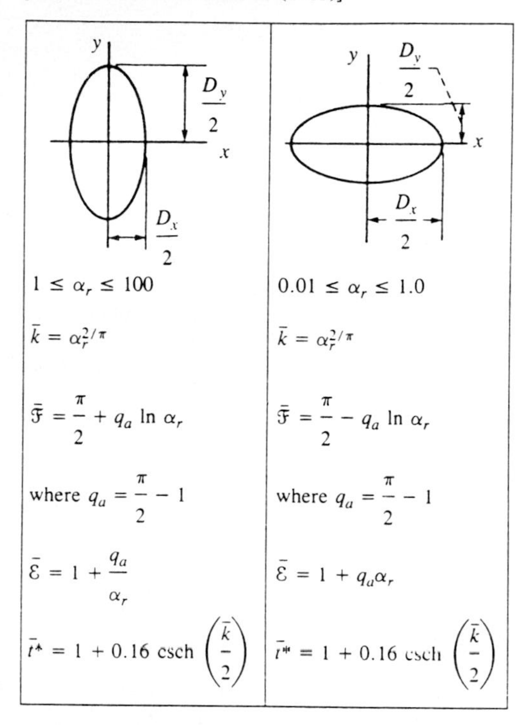

$1 \le \alpha_r \le 100$	$0.01 \le \alpha_r \le 1.0$
$\bar{k} = \alpha_r^{2/\pi}$	$\bar{k} = \alpha_r^{2/\pi}$
$\bar{\mathfrak{F}} = \frac{\pi}{2} + q_a \ln \alpha_r$	$\bar{\mathfrak{F}} = \frac{\pi}{2} - q_a \ln \alpha_r$
where $q_a = \frac{\pi}{2} - 1$	where $q_a = \frac{\pi}{2} - 1$
$\bar{\mathcal{E}} = 1 + \frac{q_a}{\alpha_r}$	$\bar{\mathcal{E}} = 1 + q_a \alpha_r$
$\bar{t}^* = 1 + 0.16 \operatorname{csch}\left(\frac{\bar{k}}{2}\right)$	$\bar{t}^* = 1 + 0.16 \operatorname{csch}\left(\frac{\bar{k}}{2}\right)$

Table 19.3, from Hamrock and Brewe (1983), gives the simplified equations for $0.01 \le \alpha_r \le 100$, which is the complete range normally experienced in practice. It is important to make the proper evaluation of α_r, since it has great significance in the outcome of the simplified equations. Table 19.3 shows that $\bar{k}$ and $\bar{t}^*$ are unaffected by the orientation of the ellipse but that the elliptical integrals of the first and second kinds ($\mathscr{F}$ and $\mathscr{E}$) are quite affected. It is important to realize that the reciprocal to α_r produces the same values of $\mathscr{F}$ and $\mathscr{E}$ as are produced by changing the orientation of the ellipse.

Figure 19.5 shows three diverse situations in which the simplified equations can be usefully applied. The locomotive wheel on a rail [Fig. 19.5(a)] illustrates an example in which the ellipticity parameter k and the radius ratio α_r are less than 1. The ball rolling against a plane [Fig. 19.5(b)] provides a pure circular contact (i.e., $\alpha_r = k = 1.00$). Figure 19.5(c) shows how the contact ellipse is formed in the ball–outer-race contact of a ball bearing. Here the semimajor axis is normal to the rolling direction, and consequently α_r and k are greater than 1. The detailed geometry and the values that can be calculated from the simplified formulas are given in Table 19.4 for each of these three configurations. In using these formulas it is important to pay attention to the

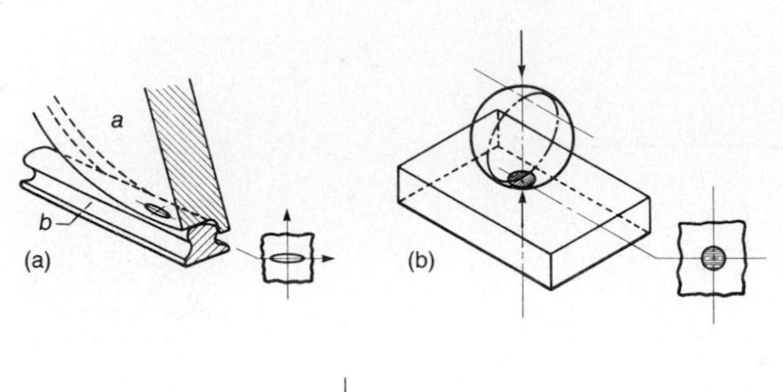

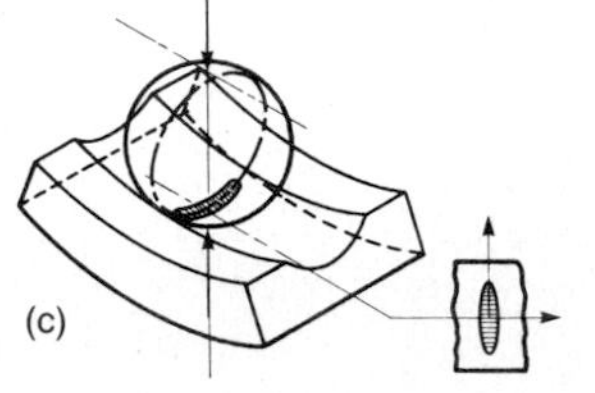

FIGURE 19.5
Three degrees of conformity. (a) Wheel on rail; (b) ball on plane; (c) ball–outer-race contact. [*From Hamrock and Brewe (1983).*]

sign of the curvatures. Note that the outer race in Fig. 19.5 is a concave surface and therefore requires a negative sign. Table 19.4 shows both the maximum pressure p_m and the maximum shear stress to be highest for the ball-on-plane configuration.

19.5 RECTANGULAR CONJUNCTIONS

For rectangular conjunctions the contact ellipse discussed throughout this chapter is of infinite width in the transverse direction ($D_y \to \infty$). This type of contact is exemplified by a cylinder loaded against a plane, a groove, or another parallel cylinder or by a roller loaded against an inner or outer race. In these situations the contact semiwidth is given by

$$b^* = R_x \left(\frac{8W'}{\pi} \right)^{1/2} \tag{19.37}$$

where the dimensionless load is

$$W' = \frac{w'}{E'R_x} \tag{19.38}$$

and w' is the load per unit width along the contact. The maximum deformation for a rectangular conjunction can be written as

$$\delta_m = \frac{2W'R_x}{\pi} \left[\ln\left(\frac{2\pi}{W'} \right) - 1 \right] \tag{19.39}$$

The maximum Hertzian pressure in a rectangular conjunction can be written as

$$p_H = E' \left(\frac{W'}{2\pi} \right)^{1/2}$$

TABLE 19.4
Practical applications for differing conformities
[From Hamrock and Brewe (1983); effective elastic modulus E', 2.197×10^{11} Pa]

Contact parameters	Wheel on rail	Ball on plane	Ball–outer-race contact
w_z, N	1.00×10^5	222.4111	222.4111
r_{ax}, m	0.5019	0.006350	0.006350
r_{ay}, m	∞	0.006350	0.006350
r_{bx}, m	∞	∞	−0.038900
r_{by}, m	0.300000	∞	−0.006600
α_r	0.5977	1.0000	22.0905
k	0.7099	1.0000	7.3649
$\bar{k}$	0.7206	1.0000	7.1738
$\mathcal{E}$	1.3526	1.5708	1.0267
$\bar{\mathcal{E}}$	1.3412	1.5708	1.0258
$\mathcal{F}$	1.8508	1.5708	3.3941
$\bar{\mathcal{F}}$	1.8645	1.5708	3.3375
D_y, m	0.010783	0.000426	0.001842
$\bar{D}_y$, m	0.010807	0.000426	0.001810
D_x, m	0.015190	0.000426	0.000250
$\bar{D}_x$, m	0.014996	0.000426	0.000252
δ_m, μm	106	7.13	3.56
$\bar{\delta}_m$, μm	108	7.13	3.57
p_m, GPa	1.166	2.34	0.922
$\bar{p}_m$, GPa	1.178	2.34	0.930
t^*	1.4354	1.2808	1.0090
$\bar{t}^*$	1.4346	1.3070	1.0089
x_0, m	±0.008862	±0.000195	±0.000096
$\bar{x}_0$, m	±0.008745	±0.000197	±0.000097
z_0, m	±0.005410	±0.000149	±0.000123
$\bar{z}_0$, m	±0.005350	±0.000145	±0.000124
τ_0, GPa	0.162	0.501	0.229
$\bar{\tau}_0$, GPa	0.164	0.494	0.232

19.6 CLOSURE

This chapter has presented an alternative approach to the classical Hertzian solution for the local stress and deformation of two elastic bodies in contact. Simplified formulas that use curve-fit analysis are given in terms of the radius ratio α_r for the ellipticity parameter k and the complete integrals $\mathcal{F}$ and $\mathcal{E}$ of the first and second kinds, respectively. Thus, their interdependence can be uncoupled and solution of the resulting transcendental equation avoided. Simplified equations were developed that permit a more direct and easier approach to the calculation of the elliptical contact deformation and the maximum

Hertzian pressure. In addition, a curve-fit analysis was used to derive a simplified formula for an auxiliary parameter t^* as a function of α_r. This eliminated having to solve a cubic equation for t^* as a function of k. A shortcut calculation could be made for the location and magnitude of the maximum subsurface shear stress. Therefore, the elliptical contact deformation and stresses presented are applicable for any contact ranging from a disk rolling on a plane ($\alpha_r = 0.03$) to a ball-on-plane contact ($\alpha_r = 1$) to a contact approaching a normal line contact ($\alpha_r \to 100$) such as a barrel-shaped roller against a plane.

19.7 PROBLEMS

19.7.1 A solid cylinder of radius 2 cm rolls around an inner ring with an internal radius of 10 cm and a large width in the axial (y) direction. What is the radius of the geometrically equivalent cylinder near a plane?

The cylinder is made of silicon nitride and the ring is made of stainless steel. If a normal applied load per unit width is 1,000 N/m determine the contact semiwidth b^*, maximum deformation δ_m, and maximum Hertzian pressure p_H. Also indicate what these values are if the silicon nitride cylinder is replaced with a stainless steel cylinder. What conclusions can you make about these results?

19.7.2 A solid sphere of radius 2 cm rolls around the inner ring with an internal radius of 10 cm and a large width in the axial (y) is quite large. What is the curvature sum?

The sphere is made of silicon nitride while the ring is made of stainless steel. If the normal applied load is 1,000 N, determine the maximum surface stress, the maximum subsurface stress, the maximum deformation, and the dimension of the contact. That is, determine b, $\mathscr{F}$, $\mathscr{E}$, t^*, D_x, D_y, δ_m, p_m τ_0, x_0, and z_0.

Also indicate what these values are if the silicon nitride sphere is replaced with a stainless steel sphere. What conclusions can be made about the results? Also compare with Prob. 19.7.1 results.

19.8 REFERENCES

Brewe, D. E., and Hamrock, B. J. (1977): Simplified Solution for Elliptical Contact Deformation Between Two Elastic Solids. *J. Lubr. Technol.*, vol. 99, no. 4, pp. 485–487.

Hamrock, B. J., and Anderson, W. J. (1973): Analysis of an Arched Outer-Race Ball Bearing Considering Centrifugal Forces. *J. Lubr. Technol.*, vol. 95, no. 3, pp. 265–276.

Hamrock, B. J., and Brewe, D. E. (1983): Simplified Solution for Stresses and Deformations. *J. Lubr. Technol.*, vol. 105, no. 2, pp. 171–177.

Hamrock, B. J., and Dowson, D. (1981): *Ball Bearing Lubrication—The Elastohydrodynamics of Elliptical Contacts*. Wiley-Interscience, New York.

Harris, T. A. (1966): *Rolling Bearing Analysis*. Wiley, New York.

Hertz, H. (1881): The Contact of Elastic Solids. *J. Reine Angew. Math.*, vol. 92, pp. 156–171.

Jones, A. B. (1946): New Departure Engineering Data; Analysis of Stresses and Deflections. Vols. I and II, General Motors, Inc., Detroit, Michigan.

Lundberg, G., and Palmgren, A. (1947): Dynamic Capacity of Rolling Bearings. *Acta Polytech., Mech. Eng. Sci.*, vol. 1, no. 3, pp. 6–9.

Timoshenko, S. P., and Goodier, J. N. (1970): *Theory of Elasticity*, 3d ed. McGraw-Hill, New York.

CHAPTER 20

GENERAL SOLUTION FOR STRESSES AND DEFORMATIONS IN DRY CONTACTS

The previous chapter described simple formulas for the maximum surface and subsurface stresses as well as the maximum deformation at the contact center. This chapter defines the stresses and the deformations in a more general manner. As in the previous chapter the contacts are assumed to be dry, or unlubricated.

When two elastic solids like those shown in Fig. 19.1 are brought into contact and subjected to a normal load, the solids deform and the nominal point of contact becomes an elliptical area. Two limiting cases can be identified from this general principle. In the first, a point contact becomes a circle, for example, when the solid surfaces are a sphere on a sphere, a sphere on a plane, or identical cylinders crossed at right angles. In the second limiting case, a nominal line contact becomes a rectangle, for example, when the solid surfaces are a cylinder on a cylinder with parallel axes, a cylinder on a plane, or a cylinder in a groove with parallel axes as in a journal. The stresses, deformations, and contact dimensions are presented here for normal loaded contacts with no tangential loading, not only for the general elliptical contact solution but also for the two limiting situations of circular and rectangular contacts.

In this chapter it is assumed that the linear size of the contact area is small relative to the radii of curvature of the contacting solids. This implies that one body can be replaced by an elastic semi-infinite space. The second assumption imposed in this chapter is that the friction forces arising between touching bodies are neglected.

20.1 GENERAL ELASTICITY THEOREMS

The behavior of an isotropic and homogeneous perfectly elastic material is generally defined by equilibrium conditions in which the body forces have been made equal to zero. Imposing these conditions yields the following equations:

$$(1+\nu)\nabla^2\sigma_x + \frac{\partial^2}{\partial x^2}(\sigma_x+\sigma_y+\sigma_z) = 0 \tag{20.1}$$

$$(1+\nu)\nabla^2\sigma_y + \frac{\partial^2}{\partial y^2}(\sigma_x+\sigma_y+\sigma_z) = 0 \tag{20.2}$$

$$(1+\nu)\nabla^2\sigma_z + \frac{\partial^2}{\partial z^2}(\sigma_x+\sigma_y+\sigma_z) = 0 \tag{20.3}$$

$$(1+\nu)\nabla^2\tau_{xy} + \frac{\partial^2}{\partial x\,\partial y}(\sigma_x+\sigma_y+\sigma_z) = 0 \tag{20.4}$$

$$(1+\nu)\nabla^2\tau_{yz} + \frac{\partial^2}{\partial y\,\partial z}(\sigma_x+\sigma_y+\sigma_z) = 0 \tag{20.5}$$

$$(1+\nu)\nabla^2\tau_{zx} + \frac{\partial^2}{\partial z\,\partial x}(\sigma_x+\sigma_y+\sigma_z) = 0 \tag{20.6}$$

where

$$\nabla^2 = \frac{\partial^2}{\partial x^2} + \frac{\partial^2}{\partial y^2} + \frac{\partial^2}{\partial z^2} \tag{20.7}$$

The solution to any elasticity problem must satisfy these conditions and the given boundary equations.

20.2 LINE LOAD SOLUTION

Figure 20.1 shows the plane polar coordinates used for a line load w'_z acting in the z direction in the $x = 0$ plane on the boundary surface of an elastic half-space. Also shown in this figure are the Cartesian coordinates. A line load

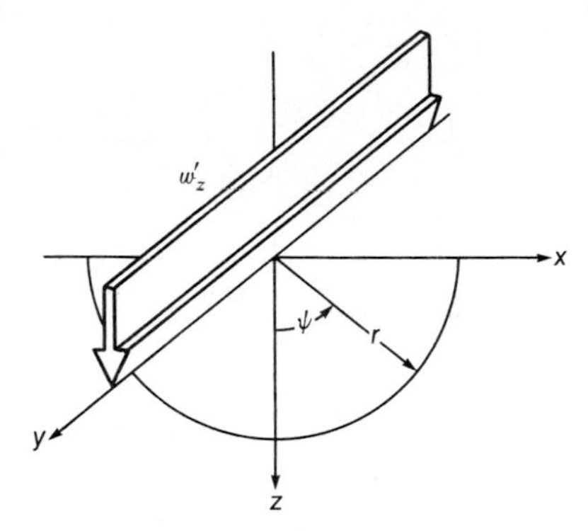

FIGURE 20.1
Plane polar coordinates used for line load w_z' acting in z direction in $x = 0$ plane on boundary surface of elastic half-space. [*From Tripp (1985).*]

condition is a two-dimensional problem, and the equilibrium conditions for this problem reduce to the following:

$$\frac{\partial \sigma_x}{\partial x} + \frac{\partial \tau_{xz}}{\partial z} = 0 \tag{20.8}$$

$$\frac{\partial \sigma_z}{\partial z} + \frac{\partial \tau_{xz}}{\partial x} = 0 \tag{20.9}$$

$$\left(\frac{\partial^2}{\partial x^2} + \frac{\partial^2}{\partial z^2}\right)(\sigma_x + \sigma_z) = 0 \tag{20.10}$$

Note that Eqs. (20.8) to (20.10) for the two-dimensional problem are much simpler than the three-dimensional equations [(20.1) to (20.6)].

Equations (20.8) to (20.10) may be solved by a "stress function approach," in which ϕ, which is a function of x and z, is introduced and expressed in terms of the stresses as

$$\sigma_x = \frac{\partial^2 \phi}{\partial z^2} \qquad \sigma_z = \frac{\partial^2 \phi}{\partial x^2} \qquad \tau_{xz} = -\frac{\partial^2 \phi}{\partial x\, \partial z} \tag{20.11}$$

The parameter ϕ is referred to as an Airy stress function. These satisfy Eqs. (20.8) and (20.9) identically, while Eq. (20.10) becomes

$$\frac{\partial^4 \phi}{\partial x^4} + 2\frac{\partial^4 \phi}{\partial x^2\, \partial z^2} + \frac{\partial^4 \phi}{\partial z^4} = 0 = \nabla^4 \phi \tag{20.12}$$

This is a biharmonic governing equation. To solve a particular problem such as the line load problem, it is necessary to find the appropriate stress function that will satisfy Eq. (20.12) and the appropriate boundary conditions. In many plane problems it is advantageous to express line load in a polar coordinate system as shown in Fig. 20.1 rather than in Cartesian coordinates. Equations (20.8) and

(20.9) become

$$\frac{\partial \sigma_r}{\partial r} + \frac{1}{r}\frac{\partial \tau_{r\psi}}{\partial \psi} + \frac{\sigma_r - \sigma_\psi}{r} = 0 \tag{20.13}$$

$$\frac{1}{r}\frac{\partial \sigma_\psi}{\partial \psi} + \frac{\partial \tau_{r\psi}}{\partial r} + \frac{2\tau_{r\psi}}{r} = 0 \tag{20.14}$$

The stress function given in Eq. (20.11) can be expressed in polar coordinates as

$$\sigma_r = \frac{1}{r}\frac{\partial \phi}{\partial r} + \frac{1}{r^2}\frac{\partial^2 \phi}{\partial \psi^2} \qquad \sigma_\psi = \frac{\partial^2 \phi}{\partial r^2} \qquad \tau_{r\psi} = -\frac{\partial}{\partial r}\left(\frac{1}{r}\frac{\partial \phi}{\partial \psi}\right) \tag{20.15}$$

The corresponding biharmonic equation in Eq. (20.12) is expressed in polar coordinates as

$$\left(\frac{\partial^2}{\partial r^2} + \frac{1}{r}\frac{\partial}{\partial r} + \frac{1}{r^2}\frac{\partial^2}{\partial \psi^2}\right)\left(\frac{\partial^2 \phi}{\partial r^2} + \frac{1}{r}\frac{\partial \phi}{\partial r} + \frac{1}{r^2}\frac{\partial^2 \phi}{\partial \psi^2}\right) = 0 \tag{20.16}$$

The line load problem shown in Fig. 20.1 is solved by using the Boussinesq stress function given as

$$\phi_b = -\frac{w'_z r\psi}{\pi} \sin \psi \tag{20.17}$$

Substituting Eq. (20.17) into Eq. (20.15) gives

$$\sigma_r = -\frac{2\, w'_z \cos \psi}{\pi r} \qquad \sigma_\psi = 0 \qquad \tau_{r\psi} = 0 \tag{20.18}$$

This indicates that the stress is radial and directed toward the line where the load is applied. Equations (20.18) are suitable for determining the stress distribution within a semi-infinite solid.

Besides the stresses the surface deformation is also of interest, especially in elastohydrodynamic lubrication studies. It is more convenient to revert to Cartesian coordinates for the deformation considerations. The stress function given in Eq. (20.17) can be expressed in Cartesian coordinates as

$$\phi_b = -\frac{w'_z x}{\pi} \tan^{-1}\frac{x}{z} \tag{20.19}$$

The stress field in Cartesian coordinates is

$$\sigma_x = -\frac{2\, w'_z x^2 z}{\pi (x^2 + z^2)^2} \tag{20.20}$$

$$\sigma_z = -\frac{2\, w'_z z^3}{\pi (x^2 + z^2)^2} \tag{20.21}$$

$$\tau_{xz} = -\frac{2\, w'_z x z^2}{\pi (x^2 + z^2)^2} \tag{20.22}$$

From Hooke's law the plane strain components in the solids can be expressed as

$$\tilde{e}_x = \frac{\partial \delta_x}{\partial x} = \frac{1}{E}\left[\sigma_x - \nu(\sigma_y + \sigma_z)\right] \tag{20.23}$$

$$\tilde{e}_y = \frac{\partial \delta_y}{\partial y} = 0 = \frac{1}{E}\left[\sigma_y - \nu(\sigma_z + \sigma_x)\right] \tag{20.24}$$

$$\tilde{e}_z = \frac{\partial \delta_z}{\partial z} = \frac{1}{E}\left[\sigma_z - \nu(\sigma_x + \sigma_y)\right] \tag{20.25}$$

$$\tilde{e}_{xz} = \frac{\partial \delta_x}{\partial z} + \frac{\partial \delta_z}{\partial x} = \frac{\tau_{xz}}{G_s} = \frac{2(1+\nu)}{E}\tau_{xz} \tag{20.26}$$

where G_s is the shear modulus of elasticity. From Eq. (20.24), the plane strain condition,

$$\sigma_y = \nu(\sigma_x + \sigma_z) \tag{20.27}$$

Substituting this into Eqs. (20.23) and (20.25) gives

$$\tilde{e}_x = \frac{\partial \delta_x}{\partial x} = \frac{1-\nu^2}{E}\sigma_x - \frac{\nu(1+\nu)}{E}\sigma_z \tag{20.28}$$

$$\tilde{e}_z = \frac{\partial \delta_z}{\partial z} = \frac{1-\nu^2}{E}\sigma_z - \frac{\nu(1+\nu)}{E}\sigma_x \tag{20.29}$$

Integrating Eqs. (20.28) and (20.29) while making use of Eqs. (20.20) and (20.21) gives

$$\delta_x = -\frac{w_z'}{\pi}\left[\frac{(1-2\nu)(1+\nu)}{E}\tan^{-1}\left(\frac{x}{z}\right) - \frac{(1+\nu)xz}{E(x^2+z^2)}\right] + \tilde{A}(z) \tag{20.30}$$

$$\delta_z = -\frac{w_z'}{\pi}\left\{\frac{1-\nu^2}{E}\left[\ln(x^2+z^2) - \frac{z^2}{x^2+z^2}\right] + \frac{\nu(1+\nu)x^2}{E(x^2+z^2)}\right\} + \tilde{B}(x) \tag{20.31}$$

By symmetry $-\delta_x(-x) = \delta_x(x)$. Hence, $\tilde{A}(z) = 0$.

Equation (20.26) must be used to determine $\tilde{B}(x)$. That is, substituting Eqs. (20.22), (20.30), and (20.31) into (20.26) gives

$$\frac{\partial \tilde{B}(x)}{\partial x} = 0 \tag{20.32}$$

This implies that $\tilde{B}$ is not dependent on the field point x but does depend on the source point $\bar{s}$, in this case chosen at the origin. The value of $\tilde{B}$ is determined by the relative positions of the source and the fixed datum. Therefore, Eqs. (20.30) and (20.31) describe the elastic deformation anywhere in the body due to a line load at the origin. The deformation can be found for *any* applied surface pressure by an integration method that uses these basic results.

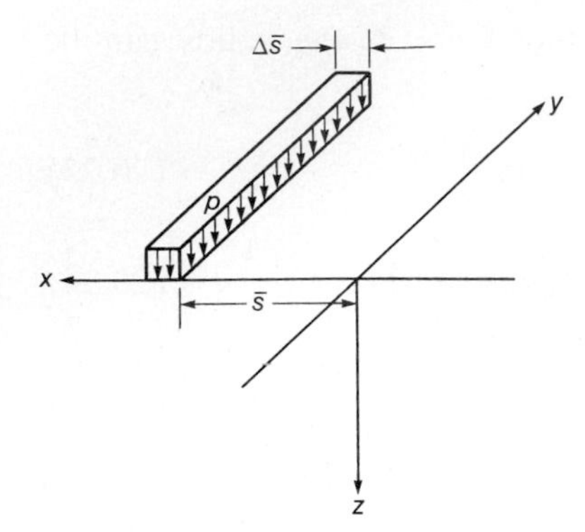

FIGURE 20.2
Strip of pressure acting as line load.

Only the normal deformation at the surface is usually of interest, but for completeness the general expression is given. From this point on in this text we will only be concerned with normal deformations so that $\delta = \delta_z$.

Sample problem 20.1 Establish what the normal deformation is for a strip of pressure p and width $d\bar{s}$ acting on a line on the surface $(z = 0)$ at a distance $\bar{s}$ from the origin $(x = \bar{s})$.

Solution. Figure 20.2 shows the strip of pressure that is acting as a line load. In Eq. (20.31) if $z = 0$, $(x - \bar{s}) \to x$, and $w'_z = p(d\bar{s})$,

$$\delta = -\frac{p(d\bar{s})}{\pi}\left\{\frac{1-\nu^2}{E}\left[\ln(x-\bar{s})^2\right] + \frac{\nu(1+\nu)}{E}\right\} + \tilde{B}$$

When displacements from the distribution of sources are superimposed, they must all be referenced to the *same* datum. Thus, an expression for $\tilde{B}$ is needed inside the integral. The result of integration is $\tilde{C}$, and it will depend on the choice of the datum.

$$\delta = \frac{-p(d\bar{s})(1-\nu^2)}{\pi E}\ln(x-\bar{s})^2 + \tilde{C}(\bar{s})$$

The displacement due to a variable pressure $p(\bar{s})$ between $x = \bar{s}_1$ and $x = \bar{s}_2$ is found by integrating the preceding equation.

$$\delta = -\frac{1-\nu^2}{\pi E}\int_{\bar{s}_2}^{\bar{s}_1}\left[p(\bar{s})\ln(x-\bar{s})^2 + \tilde{C}(\bar{s})\right]d\bar{s} \qquad (20.33)$$

When it is known how the pressure varies, this integral can be evaluated.

In general, one wants to know the displacement of the surface relative to some fixed point beneath the surface, for example, the axis of a cylinder at depth R. Since the integrand describes only the surface displacements $(z = 0)$, it cannot be used to evaluate $\tilde{B}$. Therefore, the integration will actually involve additional terms in z that can be obtained from Eq. (20.31). If, on the other hand, the datum can be taken somewhere on the surface itself, it is sufficient to evaluate the integral in Eq. (20.33) and then adjust $\tilde{C}$ by using the integrated results. Thus, the deformation of the surface $(z = 0)$ under a uniform pressure $\tilde{p}_c$ extending between

$\bar{s} = -b_2$ and $\bar{s} = b_2$ is given as

$$\delta = -\frac{(1-\nu^2)\tilde{p}_c}{\pi E}\int_{-b_2}^{b_2} \ln(x-\bar{s})^2\, d\bar{s} + \tilde{C}$$

$$= \frac{(1-\nu^2)\tilde{p}_c}{\pi E}\left[(x-b_2)\ln(x-b_2)^2 - (x+b_2)\ln(x+b_2)^2\right] + \tilde{D} \quad (20.34)$$

The preceding equation must be referred to a datum. Setting $\delta = 0$ at $x = 0$ and $z = 0$ while solving for $\tilde{D}$ gives the following:

$$\delta = \frac{2(1-\nu^2)}{\pi E}\tilde{p}_c\left(x \ln\frac{|x-b_2|}{|x+b_2|} - b_2 \ln\frac{|x^2-b_2^2|}{b_2^2}\right)$$

20.3 ELASTIC DEFORMATIONS IN RECTANGULAR CONJUNCTIONS

The general expressions for the deformations in a rectangular conjunction having been developed, these formulas will now be used in elastohydrodynamic lubrication analyses. For two surfaces having the same elasticity but made of different materials coming into contact, Eq. (20.31) gives the elastic deformation at any point x on the surface ($z = 0$) as

$$\delta = -\frac{2}{\pi E'}\int_{x_{\min}}^{x_{\text{end}}} p \ln(x-x')^2\, dx' \quad (20.35)$$

where

$$\frac{1}{E'} = \frac{1}{2}\left(\frac{1-\nu_a^2}{E_a} + \frac{1-\nu_b^2}{E_b}\right) \quad (20.36)$$

and p is pressure that is a function of x' varying from $x_{\min}$ to x_{end}. Note that since $\tilde{B}$ in Eq. (20.31) depends on the elastic constants, the expression for δ in Eq. (20.34) would have an added material-dependent term if the surfaces did not have the same elasticity. Letting

$$x = bX \qquad x' = bX' \qquad p = p_H P \qquad \delta = b^2\bar{\delta}/R_x$$

$$\frac{1}{R_x} = \frac{1}{r_{ax}} + \frac{1}{r_{bx}} \qquad \frac{R_x}{D_x} = \frac{1}{4}\left(\frac{\pi}{2W'}\right)^{1/2} \qquad W' = \frac{w_z'}{E'R_x} \quad (20.37)$$

where p_H is the maximum Hertzian pressure, causes Eq. (20.35) to become

$$\bar{\delta} = -\frac{1}{2\pi}\left[\int_{X_{\min}}^{X_{\text{end}}} P \ln(X-X')^2\, dX' + \ln\left(R_x^2\frac{8W'}{\pi}\right)\int_{X_{\min}}^{X_{\text{end}}} P\, dX'\right] \quad (20.38)$$

But the normal applied load per unit width is just the integration of the pressure from the inlet to the outlet.

$$\therefore\ w_z' = \int p\,dx$$

This implies that

$$\int P\,dX' = \frac{\pi}{2} \tag{20.39}$$

Substituting Eq. (20.39) into Eq. (20.38) gives

$$\bar{\delta} = -\frac{1}{2\pi}\int_{X_{\min}}^{X_{\text{end}}} P \ln(X - X')^2\,dX' - \frac{1}{4}\ln\left(R_x^2\frac{8W'}{\pi}\right) \tag{20.40}$$

Note that the last term on the right side of Eq. (20.40), which is a constant, depends on how X and P are made dimensionless. This term represents, in general, 80 to 90 percent of the total deformation. This grouping of the elastic deformation was discovered by Houpert and Hamrock (1986) and proved to be a useful separation in that the remaining pressure-dependent deformation is now of the same order as the film thickness at moderate loads.

Using integration by parts on the integral given in Eq. (20.40), Houpert and Hamrock (1986) found that

$$\bar{\delta} = -\frac{1}{2\pi}|PI|_{X_{\min}}^{X_{\text{end}}} + \frac{1}{2\pi}\int_{X_{\min}}^{X_{\text{end}}}\frac{dP}{dX'}I\,dX' - \frac{1}{4}\ln\left(R_x^2\frac{8W'}{\pi}\right) \tag{20.41}$$

where

$$I = \int \ln(X - X')^2\,dX' = -(X - X')\left[\ln(X - X')^2 - 2\right] \tag{20.42}$$

Since the pressure is zero at $X_{\min}$ and X_{end}, the first term on the right side of Eq. (20.41) vanishes and

$$\bar{\delta} = -\frac{1}{2\pi}\int_{X_{\min}}^{X_{\text{end}}}\frac{dP}{dX'}(X - X')\left[\ln(X - X')^2 - 2\right]dX' - \frac{1}{4}\ln\left(R_x^2\frac{8W'}{\pi}\right) \tag{20.43}$$

The integral in Eq. (20.43) can be calculated analytically by assuming that the pressure is described by a polynomial of second degree in the interval $[X_{j-1}, X_{j+1}]$. The details of the calculations are given in Appendix A. The resulting equation from their studies gives the dimensionless deformation $\bar{\delta}_i$ at node i as a function of the dimensionless pressure P_j and the influence coefficients D_{ij}:

$$\bar{\delta}_i = \sum_{j=1}^{N} D_{ij}P_j - \frac{1}{4}\ln\left(R_x^2\frac{8W'}{\pi}\right) \tag{20.44}$$

The results obtained from Houpert and Hamrock (1986) are compared in Table 20.1 with results obtained by Hamrock and Jacobson (1984) as well as by

Okamura (1982), who used simpler approaches than that used by Houpert and Hamrock (1986). Hamrock and Jacobson (1984) assumed the pressure to be constant in the interval $[X_j - \Delta X/2, X_j + \Delta X/2]$ and used an analytical expression for the integral of $\ln(X_i - X')$. Okamura (1982) did not use any analytical solutions and assumed simply that

$$\bar{\delta} = -\frac{\Delta}{2\pi}\sum_{j=1}^{N} P_j \ln\left(\left|\frac{X_{i+1}+X_i}{2} - X_j\right|\left|\frac{X_{i-1}+X_i}{2} - X_j\right|\right) \tag{20.45}$$

where $\Delta = X_{i+1} - X_i$. The three approaches can be compared by assuming a Hertzian pressure. Between $X = -1$ and $X = 1$ the film shape while assuming a Hertzian pressure should be flat, leading to $\Delta H = 0$, where ΔH is defined as

$$\Delta H = \frac{X^2}{2} + \bar{\delta} - \bar{\delta}_m \tag{20.46}$$

and $\bar{\delta}_m$ is the dimensionless maximum deformation. This can be compared with the value of $\bar{\delta}_H$ obtained by analytic integration of Eq. (20.40) with a Hertzian pressure distribution.

$$\bar{\delta}_H = -\frac{1}{4}\ln\left(R_x^2\frac{8W'}{\pi}\right) + \frac{1}{2}\ln(2) + 0.25 \tag{20.47}$$

The largest value $\Delta H_m/H_m$ of $\Delta H/H_m$ is found at $X = -1$ and $X = 1$ (because of the slope discontinuity) and is shown in Table 20.1 with the corresponding value of $\bar{\delta}_m/\bar{\delta}_H - 1$. The film thickness H_m where $dP/dX = 0$ has been chosen to be equal to 0.5; X_{min} and X_{max} define the first and last values of X; N_{max} is the number of nodes.

Table 20.1 shows that for a given mesh the best accuracy in calculating $\bar{\delta}$ and ΔH_m is obtained by using the Houpert and Hamrock (1986) approach. The value of $\bar{\delta}_m$ from Hamrock and Jacobson (1984) is in some cases less accurate than that from Okamura (1982) because they did not separate the constant as

TABLE 20.1
Three ways of calculating elastic deformations

[From Houpert and Hamrock (1986); OK denotes Okamura (1982); HJ denotes Hamrock and Jacobson (1984); HH denotes Houpert and Hamrock (1986).]

N_{max}	X_{min}	X_{end}	$(\bar{\delta}_m/\bar{\delta}_H - 1)$			$\Delta H_m/H_m$		
			OK	HJ	HH	OK	HJ	HH
51	−1.0	1.0	-1.9×10^{-3}	-2.7×10^{-3}	8.6×10^{-7}	9.4×10^{-3}	-5.5×10^{-3}	-2.7×10^{-3}
51	−3.6	1.4	-4.9×10^{-3}	-1.1×10^{-2}	1.0×10^{-5}	3.4×10^{-2}	-1.6×10^{-2}	-8.6×10^{-3}
151	−1.0	1.0	-6.3×10^{-4}	-5.1×10^{-4}	5.1×10^{-8}	4.0×10^{-3}	-1.4×10^{-3}	-6.4×10^{-4}
151	−3.6	1.4	-1.6×10^{-3}	-2.0×10^{-3}	5.4×10^{-7}	1.2×10^{-2}	-4.4×10^{-3}	-2.1×10^{-3}
301	−1.0	1.0	-3.1×10^{-4}	-1.8×10^{-4}	8.1×10^{-9}	2.2×10^{-3}	-5.6×10^{-4}	-2.5×10^{-4}
301	−3.6	1.4	-7.8×10^{-4}	-7.1×10^{-4}	9.0×10^{-8}	6.0×10^{-3}	-1.8×10^{-3}	-8.6×10^{-4}
661	−3.6	1.4	---------	2.2×10^{-6}	---------	---------	-2.6×10^{-4}	---------
51	[a]−1.0	[a]1.0	---------	---------	6.5×10^{-7}	---------	---------	-2.7×10^{-4}

[a]Nonuniform.

was done in Eq. (20.40). But $\bar{\delta}_m$ is not really significant, since any inaccuracy in its calculation can be compensated for by H_0 in the film thickness equation. The important parameter of Table 20.1 is ΔH_m, since it is a measure of the flatness of the film. This aspect is extremely important at high loads, where the elastic deformations are two or three orders of magnitude larger than the film thickness.

Also shown in Table 20.1 are the results obtained with a uniform mesh of 661 nodes by Hamrock and Jacobson (1984). Extremely small values of ΔH_m were calculated and cannot be reproduced with the new approach because of storage problems with the matrix D_{ij} and because a large system of equations would have to be solved with such a mesh. By using a nonuniform mesh with a fine grid near $X = -1$ and $X = 1$, similar results were found with a small value of N_{max} ($N_{max} = 51$), as indicated in Table 20.1. The latter case illustrates the power of the approach developed by Houpert and Hamrock (1986).

20.4 POINT LOAD SOLUTION

The stress tensor for the problem of a concentrated point load acting along the normal to the undeformed plane surface of an elastic half-space is chosen as the starting point. Because of symmetry the coordinates best suited for these considerations are the cylindrical system (r, θ, z) shown in Fig. 20.3.

Consider a point load w_z acting along the positive z axis on the boundary surface ($z = 0$) of an elastic half-space defined by $z > 0$ and having modulus of elasticity E and Poisson's ratio ν. By symmetry, polar angle θ does not appear in the stress tensor, and the shear stress components $\tau_{r\theta}$ and $\tau_{\theta z}$ vanish. The four remaining stresses can be obtained from Timoshenko and Goodier (1951) as

$$\sigma_r = \frac{w_z}{2\pi}\left\{(1-2\nu)\left[\frac{1}{r^2} - \frac{z}{r^2(r^2+z^2)^{1/2}}\right] - \frac{3zr^2}{(r^2+z^2)^{5/2}}\right\} \tag{20.48}$$

$$\sigma_z = -\frac{3w_z z^3}{2\pi(r^2+z^2)^{5/2}} \tag{20.49}$$

$$\sigma_\theta = \frac{w_z(1-2\nu)}{2\pi}\left[-\frac{1}{r^2} + \frac{z}{r^2(r^2+z^2)^{1/2}} + \frac{z}{(r^2+z^2)^{3/2}}\right] \tag{20.50}$$

$$\tau_{rz} = -\frac{3w_z rz^2}{2\pi(r^2+z^2)^{5/2}} \tag{20.51}$$

Note that $(r^2+z^2)^{1/2}$ is the distance from the point where the load is applied. The stress components given in Eqs. (20.48) to (20.51) satisfy the general requirements of mechanical equilibrium and compatibility. These equations and this physical situation can be likened to the line load solution given in Eqs. (20.20) to (20.22).

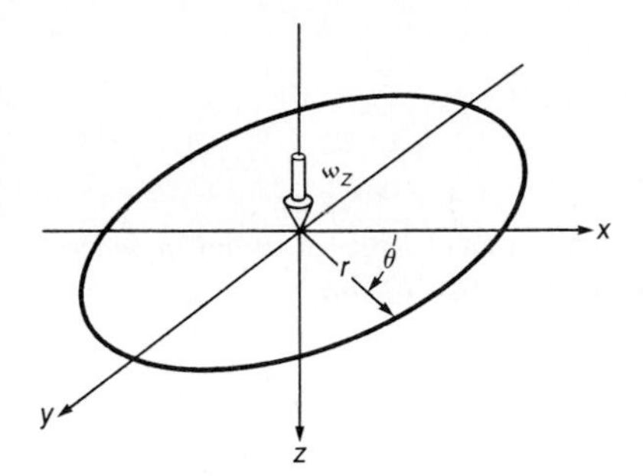

FIGURE 20.3
Cylindrical polar coordinates used for point load w_z acting in z direction at origin on bounding surface of elastic half-space. *[From Tripp (1985).]*

The deformations in the direction of increasing (r, θ, z) compatible with Eqs. (20.48) to (20.51) are

$$\delta_r = \frac{(1-2\nu)(1+\nu)\,w_z}{2\pi E r}\left[\frac{z^2}{\left(r^2+z^2\right)^{1/2}} - 1 + \frac{1}{1-2\nu}\,\frac{r^2 z}{\left(r^2+z^2\right)^{3/2}}\right] \tag{20.52}$$

$$\delta_\theta = 0 \tag{20.53}$$

$$\delta_z = \frac{w_z}{2\pi E}\left[\frac{(1+\nu)z^2}{\left(r^2+z^2\right)^{3/2}} + \frac{2(1-\nu^2)}{\left(r^2+z^2\right)^{1/2}}\right] \tag{20.54}$$

From these equations note that δ_r and δ_z are singular at the origin.

The preceding equations reduce to the following on the surface of the solids, or when $z = 0$:

$$\delta_r = -\frac{(1-2\nu)(1+\nu)\,w_z}{2\pi E r} \tag{20.55}$$

$$\delta_z = \frac{w_z(1-\nu^2)}{\pi E r} \tag{20.56}$$

Equations (20.48) to (20.56) suggest that for small values of r, infinite stresses and displacements occur. This is physically impossible, and this purely mathematical condition is avoided by assuming that the point loading is replaced by a hemispherical stress distribution that is equivalent to the load w_z.

20.5 LOADING ON A SEMI-INFINITE BODY

Consider a circular area of radius a over which a pressure acts. The deformation at point M outside the circle is shown in Fig. 20.4. From the figure

$$\text{Area of element} = (d\bar{s})(\bar{s}\,d\psi)$$

$$\text{Load on element} = p\bar{s}\,d\bar{s}\,d\psi$$

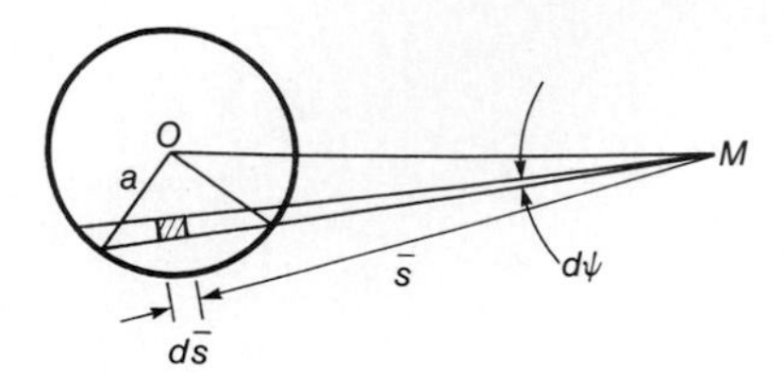

FIGURE 20.4
Deformation outside circular area of radius a where pressure p acts. *[From Tripp (1985).]*

Substituting these into Eq. (20.56) gives

$$\delta_z = p(d\bar{s})(d\psi)\frac{1-\nu^2}{\pi E}$$

Therefore, the total deflection due to all the elements of the loaded area is

$$\delta_z = \frac{1-\nu^2}{\pi E}\int\int p\, d\bar{s}\, d\psi \qquad (20.57)$$

This equation is valid regardless of the shape of the loaded area.

Figure 20.5 illustrates the situation when M lies within the loaded area. From the triangle Omn

$$mn = 2a\cos\theta$$

Also from OmM and from the law of sines, when $OM = r$

$$\frac{r}{\sin\theta} = \frac{a}{\sin\psi}$$

or

$$\sin\theta = \frac{r}{a}\sin\psi \qquad (20.58)$$

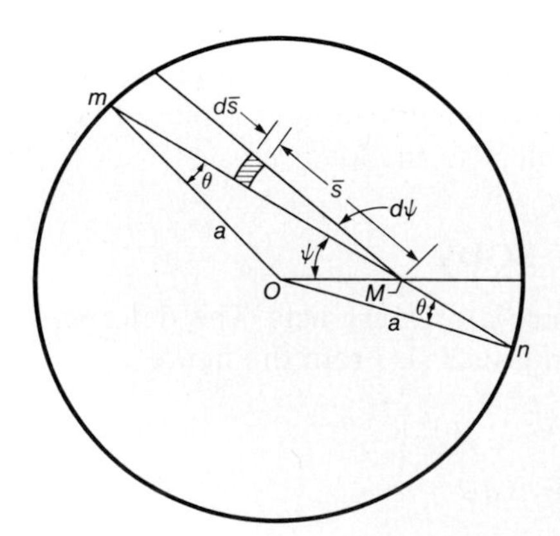

FIGURE 20.5
Deformation inside circular area of radius a where pressure p acts. *[From Tripp (1985).]*

Hence, $\int d\bar{s}$ over the length mn is

$$\int_0^{2a\cos\theta} d\bar{s} = 2a\cos\theta$$

Therefore,

$$\iint d\bar{s}\,d\psi = 2a\int_{-\pi/2}^{\pi/2}\cos\theta\,d\psi$$

From Eq. (20.58)

$$\cos\theta = (1-\sin^2\theta)^{1/2} = \left[1-\left(\frac{r}{a}\right)^2\sin^2\psi\right]^{1/2}$$

$$\therefore \iint d\bar{s}\,d\psi = 2a\int_{-\pi/2}^{\pi/2}\left[1-\left(\frac{r}{a}\right)^2\sin^2\psi\right]^{1/2} d\psi$$

Making use of Eq. (20.56) gives

$$\delta_z = \frac{4(1-\nu^2)}{\pi E}pa\int_0^{\pi/2}\left[1-\left(\frac{r}{a}\right)^2\sin^2\psi\right]^{1/2} d\psi \tag{20.59}$$

This integral can be evaluated by using tables of elliptic integrals for any particular value of r/a.

At the center of the circle when $r = 0$ ($OM = 0$), the maximum deflection is

$$(\delta_z)_{\max} = \frac{4(1-\nu^2)}{\pi E}pa\frac{\pi}{2} = \frac{2pa(1-\nu^2)}{E} \tag{20.60}$$

Note also if $r = a$, from Eq. (20.59)

$$(\delta_z)_{r=a} = \frac{4(1-\nu^2)}{\pi E}pa\int_0^{\pi/2}\cos\psi\,d\psi = \frac{4(1-\nu^2)pa}{E\pi} \tag{20.61}$$

$$\therefore (\delta_z)_{r=a} = \frac{2}{\pi}(\delta_z)_{\max} \tag{20.62}$$

Consider the stress at a point on the $\bar{s}$ axis that is produced by an element ring of pressure on the surface as shown in Fig. 20.6. From Eq. (20.49) the normal stress in the $\bar{s}$ direction can be expressed as

$$\sigma_z = -\frac{3w_z z^3}{2\pi}(r^2+\bar{s}^2)^{-5/2} \tag{20.63}$$

Now if $w_z = 2\pi rp\,dr$, this equation becomes, due to the element,

$$\sigma_z = -3pr\bar{s}^3(r^2+\bar{s}^2)^{-5/2}\,dr$$

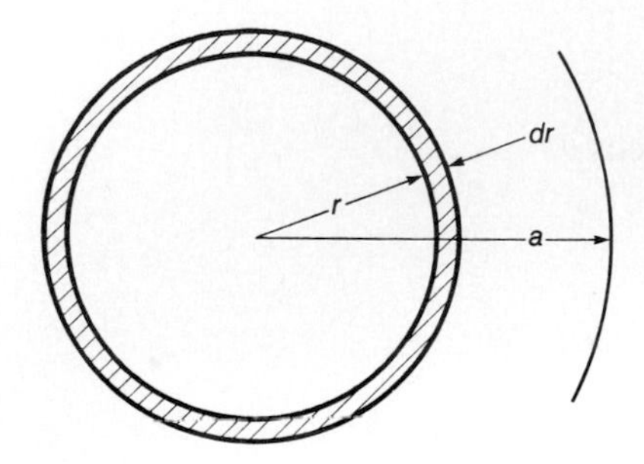

FIGURE 20.6
Ring of pressure on surface of circular area of constant pressure.

The normal stress produced by the entire distributed load is

$$\sigma_z = -3p\bar{s}^3 \int_0^a r(r^2 + \bar{s}^2)^{-5/2} \, dr = p\left[z^3(a^2 + \bar{s}^2)^{-3/2} - 1\right] \tag{20.64}$$

Note that as $\bar{s} = 0$, $\sigma_z = -p$.

The stresses σ_r and σ_θ at axial points can be calculated while transforming the stress tensor from cylindrical to Cartesian coordinates to give $\tau_{r\theta} = 0$ and

$$\sigma_r = \sigma_\theta = \frac{p}{2}\left\{-2(1+\nu) + \frac{2(1+\nu)\bar{s}}{(a^2 + \bar{s})^{1/2}} - \left[\frac{\bar{s}}{(a^2 + \bar{s}^2)^{1/2}}\right]^3\right\} \tag{20.65}$$

20.6 ELASTIC DEFORMATIONS IN ELLIPTICAL CONJUNCTIONS

Figure 20.7 shows a rectangular area of uniform pressure with the appropriate coordinate system. From Eq. (20.56) the elastic deformation at a point (x, y) of a semi-infinite solid subjected to a pressure p at the point (x_1, y_1) can be written as

$$d\delta_z = \frac{2p \, dx_1 \, dy_1}{\pi E' r}$$

The elastic deformation at a point (x, y) due to a uniform pressure over a rectangular area $2\tilde{a} \times 2\tilde{b}$ is

$$\delta_z = \frac{2P}{\pi} \int_{-\tilde{a}}^{\tilde{a}} \int_{-\tilde{b}}^{\tilde{b}} \frac{dx_1 \, dy_1}{\left[(y - y_1)^2 + (x - x_1)^2\right]^{1/2}} \tag{20.66}$$

where $P = p/E'$. Integrating gives

$$\delta_z = \frac{2}{\pi} P D^* \tag{20.67}$$

where

$$D^* = (x + \tilde{b}) \ln \frac{(y + \tilde{a}) + \left[(y + \tilde{a})^2 + (x + \tilde{b})^2\right]^{1/2}}{(y - \tilde{a}) + \left[(y - \tilde{a})^2 + (x + \tilde{b})^2\right]^{1/2}}$$

$$+ (y + \tilde{a}) \ln \frac{(x + \tilde{b}) + \left[(y + \tilde{a})^2 + (x + \tilde{b})^2\right]^{1/2}}{(x - \tilde{b}) + \left[(y + \tilde{a})^2 + (x - \tilde{b})^2\right]^{1/2}}$$

$$+ (x - \tilde{b}) \ln \frac{(y - \tilde{a}) + \left[(y - \tilde{a})^2 + (x - \tilde{b})^2\right]^{1/2}}{(y + \tilde{a}) + \left[(y + \tilde{a})^2 + (x - \tilde{b})^2\right]^{1/2}}$$

$$+ (y - \tilde{a}) \ln \frac{(x - \tilde{b}) + \left[(y - \tilde{a})^2 + (x - \tilde{b})^2\right]^{1/2}}{(x + \tilde{b}) + \left[(y - \tilde{a})^2 + (x + \tilde{b})^2\right]^{1/2}} \qquad (20.68)$$

Hamrock and Dowson (1974) used the preceding equations in their elliptical elasticity analysis as part of their treatment of elastohydrodynamic lubrication. As can be seen from Eq. (20.66), the pressure on each element can be replaced by a constant value; that is, the whole pressure distribution is replaced by blocks of uniform pressure. In this way an analytical expression for integrating the deformation is worked out, and the deformation of every node is expressed as a linear combination of the nodal pressures. In the solution of Ranger et al. (1975), a bilinear interpolating function is used to approximate the practical pressure distribution. Evans and Snidle (1982) employed a different method that was first presented by Biswas and Snidle (1977). For grid elements without singularity they directly adopted Simpson's integration rule, and for those with singularity they used a biquadratic polynomial function to express the pressure distribution approximately. In this way an analytical solution for the integration is developed without directly expressing the deformation as a linear combination of the nodal pressure. Following in these footsteps Hou et al. (1985) employed the biquadratic polynomial function for approaching the pressure distribution on any grid element. An influence coefficient matrix is introduced to reduce the amount of calculating work when repeated calculations of the elastic deformation are needed. However, a shortcoming of their work is that for a $(2n + 1)(2m + 1)$ finite difference grid, the influence matrix is composed of $(n \times m)(2n \times 2m \times 9)$ elements, which is not welcomed even with today's powerful computers.

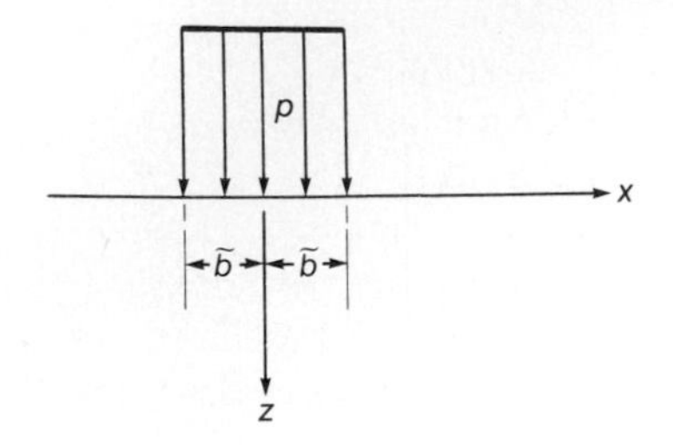

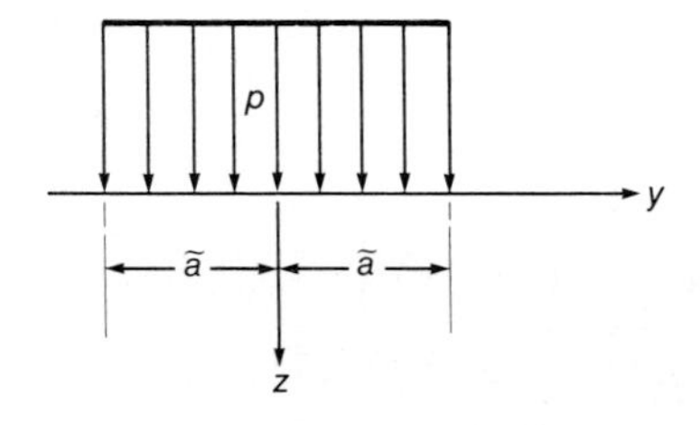

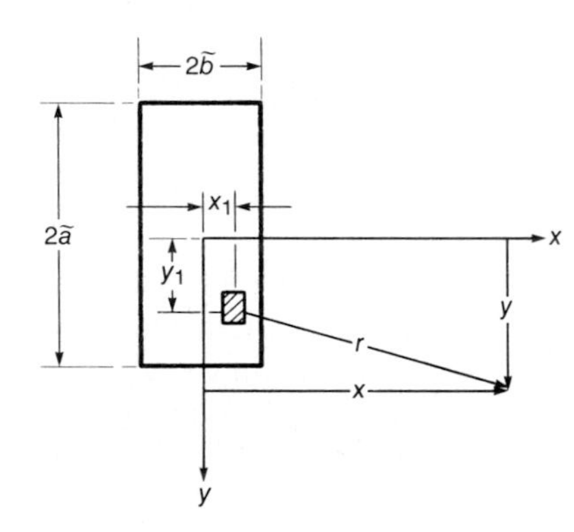

FIGURE 20.7
Surface deformation of semi-infinite body subjected to uniform pressure over rectangular area. *[From Hamrock and Dowson (1981).]*

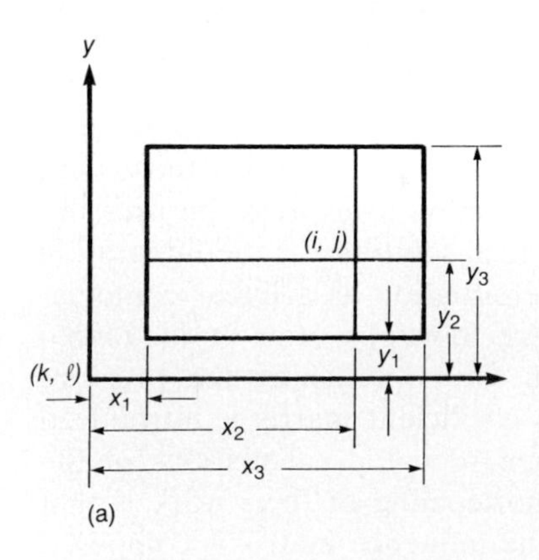

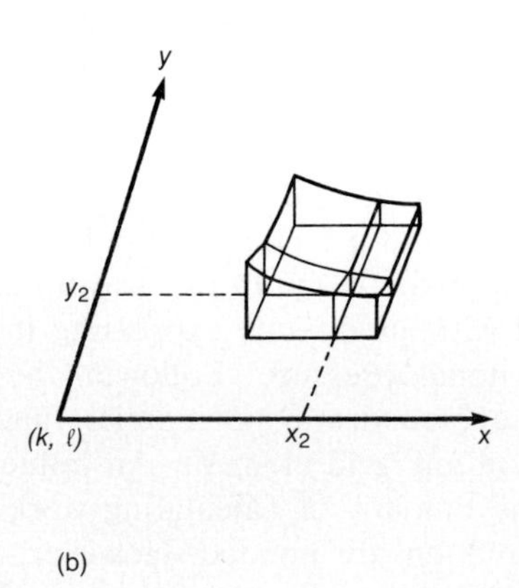

FIGURE 20.8
Coordinate system for computing element. (a) Rectangular element with nine nodes; (b) representation of pressure distribution for paraboloidal surfaces. *[From Jeng and Hamrock (1987).]*

Jeng and Hamrock (1987) used a biquadratic polynomial expressed in Lagrange form to approximate the pressure distribution on all grid elements. Figure 20.8 shows the coordinate system for the computing element. Node (k, ℓ) is the nodal point for the elastic deformation, and node (i, j) is the central point of the grid element for the pressure distribution. An influence matrix whose coefficient is dependent on the geometric factors and the distance between node (k, ℓ) and node (i, j) is introduced only to express the deformation of every node as a linear combination of the nodal pressures. In this way, only $(2n \times 2m \times 9)$ elements are needed in the influence matrix for a $(2n + 1)$ $(2m + 1)$ equidistant rectangular grid. The computational time and computer storage size for the influence coefficient matrix are greatly reduced.

20.7 CLOSURE

Stresses and deformations in nonconformal conjunctions were presented in a general manner, describing what occurs on as well as within the solids in contact. The conjunctions considered were assumed to be dry, or unlubricated, as in the preceding chapter. The material presented in this chapter is the foundation that will be used extensively in later chapters on elastohydrodynamic lubrication.

A line (two-dimensional) load situation was presented. The general stress equations were solved by a stress function approach that resulted in a biharmonic governing equation. This equation was solved by using the Boussinesq stress function. Thus, not only general expressions for the stresses but also the deformation of the surfaces could be obtained. These general line-contact formulations were then reduced to define the elastic deformation on the surfaces resulting from a rectangular contact area. An important result is that a constant term in the elastic deformation equation accounted for 80 to 90 percent of the total deformation. Isolating this term is thus important in accurately solving for the deformation of solids that have rectangular contact areas. These results will be directly applicable when dealing with elastohydrodynamic lubrication of rectangular conjunctions in Chap. 21.

A point (three-dimensional) load resulted in a stress tensor acting along the normal to the undeformed plane surface of an elastic half-space. The results indicate that for a small radius infinite stresses and displacements occur. This is physically impossible and is avoided by assuming that the point loading is replaced by a hemispherical stress distribution equivalent to the load. This general formulation of the stresses and deformations was then applied to describing the elastic deformation that results in elliptical contacts. The elastic deformation analysis assumed uniform pressure over a small rectangular area. Thus, the pressure could be placed in front of the integral equation, easing the computation considerably. The resulting equation for the elastic deformation was simply expressed as the pressure multiplied by a distance-influencing function. These results will be used when dealing with elastohydrodynamic lubrication of elliptical conjunctions in Chap. 22.

20.8 PROBLEMS

20.8.1 Explain the use of the Airy stress function in the analysis of two-dimensional problems in the theory of elasticity, with reference to both Cartesian and polar coordinates.

20.8.2 A semi-infinite uniform plate has a stress system within it defined by the stress function $\phi = Br^2\theta$. The boundary of the plate is the line $\theta = \pm\pi/2$, and the plate covers the portion of the plane where $-\pi/2 \leq 0 \leq \pi/2$ and r is positive. Find the conditions of stress that apply at the boundary and also the stress σ_x, σ_y, and τ_{xy} at any point, in terms of θ. Assume that polar and Cartesian coordinates have the same origin and that $x = r \sin\theta$ and $y = r\cos\theta$. A second stress system equal but opposite in sign is now added to the plate, but it is displaced a distance a from the first. Thus, the second system is represented by the stress function

$$\phi' = B(r')^2\theta'$$

where (r', θ') are referred to an origin at the point $(a, 0)$ in the (r, θ) coordinate system.

Show that the superposition of the boundary stresses of the two systems results in a uniform normal stress spread over the length $r = 0$ to $r = a$ and in no other stress.

20.9 REFERENCES

Biswas, S., and Snidle, R. W. (1977): Calculation of Surface Deformation in Point Contact EHD. *J. Lubr. Technol.*, vol. 99, no. 3, pp. 313–317.

Evans, H. P., and Snidle, R. W. (1982): The Elastohydrodynamic Lubrication of Point Contacts at Heavy Loads. *Proc. Roy. Soc. London, Ser. A*, vol. 382, no. 1782, July 8, pp. 183–199.

Hamrock, B. J., and Dowson, D. (1974): Numerical Evaluation of Surface Deformation of Elastic Solids Subjected to Hertzian Contact Stress. *NASA Tech. Note* D-7774.

Hamrock, B. J., and Dowson, D. (1981): *Ball Bearing Lubrication—The Elastohydrodynamics of Elliptical Contacts*. Wiley-Interscience, New York.

Hamrock, B. J., and Jacobson, B. O. (1984): Elastohydrodynamic Lubrication of Line Contacts. *ASLE Trans.*, vol. 27, no. 4, pp. 275–287.

Hou, K., Zhu, D., and Wen, S. (1985): A New Numerical Technique for Computing Surface Elastic Deformation Caused by a Given Normal Pressure Distribution. *J. Tribol.*, vol. 107, no. 1, pp. 128–131.

Houpert, L. G., and Hamrock, B, J. (1986): Fast Approach for Calculating Film Thicknesses and Pressures in Elastohydrodynamically Lubricated Contacts at High Loads. *J. Tribol.*, vol. 108, no. 3, pp. 411–420.

Jeng, Y. R., and Hamrock, B. J. (1987): The Effect of Surface Roughness on Elastohydrodynamically Lubricated Point Contacts. *ASLE Trans.*, vol. 30, no. 4, pp. 531–538.

Okamura, H. (1982): A Contribution to the Numerical Analysis of Isothermal Elastohydrodynamic Lubrication. *Tribology of Reciprocating Engines*. D. Dowson et al. (eds.). Butterworths, Guilford, England, pp. 313–320.

Ranger, A. P., Ettles, C. M. M., and Cameron, A. (1975): The Solution of the Point Contact Elastohydrodynamic Problem. *Proc. Roy. Soc. London, Ser. A*, vol. 346, no. 1645, Oct. 19, pp. 227–244.

Timoshenko, S., and Goddier, J. N. (1951): *Theory of Elasticity*, 2d ed. McGraw-Hill, New York.

Tripp, J. (1985): Hertzian Contact in Two and Three Dimensions. *NASA Tech. Paper* 2473.

CHAPTER
21
ELASTOHYDRODYNAMIC LUBRICATION OF RECTANGULAR CONJUNCTIONS

The recognition and understanding of elastohydrodynamic lubrication (EHL) represents one of the major developments in the field of tribology in the twentieth century. The revelation of a previously unsuspected lubrication film is clearly an event of some importance in tribology. In this case it not only explained the remarkable physical action responsible for the effective lubrication of many nonconformal machine elements, such as gears, rolling-element bearings, cams, and continuously variable traction drives, but also brought order to the understanding of the complete spectrum of lubrication regimes, ranging from boundary to hydrodynamic.

The development of elastohydrodynamic lubrication is fairly recent. The first notable breakthrough occurred when Grubin (1949) managed to incorporate both the elastic deformation of the solids and the viscosity-pressure characteristics of the lubricant in analyzing the inlet region of lubricated nonconformal machine elements. The most important practical significance of this work is that the film thickness equation Grubin developed yielded values one or two orders of magnitude greater than those predicted by hydrodynamic theory. These values tended to be in better agreement with experimental results for gears and rolling-element bearings. Petrusevich (1951) provided three numerical solutions to the governing elasticity and hydrodynamic equations that confirmed the essential features of the Grubin (1949) analysis and yielded additional information on the details of the film shape and pressure distribution throughout the conjunction.

Various procedures for solving the complex elastohydrodynamic lubrication problem were reported in the 1960s. Dowson and Higginson (1959) described an iterative procedure that not only yielded a wide range of solutions during the next decade, but also enabled them to derive an empirical minimum-film-thickness formula for line contacts. Crook's (1961) experiments confirmed the order of magnitude of film thickness deduced by Dowson and Higginson (1959), and he was able to produce direct evidence of the influence of load and speed on film thickness. Load was found to have an almost negligible effect on the film thickness, but speed was found to play an important role.

The major limitation of the work in the 1960s and 1970s and up to the mid-1980s was that the results were obtained for light loads and extrapolations were made for higher loads. Most nonconformal contacts, such as rolling-element bearings and gears, operate in the range of maximum Hertzian pressure from 0.5 to 3.0 GPa. Then, in 1986 Houpert and Hamrock developed an approach that enables solutions to elastohydrodynamically lubricated rectangular conjunctions to be made that have no load limitations. Successful solutions were made for a maximum Hertzian pressure of 4.8 GPa. Details of their approach as well as some simpler approaches are covered in this chapter. Furthermore, results of pressure, film thickness, and flow are given to illustrate what these features are in an elastohydrodynamically lubricated conjunction.

21.1 INCOMPRESSIBLE SOLUTION

From Eq. (7.49) the appropriate Reynolds equation for time-invariant conditions while neglecting side leakage for elastohydrodynamic lubrication is

$$\frac{d}{dx}\left(\frac{\rho h^3}{\eta}\frac{dp}{dx}\right) = 12\tilde{u}\frac{d(\rho h)}{dx} \tag{21.1}$$

where $\tilde{u} = (u_a + u_b)/2$. Assuming incompressible conditions, this equation reduces to

$$\frac{d}{dx}\left(\frac{h^3}{\eta}\frac{\partial p}{\partial x}\right) = 12\tilde{u}\frac{dh}{dx} \tag{21.2}$$

Further assuming that the pressure-viscosity effects may be expressed as Barus (1893) formulated, as discussed in Chap. 4 [Eq. (4.8)]

$$\eta = \eta_0 e^{\xi p} \tag{21.3}$$

Eq. (21.2) can be rewritten as

$$\frac{d}{dx}\left(h^3\frac{dp^*}{dx}\right) = 12\tilde{u}\eta_0\frac{dh}{dx} \tag{21.4}$$

where
$$p^* = \frac{1 - e^{-\xi p}}{\xi} \tag{21.5}$$

is the reduced pressure in pascals. The advantage of solving Eq. (21.4) over (21.2) is that, instead of both pressure and viscosity as variables, reduced pressure is the only variable. Equation (21.5) can be rewritten to express the pressure as

$$p = -\frac{1}{\xi}\ln(1 - \xi p^*) \tag{21.6}$$

From this equation observe that as $p \to 0$, $p^* \to 0$ and that as $p \to \infty$, $p^* \to 1/\xi$.

Letting

$$\begin{gathered} p^* = p_H P_r^* \qquad p_H = E'\left(\frac{W'}{2\pi}\right)^{1/2} \\ h = \frac{D_x^2}{4R_x}H_r = \frac{8R_xW'}{\pi}H_r \qquad x = \frac{D_x}{2}X_r = R_x\left(\frac{8W'}{\pi}\right)^{1/2}X_r \end{gathered} \tag{21.7}$$

Eq. (21.4) becomes

$$\frac{\partial}{\partial X_r}\left(H_r^3\frac{\partial P_r^*}{\partial X_r}\right) = \bar{K}\frac{dH_r}{dX_r} \tag{21.8}$$

where

$$\bar{K} = \text{constant} = \frac{3\pi^2U}{4(W')^2} \tag{21.9}$$

$$U = \text{dimensionless speed parameter} = \frac{\eta_0\tilde{u}}{E'R_x} \tag{21.10}$$

$$W' = \text{dimensionless load parameter} = \frac{w_z'}{E'R_x} \tag{21.11}$$

Substituting Eq. (21.7) into Eqs. (21.5) and (21.6) gives

$$P_r^* = \frac{p^*}{p_H} = \frac{1}{G}\left(\frac{2\pi}{W'}\right)^{1/2}\left\{1 - \exp\left[-G\left(\frac{W'}{2\pi}\right)^{1/2}P_r\right]\right\} \tag{21.12}$$

and

$$P_r = \frac{p}{p_H} = -\frac{1}{G}\left(\frac{2\pi}{W'}\right)^{1/2}\ln\left[1 - G\left(\frac{W'}{2\pi}\right)^{1/2}P_r^*\right] \tag{21.13}$$

where

$$G = \text{dimensionless materials parameter} = \xi E' \tag{21.14}$$

From Eqs. (21.8), (21.9), and (21.12) it can be observed that the dimensionless pressure is a function of the dimensionless speed U, load W', and materials G parameters. Note from Eq. (21.12) that $P_r \to 0$ implies $P_r^* \to 0$ and that $P_r \to \infty$ implies $P_r^* \to (2\pi/W')^{1/2}/G$.

Integrating Eq. (21.8) gives

$$H_r^3\frac{dP_r^*}{dX_r} = \bar{K}H_r + \tilde{A} \tag{21.15}$$

A boundary condition is that $dp_r/dx_r = 0$ when $h = h_m$, or in dimensionless terms $dP_r/dX_r = 0$ when $H = H_m$. To establish how this relates to the reduced pressure, let

$$\bar{B} = 1 - G\left(\frac{W'}{2\pi}\right)^{1/2} P_r^*$$

$$\frac{d\bar{B}}{dX_r} = -G\left(\frac{W'}{2\pi}\right)^{1/2} \frac{dP_r^*}{dX_r}$$

Differentiating Eq. (21.13) with respect to X_r while making use of the preceding equations gives

$$\frac{dP_r}{dX_r} = \frac{1}{1 - G(W'/2\pi)^{1/2} P_r^*} \frac{dP_r^*}{dX_r} \tag{21.16}$$

From this equation it can be concluded that the boundary condition $dP_r/dX_r = 0$ when $H = H_m$ implies that $dP_r^*/dX_r = 0$ when $H_r = H_{r,m}$. Making use of this in Eq. (21.15) gives

$$\tilde{A} = -\bar{K}H_{r,m}$$

$$\therefore \frac{dP_r^*}{dX_r} = \frac{\bar{K}(H_r - H_{r,m})}{H_r^3} \tag{21.17}$$

Often the inflection point of the pressure profile is of interest. Calculating the second derivative gives

$$\frac{d^2P_r^*}{dX_r^2} = \bar{K}\left(-\frac{2}{H_r^3} + \frac{3H_{r,m}}{H_r^4}\right)\frac{dH_r}{dX_r}$$

At the inflection point $d^2P_r^*/dX_r^2 = 0$ and $H_r = H_{r,a}$.

$$\therefore 0 = (-2H_{r,a} + 3H_{r,m})\frac{dH_r}{dX_r}$$

Since $dH_r/dX_r \neq 0$, then $-2H_{r,a} + 3H_{r,m} = 0$, or

$$H_{r,m} = 2H_{r,a}/3 \tag{21.18}$$

The expression for the film shape can be written as

$$h(x) = h_0 + S(x) + \delta(x) \tag{21.19}$$

where h_0 = constant
$S(x)$ = separation due to geometry of undeformed solids
$\delta(x)$ = elastic deformation of solids

The geometric separation, while assuming the parabolic approximation [see Eq. (13.52)], is

$$S(x) = \frac{x^2}{2R_x}$$

From this equation and (21.19) the dimensionless film shape can be written as

$$H_r = H_{r,0} + \frac{X_r^2}{2} + \frac{4R_x\delta}{D_x^2} \tag{21.20}$$

The elastic deformation can be obtained from the material presented in Sec. 20.3. The coupled solution of Eqs. (21.17) and (21.20) needs to be evaluated numerically.

21.2 COMPRESSIBLE SOLUTION

Considering compressibility effects while neglecting side leakage and assuming time-invariant conditions allows the appropriate form of the Reynolds equation to be written from Eq. (7.51) as

$$\frac{dp}{dx} = \frac{12\tilde{u}\eta(\rho h - \rho_m h_m)}{\rho h^3} \tag{7.51}$$

The equation is the integrated form of the Reynolds equation, and $\rho_m h_m$ is the corresponding value of ρh when $dp/dx = 0$. This section is extensively based on the paper of Houpert and Hamrock (1986). The analysis is presented in detail in that paper and summarized here. Letting

$$p = p_H P_r \qquad \rho = \rho_0 \bar{\rho} \qquad \eta = \eta_0 \bar{\eta} \tag{21.21}$$

and using Eq. (21.7) gives Eq. (7.51) as

$$\frac{dP_r}{dX_r} = \frac{\bar{K}\bar{\eta}(\bar{\rho}H_r - \bar{\rho}_m H_{r,m})}{\bar{\rho}H_r^3} \tag{21.22}$$

where $\bar{K}$ is the same constant developed in the previous section [Eq. (21.9)].

The basic equation that will solve for the pressure at each node within a lubricated conjunction is

$$\tilde{f}_i = H_{r,i}^3 \left(\frac{dP_r}{dX_r}\right)_i - \bar{K}\bar{\eta}_i \left(H_{r,i} - \frac{\bar{\rho}_m H_{r,m}}{\bar{\rho}_i}\right) = 0 \tag{21.23}$$

The expression for the film shape can be obtained by using the results of the previous chapter, and the expressions for the dimensionless density and viscosity can be expressed from equations developed in Chap. 4 as

$$H_{r,i} = H_{r,0} + \frac{X_{r,i}^2}{2} + \sum_{j=1}^{N} D_{ij} P_{r,j} \tag{21.24}$$

$$\bar{\rho}_i = 1 + \frac{0.6 \times 10^{-9} p_H P_{r,i}}{1 + 1.7 \times 10^{-9} p_H P_{r,i}} \tag{4.19}$$

$$\bar{\eta}_i = \exp\left\{[\ln(\eta_0) + 9.67]\left[-1 + (1 + 5.1 \times 10^{-9} p_H P_{r,i})^{Z_1}\right]\right\} \tag{4.10}$$

Note in Eq. (21.24) that $H_{r,0}$ contains the constant term $-\ln(R_x^2 8W'/\pi)/4$ introduced in Eq. (20.44). Furthermore, the density expression given in Eq.

(4.19) was obtained from Dowson and Higginson (1966), and the viscosity expression given in Eq. (4.10), from Roelands (1966). Also note that in Eqs. (4.19) and (4.10) metric units are to be used. Values of fluid properties given as Z_1 and η_0 in Eq. (4.10) may be obtained directly from Chap. 4 for the particular fluid being investigated. Observe from these equations that coupling the elasticity and rheology equations into the integrated form of the Reynolds equation proves to be a difficult task mainly for the following reasons:

1. The lubricant viscosity changes by several orders of magnitude as the lubricant travels through the conjunction ($\bar{\eta}$ can be 10^8 in the contact center and 1 at the inlet and outlet).
2. The elastic deformation of the solid surfaces can be several orders larger than the minimum film thickness (δ/h_{min} can be 10^3 or 10^4).

For both these reasons extremely accurate numerical methods are required to get successful convergence of the problem.

Once the pressure is obtained, the dimensionless normal load per unit width can be evaluated from the following equation:

$$w'_z = \int_{x_{\text{min}}}^{x_{\text{end}}} p\,dx$$

Making use of Eqs. (21.7) and (21.21) in dimensionless terms reduces this equation to

$$\int_{X_{r,\text{min}}}^{X_{r,\text{end}}} P_r\,dX = \frac{\pi}{2} \tag{21.25}$$

The unknowns in this problem are

$X_{r,\text{end}}$	outlet meniscus
N	number of nodes used
$H_{r,0}$	constant
$\bar{\rho}_m H_{r,m}$	value of $\bar{\rho} H_r$ where $dP_r/dX_r = 0$
$P_{r,j}$	pressure at node j ($j = 2, N$)

The boundary conditions are

1. $P_r = 0$ for $X_r = X_{r,\text{min}}$.
2. $P_r = dP_r/dX_r = 0$ for $X_r = X_{r,\text{end}}$.

Figure 21.1 more clearly identifies $X_{r,\text{end}}$ and N. Note that $X_{r,N}$ is the nearest node to $X_{r,\text{end}}$ such that $X_{r,N} < X_{r,\text{end}}$. A second-degree curve for $H(X)$ can be defined by using H_{N-1}, H_N, and H_{N+1}, and then X_{end} can be calculated such that

$$H_r(X_{r,\text{end}}) = \bar{\rho}_m H_{r,m} \tag{21.26}$$

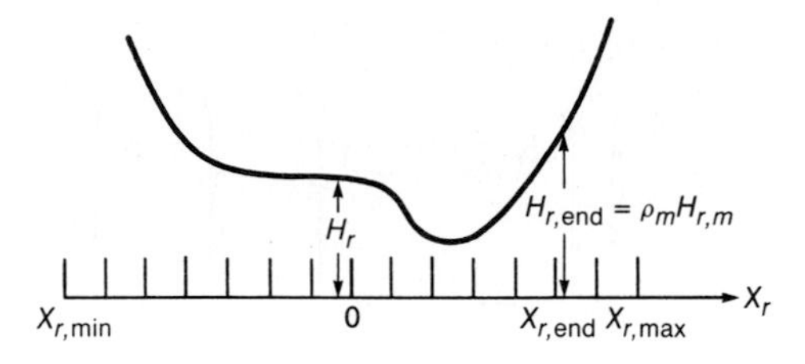

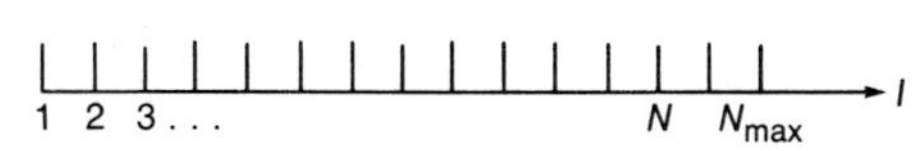

FIGURE 21.1
Sketch to illustrate calculations of $X_{r,\text{end}}$ and N. [*From Houpert and Hamrock (1986).*]

Appendix B gives the corrections to be applied to weighting factors due to X_r. Having defined $X_{r,\text{end}}$ and N, the remaining $N + 1$ unknowns, $\bar{\rho}_m H_{r,m}$, $H_{r,0}$, and P_2 to $P_{r,N}$, can be calculated by using a Newton-Raphson scheme. If the superscripts n and o are used to define the new and old values of the unknowns corresponding to two successive iterations,

$$(\bar{\rho}_m H_{r,m})^n = (\bar{\rho}_m H_{r,m})^o + [\Delta(\bar{\rho}_m H_{r,m})]^n \tag{21.27}$$

$$P_{r,j}^n = P_{r,j}^o + (\Delta P_{r,j})^n \tag{21.28}$$

$$H_{r,0}^n = H_{r,0}^o + (\Delta H_{r,0})^n \tag{21.29}$$

where $[\Delta(\bar{\rho}_m H_{r,m})]^n$, $(\Delta P_{r,j})^n$, and $(\Delta H_{r,0})^n$ are now the unknowns to the problem. They must all be small if convergence is to be obtained.

From the definition of the Newton-Raphson algorithm, for each node i

$$\left[\frac{\partial f_i}{\partial \bar{\rho}_m H_{r,m}}\right]^o [\Delta(\bar{\rho}_m H_{r,m})]^n + \sum_{j=2,\dots}^{N} \left(\frac{\partial f_i}{\partial P_{r,j}}\right)^o (\Delta P_{r,j})^n + \left(\frac{\partial f_i}{\partial H_{r,0}}\right)^o (\Delta H_{r,0})^n = -f_i^o \tag{21.30}$$

where $[\partial f_i / \partial(\bar{\rho}_m H_{r,m})]^o$, $(\partial f_i / \partial P_{r,j})^o$, and $(\partial f_i / \partial H_{r,0})^o$ are defined analytically in Appendix C.

The constant dimensionless load is taken into account by

$$\int_{X_{r,\text{min}}}^{X_{r,\text{end}}} (\Delta P_r)^n \, dX_r = \frac{\pi}{2} - \int_{X_{r,\text{min}}}^{X_{r,\text{end}}} P_r^o \, dX_r = (\Delta W')^n \tag{21.31}$$

or

$$\sum_{j=2,\dots}^{N} C_j (\Delta P_{r,j})^n = (\Delta W')^n \tag{21.32}$$

where C_j are weighting factors defined in Appendix D. Since $X_{r,N}$ does not coincide with $X_{r,\text{end}}$, minor corrections are applied on the last values of C_j.

A linear system of $N + 1$ equations is therefore to be solved.

$$\begin{bmatrix} \dfrac{\partial f_1}{\partial \bar{\rho}_m H_{r,m}} & \dfrac{\partial f_1}{\partial P_{r,2}} & \cdots & \dfrac{\partial f_1}{\partial P_{r,N}} & \dfrac{\partial f_1}{\partial H_{r,0}} \\ \dfrac{\partial f_2}{\partial \bar{\rho}_m H_{r,m}} & \dfrac{\partial f_2}{\partial P_{r,2}} & \cdots & \dfrac{\partial f_2}{\partial P_{r,N}} & \dfrac{\partial f_2}{\partial H_{r,0}} \\ \vdots & \vdots & \vdots & \vdots & \vdots \\ \dfrac{\partial f_N}{\partial \bar{\rho}_m H_{r,m}} & \dfrac{\partial f_N}{\partial P_{r,2}} & \cdots & \dfrac{\partial f_N}{\partial P_{r,N}} & \dfrac{\partial f_N}{\partial H_{r,0}} \\ 0 & C_2 & \cdots & C_N & 0 \end{bmatrix}^o \begin{Bmatrix} \Delta(\bar{\rho}_m H_{r,m}) \\ \Delta(P_{r,2}) \\ \vdots \\ \Delta(P_{r,N}) \\ \Delta(H_{r,0}) \end{Bmatrix}^n = \begin{Bmatrix} -f_1 \\ -f_2 \\ \vdots \\ -f_N \\ \Delta W' \end{Bmatrix}^o \tag{21.33}$$

A Gaussian elimination with the partial pivoting method is used to solve this linear system of equations. Iterations on the system approach are still required, as was a case in the direct method used, for example, in Hamrock and Jacobson (1984); however, much less computer time is required.

21.3 FLOW, LOADS AND CENTER OF PRESSURE

The pressure in elastohydrodynamically lubricated conjunctions having been established, it is used here to determine the mass flow rate, the various load components, and the center of pressure. In Chap. 8 these variables were found to be important in hydrodynamic lubrication; they are also important in elastohydrodynamic lubrication.

21.3.1 MASS FLOW RATE PER UNIT WIDTH. The volume flow rate per unit width for hydrodynamic lubrication was described in Eq. (7.38). The mass flow rate used in elastohydrodynamic lubrication is equivalent to the density multiplied by the volume flow rate per unit width or

$$q_m = \rho q' = \tilde{u}\rho h - \frac{\rho h^3}{12\eta}\frac{dp}{dx} \tag{21.34}$$

Making use of Eqs. (21.7) and (21.21) yields

$$Q_m = \frac{\pi q_m}{8\tilde{u}\rho_0 R_x W'} = \bar{\rho} H_r - \frac{1}{\bar{\bar{K}}}\frac{\bar{\rho} H_r^3}{\bar{\eta}}\frac{dP_r}{dX_r} \tag{21.35}$$

Making use of Eq. (21.22) reduces this equation to

$$Q_m = \bar{\rho}_m H_{r,m} \tag{21.36}$$

21.3.2 TANGENTIAL LOAD COMPONENTS. The normal load components are defined in Eq. (21.25); this section defines the tangential load components. The various load components are described in Fig. 21.2. This figure also shows the shear forces, which will be covered in the next section. These force components are acting on the two solids. Conventionally, only the z components of the normal loads per unit width acting on the solids (w'_{az} and w'_{bz}) are considered. However, it is felt that the tangential x components per unit width (w'_{ax} and w'_{bx}), shear forces per unit width (f'_a and f'_b), friction coefficient μ, and center of pressure x_{cp} should also be expressed and quantitative values obtained for each of these expressions. The tangential x component of the load per unit width acting on solid a is zero. The tangential load per unit width for solid b is not zero and can be written as

$$w'_{bx} = -\int p\,dh = -\int p\frac{dh}{dx}\,dx$$

Integrating by parts gives

$$w'_{bx} = -(ph)_{X_{\min}}^{X_{\text{end}}} + \int h\frac{dp}{dx}\,dx$$

However, the pressure at the inlet and the outlet is zero

$$\therefore\ w'_{bx} = \int h\frac{dp}{dx}\,dx \tag{21.37}$$

From Eqs. (21.7) and (21.21) this equation can be expressed in dimensionless terms as

$$W'_{bx} = \frac{w'_{bx}}{E'R_x} = 2\left(\frac{2W'}{\pi}\right)^{3/2}\int H_r\frac{dP_r}{dX_r}\,dX_r \tag{21.38}$$

The resulting normal load components per unit width can be written as

$$W'_a = \frac{w'_a}{E'R_x} = \left[(W'_{ax})^2 + (W'_{az})^2\right]^{1/2} = W'_{az} = W' \tag{21.39}$$

$$W'_b = \frac{w'_b}{E'R_x} = \left[(W'_{bx})^2 + (W'_{bz})^2\right]^{1/2} = \left[(W'_{bx})^2 + (W')^2\right]^{1/2} \tag{21.40}$$

$$\bar{\gamma} = \tan^{-1}\frac{w'_{bx}}{w'_{bz}} = \tan^{-1}\frac{W'_{bx}}{W'} \tag{21.41}$$

21.3.3 SHEAR FORCES. The shear force per unit width acting on the solid as shown in Fig. 21.2 can be written as

$$f'_a = \int(\tau)_{z=0}\,dx = \int\left(\eta\frac{du}{dz}\right)_{z=0}dx \tag{21.42}$$

From Eq. (7.32)

$$\eta\frac{du}{dz} = \frac{2z-h}{2}\frac{dp}{dx} - \frac{\eta(u_a-u_b)}{h}$$

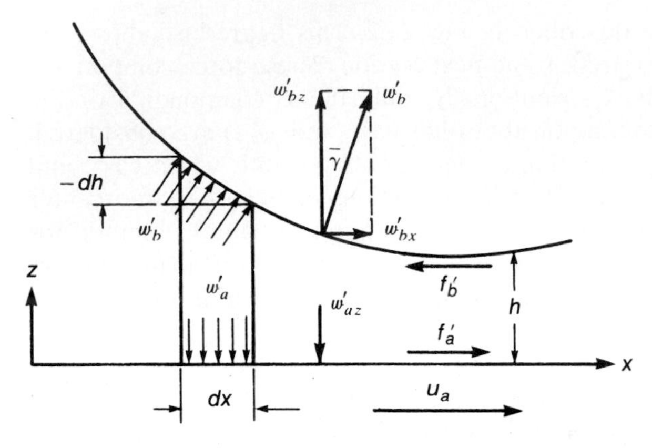

FIGURE 21.2
Load components and shear forces. [*From Hamrock and Jacobson (1984).*]

Substituting this into Eq. (21.42) gives

$$f_a' = -\int\left[\frac{h}{2}\frac{dp}{dx} + \frac{\eta(u_a - u_b)}{h}\right]dx$$

Making use of Eq. (21.33) and using Eqs. (21.7) and (21.21) to write the preceding equation in dimensionless form gives

$$F_a' = \frac{f_a'}{E'R_x} = -\frac{W_{bx}'}{2} - \left(\frac{u_a - u_b}{u_a + u_b}\right)U\left(\frac{\pi}{2W'}\right)^{1/2}\int\frac{\bar{\eta}}{H_r}\,dX_r \tag{21.43}$$

The shear force per unit length acting on solid b can be written as

$$f_b' = \int(\tau)_{z=h}\,dx = \int\left(\eta\frac{du}{dz}\right)_{z=h}dx$$

Making use of Eq. (21.33) yields the following:

$$F_b' = \frac{f_b'}{E'R_x} = \frac{W_{bx}'}{2} - \left(\frac{u_a - u_b}{u_a + u_b}\right)U\left(\frac{\pi}{2W'}\right)^{1/2}\int\frac{\bar{\eta}}{H_r}\,dX_r \tag{21.44}$$

Noe that for equilibrium to be satisfied, the following must be true (see Fig. 21.2):

$$F_a' - F_b' + W_{bx}' = 0 \tag{21.45}$$

and

$$W_{az}' - W_{bz}' = 0 \tag{21.46}$$

The coefficient of rolling friction is written as

$$\mu = -\frac{F_a'}{W'} = \frac{-F_b' + W_{bx}'}{W'} \tag{21.47}$$

21.3.4 CENTER OF PRESSURE. A useful calculation in traction studies is the location of the center of pressure. The appropriate equation is

$$x_{\mathrm{cp}} = \frac{1}{w'_z} \int px \, dx \tag{21.48}$$

From Eqs. (21.7) and (21.21) this equation in dimensionless form is

$$X_{r,\mathrm{cp}} = \frac{x_{\mathrm{cp}}}{D_x} = \frac{1}{\pi} \int P_r X_r \, dX_r \tag{21.49}$$

The location of the center of pressure indicates the position on which the resulting force is acting. The fact that the resulting force is not acting through the center of the roller creates a rolling resistance in the form of a moment. This has a significant effect on the evaluation of the resulting forces and power loss in traction devices and other machine elements.

21.4 PRESSURE SPIKE RESULTS

Because the steep pressure gradients in the outlet region of an elastohydrodynamically lubricated conjunction would otherwise increase the flow rate and clearly lead to flow continuity problems, a gap closing and an abrupt rise in pressure must occur near the outlet. This abrupt rise in pressure is defined as a "pressure spike."

Figure 21.3 gives the pressure profiles and film shapes at iterations 0, 1, and 14 as obtained from Houpert and Hamrock (1986). From the Hertzian pressure at iteration 0 the approach converges quickly. In the first iteration a pressure spike is formed that is close to the final converged pressure spike for iteration 14. The operating parameters were held fixed at $W' = 2.045 \times 10^{-5}$, $U = 1.0 \times 10^{-11}$, and $G = 5007$.

In an attempt to understand why the pressure spike exists, Hamrock et al. (1988) explored pressure spike conditions for isoviscous and viscous solutions and incompressible and compressible fluids under a wide range of loads. It was anticipated that understanding these conditions in the region of the pressure spike might lead to a better understanding of what causes the spike. The highlights of this work are given here.

The system approach developed by Houpert and Hamrock (1986) was used for the numerical evaluation and is discussed in Sec. 21.2. The operating parameters were fixed at

$$\begin{aligned} U &= \text{dimensionless speed parameter} = \frac{\eta_0 \tilde{u}}{E' R_x} = 1.0 \times 10^{-11} \\ W' &= \text{dimensionless load parameter} = \frac{w'_z}{E' R_x} = 1.3 \times 10^{-4} \\ G &= \text{dimensionless materials parameter} = \xi E' = 5007 \end{aligned} \tag{21.50}$$

These operating parameters were held fixed at these values unless otherwise stated.

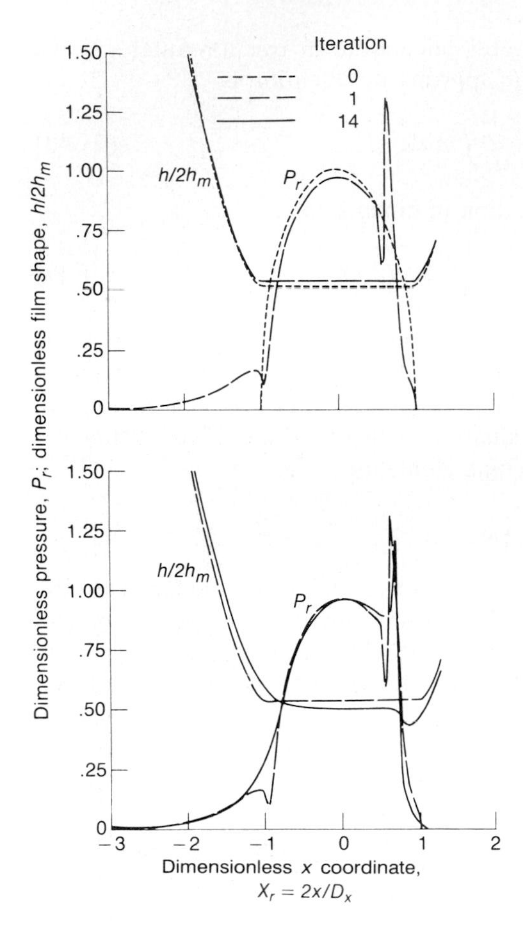

FIGURE 21.3
Pressure profiles and film shapes at iterations 0, 1, and 14 with dimensionless speed, load, and materials parameters fixed at $U = 1.0 \times 10^{-11}$, $W' = 2.045 \times 10^{-5}$, and $G = 5007$ [*From Houpert and Hamrock (1986).*]

21.4.1 ISOVISCOUS AND VISCOUS RESULTS. Figure 21.4 shows the dimensionless pressure distribution for isoviscous and viscous solutions with the operating parameters held fixed as shown in Eq. (21.50). Compressibility effects were considered. Equation (21.23) was solved numerically for the viscous case. For the isoviscous solution the only change was to replace the dimensionless viscosity $\bar{\eta}$ by 1. For this and later figures, unless otherwise stated, 400 nonuniform nodes were used. The bulk of the nodes occurred around $X_r = -1.0$ and the pressure spike. In the region of X_r between 0.9 and 1.0, 230 nodes were used. The inlet region began at $X_r = -2.0$.

Figure 21.4 shows that a pressure spike did not exist for the isoviscous solution but did exist for the viscous solution. The dimensionless viscosity $\bar{\eta}$ varied from 1 in the inlet ($X_r = -2$) and in the outlet ($X_r = 1.08$) to 0.87×10^7 at the contact center ($X_r = 0$). The pressure profiles were essentially the same except for a slight difference in the inlet, the occurrence of the pressure spike, and a slight difference in the outlet.

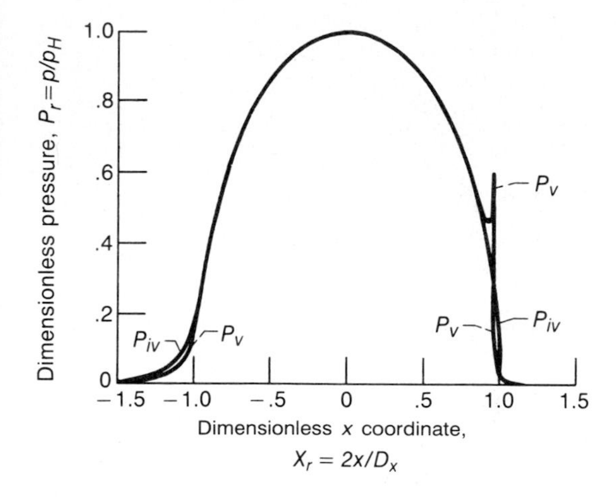

FIGURE 21.4
Dimensionless pressure profiles for isoviscous and viscous solutions. Compressibility effects were considered. [*From Hamrock et al. (1988).*]

Figure 21.5, for the same conditions as Fig. 21.4, shows the film shape for the viscous and isoviscous solutions. The film thickness in Fig. 21.5 is a dimensional value given in meters. Note that the viscous film thickness was more than three times the isoviscous film thickness in the contact region. Note also that the gap closing near the outlet of the contact region was much smaller for the isoviscous solution.

21.4.2 DETAILS OF PRESSURE SPIKE AND FILM SHAPE. Figure 21.6 shows details of the pressure spike and the minimum film thickness by considering only the range of $0.9 \leq X_r \leq 1.0$ from the results presented in Figs. 21.4 and 21.5. In this interval 230 nonuniform nodes were used to obtain the smooth profiles shown in these figures. The pressure spike [Fig. 21.6(a)] occurred at $X_r = 0.943$, but the minimum film thickness [Fig. 21.6(b)] occurred at $X_r = 0.973$. The shift in the location of the minimum film thickness relative to the location

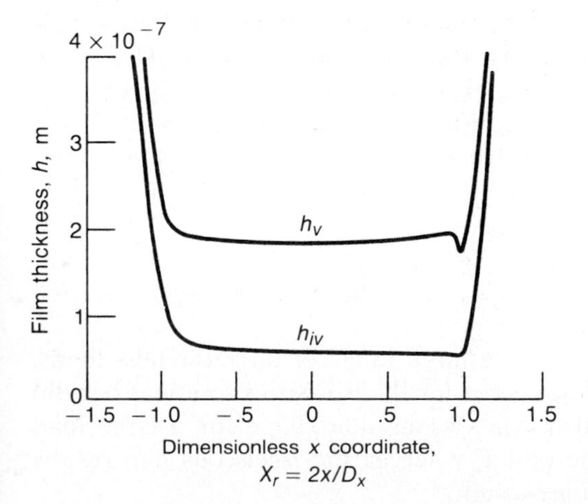

FIGURE 21.5
Film thickness profiles for isoviscous and viscous solutions. Compressibility effects were considered. [*From Hamrock et al. (1988).*]

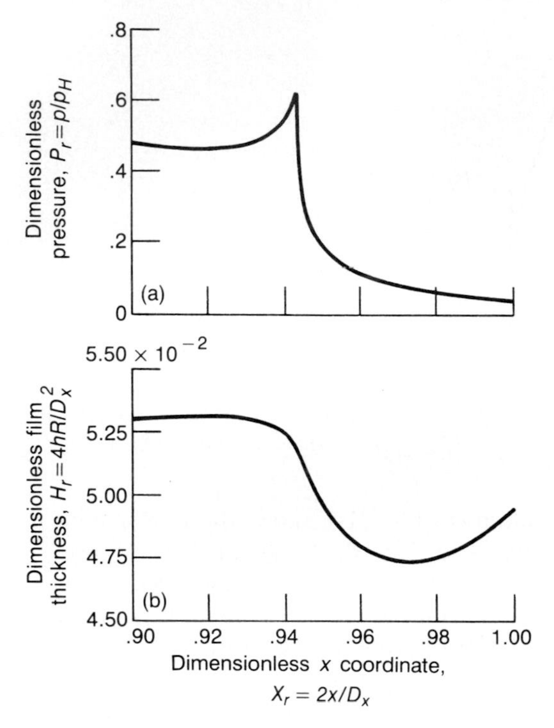

FIGURE 21.6
Pressure and film thickness profiles in region $0.9 \le X_r \le 1.0$. (a) Dimensionless pressure; (b) dimensionless film thickness [*From Hamrock et al. (1988).*]

of the pressure spike is a normal occurrence in elastohydrodynamic lubrication. Because the steep pressure gradients in the spike would otherwise increase the flow rate and clearly lead to flow continuity problems, the gap must be closed near the outlet end of the contact as shown in Fig. 21.6(b). The smoothness of the pressure spike in Fig. 21.6(a) indicates a sufficient number of nodes in the spike region.

Figure 21.7 shows the variation of the dimensionless pressure gradient dP_r/dX_r in the region $0.9 \le X_r \le 1.0$. The pressure gradient changed drastically at the spike location. Figure 21.7(b) shows the variation of $H_r - \bar{\rho}_m H_{r,m}/\bar{\rho}$ in the region $0.9 \le X_r \le 1.0$. Recall that the expression $H_r - \bar{\rho}_m H_{r,m}/\bar{\rho}$ occurs in the integrated form of the Reynolds equation given in Eq. (21.23) as does the pressure gradient plotted in Fig. 21.7(a). The value of $H_r - \bar{\rho}_m H_{r,m}/\bar{\rho}$ was near zero until the pressure spike and then varied considerably afterward. In relating dP_r/dX_r and $H_r - \bar{\rho}_m H_{r,m}/\bar{\rho}$ as they appear in Fig. 21.7 with Eq. (21.23), it should be pointed out that $\bar{\eta}$ varied from 1 in the inlet ($X_r = -2$) to 0.87×10^7 at the contact center ($X_r = 0$) to 0.55×10^5 at the pressure spike location ($X_r = 0.943$) to 1 at the outlet ($X_r = 1.08$).

Similar conclusions were drawn for a large range of dimensionless loads, namely, that the pressure spike did not occur for the isoviscous solutions but did occur for the viscous solutions. Also, the viscous film shape for a large load range shows a gap closing near the outlet, whereas the isoviscous film results had either a slight gap closing or none at all.

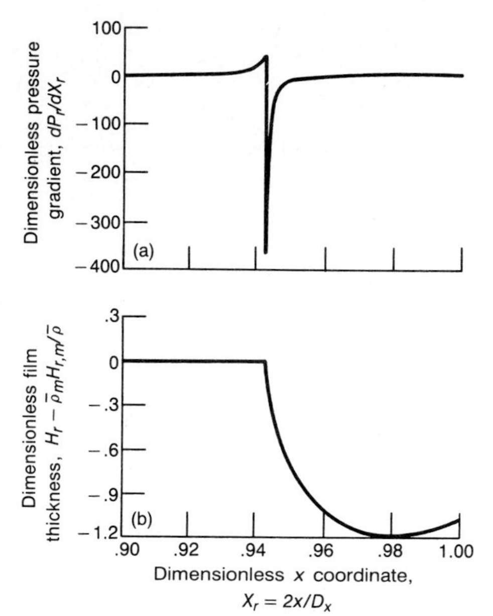

FIGURE 21.7
Pressure gradient and $H_r - \bar{\rho}_m H_{r,m}/\bar{\rho}$ profiles in region $0.9 \leq X_r \leq 1.0$. [*From Hamrock et al., (1988).*]

21.4.3 COMPRESSIBLE AND INCOMPRESSIBLE RESULTS. Figure 21.8 shows the dimensionless pressure and film profiles for an incompressible fluid. Viscous effects were considered, and the dimensionless operating parameters were fixed at the values given in Eq. (21.50). Figure 21.9 considers the same conditions as Fig. 21.8 but for a compressible fluid. The only difference in the analyses for the compressible and incompressible fluids is that for the compressible fluid $\bar{\rho}$ and $\bar{\rho}_m$ were set equal to 1 in Eq. (21.23). Also, the data given in Figs. 21.4 and 21.5 for the viscous solution are the same as those in Fig. 21.9 except that they were plotted differently to more readily compare them with the results given in Fig. 21.8.

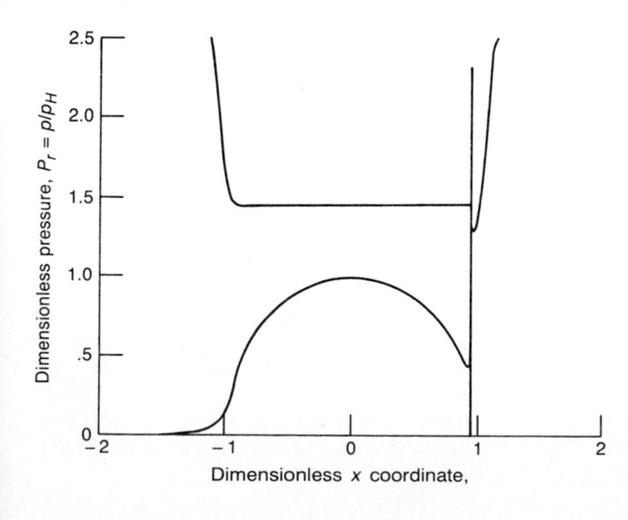

FIGURE 21.8
Dimensionless pressure and film thickness profiles for an incompressible fluid. Viscous effects were considered. [*From Hamrock et al. (1988).*]

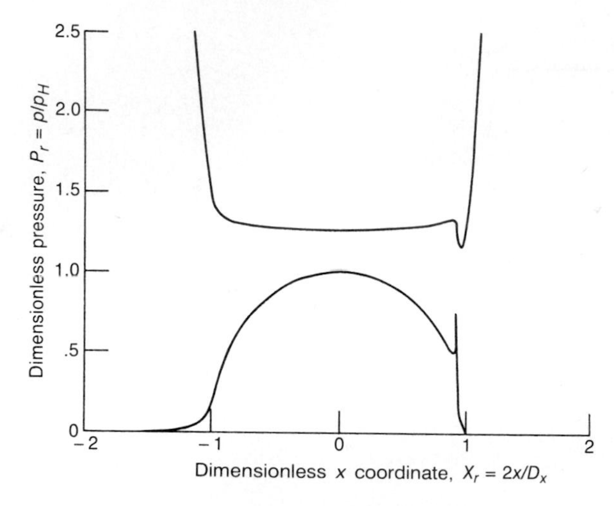

FIGURE 21.9
Dimensionless pressure and film thickness profiles for a compressible fluid. Viscous effects were considered. [*From Hamrock et al. (1988).*]

From Figs. 21.8 and 21.9 the following observations can be made: pressure spike amplitude for the incompressible fluid was 3.7 times that for the compressible fluid, and the shape of the film in the contact region was quite flat for the incompressible fluid but more rounded for the compressible fluid.

Figure 21.10 shows profiles of the dimensionless film thickness and the term $\bar{\rho}_m H_{r,m}/\bar{\rho}$ for the compressible and incompressible fluids. The range of the abscissa is limited to $0.9 \le X_r \le 1.0$, the region were the pressure spike and minimum film thickness occurred. Viscous effects were considered for both the compressible and incompressible fluids. The dimensionless operating parameters were the same as those found in Eq. (21.50).

Figure 21.10 shows that at the spike locations the film profile was sharp for the incompressible fluids but rounded for the compressible fluids. Note also that the values of $\bar{\rho}_m H_{r,m}/\bar{\rho}$ for the compressible and $H_{r,m}$ for the incompressible

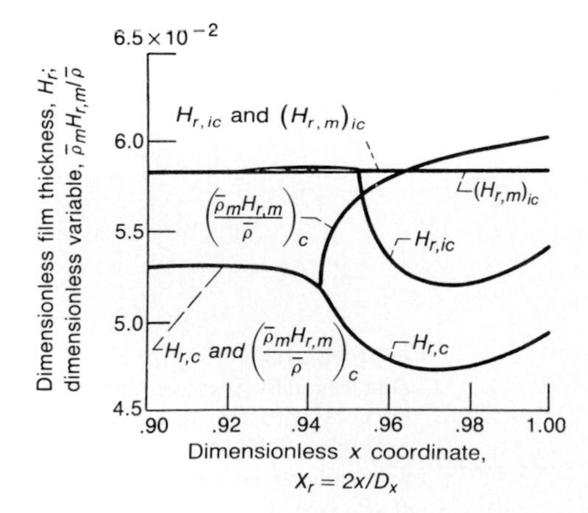

FIGURE 21.10
Dimensionless film thickness and $\bar{\rho}_m H_{r,m}/\bar{\rho}$ profiles for compressible and incompressible fluids in region $0.9 \le X_r \le 1.0$. Viscous effects were considered. [*From Hamrock et al. (1988).*]

fluids were close to the value of H_r for $X_r < X_{r,s}$, where $X_{r,s}$ is the value of X_r at the spike location. For $X_r > X_{r,s}$ the values of H_r and $\bar{\rho}_m H_{r,m}/\bar{\rho}$ and $H_{r,m}$ differ substantially. The term $H_{r,i} - \bar{\rho}_m H_{r,m}/\bar{\rho}_i$ as it appears in Eq. (21.23) is of critical importance. Note that it is multiplied by the dimensionless viscosity $\bar{\eta}_i$, which varies considerably throughout the conjunction.

21.5 USEFUL FORMULAS

Since the capability of accurately determining the amplitude and location of the pressure spike has been established, the focus of this section is to show how these quantities vary with the operating parameters U, W', and G. By quantifying these results, general formulas were developed. Formulas were also developed for the minimum film thickness, the center of pressure, and the mass flow rate, or the value of $\bar{\rho}_m H_{e,m}$, as a function of the operating parameters. These formulas were developed by curve fitting data over a wide range of operating parameters with the dimensionless load W' varying from 0.2045×10^{-4} to 5.0×10^{-4}, the dimensionless speed U varying from 0.1×10^{-11} to 5.0×10^{-11}, and values of the dimensionless materials parameter G of 2504, 5007, and 7511. The results to be presented in this section are those of Pan and Hamrock (1989).

The nondimensionalization used thus far in this chapter is different from that used in this section. The reason is that the formulas to be developed for the performance parameters must be void of the operating parameters. To accomplish this, a revised nondimensionalization was used, where

$$P_e = \frac{p}{E'} \qquad H_e = \frac{h}{R_x} \qquad X_e = \frac{x}{R_x} \tag{21.51}$$

Contrast this to the nondimensionalization used up until this section and given in Eq. (21.7). By making use of the dimensionless equations given in (21.51) the Reynolds, film shape, viscosity, and density equations were appropriately changed.

21.5.1 PRESSURE SPIKE AMPLITUDE.

The presence of a pressure spike in the outlet region of an elastohydrodynamically lubricated conjunction produces large shear stresses that are localized close to the surface. Houpert and Hamrock (1986) found that the pressure spike may halve the life of a nonconformal conjunction. For different dimensionless load, speed, and materials parameters the amplitude of the pressure spike is quite different.

21.5.1.1 Load effects.

Figure 21.11 shows the variation of dimensionless pressure in an elastohydrodynamically lubricated conjunction for six values of dimensionless load while the dimensionless speed and materials parameters are held fixed at $U = 1.0 \times 10^{-11}$ and $G = 5007$. The inlet of the conjunction is to

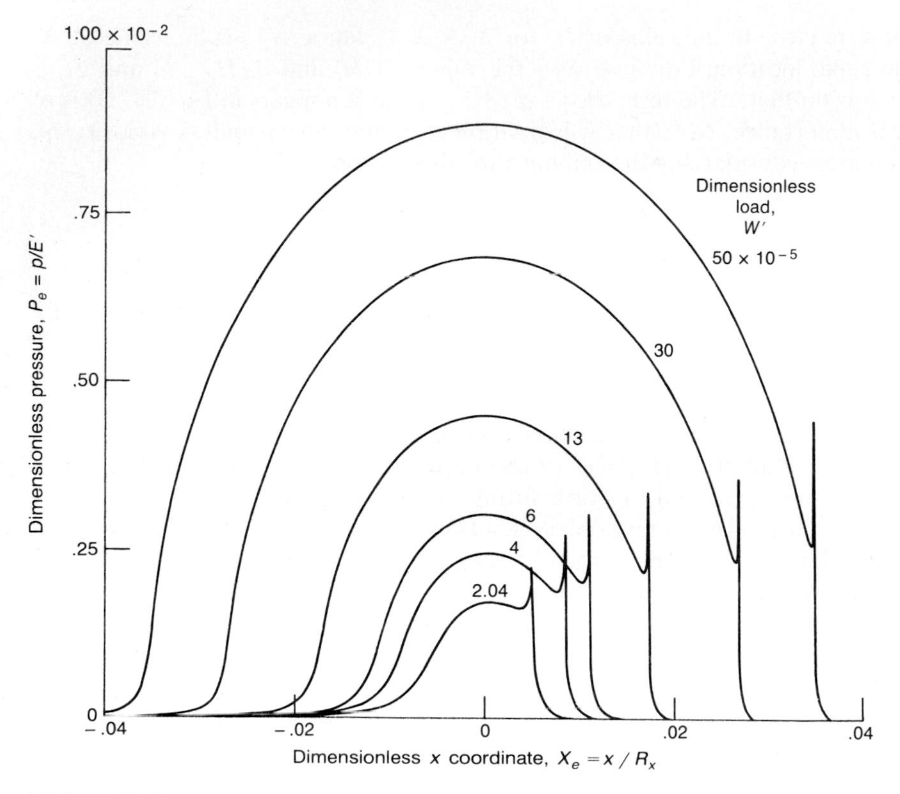

FIGURE 21.11
Variation of dimensionless pressure in elastohydrodynamically lubricated conjunction for six dimensionless loads with dimensionless speed and materials parameters held fixed at $U = 1.0 \times 10^{-11}$ and $G = 5007$. [*From Pan and Hamrock (1989).*]

the left and the outlet to the right, with the contact center occurring at $X_e = 0$. As the dimensionless load increased, the pressure spike amplitude increased.

Table 21.1 shows the effect of six values of the dimensionless load on the spike amplitude $P_{e,s}$ with the dimensionless speed and materials parameters held fixed at $U = 1.0 \times 10^{-11}$ and $G = 5007$. These values were obtained by numerically coupling the Reynolds equation with the film shape equation and the rheology equations. This is a fairly complete range of the dimensionless loads normally encountered in the physical situations existing in elastohydrodynamic lubrication. Curve fitting the pressure spike amplitude as a function of dimensionless load revealed that

$$\tilde{P}_{e,s} = \frac{\tilde{p}_{\text{sk}}}{E'} \propto (W')^{0.185} \tag{21.52}$$

The error as shown in Table 21.1 was between −5.56 and 7.88 percent.

TABLE 21.1
Effect of dimensionless load, speed, and materials parameters on dimensionless pressure spike amplitude
[From Pan and Hamrock (1989)]

Dimensionless load parameter, $W' = \frac{w'_z}{E'R_x}$	Dimensionless speed parameter, $U = \frac{\eta_0\tilde{u}}{E'R_x}$	Dimensionless materials parameter, $G = \xi E'$	Dimensionless pressure spike amplitude, $P_{e,s} = \frac{p_{sk}}{E'}$	Curve-fit dimensionless pressure spike amplitude, $\tilde{P}_{e,s} = \frac{\tilde{p}_{sk}}{E'}$	Error, $\left(\frac{P_{e,s} - \tilde{P}_{e,s}}{P_{e,s}}\right)100$, percent	Results
0.2045×10^{-4}	1.0×10^{-11}	5007	0.22781×10^{-2}	0.23218×10^{-2}	−1.9161	Load
.4	↓	↓	.27653	.26285	4.9473	
.6			.30755	.28332	7.8771	
1.3			.33890	.32689	3.5431	
3.0			.36150	.38159	−5.5564	
5.0			.44791	.41941	6.3637	
1.3×10^{-4}	0.1×10^{-11}	5007	0.17489×10^{-2}	0.17354×10^{-2}	0.7707	Speed
↓	.25	↓	.20763	.22327	−7.5346	
	.5		.24706	.27016	−9.3501	
	.75		.28725	.30203	−5.1445	
	1.0		.33890	.32689	3.5431	
	3.0		.42776	.44219	−3.3744	
	5.0		.49218	.50889	−3.3946	
2.6×10^{-4}	2.0×10^{-11}	2504	0.33549×10^{-2}	0.34291×10^{-2}	−2.2107	Materials
1.3	1.0	5007	.33890	.32689	3.5431	
.8667	.6667	7511	.30760	.31788	−3.3421	

21.5.1.2 Speed effects. Figure 21.12 shows the variation of dimensionless pressure for three values of dimensionless speed while the dimensionless load and materials parameters were held fixed at $W' = 1.3 \times 10^{-4}$ and $G = 5007$. As the dimensionless speed increased, the amplitude of the dimensionless pressure spike increased significantly.

Table 21.1 shows the effect of seven values of dimensionless speed on the pressure spike amplitude at a dimensionless load of 1.3×10^{-4} and a dimensionless materials parameter of 5007. For the seven sets of data a curve fit was applied that resulted in

$$\tilde{P}_{e,s} = \frac{\tilde{p}_{sk}}{E'} \propto U^{0.275} \tag{21.53}$$

The error as it related to the speed data was between −9.35 and 3.54 percent.

21.5.1.3 Materials effects. A study of the influence of the dimensionless materials parameter G on the pressure spike amplitude has to be approached with caution, since in practice it is not possible to change the physical properties of the materials, and hence the value of G, without influencing the other

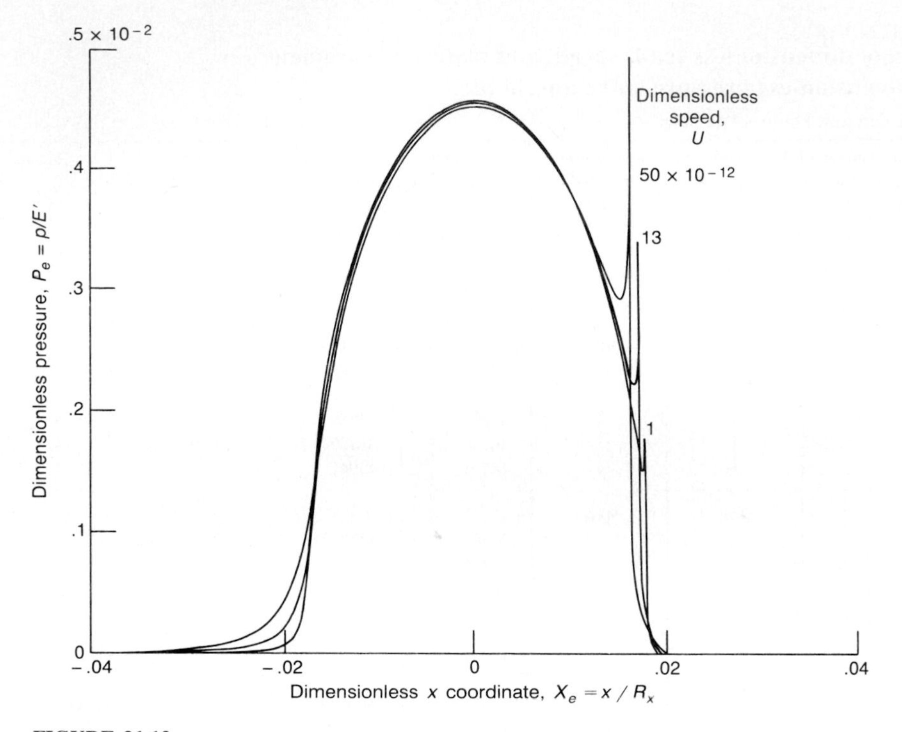

FIGURE 21.12
Variation of dimensionless parameter in elastohydrodynamically lubricated conjunction for three dimensionless speeds with dimensionless load and materials parameters held fixed at $W' = 1.3 \times 10^{-4}$ and $G = 5007$. [*From Pan and Hamrock (1989).*]

operating parameters (U and W'). The results obtained from calculations performed for three values of the dimensionless materials parameter are shown in Table 21.1. The general form of these results, showing how the spike amplitude is a function of the dimensionless materials parameter, is written as

$$\tilde{C} = C_7 G^{C_8} \tag{21.54}$$

where

$$\tilde{C} = \frac{P_{e,s}}{(W')^{0.185} U^{0.275}} \tag{21.55}$$

By applying a least-squares power fit to the three pairs of data

$$\tilde{P}_{e,s} = \frac{\tilde{p}_{\text{sk}}}{E'} \propto G^{0.391} \tag{21.56}$$

The proportionality equations (21.52), (21.53), and (21.56) have established how the pressure spike amplitude varies with the load, speed, and materials parameters, respectively. This enables a composite pressure spike amplitude to be expressed as

$$\tilde{P}_{e,s} = \frac{\tilde{p}_{sk}}{E'} = 0.648(W')^{0.185} U^{0.275} G^{0.391} \tag{21.57}$$

It is evident that the materials parameter has the largest effect on the pressure spike amplitude, followed by speed and load.

21.5.2 PRESSURE SPIKE LOCATION. As shown in Fig. 21.11, as the dimensionless load increased, the dimensionless pressure spike location $X_{e,s}$ moved toward the outlet. These results are quantified in Table 21.2 for six dimensionless loads with the speed and materials parameters held fixed at $U = 1.0 \times 10^{-11}$ and $G = 5007$. From Fig. 21.12, the location of the pressure spike also moved toward the outlet as the speed was decreased. These results are also quantified in Table 21.2. Also shown in Table 21.2 are results for three materials parameters. Making use of the data in Table 21.2 by following the procedure used in the previous section allows an equation for the dimensionless pressure spike

TABLE 21.2
Effect of dimensionless load, speed, and materials parameters on dimensionless pressure spike location
[From Pan and Hamrock (1989)]

Dimensionless load parameter, $W' = \frac{w'_z}{E'R_x}$	Dimensionless speed parameter, $U = \frac{\eta_0 \tilde{u}}{E'R_x}$	Dimensionless materials parameter, $G = \xi E'$	Dimensionless pressure spike location, $X_{e,s} = \frac{x_s}{R_x}$	Curve-fit dimensionless pressure spike location, $\tilde{X}_{e,s} = \frac{\tilde{x}_s}{R_x}$	Error, $\left(\frac{X_{e,s} - \tilde{X}_{e,s}}{X_{e,s}}\right) 100$, percent	Results
0.2045×10^{-4}	1.0×10^{-11}	5007	0.48561×10^{-2}	0.52467×10^{-2}	−8.0429	Load
.4	↓	↓	.82948	0.78778	5.0254	
.6			1.08220	1.00722	6.9284	
1.3			1.70980	1.60922	5.8827	
3.0			2.68600	2.67116	.5524	
5.0			3.49820	3.64033	−4.0630	
1.3×10^{-4}	0.1×10^{-11}	5007	1.76520×10^{-2}	1.68895×10^{-2}	4.3201	Speed
↓	.25	↓	1.74670	1.65675	5.1495	
	.5		1.73000	1.63281	5.6178	
	.75		1.71620	1.61896	5.6655	
	1.0		1.70980	1.60922	5.8827	
	3.0		1.65310	1.57251	4.8747	
	5.0		1.62860	1.55574	4.4739	
2.6×10^{-4}	2.0×10^{-11}	2504	2.42060×10^{-2}	2.28843×10^{-2}	5.4601	Materials
1.3	1.0	5007	1.70980	1.60922	5.8827	
.8667	.6667	7511	1.38400	1.30969	5.3689	

location to be written in terms of the operating parameters as

$$\tilde{X}_{e,s} = \frac{\tilde{x}_s}{R_x} = 1.111(W')^{0.606}U^{-0.021}G^{0.077} \qquad (21.58)$$

This equation shows the spike location to be highly affected by the dimensionless load but only slightly affected by the dimensionless speed and materials parameters. Knowing the location of the pressure spike in terms of the operating parameters is extremely valuable in setting up a nodal structure that will produce convergent results. The percentage of error (Table 21.2) was between −8.04 and 6.93 percent.

21.5.3 MINIMUM AND CENTRAL FILM THICKNESSES. Figure 21.13 shows the effect of dimensionless load on film shape for exactly the same conditions as presented in Fig. 21.11. The dimensionless speed and materials parameters were held fixed at $U = 1.0 \times 10^{-11}$ and $G = 5007$. Figure 21.13 shows that the minimum film thickness decreased as the load was increased except for the

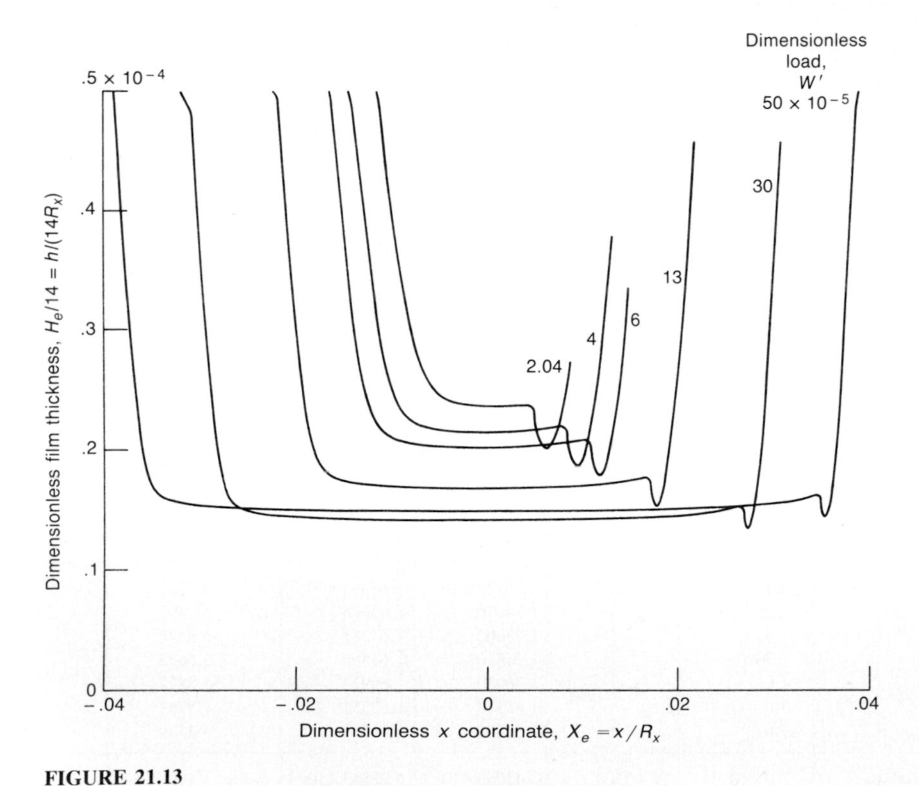

FIGURE 21.13
Variation of dimensionless film shape in elastohydrodynamically lubricated conjunction for six dimensionless loads with dimensionless speed and materials parameters held fixed at $U = 1.0 \times 10^{-1}$ and $G = 5007$. [*From Pan and Hamrock (1989).*]

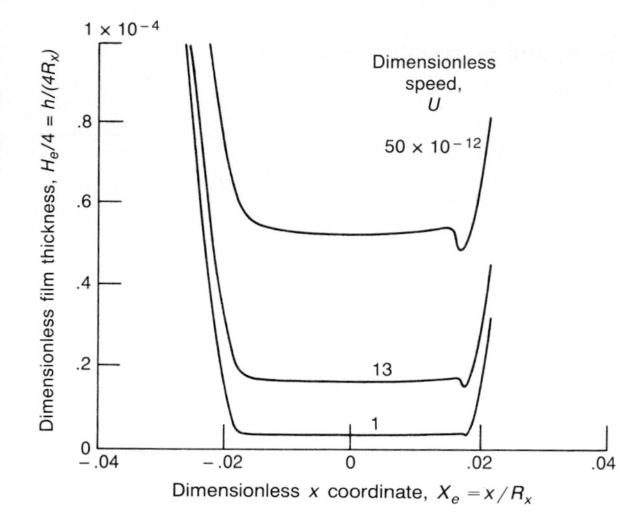

FIGURE 21.14
Variation of dimensionless film shape for three dimensionless speeds with dimensionless load and materials parameters fixed at $W' = 1.3 \times 10^{-4}$ and $G = 5007$. [*From Pan and Hamrock (1989).*]

highest load. The discrepancy of the highest-load case from the other results is not understood. As shown in Fig. 21.13 the location of the minimum film thickness moved considerably toward the outlet as the dimensionless load was increased.

Figure 21.14 shows the variation of dimensionless film shape for three dimensionless speed parameters with the dimensionless load and materials parameters held fixed at $W' = 1.3 \times 10^{-4}$ and $G - 5007$. The minimum film thickness decreased significantly as the dimensionless speed decreased.

Table 21.3 shows the effect of the dimensionless load, speed, and materials parameters on dimensionless minimum film thickness. The formula developed from curving fitting the 16 groups of data is

$$\tilde{H}_{e,\min} = \frac{\tilde{h}_{\min}}{R_x} = 1.714(W')^{-0.128}U^{0.694}G^{0.568} \tag{21.59}$$

This curve-fit formula fits the numerical data within −7.72 and 6.69 percent.

By making use of Eqs. (21.10), (21.11), and (21.14), Eq. (21.59) can be expressed in dimensional form as

$$\tilde{h}_{\min} = 1.714(E')^{0.002}(w_z')^{-0.128}(h_0\tilde{u})^{0.694}\xi^{0.568}R_x^{0.434} \tag{21.60}$$

Three things should be noted from Eq. (21.60):

1. The minimum film thickness is virtually independent of the effective modulus of elasticity.
2. The minimum film thickness is only slightly dependent on the normal applied load.
3. The minimum film thickness is most affected by the speed and the viscosity at atmospheric condition.

TABLE 21.3
Effect of dimensionless load, speed, and materials parameters on dimensionless minimum film thickness
[From Pan and Hamrock (1989)]

Dimensionless load parameter, $W' = \frac{w'_z}{E'R_x}$	Dimensionless speed parameter, $U = \frac{\eta_0 \tilde{u}}{E'R_x}$	Dimensionless materials parameter, $G = \xi E'$	Dimensionless minimum film thickness, $H_{e,\min} = \frac{h_{\min}}{R_x}$	Curve-fit dimensionless minimum film thickness, $\tilde{H}_{e,\min} = \frac{\tilde{h}_{\min}}{R_x}$	Error, $\left(\frac{H_{e,\min} - \tilde{H}_{e,\min}}{H_{e,\min}}\right) 100$, percent	Res
0.20453×10^{-4}	1.0×10^{-11}	5007	0.19894×10^{-4}	0.20030×10^{-4}	−0.6840	Load
.4	↓	↓	.18404	.18382	.1189	
.6			.17558	.17452	.6012	
1.3			.15093	.15808	−4.7367	
3.0			.13185	.14203	−7.7228	
5.0			.14067	.13304	5.4221	
1.3×10^{-4}	0.1×10^{-11}	5007	0.03152×10^{-4}	0.03198×10^{-4}	−1.4585	Spee
↓	.25	↓	.06350	.06040	4.8801	
	.5		.10472	.09771	6.6896	
	.75		.13791	.12947	6.1204	
	1.0		.15093	.15808	−4.7367	
	3.0		.34963	.33884	3.0857	
	5.0		.48807	.48301	1.0359	
2.6×10^{-4}	2.0×10^{-11}	2504	0.16282×10^{-4}	0.15786×10^{-4}	3.0462	Mate
1.3	1.0	5007	.15093	.15808	−4.7367	
.8667	.6667	7511	.16573	.15821	4.5362	

If in a particular application the minimum film thickness is found to be too small relative to the composite surface roughness, the recognition of the influence of the various dimensional operating parameters on minimum film thickness is important.

In practical situations there is considerable interest in the central as well as the minimum film thickness in an elastohydrodynamically lubricated conjunction. This is particularly true when traction is considered, since the surfaces in relative motion are separated by a film of almost constant thickness that is well represented by the central value over much of the Hertzian contact zone. Table 21.4 shows the effect of the dimensionless load, speed, and materials parameters on dimensionless central film thickness. The formula developed from curve fitting the data is

$$\tilde{H}_{e,c} = \frac{\tilde{h}_c}{R_x} = 2.922(W')^{-0.166} U^{0.692} G^{0.470} \tag{21.61}$$

The percentage of error was between −7.46 and 6.49 percent.

21.5.4 LOCATION OF MINIMUM FILM THICKNESS. Figure 21.13 shows the effect of dimensionless load on the location of the minimum film thickness for

TABLE 21.4
Effect of dimensionless load, speed, and materials parameters on dimensionless central film thickness
[From Pan and Hamrock (1989)]

Dimensionless load parameter, $W' = \frac{w_z'}{E'R_x}$	Dimensionless speed parameter, $U = \frac{\eta_0 \tilde{u}}{E'R_x}$	Dimensionless materials parameter, $G = \xi E'$	Dimensionless central film thickness, $H_{e,c} = \frac{h_c}{R_x}$	Curve-fit dimensionless central film thickness, $\tilde{H}_{e,c} = \frac{\tilde{h}_c}{R_x}$	Error, $\left(\frac{H_{e,c} - \tilde{H}_{e,c}}{H_{e,c}}\right)100$, percent	Results
0.2045×10^{-4}	1.0×10^{-11}	5007	0.23436×10^{-4}	0.23490×10^{-4}	−0.2318	Load
.4	↓	↓	.21242	.21015	1.0679	
.6	↓	↓	.20003	.19647	1.7785	
1.3	↓	↓	.16607	.17281	−4.0564	
3.0	↓	↓	.13997	.15041	−7.4577	
5.0	↓	↓	.14777	.13818	6.4897	
1.3×10^{-4}	0.1×10^{-11}	5007	0.03508×10^{-4}	0.03512×10^{-4}	−0.1153	Speed
↓	.25	↓	.06796	.06621	2.5726	
↓	.5	↓	.11135	.10697	3.9368	
↓	.75	↓	.14810	.14161	4.3803	
↓	1.0	↓	.16607	.17281	−4.0564	
↓	3.0	↓	.38079	.36960	2.9395	
↓	5.0	↓	.53012	.52632	.7174	
2.6×10^{-4}	2.0×10^{-11}	2504	0.18627×10^{-4}	0.17965×10^{-4}	3.5561	Materials
1.3	1.0	5007	.16607	.17281	−4.0564	
.8667	.6667	7511	.17767	.16893	4.9183	

exactly the same conditions as presented in Fig. 21.11. The location of the minimum film thickness moved toward the outlet as the load was increased. Figure 21.14 shows the effect of dimensionless speed on the location of the minimum film thickness. The location of the minimum film thickness did not change much as the speed was increased.

Table 21.5 shows the effect of the dimensionless load, speed, and materials parameters on the dimensionless location of the minimum film thickness. The formula developed from curve fitting the data is

$$\tilde{X}_{e,\min} = \frac{\tilde{x}_{\min}}{R_x} = 1.439(W')^{0.548}U^{-0.011}G^{0.026} \tag{21.62}$$

The curve-fit formula fits the numerically obtained data with an error between −2.64 and 2.79 percent.

21.5.5 CENTER OF PRESSURE. The dimensionless center of pressure is expressed in Eq. (21.49). Table 21.6 shows how the dimensionless load, speed, and materials parameters affect the dimensionless center of pressure. A least-squares

TABLE 21.5
Effect of dimensionless load, speed, and materials parameters on dimensionless location of minimum film thickness
[From Pan and Hamrock (1989)]

Dimensionless load parameter, $W' = \frac{w'_z}{E'R_x}$	Dimensionless speed parameter, $U = \frac{\eta_0 \tilde{u}}{E'R_x}$	Dimensionless materials parameter, $G = \xi E'$	Dimensionless location of minimum film thickness, $X_{e,\min} = \frac{x_{\min}}{R_x}$	Curve-fit dimensionless location of minimum film thickness, $\tilde{X}_{e,\min} = \frac{\tilde{x}_{\min}}{R_x}$	Error, $\left(\frac{X_{e,\min} - \tilde{X}_{e,\min}}{X_{e,\min}}\right) 100$, percent	Re
0.2045×10^{-4}	1.0×10^{-11}	5007	0.60900×10^{-2}	0.62598×10^{-2}	2.7883	Load
.4			.92850	.90398	−2.6410	
.6			1.11681	1.12884	1.0774	
1.3			1.76850	1.72427	−2.5011	
3.0			2.72800	2.72634	−.0609	
5.0			3.53610	3.60682	2.0000	
1.3×10^{-4}	0.1×10^{-11}	5007	1.79940×10^{-2}	1.76874×10^{-2}	−1.7038	Spee
	.25		1.78670	1.75091	−2.0032	
	.5		1.77760	1.73754	−2.2538	
	.75		1.77030	1.72976	−2.2898	
	1.0		1.76850	1.72427	−2.5011	
	3.0		1.73760	1.70344	−1.9657	
	5.0		1.72300	1.69385	−1.6920	
2.6×10^{-4}	2.0×10^{-11}	2504	2.5139×10^{-2}	2.45687×10^{-2}	−2.2686	Mate
1.3	1.0	5007	1.76850	1.72427	−2.5011	
.8667	.6667	7511	1.43360	1.40172	−2.2241	

fit of the 16 data sets resulted in the following formula:

$$\tilde{X}_{e,\,\mathrm{cp}} = \frac{\tilde{x}_{\mathrm{cp}}}{R_x} = -3.595(W')^{-1.019} U^{0.638} G^{-0.358} \tag{21.63}$$

The percentage of error was between −11.91 and 9.62 percent.

21.5.6 MASS FLOW RATE. From Eq. (21.36) the dimensionless mass flow rate Q_m is equal to $\bar{\rho}_m H_{r,m}$. Recall from the statement of the elastohydrodynamic rectangular conjunction problem that $\bar{\rho}_m H_{r,m}$ is an unknown and will need to be guessed at initially. Therefore, a formula that would provide an initial guess for given operating parameters (W', U, and G) would help in the convergence of the problem.

Table 21.7 shows the effect of a wide range of operating parameters (W', U, and G) on the dimensionless mass flow rate. A least-squares fit to the data enabled the following equation to be obtained:

$$\tilde{Q}_m = \frac{\tilde{\bar{\rho}}_m \tilde{h}_m}{\rho_0 R_x} = 2.698(W')^{-0.131} U^{0.692} G^{0.539} \tag{21.64}$$

TABLE 21.6
Effect of dimensionless load, speed, and materials parameters on dimensionless center of pressure [From Pan and Hamrock (1989)]

Dimensionless load parameter, $W' = \frac{w'_z}{E'R_x}$	Dimensionless speed parameter, $U = \frac{\eta_0 \tilde{u}}{E'R_x}$	Dimensionless materials parameter, $G = \xi E'$	Dimensionless center of pressure, $X_{e,cp} = \frac{x_{cp}}{R_x}$	Curve-fit dimensionless center of pressure, $\tilde{X}_{e,cp} = \frac{\tilde{x}_{cp}}{R_x}$	Error, $\left(\frac{X_{e,cp} - \tilde{X}_{e,cp}}{X_{e,cp}}\right)$ 100, percent	Results
0.20453×10^{-4}	1.0×10^{-11}	5007	-1.00670×10^{-5}	-1.02309×10^{-5}	−1.6284	Load
.4			−.57148	−.51651	9.6191	
.6			−.33977	−.34169	−.5670	
1.3			−.14402	−.15541	−7.9060	
3.0			−.06233	−.06628	−6.3391	
5.0	↓	↓	−.04330	−.03939	9.0425	
1.3×10^{-4}	0.1×10^{-11}	5007	-0.03231×10^{-5}	-0.03577×10^{-5}	−10.6959	Speed
	.25		−.05883	−.06417	−9.0826	
	.5		−.09300	−.09987	−7.3812	
	.75		−.12202	−.12935	−6.0051	
	1.0		−.14402	−.15540	−7.9060	
	3.0		−.29264	−.31324	−7.0380	
↓	5.0	↓	−.38774	−.43392	−11.9105	
2.6×10^{-4}	2.0×10^{-11}	2504	-0.14662×10^{-5}	-0.15295×10^{-5}	−4.3162	Materials
1.3	1.0	5007	−.14402	−.15541	−7.9060	
.8667	.6667	7511	−.15127	−.15686	−3.6961	

TABLE 21.7
Effect of dimensionless load, speed, and materials parameters on dimensionless mass flow rate [From Pan and Hamrock (1989)]

Dimensionless load parameter, $W' = \frac{w'_z}{E'R_x}$	Dimensionless speed parameter, $U = \frac{\eta_0 \tilde{u}}{E'R_x}$	Dimensionless materials parameter, $G = \xi E'$	$\bar{\rho}_m H_m$	Curve-fit $\tilde{\bar{\rho}}_m \tilde{H}_m$	Error, $\left(\frac{\bar{\rho}_m H_m - \tilde{\bar{\rho}}_m \tilde{H}_m}{\bar{\rho}_m H_m}\right)$ 100, percent	Results
0.20453×10^{-4}	1.0×10^{-11}	5007	0.26694×10^{-4}	0.26760×10^{-4}	−0.2445	Load
.4			.24844	.24508	1.3511	
.6			.23767	.23241	2.2152	
1.3			.20291	.21002	−3.5031	
3.0			.17557	.18823	−7.2087	
5.0	↓	↓	.18788	.17604	6.3004	
1.3×10^{-4}	0.1×10^{-11}	5007	0.04288×10^{-4}	0.04268×10^{-4}	−0.4590	Speed
	.25		.08350	.08047	3.1072	
	.5		.13607	.13000	4.4608	
	.75		.18096	.17211	4.8921	
	1.0		.20291	.21002	−3.5031	
	3.0		.46515	.44918	3.4323	
↓	5.0	↓	.64744	.63965	1.2028	
2.6×10^{-4}	2.0×10^{-11}	2504	0.22212×10^{-4}	0.21325×10^{-4}	3.9954	Materials
1.3	1.0	5007	.20291	.21002	−3.5031	
.8667	.6667	7511	.21995	.20816	5.3607	

21.6 CLOSURE

Methods of elastohydrodynamically calculating the film thickness and pressure in a rectangular conjunction were presented for both incompressible and compressible fluids. Recent developments have had a profound effect in that elastohydrodynamically lubricated conjunctions can now be numerically evaluated in the load range that is normally experienced by nonconformal contacts such as rolling-element bearings and gears. A new approach in which there are no load limitations was described, and successful results were presented for conditions corresponding to a maximum Hertzian pressure of 4.8 GPa. These results were obtained with little computer time.

The focus of the chapter then turned to getting a better understanding of why pressure spikes occur in elastohydrodynamically lubricated conjunctions. Various combinations of viscous or isoviscous and compressible or incompressible situations were studied for a wide range of load conditions. Comparing the results of the isoviscous and viscous solutions showed that a pressure spike did not occur for the isoviscous solution but did occur for the viscous solution. These results indicated also that the film thickness in the contact region was more than three times greater for the viscous solution than for the isoviscous solution. The viscous film shape results also showed a gap closing near the outlet, whereas the isoviscous film shape results had either none or just a slight gap closing. Similar conclusions were obtained for a number of different loads. Therefore, it was concluded that the pressure spike and gap closing at the outlet were viscosity driven. Studies with compressible and incompressible fluids showed that the amplitude of the pressure spike for the incompressible fluid was 3.7 times larger.

Having explained the details of the pressure and film shape, the chapter turned to studying the influence of the operating parameters on the performance parameters and to developing simple formulas that relate the two. The operating parameters that were studied were the dimensionless load, speed, and materials parameters. The dimensionless load parameter W' was varied from 0.2045×10^{-4} to 5.0×10^{-4}. The dimensionless speed parameter U was varied from 0.1×10^{-11} to 5.0×10^{-11}. And the dimensionless materials parameter G had values of 2504, 5007, and 7511, corresponding to solid materials of bronze, steel, and silicon nitride, respectively. Fourteen cases were investigated, covering a complete range of operating parameters normally experienced in elastohydrodynamic lubrication. Formulas were obtained for the following performance parameters: pressure spike amplitude and location, minimum and central film thicknesses, location of the minimum film thickness, location of the center of pressure, and mass flow rate.

21.7 PROBLEM

21.7.1 Two disks 35 mm wide (b = 35 mm) and 180 and 240 mm in radius (r_a = 180 mm and r_b = 240 mm) are each rotating at 580 rpm in a liquid bath in the directions shown in Fig. 21.7.1. If the disks are subjected to a normal applied load of 2,000

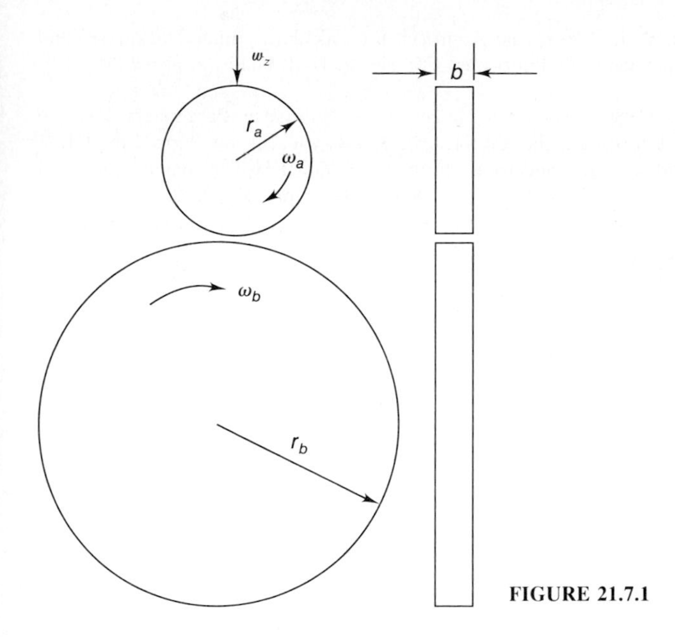

FIGURE 21.7.1

N, determine the dimensions of the contact zone, the maximum normal stress, and the minimum film thickness for

(a) *Water* being the lubricant (viscosity, 0.01 P; ξ, 6.68×10^{-10} m^2/N)

(b) *Oil* being the lubricant (viscosity, 0.36 P; ξ, 2.2×10^{-8} m^2/N)

Assume the disks are made of steel with a modulus of elasticity of 2.1×10^{11} N/m^2 and Poisson's ratio of 0.3.

21.8 REFERENCES

Barus, C. (1893): Isothermals, Isopiestics, and Isometrics Relative to Viscosity. *Am. J. Sci.*, vol. 45, pp. 87–96.

Crook, A. W. (1961): Elasto-Hydrodynamic Lubrication of Rollers. *Nature*, vol. 190, p. 1182.

Dowson, D., and Higginson, G. R. (1959): A Numerical Solution to the Elastohydrodynamic Problem. *J. Mech. Eng. Sci.*, vol. 1, no. 1, pp. 6–15.

Dowson, D., and Higginson, G. R. (1966): *Elastohydrodynamic Lubrication, The Fundamentals of Roller and Gear Lubrication*. Pergamon, Oxford.

Grubin, A. N. (1949): Fundamentals of the Hydrodynamic Theory of Lubrication of Heavily Loaded Cylindrical Surfaces. *Investigation of the Contact Machine Components*. Kh. F. Ketova, (Ed.). Translation of Russian Book No. 30, Central Scientific Institute for Technology and Mechanical Engineering, Moscow, chap. 2. (Available From Department of Scientific and Industrial Research, Great Britain, Trans. CTS–235 and Special Libraries Association, Trans. R–3554).

Hamrock, B. J., and Jacobson, B. O. (1984): Elastohydrodynamic Lubrication of Line Contacts. *ASLE Trans.*, vol. 24, no. 4, pp. 275–287.

Hamrock, B. J., Pan, P., and Lee, R. T. (1988): Pressure Spikes in Elastohydrodynamically Lubricated Conjunctions. *J. Tribol.*, vol. 110, no. 2, pp. 279–284.

Houpert, L. G., and Hamrock, B. J. (1986): Fast Approach for Calculating Film Thicknesses and Pressures in Elastohydrodynamically Lubricated Contacts at High Loads. *J. Tribol.*, vol. 108, no. 3, pp. 411–420.

Pan, P., and Hamrock, B. J. (1989): Simple Formulae for Performance Parameters Used in Elastohydrodynamically Lubricated Line Contacts. *J. Tribol.*, vol. 111, no. 2, pp. 246–251.

Petrusevich, A. I. (1951): Fundamental Conclusions From the Contact-Hydrodynamic Theory of Lubrication. *Izv. Akad. Nauk, SSSR, Otd. Tekh. Nauk.*, vol. 2, pp. 209–233.

Roelands, C. J. A. (1966): Correlational Aspects of the Viscosity-Temperature-Pressure Relationships of Lubrication Oils, Druk. V. R. B., Groningen, Netherlands.

CHAPTER 22

ELASTOHYDRODYNAMIC LUBRICATION OF ELLIPTICAL CONJUNCTIONS

In the 1950s and 1960s—and again in the 1980s, when better numerical approaches were developed—elastohydrodynamic lubrication focused primarily on rectangular conjunctions. However, in the 1970s theoretical EHL studies changed direction dramatically, switching from rectangular to elliptical conjunctions. The change began in 1970 when Cheng developed a Grubin type of inlet analysis applicable to elliptical Hertzian contact areas. Later in the 1970s a numerical solution for the isothermal elastohydrodynamic lubrication of elliptical conjunctions successfully emerged. In the years 1974–79, Hamrock and Dowson published nine papers on elastohydrodynamic lubrication (Hamrock and Dowson, 1974; Dowson and Hamrock, 1976; and Hamrock and Dowson, 1976a, b, 1977a, b, 1978, 1979a, b). Many of the essential theoretical features of elastohydrodynamically lubricated elliptical conjunctions have been shown to be in overall agreement with experiments (e.g., see Koye and Winer, 1980). Hamrock and Dowson's collective efforts appear in book form (Hamrock and Dowson, 1981), where they also apply elastohydrodynamic lubrication to rolling-element bearings. This chapter covers the highlights of the Hamrock and Dowson research.

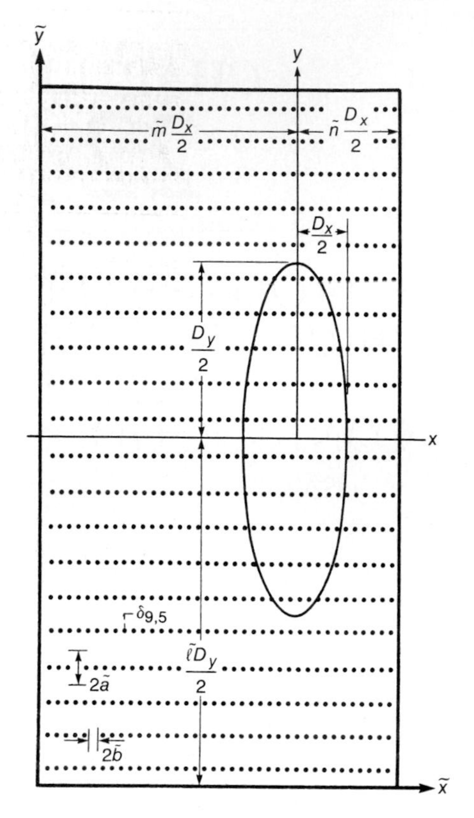

FIGURE 22.1
Example of dividing area in and around contact zone into equal rectangular areas. [*From Hamrock and Dowson (1976a).*]

22.1 RELEVANT EQUATIONS

Figure 22.1 shows the coordinate system to used as well as an example of how to divide the area in and around the contact zone into equal rectangular areas. From Eq. (7.58) the general Reynolds equation can be expressed as

$$\frac{\partial}{\partial x}\left(\frac{\rho h^3}{12\eta}\frac{\partial p}{\partial x}\right)+\frac{\partial}{\partial y}\left(\frac{\rho h^3}{12\eta}\frac{\partial p}{\partial y}\right)$$

$$=\frac{\partial}{\partial x}\left[\frac{\rho h(u_a+u_b)}{2}\right]+\frac{\partial}{\partial y}\left[\frac{\rho h(v_a+v_b)}{2}\right]+\frac{\partial(\rho h)}{\partial t} \tag{7.58}$$

Assuming steady-state operation and that the velocity components are not a function of x or y, this equation reduces to

$$\frac{\partial}{\partial x}\left(\frac{\rho h^3}{\eta}\frac{\partial p}{\partial x}\right)+\frac{\partial}{\partial y}\left(\frac{\rho h^3}{\eta}\frac{\partial p}{\partial y}\right)=12\tilde{u}\frac{\partial(\rho h)}{\partial x}+12\tilde{v}\frac{\partial(\rho h)}{\partial y} \tag{22.1}$$

where $\tilde{u}$ = mean velocity in x direction = $(u_a+u_b)/2$
$\tilde{v}$ = mean velocity in y direction = $(v_a+v_b)/2$

Letting

$$X_r = \frac{2\tilde{x}}{D_x} \qquad Y = \frac{2\tilde{y}}{D_y} \qquad \bar{\rho} = \frac{\rho}{\rho_0} \qquad \bar{\eta} = \frac{\eta}{\eta_0} \qquad H_e = \frac{h}{R_x}$$

$$\frac{1}{R_x} = \frac{1}{r_{ax}} + \frac{1}{r_{bx}} \qquad P_e = \frac{p}{E'} \tag{22.2}$$

$$V = (\tilde{u}^2 + \tilde{v}^2)^{1/2} \qquad \theta = \tan^{-1}\frac{\tilde{v}}{\tilde{u}}$$

causes Eq. (22.1) to become

$$\frac{\partial}{\partial X_r}\left(\frac{\bar{\rho}H_e^3}{\bar{\eta}}\frac{\partial P_e}{\partial X_r}\right) + \frac{1}{k^2}\frac{\partial}{\partial Y}\left(\frac{\bar{\rho}H_e^3}{\bar{\eta}}\frac{\partial P_e}{\partial Y}\right)$$

$$= 12U\left(\frac{D_x}{2R_x}\right)\left[\cos\theta\frac{\partial(\bar{\rho}H_e)}{\partial X_r} + \frac{\sin\theta}{k}\frac{\partial(\rho H_e)}{\partial Y}\right] \tag{22.3}$$

where $$k = \text{ellipicity parameter} = \frac{D_y}{D_x} \approx \alpha_r^{2/\pi} \tag{19.29}$$

and

$$\alpha_r = \frac{R_y}{R_x} \tag{18.57}$$

where $$\frac{1}{R_y} = \frac{1}{r_{ay}} + \frac{1}{r_{by}}$$

$$U_v = \text{dimensionless speed parameter} = \frac{V\eta_0}{E'R_x}$$

Equation (22.3) is a nonhomogeneous partial differential equation in two variables. This is generally a difficult equation to solve. The situation is worsened by the facts, as found for rectangular conjunctions in the last chapter, that the viscosity varies by several orders of magnitude throughout the conjunction and that the elastic deformation of the solid surfaces can be three or four orders of magnitude larger than the minimum film thickness within the conjunction. The severity of these conditions enables solutions to be obtained numerically only when extreme care is taken.

The dimensionless density, viscosity, and film shape appear in Eq. (22.3) and therefore need to be defined. Note that in Eq. (22.2) the pressure was made dimensionless in a different manner than was done for rectangular conjunctions in Chap. 21 [Eq. (21.21)]. Therefore, the expressions for the density, viscosity, and film shape will be different. Density is expressed as

$$\bar{\rho} = 1 + \frac{0.6E'P_e}{1 + 1.7E'P_e} \tag{4.20}$$

where the effective elastic modulus is

$$E' = \frac{2}{\left(1 - \nu_a^2\right)/E_a + \left(1 - \nu_b^2\right)/E_b} \qquad \text{GPa}$$

and viscosity is expressed as

$$\bar{\eta} = \left(\frac{\eta_\infty}{\eta_0}\right)^{[1-(1+p/c_p)^{Z_1}]} \tag{4.10}$$

where $\eta_\infty = 6.31 \times 10^{-5}\text{Pa} \cdot \text{s}\ (9.15 \times 10^{-9}\ \text{lbf} \cdot \text{s/in}^2)$
$c_p = 1.96 \times 10^8\text{Pa}\ (28{,}440\ \text{lbf/in}^2)$
Z_1 = viscosity-pressure index, a dimensionless constant

In Eqs. (4.10) and (4.20) it is important that the same dimensions are used in defining the constants.

The film shape can be written simply as

$$h(\tilde{x}, \tilde{y}) = h_0 + S(\tilde{x}, \tilde{y}) + \delta(\tilde{x}, \tilde{y}) \tag{22.4}$$

where h_0 = constant
$S(\tilde{x}, \tilde{y})$ = separation due to geometry of undeformed ellipsoidal solids
$\delta(\tilde{x}, \tilde{y})$ = elastic deformation of solid surfaces

The separation due to the geometry of the two undeformed ellipsoidal solids is shown in Fig. 19.1 and can be described by an equivalent ellipsoidal solid near a plane. The geometrical requirement is that the separation of the two ellipsoidal solids in the initial and equivalent situations should be the same at equal values of $\tilde{x}$ and $\tilde{y}$. Therefore, from Fig. 22.1 the separation due to the undeformed geometry of two ellipsoids can be written as

$$S(\tilde{x}, \tilde{y}) = \frac{\left(\tilde{x} - \tilde{m}D_x/2\right)^2}{2R_x} + \frac{\left(\tilde{y} - \tilde{b}D_y/2\right)^2}{2R_y} \tag{22.5}$$

where $\tilde{m}$ = constant used to determine length of inlet region
$\tilde{b}$ = constant used to determine side-leakage region

For illustration, the mesh described in Fig. 22.1 uses an $\tilde{m} = 4$ and an $\tilde{b} = 2$.

Figure 22.2 illustrates the film thickness and its components for an ellipsoidal solid near a plane. Note that δ is a maximum and S is a minimum at the contact center. Also h_0 assumes a large negative constant value. Substituting Eq. (22.5) into (22.4) while using Eq. (22.2) to make it dimensionless gives

$$H_e = \frac{h}{R_x} = H_{e,0} + \frac{D_x^2\left(X_r - \tilde{m}\right)^2}{8R_x^2} + \frac{D_y^2\left(Y - \tilde{b}\right)^2}{8R_xR_y} + \frac{\delta(X_r, Y)}{R_x} \tag{22.6}$$

where $H_{e,0} = h_0/R_x$ is a constant that is initially estimated.

The simple way of calculating the elastic deformation in a constant rectangular area as described by Hamrock and Dowson (1974) is used here. This method was covered in Sec. 20.6 so that only the conclusions are stated here. Therefore, the elastic deformation can be written as

$$\delta_{\alpha,\beta}(X_r, Y) = \frac{2}{\pi} \sum_{j=1,2,\ldots}^{2\tilde{b}\tilde{c}} \sum_{i=1,2,\ldots}^{(\tilde{m}+\tilde{n})\tilde{d}} P_{e,ij} D_{i^*j^*} \tag{22.7}$$

where $\tilde{c}$ = number of divisions in semidiameter of contact ellipse in y direction
$\tilde{n}$ = constant used to determine length of outlet region
$\tilde{d}$ = number of divisions in semidiameter of contact ellipse in x direction

$$i^* = |\alpha - i| + 1$$

$$j^* = |\beta - j| + 1$$

$$\begin{aligned}
D = {} & \frac{D_x}{2}\left(X_r + \frac{1}{2\tilde{d}}\right) \ln \frac{k\left(Y + \frac{1}{2\tilde{c}}\right) + \left[k^2\left(Y + \frac{1}{2\tilde{c}}\right)^2 + \left(X_r + \frac{1}{2\tilde{d}}\right)^2\right]^{1/2}}{k\left(Y - \frac{1}{2\tilde{c}}\right) + \left[k^2\left(Y - \frac{1}{2\tilde{c}}\right)^2 + \left(X_r + \frac{1}{2\tilde{d}}\right)^2\right]^{1/2}} \\
& + \frac{D_y}{2}\left(Y + \frac{1}{2\tilde{c}}\right) \ln \frac{\left(X_r + \frac{1}{2\tilde{d}}\right) + \left[k^2\left(Y + \frac{1}{2\tilde{c}}\right)^2 + \left(X_r + \frac{1}{2\tilde{d}}\right)^2\right]^{1/2}}{\left(X_r - \frac{1}{2\tilde{d}}\right) + \left[k^2\left(Y + \frac{1}{2\tilde{c}}\right)^2 + \left(X_r - \frac{1}{2\tilde{d}}\right)^2\right]^{1/2}} \\
& + \frac{D_x}{2}\left(X_r - \frac{1}{2\tilde{d}}\right) \ln \frac{k\left(Y - \frac{1}{2\tilde{c}}\right) + \left[k^2\left(Y - \frac{1}{2\tilde{c}}\right)^2 + \left(X_r - \frac{1}{2\tilde{d}}\right)^2\right]^{1/2}}{k\left(Y + \frac{1}{2\tilde{c}}\right) + \left[k^2\left(Y + \frac{1}{2\tilde{c}}\right)^2 + \left(X_r - \frac{1}{2\tilde{d}}\right)^2\right]^{1/2}} \\
& + \frac{D_y}{2}\left(Y - \frac{1}{2\tilde{c}}\right) \ln \frac{\left(X_r - \frac{1}{2\tilde{d}}\right) + \left[k^2\left(Y - \frac{1}{2\tilde{c}}\right)^2 + \left(X_r - \frac{1}{2\tilde{d}}\right)^2\right]^{1/2}}{\left(X_r + \frac{1}{2\tilde{d}}\right) + \left[k^2\left(Y - \frac{1}{2\tilde{c}}\right)^2 + \left(X_r + \frac{1}{2\tilde{d}}\right)^2\right]^{1/2}}
\end{aligned} \tag{22.8}$$

Equation (22.8) and Fig. 22.1 clarify the meaning of Eq. (22.7). The elastic deformation at the center of the rectangular area $\delta_{9,5}$ (Fig. 22.1) caused by the

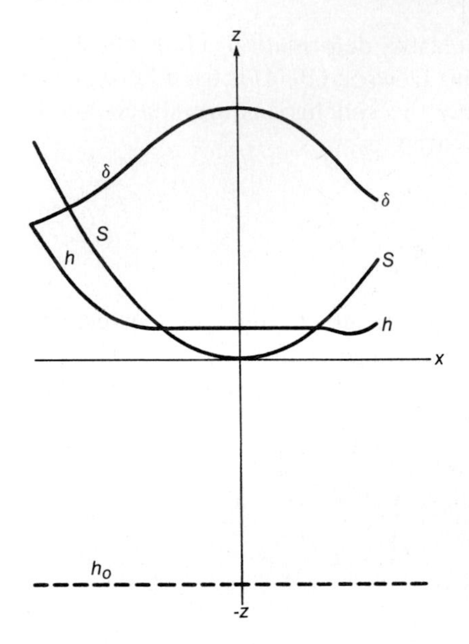

FIGURE 22.2
Components of film thickness for ellipsoidal solid near plane. [*From Hamrock and Dowson (1981).*]

pressure on the various rectangular areas in and around the contact ellipse can be written as

$$\delta_{9,5} = \frac{2}{\pi} \left(\begin{array}{llll} P_{e,1,1} D_{9,5} & + P_{e,2,1} D_{8,5} & + \cdots & + P_{e,35,1} D_{27,5} \\ P_{e,1,2} D_{9,4} & + P_{e,2,2} D_{8,4} & + \cdots & + P_{e,35,2} D_{27,4} \\ \hline P_{e,1,20} D_{9,16} & + P_{e,2,20} D_{8,16} & + \cdots & + P_{e,35,20} D_{27,16} \end{array} \right) \tag{22.9}$$

22.2 DIMENSIONLESS GROUPINGS

The variables resulting from the analysis of elastohydrodynamically lubricated elliptical conjunctions are

R_x	effective radius in x direction, m
R_y	effective radius in y direction, m
E'	effective modulus of elasticity, $2\left(\frac{1-\nu_a^2}{E_a} + \frac{1-\nu_b^2}{E_b}\right)^{-1}$, Pa
h	film thickness, m
$\tilde{u}$	mean surface velocity in x direction, m/s
$\tilde{v}$	mean surface velocity in y direction, m/s
η_0	absolute viscosity at $p = 0$ and constant temperature, $N \cdot s/m^2$
Z_1	viscosity-pressure index
w_z	normal load component, N

From the nine variables six dimensionless groupings can be written:

Dimensionless film thickness:

$$H_e = \frac{h}{R_x} \tag{22.10}$$

where

$$\frac{1}{R_x} = \frac{1}{r_{ax}} + \frac{1}{r_{bx}} \tag{22.11}$$

The radii of curvature defined in Eq. (22.11) are shown in Fig. 19.1. As stated in Chap. 19 it is assumed that convex surfaces, as shown in Fig. 19.1, have a positive value and concave surfaces, a negative value.

Dimensionless load parameter:

$$W = \frac{w_z}{E'R_x^2} \tag{22.12}$$

Dimensionless speed parameter:

$$U_v = \frac{V\eta_0}{E'R_x} \tag{22.13}$$

where

$$V = (\tilde{u}^2 + \tilde{v}^2)^{1/2} \tag{22.14}$$

Dimensionless materials parameter:

$$G = \frac{E'}{p_{iv,as}} = \xi E' \tag{22.15}$$

where $p_{iv,as}$ is the asymptotic isoviscous pressure obtained from Roelands (1966) [see Eq. (4.9)].

Ellipticity parameter:

$$k = \frac{D_y}{D_x} \approx \alpha_r^{2/\pi} \tag{19.29}$$

where

$$\alpha_r = \frac{R_y}{R_x} \tag{18.57}$$

Slide-roll ratio:

$$\theta = \tan^{-1} \frac{\tilde{v}}{\tilde{u}} \tag{22.16}$$

Therefore, the dimensionless minimum-film-thickness equation can be written as

$$H_{\min} = f(W, U_v, G, k, \theta) \tag{22.17}$$

The results presented in this chapter assume pure rolling ($\tilde{v} = 0$ or $\theta = 0$).

The set of dimensionless groups $\{H_{\min}, k, U, W,$ and $G\}$ is a useful collection of parameters for evaluating the results presented in this chapter. It is exactly the same set of dimensionless parameters used in Chap. 21 for rectangular conjunctions with the exception that the ellipticity parameter k has been added to handle the side-leakage effects in this chapter. However, a number of authors (e.g., Moes, 1965–66; and Theyse, 1966) have noted that this set of dimensionless groups can be reduced by one parameter without any loss of generality. This approach is followed in the next chapter, where the film thicknesses to be expected in each of the four regimes of fluid film lubrication are presented graphically and it is necessary to reduce the number of dimensionless groups. The disadvantage of reducing the number of parameters by one is that the new dimensionless groups do not have the physical significance which exists for the original grouping.

22.3 HARD-EHL RESULTS

Results in this section are to be applied to materials of high elastic modulus lubricated elastohydrodynamically, a situation often referred to as "hard EHL." By using the numerical procedures outlined in Hamrock and Dowson (1976a), the influence of the ellipticity parameter and the dimensionless speed, load, and materials parameters on minimum film thickness has been investigated for fully flooded hard-EHL contacts by Hamrock and Dowson (1977a). The ellipticity parameter k wwas varied from 1 (a ball-on-plane configuration) to 8 (a configuration approaching a rectangular conjunction). The dimensionless speed parameter U was varied over a range of nearly two orders of magnitude, and the dimensionless load parameter W, over one order of magnitude. Situations equivalent to using solid materials of bronze, steel, and silicon nitride and lubricants of paraffinic and naphthenic oils were considered in an investigation of the role of the dimensionless materials parameter G on minimum film thickness. The 34 cases used to generate the minimum-film-thickness formula are given in Table 22.1. In the table $\tilde{H}_{e,\min}$ corresponds to the minimum film thickness obtained from solving the coupled Reynolds, rheology, and elasticity equations for an elastohydrodynamically lubricated elliptical conjunction by the approach given in Hamrock and Dowson (1976a). The minimum-film-thickness formula as obtained by Hamrock and Dowson (1977a) from a least-squares fit of the $H_{e,\min}$ data given in the table is

$$\tilde{H}_{e,\min} = \frac{\tilde{h}_{\min}}{R_x} = 3.63 U^{0.68} G^{0.49} W^{-0.073} (1 - e^{-0.68k}) \tag{22.18}$$

In Table 22.1, $\tilde{H}_{e,\min}$ is the minimum film thickness obtained from Eq. (22.18).

TABLE 22.1

Effect of ellipticity, load, speed, and materials parameters on minimum film thickness for hard-EHL contacts [From Hamrock and Dowson (1977a)]

se	Ellipticity parameter, k	Dimensionless load parameter, W	Dimensionless speed parameter, U	Dimensionless materials parameter, G	Minimum film thickness		Difference between $H_{e,\min}$ and $\tilde{H}_{e,\min}$, D_1, percent	Results
					Obtained from EHL elliptical contact theory, $H_{e,\min}$	Obtained from least-squares fit, $\tilde{H}_{e,\min}$		
	1	0.1106×10^{-6}	0.1683×10^{-11}	4522	3.367×10^{-6}	3.514×10^{-6}	4.37	Ellipticity
	1.25	↓	↓	↓	4.105	4.078	−.66	
	1.5				4.565	4.554	−.24	
	1.75				4.907	4.955	.98	
	2				5.255	5.294	.74	
	2.5				5.755	5.821	1.15	
	3				6.091	6.196	−1.72	
	4				6.636	6.662	−.24	
	6				6.969	7.001	.46	
	8				7.048	7.091	.61	
	6	0.2211×10^{-6}	0.1683×10^{-11}	4522	6.492×10^{-6}	6.656×10^{-6}	2.53	Load plus case 9
	↓	.3686	↓	↓	6.317	6.412	1.50	
		.5528			6.268	6.225	−.69	
		.7371			6.156	6.095	−.99	
		.9214			6.085	5.997	−1.45	
		1.106			5.811	5.918	1.84	
		1.290			5.657	5.851	3.43	
	6	0.7371×10^{-6}	0.08416×10^{-11}	4522	3.926×10^{-6}	3.805×10^{-6}	−3.08	Speed plus case 14
	↓	↓	.2525	↓	8.372	8.032	−4.06	
			.3367		9.995	9.769	−2.26	
			.4208		11.61	11.37	−2.07	
			.5892		14.39	14.29	−.69	
			.8416		18.34	18.21	−.71	
			1.263		24.47	24.00	−1.92	
			1.683		29.75	29.18	−1.92	
			2.104		34.58	33.96	−1.79	
			2.525		39.73	38.44	−3.25	
			2.946		43.47	42.69	−1.79	
			3.367		47.32	46.76	−1.18	
			4.208		54.57	54.41	−.29	
			5.050		61.32	61.59	.44	
	6	0.7216×10^{-6}	0.3296×10^{-11}	2310	6.931×10^{-6}	6.938×10^{-6}	0.10	Materials plus case 9
	6	.7216	.9422	3491	17.19	17.59	2.33	
	6	.2456	.1122	6785	6.080	6.116	.59	

The percentage difference between $H_{e,\min}$ and $\tilde{H}_{e,\min}$ is expressed by

$$D_1 = \frac{\left(\tilde{H}_{e,\min} - H_{e,\min}\right)100}{H_{e,\min}} \tag{22.19}$$

In table 22.1 the values of D_1 are within ± 5 percent.

It is interesting to compare the minimum-film-thickness formula [Eq. (22.18)] for elliptical conjunctions with that for rectangular conjunctions given in the last chapter.

$$\tilde{H}_{e,\min} = 1.714(W')^{-0.128} U^{0.694} G^{0.568} \tag{21.59}$$

where

$$W' = \frac{w'_z}{E' R_x}$$

and w'_z is the normal load per unit length. The powers on U, W, and G in Eq. (22.18) are quite similar to the powers on U, W', and G in Eq. (21.59).

In practical situations there is considerable interest in the central as well as the minimum film thickness in elastohydrodynamically lubricated conjunctions. This is particularly true when traction is considered, since the surfaces in relative motion are separated by a film of almost constant thickness that is well represented by the central value over much of the Hertzian contact zone. The procedure used in obtaining the central film thickness was the same as that used in obtaining the minimum film thickness. From Hamrock and Dowson (1981) the central-film-thickness formula for hard-EHL conjunctions is

$$\tilde{H}_{e,c} = 2.69 U^{0.67} G^{0.53} W^{-0.067} (1 - 0.61 e^{-0.73k}) \tag{22.20}$$

A representative contour plot of dimensionless pressure is shown in Fig. 22.3 for $k = 1.25$, $U = 0.168 \times 10^{-11}$, $W = 0.111 \times 10^{-6}$, and $G = 4522$ as obtained from Hamrock and Dowson (1977a). This is one of the 34 cases presented in Table 22.1. In Fig. 22.3 as well as in Fig. 22.4, the + symbol indicates the center of the Hertzian contact zone. The dimensionless representation of the X and Y coordinates causes the actual Hertzian contact ellipse to be a circle regardless of the value of the ellipticity parameter. The Hertzian contact circle is shown in asterisks. On the figure a key shows the contour labels and each corresponding value of the dimensionless pressure. The inlet region is to the left, and the outlet region is to the right. The pressure gradient at the outlet is much larger than that at the inlet. In Fig. 22.3 a pressure spike is visible at the outlet.

Figure 22.4 shows contour plots of film thickness for the same case shown in Fig. 22.3 for the pressure profile ($k = 1.25$, $U = 0.168 \times 10^{-11}$, $W = 0.111 \times 10^{-6}$, and $G = 4522$). In this figure two minimum-film-thickness regions occur

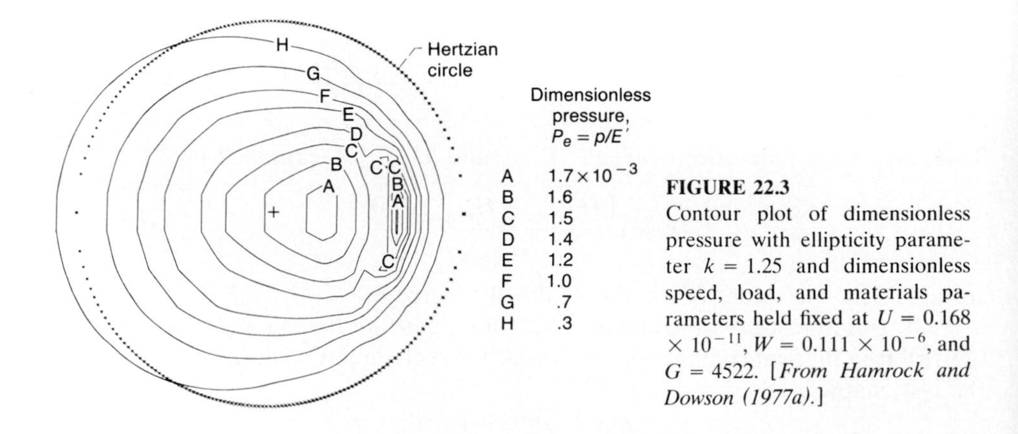

FIGURE 22.3
Contour plot of dimensionless pressure with ellipticity parameter $k = 1.25$ and dimensionless speed, load, and materials parameters held fixed at $U = 0.168 \times 10^{-11}$, $W = 0.111 \times 10^{-6}$, and $G = 4522$. [*From Hamrock and Dowson (1977a).*]

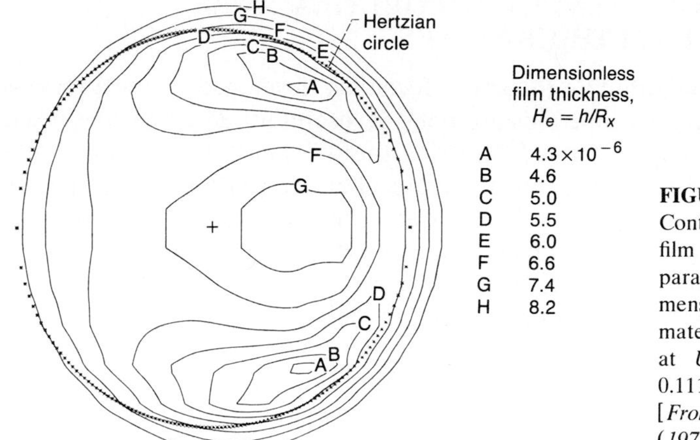

FIGURE 22.4
Contour plot of dimensionless film thickness with ellipticity parameter $k = 1.25$ and dimensionless speed, load, and materials parameters held fixed at $U = 0.168 \times 10^{-11}$, $W = 0.111 \times 10^{-6}$, and $G = 4522$. [*From Hamrock and Dowson (1977a).*]

in well-defined side lobes that follow, and are close to, the edge of the Hertzian contact ellipse. These results reproduce all the essential features of previously reported experimental observations based on optical interferometry as found, for example, in Cameron and Gohar (1966).

The more general consideration that allows the velocity vector to be in any direction, as discussed at the beginning of this chapter, has been recently considered by Chittenden et al. (1985). They used the Hamrock and Dowson (1977a) formulas but added a factor to account for more general consideration of the velocity vector. The film formula from Chittenden et al. (1985) is

$$H_{\min} = 3.68 U_v^{0.68} G^{0.49} W^{-0.073} \left\{ 1 - \exp\left[-0.68 \left(\frac{R_s}{R_e} \right)^{2/3} \right] \right\} \tag{22.21}$$

where

$$\frac{1}{R_e} = \frac{\cos^2\theta}{R_x} + \frac{\sin^2\theta}{R_y}$$

$$\frac{1}{R_s} = \frac{\sin^2\theta}{R_x} + \frac{\cos^2\theta}{R_y}$$

$$U_v = \frac{V\eta_0}{E'R_x}$$

$$V = (\tilde{u}^2 + \tilde{v}^2)^{1/2}$$

Note in the preceding equations that if pure rolling or pure sliding exists, $\theta = 0$ and $\tilde{v} = 0$ and Eq. (22.21) is exactly Eq. (22.20).

22.4 COMPARISON BETWEEN THEORETICAL AND EXPERIMENTAL FILM THICKNESSES

The minimum- and central-film-thickness formulas developed in the previous section for fully flooded conjunctions are not only useful for design purposes but also provide a convenient means of assessing the influence of various parameters on the elastohydrodynamic film thickness. For the purpose of comparing theoretical film thicknesses with those found in actual elastohydrodynamic contacts, Table 22.2 was obtained from Kunz and Winer (1977). The experimental apparatus consisted of a steel ball rolling and sliding on a sapphire plate. This generated a circular conjunction, or an ellipticity parameter of unity. Measurements were made by using the technique of optical interferometry. Table 22.2 shows the results of both calculations and measurements for three lubricants in pure sliding, each under two different loads and three speeds. The notation $\tilde{H}_{e,\min}$ and $\tilde{H}_{e,c}$ is used to denote the dimensionless film thicknesses calculated from Eqs. (22.18) and (22.20), respectively. The measured minimum and central film thicknesses obtained from Kunz and Winer (1977) are denoted by $H_{e,\min}$ and $H_{e,c}$. Figure 22.5 shows comparisons between the calculated and measured film thicknesses for the two loads shown in Table 22.2.

For the smaller load ($W = 0.1238 \times 10^{-6}$) the results shown in Fig. 22.5(a) compare remarkably well if one bears in mind the difficulties associated with the experimental determination of such small quantities under arduous conditions and the error associated with the complex numerical evaluations of elastohydrodynamic conjunctions. The ratios between the central and minimum film thicknesses are similar for the calculations and the measurements, and the dependence of film thickness on speed thus appears to be well represented by Eqs. (22.18) and (22.20).

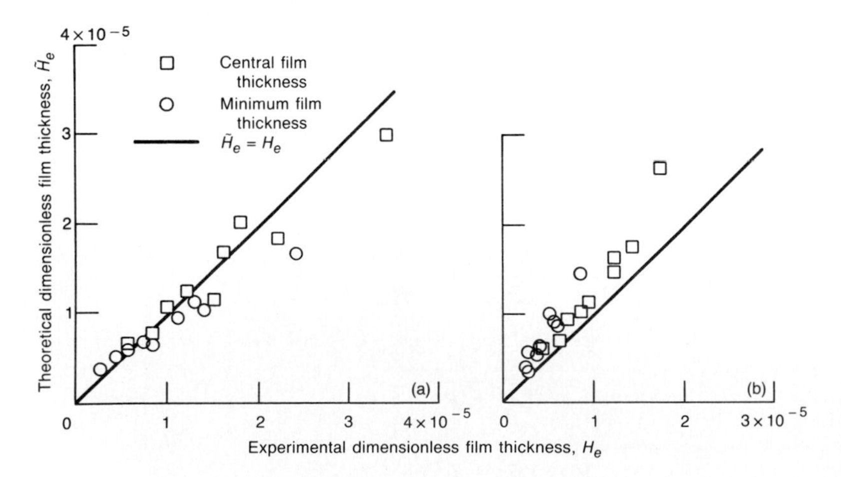

FIGURE 22.5
Theoretical and experimental central and minimum film thicknesses for pure sliding for two dimensionless load parameters W. (a) $W = 0.1238 \times 10^{-6}$; (b) $W = 0.9287 \times 10^{-6}$. [*From Kunz and Winer (1977).*]

For the larger load ($W = 0.9287 \times 10^{-6}$) the agreement shown in Fig. 22.5(b) is not quite so good, with the theoretical predictions of film thickness being consistently larger than the measured values. This discrepancy is sometimes attributed to viscous heating, as discussed by Greenwood and Kauzlarich (1973), or to non-Newtonian behavior, as discussed by Moore (1973). Viscous heating appears to enjoy the most support. Since the measurements were made during pure sliding, thermal effects might well be significant, particularly at the larger load. The value of viscosity used in the calculations reported in Table

TABLE 22.2
Theoretical and experimental film thicknesses [From Kunz and Winer (1977)]

Lubricant[a]	Dimensionless speed parameter, U	Central film thickness		Minimum film thickness	
		Theoretical, eq. (22-20), $\tilde{H}_{e,c}$	Experimental, $H_{e,c}$	Theoretical, eq. (22-18), $\tilde{H}_{e,\min}$	Experimental, $H_{e,\min}$
Dimensionless load parameter, W, 0.1238×10^{-6}					
A	0.1963×10^{-11}	6.84×10^{-6}	5.7×10^{-6}	3.87×10^{-6}	2.8×10^{-6}
	.3926	10.90	9.9	6.20	5.7
	.7866	17.30	16.0	9.30	11.0
B	0.2637×10^{-11}	11.00×10^{-6}	15.0×10^{-6}	6.58×10^{-6}	8.4×10^{-6}
	.5274	18.90	22.0	10.5	14.0
	1.0570	30.20	34.0	17.0	24.0
C	0.2268×10^{-11}	8.04×10^{-6}	8.2×10^{-6}	4.53×10^{-6}	5.0×10^{-6}
	.4536	12.80	12.0	7.27	7.5
	.9089	20.60	18.0	11.60	13.0
Dimensionless load parameter, W, 0.9287×10^{-6}					
A	0.1963×10^{-11}	5.96×10^{-6}	4.3×10^{-6}	3.33×10^{-6}	2.8×10^{-6}
	.3926	9.50	7.1	5.35	4.3
	.7866	15.10	12.0	8.58	5.7
B	0.2637×10^{-11}	10.40×10^{-6}	8.4×10^{-6}	5.68×10^{-6}	2.8×10^{-6}
	.5274	16.60	12.0	9.11	5.6
	1.0570	26.40	17.0	14.60	8.4
C	0.2258×10^{-11}	7.01×10^{-6}	6.3×10^{-6}	3.92×10^{-6}	2.6×10^{-6}
	.4536	11.20	9.4	6.27	3.8
	.9089	17.80	14.0	10.10	5.0

[a] Lubricant A is polyalkyl aromatic ($\xi = 1.58 \times 10^{-8}$ m^2/N; $\eta_0 = 0.0255$ N s/m^2; $G = 4507$); lubricant B is synthetic hydrocarbon ($\xi = 3.11 \times 10^{-8}$ m^2/N; $\eta_0 = 0.0343$ N s/m^2; $G = 8874$); lubricant C is modified polyphenyl ether ($\xi = 1.79 \times 10^{-8}$ m^2/N; $\eta_0 = 0.0294$ N s/m^2; $G = 5107$), where ξ is the pressure-viscosity constant. η_0 is the absolute viscosity at $p = 0$ and constant temperature, and G is the dimensionless materials parameter, $\xi E'$.

22.2 corresponded to the temperature of the lubricant bath. If thermal effects become important, the isothermal assumption used in deriving Eqs. (22.18) and (22.20) is violated. But, although the value of the viscosity used in these expressions is necessarily somewhat arbitrary, there is evidence from previous studies of line or rectangular elastohydrodynamic conjunctions that the film thickness is determined by effective viscosity in the inlet region. If this viscosity is known with reasonable accuracy, the predicted film thicknesses are quite reliable. In addition, at the more severe conditions imposed by the larger load the lubricant may no longer behave like a Newtonian fluid. This would violate the assumptions used in deriving Eqs. (22.18) and (22.20).

The results of Kunz and Winer (1977) presented in Table 22.2 and Fig. 22.5 suggest that at large loads lubricant film thickness changes more rapidly than would be predicted by the isothermal elastohydrodynamic theory for elliptical conjunctions as presented in Eqs. (22.18) and (22.20). This view is supported by the experimental results based on the x-ray technique reported by Parker and Kannel (1971) and the results of optical interferometry presented by Lee et al. (1973). This observation, however, contradicts the results of Johnson and Roberts (1974), who used the spring dynamometer method to estimate elastohydrodynamic film thickness. Johnson and Roberts found that, in spite of the approximate nature of their method, the results indicated strongly that elastohydrodynamic film thickness as predicted by Eqs. (22.18) and (22.20) is maintained up to the highest contact pressures that are practical with ball bearing steel. Their results support the more precise measurements of Gentle and Cameron (1973) and extend the range of operating conditions to higher rolling speeds. The discrepancy between these investigations has yet to be explained.

Another comparison between experimental findings and theoretical predictions can be based on the experimental results provided by Dalmaz and Godet (1978). They measured film thickness optically in a pure-sliding, circular-contact apparatus for different fire-resistant fluids. An example of the good correlation between the theoretical predictions based on the formulas developed in the previous section and these experimental results is shown in Fig. 22.6. The agreement between the experimentally determined variation of central film thickness with speed for mineral oil and water-glycol lubricants of similar viscosity and the theoretical predictions is most encouraging. The ellipticity parameter was unity and applied load was 2.6 N for these experiments. Figure 22.6 also shows that for water-glycol the film thickness generated was barely one-third of that developed for mineral oil of similar viscosity. This drastic reduction in film thickness is attributed to the pressure-viscosity coefficient of water-glycol, which is only about one-fifth that of mineral oil.

Another important comparison between the theoretical film thickness equations presented in this chapter and experimental measurements of film thickness in elliptical elastohydrodynamic conjunctions was presented by Koye and Winer (1980). They presented experimental values of film thickness for ellipticity ratios ranging from 3.7 to 0.117. The pressure-viscosity properties of the lubricant and the effective elasticity of the crowned rollers and the sapphire

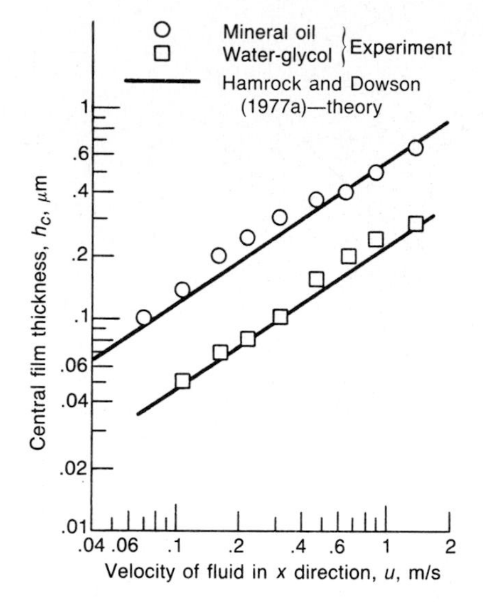

FIGURE 22.6
Effect of velocity on central film thickness at constant load (2.6 N) for mineral oil and water-glycol lubricants of similar viscosity. [*From Dalmaz and Godet (1978).*]

disk yielded an average value of the dimensionless materials parameter G of 10, 451. The dimensionless speeds U and loads W were varied by changing the rolling velocities and the applied loads such that

$$2.14 \times 10^{-11} < U < 8.90 \times 10^{-11}$$

$$0.038 \times 10^{-6} < W < 5.32 \times 10^{-6}$$

A statistical analysis of the 57 acceptable results showed that the experimental values of minimum film thickness were about 23 percent greater than the corresponding predictions of Hamrock and Dowson (1977a) given in Eq. (22.18) for values of $k \geq 1$. Koye and Winer (1980) found that their results suggested that both the dimensionless speed U and load W had a slightly more dominant influence on minimum film thickness than the theory given in Eq. (22.18) predicted. A summary of the results, in which the experimental dimensionless minimum film thicknesses are plotted against the theoretical predictions, is shown in Fig. 22.7.

There are two significant features of the findings from Koye and Winer's (1980) experiments:

1. The theoretical minimum-film-thickness formula given in Eq. (22.18) tends to underestimate the actual minimum film thickness.
2. The experimental findings suggest that the theoretical minimum-film-thickness formula given in Eq. (22.18) is just as valid for ellipticity ratios less than unity as it is for those greater than unity.

The first conclusion appears to contradict the earlier finding of Kunz and Winer (1977), although it must be said that the overall agreement between

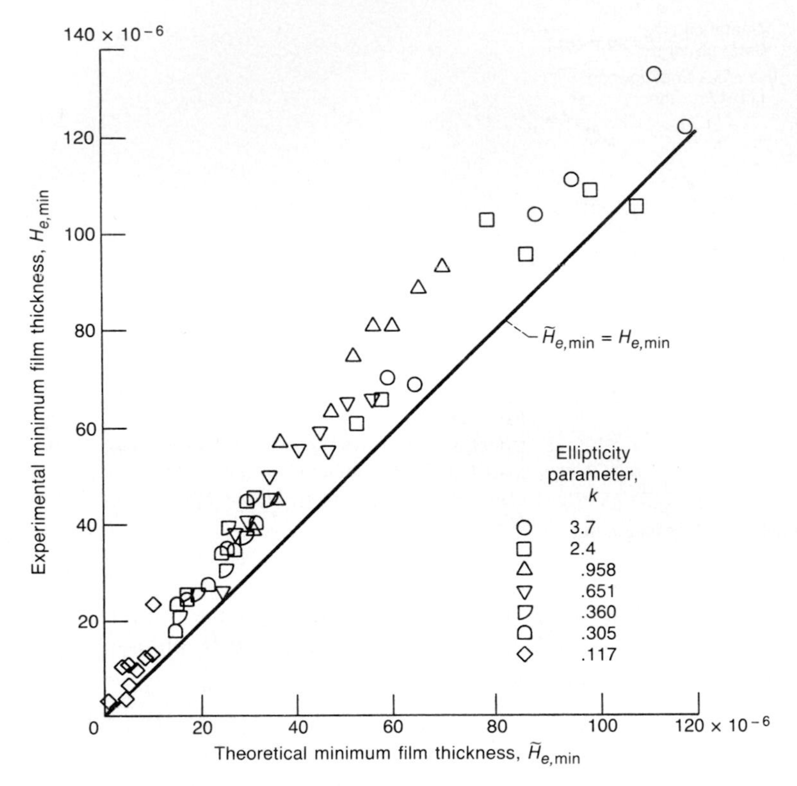

FIGURE 22.7
Predicted dimensionless minimum film thickness from Hamrock and Dowson [1977a, Eq. (22.18)] compared with experimental results obtained by Koye and Winer. [*From Koye and Winer (1980).*]

theory and experiment in this difficult field is encouraging. If the finding that the minimum-film-thickness formula tends to underestimate the actual film thickness is confirmed, the theoretical predictions will at least possess the merit of being conservative.

Perhaps the most remarkable and valuable feature of the Koye and Winer (1980) study is that it has confirmed the utility of the minimum-film-thickness formula at values of k considerably less than unity. This is well outside the range of ellipticity ratios considered by Hamrock and Dowson (1977a) in their numerical solutions to the elastohydrodynamic problem, and it engenders confidence in the application of the film thickness equation to machine elements that normally produce a long, elliptical contact in the direction of motion. This field of applications includes the roller-flange conjunction in axially loaded roller bearings, Novikov gear contacts, and certain situations involving microelastohydrodynamic action on surface asperities.

Various theoretical predictions are compared with the experimental results of Archard and Kirk (1964) in Fig. 22.8. These results are presented in a

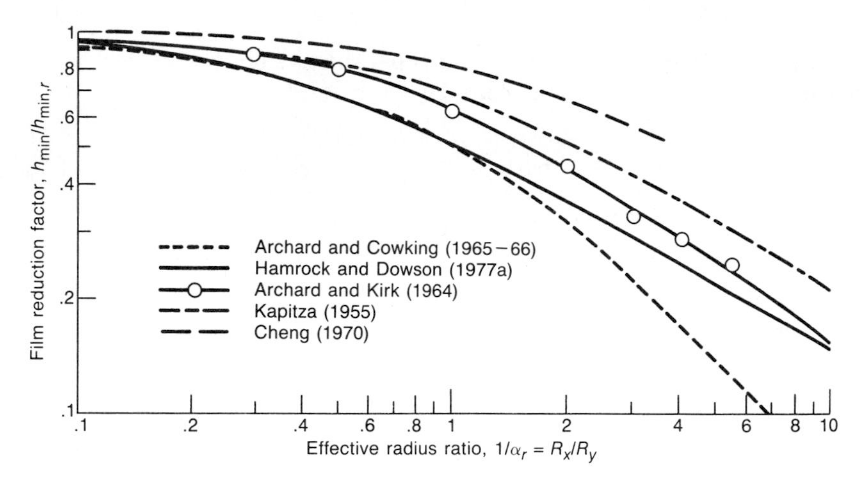

FIGURE 22.8
Side-leakage factor for elliptical conjunctions in hard EHL. [*From Hamrock and Dowson (1981).*]

form that indicates the influence of side leakage on film thickness in elliptical conjunctions by plotting a film reduction factor $h_{\text{min}}/h_{\text{min},r}$ against the effective radius ratio R_x/R_y. The film reduction factor is defined as the ratio of the minimum film thickness achieved in an elliptical conjunction to that achieved in a rectangular conjunction formed between two cylinders in nominal line contact. The theoretical results of Archard and Cowking (1965–66), Cheng (1970), and Hamrock and Dowson (1977a), along with the theoretical solution for rigid solids by Kapitza (1955), are presented in this figure. Archard and Cowking (1965–66) adopted an approach for elliptical conjunctions similar to that used by Grubin (1949) for a rectangular conjunction condition. The Hertzian contact zone is assumed to form a parallel film region, and the generation of high pressure in the approach to the Hertzian zone is considered.

Cheng (1970) solved the coupled system of equations separately by first calculating the deformations from the Hertz equation. Cheng then applied the Reynolds equation to this geometry. He did not consider a change in the lubricant density and assumed an exponential law for the viscosity change due to pressure. Hamrock and Dowson (1977a) developed a procedure for the numerical solution of the complete isothermal elastohydrodynamic lubrication of elliptical conjunctions.

For $0.1 < R_x/R_y < 1$ the Archard and Cowking (1965–66) and Hamrock and Dowson (1977a) predictions are in close agreement, and both overestimate the film thickness reduction evident in the experimental results of Archard and Kirk (1964). For $1 < R_x/R_y < 10$ the Hamrock and Dowson (1977a) film thickness predictions exceed and gradually diverge from those of Archard and Cowking (1965–66). They are also in better agreement with the experimental results of Archard and Kirk (1964).

TABLE 22.3
Effect of ellipticity, load, speed, and materials parameters on minimum film thickness for soft-EHL contacts
[From Hamrock and Dowson (1978)]

Case	Ellipticity parameter, k	Dimensionless load parameter, W	Dimensionless speed parameter, U	Dimensionless materials parameter, G	Minimum film thickness		Difference between $H_{e,min}$ and $\tilde{H}_{e,min}$, D_1, percent	Results
					Obtained from EHL elliptical contact theory, $H_{e,min}$	Obtained from least-squares fit, $\tilde{H}_{e,min}$		
1	1	0.4405×10^{-3}	0.1028×10^{-7}	0.4276	88.51×10^{-6}	91.08×10^{-6}	2.90	Ellipticity
2	2	↓	↓	↓	142.5	131.2	−7.93	
3	3				170.4	160.8	−5.63	
4	4				186.7	182.4	−2.30	
5	6				206.2	209.8	1.75	
6	8				219.7	224.6	2.23	
7	12				235.2	236.0	.34	
8	6	0.4405×10^{-3}	0.5139×10^{-7}	0.4276	131.8×10^{-6}	133.7×10^{-6}	1.44	Speed plus case 5
9	↓	↓	.1542	↓	268.1	273.1	1.86	
10			.2570		381.6	380.7	−.24	
11			.05139		584.7	597.3	2.15	
12	6	0.2202×10^{-3}	0.1028×10^{-7}	0.4276	241.8×10^{-6}	242.7×10^{-6}	0.37	Load plus case 5
13	↓	.6607	↓	↓	190.7	192.7	1.05	
14		1.101			170.5	173.1	1.52	
15		1.542			160.4	161.3	.56	
16		2.202			149.8	149.7	−.07	
17	6	0.1762×10^{-3}	0.06169×10^{-7}	1.069	181.8×10^{-6}	182.5×10^{-6}	0.39	Material

22.5 SOFT-EHL RESULTS

The earlier studies of elastohydrodynamic lubrication of elliptical conjunctions are applied to the particular and interesting situation exhibited by low-elastic-modulus materials (soft EHL). The procedure used in obtaining the soft-EHL results is given in Hamrock and Dowson (1978). The ellipticity parameter was varied from 1 (a ball-on-plane configuration) to 12 (a configuration approaching a nominal line or rectangular contact). The dimensionless speed and load parameters were varied by one order of magnitude. Seventeen cases used to generate the minimum-film-thickness formula are given in Table 22.3. In the table $H_{e,\min}$ corresponds to the minimum film thickness obtained from applying the EHL elliptical contact theory given in Hamrock and Dowson (1976a) to the soft-EHL contacts. The minimum-film-thickness formula obtained from a least-squares fit of the data was first given in Hamrock and Dowson (1978) as

$$\tilde{H}_{e,\min} = 7.43U^{0.65}W^{-0.21}(1 - 0.85e^{-0.31k}) \tag{22.22}$$

In Table 22.3, $\tilde{H}_{e,\min}$ is the minimum film thickness obtained from Eq. (22.22). The percentage difference between $H_{e,\min}$ and $\tilde{H}_{e,\min}$ is expressed by D_1, given in Eq. (22.19). The values of D_1 in Table 22.3 are within the range -7.9 to 2.9 percent.

It is interesting to compare the equation for low-elastic-modulus materials [soft EHL, Eq. (22.22)] with the corresponding equation for high-elastic-modulus materials [hard EHL, Eq. (22.18)]. The powers of U in Eqs. (22.22) and (22.18) are similar, but the power of W is much more significant for low-elastic-modulus materials. The expression showing the effect of the ellipticity parameter is of exponential form in both equations, but with different constants.

A major difference between Eqs. (22.22) and (22.18) is the absence of a materials parameter G in the expression for low-elastic-modulus materials. There are two reasons for this. One is the negligible effect of pressure on the viscosity of the lubricating fluid, and the other is the way in which the role of elasticity is simply and automatically incorporated into the prediction of conjunction behavior through an increase in the size of the Hertzian contact zone corresponding to load changes. As a check on the validity of this absence, case 9 of Table 22.3 was repeated with the material changed from nitrile to silicon rubber. The results of this change are recorded as case 17 in Table 22.3. The dimensionless minimum film thickness calculated from the full numerical solution to the elastohydrodynamic contact theory was 181.8×10^{-6}, and the dimensionless minimum film thickness predicted from Eq. (22.22) turned out to be 182.5×10^{-6}. This clearly indicates that the minimum film thickness for low-elastic-modulus materials does not depend on the materials parameter.

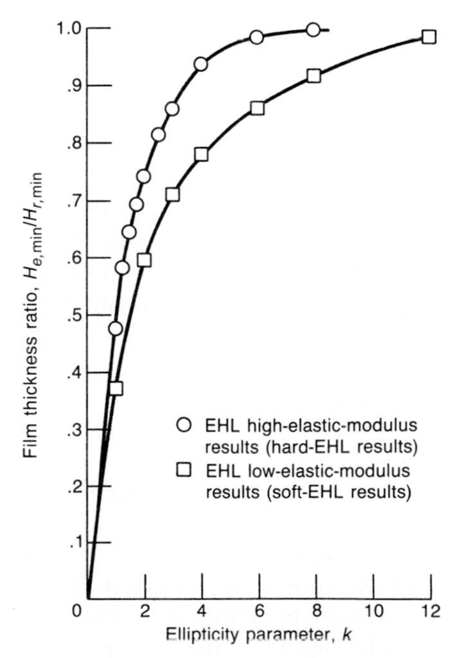

FIGURE 22.9
Effect of ellipticity parameter on ratio of dimensionless minimum film thickness to line-contact dimensionless minimum film thickness, for EHL high- and low-elastic-modulus analyses. [*From Hamrock and Dowson (1978).*]

There is interest in knowing the central film thickness, in addition to the minimum film thickness, in elastohydrodynamic conjunctions. The procedure used to obtain an expression for the central film thickness was the same as that used to obtain the minimum-film-thickness expression. The central-film-thickness formula for low-elastic-modulus materials (soft EHL) as obtained from Hamrock and Dowson (1978) is

$$\tilde{H}_{e,c} = 7.32(1 - 0.72e^{-0.28k})U^{0.64}W^{-0.22} \tag{22.23}$$

Comparing the central- and minimum-film-thickness equations, (22.23) and (22.22), reveals only slight differences. The ratio of minimum to central film thicknesses evident in the computed values given in Hamrock and Dowson (1978) ranges from 70 to 83 percent, the average being 77 percent.

Figure 22.9 shows the variation of the ratio $h_{\min}/h_{\min,r}$ where $h_{\min,r}$ is the minimum film thickness for rectangular conjunctions, with the ellipticity parameter k for both high- and low-elastic-modulus materials. If it is assumed that the minimum film thickness obtained from the elastohydrodynamic analysis of elliptical conjunctions can only be accurate to 3 percent, the ratio $h_{\min}/h_{\min,r}$ approaches the limiting value at $k = 5$ for high-elastic-modulus materials. For low-elastic-modulus materials the ratio approaches the limiting value more slowly, but it is reasonable to state that the rectangular conjunction solution will give a good prediction of the minimum film thickness for conjunctions in which k exceeds about 11.

22.6 STARVATION RESULTS

It was not until the late 1960s and early 1970s that the influence of lubricant starvation on elastohydrodynamic behavior received serious consideration. Before this time it was assumed that inlets were always fully flooded. This assumption seemed to be entirely reasonable in view of the minute quantities of lubricant required to provide an adequate film. However, in due course it was recognized that some machine elements suffered from lubricant starvation.

How partial filling of the inlet to an elastohydrodynamic conjunction influences pressure and film thickness can readily be explored theoretically by adopting different starting points for the inlet pressure boundary. Orcutt and Cheng (1966) appear to have been the first to proceed in this way for a specific case corresponding to a particular experimental situation. Their results showed that lubricant starvation lessened the film thickness. Wolveridge et al. (1971) used a Grubin (1949) approach in analyzing starved, elastohydrodynamically lubricated line contacts. Wedeven et al. (1971) analyzed a starved condition in a ball-on-plane geometry. Castle and Dowson (1972) presented a range of numerical solutions for the starved line-contact elastohydrodynamic situation. These analyses yielded the proportional reductions in film thickness from the fully flooded condition in terms of a dimensionless inlet boundary parameter.

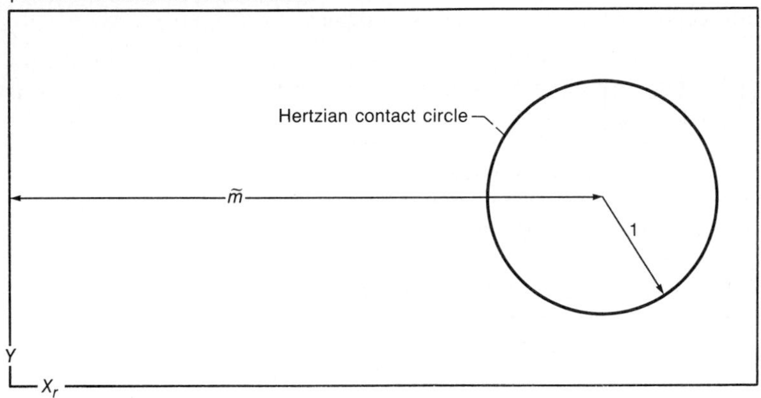

FIGURE 22.10
Computing area in and around Hertzian contact zone. [*From Hamrock and Dowson (1977b).*]

22.6.1 FULLY FLOODED–STARVED BOUNDARY. Figure 22.10 shows the computing area in and around the Hertzian contact. In this figure the coordinate X_r is made dimensionless with respect to the semiminor axis $D_x/2$ of the contact ellipse, and the coordinate Y is made dimensionless with respect to its semimajor axis $D_y/2$. The ellipticity parameter k is defined as the semimajor axis divided by the semiminor axis ($k = D_y/D_x$). Because the coordinates X_r and Y are dimensionless, the Hertzian contact ellipse becomes a Hertzian circle regardless of the value of k. This Hertzian contact circle is shown in Fig. 22.10 with a radius of unity. The edges of the computing area, where the pressure is assumed to be ambient, are also denoted. Also shown is the dimensionless inlet distance $\tilde{m}$, which is equal to the dimensionless distance from the center of the Hertzian contact zone to the inlet edge of the computing area.

Lubricant starvation can be studied by simply changing $\tilde{m}$. A fully flooded condition is said to exist when $\tilde{m}$ ceases to influence the minimum film thickness to any significant extent. The value at which the minimum film thickness first starts to change when $\tilde{m}$ is gradually reduced from a fully flooded condition is called the "fully flooded–starved boundary position" and is denoted by m^*. Therefore, lubricant starvation was studied by using the basic EHL elliptical contact theory developed in Sec. 22.1 and by observing how reducing the dimensionless inlet distance affected the basic features of the conjunction.

22.6.2 HARD-EHL RESULTS. Table 22.4 shows how changing the dimensionless inlet distance affected the dimensionless minimum film thickness for three groups of dimensionless load and speed parameters. All the data presented in this section are for hard-EHL contacts that have a materials parameter G fixed at 4522 and an ellipticity parameter k fixed at 6. It can be seen from Table 22.4

TABLE 22.4
Effect of starvation on minimum film thickness for hard-EHL contacts [From Hamrock and Dowson (1977b)]

Dimensionless inlet distance, $\tilde{m}$	Group		
	1	2	3
	Dimensionless load parameter, W		
	0.3686×10^{-6}	0.7371×10^{-6}	0.7371×10^{-6}
	Dimensionless speed parameter, U		
	0.1683×10^{-11}	1.683×10^{-11}	5.050×10^{-11}
	Dimensionless minimum film thickness, $H_{e,\min}$		
6	------------	29.75×10^{-6}	61.32×10^{-6}
4	6.317×10^{-6}	29.27	57.50
3	6.261	27.84	51.70
2.5	------------	26.38	46.89
2	5.997	23.46	39.91
1.75	------------	21.02	34.61
1.5	5.236	------------	27.90
1.25	3.945	------------	-----------

that as $\tilde{m}$ decreased, the dimensionless minimum film thickness $H_{e,\min}$ also decreased.

Table 22.5 shows how the three groups of dimensionless speed and load parameters considered affected the location of the fully flooded–starved boundary m^*. Also given in this table are the corresponding values of dimensionless central and minimum film thicknesses for the fully flooded condition as obtained by interpolating the numerical values. The values of m^* shown in Table 22.5 were obtained by using the data from Table 22.4 when the following

TABLE 22.5
Effect of dimensionless speed and load parameters on fully flooded–starved boundary for hard-EHL contacts
[From Hamrock and Dowson (1977b)]

Group (from table 22-4)	$2R_x/D_x$	Fully flooded film thickness		Fully flooded–starved boundary, m^*
		Central, $H_{e,c}$	Minimum, $H_{e,\min}$	
1	205.9	7.480×10^{-6}	5.211×10^{-6}	2.62
2	163.5	33.55	29.29	3.71
3	163.5	70.67	60.92	5.57

equation was satisfied:

$$\frac{H_{e,\min} - (H_{e,\min})_{\tilde{m}=m^*}}{H_{e,\min}} = 0.03 \qquad (22.24)$$

The value of 0.03 was used in Eq. (22.24) because it was ascertained that the data in Table 22.4 were accurate to only ± 3 percent.

The general form of the equation that describes how m^* varies with the geometry and central film thickness of an elliptical elastohydrodynamic conjunction is given as

$$m^* - 1 = A^* \left[\left(\frac{2R_x}{D_x} \right)^2 H_{e,c} \right]^{B^*} \qquad (22.25)$$

The right side of Eq. (22.25) is similar in form to the equations given by Wolveridge et al. (1971) and Wedeven et al. (1971). Applying a least-squares power fit to the data obtained from Hamrock and Dowson (1977b) shown in Table 22.4 gives

$$m^* = 1 + 3.06 \left[\left(\frac{2R_x}{D_x} \right)^2 H_{e,c} \right]^{0.58} \qquad (22.26)$$

A fully flooded condition exists when $\tilde{m} \geq m^*$, and a starved condition exists when $\tilde{m} < m^*$. The coefficient of determination for these results is 0.9902, which is entirely satisfactory.

If in Eq. (22.25) the dimensionless minimum film thickness is used instead of the central film thickness,

$$m^* = 1 + 3.34 \left[\left(\frac{2R_x}{D_x} \right)^2 H_{e,\min} \right]^{0.56} \qquad (22.27)$$

The coefficient of determination for these results is 0.9869, which is again extremely good.

Once the limiting location of the inlet boundary for the fully flooded conditions [Eqs. (22.26) and (22.27)] has been clearly established, and equation can be developed that defines the dimensionless film thickness for elliptical conjunctions operating under starved lubrication conditions. The ratio between the dimensionless central film thicknesses under starved and fully flooded conditions can be expressed in general form as

$$\frac{H_{e,c,s}}{H_{e,c}} = C^* \left(\frac{\tilde{m} - 1}{m^* - 1} \right)^{D^*} \qquad (22.28)$$

Table 22.6 shows how the ratio of the dimensionless inlet distance to the fully flooded–starved boundary $(\tilde{m} - 1)/(m^* - 1)$ affects the ratio of central film thicknesses under starved and fully flooded conditions $H_{e,c,s}/H_{e,c}$. A least-squares power curve fit to the 16 pairs of data points

TABLE 22.6
Effect of dimensionless inlet distance on central- and minimum-film-thickness ratios for hard-EHL contacts
[From Hamrock and Dowson (1977b)]

Group (from table 22-4)	Dimensionless inlet distance, $\tilde{m}$	Film thickness ratios for starved and flooded conditions		Inlet boundary parameters	
		Central, $H_{e,c,s}/H_{e,c}$	Minimum, $H_{e,\min,s}/H_{e,\min}$	Critical, $(\tilde{m}-1)/(m^*-1)$	Wedeven et al. (1971), $(\tilde{m}-1)/(m_W-1)$
1	2.62	1	1	1	0.9895
	2	.9430	.9640	.6173	.6108
	1.5	.7697	.8417	.3086	.3054
	1.25	.5689	.6341	.1543	.1527
2	3.71	1	1	1	0.8281
	2	.9574	.9534	.7380	.6111
	2.5	.8870	.9034	.5525	.4584
	2	.7705	.8034	.3690	.3056
	1.75	.7151	.7199	.2768	.2292
3	5.57	1	1	1	0.8498
	4	.9348	.9439	.6565	.5579
	3	.8330	.8487	.4376	.3719
	2.5	.7440	.7697	.3282	.2789
	2	.5223	.6551	.2188	.1860
	1.75	.5309	.5681	.1641	.1395
	1.5	.4155	.4580	.1094	.0930

$$\left[\left(\frac{H_{e,c,s}}{H_{e,c}}\right)_i, \left(\frac{\tilde{m}-1}{m^*-1}\right)_i\right] \qquad i = 1, 2, \ldots, 16$$

was used in obtaining values for C^* and D^* in Eq. (22.28). For these values the dimensionless central film thickness for a starved condition can be written as

$$H_{e,c,s} = H_{e,c}\left(\frac{\tilde{m}-1}{m^*-1}\right)^{0.29} \tag{22.29}$$

By using a similar approach and the data in Table 22.6 the dimensionless minimum film thickness for a starved condition can be written as

$$H_{e,\min,s} = H_{e,\min}\left(\frac{\tilde{m}-1}{m^*-1}\right)^{0.25} \tag{22.30}$$

Therefore, whenever $\tilde{m} < m^*$, where m^* is defined by either Eq. (22.26) or (22.27), a starved lubrication condition exists. When this is true, the dimensionless central film thickness is expressed by Eq. (22.29), and the dimensionless minimum film thickness is expressed by Eq. (22.30). If $\tilde{m} \geq m^*$, a fully flooded condition exists. Expressions for the dimensionless central and minimum film thicknesses for a fully flooded condition ($H_{e,c}$ and $H_{e,\min}$) were developed in Sec. 22.5.

Figures 22.11 and 22.12 explain more fully what happens in going from a fully flooded to a starved lubrication condition. As in Sec. 22.2 the + symbol indicates the center of the Hertzian contact, and the asterisks indicate the Hertzian contact circle. The contour labels and each corresponding value are given on each figure.

In Fig. 22.11(a), (b), and (c) contour plots of the dimensionless pressure ($P_e = p/E'$) are given for group 1 of Table 22.5 and for dimensionless inlet distances $\tilde{m}$ of 4, 2, and 1.25, respectively. In this figure the contour values are the same in each plot. The pressure spikes are evident in Figs. 22.11(a) and (b), but no pressure spike occurs in Fig. 22.11(c). This implies that as $\tilde{m}$ decreases, or as the severity of lubricant starvation increases, the pressure spike is suppressed. Figure 22.11(*a*), with $\tilde{m} = 4$, corresponds to a fully flooded condition; Fig. 22.11(b), with $\tilde{m} = 2$, to a starved condition; and Fig. 22.11(c), with $m = 1.25$, to even more severe starvation. Once lubricant starvation occurs, the severity of the situation within the conjunction increases rapidly as $\tilde{m}$ is decreased and dry contact conditions are approached.

Contour plots of the dimensionless film thickness ($H_e = h/R_x$) for the results shown in group 1 of Table 22.5 and for conditions corresponding to the three pressure distributions shown in Fig. 22.11 are reproduced in Fig. 22.12. It is clear that the film contours in the central region of the elastohydrodynamic conjunction become more parallel as lubricant starvation increases and that the region occupied by the minimum film thickness becomes more concentrated. Note also that the values attached to the film thickness contours for the severely starved condition [Fig. 22.12(c)] are much smaller than those for the fully flooded condition [Fig. 22.12(a)].

22.6.3 SOFT-EHL RESULTS. The theory and numerical procedure mentioned in Sec. 22.5 can be used to investigate how lubricant starvation influences minimum film thickness in starved, elliptical, elastohydrodynamic conjunctions for low-elastic-modulus materials (soft EHL). Lubricant starvation was studied by Hamrock and Dowson (1979a) by simply moving the inlet boundary closer to the contact center, as described in the previous section.

Table 22.7 shows how the dimensionless inlet distance $\tilde{m}$ affects the dimensionless minimum film thickness $H_{e,\min}$ for three groups of dimensionless load and speed parameters. For all the results presented in this section, the dimensionless materials parameter G was fixed at 0.4276 and the ellipticity parameter k was fixed at 6. The results shown in Table 22.7 clearly indicate the adverse effect of lubricant starvation in the sense that as $\tilde{m}$ decreases, $H_{e,\min}$ also decreases.

Table 22.8 shows how the three groups of dimensionless speed and load parameters affect the limiting location of the fully flooded–starved boundary m^*. Also given in this table are corresponding values of the dimensionless

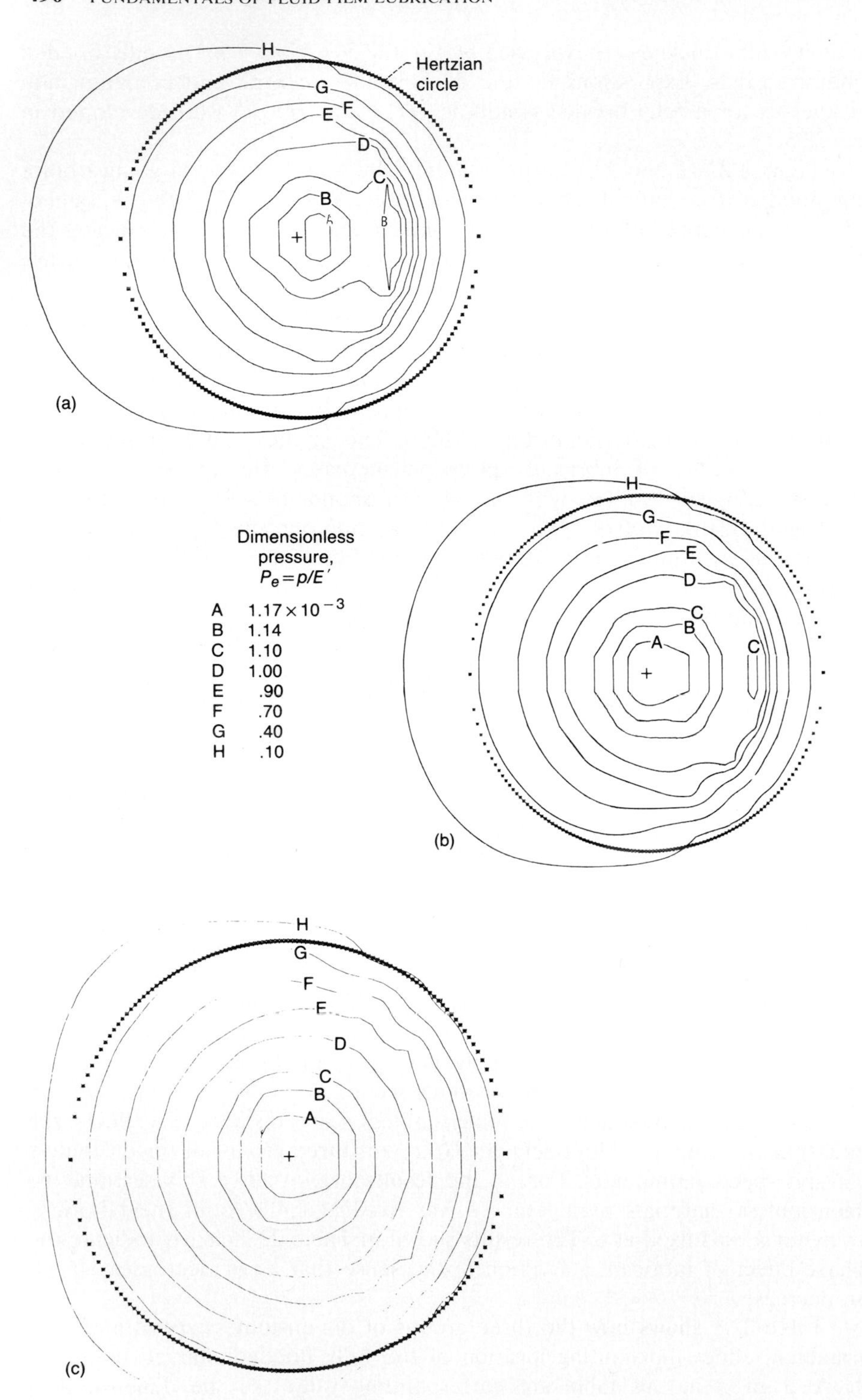

FIGURE 22.11
Contour plots of dimensionless pressure for dimensionless inlet distances $\tilde{m}$ of (a) 4, (b) 2, and (c) 1.25 and for group 1 of Table 22.5. [*From Hamrock and Dowson (1977b).*]

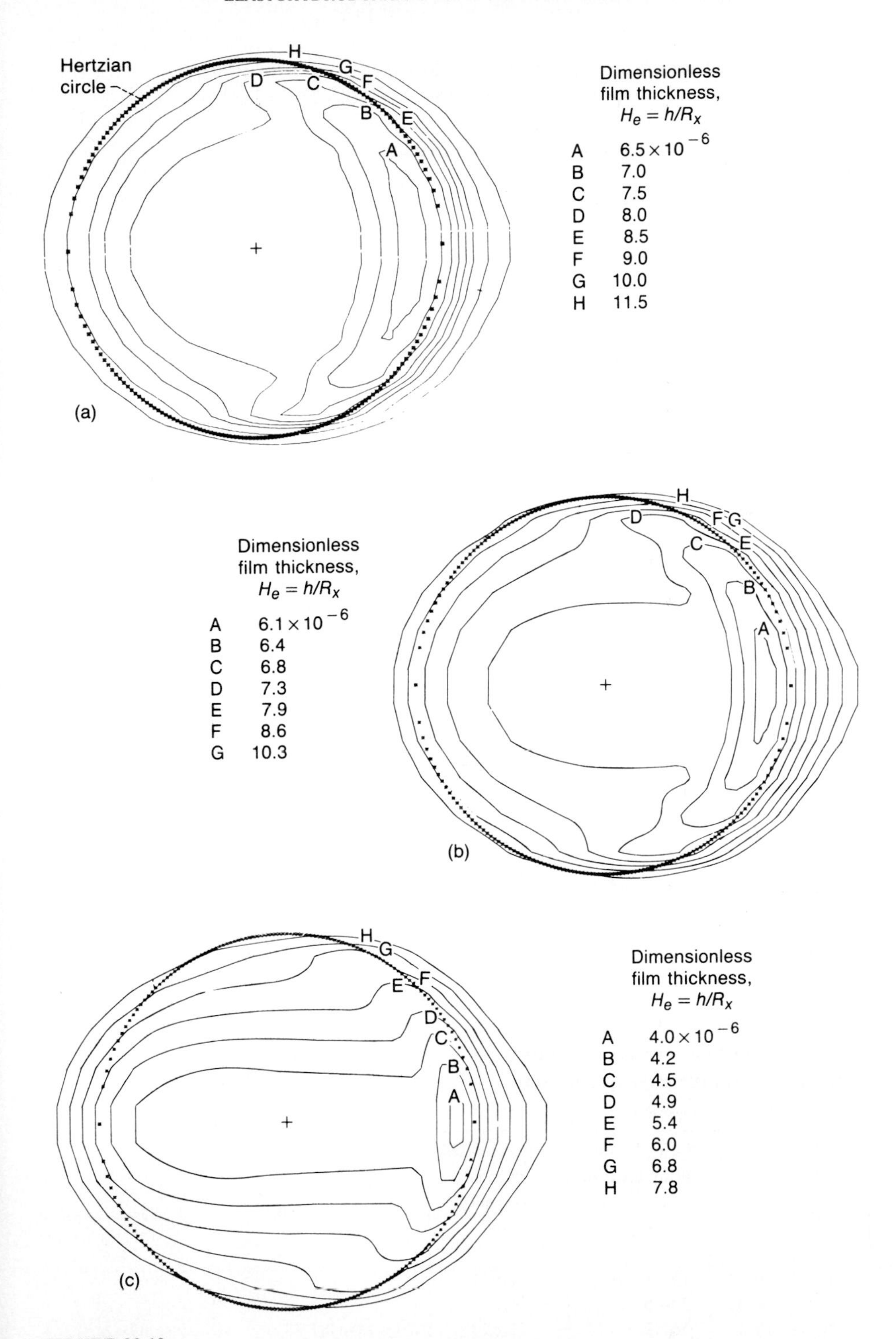

FIGURE 22.12
Contour plots of dimensionless film thickness for dimensionless inlet distances $\tilde{m}$ of (a) 4, (b) 2, and (c) 1.25 and for group 1 of Table 22.5. [*From Hamrock and Dowson (1977b).*]

TABLE 22.7
Effect of starvation on minimum film thickness for soft-EHL contacts [From Hamrock and Dowson (1979a)]

Dimensionless inlet distance, $\tilde{m}$	Group		
	1	2	3
	Dimensionless load parameter, W		
	0.4405×10^{-3}	0.2202×10^{-3}	0.4405×10^{-3}
	Dimensionless speed parameter, U		
	0.5139×10^{-8}	1.027×10^{-8}	5.139×10^{-8}
	Dimensionless minimum film thickness, $H_{e,\min}$		
1.967	131.8×10^{-6}	241.8×10^{-6}	584.7×10^{-6}
1.833	131.2	238.6	572.0
1.667	129.7	230.8	543.1
1.500	125.6	217.2	503.0
1.333	115.9	199.3	444.9
1.167	98.11	120.8	272.3
1.033	71.80	120.8	272.3

minimum film thickness for the fully flooded condition, as obtained by interpolating the numerical values. Making use of Table 22.7 and following the procedure outlined in the previous section yield the critical boundary m^* at which starvation becomes important for low-elastic-modulus materials:

$$m^* = 1 + 1.07\left[\left(\frac{2R_x}{D_x}\right)^2 \tilde{H}_{e,\min}\right]^{0.16} \tag{22.31}$$

where $\tilde{H}_{e,\min}$ is obtained from the fully flooded soft-EHL results in Sec. 22.5.

TABLE 22.8
Effect of inlet distance on minimum film thickness for soft-EHL contacts
[From Hamrock and Dowson (1979a)]

Group (from table 22-7)	$2R_x/D_x$	Fully flooded minimum film thickness, $H_{e,\min}$	Fully flooded-starved boundary, m^*
1	19.41	127.8×10^{-6}	1.661
2	24.45	234.5	1.757
3	19.41	567.2	1.850

TABLE 22.9
Effect of inlet distance on minimum-film-thickness ratio for soft-EHL contacts [From Hamrock and Dowson (1979a)]

Group (from table 22.7)	Dimensionless inlet distance, $\tilde{m}$	Ratio of minimum film thicknesses for starved and flooded conditions, $H_{e,\min,s}/H_{e,\min}$	Critical inlet boundary parameter, $(\tilde{m}-1)/(m^*-1)$
1	1.661	1	1
	1.500	.9828	.7564
	1.333	.9069	.5038
	1.167	.7677	.2526
	1.033	.5618	.0499
2	1.757	1	1
	1.667	.9842	.8811
	1.500	.9262	.6605
	1.333	.8499	.4399
	1.167	.7267	.2206
	1.033	.5151	.0436
3	1.850	1	1
	1.667	.9575	.7847
	1.500	.8868	.5882
	1.333	.7844	.3918
	1.167	.6761	.1965
	1.033	.4801	.0388

Table 22.9 shows how $\tilde{m}$ affects the ratio of minimum film thicknesses for the starved and fully flooded conditions $H_{e,\min,s}/H_{e,\min}$. The dimensionless minimum film thickness for a starved condition for low-elastic-modulus materials can thus be written as

$$\tilde{H}_{e,\min,s} = H_{e,\min}\left(\frac{\tilde{m}-1}{m^*-1}\right)^{0.22} \tag{22.32}$$

Therefore, whenever $\tilde{m} < m^*$, where m^* is defined by Eq. (22.31), a lubricant starvation condition exists. When this is true, the dimensionless minimum film thickness is expressed by Eq. (22.32). If $\tilde{m} \geq m^*$, a fully flooded condition exists. The expression for the dimensionless minimum film thickness for a fully flooded condition $H_{e,\min}$ for materials of low elastic modulus was developed in Sec. 22.5.

In Fig. 22.13 contour plots of the dimensionless pressure ($P_e = p/E'$) are shown for the third set of conditions recorded in Table 22.7 and for $\tilde{m}$ of 1.967, 1.333, and 1.033. Note that the contour levels and intervals are identical in all parts of Fig. 22.13. In Fig. 22.13(a), with $\tilde{m}$ = 1.967, an essentially fully flooded condition exists. The contours are almost circular and extend farther into the

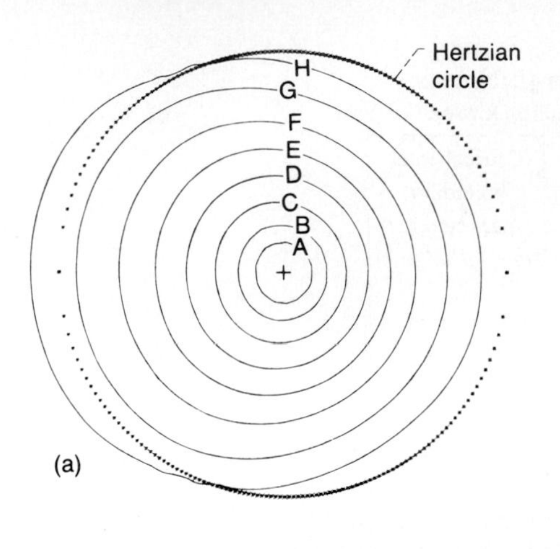

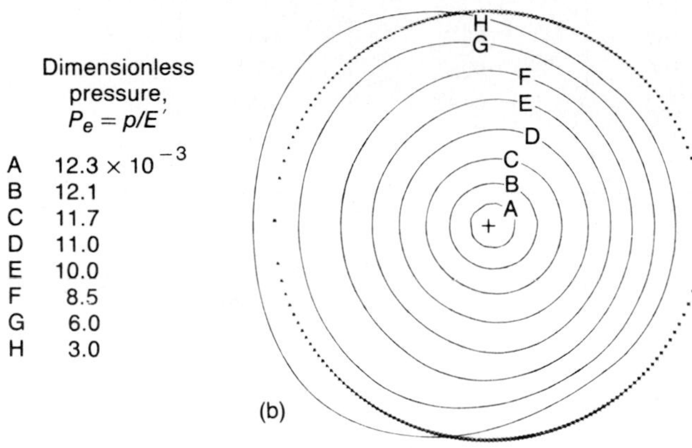

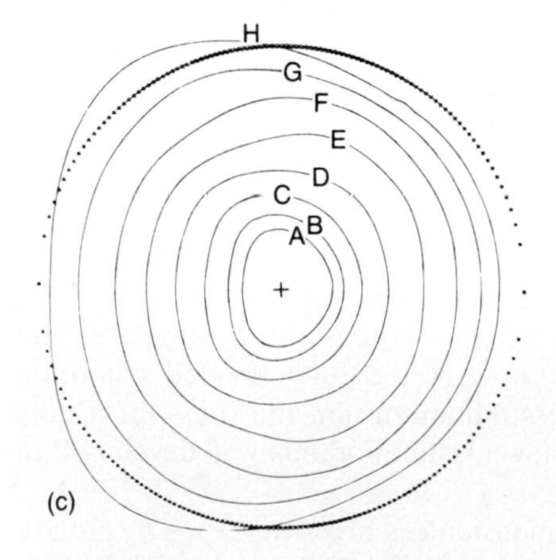

FIGURE 22.13
Contour plots of dimensionless pressure for dimensionless inlet distances $\tilde{m}$ of (a) 1.967, (b) 1.333, and (c) 1.033 and for group 3 of Table 22.7. [*From Hamrock and Dowson (1979a).*]

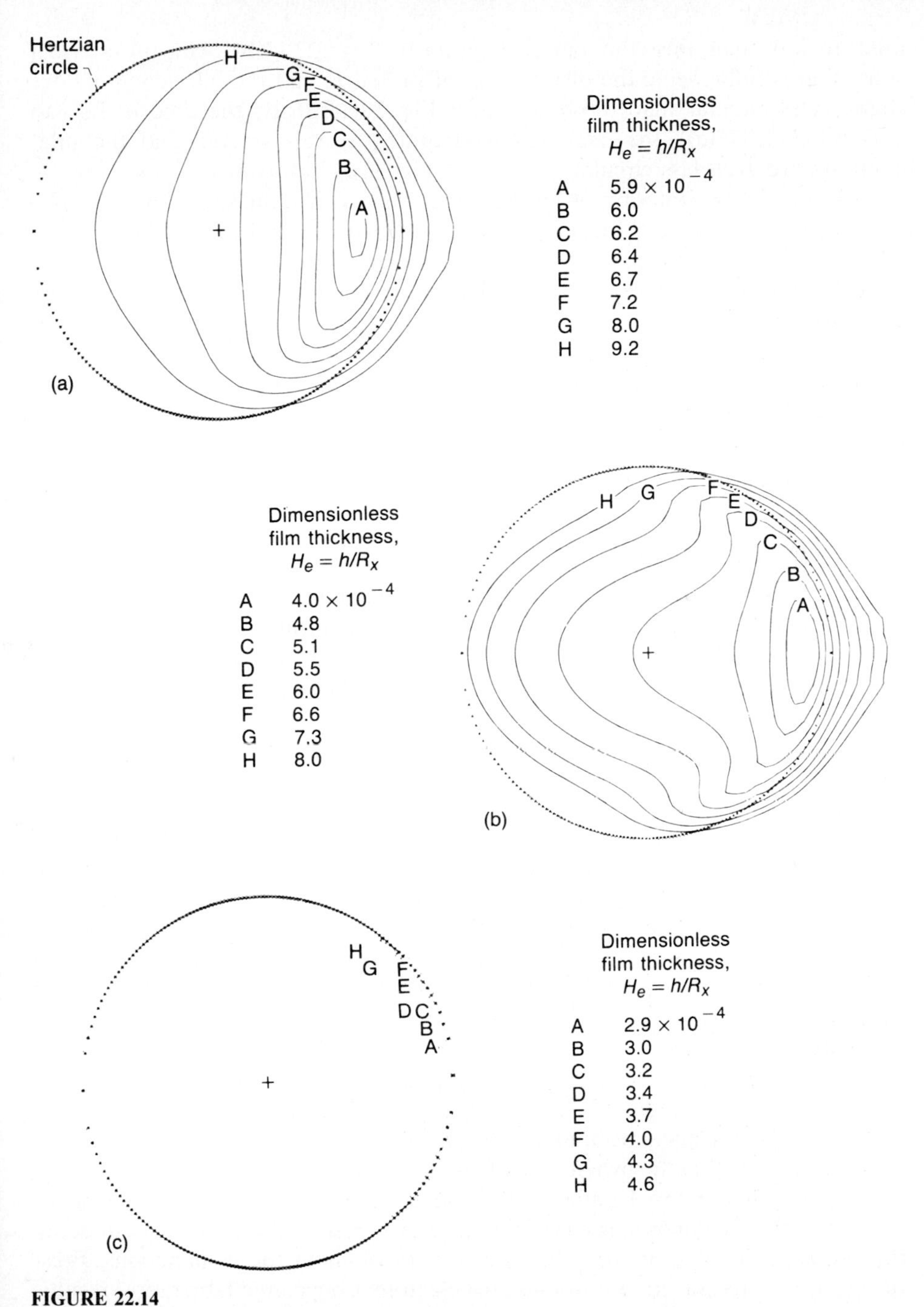

FIGURE 22.14
Contour plots of dimensionless film thickness for dimensionless inlet distances $\tilde{m}$ of (a) 1.967, (b) 1.333, and (c) 1.033 and for group 3 of Table 22.7. [*From Hamrock and Dowson (1979a).*]

inlet region than into the outlet region. In Fig. 22.13(b), with $\tilde{m} = 1.333$, starvation is influencing the distribution of pressure, and the inlet contours are slightly less circular than those shown in Fig. 22.13(a). By the time $\tilde{m}$ falls to 1.033 [Fig. 22.13(c)], the conjunction is quite severely starved and the inlet contours are even less circular.

In Fig. 22.14 contour plots of the dimensionless film thickness ($H_e = h/R_x$) are shown, also for the third set of conditions recorded in Table 22.7 and for $\tilde{m}$ of 1.967, 1.333, and 1.033. These film thickness results correspond to the pressure results shown in Fig. 22.13. The central portions of the film thickness contours become more parallel as starvation increases and the minimum-film-thickness area moves to the outlet region. The values of the film thickness contours for the severely starved condition [Fig. 22.14(c)] are much lower than those for the fully flooded condition [Fig. 22.14(a)].

22.7 CLOSURE

This chapter focused on the transition from rectangular conjunctions covered in the last chapter to elliptical conjunctions. The numerical analysis and computational technique represent a substantial extension. A procedure for the numerical solution of isothermal elastohydrodynamically lubricated elliptical conjunctions was outlined. The influence of the ellipticity parameter and the dimensionless speed, load, and materials parameters on pressure and film shape was investigated in this chapter for elastohydrodynamically lubricated elliptical conjunctions. The ellipticity parameter was varied from 1 (a ball-on-plane configuration) to 8 (a configuration approaching a line contact). The dimensionless speed parameter was varied over a range of nearly two orders of magnitude. The dimensionless load was varied over a range of one order of magnitude. Situations equivalent to using solid materials of bronze, steel, and silicon nitride, and lubricants of paraffinic and naphthenic mineral oils were considered in an investigation of the role of the dimensionless materials parameter. Thirty-four cases were used to generate the minimum-film-thickness formula given below for elastohydrodynamically lubricated elliptical conjunctions while assuming that the solid surface materials are hard and that fully flooded lubrication conditions exist:

$$\tilde{H}_{e,\min} = 3.63 U^{0.68} G^{0.49} W^{-0.073} (1 - e^{-0.68k})$$

Minimum-film-thickness formulas were also obtained for starved lubricating conditions as well as for when one of the solid surfaces is made of a soft (low elastic modulus) material under fully flooded or starved lubricating conditions. For each of these conditions contour plots were presented that indicate in detail the pressure distribution and film thickness throughout the conjunctions. Pressure spikes were in evidence in the hard elastohydrodynamic lubrication results, as was found in the last chapter for rectangular conjunctions. The theoretical solutions of film thickness have all the essential features of previously reported experimental observations based on optical interferometry.

22.8 PROBLEM

22.8.1 Describe at least six practical situations that illustrate the importance of elastohydrodynamic lubrication. Also indicate whether in each of these applications you might anticipate side leakage or starvation to play a significant roll in determining minimum film thickness.

22.9 REFERENCES

Archard, J. F., and Cowking, E. W. (1965–66): Elastohydrodynamic Lubrication at Point Contacts. *Elastohydrodynamic Lubrication. Proc. Inst. Mech. Eng. (London)*, vol. 180, pt. 3B, pp. 17–26.

Archard, J. F., and Kirk, M. T. (1964): Film Thickness for a Range of Lubricants Under Severe Stress. *J. Mech. Eng. Sci.*, vol. 6, pp. 101–102.

Cameron, A., and Gohar, R. (1966): Theoretical and Experimental Studies of the Oil Film in Lubricated Point Contact. *Proc. Roy. Soc. London, Ser. A*, vol. 291, no. 1427, Apr. 26, pp. 520–536.

Castle, P., and Dowson, D. (1972): A Theoretical Analysis of the Starved Elastohydrodynamic Lubrication Problem for Cylinders in Line Contact. Proceedings of Second Symposium on Elastohydrodynamic Lubrication. Institution of Mechanical Engineers, London, pp. 131–137.

Cheng, H. S. (1970): A Numerical Solution to the Elastohydrodynamic Film Thickness in an Elliptical Contact. *J. Lubr. Technol.*, vol. 92, no. 1, pp. 155–162.

Chittenden, R. J. et al. (1985): Theoretical Analysis of Isothermal EHL Concentrated Contacts: Parts I and II. *Proc. Roy. Soc. London, Ser. A*, vol. 387, pp. 245–294.

Dalmaz, G., and Godet, M. (1978): Film Thickness and Effective Viscosity of Some Fire Resistant Fluids in Sliding Point Contacts. *J. Lubr. Technol.*, vol. 100, no. 2, pp. 304–308.

Dowson, D., and Hamrock, B. J. (1976): Numerical Evaluation of the Surface Deformation of Elastic Solids Subjected to a Hertzian Contact Stress. *ASLE Trans.*, vol. 19, no. 4, pp. 279–286.

Gentle, C. R., and Cameron, A. (1973): Optical Elastohydrodynamics at Extreme Pressure. *Nature*, vol. 246, no. 5434, pp. 478–479.

Greenwood, J. A., and Kauzlarich, J. J. (1973): Inlet Shear Heating in Elastohydrodynamic Lubrication. *J. Lubr. Technol.*, vol. 95, no. 4, pp. 417–426.

Grubin, A. N. (1949): Fundamentals of the Hydrodynamic Theory of Lubrication of Heavily Loaded Cylindrical Surfaces. *Investigation of the Contact Machine Components*. Kh. F. Ketova, (ed.). Translation of Russian Book No. 30, Central Scientific Institute for Technology and Mechanical Engineering, Moscow, chap. 2. (Available from Department of Scientific and Industrial Research, Great Britain, Trans. CTS-235 and Special Libraries Association, Trans. R-3554).

Hamrock, B. J., and Dowson, D. (1974): Numerical Evaluation of Surface Deformation of Elastic Solids Subjected to Hertzian Contact Stress. *NASA Tech. Note* D-7774.

Hamrock, B. J., and Dowson, D. (1976a): Isothermal Elastohydrodynamic Lubrication of Point Contacts, Part I—Theoretical Formulation. *J. Lubr. Technol.*, vol. 98, no. 2, pp. 223–229.

Hamrock, B. J., and Dowson, D. (1976b): Isothermal Elastohydrodynamic Lubrication of Point Contacts, Part II—Ellipticity Parameter Results. *J. Lubr. Technol.*, vol. 98, no. 3, pp. 375–383.

Hamrock, B. J., and Dowson, D. (1977a): Isothermal Elastohydrodynamic Lubrication of Point Contacts, Part III—Fully Flooded Results. *J. Lubr. Technol.*, vol. 99, no. 2, pp. 264–276.

Hamrock, B. J., and Dowson, D. (1977b): Isothermal Elastohydrodynamic Lubrication of Point Contacts, Part IV—Starvation Results. *J. Lubr. Technol.*, vol. 99, no. 1, pp. 15–23.

Hamrock, B. J., and Dowson, D. (1978): Elastohydrodynamic Lubrication of Elliptical Contacts for Materials of Low Elastic Modulus, Part I—Fully Flooded Conjunctions. *J. Lubr. Technol.*, vol. 100, no. 2, pp. 236–245.

Hamrock, B. J., and Dowson, D. (1979a): Elastohydrodynamic Lubrication of Elliptical Contacts for Materials of Low Elastic Modulus, Part II—Starved Conjunctions. *J. Lubr. Technol.*, vol. 101, no. 1, pp. 92–98.

Hamrock, B. J., and Dowson, D. (1979b): Minimum Film Thickness in Elliptical Contacts for Different Regimes of Fluid-Film Lubrication. *Elastohydrodynamics and Related Topics.* Proceedings of the 5th Leeds-Lyon Symposium on Tribology, D. Dowson et al. (eds.), Mechanical Engineering Publications Ltd., England, pp. 22–27.

Hamrock, B. J., and Dowson, D. (1981): *Ball Bearing Lubrication—The Elastohydrodynamics of Elliptical Contacts.* Wiley-Interscience, New York.

Johnson, K. L., and Roberts, A. D. (1974): Observation of Viscoelastic Behavior of an Elastohydrodynamic Lubricant Film. *Proc. Roy. Soc. London, Ser. A*, vol. 337, no. 1609, Mar. 19, pp. 217–242.

Kapitza, P. L. (1955): Hydrodynamic Theory of Lubrication During Rolling. *Zh. Tekh. Fiz.*, vol. 25, no. 4, pp. 747–762.

Koye, K. A., and Winer, W. O. (1980): An Experimental Evaluation of the Hamrock and Dowson Minimum Film Thickness Equations for Fully Flooded EHD Point Contacts. *J. Lubr. Technol.*, vol. 103, no. 2, pp. 284–294.

Kunz, R. K., and Winer, W. O. (1977): Discussion on pp. 275–276 of Hamrock, B. J., and Dowson, D., Isothermal Elastohydrodynamic Lubrication of Point Contacts, Part III—Fully Flooded Results. *J. Lubr. Technol.*, vol. 99, no. 2, pp. 264–275.

Lee, D., Sanborn, D. M., and Winer, W. O. (1973): Some Observations of the Relationship Between Film Thickness and Load in High Hertz Pressure Sliding Elastohydrodynamic Contacts. *J. Lubr. Technol.*, vol. 95, no. 3, pp. 386–390.

Moes, H. (1965–66): Discussion of Dowson, D., Elastohydrodynamic Lubrication: An Introduction and Review of Theoretical Studies. *Elastohydrodynamic Lubrication, Proc. Inst. Mech. Eng. (London)*, vol. 180, pt. 3B, pp. 244–245.

Moore, A. J. (1973): "Non-Newtonian Behavior in Elastohydrodynamic Lubrication." Ph.D. thesis. University of Reading.

Orcutt, F. K., and Cheng, H. S. (1966): Lubrication of Rolling-Contact Instrument Bearings. Proceedings of the Gyro-Spin Axis Hydrodynamic Bearing Symposium, Vol. 2, Ball Bearings, Massachusetts Institute of Technology, Cambridge, pp. 1–25.

Parker, R. J., and Kannel, J. W. (1971): Elastohydrodynamic Film Thickness Between Rolling Disks With a Synthetic Paraffinic Oil to 589 K (600 °F). *NASA Tech. Note* D–6411.

Roelands, C. J. A. (1966): *Correlational Aspects of the Viscosity-Temperature-Pressure Relationship of Lubricating Oils.* Druk. V. R. B., Groningen, Netherlands.

Theyse, F. H. (1966): Some Aspects of the Influence of Hydrodynamic Film Formation on the Contact Between Rolling/Sliding Surfaces. *Wear*, vol. 9, pp. 41–59.

Wedeven, L. E., Evans, D., and Cameron, A. (1971): Optical Analysis of Ball Bearing Starvation. *J. Lubr. Technol.*, vol. 93, no. 3, pp. 349–363.

Wolveridge, P. E., Baglin, K. P., and Archard, J. F. (1971): The Starved Lubrication of Cylinders in Line Contact. *Proc. Inst. Mech. Eng. (London)*, vol, 185, no. 81/71, pp. 1159–1169.

CHAPTER
23

FILM THICKNESSES FOR DIFFERENT REGIMES OF FLUID FILM LUBRICATION

The type of lubrication taking place in nonconformal conjunctions is influenced by two major physical effects: the elastic deformation of the solids under an applied load and the increase in fluid viscosity with pressure. There are four main regimes of fluid film lubrication, depending on the magnitude of these effects and on their importance. These four regimes are defined as follows:

1. *Isoviscous-rigid*: In this regime the amount of elastic deformation of the surfaces is insignificant relative to the thickness of the fluid film separating them, and the maximum contact pressure is too low to increase fluid viscosity significantly. This form of lubrication is typically encountered in circular-arc thrust bearing pads and in industrial coating processes in which paint, emulsion, or protective coatings are applied to sheet or film materials passing between rollers.
2. *Viscous-rigid*: If the contact pressure is sufficiently high to significantly increase the fluid viscosity within the conjunction, it may be necessary to consider the pressure-viscosity characteristics of the lubricant while assuming

that the solids remain rigid. For the latter part of this assumption to be valid the deformation of the surfaces must remain insignificant relative to the fluid film thickness. This form of lubrication may be encountered on roller end-guide flanges, in contacts in moderately loaded cylindrical tapered rollers, and between some piston rings and cylinder liners.

3. *Isoviscous-elastic*: In this regime the elastic deformation of the solids is significant relative to the thickness of the fluid film separating them, but the contact pressure is quite low and insufficient to cause any substantial increase in viscosity. This situation arises with low-elastic-modulus materials (soft EHL) and may be encountered in seals, human joints, tires, and elastomeric-material machine elements.
4. *Viscous-elastic*: In fully developed elastohydrodynamic lubrication the elastic deformation of the solids is often significant relative to the thickness of the fluids film separating them, and the contact pressure is high enough to significantly increase the lubricant viscosity within the conjunction. This form of lubrication is typically encountered in ball and roller bearings, gears, and cams.

Several authors—Moes (1965–66), Theyse (1966), Archard (1968), Greenwood (1969), Johnson (1970), and Hooke (1977)—have contributed solutions for the film thickness in the four lubrication regimes, but their results have been confined largely to rectangular conjunctions. The essential difference between these contributions is the way in which the parameters were made dimensionless. In this chapter the film thickness is defined for the four fluid film lubrication regimes just described for conjunctions ranging from circular to rectangular. The film thickness equations for the respective lubrication regimes come from my theoretical studies on elastohydrodynamic and hydrodynamic lubrication of elliptical conjunctions. The results are valid for isothermal, fully flooded conjunctions. In addition to the film thickness equations for the various conditions, maps are presented of the lubrication regimes, with film thickness contours represented on a log-log grid of the viscosity and elasticity parameters for three values of the ellipticity parameter. This chapter draws extensively from Hamrock and Dowson (1979) and Esfahanian and Hamrock (1991).

23.1 DIMENSIONLESS GROUPING

Chapter 22 presented elastohydrodynamic theory for elliptical conjunctions in terms of the dimensionless groups (H_e, U, W, G, k). This has been particularly helpful, since it allows the physical explanation of conjunction behavior to be readily associated with each set of numerical results. However, several authors have noted that this dimensionless group can be reduced by one parameter—without any loss of generality—by using dimensional analysis. The film thickness contours for the four fluid film lubrication regimes can be conveniently represented graphically by the fewest parameters, even though the physical meaning of each composite parameter requires careful consideration.

Johnson (1970) has pointed out that the behavior distinguishing the four lubrication regimes can be characterized by three dimensionless quantities, each having the dimensions of pressure:

1. The reduced pressure parameter p^*, a measure of the fluid pressure generated by an isoviscous lubricant when elastic deformation is neglected
2. The inverse pressure-viscosity coefficient $1/\xi$, a measure of viscosity change with pressure
3. The maximum Hertzian pressure p_H, the maximum pressure of a dry elastic contact

Although Johnson (1970) does not consider elliptical conjunctions, he does state what the dimensionless parameters for such configurations should be.

Dimensionless film thickness parameter:

$$\hat{H} = H_e \left(\frac{W}{U} \right)^2 \tag{23.1}$$

Dimensionless viscosity parameter:

$$g_V = \frac{GW^3}{U^2} \tag{23.2}$$

Dimensionless elasticity parameter:

$$g_E = \frac{W^{8/3}}{U^2} \tag{23.3}$$

The ellipticity parameter k remains as discussed in Chap. 22. Therefore, the reduced dimensionless group is $(\hat{H}, g_V, g_E, k)$.

23.2 ISOVISCOUS-RIGID REGIME

How the conjunction geometry influences the isothermal hydrodynamic film separating two rigid solids was investigated by Brewe et al. (1979) for fully flooded, isoviscous conditions and covered in Chap. 18. Brewe et al. varied the radius ratio R_y/R_x from 1 (a circular configuration) to 36 (a configuration approaching a rectangular conjunction) and varied the film thickness over two orders of magnitude for conditions representing steel solids separated by a paraffinic mineral oil. They found that the computed minimum film thickness had the same speed, viscosity, and load dependence as the classical Kapitza (1955) solution. However, when they introduced the Reynolds cavitation boundary condition ($\partial p/\partial x = 0$ and $p = 0$ at the cavitation boundary), an additional geometrical effect emerged. Therefore, from Brewe et al. (1979) the dimensionless minimum (or central) film thickness parameter for the isoviscous-rigid lubrication regime can be written from Eqs. (18.75) and (18.76) in terms of the new dimensionalization as

$$\left(\hat{H}_{\min}\right)_{\mathrm{IR}} = \left(\hat{H}_c\right)_{\mathrm{IR}} = 128\alpha_r\left[0.131\tan^{-1}\left(\frac{\alpha_r}{2}\right) + 1.683\right]^2 \varphi^2 \tag{23.4}$$

where

$$\alpha_r = \frac{R_y}{R_x} = k^{\pi/2} \tag{19.29}$$

$$\varphi = \left(1 + \frac{2}{3\alpha_r}\right)^{-1} \tag{18.64}$$

In Eq. (23.4) the dimensionless film thickness parameter $\hat{H}$ is shown to be strictly a function of the contact geometry R_y/R_x.

23.3 VISCOUS-RIGID REGIME

Jeng et al. (1987) have shown that the minimum (or central) film thickness for the viscous-rigid regime in a elliptical conjunction can be expressed as

$$H_{\min,e} = H_{c,e} = 178G^{0.386}U^{1.266}W^{-0.880}\left(1 - e^{-0.0387\alpha_r}\right) \tag{23.5}$$

Forty-one cases were used in obtaining this formula. Note that, in the past, several formulas have been proposed for the lubricant film thickness in this lubrication regime. However, either the load parameter W, which has a strong effect on film thickness, was not included or the film thickness was overestimated by using the Barus (1893) formula for pressure-viscosity characteristics. The Roelands (1966) formula was used for the pressure-viscosity relationship in the analysis that produced Eq. (23.5).

Rewriting Eq. (23.5) in terms of the dimensionless groupings given in Eqs. (23.1) to (23.3) gives

$$\left(\hat{H}_{\min}\right)_{\mathrm{VR}} = \left(\hat{H}_c\right)_{\mathrm{VR}} = 141g_V^{0.375}\left(1 - e^{-0.0387\alpha_r}\right) \tag{23.6}$$

Note that before the Jeng et al. (1987) work a modified version of the Blok (1952) work had been used to represent the film thickness in the viscous-rigid regime. For example, in Hamrock and Dowson (1979) the following modified Blok equation was used:

$$\left(\hat{H}_{\min}\right)_{\mathrm{VR}} = \left(\hat{H}_c\right)_{\mathrm{VR}} = 1.66g_V^{0.667}\left(1 - e^{-0.68\alpha_r^{2/\pi}}\right) \tag{23.7}$$

Two differences should be observed in Eqs. (23.6) and (23.7). The first is that the exponent on g_V in (23.6) is almost half that given in (23.7). In deriving Eq. (23.7) it was assumed, since no other equation was available at the time, that the geometry effect for the viscous-rigid regime was the same as that for the viscous-elastic regime. This assumption turns out not to be true as demonstrated by the difference in the geometry effect shown in Eqs. (23.6) and (23.7). Figure

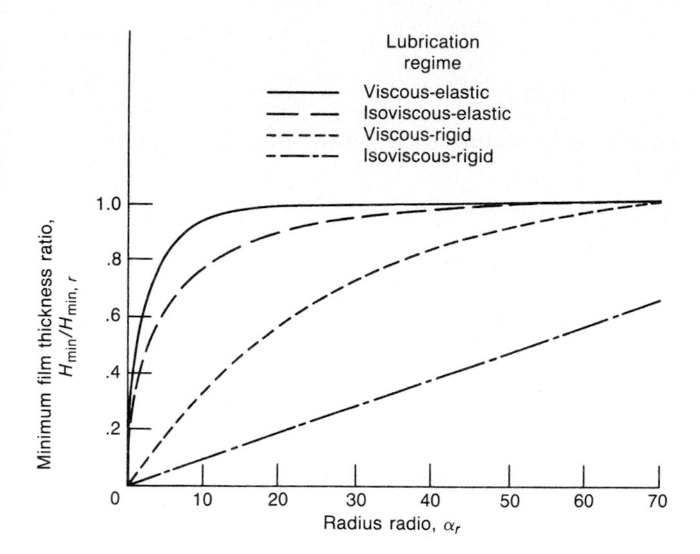

FIGURE 23.1
Comparison of geometry effects in four lubrication regimes. [*From Jeng et al. (1987).*]

23.1 is presented to further illustrate this point. It is assumed that when $\alpha_r = 150$ a rectangular conjunction is obtained. From Fig. 23.1 it is observed that $H_{min}/H_{min,r}$ approaches the limiting value quickly for the viscous-elastic regime, and most slowly for the isoviscous-rigid regime. It is further observed that the assumption that the geometry effects for the viscous-rigid regime are similar to those for the viscous-elastic regime is not true. Therefore, both the power on g_V and the difference in geometry effect should appreciably change the results given in Hamrock and Dowson (1979).

23.4 ISOVISCOUS-ELASTIC REGIME

The influence of the ellipticity parameter k and the dimensionless speed U, load W, and materials G parameters on the minimum (or central) film thickness was investigated theoretically for the isoviscous-elastic (soft EHL) regime. The results were presented in Sec. 22.5. The ellipticity parameter was varied from 1 (a circular configuration) to 12 (a configuration approaching a rectangular conjunction). The dimensionless speed and load parameters were each varied by one order of magnitude. Seventeen cases were considered in obtaining the dimensionless minimum-film-thickness equation

$$\tilde{H}_{e,\,min} = 7.43U^{0.65}W^{-0.21}(1 - 0.85e^{-0.31k}) \tag{22.22}$$

From Eqs. (23.1) and (23.3) the general form of the dimensionless minimum-film-thickness parameter for the isoviscous-elastic lubrication regime can be expressed as

$$\hat{H}_{\min} = A g_E^c (1 - 0.85 e^{-0.31k}) \tag{23.8}$$

where A and c are constants to be determined. From Eqs. (23.1) and (23.3), Eq. (23.8) can be written as

$$H_{e,\min} = A U^{2-2c} W^{(8/3c)-2} (1 - 0.85 e^{-0.31k}) \tag{23.9}$$

Comparing Eq. (22.22) with (23.9) gives $c = 0.67$. Substituting this into Eq. (23.8) while solving for A gives

$$A = \frac{\hat{H}_{\min}}{g_E^{0.67}(1 - 0.85 e^{-0.31k})} \tag{23.10}$$

The arithmetic mean for A based on the 17 cases considered in Sec. 22.5 is 8.70, with a standard deviation of ± 0.05. Therefore, the dimensionless minimum-film-thickness parameter for the isoviscous-elastic lubrication regime can be written as

$$\left(\hat{H}_{\min}\right)_{\mathrm{IE}} = 8.70 g_E^{0.67} (1 - 0.85 e^{-0.31k}) \tag{23.11}$$

With a similar approach the dimensionless central-film-thickness parameter for the isoviscous-elastic lubrication regime can be written as

$$\left(\hat{H}_c\right)_{\mathrm{IE}} = 11.15 g_E^{0.67} (1 - 0.72 e^{-0.28k}) \tag{23.12}$$

23.5 VISCOUS-ELASTIC REGIME

In Sec. 22.3 the influence of the ellipticity parameter and the dimensionless speed, load, and materials parameters on the minimum (or central) film thicknesses of hard-EHL contacts was investigated theoretically for the viscous-elastic regime. The ellipticity parameter was varied from 1 to 8, the dimensionless speed parameter was varied over nearly two orders of magnitude, and the dimensionless load parameter was varied over one order of magnitude. Conditions corresponding to the use of bronze, steel, and silicon nitride and paraffinic and naphthenic oils were considered in obtaining the exponent on the dimensionless materials parameter. Thirty-four cases were used in obtaining the following dimensionless minimum-film-thickness formula:

$$\tilde{H}_{e,\min} = 3.63 U^{0.68} G^{0.49} W^{-0.073} (1 - e^{-0.68k}) \tag{22.18}$$

The general form of the dimensionless minimum-film-thickness parameter for the viscous-elastic lubrication regime can be written as

$$\hat{H}_{\min} = B g_V^d g_E^f (1 - e^{-0.68k}) \tag{23.13}$$

where B, d, and f are constants to be determined. From Eqs. (23.1) to (23.3), Eq. (23.13) can be written as

$$H_{e,\min} = B G^d U^{2-2d-2f} W^{-2+3d+(8f/3)} (1 - e^{-0.68k}) \tag{23.14}$$

Comparing Eq. (22.18) with (23.14) gives $d = 0.49$ and $f = 0.17$. Substituting these values into Eq. (23.13) while solving for B gives

$$B = \frac{\hat{H}_{\min}}{g_V^{0.49} g_E^{0.17}(1 - e^{-0.68k})} \tag{23.15}$$

For the 34 cases considered in Table 22.1 for the derivation of Eq. (22.18), the arithmetic mean for B was 3.42, with a standard deviation of ± 0.03. Therefore, the dimensionless minimum-film-thickness parameter for the viscous-elastic lubrication regime can be written as

$$\left(\hat{H}_{\min}\right)_{\mathrm{VE}} = 3.42 g_V^{0.49} g_E^{0.17}(1 - e^{-0.68k}) \tag{23.16}$$

An interesting observation to make in comparing Eqs. (23.11) and (23.16) is that the sum of the exponents on g_V and g_E is close to $\frac{2}{3}$ for the isoviscous-elastic and viscous-elastic cases.

By adopting a similar approach to that outlined here, the dimensionless central-film-thickness parameter for the viscous-elastic lubrication regime can be written as

$$\left(\hat{H}_c\right)_{\mathrm{VE}} = 3.61 g_V^{0.53} g_E^{0.13}(1 - 0.61 e^{-0.73k}) \tag{23.17}$$

23.6 PROCEDURE FOR MAPPING THE DIFFERENT LUBRICATION REGIMES

The dimensionless minimum-film-thickness parameters for the four fluid film lubrication regimes expressed in Eqs. (23.4), (23.6), (23.11), and (23.16) were used to develop a map of the lubrication regimes in the form of dimensionless minimum-film-thickness-parameter contours. These maps are shown in Fig. 23.2 on a log-log grid of the dimensionless viscosity and elasticity parameters for ellipticity parameters k of $\frac{1}{2}$, 1, 3, and 6, respectively. The procedure used to obtain these figures was as follows:

1. For a given value of the ellipticity parameter, $(\hat{H}_{\min})_{\mathrm{IR}}$ was calculated from Eq. (23.4).
2. For a value of $\hat{H}_{\min} > (\hat{H}_{\min})_{\mathrm{IR}}$ and the value of k chosen in step 1, the dimensionless viscosity parameter was calculated from Eq. (23.6) as

$$g_V = \left[\frac{\hat{H}_{\min}}{141(1 - e^{-0.0387\alpha_r})}\right]^{1/0.375} \tag{23.18}$$

   This established the dimensionless minimum-film-thickness-parameter contours $\hat{H}_{\min}$ as a function of g_V for a given value of k in the viscous-rigid regime.
3. For the values of k selected in step 1, $\hat{H}_{\min}$ selected in step 2, and g_V obtained from Eq. (23.18), the dimensionless elasticity parameter was calcu-

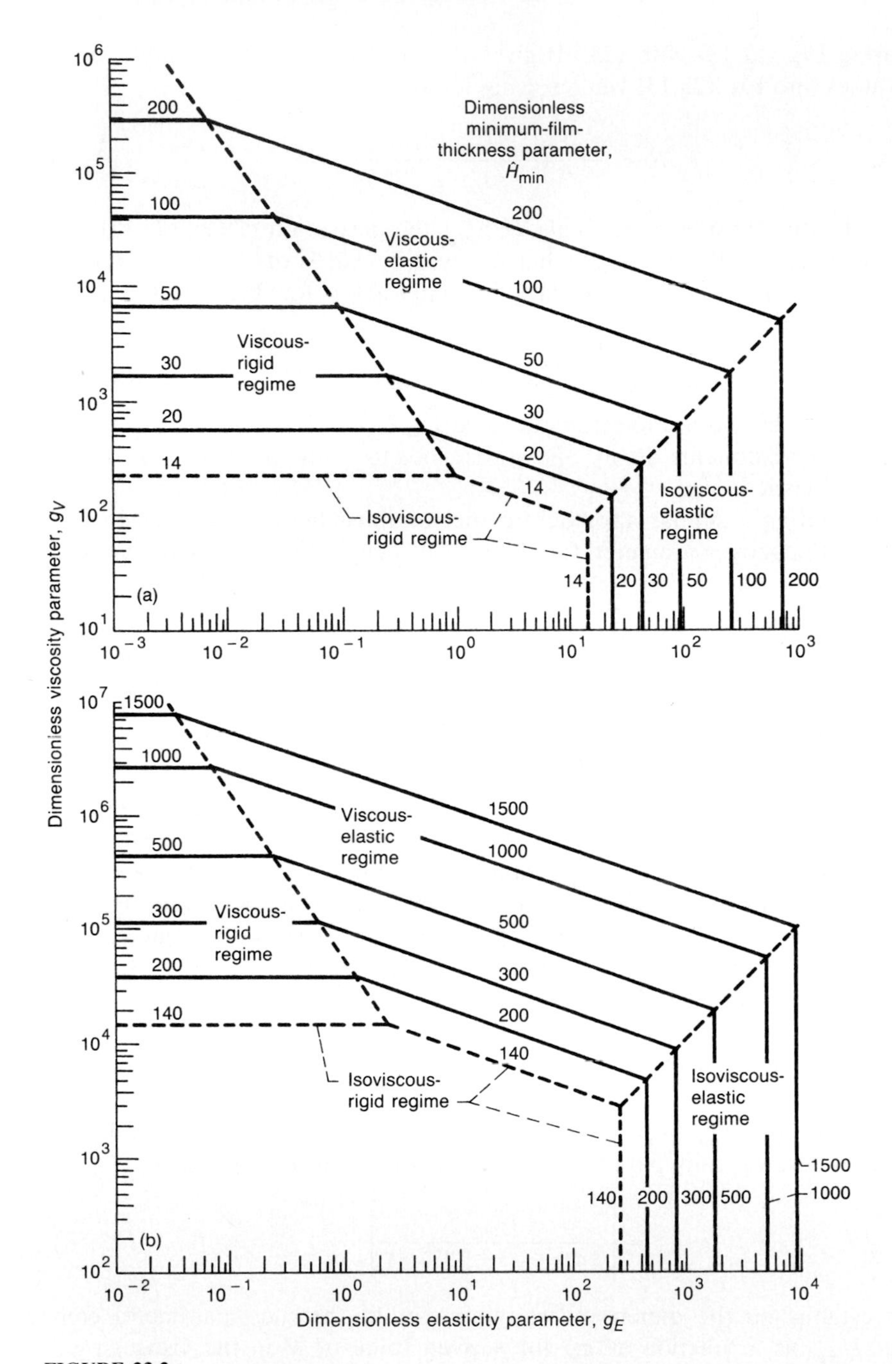

FIGURE 23.2
Maps of lubrication regimes for four values of ellipticity parameter k. (a) $k = \frac{1}{2}$; (b) $k = 1$. [*From Esfahanian and Hamrock (1991).*]

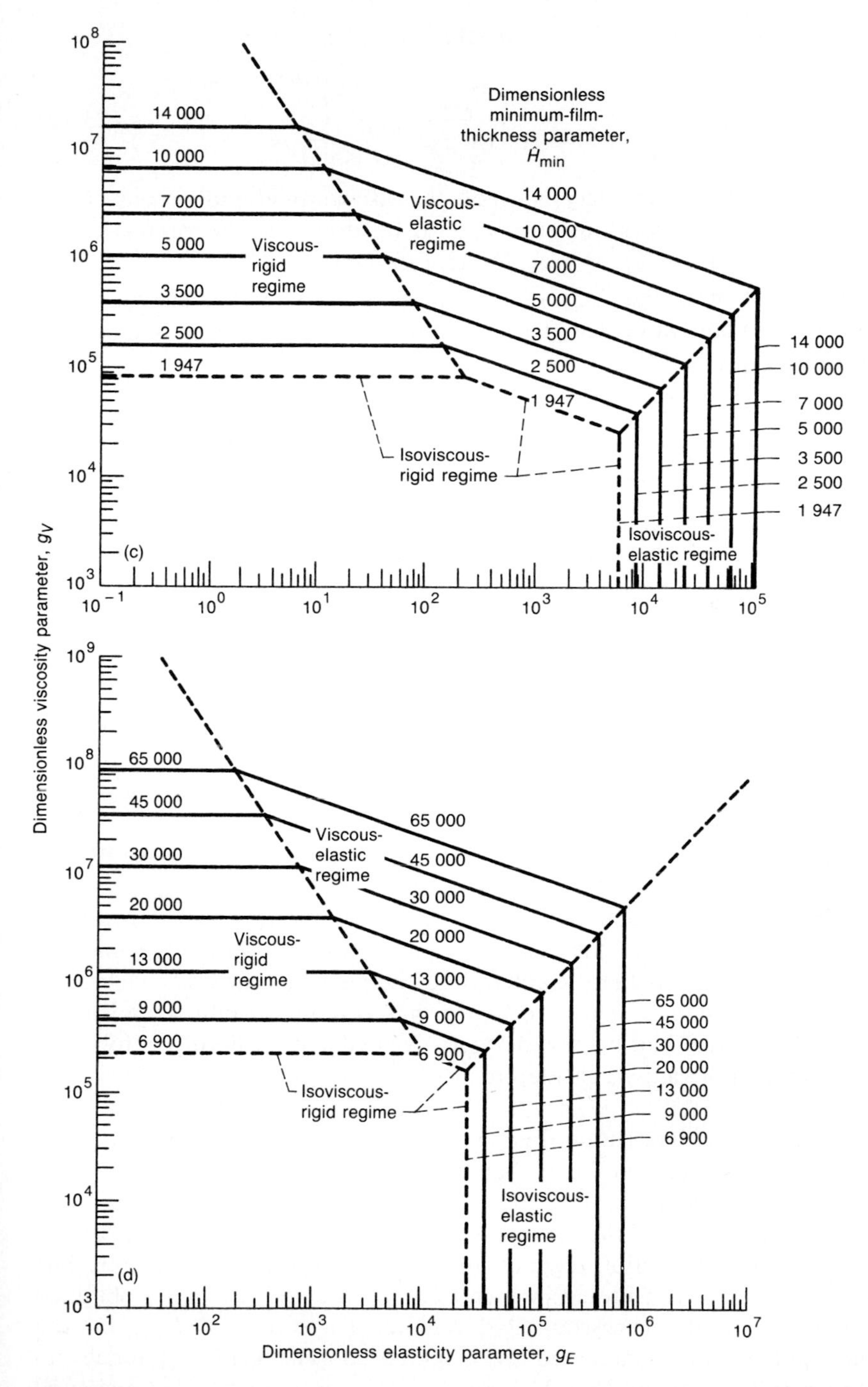

FIGURE 23.2
Concluded. (c) $k = 3$; (d) $k = 6$.

lated from the following equation, derived from Eq. (23.16):

$$g_E = \left[\frac{\hat{H}_{\min}}{3.42 g_V^{0.49}(1 - e^{-0.68k})} \right]^{1/0.17} \tag{23.19}$$

This established the boundary between the viscous-rigid and viscous-elastic regimes and enabled contours of $\hat{H}_{\min}$ to be drawn in the viscous-elastic regime as functions of g_V and g_E for given values of k.

4. For the values of k and $\hat{H}_{\min}$ chosen in steps 1 and 2, the dimensionless elasticity parameter was calculated from the following equation, obtained by rearranging Eq. (23.11):

$$g_E = \left[\frac{\hat{H}_{\min}}{8.70(1 - 0.85 e^{-0.31k})} \right]^{1/0.67} \tag{23.20}$$

This established the dimensionless minimum-film-thickness-parameter contours $\hat{H}_{\min}$ as a function of g_E for a given value of k in the isoviscous-elastic lubrication regime.

5. For the values of k and $\hat{H}_{\min}$ selected in steps 1 and 2 and the value of g_E obtained from Eq. (23.20), the viscosity parameter was calculated from the following equation:

$$g_V = \left[\frac{\hat{H}_{\min}}{3.42 g_E^{0.17}(1 - e^{-0.68k})} \right]^{1/0.49} \tag{23.21}$$

This established the isoviscous-elastic and viscous-elastic boundaries for the particular values of k and $\hat{H}_{\min}$ chosen in steps 1 and 2.

6. At this point, for particular values of k and $\hat{H}_{\min}$, the contours were drawn through the viscous-rigid, viscous-elastic, and isoviscous-elastic regimes. A new value of $\hat{H}_{\min}$ was then selected, and the new contour was constructed by returning to step 2. This procedure was continued until an adequate number of contours had been generated. A similar procedure was followed for the range of k values considered.

The maps of the lubrication regimes shown in Fig. 23.2 were generated by following the procedure outlined here. The dimensionless minimum-film-thickness-parameter contours were plotted on a log-log grid of the dimensionless viscosity parameter and the dimensionless elasticity parameter for k of $\frac{1}{2}$, 1, 3, and 6. The four lubrication regimes are clearly shown in these figures. The smallest $\hat{H}_{\min}$ contour considered in each case represented the values obtained from Eq. (23.4) in the isoviscous-rigid regime. The value of $\hat{H}_{\min}$ on the isoviscous-rigid boundary increased as k increased. The main difference between the results given in Fig. 23.2 and those presented by Hamrock and Dowson (1979) is that the viscous-elastic regime is much larger in Fig. 23.2 than in the Hamrock and Dowson results. This is entirely due to the use of the Jeng et al. (1987) formula in the viscous-rigid regime instead of the modified Blok (1952) formula.

By using Fig. 23.2 for given values of the parameters k, g_V, and g_E, the fluid film lubrication regime in which any elliptical conjunction is operating can be ascertained and the approximate value of $\hat{H}_{\min}$ determined. When the lubrication regime is known, a more accurate value of $\hat{H}_{\min}$ can be obtained by using the appropriate dimensionless minimum-film-thickness-parameter equation.

23.7 THERMAL CORRECTION FACTOR

The relevant film thickness formulas for the various lubrication regimes having been described, the next task is to look at secondary effects that alter these formulas. The first of these correction factors has already been considered under starvation effects. That is, the starvation effects considered in Sec. 18.5 are applicable for both the isoviscous-rigid and viscous-rigid regimes. For the viscous-elastic regime the starvation effects covered in Sec. 22.6.2 are applicable; for the isoviscous-elastic regime Sec. 22.6.3 is applicable. The film thickness reduction due to starvation needs to be properly taken care of while applying the information to the specified section of the fully flooded film equation.

The film thickness predictions given elsewhere in this chapter require knowledge of the lubricant temperature in the conjunction inlet, which can be used in determining the absolute viscosity at $p = 0$ and constant temperature η_0. A method of estimating lubricant film thickness when considering shear heating effects is described in this section.

The film thickness reduction due to viscous heating of the lubricant at the conjunction inlet was examined by Cheng (1967) and by Sheu and Wilson (1982). These data were recently consolidated and correlated by Gupta et al. (1991) to give the following integrated empirical formula for calculating the percentage of film thickness reduction due to inlet heating, or the thermal correction factor C_t.

$$C_t = \frac{1 - 13.2(p_H/E')(L^*)^{0.42}}{1 + 0.213(1 + 2.23A_c^{0.83})(L^*)^{0.64}} \tag{23.22}$$

where

$$L^* = \left(-\frac{\partial \eta}{\partial t_m}\right)\frac{(\tilde{u})^2}{K_f} \tag{23.23}$$

and $C_t = h_{\text{thermal}}/h_{\text{isothermal}}$
p_H = maximum Hertzian pressure, N/m^2
E' = see Eq. (20.36), N/m^2
η = absolute viscosity, $\text{N} \cdot \text{s/m}^2$
t_m = lubricant temperature, °C
$\tilde{u}$ = mean surface velocity, m/s
K_f = lubricant thermal conductivity, W/(m · °C)
A_c = slide-roll ratio, $2(u_b - u_a)/(u_a + u_b)$

Thus, C_t calculated from Eq. (23.22) must be multiplied by $\hat{H}_{\min}$ to get the effect of the thermal correction factor. It is suggested that Eq. (23.22) be applied to all the fluid film regimes (IR, VR, IE, and VE). Note that if pure rolling exists, A_c is equal to zero and Eq. (23.22) is still valid. Substantial reduction in film thickness can result if thermal effects are considered.

Sheu and Wilson (1982) discovered no conditions under which thermal action leads to a catastrophic breakdown in lubrication. They concluded that no conditions have been found under which a lubricant film cannot be formed nor do the forms of the relationships given in Eq. (23.22) suggest that such a condition exists. Thus, the analysis seems to indicate that lubricant failure due to thermal action is not a direct result of viscosity reduction due to temperature variations across the film. If thermally induced failure exists, it must be due to either surface temperature rises or changes in some property other than viscosity.

The effects of temperature on the film thickness, pressure, and surface shear stress distributions within the entire conjunction were examined early on by Cheng (1965) and Dowson and Whitaker (1965). They discovered that although temperature is a major influence on friction, it does not alter significantly the minimum film thickness at the outlet of the conjunction, at least for moderately loaded cases and slide-roll ratios up to 25 percent. Recently, Sadeghi and Sui (1990) extended these results to much higher loads and slide-roll ratios up to 30 percent. A somewhat higher reduction in minimum film thickness, up to 15 percent, was found in their calculations.

Sliding friction is affected significantly by heat not only in thermal EHL theories based on the Newtonian model, as alluded to earlier, but also in later analyses based on non-Newtonian models, such as that by Bair and Winer (1979).

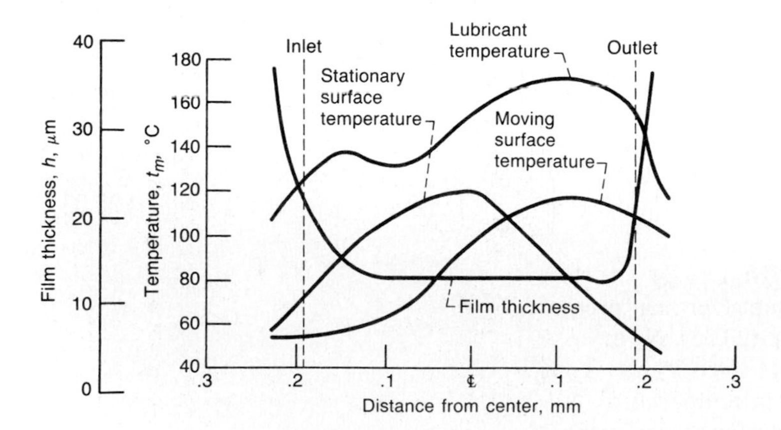

FIGURE 23.3
Lubricant and ball surface temperatures along contact centerline. Sliding speed, 1.40 m/s; load, 67 N. [*From Winer (1983)*.]

Experimental verification of film temperature rise has proven to be extremely difficult. Winer (1983) employed infrared thermometry and was successful in measuring the temperature rise in the film between a steel ball sliding against a sapphire disk. Figure 23.3 shows the film thickness, the film temperature rise, and both surface temperatures with $\sigma_{max} = 1.82$ GPa and a bath temperature of 40°C. Even though these measured film temperature profiles cannot be directly compared with those calculated recently by Sadeghi and Sui (1990) because of the difference in line and point contacts, we can still observe similar features in the measured and calculated profiles.

23.8 SURFACE ROUGHNESS CORRECTION FACTOR

Surface roughness refers to any departure of the actual surface height from a reference level as discussed in Chap. 3. It is due to the physical and chemical methods used in preparing the surface as well as to the microstructure of the material. Surface roughness can appreciably affect the lubrication of surfaces if the film thickness is of the same order as the roughness. For example, endurance tests of ball bearings, as reported by Tallian et al. (1965), have demonstrated that when the lubricant film is thick enough to separate the contacting bodies, the fatigue life of the bearing is greatly extended. Conversely, when the film is not thick enough to provide full separation between the asperities in the contact zone, the bearing life is adversely affected by the high shear resulting from direct metal-to-metal contact.

A surface roughness correction factor C_r for the minimum film thickness is defined as a function of the ratio of minimum film thickness to composite surface roughness Λ as described in Chap. 3. The mathematical expression for Λ is

$$\Lambda = \frac{h_{min}}{R_q} \qquad (3.20)$$

where

$$R_q = \left(R_{q,a}^2 + R_{q,b}^2\right)^{1/2} \qquad (23.24)$$

and $R_{q,a}$ and $R_{q,b}$ are the rms surface finishes of solids a and b, respectively (see Sec. 3.5 for more details).

Another parameter used to define surface texture is the surface pattern parameter. Many engineering surfaces have roughnesses that are directionally oriented and can be described by the surface pattern parameter γ_c, first introduced by Peklenik (1968) as

$$\gamma_c = \frac{\lambda_x}{\lambda_y} \qquad (23.25)$$

where λ_x and λ_y are autocorrelation lengths in the x and y directions. The surface pattern parameter γ_c may be interpreted as the length-to-width ratio of a representative asperity contact. Purely transverse, isotropic, and purely longitudinal patterns have γ_c of 0, 1, and ∞, respectively. Surfaces with $\gamma_c > 1$ are longitudinally oriented.

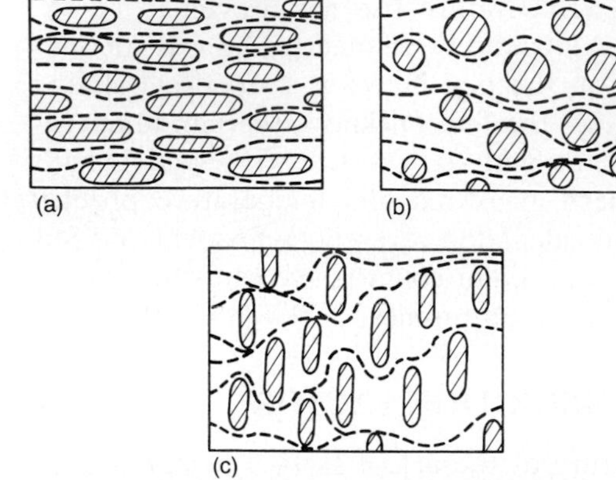

FIGURE 23.4
Typical contact areas for (a) longitudinally oriented ($\gamma_c = 6$), (b) isotropic ($\gamma_c = 1$), and (c) transversely oriented ($\gamma_c = \frac{1}{6}$) rough surfaces. [*From Patir and Cheng (1978).*]

Cheng and Dyson (1978) showed the effect of roughness on the pressure buildup in the inlet region, up to the center of the contact. They used a stochastic Reynolds equation for longitudinal striations and a force compliance relationship to account for the mean contact pressure between the asperities.

In a later publication, Patir and Cheng (1978) made the solution more general by replacing the stochastic Reynolds equation, used by Cheng and Dyson, with an average Reynolds equation. They developed a method known as the "average flow model" to handle surface roughnesses at any arbitrary surface pattern parameter γ_c. The flow factors in the average Reynolds equation are defined by numerical flow simulation. This method enables one not only to investigate the effect of more practical surface roughness textures but also to

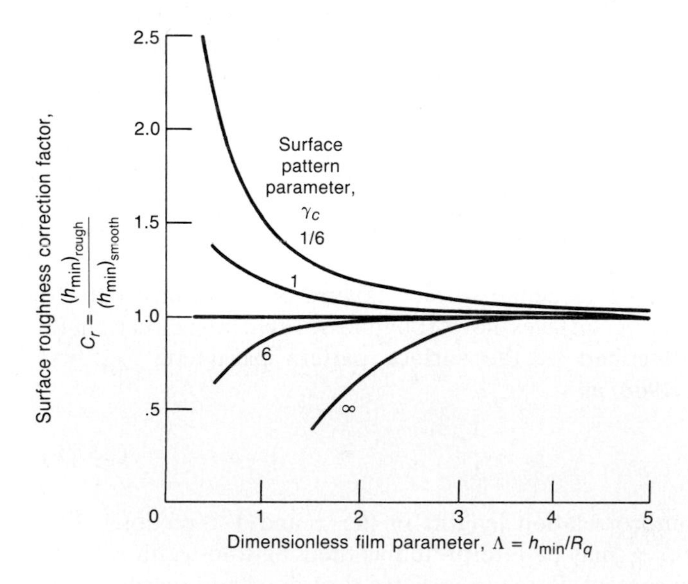

FIGURE 23.5
Effect of surface roughness on average film thickness of EHL contacts. [*From Patir and Cheng (1978).*]

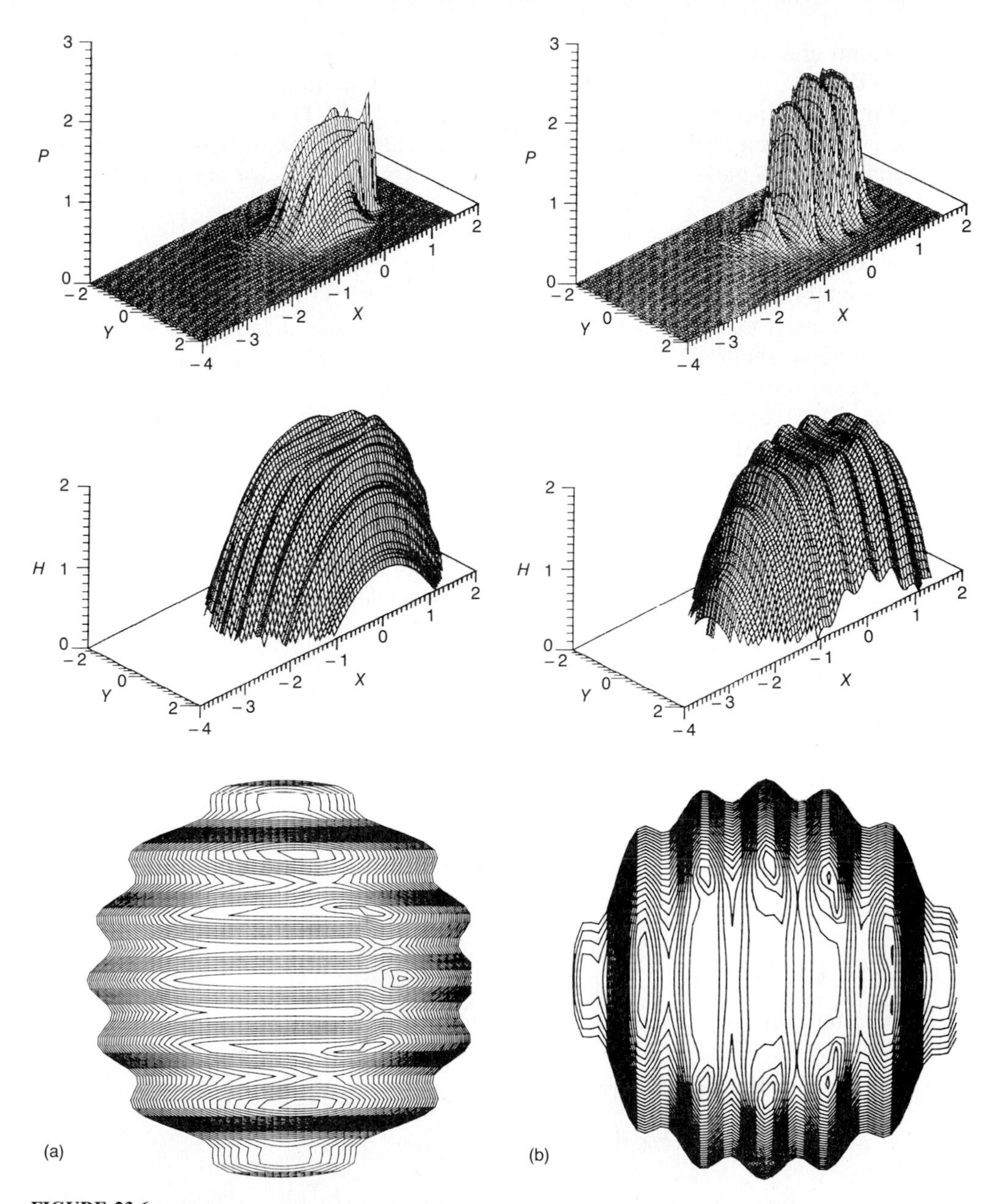

FIGURE 23.6
Pressure profiles, film thickness profiles, and film thickness contour plots for longitudinal and transverse roughness. (a) Longitudinal roughness. Dimensionless film thickness where $dP/dX = 0$, $H_m = 0.130$. (b) Transverse roughness. $H_m = 0.155$. Dimensionless viscosity index $\tilde{\alpha}$, 10; dimensionless velocity parameter λ, 0.1; dimensionless amplitude of asperity A_a, 0.1; dimensionless radius of asperity R_b, 0.125. [*From Lubrecht et al. (1988).*]

extend the results to $\Lambda < 3$, where part of the load is shared by asperity contacts. Figure 23.4 shows the flow pattern for longitudinally oriented ($\gamma_c = 6$), transversely oriented ($\gamma_c = 1/6$), and isotropic ($\gamma_c = 1$) roughness. A simplified way of obtaining the value of γ_c is to calculate the ratio between high spot counts (HSC) measured in the lateral direction y and HSC measured in the rolling direction x.

Figure 23.5, obtained from Patir and Cheng (1978), shows C_r plotted against the film parameter Λ. A conclusion reached from Fig. 23.5 is that transverse roughness orientation leads to a thicker mean film than isotropic roughness orientation, and isotropic to a thicker mean film than longitudinal roughness orientation. For $\Lambda \sim 2$, $h_{\text{rough}} \approx h_{\text{smooth}}$. The surface roughness correction factor C_r can be estimated from Fig. 23.5 and then multiplied by $(h_{\min})_{\text{smooth}}$ to get $(h_{\min})_{\text{rough}}$.

Results that include the effects of surface roughness have been recently obtained for point contacts by Zhu and Cheng (1988) using the average flow factors provided by Patir and Cheng. They show the same features found in line contacts.

With the recent dramatic improvements in computer memory and speed and with the recent development of more efficient numerical schemes in EHL, such as the multigrid method by Lubrecht et al. (1988), it is now possible to calculate the detailed pressure and deformation around idealized asperities in a circular or elliptical Hertzian conjunction. Efforts in this area are now known as "microelastohydrodynamic lubrication." Representative works in micro-EHL include those on longitudinal and transverse roughness by Lubrecht et al. (1988), as shown in Fig. 23.6, by Kweh et al. (1989), and by Barragan de Ling et al. (1989). Their powerful methods offer considerable promise for verifying the average quantities obtained from the Reynolds equation and for determining asperity film thickness, peak pressure, temperature, shear stress, and near-surface stress field.

23.9 CLOSURE

Relationships for the dimensionless minimum-film-thickness-parameter equations for the four lubrication regimes found in elliptical conjunctions have been developed and expressed as

Isoviscous-rigid regime:

$$\left(\hat{H}_{\min}\right)_{\text{IR}} = \left(\hat{H}_c\right)_{\text{IR}} = 128\alpha_r\left[0.131\tan^{-1}\left(\frac{\alpha_r}{2}\right) + 1.683\right]^2\varphi^2$$

$$\alpha_r = \frac{R_y}{R_x} = k^{\pi/2}$$

$$\varphi = \left(1 + \frac{2}{3\alpha_r}\right)^{-1}$$

Viscous-rigid regime:

$$\left(\hat{H}_{\min}\right)_{\mathrm{VR}} = 141 g_V^{0.375}\left(1 - e^{-0.0387\alpha_r}\right)$$

Isoviscous-elastic regime:

$$\left(\hat{H}_{\min}\right)_{\mathrm{IE}} = 8.70 g_E^{0.67}\left(1 - 0.85 e^{-0.31k}\right)$$

Viscous-elastic regime:

$$\left(\hat{H}_{\min}\right)_{\mathrm{VE}} = 3.42 g_V^{0.49} g_E^{0.17}\left(1 - e^{-0.68k}\right)$$

The relative importance of the influence of pressure on elastic distortion and lubricant viscosity is the factor that distinguishes these regimes for a given conjunction geometry.

These equations were used to develop maps of the lubrication regimes by plotting film thickness contours on a log-log grid of the viscosity and elasticity parameters for four values of the ellipticity parameter. These results present a complete theoretical minimum-film-thickness-parameter solution for elliptical conjunctions in the four lubrication regimes. The results are particularly useful in initial investigations of many practical lubrication problems involving elliptical conjunctions. Correction factors taking into account lubricant supply, thermal effects, and surface roughness effects were also presented.

23.10 PROBLEM

23.10.1 In Problem 21.7.1 we assumed that the viscous-elastic regime was the proper lubrication regime while considering water or oil as a lubricant. Making use of the fluid film regime charts of this chapter, indicate if this is a valid assumption for both lubricants. Also indicate how the minimum film thickness obtained for rectangular conjunction given in Prob. 21.7.1 compares with the elliptical conjunctions results found in this problem.

23.11 REFERENCES

Archard, J. F. (1968): Non-Dimensional Parameters in Isothermal Theories of Elastohydrodynamic Lubrication. *J. Mech. Eng. Sci.*, vol. 10, no. 2, pp. 165–167.

Bair, S., and Winer, W. O. (1979): Shear Strength Measurements of Lubricants at High Pressures, *J. Lubr. Technol.*, vol. 101, no. 3, pp. 251–257.

Barragan de Ling, FdM., Evans, H. P., and Snidle, R. W. (1989): Microelastohydrodynamic Lubrication of Circumferentially Finished Rollers: The Influence of Temperature and Roughness, *J. Tribol*, vol. 111, no. 4, pp. 730–736.

Barus, C. (1893): Isotherms, Isopiestics, and Isometrics Relative to Viscosity. *Am. J. Sci.*, vol. 45, pp. 87–96.

Blok, H. (1952): Discussion of paper by E. McEwen, The Effect of Variation of Viscosity With Pressure on the Load-Carrying Capacity of the Oil Film Between Gear Teeth. *J. Inst. Petrol.*, vol. 38, no. 344, pp. 673–683.

Brewe, D. E., Hamrock, B. J., and Taylor, C. M. (1979): Effects of Geometry on Hydrodynamic Film Thickness. *J. Lubr. Technol.*, vol. 101, no. 2, pp. 231–239.

Cheng, H. S. (1965): A Refined Solution to the Thermal Elastohydrodynamic Lubrication of Rolling-Sliding Cylinders, *Trans. ASLE*, vol. 8, no. 4, pp. 397–410.

Cheng, H. S. (1967): Calculation of Elastohydrodynamic Film Thickness in High Speed Rolling and Sliding Contacts, Mechanical Technical Report 67–TR–24, Mechanical Technology Inc., Latham, New York. (Avail. NTIS, AD–652924.)

Cheng, H. S., and Dyson, A. (1978): Elastohydrodynamic Lubrication of Circumferentially Ground Rough Disks, *Trans. ASLE*, vol. 21, no. 1, pp. 25–40.

Dowson, D., and Whitaker, A. V. (1965): A Numerical Procedure for the Solution of the Elastohydrodynamic Problem of Rolling and Sliding Contacts Lubricated by a Newtonian Fluid, *Elastohydrodynamic Lubrication. Proc. Inst. Mech. Eng. (London)*, vol. 180, pp. 57–71.

Esfahanian, M., and Hamrock, B. J. (1991): Fluid-Film Lubrication Regimes Revisited, *STLE Tribol. Trans.*, vol. 34, no. 4, pp. 618–632.

Greenwood, J. A. (1969): Presentation of Elastohydrodynamic Film-Thickness Results. *J. Mech. Eng. Sci.*, vol. 11, no. 2, pp. 128–132.

Gupta, P. K. et al. (1991): Visco-Elastic Effects in Mil-L-7808 Type Lubricant, Part I: Analytical Formulation. *STLE Tribol. Trans.*, vol. 34, no. 4, pp. 608–617.

Hamrock, B. J., and Dowson, D. (1979): Minimum film thickness in Elliptical Contacts for Different Regimes of Fluid-Film Lubrication. *Elastohydrodynamics and Related Topics*. Proceedings of the 5th Leeds-Lyon Symposium on Tribology, D. Dowson et al. (eds.), Mechanical Engineering Publications Ltd., England, pp. 22–27.

Hooke, C. J. (1977): The Elastohydrodynamic Lubrication of Heavily Loaded Contacts. *J. Mech. Eng. Sci.*, vol. 19, no. 4, pp. 149–156.

Jeng, Y. R., Hamrock, B. J., and Brewe, D. E. (1987): Piezoviscous Effects in Nonconformal Contacts Lubricated Hydrodynamically. *ASLE Trans.*, vol. 30, no. 4, pp. 452–464.

Johnson, K. L. (1970): Regimes of Elastohydrodynamic Lubrication. *J. Mech. Eng. Sci.*, vol. 12, no. 1, pp. 9–16.

Kapitza, P. L. (1955): Hydrodynamic Theory of Lubrication During Rolling. *Zh. Tekh. Fiz.*, vol. 25, no. 4, pp. 747–762.

Kweh, C. C., Evans, H. P., and Snidle, R. W. (1989): Micro-Elastohydrodynamic Lubrication of an Elliptical Contact With Transverse and Three-Dimensional Sinusoidal Roughness. *J. Tribol.*, vol. 111, no. 4, pp. 577–584.

Lubrecht, A. A., Ten Napel, W. E., and Bosma, R. (1988): The Influence of Longitudinal and Transverse Roughness on the Elastohydrodynamic Lubrication of Circular Contacts, *J. Tribol.*, vol. 110, no. 3, pp. 421–426.

Moes, H. (1965–66): Discussion of Dowson, D., Elastohydrodynamic Lubrication: An Introduction and Review of Theoretical Studies. *Elastohydrodynamic Lubrication. Proc. Inst. Mech. Eng. (London)*, vol. 180, pt. 3B, pp. 244–245.

Patir, N., and Cheng, H. S. (1978): Effect of Surface Roughness on the Central Film Thickness in EHD Contacts. *Elastohydrodynamic and Related Topics*. Proceedings of the Fifth Leeds-Lyon Symposium on Tribology, Institution of Mechanical Engineers, London, pp. 15–21.

Peklenik, J. (1968): New Developments in Surface Characterization and Measurement by Means of Random Process Analysis. *Properties and Metrology of Surfaces. Proc. Inst. Mech. Eng. (London)*, vol. 182, part 3K, pp. 108–114.

Roelands, C. J. A. (1966): Correlational Aspects of the Viscosity-Temperature-Pressure Relationship of Lubricating Oils. Druk. V. R. B., Groningen, Netherlands.

Sadeghi, F., and Sui, P. (1990): Thermal Elastohydrodynamic Lubrication of Rolling-Sliding Contacts. *J. Tribol.*, vol. 112, no. 2, pp. 189–195.

Sheu, S., and Wilson, W. R. D. (1982): Viscoplastic Lubrication of Asperities. *J. Lubr. Technol.*, vol. 104, no. 4, pp. 568–574.

Tallian, T., Sibley, L., and Valori, R. (1965): Elastohydrodynamic Film Effects on the Load-Life Behavior of Rolling Contacts. *ASME Paper* 65–LUB–11.

Theyse, F. H. (1966): Some Aspects of the Influence of Hydrodynamic Film Formulation on the Contact between Rolling/Sliding Surfaces. *Wear*, vol. 9, pp. 41–59.

Winer, W. O. (1983): Temperature Effects in Elastohydrodynamically Lubricated Contacts. *Tribology in the '80's. NASA CP*–2300, vol. II, pp. 533–543.

Zhu, D., and Cheng, H. S. (1988): Effect of Surface Roughness on the Point Contact EHL. *J. Tribol.*, vol. 110, no. 1, pp. 32–37.

CHAPTER 24

ROLLING-ELEMENT BEARINGS

Since Chap. 18 the focus has been on nonconformal lubricated conjunctions and the development of elastohydrodynamic lubrication theory. In this and the following chapter the principles developed in those chapters are applied to specific machine elements. This chapter focuses on rolling-element bearings, which are a precise, yet simple, machine element of great utility, and draws together the current understanding of these bearings. First, the history of rolling-element bearings is briefly reviewed. Subsequent sections are devoted to describing the types of rolling-element bearing, their geometry and kinematics, the materials they are made from, and the manufacturing processes they involve. The organization of this chapter is such that unloaded and unlubricated rolling-element bearings are considered in the first five sections, loaded but unlubricated rolling-element bearings in Secs. 24.6 and 24.7 (as well as Chap. 19), and loaded and lubricated rolling-element bearings in Secs. 24.8 to 24.11 (as well as Chaps. 21 to 23).

24.1 HISTORICAL OVERVIEW

The purpose of a bearing is to provide relative positioning and rotational freedom while transmitting a load between two structures, usually a shaft and a housing. The basic form and concept of the rolling-element bearing are simple. If loads are to be transmitted between surfaces in relative motion in a machine, the action can be facilitated in a most effective manner if rolling elements are interposed between the sliding members. The frictional resistance encountered in sliding is then largely replaced by the much smaller resistance associated with rolling, although the arrangement is inevitably afflicted with high stresses in the restricted regions of effective load transmission.

The precision rolling-element bearing of the twentieth century is a product of exacting technology and sophisticated science. It is simple in form and concept and yet extremely effective in reducing friction and wear in a wide range of machinery. The spectacular development of numerous forms of rolling-element bearing in the twentieth century is well known and documented, but it is possible to trace the origins and development of these vital machine elements to periods long before there was a large industrial demand for such devices and certainly long before adequate machine tools for their effective manufacture existed in large quantities. A complete historical development of rolling-element bearings is given in Hamrock and Dowson (1981), and therefore only a brief overview is presented here.

The influence of general technological progress on the development of rolling-element bearings, particularly those concerned with the movement of heavy stone building blocks and carvings, road vehicles, precision instruments, water-raising equipment, and windmills is discussed in Hamrock and Dowson (1981). The concept of rolling-element bearings emerged in embryo form in Roman times, faded from the scene during the Middle Ages, was revived during the Renaissance, developed steadily in the seventeenth and eighteenth centuries for various applications, and was firmly established for individual road carriage bearings during the Industrial Revolution. Toward the end of the nineteenth century, the great merit of ball bearings for bicycles promoted interest in the manufacture of accurate steel balls. Initially, the balls were turned from bars on special lathes, with individual general machine manufacturing companies making their own bearings. Growing demand for both ball and roller bearings encouraged the formation of specialist bearing manufacturing companies at the turn of the century and thus laid the foundations of a great industry. The advent of precision grinding techniques and the availability of improved materials did much to confirm the future of the new industry.

The essential features of most forms of modern rolling-element bearing were therefore established by the second half of the nineteenth century, but it was the formation of specialist precision-manufacturing companies in the early years of the twentieth century that finally established the rolling-element bearing as a most valuable, high-quality, readily available machine component. The availability of ball and roller bearings in standard sizes has had a tremendous impact on machine design throughout the twentieth century. Such bearings still provide a challenging field for further research and development, and many engineers and scientists are currently engaged in exciting and demanding research projects in this area. In many cases, new improved materials or enlightened design concepts have extended the life and range of application of rolling-element bearings, yet in other respects, much remains to be done in explaining the extraordinary operating characteristics of bearings that have served our technology so well for almost a century. A recent development, covered in Chaps. 21 to 23, is the understanding and analysis of one important aspect of rolling-element performance—the lubrication mechanism in small, highly stressed conjunctions between the rolling element and the rings, or races.

24.2 BEARING TYPES

Ball and roller bearings are available to the designer in a great variety of designs and size ranges. The intent of this chapter is not to duplicate the complete descriptions given in manufacturers' catalogs but rather to present a guide to representative bearing types along with the approximate range of sizes available. Tables 24.1 to 24.9 illustrate some of the more widely used bearing types. In addition, numerous types of specialty bearing are available; space does not permit a complete cataloging of all available bearings. Size ranges are generally given in metric units. Traditionally, most rolling-element bearings have been manufactured to metric dimensions, predating the efforts toward a metric standard.

In addition to bearing types and approximate size ranges available, Tables 24.1 to 24.9 also list approximate relative load-carrying capacities, both radial and thrust, and, where relevant, approximate tolerances to misalignment.

Rolling-element bearings are an assembly of several parts: an inner race, an outer race, a set of balls or rollers, and a cage or separator. The cage or separator maintains even spacing of the rolling elements. A cageless bearing, in which the annulus is packed with the maximum rolling-element complement, is called a "full-complement bearing." Full-complement bearings have high load-carrying capacity but lower speed limits than bearings equipped with cages. Tapered-roller bearings are an assembly of a cup, a cone, a set of tapered rollers, and a cage.

24.2.1 BALL BEARINGS. Ball bearings are used in greater quantity than any other type of rolling-element bearing. For an application where the load is primarily radial with some thrust load present, one of the types in Table 24.1 can be chosen. A Conrad, or deep-groove, bearing has a ball complement limited by the number of balls that can be packed into the annulus between the inner and outer races with the inner race resting against the inside diameter of the outer race. A stamped and riveted two-piece cage, piloted on the ball set, or a machined two-piece cage, ball-piloted or race-piloted, is almost always used in a Conrad bearing. The only exception is a one-piece cage with open-sided pockets that is snapped into place. A filling-notch bearing most often has both inner and outer races notched so that a ball complement limited only by the annular space between the races can be used. It has low thrust capacity because of the filling notch.

The self-aligning internal bearing shown in Table 24.1 has an outer-race ball path ground in a spherical shape so that it can accept high levels of misalignment. The self-aligning external bearing has a multipiece outer race with a spherical interrace. It too can accept high misalignment and has higher

TABLE 24.1
Characteristics of representative radial ball bearings [From Hamrock and Anderson (1983)]

Type	Approximate range of bore sizes, mm		Relative capacity		Limiting speed factor	Tolerance to mis-alignment
	Mini-mum	Maxi-mum	Radial	Thrust		
Conrad or deep groove	3	1060	1.00	[a]0.7	1.0	±0°15′
Maximum capacity or filling notch	10	130	1.2–1.4	[a]0.2	1.0	±0°3′
Magneto or counterbored outer	3	200	0.9–1.3	[b]0.5–0.9	1.0	±0°5′
Airframe or aircraft control	4.826	31.75	High static capacity	[a]0.5	0.2	0°
Self-aligning, internal	5	120	0.7	[b]0.2	1.0	±2°30′
Self-aligning, external	------	------	1.0	[a]0.7	1.0	High
Double row, maximum	6	110	1.5	[a]0.2	1.0	±0°3′
Double row, deep groove	6	110	1.5	[a]1.4	1.0	0°

[a]Two directions. [b]One direction.

capacity than the self-aligning internal bearing. However, the self-aligning external bearing is somewhat less self-aligning than its internal counterpart because of friction in the multipiece outer race.

Representative angular-contact ball bearings are illustrated in Table 24.2. An angular-contact ball bearing has a two-shouldered ball groove in one race and a single-shouldered ball groove in the other race. Thus, it is capable of supporting only a unidirectional thrust load. The cutaway shoulder allows assembly of the bearing by snapping the inner race over the ball set after it is positioned in the cage and outer race. This also permits use of a one-piece,

TABLE 24.2
Characteristics of representative angular-contact ball bearings
[Minimum bore size, 10 mm. From Hamrock and Anderson (1983)]

Type	Approximate maximum bore size, mm	Relative capacity		Limiting speed factor	Tolerance to mis-alignment
		Radial	Thrust		
One-directional thrust	320	[b]1.00–1.15	[a,b]1.5–2.3	[b]1.1–3.0	±0°2′
Duplex, back to back	320	1.85	[c]1.5	3.0	0°
Duplex, face to face	320	1.85	[c]1.5	3.0	0°
Duplex, tandem	320	1.85	[a]2.4	3.0	0°
Two directional or split ring	110	1.15	[c]1.5	3.0	±0°2′
Double row	140	1.5	[c]1.85	0.8	0°
Double row, maximum	110	1.65	[a]0.5 [d]1.5	0.7 [d]1.5	0°

[a]One direction.
[b]Depends on contact angle.
[c]Two directions.
[d]In other direction.

machined, race-piloted cage that can be balanced for high-speed operation. Typical contact angles vary from 15 to 25°.

Angular-contact ball bearings are used in duplex pairs mounted either back to back or face to face as shown in Table 24.2. Duplex bearing pairs are manufactured so that they "preload" each other when clamped together in the housing and on the shaft. The use of preloading provides stiffer shaft support and helps prevent bearing skidding at light loads. Proper levels of preload can be obtained from the manufacturer. A duplex pair can support bidirectional thrust load. The back-to-back arrangement offers more resistance to moment or overturning loads than does the face-to-face arrangement.

Where thrust loads exceed the capability of a simple bearing, two bearings can be used in tandem, with both bearings supporting part of the thrust load. Three or more bearings are occasionally used in tandem, but this is discouraged because of the difficulty in achieving good load sharing. Even slight differences in operating temperature will cause a maldistribution of load sharing.

TABLE 24.3
Characteristics of representative thrust ball bearings
[Relative radial capacity, 0. From Hamrock and Anderson (1983)]

Type	Approximate range of bore sizes, mm		Relative thrust capacity	Limiting speed factor	Tolerance to misalignment
	Minimum	Maximum			
One directional, flat race	6.45	88.9	[a]0.7	0.10	[b]0°
One directional, grooved race	6.45	1180	[a]1.5	0.30	0°
Two directional, grooved race	15	220	[c]1.5	0.30	0°

[a]One direction.
[b]Accepts eccentricity.
[c]Two directions.

The split-ring bearing shown in Table 24.2 offers several advantages. The split ring (usually the inner) has it ball groove ground as a circular arc with a shim between the ring halves. The shim is then removed when the bearing is assembled so that the split-ring ball groove has the shape of a gothic arch. This reduces the axial play for a given radial play and results in more accurate axial positioning of the shaft. The bearing can support bidirectional thrust loads but must not be operated for prolonged periods of time at predominantly radial loads. This restriction results in three-point ball-race contact and relatively high frictional losses. As with the conventional angular-contact bearing, a one-piece precision-machined cage is used.

Ball thrust bearings (90° contact angle), Table 24.3, are used almost exclusively for machinery with vertically oriented shafts. The flat-race bearing allows eccentricity of the fixed and rotating members. An additional bearing must be used for radial positioning. It has low load-carrying capacity because of its small ball-race contact and consequent high Hertzian stress. Grooved-race bearings have higher load-carrying capacities and are capable of supporting low-magnitude radial loads. All the pure thrust ball bearings have modest speed capability because of the 90° contact angle and the consequent high level of ball spinning and frictional losses.

24.2.2 ROLLER BEARINGS. Cylindrical roller bearings, Table 24.4, provide purely radial load support in most applications. An N or U type of bearing will

TABLE 24.4
Characteristics of representative cylindrical roller bearings
[From Hamrock and Anderson (1983)]

Type	Approximate range of bore sizes, mm		Relative capacity		Limiting speed factor	Tolerance to mis-alignment
	Mini-mum	Maxi-mum	Radial	Thrust		
Separable outer ring, nonlocating (RN, RIN)	10	320	1.55	0	1.20	±0°5′
Separable inner ring, nonlocating (RU, RIU)	12	500	1.55	0	1.20	±0°5′
Separable outer ring, one-direction locating (RF, RIF)	40	177.8	1.55	[a]Locating	1.15	±0°5′
Separable inner ring, one-direction locating (RJ, RIJ)	12	320	1.55	[a]Locating	1.15	±0°5′
Self-contained, two-direction locating	12	100	1.35	[b]Locating	1.15	±0°5′
Separable inner ring, two-direction locating (RT, RIT)	20	320	1.55	[b]Locating	1.15	±0°5′
Nonlocating, full complement (RK, RIK)	17	75	2.10	0	0.20	±0°5′
Double row, separable outer ring, nonlocating (RD)	30	1060	1.85	0	1.00	0°
Double row, separable inner ring, nonlocating	70	1060	1.85	0	1.00	0°

[a]One direction.
[b]Two directions.

TABLE 24.5
Characteristics of representative spherical roller bearings
[From Hamrock and Anderson (1983)]

Type	Approximate range of bore sizes, mm		Relative capacity		Limiting speed factor	Tolerance to misalignment
	Minimum	Maximum	Radial	Thrust		
Single row, barrel or convex	20	320	2.10	0.20	0.50	±2°
Double row, barrel or convex	25	1250	2.40	0.70	0.50	±1°30′
Thrust	85	360	[a]0.10 [b]0.10	[a]1.80 [b]2.40	0.35–0.50	±3°
Double row, concave	50	130	2.40	0.70	0.50	±1°30′

[a]Symmetric rollers. [b]Asymmetric rollers.

allow free axial movement of the shaft relative to the housing to accommodate differences in thermal growth. An F or J type of bearing will support a light thrust load in one direction; and a T type of bearing, a light bidirectional thrust load. Cylindrical roller bearings have moderately high radial load-carrying capacity as well as high speed capability. Their speed capability exceeds that of either spherical or tapered-roller bearings. A commonly used bearing combination for support of a high-speed rotor is an angular-contact ball bearing or duplex pair and a cylindrical roller bearing. As explained in the following section on bearing geometry, the rollers in cylindrical roller bearings are seldom pure cylinders. They are crowned, or made slightly barrel shaped, to relieve stress concentrations on the roller ends when any misalignment of the shaft and housing is present. Cylindrical roller bearings may be equipped with one- or two-piece cages, usually race piloted. For greater load-carrying capacity, full-complement bearings can be used, but at a significant sacrifice in speed capability.

Spherical roller bearings, Tables 24.5 to 24.7, are made as either single- or double-row bearings. The more popular bearing design uses barrel-shaped rollers. An alternative design employs hourglass-shaped rollers. Spherical roller bearings combine extremely high radial load-carrying capacity with modest thrust load-carrying capacity (with the exception of the thrust type) and excellent tolerance to misalignment. They find widespread use in heavy-duty rolling

TABLE 24.6
Characteristics of standardized double-row, spherical roller bearings
[From Hamrock and Anderson (1983)]

Type	Roller design	Retainer design	Roller guidance	Roller-race contact
SLB	Symmetric	Machined, roller piloted	Retainer pockets	Modified line, both races
SC	Symmetric	Stamped, race piloted	Floating guide ring	Modified line, both races
SD	Asymmetric	Machined, race piloted	Inner-ring center rib	Line contact, outer; point contact, inner

mill and industrial gear drives, where all these bearings characteristics are requisite.

Tapered-roller bearings, Table 24.8, are also made as single- or double-row bearings with combinations of one- or two-piece cups and cones. A four-row bearing assembly with two- or three-piece cups and cones is also available. Bearings are made with either a standard angle for applications in which moderate thrust loads are present or with a steep angle for high thrust capacity.

TABLE 24.7
Characteristics of spherical roller bearings
[From Hamrock and Anderson (1983)]

Series	Types	Approximate range of bore sizes, mm		Approximate relative capacity[a]		Limiting speed factor
		Minimum	Maximum	Radial	Thrust	
202	Single-row barrel	20	320	1.0	0.11	0.5
203	Single-row barrel	20	240	1.7	.18	.5
204	Single-row barrel	25	110	2.1	.22	.4
212	SLB	35	75	1.0	.26	.6
213	SLB	30	70	1.7	.53	↓
22, 22K	SLB, SC, SD	30	320	1.7	.46	↓
23, 23K	↓	40	280	2.7	1.0	↓
30, 30K	↓	120	1250	1.2	.29	.7
31, 31K	↓	110	1250	1.7	.54	.6
32, 32K	↓	100	850	2.1	.78	.6
39, 39K	SD	120	1250	.7	.18	.7
40, 40K	SD	180	250	1.5	----	.7

[a]Load capacities are comparative within the various series of spherical roller bearings only. For a given envelope size, a spherical roller bearing has a radial capacity approximately equal to that of a cylindrical roller bearing.

TABLE 24.8
Characteristics of representative tapered-roller bearings
[From Hamrock and Anderson (1983)]

Type	Subtype	Approximate range of bore sizes, mm	
		Minimum	Maximum
Single row (TS)	TS—straight bore	8	1690
	TST—tapered bore	24	430
	TSS—steep angle	16	1270
	TSE, TSK—keyway cone	12	380
	TSF, TSSF—flanged cup	8	1070
Two row, double cone, single cup (TDI)	TDI—straight bore	30	1200
	TDIK, TDIT, TDITP—tapered bore	30	860
	TDIE, TDIKE—slotted double cone	24	690
	TDIS—steep angle	55	520
Two row, double cup, single cone, adjustable (TDO)	TDO	8	1830
	TDOS—steep angle	20	1430
Two row, double cup, single cone, nonadjustable (TNA)	TNA—straight bore	20	60
	TNASW—slotted cone	30	260
	TNASWE—extended cone rib	20	305
	TNASWH—slotted cone, sealed	8	70
Four row, cup adjusted (TQO)	TQO—straight bore	70	1500
	TQOT—tapered bore	250	1500

Standard and special cages are available to suit the application requirements. Single-row tapered-roller bearings must be used in pairs because a radially loaded bearing generates a thrust reaction that must be taken up by a second bearing. Tapered-roller bearings are normally set up with spacers designed so that they operate with some internal play. Manufacturers' engineering journals should be consulted for proper setup procedures.

TABLE 24.9
Characteristics of representative needle roller bearings
[From Hamrock and Anderson (1983)]

Type	Bore size, mm		Relative load capacity		Limiting speed factor	Misal tole
	Minimum	Maximum	Dynamic	Static		
Drawn cup, needle Open end Closed end	3	185	High	Moderate	0.3	Low
Drawn cup, needle greese retained	4	25	High	Moderate	0.3	Low
Drawn cup, roller Open end Closed end	5	70	Moderate	Moderate	0.9	Moder
Heavy-duty roller	16	235	Very high	Moderate	1.0	Moder
Caged roller	12	100	Very high	High	1.0	Moder
Cam follower	12	150	Moderate to high	Moderate to high	0.3–0.9	Low
Needle thrust	6	105	Very high	Very high	0.7	Low

Needle roller bearings, Table 24.9, are characterized by compactness in the radial direction and are frequently used without an inner race. In the latter case the shaft is hardened and ground to serve as the inner race. Drawn cups, both open and closed end, are frequently used for grease retention. Drawn cups are thin walled and require substantial support from the housing. Heavy-duty roller bearings have relatively rigid races and are more akin to cylindrical roller bearings with long-width-to-diameter-ratio rollers.

Needle roller bearings are more speed limited than cylindrical roller bearings because of roller skewing at high speeds. A high percentage of needle roller bearings are full-complement bearings. They have higher load-carrying capacity than a caged needle bearing, but lower speed capability.

Many types of specialty bearing are available other than those discussed here. Aircraft bearings for control systems, thin-section bearings, and fractured-ring bearings are some of the more widely used bearings among the many types manufactured. Complete coverage of all bearing types is beyond the scope of this chapter.

Angular-contact ball bearings and cylindrical roller bearings are generally considered to have the highest speed capabilities. Speed limits of roller bearings are discussed in conjunction with lubrication methods. The lubrication system employed has as great an influence on bearing limiting speed as the bearing design.

24.3 GEOMETRY

The operating characteristics of a rolling-element bearing depend greatly on its diametral clearance. This clearance varies for the different types of bearing discussed in the preceding section. In this section the principal geometrical relationships governing the operation of unloaded rolling-element bearings are developed. This information will be of vital interest when such quantities as stress, deflection, load-carrying capacity, and life are considered in subsequent sections. Although bearings rarely operate in the unloaded state, an understanding of this section is essential to appreciation of the remaining sections.

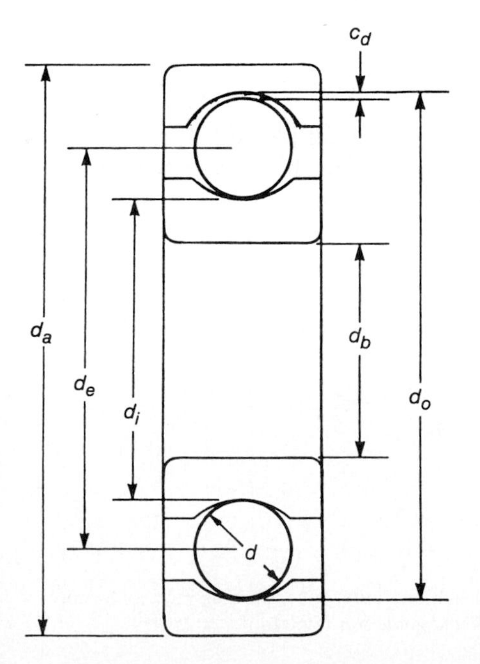

FIGURE 24.1
Cross section through radial single-row ball bearings. [*From Hamrock and Anderson (1983).*]

24.3.1 GEOMETRY OF BALL BEARINGS. The geometry of ball bearings involves the elements of pitch diameter and clearance, race conformity, contact angle, endplay, shoulder height, and curvature sum and difference.

24.3.1.1 Pitch diameter and clearance. The cross section through a radial single-row ball bearing shown in Fig. 24.1 depicts the radial clearance and various diameters. The pitch diameter d_e is the mean of the inner- and outer-race contact diameters and is given by

$$d_e = d_i + \frac{1}{2}(d_o - d_i)$$

or

$$d_e = \frac{1}{2}(d_o + d_i) \tag{24.1}$$

Also from Fig. 24.1 the diametral clearance, denoted by c_d, can be written as

$$c_d = d_o - d_i - 2d \tag{24.2}$$

Diametral clearance may therefore be thought of as the maximum distance that one race can move diametrally with respect to the other when no measurable force is applied and both races lie in the same plane. Although diametral clearance is generally used in connection with single-row radial bearings, Eq. (24.2) is also applicable to angular-contact bearings.

24.3.1.2 Race conformity. Race conformity is a measure of the geometrical conformity of the race and the ball in a plane passing through the bearing axis, which is a line passing through the center of the bearing perpendicular to its plane and transverse to the race. Figure 24.2 is a cross section of a ball bearing showing race conformity, expressed as

$$R_r = \frac{r}{d} \tag{24.3}$$

For perfect conformity, where the race radius is equal to the ball radius, R_r is equal to $\frac{1}{2}$. The closer the race conforms to the ball, the greater the frictional heat within the contact. On the other hand, open-race curvature and reduced geometrical conformity, which reduce friction, also increase the maximum contact stresses and consequently reduce the bearing fatigue life. For this reason most ball bearings made today have race conformity ratios in the range $0.51 \le R_r \le 0.54$, with $R_r = 0.52$ being the most common value. The race

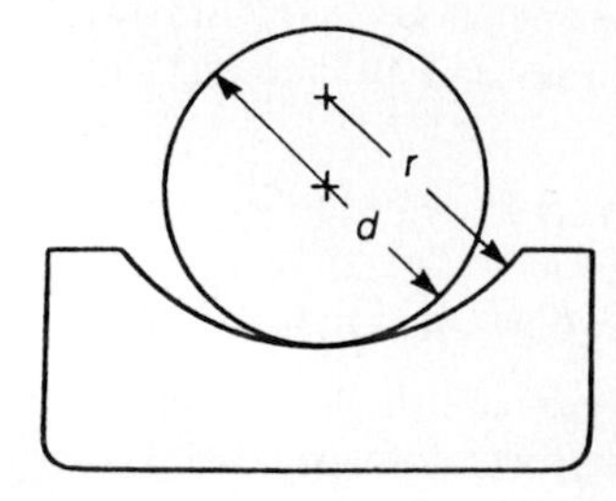

FIGURE 24.2
Cross section of ball and outer race, showing race conformity. [*From Hamrock and Anderson (1983).*]

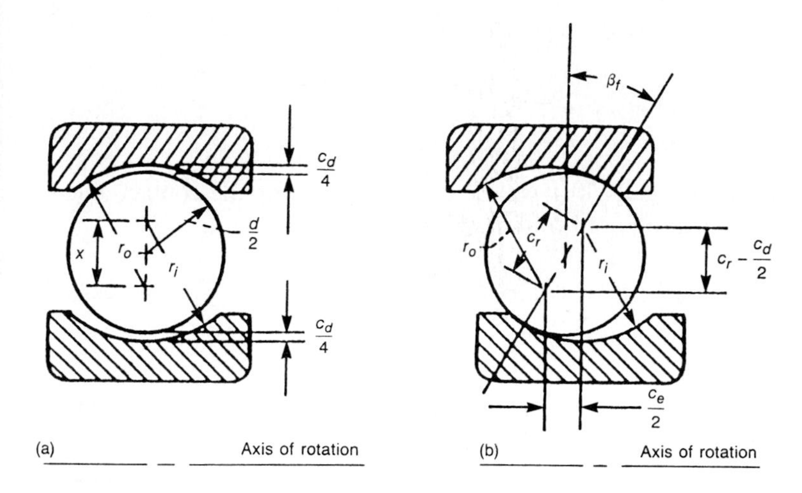

FIGURE 24.3
Cross section of radial bearing, showing ball-race contact due to axial shift of inner and outer races. (a) Initial position; (b) shifted position. [*From Hamrock and Anderson (1983)*.]

conformity ratio for the outer race is usually made slightly larger than that for the inner race to compensate for the closer conformity in the plane of the bearing between the outer race and the ball than between the inner race and the ball. This tends to equalize the contact stresses at the inner- and outer-race contacts. The difference in race conformities does not normally exceed 0.02.

24.3.1.3 Contact angle. Radial bearings have some axial play, since they are generally designed to have a diametral clearance, as shown in Fig. 24.3. This implies a free contact angle different from zero. Angular-contact bearings are specifically designed to operate under thrust loads. The clearance built into the unloaded bearing, along with the race conformity ratio, determines the bearing free contact angle. Figure 24.3 shows a radial bearing with contact due to the axial shift of the inner and outer races when no measurable force is applied.

Before the free contact angle is discussed, it is important to define the distance between the centers of curvature of the two races in line with the center of the ball in Fig. 24.3(a) and (b). This distance—denoted by x in Fig. 24.3(a) and by c_r in Fig. 24.3(b)—depends on race radius and ball diameter. When quantities referred to the inner and outer races are denoted by subscripts i and o, respectively, Fig. 24.3(a) and (b) shows that

$$\frac{c_d}{4} + d + \frac{c_d}{4} = r_o - x + r_i$$

or

$$x = r_o + r_i - d - \frac{c_d}{2}$$

and

$$d = r_o - c_r + r_i$$

or

$$c_r = r_o + r_i - d \tag{24.4}$$

From these equations

$$x = c_r - \frac{c_d}{2}$$

This distance, shown in Fig. 24.3(b), will be useful in defining the contact angle.
Equation (24.3) can be used to write Eq. (24.4) as

$$c_r = Bd \tag{24.5}$$

where

$$B = R_{r,o} + R_{r,i} - 1 \tag{24.6}$$

The quantity B in Eq. (24.6) is known as the "total conformity ratio" and is a measure of the combined conformity of both the outer and inner races to the ball. Calculations of bearing deflection in later sections depend on the quantity B.

The free contact angle β_f [Fig. 24.3(b)] is defined as the angle made by a line through the points where the ball contacts both races and a plane perpendicular to the bearing axis of rotation when no measurable force is applied. Note that the centers of curvature of both the outer and inner races lie on the line defining the free contact angle. From Fig. 24.3(b) the expression for the free contact angle can be written as

$$\cos \beta_f = 1 - \frac{c_d}{2c_r} \tag{24.7}$$

Equations (24.2) and (24.4) can be used to write Eq. (24.7) as

$$\beta_f = \cos^{-1} \frac{r_o + r_i - \frac{1}{2}(d_o - d_i)}{r_o + r_i - d} \tag{24.8}$$

Equation (24.8) shows that if the size of the balls is increased and everything else remains constant, the free contact angle is decreased. Similarly, if the ball size is decreased, the free contact angle is increased.

From Eq. (24.7) the diametral clearance c_d can be written as

$$c_d = 2c_r(1 - \cos \beta_f) \tag{24.9}$$

This is an alternative definition of the diametral clearance given in Eq. (24.2).

24.3.1.4 Endplay. Free endplay c_e is the maximum axial movement of the inner race with respect to the outer when both races are coaxially centered and no measurable force is applied. Free endplay depends on total curvature and

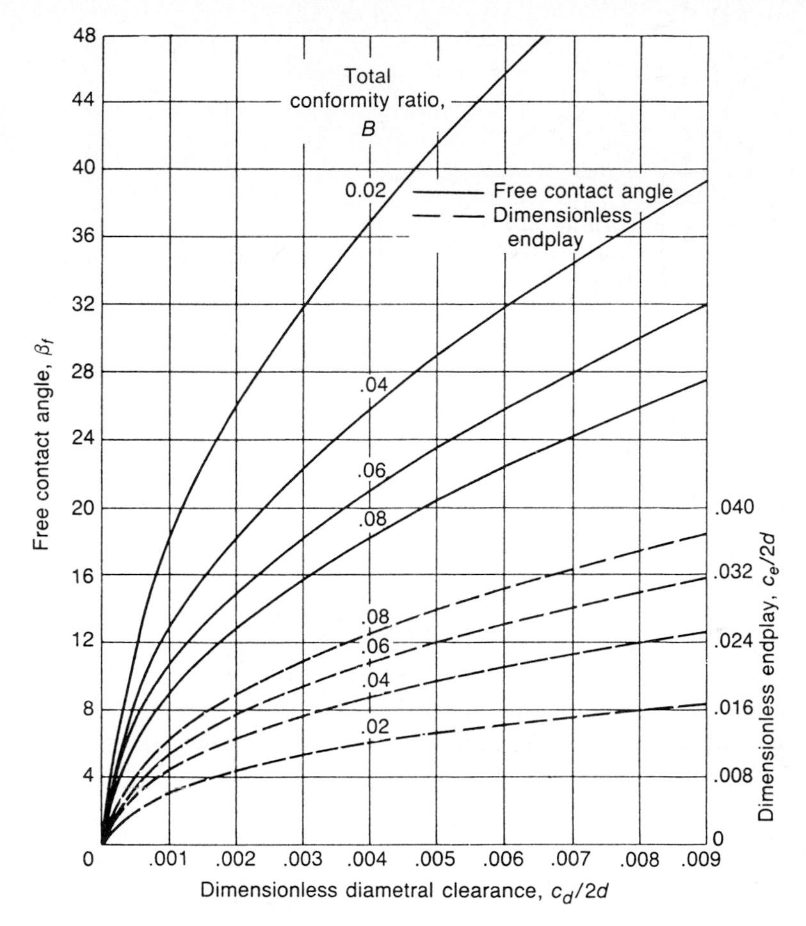

FIGURE 24.4
Free contact angle and endplay as function of $c_d/2d$ for four values of total conformity. [*From Hamrock and Anderson (1983).*]

contact angle, as shown in Fig. 24.3(b), and can be written as

$$c_e = 2c_r \sin \beta_f \tag{24.10}$$

The variation of free contact angle and endplay with the ratio $c_d/2d$ is shown in Fig. 24.4 for four values of the total conformity ratio normally found in single-row ball bearings. Eliminating β_f in Eqs. (24.9) and (24.10) enables the following relationships between free endplay and diametral clearance to be established:

$$c_d = 2c_r - \left[(2c_r)^2 - c_e^2\right]^{1/2}$$

$$c_e = \left(4c_r c_d - c_d^2\right)^{1/2}$$

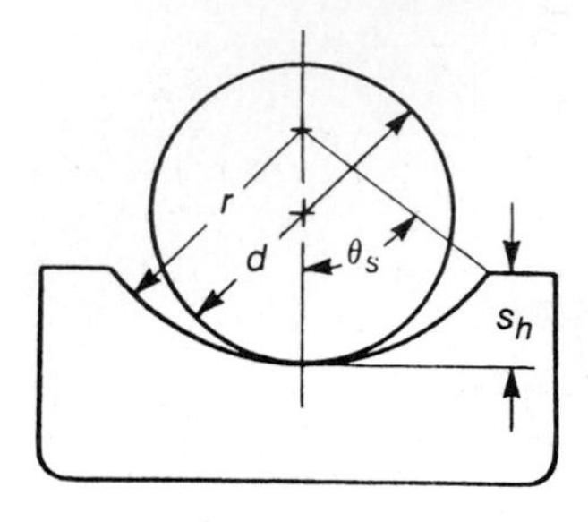

FIGURE 24.5
Shoulder height in ball bearing. [*From Hamrock and Anderson (1983).*]

24.3.1.5 Shoulder height. The shoulder height of ball bearings is illustrated in Fig. 24.5. Shoulder height, or race depth, is the depth of the race groove measured from the shoulder to the bottom of the groove and is denoted by s_h in Fig. 24.5. From this figure the equation defining the shoulder height can be written as

$$s_h = r(1 - \cos \theta_s) \tag{24.11}$$

The maximum possible diametral clearance for complete retention of the ball-race contact within the race under zero thrust load is given by the condition $(\beta_f)_{\max} = \theta_s$. Making use of Eqs. (24.9) and (24.11) gives

$$(c_d)_{\max} = \frac{2c_r s_h}{r}$$

24.3.1.6 Curvature sum and difference. The undeformed geometry of contacting solids in a ball bearing can be represented by two ellipsoids as discussed in Sec. 19.1. This section will apply the information provided there to a ball bearing.

A cross section of a ball bearing operating at a contact angle β is shown in Fig. 24.6. Equivalent radii of curvature for both inner- and outer-race contacts in, and normal to, the direction of rolling can be calculated from this figure. The radii of curvature for the *ball–inner-race contact* are

$$r_{ax} = r_{ay} = \frac{d}{2} \tag{24.12}$$

$$r_{bx} = \frac{d_e - d \cos \beta}{2 \cos \beta} \tag{24.13}$$

$$r_{by} = -R_{r,i} d = -r_i \tag{24.14}$$

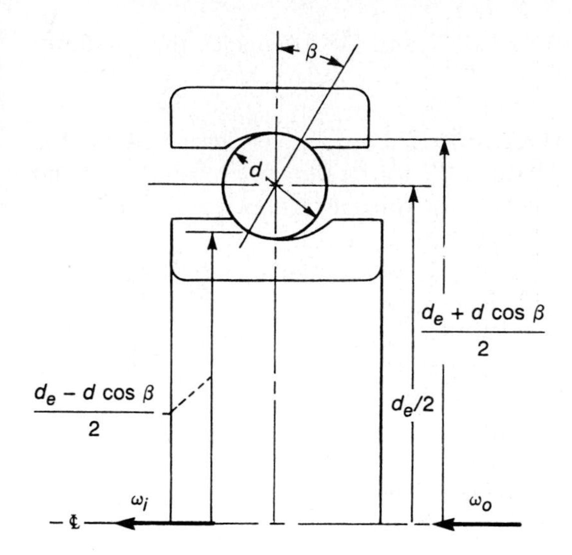

FIGURE 24.6
Cross section of ball bearing. [*From Hamrock and Anderson (1983).*]

The radii of curvature for the *ball–outer-race contact* are

$$r_{ax} = r_{ay} = \frac{d}{2} \tag{24.15}$$

$$r_{bx} = -\frac{d_e + d \cos \beta}{2 \cos \beta} \tag{24.16}$$

$$r_{by} = -R_{r,o} d = -r_o \tag{24.17}$$

In Eqs. (24.13) and (24.16), β is used instead of β_f because these equations are also valid when a load is applied to the contact. By setting $\beta = 0°$, Eqs. (24.12) to (24.17) are equally valid for radial ball bearings. For thrust ball bearings $r_{bx} = \infty$, and the other radii are defined as given in the preceding equations.

From the preceding radius-of-curvature expressions and Eq. (19.4) and (19.5) for the *ball–inner-race contact*

$$R_{x,i} = \frac{d(d_e - d \cos \beta)}{2d_e} \tag{24.18}$$

$$R_{y,i} = \frac{R_{r,i} d}{2R_{r,i} - 1} \tag{24.19}$$

and for the *ball–outer-race contact*

$$R_{x,o} = \frac{d(d_e + d \cos \beta)}{2d_e} \tag{24.20}$$

$$R_{y,o} = \frac{R_{r,o} d}{2R_{r,o} - 1} \tag{24.21}$$

Substituting these equations into Eqs. (19.2) and (19.3) enables the curvature sum and difference to be obtained.

24.3.2 GEOMETRY OF ROLLER BEARINGS. The equations developed for the pitch diameter d_e and diametral clearance c_d for ball bearings, Eqs. (24.1) and (24.2), respectively, are directly applicable for roller bearings.

24.3.2.1 Crowning. High stresses at the edges of the rollers in cylindrical roller bearings are usually prevented by crowning the rollers as shown in Fig. 24.7. A fully crowned roller is shown in Fig. 24.7(a) and a partially crowned roller in Fig. 24.7(b). The crown curvature is greatly exaggerated for clarity. The crowning of rollers also gives the bearing protection against the effects of slight misalignment. For cylindrical rollers $r_r/d \approx 10^2$. In contrast, for spherical rollers in spherical roller bearings, as shown in Fig. 24.7(a), $r_r/d \approx 4$. Observe in Fig. 24.7 that the roller effective length ℓ_l is the length presumed to be in contact with the races under loading. Generally, the roller effective length can be written as

$$\ell_l = \ell_t - 2r_c$$

where r_c is the roller corner radius or the grinding undercut, whichever is larger.

24.3.2.2 Race conformity. Race conformity applies to roller bearings much as it applies to ball bearings. It is a measure of the geometrical conformity of the

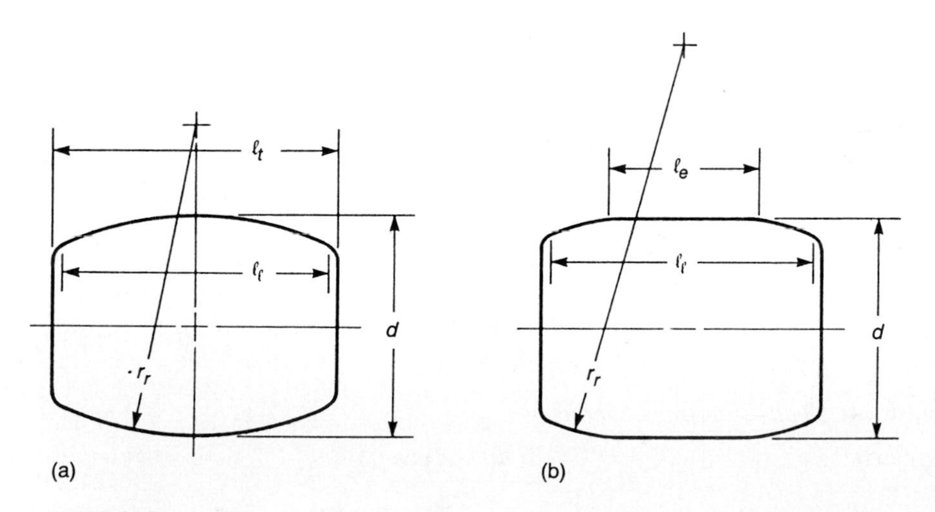

FIGURE 24.7
(a) Spherical roller (fully crowned) and (b) cylindrical roller (partially crowned). [*From Hamrock and Anderson (1983).*]

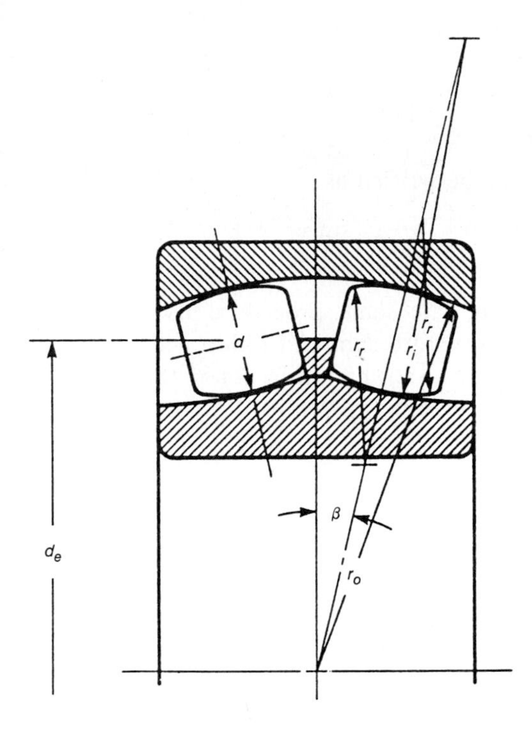

FIGURE 24.8
Geometry of spherical roller bearing. [*From Hamrock and Anderson (1983).*]

race and the roller. Figure 24.8 shows a cross section of a spherical roller bearing. From this figure the race conformity can be written as

$$R_r = \frac{r}{2r_r}$$

In this equation if R_r and r are subscripted with i or o, the race conformity values are for the inner- or outer-race contacts.

24.3.2.3 Free endplay and contact angle. Cylindrical roller bearings have a contact angle of zero and may take thrust load only by virtue of axial flanges. Tapered-roller bearings must be subjected to a thrust load or the inner and outer races (the cone and the cup) will not remain assembled; therefore, tapered-roller bearings do not exhibit free diametral play. Radial spherical roller bearings are, however, normally assembled with free diametral play and hence exhibit free endplay. The diametral play c_d for a spherical roller bearing is the same as that obtained for ball bearings as expressed in Eq. (24.2). This diametral play as well as endplay is shown in Fig. 24.9 for a spherical roller bearing. From this figure

$$r_o \cos \beta = \left(r_o - \frac{c_d}{2}\right) \cos \gamma_d$$

or

$$\beta = \cos^{-1}\left[\left(1 - \frac{c_d}{2r_o}\right)\cos\gamma_d\right]$$

Also from Fig. 24.9 the free endplay can be written as

$$c_e = 2r_o(\sin\beta - \sin\gamma_d) + c_d \sin\gamma_d$$

24.3.2.4 Curvature sum and difference. The same procedure used for ball bearings will be used for defining the curvature sum and difference for roller bearings. For spherical roller bearings, as shown in Fig. 24.8, the radii of curvature for the *roller–inner-race contact* can be written as

$$r_{ax} = \frac{d}{2}$$

$$r_{ay} = \frac{r_i}{2R_{r,i}} = r_r$$

$$r_{bx} = \frac{d_e - d\cos\beta}{2\cos\beta}$$

$$r_{by} = -2R_{r,i}r_r = -r_i$$

and the radii of curvature for the *roller–outer-race contact* can be written as

$$r_{ax} = \frac{d}{2}$$

$$r_{ay} = \frac{r_o}{2R_{r,o}} = r_r$$

$$r_{bx} = -\frac{d_e + d\cos\beta}{2\cos\beta}$$

$$r_{by} = -2R_{r,o}r_r = -r_o$$

Once the radii of curvature for the respective contact conditions are known, the curvature sum and difference can be written directly from Eqs. (19.2) and (19.3). Furthermore, the radius-of-curvature expressions R_x and R_y for spherical roller bearings can be written for the *roller–inner-race contact* as

$$R_x = \frac{d(d_e - d\cos\beta)}{2d_e} \tag{24.22}$$

$$R_y = \frac{r_i r_r}{r_i - r_r} \tag{24.23}$$

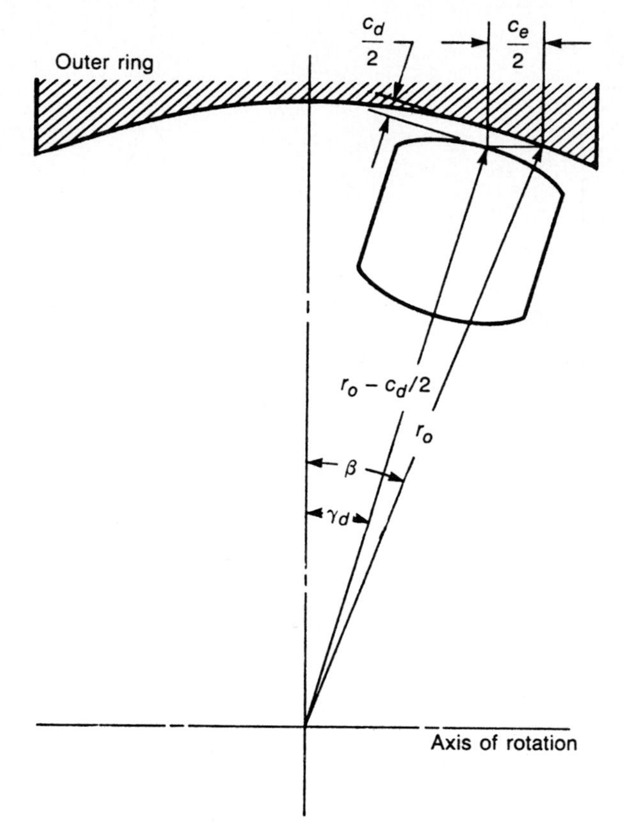

FIGURE 24.9
Schematic diagram of spherical roller bearing, showing diametral play and endplay. [*From Hamrock and Anderson (1983).*]

and for the *roller–outer-race contact* as

$$R_x = \frac{d(d_e + d\cos\beta)}{2d_e} \tag{24.24}$$

$$R_y = \frac{r_o r_r}{r_o - r_r} \tag{24.25}$$

Substituting these equations into Eqs. (19.2) and (19.3) enables the curvature sum and difference to be obtained.

24.4 KINEMATICS

The relative motions of the separator, the balls or rollers, and the races of rolling-element bearings are important to understanding their performance. The relative velocities in a ball bearing are somewhat more complex than those in

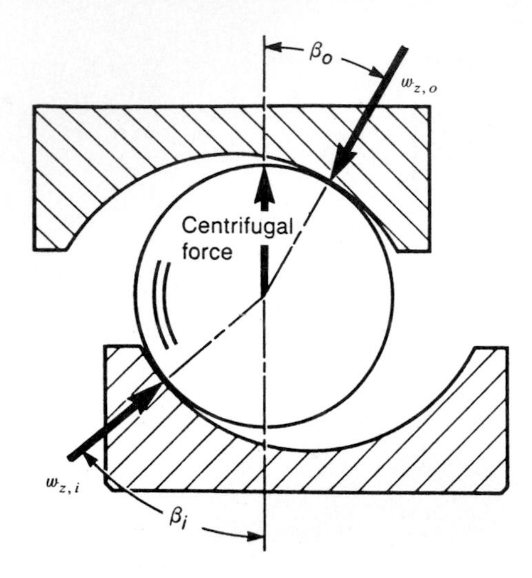

FIGURE 24.10
Contact angles in ball bearing at appreciable speeds. [*From Hamrock and Anderson (1983).*]

roller bearings, the latter being analogous to the specialized case of a zero- or fixed-contact-angle ball bearing. For that reason the ball bearing is used as an example here to develop approximate expressions for relative velocities. These expressions are useful for rapid but reasonably accurate calculation of elastohydrodynamic film thickness, which can be used with surface roughnesses to calculate the lubrication life factor.

The precise calculation of relative velocities in a ball bearing in which speed or centrifugal force effects, contact deformations, and elastohydrodynamic traction effects are considered requires a large computer to numerically solve the relevant equations. Refer to the growing body of computer codes discussed in Sec. 24.10 for precise calculations of bearing performance. Such a treatment is beyond the scope of this section. However, approximate expressions that yield answers with accuracies satisfactory for many situations are available.

When a ball bearing operates at high speeds, the centrifugal force acting on the ball creates a divergency of the inner- and outer-race contact angles, as shown in Fig. 24.10, in order to maintain force equilibrium on the ball. For the most general case of rolling and spinning at both inner- and outer-race contacts, the rolling and spinning velocities of the ball are as shown in Fig. 24.11.

The equations for ball and separator angular velocity for all combinations of inner- and outer-race rotation were developed by Jones (1964). Without introducing additional relationships to describe the elastohydrodynamic conditions at both ball-race contacts, however, the ball–spin axis orientation angle ϕ_s cannot be obtained. As mentioned, this requires a lengthy numerical solution except for the two extreme cases of outer- or inner-race control. These are illustrated in Fig. 24.12.

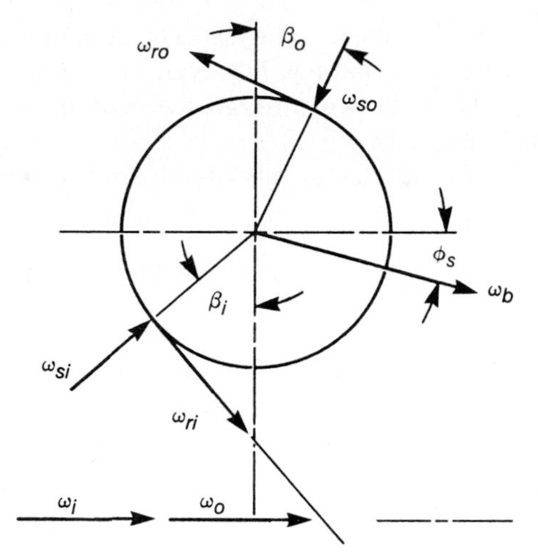

FIGURE 24.11
Angular velocities of ball. [*From Hamrock and Anderson (1983).*]

Race control assumes that pure rolling occurs at the controlling race, with all the ball spin occurring at the other race contact. The orientation of the ball rotation axis can then be easily determined from bearing geometry. Race control probably occurs only in dry bearings or dry-film-lubricated bearings where Coulomb friction conditions exist in the ball-race contact ellipses. The spin-resisting moment will always be greater at one of the race contacts. Pure rolling will occur at the race contact with the higher magnitude spin-resisting moment. This is usually the inner race at low speeds and the outer race at high speeds.

In oil-lubricated bearings in which elastohydrodynamic films exist in both ball-race contacts, rolling with spin occurs at both contacts. Therefore, precise

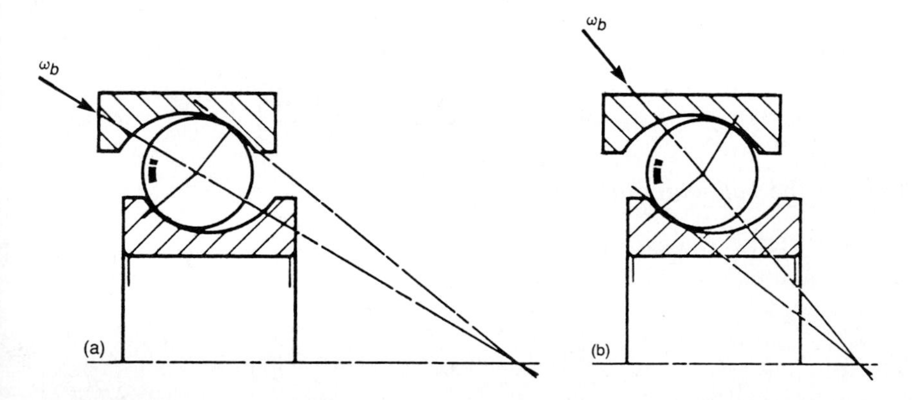

FIGURE 24.12
Ball spin axis orientations for (a) outer- and (b) inner-race control. [*From Hamrock and Anderson (1983).*]

ball motions can only be determined through a computer analysis. The situation can be approximated with a reasonable degree of accuracy, however, by assuming that the ball rolling axis is normal to the line drawn through the centers of the two ball-race contacts. This is shown in Fig. 24.6.

The angular velocity of the separator or ball set ω_c about the shaft axis can be shown to be (Anderson, 1970)

$$\omega_c = \frac{(v_i + v_o)/2}{d_e/2} = \frac{1}{2}\left[\omega_i\left(1 - \frac{d\cos\beta}{d_e}\right) + \omega_o\left(1 + \frac{d\cos\beta}{d_e}\right)\right] \tag{24.26}$$

where v_i and v_o are the linear velocities of the inner and outer contacts. The angular velocity of a ball about its own axis ω_b, assuming no spin, is

$$\omega_b = \frac{v_i - v_o}{d} = \frac{d_e}{2d}\left[\omega_i\left(1 - \frac{d\cos\beta}{d_e}\right) - \omega_o\left(1 + \frac{d\cos\beta}{d_e}\right)\right] \tag{24.27}$$

It is convenient, for calculating the velocities of the ball-race contacts, which are required for calculating elastohydrodynamic film thicknesses, to use a coordinate system that rotates at ω_c. This fixes the ball-race contacts relative to the observer. In the rotating coordinate system the angular velocities of the inner and outer races become

$$\omega_{ri} = \omega_i - \omega_c = \frac{\omega_i - \omega_o}{2}\left(1 + \frac{d\cos\beta}{d_e}\right)$$

$$\omega_{ro} = \omega_o - \omega_c = \frac{\omega_o - \omega_i}{2}\left(1 - \frac{d\cos\beta}{d_e}\right)$$

The surface velocities entering the *ball–inner-race contact* for pure rolling are

$$u_{ai} = u_{bi} = \frac{\omega_{ri}(d_e - d\cos\beta)}{2} \tag{24.28}$$

or

$$u_{ai} = u_{bi} = \frac{d_e(\omega_i - \omega_o)}{4}\left(1 - \frac{d^2\cos^2\beta}{d_e^2}\right) \tag{24.29}$$

and those at the *ball–outer-race contact* are

$$u_{ao} = u_{bo} = \frac{\omega_{ro}(d_e + d\cos\beta)}{2}$$

or

$$u_{ao} = u_{bo} = \frac{d_e(\omega_o - \omega_i)}{4}\left(1 - \frac{d^2\cos^2\beta}{d_e^2}\right) \tag{24.30}$$

Thus,

$$|u_{ai}| = |u_{ao}|$$

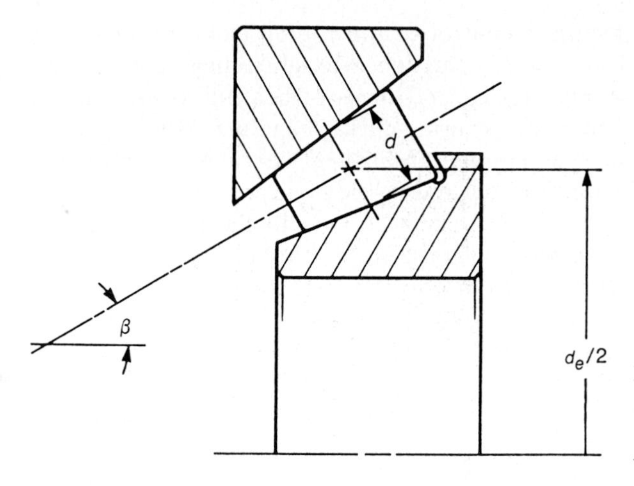

FIGURE 24.13
Simplified geometry for tapered-roller bearing. [*From Hamrock and Anderson (1983).*]

For a cylindrical roller bearing $\beta = 0°$, and Eqs. (24.26), (24.27), (24.29), and (24.30) become, if d is roller diameter,

$$\omega_c = \frac{1}{2}\left[\omega_i\left(1 - \frac{d}{d_e}\right) + \omega_o\left(1 + \frac{d}{d_e}\right)\right]$$

$$\omega_b = \frac{d_e}{2d}\left[\omega_i\left(1 - \frac{d}{d_e}\right) + \omega_o\left(1 + \frac{d}{d_e}\right)\right]$$

$$u_{ai} = u_{bi} = \frac{d_e(\omega_i - \omega_o)}{4}\left(1 - \frac{d^2}{d_e^2}\right)$$

$$u_{ao} = u_{bo} = \frac{d_e(\omega_o - \omega_i)}{4}\left(1 - \frac{d^2}{d_e^2}\right) \qquad (24.31)$$

Equations directly analogous to those for a ball bearing can be used for a tapered-roller bearing if d is the average diameter of the tapered roller, d_e is the diameter at which the geometric center of the rollers is located, and β is the angle as shown in Fig. 24.13.

24.5 SEPARATORS

Ball and rolling-element bearing separators (sometimes called "cages" or "retainers") are bearing components that, although they never carry load, are capable of exerting a vital influence on bearing efficiency. In a bearing without a separator the rolling elements contact each other during operation and in so doing experience severe sliding and friction. The primary functions of a separator are to maintain the proper distance between the rolling elements and to ensure proper load distribution and balance within the bearing. Another function of the separator is to maintain control of the rolling elements in such a manner as to produce the least possible friction through sliding contact. Furthermore, a separator is necessary for several types of bearing to prevent the rolling elements from falling out of the bearing during handling. Most separator troubles occur from improper mounting, misaligned bearings, or improper (inadequate or excessive) clearance in the rolling-element pocket.

The materials used for separators vary according to the type of bearing and the application. In ball bearings and some sizes of roller bearing the most common type of separator is made from two strips of carbon steel that are pressed and riveted together. Called "ribbon separators," they are the least expensive separators to manufacture and are entirely suitable for many applications. They are also lightweight and usually require little space.

The design and construction of angular-contact ball bearings allow the use of a one-piece separator. The simplicity and inherent strength of one-piece separators permit their fabrication from many desirable materials. Reinforced phenolic and bronze are the two most commonly used materials. Bronze separators offer strength and low-friction characteristics and can be operated at temperatures to 230°C (450°F). Machined, silver-plated ferrous alloy separators are used in many demanding applications. Because reinforced cotton-base phenolic separators combine the advantages of low weight, strength, and nongalling properties, they are used for such high-speed applications as gyro bearings. Lightness and strength are particularly desirable in high-speed bearings, since the stresses increase with speed but may be greatly minimized by reducing separator weight. A limitation of phenolic separators, however, is that they have an allowable maximum temperature of about 135°C (275°F).

24.6 STATIC LOAD DISTRIBUTION

Since a simple analytical expression for the deformation in terms of load was defined in Sec. 19.4, it is possible to consider how the bearing load is distributed among the elements. Most rolling-element bearing applications involve steady-state rotation of either the inner or outer race or both; however, the rotational speeds are usually not so great as to cause ball or roller centrifugal forces or gyroscopic moments of significant magnitudes. In analyzing the loading distribution on the rolling elements, it is usually satisfactory to ignore these effects in most applications. In this section the load deflection relationships for ball and roller bearings are given, along with radial and thrust load distributions of statically loaded rolling elements.

24.6.1 LOAD DEFLECTION RELATIONSHIPS. For an elliptical conjunction the load deflection relationship given in Eq. (19.16) can be written as

$$w_z = K_{1.5}\delta_m^{3/2} \tag{24.32}$$

where

$$K_{1.5} = \pi k E' \left(\frac{2\mathscr{E} R}{9\mathscr{F}^3} \right)^{1/2} \tag{24.33}$$

Similarly, for a rectangular conjunction from Eq. (19.39) we get

$$w_z = K_1 \delta_m$$

where

$$K_1 = \frac{\pi \ell E'}{2\left[\frac{2}{3} + \ln(4r_{ax}/b) + \ln(4r_{bx}/b)\right]} \tag{24.34}$$

and ℓ is the length of the rolling element. In general then,

$$w_z = K_j \delta_m^j \tag{24.35}$$

in which $j = 1.5$ for ball bearings and 1.0 for roller bearings. The total normal approach between two races separated by a rolling element is the sum of the deformations under load between the rolling element and both races. Therefore,

$$\delta_m = \delta_{mo} + \delta_{mi} \tag{24.36}$$

where

$$\delta_{mo} = \left[\frac{w_z}{(K_j)_o} \right]^{1/j} \tag{24.37}$$

$$\delta_{mi} = \left[\frac{w_z}{(K_j)_i} \right]^{1/j} \tag{24.38}$$

Substituting Eqs. (24.36) to (24.38) into Eq. (24.35) gives

$$K_j = \frac{1}{\left\{ \left[1/(K_j)_o \right]^{1/j} + \left[1/(K_j)_i \right]^{1/j} \right\}^j} \tag{24.39}$$

Recall that $(K_j)_o$ and $(K_j)_i$ are defined by Eq. (24.33) and (24.34) for an elliptical and a rectangular conjunction, respectively. These equations show that $(K_j)_o$ and $(K_j)_i$ are functions only of the contact geometry and the material properties. The radial and thrust load analyses are presented in the following two sections and are directly applicable for radially loaded ball and roller bearings and thrust-loaded ball bearings.

24.6.2 RADIALLY LOADED BALL AND ROLLER BEARINGS. A radially loaded rolling element with radial clearance $c_d/2$ is shown in Fig. 24.14. In the concentric position shown in Fig. 24.14(a) a uniform radial clearance between the rolling element and the races of $c_d/2$ is evident. The application of an arbitrarily small radial load to the shaft causes the inner race to move a distance $c_d/2$ before contact is made between a rolling element located on the load line

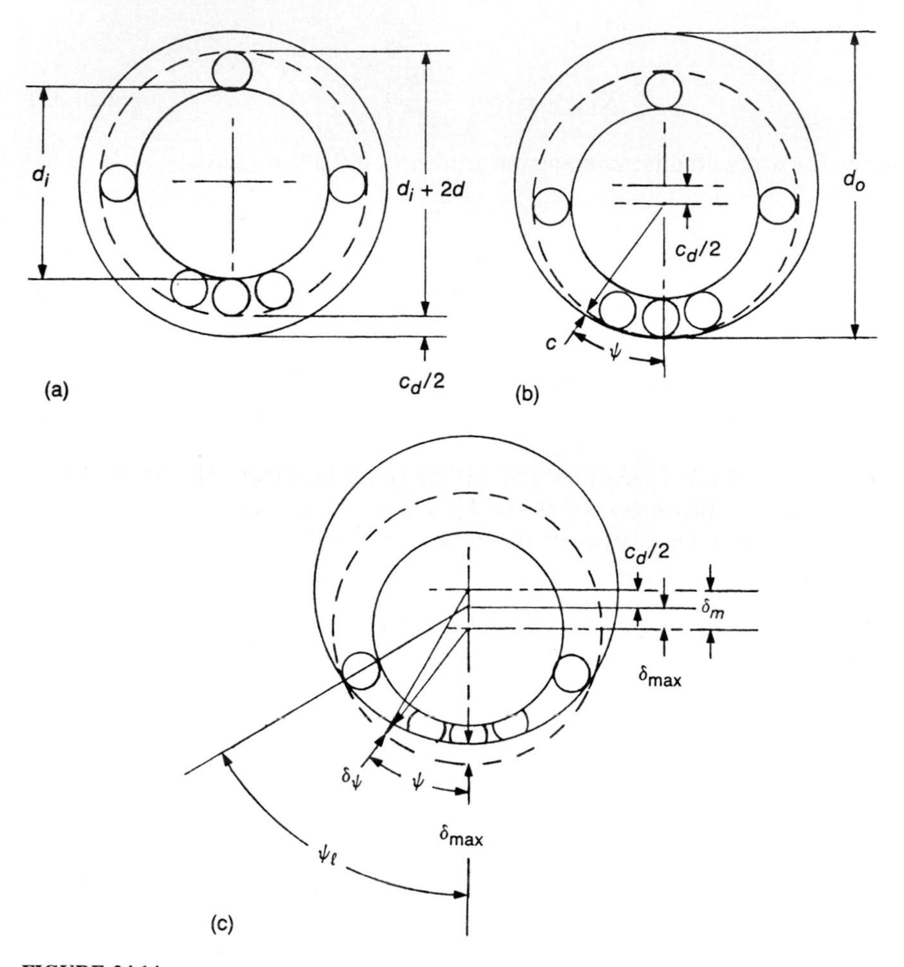

FIGURE 24.14
Radially loaded rolling-element bearing. (a) Concentric arrangement; (b) initial contact; (c) interference. [*From Hamrock and Anderson (1983).*]

and the inner and outer races. At any angle there will still be a radial clearance c that, if c_d is small in relation to the radius of the tracks, can be expressed with adequate accuracy by

$$c = (1 - \cos \psi)\frac{c_d}{2}$$

On the load line where $\psi = 0°$, the clearance is zero; but when $\psi = 90°$, the clearance retains its initial value of $c_d/2$.

The application of further load will cause elastic deformation of the balls and elimination of clearance around an arc $2\psi_\ell$. If the interference or total elastic compression on the load line is δ_{max}, the corresponding elastic compression of the ball δ_ψ along a radius at angle ψ to the load line will be given by

$$\delta_\psi = (\delta_{\max} \cos\psi - c) = \left(\delta_{\max} + \frac{c_d}{2}\right)\cos\psi - \frac{c_d}{2}$$

This assumes that the races are rigid. Now it is clear from Fig. 24.14(c) that $\delta_{\max} + c_d/2$ represents the total relative radial displacements of the inner and outer races. Hence,

$$\delta_\psi = \delta_m \cos\psi - \frac{c_d}{2} \tag{24.40}$$

The relationship between load and elastic compression along the radius at angle ψ to the load vector is given by Eq. (24.35) as

$$w_\psi = K_j \delta_\psi^j$$

Substituting Eq. (24.40) into this equation gives

$$w_\psi = K_j\left(\delta_m \cos\psi - \frac{c_d}{2}\right)^j$$

For static equilibrium the applied load must equal the sum of the components of the rolling-element loads parallel to the direction of the applied load.

$$w_t = \Sigma\, w_\psi \cos\psi$$

Therefore,

$$w_t = K_j \Sigma\left(\delta_m \cos\psi - \frac{c_d}{2}\right)^j \cos\psi \tag{24.41}$$

The angular extent of the bearing arc $2\psi_\ell$ in which the rolling elements are loaded is obtained by setting the root expression in Eq. (24.41) equal to zero and solving for ψ.

$$\psi_\ell = \cos^{-1}\frac{c_d}{2\delta_m}$$

The summation in Eq. (24.41) applies only to the angular extent of the loaded region. This equation can be written for a roller bearing as

$$w_t = \left(\psi_\ell - \frac{c_d}{2\delta}\sin\psi_\ell\right)\frac{nK_1\delta_m}{2\pi} \tag{24.42}$$

and similarly in integral form for a ball bearing as

$$w_t = \frac{n}{P}K_{1.5}\delta_m^{3/2}\int_0^{\psi_\ell}\left(\cos\psi - \frac{c_d}{2\delta}\right)^{3/2}\cos\psi\, d\psi$$

The integral in the equation can be reduced to a standard elliptic integral by the hypergeometric series and the beta function. If the integral is numerically evaluated directly, the following approximate expression is derived:

$$\int_0^{\psi_\ell}\left(\cos\psi - \frac{c_d}{2\delta_m}\right)^{3/2}\cos\psi\,d\psi = 2.491\left\{\left[1+\left(\frac{c_d/2\delta_m - 1}{1.23}\right)^2\right]^{1/2} - 1\right\}$$

This approximate expression fits the exact numerical solution to within ± 2 percent for a complete range of $c_d/2\delta_m$.

The load carried by the most heavily loaded rolling element is obtained by substituting $\psi = 0°$ in Eq. (24.41) and dropping the summation sign.

$$(w_z)_{\max} = K_j\delta_m^j\left(1 - \frac{c_d}{2\delta_m}\right)^j \tag{24.43}$$

Dividing this maximum load by the total radial load for a roller bearing [Eq. (24.42)] gives

$$w_z = \frac{\left(\psi_\ell - \dfrac{c_d}{2\delta_m}\sin\psi_\ell\right)\dfrac{n(w_z)_{\max}}{2\pi}}{1 - c_d/2\delta_m} \tag{24.44}$$

and similarly for a ball bearing

$$w_z = \frac{n(w_z)_{\max}}{Z_w} \tag{24.45}$$

where

$$Z_w = \frac{\pi(1 - c_d/2\delta_m)^{3/2}}{2.491\left\{\left[1+\left(\dfrac{1 - c_d/2\delta_m}{1.23}\right)^2\right]^{1/2} - 1\right\}} \tag{24.46}$$

For roller bearings when the diametral clearance c_d is zero, Eq. (24.44) gives

$$w_z = \frac{n(w_z)_{\max}}{4} \tag{24.47}$$

For ball bearings when the diametral clearance c_d is zero, the value of Z_w in Eq. (24.45) becomes 4.37. This is the value derived by Stribeck (1901) for ball bearings of zero diametral clearance. The approach used by Stribeck was to evaluate the finite summation for various numbers of balls. He then derived the celebrated Stribeck equation for static load-carrying capacity by writing the more conservative value of 5 for the theoretical value of 4.37:

$$w_z = \frac{n(w_z)_{\max}}{5} \tag{24.48}$$

In using Eq. (24.48) remember that Z_w was considered to be a constant and that the effects of clearance and applied load on load distribution were not taken into account. These effects were, however, considered in obtaining Eq. (24.45).

24.6.3 THRUST-LOADED BALL BEARING. The static thrust load-carrying capacity of a ball bearing may be defined as the maximum thrust load that the bearing can endure before the contact ellipse approaches a race shoulder, as shown in Fig. 24.15, or as the load at which the allowable mean compressive stress is reached, whichever is smaller. Both the limiting shoulder height and the mean compressive stress must be calculated to find the static thrust load-carrying capacity.

Each ball is subjected to an identical thrust component w_t/n, where w_t is the total thrust load. The initial contact angle before the application of a thrust load is denoted by β_f. Under load the normal ball thrust load w_t acts at the contact angle β and is written as

$$w = \frac{w_t}{n \sin \beta} \tag{24.49}$$

A cross section through an angular-contact bearing under a thrust load w_t is shown in Fig. 24.16. From this figure the contact angle after the thrust load has been applied can be written as

$$\beta = \cos^{-1} \frac{c_r - c_d/2}{c_r + \delta_m} \tag{24.50}$$

The initial contact angle was given in Eq. (24.7). Using that equation and rearranging terms in Eq. (24.50) give, solely from geometry (Fig. 24.16),

$$\delta_m = c_r \left(\frac{\cos \beta_f}{\cos \beta} - 1 \right)$$

$$\delta_m = \delta_{mo} + \delta_{mi}$$

$$\delta_m = \left[\frac{w}{(K_j)_o} \right]^{1/j} + \left[\frac{w}{(K_j)_i} \right]^{1/j}$$

$$K_j = \frac{1}{\left\{ \left[1/(K_j)_o \right]^{1/j} + \left[1/(K_j)_i \right]^{1/j} \right\}^j}$$

$$K_{1.5} = \frac{1}{\left\{ \left[\dfrac{(4.5\mathscr{F}_o^3)^{1/2}}{\pi k_o E_o' (R_o \mathscr{E}_o)^{1/2}} \right]^{2/3} + \left[\dfrac{(4.5\mathscr{F}_i^3)^{1/2}}{\pi k_i E_i' (R_i \mathscr{E}_i)^{1/2}} \right]^{2/3} \right\}^{3/2}} \tag{24.51}$$

$$w = K_{1.5} c_r^{3/2} \left(\frac{\cos \beta_f}{\cos \beta} - 1 \right)^{3/2} \tag{24.52}$$

where Eq. (24.32) for $K_{1.5}$ is replaced by Eq. (24.51) and $k, \mathscr{E}, \mathscr{F}$ are given by Eq. (19.29), (19.30), and (19.32), respectively.

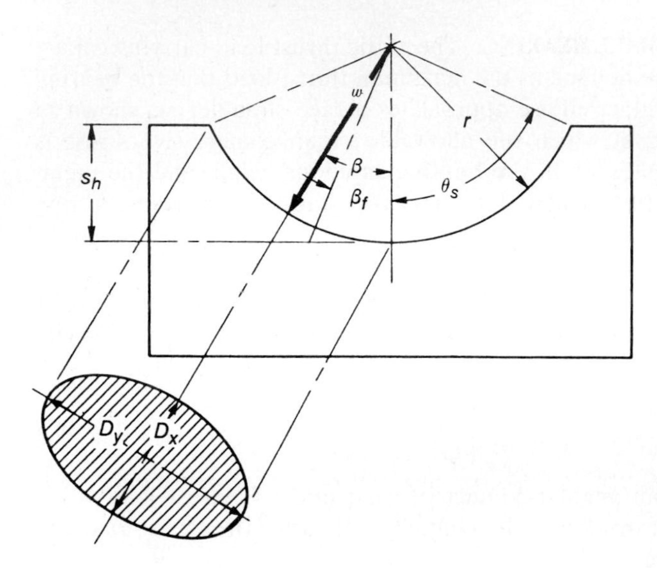

FIGURE 24.15
Contact ellipse in bearing race under load. [*From Hamrock and Anderson (1983).*]

From Eq. (24.49) and (24.52)

$$\frac{w_t}{n \sin \beta} = w$$

or

$$\frac{w_t}{nK_j c_r^{3/2}} = \sin \beta \left(\frac{\cos \beta_f}{\cos \beta} - 1 \right)^{3/2} \tag{24.53}$$

This equation can be solved numerically by the Newton-Raphson method. The iterative equation to be satisfied is

$$\beta' - \beta = \frac{\dfrac{w_t}{nK_{1.5} c_r^{3/2}} - \sin \beta \left(\dfrac{\cos \beta_f}{\cos \beta} - 1 \right)^{3/2}}{\cos \beta \left(\dfrac{\cos \beta_f}{\cos \beta} - 1 \right)^{3/2} + \dfrac{3}{2} \cos \beta_f \tan^2 \beta \left(\dfrac{\cos \beta_f}{\cos \beta} - 1 \right)^{1/2}} \tag{24.54}$$

In this equation convergence is satisfied when $\beta' - \beta$ becomes essentially zero.

When a thrust load is applied, the shoulder height limits the axial deformation, which can occur before the pressure-contact ellipse reaches the shoulder. As long as the following inequality is satisfied, the contact ellipse will not exceed this limit:

$$\theta_s > \beta + \sin^{-1} \frac{D_y}{R_r d}$$

From Fig. 24.5 and Eq. (24.11) the angle θ_s used to define the shoulder height

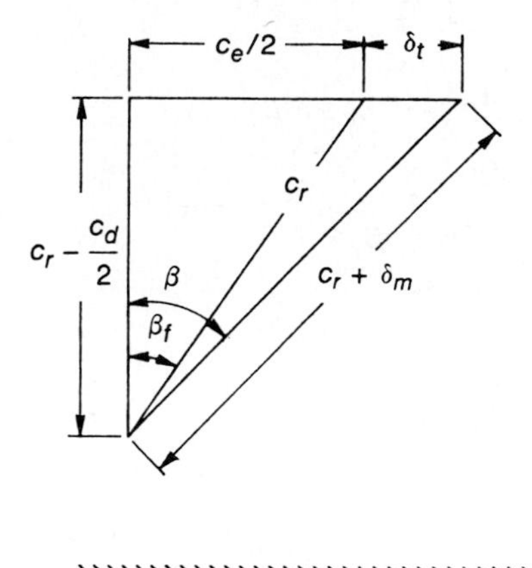

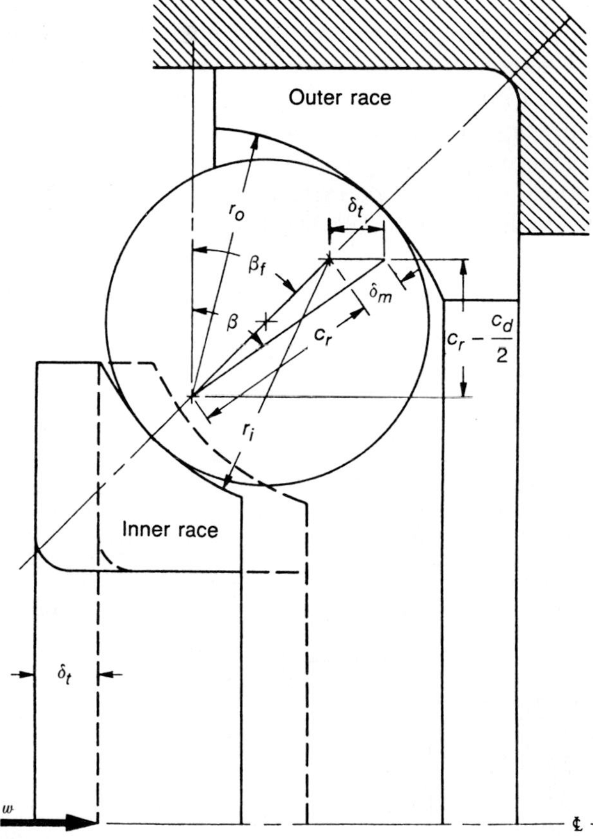

FIGURE 24.16
Angular-contact ball bearing under thrust load. [*From Hamrock and Anderson (1983).*]

can be written as

$$\theta_s = \cos^{-1}\left(1 - \frac{s_h}{R_r d}\right)$$

From Fig. 24.3 the axial deflection δ_t corresponding to a thrust load can be written as

$$\delta_t = (c_r + \delta) \sin \beta - c_r \sin \beta_f \tag{24.55}$$

Substituting Eq. (24.51) into (24.55) gives

$$\delta_t = \frac{c_r \sin (\beta - \beta_f)}{\cos \beta}$$

Once β has been determined from Eq. (24.54) and β_f from Eq. (24.7), the relationship for δ_t can be easily evaluated.

24.6.4 PRELOADING. The use of angular-contact bearings as duplex pairs preloaded against each other is discussed in Sec. 24.2.1. As shown in Table 24.2, duplex bearing pairs are used in either back-to-back or face-to-face arrangements. Such bearings are usually preloaded against each other by providing what is called "stickout" in the manufacture of the bearing. This is illustrated in Fig. 24.17 for a bearing pair used in a back-to-back arrangement. The magnitude of the stickout and the bearing design determine the level of preload on each bearing when the bearings are clamped together as in Fig. 24.17. The magnitude of preload and the load deflection characteristics for a given bearing pair can be calculated by using Eqs. (24.7), (24.32), (24.49), and (24.51) to (24.53).

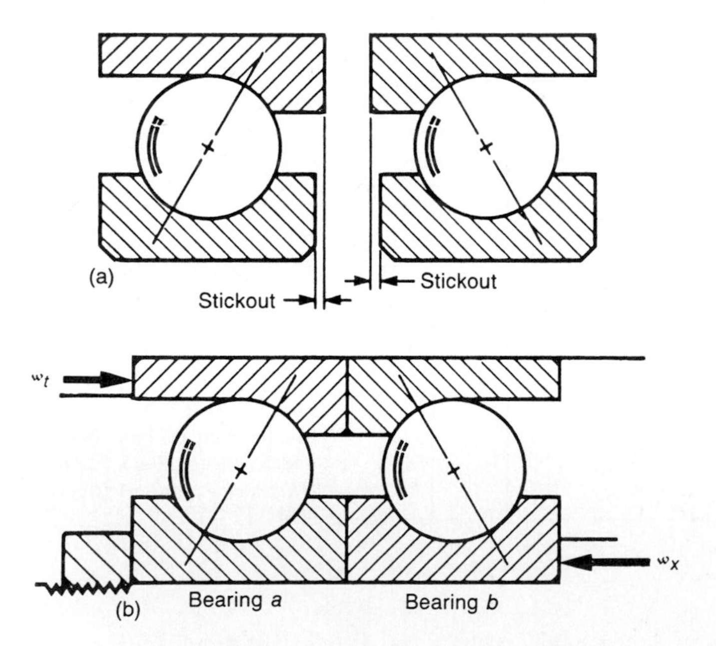

FIGURE 24.17
Angular-contact bearings in back-to-back arrangement, shown (a) individually as manufactured and (b) as mounted with preload. [*From Hamrock and Anderson (1983).*]

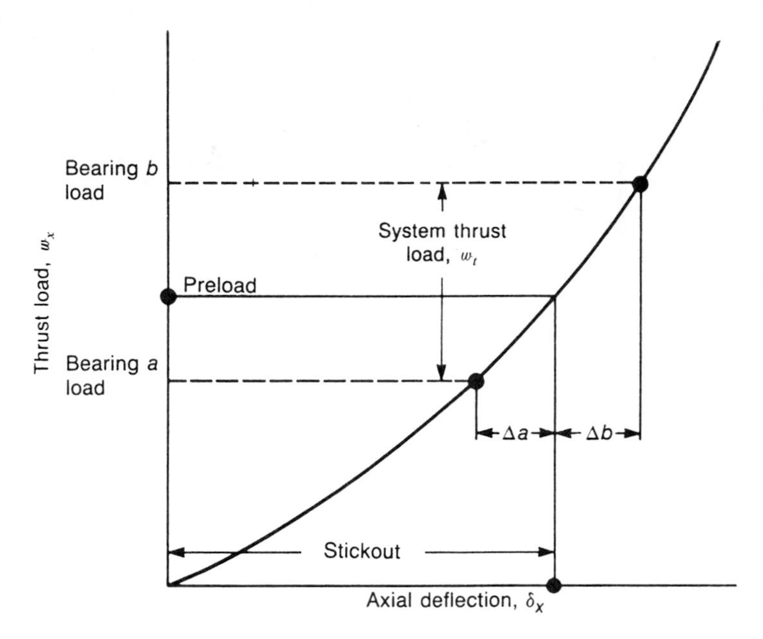

FIGURE 24.18
Thrust load–axial deflection curve for typical ball bearing. [*From Hamrock and Anderson (1983).*]

The relationship of initial preload, system load, and final load for bearings a and b is shown in Fig. 24.18. The load deflection curve follows the relationship $\delta_m = K w^{2/3}$. When a system thrust load w_t is imposed on the bearing pairs, the magnitude of load on bearing b increases while that on bearing a decreases until the difference equals the system load. The physical situation demands that the change in each bearing deflection be the same ($\Delta a = \Delta b$ in Fig. 24.18). The increments in bearing load, however, are not the same. This is important because it always requires a system thrust load far greater than twice the preload before one bearing becomes unloaded. Prevention of bearing unloading, which can result in skidding and early failure, is an objective of preloading.

24.7 ROLLING FRICTION AND FRICTION LOSSES

24.7.1 ROLLING FRICTION. The concepts of rolling friction are important generally in understanding the behavior of machine elements in rolling contact and particularly because rolling friction influences the overall behavior of rolling-element bearings. The theories of Reynolds (1876) and Heathcote (1921) attempted to explain rolling friction in terms of the energy required to overcome the interfacial slip that occurs because of the curved shape of the contact area. As shown in Fig. 24.19, the ball rolls about the Y axis and makes contact

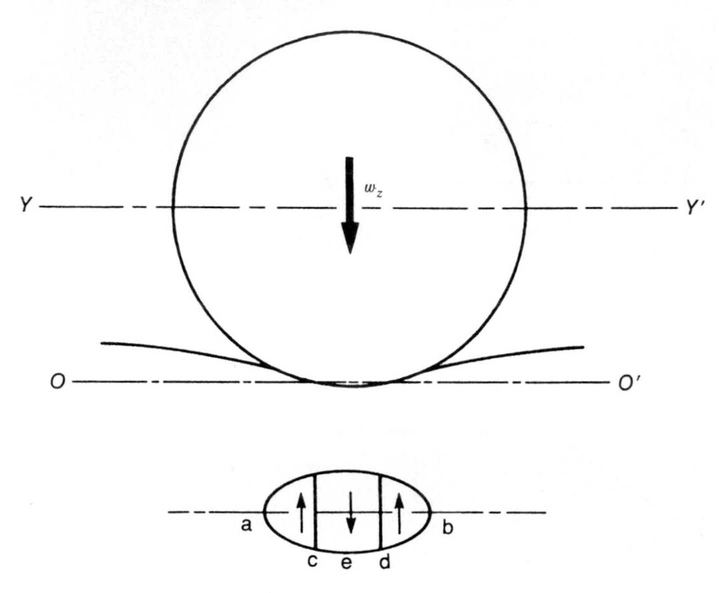

FIGURE 24.19
Differential slip due to curvature of contact ellipse. [*From Hamrock and Anderson (1983).*]

with the groove from a to b. If the groove is fixed, then for zero slip over the contact area no point within the area should have a velocity in the rolling direction. The surface of the contact area is curved, however, so that points a and b are at different radii from the Y axis than are points c and d. For a rigid ball, points a and b must have different velocities with respect to the Y axis than do points c and d because the velocity of any point on the ball relative to the Y axis equals the angular velocity times the radius from the Y axis. Slip must occur at various points over the contact area unless the body is so elastic that yielding can take place in the contact area to prevent this interfacial slip. The theories of Reynolds and later Heathcote assumed that this interfacial slip took place and that the forces required to make a ball roll were those required to overcome the friction due to this interfacial slip. In the contact area rolling without slip will occur at a specific radius from the Y axis. Where the radius is greater than this radius to the rolling point, slip will occur in one direction; where it is less, slip will occur in the other direction. In Fig. 24.19 the lines at points c and d represent the approximate location of the rolling bands, and the arrows shown in the three portions of the contact area represent the directions of interfacial slip when the ball is rolling into the paper.

The location of the two rolling bands relative to the axis of the contact ellipse can be obtained by summing the loads acting on the ball in the rolling direction. In Fig. 24.20 these are

$$2\,w_{x,b} - w_{x,a} = \mu_r\, w_z$$

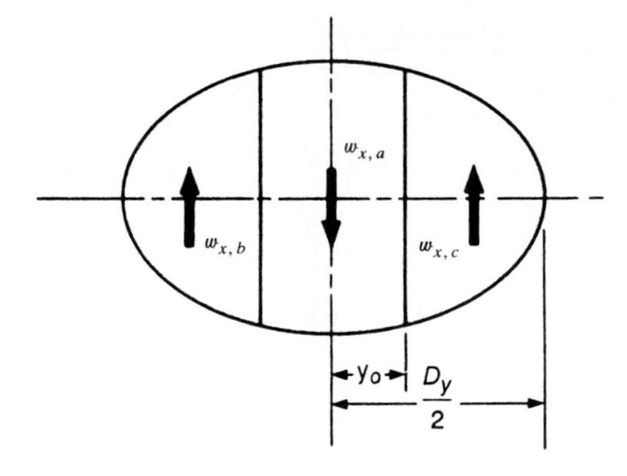

FIGURE 24.20
Load components in contact ellipse. [*From Hamrock and Anderson (1983)*.]

where μ_r is the coefficient of rolling friction. If a Hertzian ellipsoidal pressure distribution is assumed, the location of the rolling bands can be determined (Bisson and Anderson, 1964). For rolling with zero traction ($\mu_r = 0$) the result is

$$\frac{y_o}{D_y} \approx 0.174$$

where D_y is the diameter of the contact ellipse along the Y axis [Eq. (19.14)].

In reality, all materials are elastic so that areas of no slip as well as areas of microslip exist within the contact as pointed out by Johnson and Tabor (1967–68). Differential strains in the materials in contact will cause slip unless it is prevented by friction. In high-conformity contacts slip is likely to occur over most of the contact region, as can be seen in Fig. 24.21, which shows the nondimensional friction force as a function of the conformity parameter

$$\lambda_f = \frac{\pi D_y^2 E'}{4\mu r_{ax}^2 p_{\text{mean}}}$$

where μ is the coefficient of sliding friction and p_{mean} is the mean pressure of the contact ellipse. The limiting value of friction force w_x is

$$w_x = \frac{0.02\mu D_y^2 w_z}{r_{ax}^2}$$

A fraction λ_h of the elastic energy of compression in rolling is always lost because of hysteresis. The effect of hysteresis losses on rolling resistance has been studied by Tabor (1955). Tabor developed the following expression:

$$w_x = c_4 \lambda_h \frac{w_z D_y}{r_{ax}} \tag{24.56}$$

where c_4 is $1/3\pi$ for rectangular conjunctions and $3/32$ for elliptical conjunctions. Two hard-steel surfaces show a λ_h of about 1 percent.

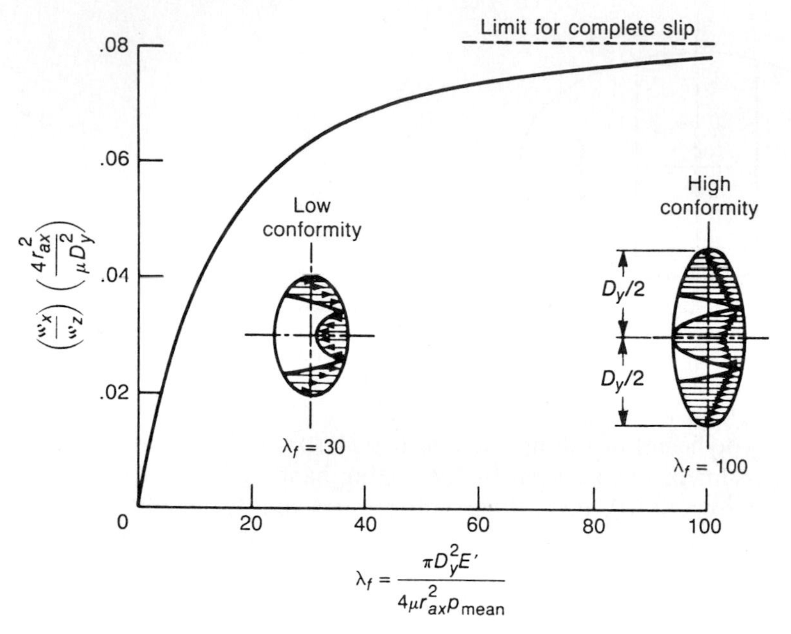

FIGURE 24.21
Frictional resistance of ball in conforming groove. [*From Hamrock and Anderson (1983).*]

Of far greater importance in contributing to frictional losses in rolling-element bearings, especially at high speeds, is ball spinning. Spinning as well as rolling takes place in one or both of the ball-race contacts of a ball bearing. The situation is shown schematically in Fig. 24.11.

High speeds cause a divergency of the contact angles β_i and β_o. In Fig. 24.12(b) the ball is rolling with inner-race control so that approximately pure rolling takes place at the inner-race contact. The rolling and spinning vectors ω_r and ω_s at the outer-race contact are shown in Fig. 24.11. The higher the ratio ω_s/ω_r, the higher the friction losses.

24.7.2 FRICTION LOSSES. Some factors that affect the magnitude of friction losses in rolling-element bearings are

1. Bearing size
2. Bearing type
3. Bearing design
4. Load (magnitude and type, either thrust or radial)
5. Speed
6. Oil viscosity
7. Oil flow

Friction losses in a specific bearing consist of the following:

1. Sliding friction losses in the contacts between the rolling elements and the races. These losses include differential slip and slip due to ball spinning. They are complicated by the presence of elastohydrodynamic lubrication films. The shearing of these films, which are extremely thin and contain oil whose viscosity is increased by orders of magnitude above its atmospheric pressure value, accounts for a significant fraction of the friction losses in rolling-element bearings.
2. Hysteresis losses due to the damping capacity of the race and ball materials.
3. Sliding friction losses between the separator and its locating race surface and between the separator pockets and the rolling elements.
4. Shearing of oil films between the bearing parts and oil churning losses caused by excess lubricant within the bearing.
5. Flinging of oil off the rotating parts of the bearing.

Because of the dominant role in bearing frictional losses played by the lubrication method, there are no quick and easy formulas for computing rolling-element bearing power loss. Friction coefficients for a particular bearing can vary by a factor of 5 depending on lubrication. A flood-lubricated bearing may consume five times the power of one that is merely wetted by oil-air mist lubrication.

The friction coefficients given by Palmgren (1959) are a rough but useful guide. These values were computed at a bearing load that will give a life of 1×10^9 revolutions for the the respective bearings. The friction coefficients for several different bearings are shown here:

Self-aligning ball	0.0010
Cylindrical roller with flange-guided short rollers	0.0011
Thrust ball	0.0013
Single-row, deep-groove ball	0.0015
Tapered and spherical roller with flange-guided rollers	0.0018
Needle roller	0.0045

All these friction coefficients are referenced to the bearing bore.

More accurate estimates of bearing power loss and temperature rise can be obtained by using one or more of the available computer codes that represent the basis for the current design methodology for rolling-element bearings. These are discussed in Anderson (1979) and Pirvics (1980) and briefly reviewed in Sec. 24.10.

24.8 LUBRICATION SYSTEMS

A liquid lubricant has several functions in a rolling-element bearing. It provides separating films between the bearing parts (elastohydrodynamic lubrication between the races and the rolling elements and hydrodynamic between the cage or separator and its locating surface). It serves as a coolant if either circulated through the bearing to an external heat exchanger or simply brought into contact with the bearing housing and the machine casing. A circulating lubricant also serves to flush out wear debris and carry it to a filter where it can be removed from the system. Finally, it provides corrosion protection. The different methods of providing liquid lubricant to a bearing are each discussed here. Note that Chap. 4 presents lubricant properties, whereas this section explores effective means of lubrication.

24.8.1 SOLID LUBRICATION. An increasing number of rolling-element bearings are lubricated with solid-film lubricants, usually in applications, such as extreme temperature or the vacuum of space, where conventional liquid lubricants are not suitable.

Success in cryogenic applications, where the bearing is cooled by the cryogenic fluid (liquid oxygen or hydrogen), has been achieved with transfer films of polytetrafluoroethylene (Scibbe, 1968). Bonded films of soft metals such as silver, gold, and lead applied by ion plating as extremely thin films (0.2 to 0.3 mm) have also been used (Todd and Bentall, 1978). Silver and lead in particular have found use in bearings that support the rotating anode in x-ray tubes. Extremely thin films are required in rolling-element bearings in order not to significantly alter the bearing internal geometry and in order to retain the basic mechanical properties of the substrate materials in the Hertzian contacts.

24.8.2 LIQUID LUBRICATION. The great majority of rolling-element bearings are lubricated by liquids. The liquid in greases can be used, or liquid can be supplied to the bearing from either noncirculating or circulating systems.

24.8.2.1 Greases. The most common and probably least expensive mode of lubrication is grease lubrication. In the strictest sense a grease is not a liquid, but the liquid or fluid constituent in the grease is the lubricant. Greases consist of a fluid phase of either a petroleum or synthetic oil and a thickener. The most common thickeners are sodium-, calcium-, or lithium-based soaps, although thickeners of inorganic materials such as bentonite clay have been used in synthetic greases. Some discussion of the characteristics and temperature limits of greases is given in Bisson and Anderson (1964) and McCarthy (1973).

Greases are usually retained within the bearing by shields or seals that are an integral part of the assembled bearing. Since there is no recirculating fluid, grease-lubricated bearings must reject heat by conduction and convection and are therefore limited to maximum $d_b N_a$ (ball bearing bore diameter in millimeters times rotational speed in revolutions per minute) values of 0.25 to 0.4 million.

The proper grease for a particular application depends on the temperature, speed, and ambient pressure environment to which the bearing is exposed. McCarthy (1973) presents a comprehensive discussion useful in the selection of a grease; the bearing manufacturer can also recommend the most suitable grease and bearing type.

24.8.2.2 Nonrecirculating liquid lubrication systems. At low to moderate speeds, where the use of grease lubrication is not suitable, other methods of supplying lubricant to the bearing can be used. These include splash or bath, wick, oil-ring, and oil-air mist lubrication. Felt wicks can be used to transport oil by capillary action from a nearby reservoir. Oil rings, which are driven by frictional contact with the rotating shaft, run partially immersed in an oil reservoir and feed oil mechanically to the shaft, which is adjacent to the bearing. The bearing may itself be partially immersed in an oil reservoir to splash-lubricate itself. All these methods require modest ambient temperatures and thermal conditions as well as speed conditions equivalent to a maximum $d_b N_a$ of about 0.5 million. The machinery must also remain in a fixed-gravity orientation.

Oil-air mist lubrication supplies atomized oil in an airstream to the bearing, where a reclassifier increases the droplet size, allowing it to condense on the bearing surfaces. Feed rates are low, and a portion of the oil flow escapes with the feed air to the atmosphere. Commercial oil-air mist generators are available for systems ranging from a single bearing to hundreds of bearings. Bearing friction losses and heat generation are low with mist lubrication, but ambient temperatures and cooling requirements must be moderate because oil-air mist systems provide minimal cooling. Many bearings, especially small-bore bearings, are successfully operated at high speeds ($d_b N_a$ values to greater than 1 million) with oil-air mist lubrication.

24.8.2.3 Jet lubrication. In applications where speed or heat rejection requirements are too high, jet lubrication is frequently used both to lubricate and to control bearing temperatures. A number of variables are critical to achieving not only satisfactory but near optimal performance and bearing operation. These include the placement of the nozzles, the number of nozzles, the jet velocity, the lubricant flow rate, and the scavenging of the lubricant from the bearing and its immediate vicinity. The importance of proper jet lubricating system design is shown by Matt and Gianotti (1966). Their results are summarized in Fig. 24.22.

Proper placement of the jets should take advantage of any natural pumping ability of the bearings. Figure 24.23 (Parker, 1980) illustrates jet lubrication of ball and tapered-roller bearings. Centrifugal forces aid in moving the oil through the bearing to cool and lubricate the elements. Directing jets at the radial gaps between the cage and the races achieves maximum oil penetration into the bearing interior. Miyakawa et al. (1972), Anderson et al. (1954), Zaretsky et al. (1976), and Parker and Signer (1978) present useful data on how jet placement and velocity influence the lubrication of several types of bearing.

24.8.2.4 Underrace lubrication. As bearing speeds increase, centrifugal effects become more predominant, making it increasingly difficult to lubricate and cool a bearing effectively. The jetted oil is thrown off the sides of the bearing rather than penetrating to the interior. At extremely high $d_b N_a$ values (2.4 million and

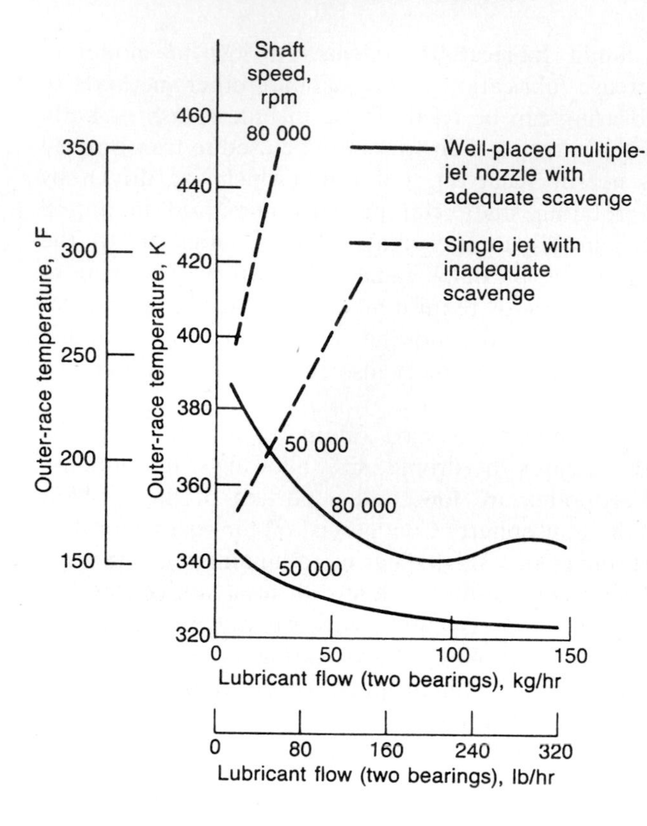

FIGURE 24.22
Effectiveness of proper jet lubrication. Test bearings, 20-mm-bore angular-contact ball bearings; thrust load, 222 N (50 lbf). [*From Matt and Gianotti (1966).*]

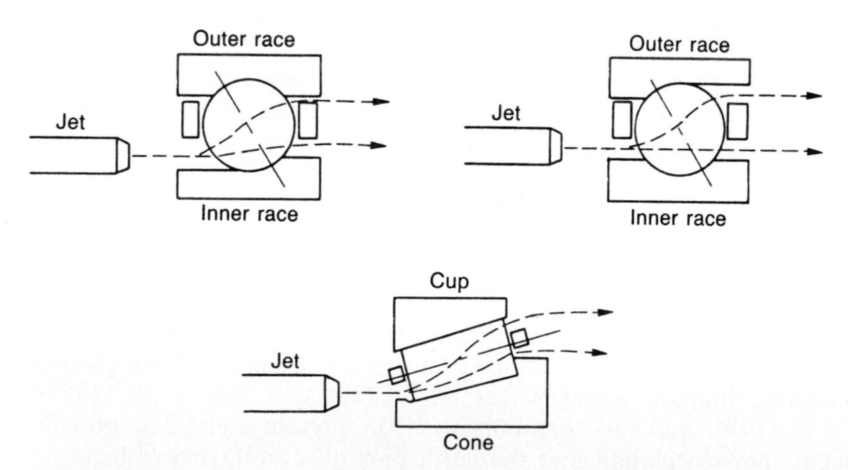

FIGURE 24.23
Placement of jets for ball bearings with relieved rings and tapered-roller bearings. [*From Parker (1980).*]

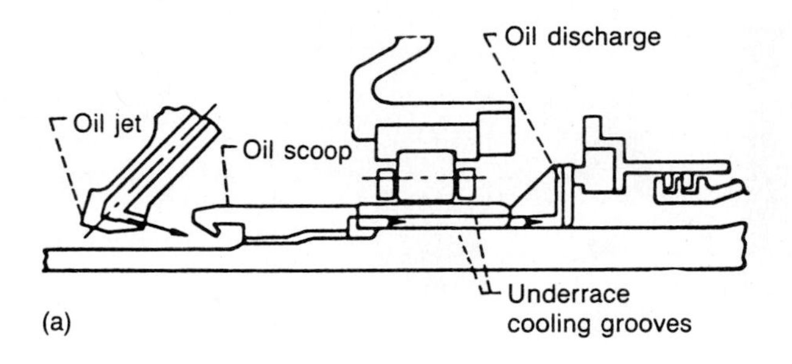

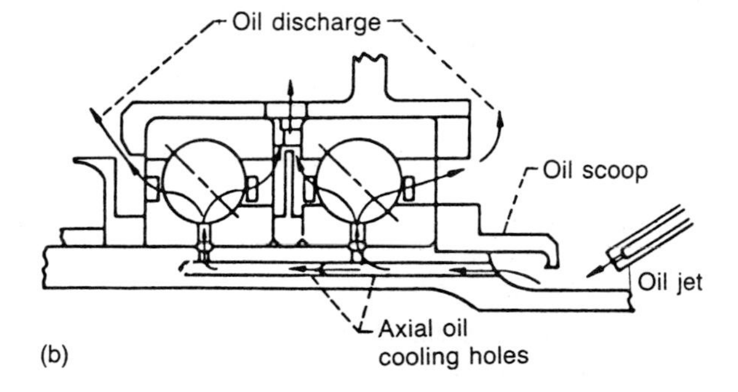

FIGURE 24.24
Underrace oiling system for main shaft bearings on turbofan engine. (a) Cylindrical roller bearing; (b) ball thrust bearing. [*From Brown (1970).*]

higher), jet lubrication becomes ineffective. Increasing the flow rate only adds to heat generation through increased churning losses. Brown (1970) describes an underrace oiling system used in a turbofan engine for both ball and cylindrical roller bearings. Figure 24.24 illustrates the technique. The lubricant is directed radially under the bearings. Centrifugal effects assist in pumping oil out through the bearings via suitable slots and holes, which are made a part of both the shaft and the mounting system and the bearings themselves. Holes and slots are provided within the bearing to feed oil directly to the ball-race and cage-race contacts.

This lubrication technique has been thoroughly tested for large-bore ball and roller bearings up to 3 million $d_b N_a$. Pertinent data are reported by Signer et al. (1974), Brown et al. (1977), Schuller (1979), and Signer and Schuller (1982).

24.9 FATIGUE LIFE

24.9.1 CONTACT FATIGUE THEORY. Rolling fatigue is a material failure caused by the application of repeated stresses to a small volume of material. It is a unique failure type—essentially a process of seeking out the weakest point at which the first failure will occur. A typical spall is shown in Fig. 24.25. On a microscale there will probably be a wide dispersion in material strength or resistance to fatigue because of inhomogeneities in the material. Because

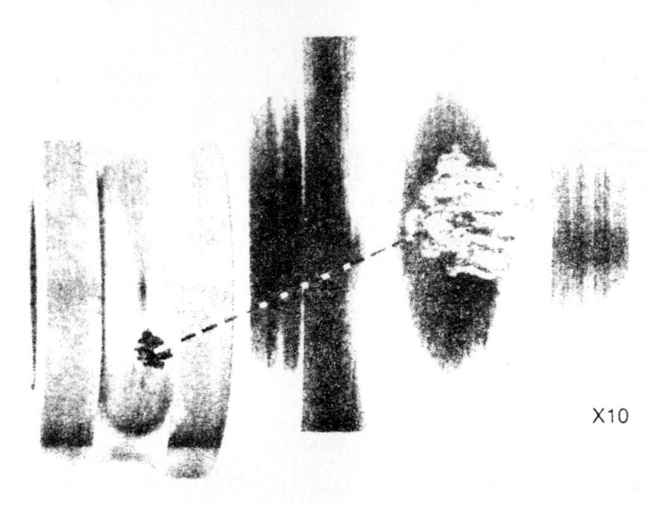

FIGURE 24.25
Typical fatigue spall. [*From Hamrock and Anderson (1983).*]

bearing materials are complex alloys, they are not homogeneous or equally resistant to failure at all points. Therefore, the fatigue process can be expected to be one in which a group of supposedly identical specimens exhibit wide variations in failure time when stressed in the same way. For this reason it is necessary to treat the fatigue process statistically.

Predicting how long a particular bearing will run under a specific load requires the following two essential pieces of information:

1. An accurate, quantitative estimate of the life dispersion or scatter
2. The life at a given survival rate or reliability level

This translates into an expression for the "load-carrying capacity," or the ability of the bearing to endure a given load for a stipulated number of stress cycles or revolutions. If a group of supposedly identical bearings is tested at a specific load and speed, the distribution in bearing lives shown in Fig. 24.26 will occur.

24.9.2 WEIBULL DISTRIBUTION. Weibull (1949) postulates that the fatigue lives of a homogeneous group of rolling-element bearings are dispersed according to the following relation:

$$\ln \ln \frac{1}{\tilde{S}} = e_1 \ln \frac{\tilde{L}}{\tilde{A}} \tag{24.57}$$

where $\tilde{S}$ is the probability of survival, $\tilde{L}$ is the fatigue life, and e_1 and $\tilde{A}$ are constants. The Weibull distribution results from a statistical theory of strength based on probability theory, where the dependence of strength on volume is

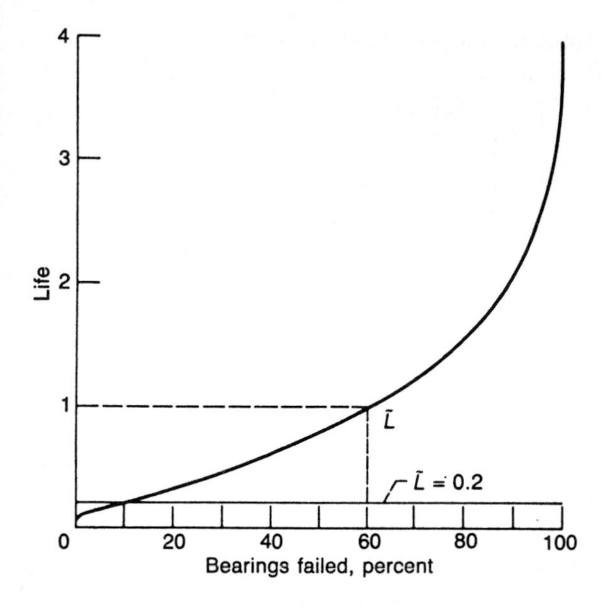

FIGURE 24.26
Distribution of bearing fatigue failures. [*From Hamrock and Anderson (1983).*]

explained by the dispersion in material strength. This is the "weakest link" theory.

Consider a volume being stressed that is broken up into m similar volumes

$$\tilde{S}_1 = 1 - \tilde{M}_1 \qquad \tilde{S}_2 = 1 - \tilde{M}_2 \qquad \tilde{S}_3 = 1 - \tilde{M}_3 \qquad \cdots \qquad \tilde{S}_m = 1 - \tilde{M}_m$$

The $\tilde{M}$'s represent the probability of failure and the $\tilde{S}$'s, the probability of survival. For the entire volume

$$\tilde{S} = \tilde{S}_1 \cdot \tilde{S}_2 \cdot \tilde{S}_3 \cdot \cdots \cdot \tilde{S}_m$$

Then

$$1 - \tilde{M} = (1 - \tilde{M}_1)(1 - \tilde{M}_2)(1 - \tilde{M}_3) \cdots (1 - \tilde{M}_m)$$

$$1 - \tilde{M} = \prod_{i=1}^{m} (1 - \tilde{M}_i)$$

$$\tilde{S} = \prod_{i=1}^{m} (1 - \tilde{M}_i) \qquad (24.58)$$

The probability of a crack starting in the ith volume is

$$\tilde{M}_i = f(x)\tilde{V}_i \qquad (24.59)$$

where $f(x)$ is a function of the stress level, the number of stress cycles, and the depth into the material where the maximum stress occurs and $\tilde{V}_i$ is the elementary volume. Therefore, substituting Eq. (24.59) into (24.58) gives

$$\tilde{S} = \prod_{i=1}^{m} \left[1 - f(x)\tilde{V}_i\right]$$

$$\ln \tilde{S} = \sum_{i=1}^{m} \ln\left[1 - f(x)\tilde{V}_i\right]$$

Now if $f(x)\tilde{V}_i \ll 1$, then $\ln[1 - f(x)\tilde{V}_i] = -f(x)\tilde{V}_i$ and

$$\ln \tilde{S} = -\sum_{i=1}^{m} f(x)\tilde{V}_i \tag{24.60}$$

Let $\tilde{V}_i \to 0$; then

$$\sum_{i=1}^{m} f(x)\tilde{V}_i = \int f(x)\, d\tilde{V} = \bar{f}(x)\tilde{V} \tag{24.61}$$

where $\bar{f}(x)$ is a volume-average value of $f(x)$.

Lundberg and Palmgren (1947) assumed that $f(x)$ could be expressed as a power function of shear stress τ_0, number of stress cycles $\tilde{J}$, and depth of the maximum shear stress z_0.

$$\bar{f}(x) = \frac{\tau_0^{c_1}\tilde{J}^{c_2}}{z_0^{c_3}} \tag{24.62}$$

They also chose as the stressed volume

$$\tilde{V} = D_y z_0 \ell_v \tag{24.63}$$

Substituting Eqs. (24.61) to (24.63) into (24.60) gives

$$\ln \tilde{S} = -\frac{\tau_0^{c_1}\tilde{J}^{c_2} D_y \ell_v}{z_0^{c_3 - 1}}$$

or

$$\ln \frac{1}{\tilde{S}} = \frac{\tau_0^{c_1}\tilde{J}^{c_2} D_y \ell_v}{z_0^{c_3 - 1}} \tag{24.64}$$

For a specific bearing and load (e.g., stress), τ_0, D_y, ℓ_v, and z_0 are all constant so that

$$\ln \frac{1}{\tilde{S}} \approx \tilde{J}^{c_2}$$

Designating $\tilde{J}$ as life $\tilde{L}$ in stress cycles gives

$$\ln \frac{1}{\tilde{S}} = \left(\frac{\tilde{L}}{\tilde{A}}\right)^{c_2}$$

or

$$\ln \ln \frac{1}{\tilde{S}} = c_2 \ln \frac{\tilde{L}}{\tilde{A}} \tag{24.65}$$

This is the Weibull distribution, which relates probability of survival and life. It has two principal functions. First, bearing fatigue lives plot as a straight line on Weibull coordinates (log-log versus log) so that the life at any reliability level can be determined. Of most interest are the $\tilde{L}_{10}$ life ($\tilde{S} = 0.9$) and the $\tilde{L}_{50}$ life ($\tilde{S} = 0.5$). Bearing load ratings are based on the $\tilde{L}_{30}$ life. Second, Eq. (24.65) can be used to determine what the $\tilde{L}_{10}$ life must be to obtain a required life at any reliability level. The $\tilde{L}_{10}$ life is calculated, from the load on the bearing and the bearing dynamic capacity or the load rating given in manufacturers' catalogs and engineering journals, by using the equation

$$\tilde{L} = \left(\frac{\overline{C}}{w_e} \right)^{m_k} \tag{24.66}$$

where $\overline{C}$ = basic dynamic capacity or load rating, N
w_e = equivalent bearing load, N
m_k = load-life exponent; 3 for elliptical contacts and $\frac{10}{3}$ for rectangular contacts

A typical Weibull plot is shown in Fig. 24.27.

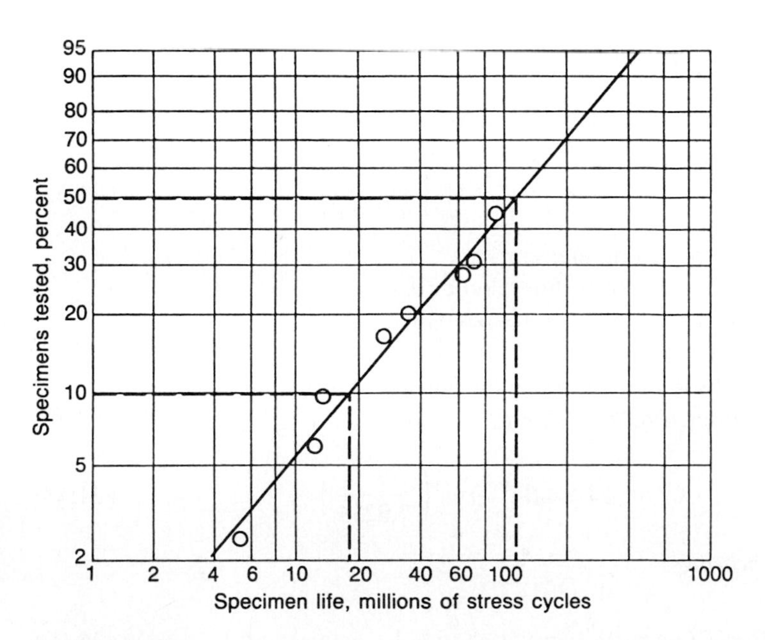

FIGURE 24.27
Typical Weibull plot of bearing fatigue failures. [*From Hamrock and Anderson (1983).*]

24.9.3 LUNDBERG-PALMGREN THEORY. The Lundberg-Palmgren theory, on which bearing ratings are based, is expressed by Eq. (24.64). The exponents in this equation are determined experimentally from the dispersion of bearing lives and the dependence of life on load, geometry, and bearing size. As a standard of reference, all bearing load ratings are expressed in terms of the specific dynamic capacity $\overline{C}$, which, by definition, is the load that a bearing can carry for 1 million inner-race revolutions with a 90 percent chance of survival.

Factors on which specific dynamic capacity and bearing life depend are

1. Size of rolling element
2. Number of rolling elements per row
3. Number of rows of rolling elements
4. Conformity between rolling elements and races
5. Contact angle under load
6. Material properties
7. Lubricant properties
8. Operating temperature
9. Operating speed

Only factors 1 to 5 are incorporated in bearing dynamic capacities developed from the Lundberg-Palmgren theory. The remaining factors must be taken into account in the life adjustment factors discussed in Sec. 24.9.5.

The formulas for specific dynamic capacity as developed by Lundberg and Palmgren (1947, 1952) are as follows:

For radial ball bearings with $d < 25$ mm:

$$\overline{C} = f_c(i \cos \beta)^{0.7} n^{2/3} \left(\frac{d}{0.0254} \right)^{1.8} \tag{24.67}$$

where d = diameter of rolling element, m
i = number of rows of rolling elements
n = number of rolling elements per row
β = contact angle
f_c = coefficient dependent on material and bearing type

For radial ball bearings with $d \geq 25$ mm:

$$\overline{C} = f_c(i \cos \beta)^{0.7} n^{2/3} \left(\frac{d}{0.0254} \right)^{1.4} \tag{24.68}$$

For radial roller bearings:

$$\overline{C} = f_c(i \cos \beta)^{0.78} n^{3/4} \left(\frac{d}{0.0254} \right)^{1.07} \left(\frac{\ell_t}{0.0254} \right)^{0.78} \tag{24.69}$$

where ℓ_t is roller length in meters.

For thrust ball bearings with $\beta \neq 90°$:

$$\bar{C} = f_c(i \cos \beta)^{0.7}(\tan \beta) n^{2/3}\left(\frac{d}{0.0254}\right)^{1.8} \tag{24.70}$$

For thrust roller bearings with $\beta \neq 90°$:

$$\bar{C} = f_c(i \cos \beta)^{0.78}(\tan \beta) n^{3/4}\left(\frac{\ell_t}{0.0254}\right)^{0.78} \tag{24.71}$$

For thrust ball bearings with $\beta = 90°$:

$$\bar{C} = f_c i^{0.7} n^{2/3}\left(\frac{d}{0.0254}\right)^{1.8} \tag{24.72}$$

For thrust roller bearings with $\beta = 90°$:

$$\bar{C} = f_c i^{0.78} n^{3/4}\left(\frac{d}{0.0254}\right)^{1.07}\left(\frac{\ell_t}{0.0254}\right)^{0.78} \tag{24.73}$$

For ordinary bearing steels such as SAE 52100 with mineral oil lubrication, f_c can be evaluated by using Tables 24.10 and 24.11, but a more convenient method is to use tabulated values from the most recent Antifriction Bearing Manufacturers Association (AFBMA) documents on dynamic load ratings and life (ISO, 1976). The values of $\bar{C}$ appear in bearing manufacturers' catalogs, together with factors $\tilde{X}$ and $\tilde{Y}$ used to determine the equivalent load w_e. The equivalent load can be calculated from the equation

$$w_e = \tilde{X} w_z + \tilde{Y} w_x \tag{24.74}$$

In addition to specific dynamic capacity $\bar{C}$, every bearing has a specific static capacity, usually designated as $\bar{C}_0$. Specific static capacity is defined as the load that under static conditions will result in a permanent deformation of 0.0001 times the rolling-element diameter. For some bearings $\bar{C}_0$ is less than $\bar{C}$; therefore it is important to avoid exposing a bearing to a static load that exceeds $\bar{C}_0$. Values of $\bar{C}_0$ are also given in bearing manufacturers' catalogs.

24.9.4 AFBMA METHODS. Shortly after publication of the Lundberg-Palmgren theory the AFBMA began efforts to standardize methods for establishing bearing load ratings and making life predictions. Standardized methods of establishing load ratings for ball bearings (AFBMA, 1960b) and roller bearings (AFBMA, 1960a) were devised, based essentially on the Lundberg-Palmgren theory. These early standards are published in their entirety in Jones (1964). In recent years significant advances have been made in rolling-element bearing material quality and in our understanding of the role of lubrication in bearing life through the development of elastohydrodynamic theory. Therefore, the original AFBMA standards in AFBMA (1960a, b) have been updated with life adjustment factors. These factors have been incorporated into ISO (1976), which is discussed in the following section.

TABLE 24.10
Capacity formulas for rectangular and elliptical conjunctions
[Units in kilograms and millimeters.]

Function	Elliptical contact of ball bearings	Rectangular contact of roller bearings
$\bar{C}$	$f_c f_a i^{0.7} n^{2/3} d^{1.8}$	$f_c f_a i^{7/9} n^{3/4} d^{29/27} \ell_{t,i}^{7/9}$
f_c	$g_c f_1 f_2 \left(\dfrac{d_i}{d_i - d}\right)^{0.41}$	$g_c f_1 f_2$
g_c	$\left[1 + \left(\dfrac{\bar{C}_i}{\bar{C}_o}\right)^{10/8}\right]^{-0.8}$	$\left[1 + \left(\dfrac{\bar{C}_i}{\bar{C}_o}\right)^{9/2}\right]^{-2/9}$
$\bar{C}_i / \bar{C}_o$	$f_3 \left[\dfrac{d_i (d_o - d)}{d_o (d_i - d)}\right]^{0.41}$	$f_3 \left(\dfrac{\ell_{t,i}}{\ell_{t,o}}\right)^{7/9}$

Function	Radial	Thrust		Radial	Thrust	
		$\beta \neq 90°$	$\beta = 90°$		$\beta \neq 90°$	$\beta = 90°$
γ_n	$\dfrac{d \cos \beta}{d_e}$		$\dfrac{d}{d_e}$	$\dfrac{d \cos \beta}{d_e}$		$\dfrac{d}{d_e}$
f_a	$(\cos \beta)^{0.7}$	$(\cos \beta)^{0.7} \tan \beta$	1	$(\cos \beta)^{7/9}$	$(\cos \beta)^{7/9} \tan \beta$	1
f_1	3.7–4.1	6–10		18–25	36–60	
f_2	$\dfrac{\gamma_n^{0.3}(1-\gamma_n)^{1.39}}{(1+\gamma_n)^{1/3}}$		$\gamma_n^{0.3}$	$\dfrac{\gamma_n^{2/9}(1-\gamma_n)^{29/27}}{(1+\gamma_n)^{1/3}}$		$\gamma_n^{2/9}$
f_3	$1.04 f_4$	f_4	1	$1.14 f_4$	f_4	1
f_4	$\left(\dfrac{1-\gamma_n}{1+\gamma_n}\right)^{1.72}$			$\left(\dfrac{1-\gamma_n}{1+\gamma_n}\right)^{38/37}$		

24.9.5 LIFE ADJUSTMENT FACTORS. A comprehensive study of the factors affecting the fatigue life of bearings, which were not taken into account in the Lundberg-Palmgren theory, is reported in Bamberger (1971). In that reference it is assumed that the various environmental or bearing design factors are

TABLE 24.11
Capacity formulas for mixed rectangular and elliptical conjunctions

[$\bar{C} = \bar{C}_i[1 + (\bar{C}_i/\bar{C}_o)^4]^{1/4}$; units in kilograms and millimeters.]

Function	Radial bearing	Thrust bearing		Radial bearing	Thrust bearing	
		$\beta \neq 90°$	$\beta = 90°$		$\beta \neq 90°$	$\beta = 90°$
	Inner race			Outer race		
γ_n	$\dfrac{d \cos \beta}{d_e}$		$\dfrac{d}{d_e}$	$\dfrac{d \cos \beta}{d_e}$		$\dfrac{d}{d_e}$
Function	Rectangular conjunction ($\bar{C}_i$)			Elliptical conjunction ($\bar{C}_o$)		
$\bar{C}_i$ or $\bar{C}_o$	$f_1 f_2 f_a i^{7/9} n^{3/4} d^{29/27} \ell_{t,i}^{7/9}$			$f_1 f_2 f_a \left(\dfrac{2R}{D} \dfrac{r_o}{r_o - R}\right)^{0.41} i^{0.7} n^{2/3} d^{1.8}$		
f_a	$(\cos \beta)^{7/9}$	$(\cos \beta)^{7/9} \tan \beta$	1	$(\cos \beta)^{0.7}$	$(\cos \beta)^{0.7} \tan \beta$	1
f_1	18–25	36–60		3.5–3.9	6–10	
f_2	$\dfrac{\gamma_n^{2/9}(1-\gamma_n)^{29/27}}{(1+\gamma_n)^{1/3}}$		$\gamma_n^{3/9}$	$\dfrac{\gamma_n^{0.3}(1+\gamma_n)^{1.39}}{(1-\gamma_n)^{1/3}}$		$\gamma_n^{0.3}$
Function	Point contact ($\bar{C}_i$)			Line contact ($\bar{C}_o$)		
$\bar{C}_i$ or $\bar{C}_o$	$f_1 f_2 f_a \left(\dfrac{2R}{D} \dfrac{r_i}{r_i - R}\right)^{0.41} i^{0.7} n^{2/3} d^{1.8}$			$f_1 f_2 f_a i^{7/9} n^{3/4} d^{29/27} \ell_{t,o}^{7/9}$		
f_a	$(\cos \beta)^{0.7}$	$(\cos \beta)^{0.7} \tan \beta$	1	$(\cos \beta)^{7/9}$	$(\cos \beta)^{7/9} \tan \beta$	1
f_1	3.7–4.1	6–10		15–22	36–60	
f_2	$\dfrac{\gamma_n^{0.3}(1-\gamma_n)^{1.39}}{(1+\gamma_n)^{1/3}}$		$\gamma_n^{0.3}$	$\dfrac{\gamma_n^{2/9}(1+\gamma_n)^{29/27}}{(1-\gamma_n)^{1/3}}$		$\gamma_n^{2/9}$

multiplicative in their effect on bearing life. The following equation results:

$$L_A = (\bar{D})(\bar{E})(\bar{F}_\ell)(\bar{G})(\bar{H}_m)\bar{L}_{10} \tag{24.75}$$

or

$$L_A = (\bar{D})(\bar{E})(\bar{F}_\ell)(\bar{G})(\bar{H}_m)\left(\frac{\bar{C}}{w_e}\right)^{m_k} \tag{24.76}$$

where $\overline{D}$ = material factor
$\overline{E}$ = metallurgical processing factor
$\overline{F}_\ell$ = lubrication factor
$\overline{G}$ = speed effect factor
$\overline{H}_m$ = misalignment factor
w_e = bearing equivalent load
m_k = load-life exponent; 3 for ball bearings or $\frac{10}{3}$ for roller bearings

Factors $\overline{D}$, $\overline{E}$, and $\overline{F}_\ell$ are reviewed briefly here. Refer to Bamberger (1971) for a complete discussion of all five life adjustment factors.

24.9.5.1 Materials factors $\overline{D}$ and $\overline{E}$. For over a century AISI 52100 steel has been the predominant material for rolling-element bearings. In fact, the basic dynamic capacity as defined by AFBMA in 1949 is based on an air-melted 52100 steel, hardened to at least Rockwell C 58. Since that time, as discussed in Sec. 5.5, better control of air-melting processes and the introduction of vacuum-remelting processes have resulted in more homogeneous steels with fewer impurities. Such steels have extended rolling-element bearing fatigue lives to several times the AFBMA (or catalog) life. Life extensions of three to eight times are not uncommon. Other steel compositions, such as AISI M–1 and AISI M–50, chosen for their higher temperature capabilities and resistance to corrosion, also have shown greater resistance to fatigue pitting when vacuum-melting techniques are employed. Case-hardened materials, such as AISI 4620, AISI 4118, and AISI 8620, used primarily for rolling-element bearings, have the advantage of a tough, ductile steel core with a hard, fatigue-resistant surface.

The recommended material factors $\overline{D}$ for various alloys processed by air melting are shown in Table 24.12. Insufficient definitive life data were found for

TABLE 24.12
Material factors for through-hardened bearing materials
[From Bamberger (1971); air-melted materials assumed.]

Material	Material factor, $\overline{D}$
52100	2.0
M-1	.6
M-2	.6
M-10	2.0
M-50	2.0
T-1	.6
Halmo	2.0
M-42	.2
WB 49	.6
440C	0.6–0.8

case-hardened materials to recommend values of $\overline{D}$ for them. Refer to the bearing manufacturer for the choice of a specific case-hardened material.

The metallurgical processing variables considered in the development of the metallurgical processing factor $\overline{E}$ included melting practice (air and vacuum melting) and metalworking (thermomechanical working). Thermomechanical working of M–50 has also been shown to lengthen life, but in a practical sense it is costly and still not fully developed as a processing technique. Bamberger (1971) recommends an $\overline{E}$ of 3 for consumable-electrode-vacuum-melted materials.

The translation of these factors into a standard (ISO, 1976) is discussed later.

24.9.5.2 Lubrication factor $\overline{F}_\ell$. Until approximately 1960 the role of the lubricant between surfaces in rolling contact was not fully appreciated. Metal-to-metal contact was presumed to occur in all applications with attendant required boundary lubrication. The development of elastohydrodynamic lubrication theory showed that lubricant films with thicknesses on the order of microinches and tenths of microinches occur in rolling contact. Since surface finishes are of the same order of magnitude as the lubricant film thicknesses, the significance of rolling-element bearing surface roughnesses to bearing performance became apparent. Tallian (1967) was the first to report on the importance to bearing life of the ratio of elastohydrodynamic lubrication film thickness to surface roughness. Figure 24.28 shows calculated $\tilde{L}_{10}$ life as a function of the dimensionless film parameter Λ, which was introduced in

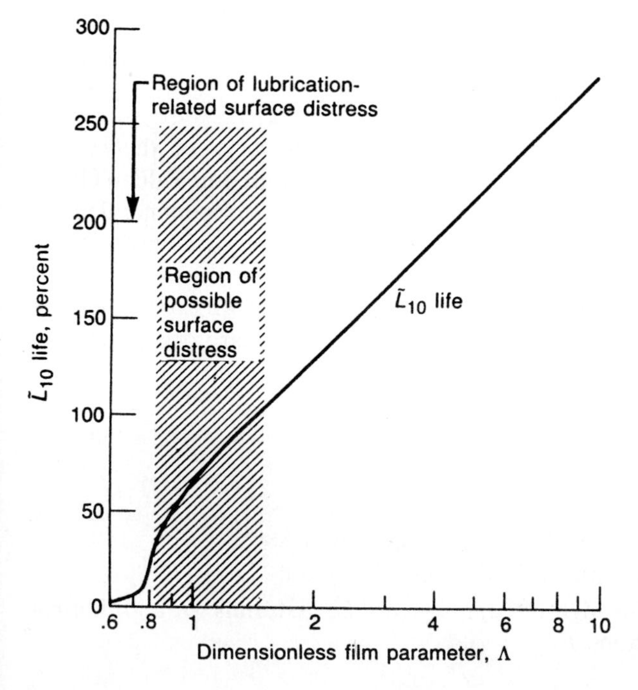

FIGURE 24.28
Group fatigue life $\tilde{L}_{10}$ as function of dimensionless film parameter. [*From Tallian (1967).*]

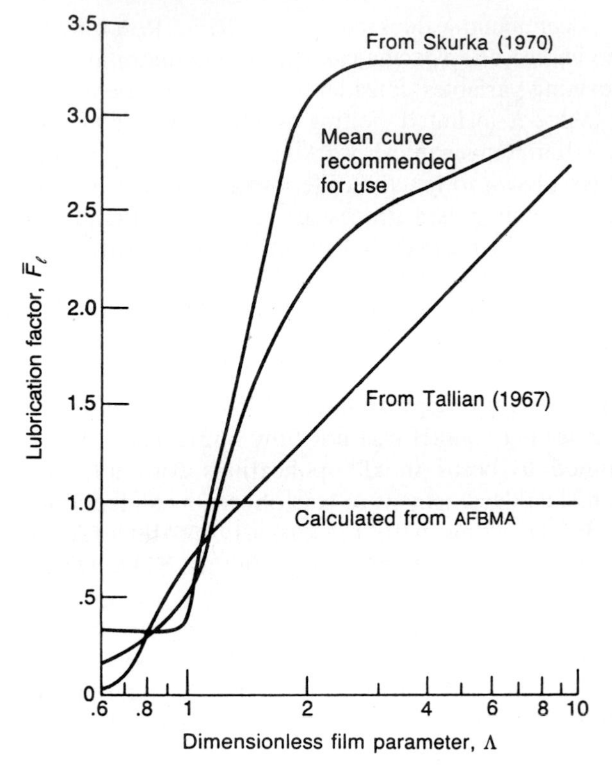

FIGURE 24.29
Lubrication factor as function of dimensionless film parameter. [*From Bamberger (1971).*]

Chap. 3, where

$$\Lambda = \frac{h_{\min}}{\left(R_{q,a}^2 + R_{q,b}^2\right)^{1/2}} \tag{3.22}$$

Figure 24.29, from Bamberger (1971), presents a curve of the recommended $\bar{F}_\ell$ as a function of Λ. A mean of the curves presented in Tallian (1967) for ball bearings and in Skurka (1970) for roller bearings is recommended for use. A formula for calculating the minimum film thickness $h_{\min}$ is given in Eq. (22.18).

The results of Bamberger (1971) have not been fully accepted into the current AFBMA standard represented by ISO (1976). The standard does, however, include the following:

1. Life and dynamic load rating formulas for radial and thrust ball bearings and radial and thrust roller bearings
2. Tables of f_c for all cases
3. Tables of $\tilde{X}$ and $\tilde{Y}$ factors for calculating equivalent loads
4. Load rating formulas for multirow bearings
5. Life correction factors for high-reliability levels $\bar{E}$, materials $\bar{D}$, and lubrication or operating conditions $\bar{F}_\ell$

Procedures for calculating $\bar{D}$ and $\bar{F}_\ell$ are unfortunately at present less than definitive, reflecting the need for additional research, life data, and operating experience.

24.10 DYNAMIC ANALYSES AND COMPUTER CODES

As has been stated, rolling-element bearing kinematics, stresses, deflections, and life can be analyzed precisely only by using a large-scale computer code. Presented here is a brief discussion of some of the significant analyses that have led to the formulation of modern bearing analytical computer codes. Refer to COSMIC, University of Georgia, Athens, GA 30601, for the public availability of program listings.

24.10.1 QUASI-STATIC ANALYSES. Rolling-element bearing analysis began with the work of Jones on ball bearings (Jones, 1959 and 1960). Jones did his work before there was a general awareness of elastohydrodynamic lubrication, and he assumed Coulomb friction in the race contacts. This led to the commonly known "race control" theory, which assumes that pure rolling (except for Heathcote interfacial slip) can occur at one of the ball-race contacts. All the spinning required for dynamic equilibrium of the balls would then take place at the other, or "noncontrolling," race contact. Jones' analysis proved to be quite effective for predicting fatigue life but less useful for predicting cage slip, which usually occurs at high speeds and light loads. Harris (1971b) extended Jones' analysis, retaining the assumption of Coulomb friction but allowing a frictional resistance to gyroscopic moments at the noncontrolling as well as the controlling race contact. Harris' analysis (1971b) is adequate for predicting bearing performance under conditions of dry-film lubrication or whenever there is a complete absence of any elastohydrodynamic film.

Harris (1971a) first incorporated elastohydrodynamic relationships into a ball bearing analysis. A revised version of Harris' computer program, called SHABERTH, was developed that incorporated actual traction data from a disk machine. The program SHABERTH has been expanded until today it encompasses ball, cylindrical roller, and tapered-roller bearings.

Harris (1966) first introduced elastohydrodynamics into a cylindrical roller bearing analysis. His initial analysis has been augmented with more precise viscosity-pressure and temperature relationships and traction data for the lubricant. This augmented analysis has evolved into the program CYBEAN and more recently has been incorporated into SHABERTH. Parallel efforts by Harris' associates have resulted in SPHERBEAN, a program that can be used to predict the performance of spherical roller bearings. These analyses can range from relatively simple force balance and life analyses through a complete thermal analysis of a shaft bearing system in several steps of varying complexity.

24.10.2 DYNAMIC ANALYSES. The work of Jones and Harris is categorized as quasi-static because it applies only when steady-state conditions prevail. Under highly transient conditions, such as accelerations or decelerations, only a true dynamic analysis will suffice. Walters (1971) made the first attempt at a dynamic analysis to explain cage dynamics in gyro-spin-axis ball bearings. Gupta (1975) solved the generalized differential equations of ball motion in an angular-contact ball bearing. Gupta continued his dynamic analyses for both cylindrical roller bearings (1979a, b) and ball bearings (1979c, d). Gupta's work is available in the programs DREB and RAPIDREB. Thus, the development of codes with real-time dynamic simulations that integrate the classical differential equations of motion for each bearing element has followed in the wake of the quasi-static codes. Contributors to dynamic code development are Walters (1971), Gupta (1979a–d, 1984, 1985, 1986, 1990), and Meeks and Ng (1985a, b). They have produced powerful computational tools that are useful in analyzing transient dynamics.

24.11 IOANNIDES-HARRIS THEORY

Ioannides and Harris (1985) extended the Lundberg-Palmgren theory on the basis of an elemental calculation of the risk of fatigue and the use of a material fatigue limit, akin to an endurance limit below which fatigue will not occur. They obtained the survival probability $\Delta\tilde{S}$ to N_c cycles of a volume element ΔV of material under known conditions of stress to be

$$-\ln \Delta\tilde{S} \sim N_c^{c_1}(\tau - \tau_u)^{c_2}(z')^{-c_3}\,\Delta V \tag{24.77}$$

This resembles the classic Lundberg-Palmgren method covered earlier with exponents c_1, c_2, and c_3, but Eq. (24.77) incorporates a local Weibull endurance strength or threshold stress τ_u and a stress-weighted mean depth z' to replace the maximum shear stress depth z_0. Equation (24.77) is not built on a detailed model of damage (e.g., the crack propagation and spall formation mechanisms treated by Tallian et al., 1978). Instead, by considering the stress distribution throughout the whole material, the need for an asperity model and the problem of distinguishing competing failure modes originating in the surface and subsurface are avoided. The characteristic stress σ may be selected from among several criteria, such as a single shear stress component, maximum shear stress, or von Mises.

Integrating Eq. (24.77) over the region in which the shear stress τ exceeds the threshold stress and setting $\tilde{S} = 0.9$ as appropriate to $\tilde{L}_{10}$ life, the following expression for life is obtained:

$$\tilde{L}_{10} = A_{\text{av}}\left[\frac{(\tau - \tau_u)^{c_2}}{(z')^{c_3}}\,dV\right]^{-1/c_1} \tag{24.78}$$

The coefficient A_{av} contains information on average failure risk that is only partially accessible, such as local material property variations or lubricant

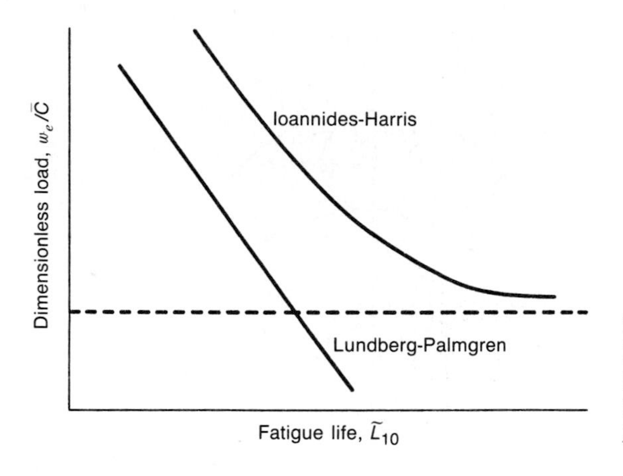

FIGURE 24.30
Dimensionless load-life comparison for Lundberg-Palmgren and Ioannides-Harris theories. [*From Ioannides and Harris (1985).*]

effects. Fig. 24.30 gives a load-life comparison for the Lundberg-Palmgren and Ioannides-Harris theories. The impact of the different theories is easily seen from this figure.

24.12 APPLICATIONS

In this section two applications of the film thickness equations developed throughout this chapter are presented to illustrate how the fluid film lubrication conditions in machine elements can be analyzed. Specifically, a typical roller and a typical ball bearing problem are considered.

24.12.1 CYLINDRICAL ROLLER BEARING PROBLEM. The equations for elastohydrodynamic film thickness developed in Chap. 22 relate primarily to elliptical conjunctions, but they are sufficiently general to allow them to be used with adequate accuracy in line-contact problems, as would be found in a cylindrical roller bearing. Therefore, the minimum elastohydrodynamic film thicknesses are calculated on the inner and outer races of a cylindrical roller bearing with the following dimensions—for both the elliptical conjunction of Chap. 22 and the rectangular conjunction of Chap. 21:

Inner-race diameter d_i, mm	64
Outer-race diameter d_o, mm	96
Diameter of cylindrical rollers d, mm	16
Axial length of cylindrical rollers ℓ_l, mm	16
Number of rollers in complete bearing n	9

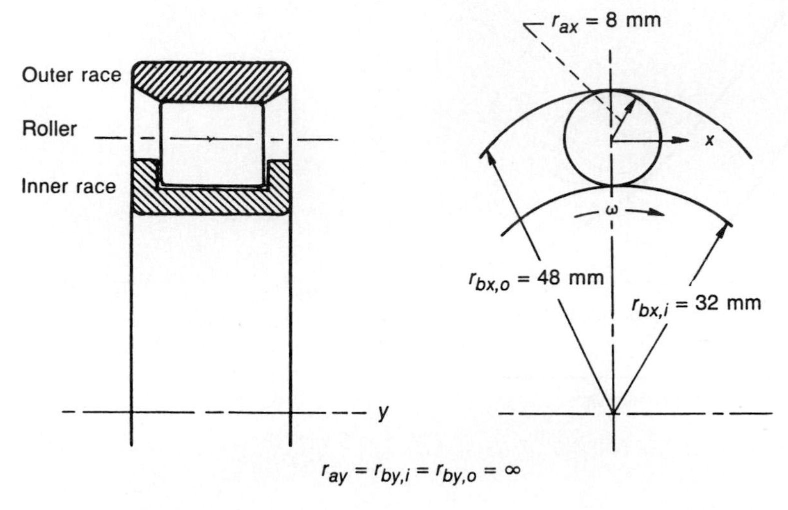

FIGURE 24.31
Roller bearing example. [*From Hamrock and Anderson (1983).*]

A bearing of this kind might well experience the following operating conditions:

Radial load w_z, N	10,800
Inner-race angular velocity ω_i, rad/s	524
Outer-race angular velocity ω_o, rad/s	0
Absolute viscosity at $p = 0$ and bearing	
Effective operating temperature η_0, $\text{N} \cdot \text{s/m}^2$	0.01
Viscosity-pressure coefficient ξ, m^2/N	2.2×10^{-8}
Modulus of elasticity for both rollers and races E, N/m^2	2.075×10^{11}
Poisson's ratio ν	0.3

Since the diametral clearance c_d is zero, from Eq. (24.47) the most heavily loaded roller can be expressed as

$$(w_z)_{\max} = \frac{4w_z}{n} = \frac{4(10{,}800\ \text{N})}{9} = 4800\ \text{N} \tag{24.47}$$

Therefore, the radial load per unit length on the most heavily loaded roller is

$$(w'_z)_{\max} = \frac{4800\ \text{N}}{0.016\ \text{m}} = 0.3\ \text{MN/m}$$

From Fig. 24.31 the radii of curvature are

$$r_{ax} = 0.008 \text{ m} \qquad r_{ay} = \infty$$

$$r_{bx,i} = 0.032 \text{ m} \qquad r_{by,i} = \infty$$

$$r_{bx,o} = -0.048 \text{ m} \qquad r_{by,o} = \infty$$

Then

$$\frac{1}{R_{x,i}} = \frac{1}{0.008} + \frac{1}{0.032} = \frac{5}{0.032} \tag{19.4}$$

giving $R_{x,i} = 0.0064$ m,

$$\frac{1}{R_{x,o}} = \frac{1}{0.008} - \frac{1}{0.048} = \frac{5}{0.048} \tag{19.4}$$

giving $R_{x,o} = 0.0096$ m, and

$$\frac{1}{R_{y,i}} = \frac{1}{R_{y,o}} = \frac{1}{\infty} + \frac{1}{\infty} = 0 \tag{19.5}$$

giving $R_{y,i} = R_{y,o} = \infty$.

From the input information the effective modulus of elasticity can be written as

$$E' = \frac{2}{\left(1 - \nu_a^2\right)/E_a + \left(1 - \nu_b^2\right)/E_b} = 2.28 \times 10^{11} \text{ N/m}^2 \tag{19.17}$$

For pure rolling the surface velocity $\tilde{u}$ relative to the lubricated conjunctions for a cylindrical roller is

$$\tilde{u} = \frac{|\omega_i - \omega_o|\left(d_e^2 - d^2\right)}{4d_e} \tag{24.29}$$

where d_e is the pitch diameter and d is the roller diameter.

$$d_e = \frac{d_o + d_i}{2} = \frac{0.096 + 0.064}{2} = 0.08 \text{ m} \tag{24.1}$$

Hence,

$$\tilde{u} = \frac{0.08^2 - 0.16^2}{4 \times 0.08}|524 - 0| = 10.061 \text{ m/s} \tag{24.28}$$

The dimensionless speed, materials, and load parameters for the inner- and outer-race conjunctions thus become, when the velocity is only in the rolling direction ($\tilde{v} = 0$),

$$U_i = \frac{\eta_0 \tilde{u}}{E' R_{x,i}} = \frac{(0.01)(10.061)}{(2.28 \times 10^{11})(0.0064)} = 6.895 \times 10^{-11} \quad (22.13)$$

$$G_i = \xi E' = 5016 \quad (22.15)$$

$$W_i = \frac{w_z}{E'(R_{x,i})^2} = \frac{4800}{(2.28 \times 10^{11})(0.0064)^2} = 5.140 \times 10^{-4} \quad (22.12)$$

$$U_o = \frac{\eta_0 \tilde{u}}{E' R_{x,o}} = \frac{(0.01)(10.061)}{(2.28 \times 10^{11})(0.0096)} = 4.597 \times 10^{-11} \quad (22.13)$$

$$G_o = \xi E' = 5016 \quad (22.15)$$

$$W_o = \frac{w_z}{E'(R_{x,o})^2} = \frac{4800}{(2.28 \times 10^{11})(0.0096)^2} = 2.284 \times 10^{-4} \quad (22.12)$$

The appropriate elastohydrodynamic film thickness equation for a fully flooded elliptical conjunction is developed in Chap. 22 and recorded as Eq. (22.18):

$$\tilde{H}_{e,\min} = \frac{\tilde{h}_{\min}}{R_x} = 3.63 U^{0.68} G^{0.49} W^{-0.073} (1 - e^{-0.68k}) \quad (22.18)$$

In the case of a roller bearing $k = \infty$, and this equation reduces to

$$\tilde{H}_{e,\min} = 3.63 U^{0.68} G^{0.49} W^{-0.073}$$

The dimensionless film thickness for the roller–inner-race conjunction is

$$\tilde{H}_{e,\min} = \frac{\tilde{h}_{\min}}{R_{x,i}} = (3.63)(1.231 \times 10^{-7})(65.04)(1.738) = 50.5 \times 10^{-6}$$

and hence

$$h_{\min} = (0.0064)(50.5 \times 10^{-6}) = 0.32\ \mu\text{m}$$

The dimensionless film thickness for the roller–outer-race conjunction is

$$\tilde{H}_{e,\min} = \frac{\tilde{h}_{\min}}{R_{x,o}} = (3.63)(9.343 \times 10^{-8})(65.04)(1.844) = 40.7 \times 10^{-6}$$

and hence

$$\tilde{h}_{\min} = (0.0096)(40.7 \times 10^{-6}) = 0.39\ \mu\text{m}$$

It is clear from these calculations that the smaller minimum film thickness in the bearing occurs at the roller–inner-race conjunction, where the geometrical conformity is less favorable. It was found that if the ratio of minimum film thickness to composite surface roughness is greater than 3, an adequate elastohydrodynamic film is maintained. This implies that a composite surface roughness of less than 0.1 μm is needed to ensure that an elastohydrodynamic film is maintained.

Now using, instead of the elliptical conjunction results of Chap. 22, the rectangular conjunction results of Chap. 21 gives

$$\tilde{H}_{e,\min} = \frac{\tilde{h}_{\min}}{R_x} = 1.714(W')^{-0.128} U^{0.694} G^{0.568} \tag{21.59}$$

where

$$W' = \frac{w'_z}{E'R_x}$$

For the roller–inner-race conjunction

$$W' = \frac{w'_z}{E'R_x} = \frac{0.3 \times 10^6}{(2.28 \times 10^{11})(0.0064)} = 2.056 \times 10^{-4}$$

$$\tilde{H}_{e,\min} = (1.714)(2.056 \times 10^{-4})^{-0.128}(6.895 \times 10^{-11})^{0.694}(5016)^{0.568}$$

$$= 39.03 \times 10^{-6}$$

For the roller–outer-race conjunction

$$W' = \frac{w'_z}{E'R_x} = \frac{0.3 \times 10^6}{(2.28 \times 10^{11})(0.0096)} = 1.371 \times 10^{-4}$$

$$\tilde{H}_{e,\min} = (1.714)(1.371 \times 10^{-4})^{-0.128}(4.597 \times 10^{-11})^{0.694}(5016)^{0.568}$$

$$= 31.02 \times 10^{-6}$$

The film thickness obtained from Chap. 21, the rectangular conjunction results, produces approximately a 23 percent decrease from the elliptical conjunction results obtained from Chap. 22. Recall from Chap. 22 that the theoretical results for the elliptical conjunction were about 20 percent larger than the Kunz and Winer (1977) experimental results. The Kunz and Winer (1977) experimental results appear to be in better agreement with the rectangular conjunction results presented in Chap. 21. Therefore, it is recommended that the results of Chap. 21 [Eq. (21.59)] be used to evaluate the minimum film thickness for a roller bearing.

24.12.2 RADIAL BALL BEARING PROBLEM. Consider a single-row, radial, deep-groove ball bearing with the following dimensions:

Inner-race diameter d_i, m	0.052291
Outer-race diameter d_o, m	0.077706
Ball diameter d, m	0.012700
Number of balls in complete bearing n	9
Inner-groove radius r_i, m	0.006604

Outer-groove radius r_o, m	0.006604
Contact angle β, deg	0
rms surface finish of balls $R_{q,b}$, μm	0.0625
rms surface finish of races $R_{q,a}$, μm	0.175

A bearing of this kind might well experience the following operating conditions:

Radial load w_z, N	8900
Inner-race angular velocity ω_i, rad/s	400
Outer-race angular velocity ω_o, rad/s	0
Absolute viscosity at $p = 0$ and bearing effective operating temperature η_0, $N \cdot s/m^2$	0.04
Viscosity-pressure coefficient ξ, m^2/N	2.3×10^{-8}
Modulus of elasticity for both balls and races E, N/m^2	2×10^{11}
Poisson's ratio for both balls and races ν	0.3

The essential features of the geometry of the inner and outer conjunctions (Figs. 24.1 and 24.6) can be ascertained as follows:

Pitch diameter:

$$d_e = \tfrac{1}{2}(d_o + d_i) = 0.065 \text{ m} \tag{24.1}$$

Diametral clearance:

$$c_d = d_o - d_i - 2d = 1.5 \times 10^{-5} \text{ m} \tag{24.2}$$

Race conformity:

$$R_{r,i} = R_{r,o} = \frac{r}{d} = 0.52 \tag{24.3}$$

Equivalent radii:

$$R_{x,i} = \frac{d(d_e - d\cos\beta)}{2d_e} = 0.00511 \text{ m} \tag{24.18}$$

$$R_{x,o} = \frac{d(d_e + d\cos\beta)}{2d_e} = 0.00759 \text{ m} \tag{24.20}$$

$$R_{y,i} = \frac{R_{r,i}d}{2R_{r,i} - 1} = 0.165 \text{ m} \tag{24.19}$$

$$R_{y,o} = \frac{R_{r,o}d}{2R_{r,o} - 1} = 0.165 \text{ m} \tag{24.21}$$

The curvature sum of inner-ring and ball contact

$$\frac{1}{R_i} = \frac{1}{R_{x,i}} + \frac{1}{R_{y,i}} = 201.76 \qquad (19.2)$$

gives $R_i = 4.956 \times 10^{-3}$ m, and the curvature sum of outer-ring and ball contact

$$\frac{1}{R_o} = \frac{1}{R_{x,o}} + \frac{1}{R_{y,o}} = 137.81 \qquad (19.2)$$

gives $R_o = 7.256 \times 10^{-3}$ m. Also, $\alpha_{r,i} = R_{y,i}/R_{x,i} = 32.35$ and $\alpha_{r,o} = R_{y,o}/R_{x,o} = 21.74$.

The nature of the Hertzian contact conditions can now be assessed.

Ellipticity parameters:

$$\bar{k}_i = \alpha_{r,i}^{2/\pi} = 9.15$$

$$\bar{k}_o = \alpha_{r,o}^{2/\pi} = 7.09 \qquad (19.29)$$

$$q_a = \frac{\pi}{2} - 1$$

Elliptic integrals:

$$\bar{\mathscr{E}}_i = 1 + \frac{q_a}{\alpha_{r,i}} = 1.0176$$

$$\bar{\mathscr{E}}_o = 1 + \frac{q_a}{\alpha_{r,o}} = 1.0263 \qquad (19.30)$$

$$\bar{\mathscr{F}}_i = \frac{\pi}{2} + q_a \ln \alpha_{r,i} = 3.555$$

$$\bar{\mathscr{F}}_o = \frac{\pi}{2} + q_a \ln \alpha_{r,o} = 3.3284 \qquad (19.32)$$

The effective elastic modulus E' is given by

$$E' = \frac{2}{(1 - \nu_a^2)/E_a + (1 - \nu_b^2)/E_b} = 2.198 \times 10^{11} \text{ N/m}^2$$

To determine the load carried by the most heavily loaded ball in the bearing, it is necessary to adopt an iterative procedure based on the calculation of local static compression and the analysis presented in Sec. 24.6.2. Stribeck (1901) found that the value of Z_w was about 4.37 in the expression

$$(w_z)_{max} = \frac{Z_w w_z}{n} \tag{24.45}$$

where $(w_z)_{max}$ = load on most heavily loaded ball
w_z = radial load on bearing
n = number of balls

However, it is customary to adopt a value of $Z_w = 5$ in simple calculations to produce a conservative design, and this value will be used to begin the iterative procedure.

Stage 1. *Assume* $Z_w = 5$. *Then*,

$$(w_z)_{max} = \frac{5w_z}{9} = \frac{5}{9} \times 8900 = 4944 \text{ N} \tag{24.45}$$

The maximum local elastic compression is

$$\delta_i = \mathscr{F}_i \left[\frac{9}{2\mathscr{E}_i R_i} \left(\frac{(w_z)_{max}}{\pi k_i E'} \right)^2 \right]^{1/3} = 2.906 \times 10^{-5} \text{ m}$$

$$\delta_o = \mathscr{F}_o \left[\frac{9}{2\mathscr{E}_o R_o} \left(\frac{(w_z)_{max}}{\pi k_o E'} \right)^2 \right]^{1/3} = 2.832 \times 10^{-5} \text{ m} \tag{19.16}$$

The sum of the local compressions on the inner and outer races is

$$\delta = \delta_i + \delta_o = 5.738 \times 10^{-5} \text{ m}$$

A better value for Z_w can now be obtained from

$$Z_w = \frac{\pi(1 - c_d/2\delta)^{3/2}}{2.491 \left\{ \left[1 + \left(\frac{1 - c_d/2\delta}{1.23} \right)^2 \right]^{1/2} - 1 \right\}}$$

since $c_d/2\delta = (1.5 \times 10^{-5})/(5.779 \times 10^{-5}) = 0.1298$. Thus, from Eq. (24.46)

$$Z_w = 4.551$$

Stage 2.

$$Z_w = 4.551$$

$$(w_z)_{max} = \frac{4.551 \times 8900}{9} = 4500 \text{ N}$$

$$\delta_i = 2.725 \times 10^{-5} \text{ m}$$

$$\delta_o = 2.702 \times 10^{-5} \text{ m}$$

$$\delta = 5.427 \times 10^{-5} \text{ m}$$

$$\frac{c_d}{2\delta} = 0.1382$$

Thus

$$Z_w = 4.565$$

Stage 3.

$$Z_w = 4.565$$

$$(w_z)_{max} = \frac{4.565 \times 8900}{9} = 4514 \text{ N}$$

$$\delta_i = 2.731 \times 10^{-5} \text{ m}$$

$$\delta_o = 2.708 \times 10^{-5} \text{ m}$$

$$\delta = 5.439 \times 10^{-5} \text{ m}$$

$$\frac{c_d}{2\delta} = 0.1379$$

and hence

$$Z_w = 4.564$$

This value is close to the previous value from stage 2 of 4.565, and a further iteration confirms its accuracy.

Stage 4.

$$Z_w = 4.564$$

$$(w_z)_{max} = \frac{4.564 \times 8900}{9} = 4513 \text{ N}$$

$$\delta_i = 2.731 \times 10^{-5} \text{ m}$$

$$\delta_o = 2.707 \times 10^{-5} \text{ m}$$

$$\delta = 5.438 \times 10^{-5} \text{ m}$$

$$\frac{c_d}{2\delta} = 0.1379$$

and hence

$$Z_w = 4.564$$

The load on the most heavily loaded ball is thus 4513 N.

The elastohydrodynamic minimum film thickness is calculated as follows: For pure rolling

$$\tilde{u} = \frac{|\omega_o - \omega_i|\left(d_e^2 - d^2\right)}{4d_e} = 6.252 \text{ m/s} \tag{24.29}$$

The dimensionless speed, materials, and load parameters for the inner- and outer-race conjunctions thus become, when the velocity is only in the rolling direction ($\tilde{v} = 0$),

$$U_i = \frac{\eta_0 \tilde{u}}{E' R_{x,i}} = \frac{0.04 \times 6.252}{(2.198 \times 10^{11})(5.11 \times 10^{-3})} = 2.227 \times 10^{-10} \tag{22.13}$$

$$G_i = \xi E' = (2.3 \times 10^{-8})(2.198 \times 10^{11}) = 5055 \tag{22.15}$$

$$W_i = \frac{w_z}{E'(R_{x,i})^2} = \frac{4513}{(2.198 \times 10^{11})(5.11^2 \times 10^{-6})} = 7.863 \times 10^{-4} \tag{22.12}$$

$$U_o = \frac{\eta_0 \tilde{u}}{E' R_{x,o}} = \frac{0.04 \times 6.252}{(2.198 \times 10^{11})(7.59 \times 10^{-3})} = 1.499 \times 10^{-10} \tag{22.13}$$

$$G_o = \xi E' = (2.3 \times 10^{-8})(2.198 \times 10^{11}) = 5055 \tag{22.15}$$

$$W_o = \frac{w_z}{E'(R_{x,o})^2} = \frac{4513}{(2.198 \times 10^{11})(7.59^2 \times 10^{-6})} = 3.564 \times 10^{-4} \tag{22.12}$$

The dimensionless minimum elastohydrodynamic film thickness in a fully flooded elliptical conjunction is again obtained from Eq. (22.18).

$$\tilde{H}_{e,\min} = \frac{\tilde{h}_{\min}}{R_x} = 3.63 U^{0.68} G^{0.49} W^{-0.073}\left(1 - e^{-0.68k}\right) \tag{22.18}$$

Ball–inner-race conjunction:

$$\left(\tilde{H}_{e,\min}\right)_i = (3.63)(2.732 \times 10^{-7})(65.29)(1.685)(0.9983)$$

$$= 1.09 \times 10^{-4} \tag{22.18}$$

Thus,

$$\left(\tilde{h}_{\min}\right)_i = 1.09 \times 10^{-4} R_{x,i} = 0.557\ \mu\text{m}$$

The dimensionless film parameter Λ discussed in Sec. 3.8 was found to play a significant role in determining the fatigue life of rolling-element bearings. In this case

$$\Lambda_i = \frac{(\tilde{h}_{\min})_i}{\left(R_{q,a}^2 + R_{q,b}^2\right)^{1/2}} = \frac{0.557 \times 10^{-6}}{\left[(0.175)^2 + (0.0625)^2\right]^{1/2} \times 10^{-6}} = 3.00 \quad (3.22)$$

Ball–outer-race conjunction

$$(\tilde{H}_{\min})_o = \frac{(\tilde{h}_{\min})_o}{R_{x,o}} = 3.63 U_o^{0.68} G^{0.49} W^{-0.073}(1 - e^{-0.68 k_o})$$

$$= (3.63)(2.087 \times 10^{-7})(65.29)(1.785)(0.9919)$$

$$= 0.876 \times 10^{-4} \quad (22.18)$$

Thus,

$$(h_{\min})_o = 0.876 \times 10^{-4} R_{x,o} = 0.665\ \mu\text{m}$$

In this case the dimensionless film parameter is given by

$$\Lambda_o = \frac{0.665 \times 10^{-6}}{\left[(0.175)^2 + (0.0625)^2\right]^{1/2} \times 10^{-6}} = 3.58 \quad (3.22)$$

It is evident that the smaller minimum film thickness occurs between the most heavily loaded ball and the inner race. However, in this case the minimum elastohydrodynamic film thickness is about three times the composite surface roughness, and the bearing lubrication can be deemed to be entirely satisfactory. Indeed, it is clear from Fig. 24.29 that little improvement in the lubrication factor $\overline{F}_\ell$ and thus in the fatigue life of the bearing could be achieved by further improving the minimum film thickness and hence the dimensionless film parameter.

24.13 CLOSURE

Rolling-element bearings are precise, yet simple, machine elements of great utility. This chapter drew together the current understanding of rolling-element bearings and attempted to present it in a concise manner. The material presented the operation of rolling-element bearings; the detailed precise calculations of bearing performance were only summarized and appropriate references given. The history of rolling-element bearings was briefly reviewed, and subsequent sections described the types of rolling-element bearing, their geometry, and their kinematics. Having defined conditions of a ball bearing under unloaded and unlubricated conditions, the chapter then focused on static loading of rolling-element bearings. Most rolling-element bearing applications involve steady-state rotation of either the inner or outer race or both; however, the rotational speeds are usually not so great as to cause ball or roller centrifugal forces or gyroscopic moments of significant magnitudes. Thus, these were neglected. Radial, thrust, and preloaded bearings that are statically loaded were considered. Rolling friction and friction in bearings concluded the second major thrust of the chapter, which was on loaded but unlubricated rolling-element bearings.

The last major thrust of the chapter dealt with loaded and lubricated rolling-element bearings. Topics covered were lubrication systems, fatigue life, dynamic analyses, and computer codes. The chapter concluded by applying the knowledge of this and previous chapters to roller and ball bearing applications. The use of the elastohydrodynamic lubrication film thickness developed in Chaps. 21 and 22 was integrated with the rolling-element bearing ideas developed in this chapter. It was found that the most critical conjunctions of both ball and roller bearings occurred between the rolling elements and the inner races.

24.14 PROBLEM

24.14.1 Outline the considerations that govern the following aspects of the performance of rolling-element bearings:

(*a*) Load capacity
(*b*) Rotational speed
(*c*) Life
(*d*) Lubrication requirements

24.15 REFERENCES

AFBMA (1960a): Method of Evaluating Load Ratings for Roller Bearings. AFBMA Standard System No. 11. Anti-Friction Bearings Manufacturers Association, Inc., New York.

AFBMA (1960b): Method of Evaluating Load Ratings for Ball Bearings. AFBMA Standard System No. 9, Revision No. 4. Anti-Friction Bearing Manufacturers Association, Inc., New York.

Anderson, W. J. (1970): Elastohydrodynamic Lubrication Theory as a Design Parameter for Rolling Element Bearings. *ASME*, *Pap*. 70–DE–19.

Anderson, W. J. (1979): Practical Impact of Elastohydrodynamic Lubrication. *Elastohydrodynamics and Related Topics*. Proceedings of Fifth Leeds-Lyon Symposium in Tribology, D. Dowson et al. (eds.). Mechanical Engineering Publications, Bury St. Edmunds, Suffolk, England, pp. 217–226.

Anderson, W. J., Macks, E. F., and Nemeth, Z. N. (1954): Comparison of Performance of Experimental and Conventional Cage Designs and Materials for 75-Millimeter-Bore Cylindrical Roller Bearings at High Speeds. *NACA Rep*. 1177.

Bamberger, E. N. (1971): *Life Adjustment Factors for Ball and Roller Bearings—An Engineering Design Guide*. American Society for Mechanical Engineers, New York.

Bisson, E. E., and Anderson, W. J. (1964): Advanced Bearing Technology. *NASA Spec. Publ.* 38.

Brown, P. F. (1970): Bearings and Dampers for Advanced Jet Engines. *SAE Pap*. 700318.

Brown, P. F., et al. (1977): Development of Mainshaft High Speed Cylindrical Roller Bearings for Gas Turbine Engines. PWA–FR–8615, Pratt & Whitney Group, West Palm Beach, Florida. (Avail. NTIS, AD–A052351.)

Gupta, P. K. (1975): Transient Ball Motion and Skid in Ball Bearings. *J. Lubr. Technol.*, vol. 97, no. 2, pp. 261–269.

Gupta, P. K. (1979a): Dynamics of Rolling Element Bearings—Part I, Cylindrical Roller Bearing Analysis. *J. Lubr. Technol.*, vol. 101, no. 3, pp. 293–304.

Gupta, P. K. (1979b): Dynamics of Rolling Element Bearings—Part II, Cylindrical Roller Bearing Results. *J. Lubr. Technol.*, vol. 101, no. 3, pp. 305–311.

Gupta, P. K. (1979c): Dynamics of Rolling Element Bearings—Part III, Ball Bearing Analysis. *J. Lubr. Technol.*, vol. 101, no. 3, pp. 312–318.

Gupta, P. K. (1979d): Dynamics of Rolling Element Bearings—Part IV, Ball Bearing Results. *J. Lubr. Technol.*, vol. 101, no. 3, pp. 319–326.

Gupta, P. K. (1984): *Advanced Dynamics of Rolling Elements*, Springer-Verlag, New York.

Gupta, P. K., Dill, J. F., and Bandow, H. E. (1985): Dynamics of Rolling Element Bearings—Experimental Validation of the DREB and RAPIDREB Computer Programs. *J. Tribol.*, vol. 107, no. 1, pp. 132–137.

Gupta, P. K. et al. (1986): Ball Bearing Response to Cage Unbalance. *J. Tribol.*, vol. 108, no. 3, pp. 462–467.

Gupta, P. K., and Tallian, T. E. (1990): Rolling Bearing Life Prediction—Correction for Materials and Operating Conditions—Part III: Implementation in Bearing Dynamic Computer Codes. *J. Tribol.*, vol. 112, no. 1, pp. 23–26.

Hamrock, B. J., and Anderson, W. J. (1983): Rolling-Element Bearings. *NASA Ref. Publ.*-1105.

Hamrock, B. J., and Dowson, D. (1981): *Ball Bearing Lubrication—The Elastohydrodynamics of Elliptical Contacts*. Wiley-Interscience, New York.

Harris, T. A. (1966): An Analytical Method to Predict Skidding in High Speed Roller Bearings. *ASLE Trans.*, vol. 9, pp. 229–241.

Harris, T. A. (1971a): An Analytical Method to Predict Skidding in Thrust-Loaded, Angular-Contact Ball Bearings. *J. Lubr. Technol.*, vol. 93, no. 1, pp. 17–24.

Harris, T. A. (1971b): Ball Motion in Thrust-Loaded, Angular Contact Bearings With Coulomb Friction. *J. Lubr. Technol.*, vol. 93, no. 1, pp. 32–38.

Heathcote, H. L. (1921): The Ball Bearing: In the Making, Under Test and in Service. *Proc. Inst. Auto. Eng. (London)*, vol. 15, pp. 569–702.

Ioannides, E., and Harris, T. A. (1985): A New Fatigue Life Model for Rolling Bearings. *J. Tribol.*, vol. 107, no. 3, pp. 367–378.

ISO (1976): Rolling Bearings, Dynamic Load Ratings and Rating Life. ISO/TC4/JC8, Revision of ISOR281. Issued by International Organization for Standardization, Technical Committee ISO/TC4.

Johnson, K. L., and Tabor, D. (1967–68): Rolling Friction. Lubrication and Wear: Fundamentals and Application to Design, *Proc. Inst. Mech. Eng.*, vol. 182, pt. 3A, pp. 168–172.

Jones, A. B. (1959): Ball Motion and Sliding Friction in Ball Bearings. *J. Basic Eng.*, vol. 81, no. 1, pp. 1–12.

Jones, A. B. (1960): A General Theory for Elastically Constrained Ball and Radial Roller Bearings Under Arbitrary Load and Speed Conditions. *J. Basic Eng.*, vol. 82, no. 2, pp. 309–320.

Jones, A. B. (1964): The Mathematical Theory of Rolling Element Bearings. *Mechanical Design and Systems Handbook*. H. A. Rothbart (ed.). McGraw-Hill, New York, pp. 13–1 to 13–76.

Kunz, R. K., and Winer, W. O. (1977): Discussion on pp. 275–276 of Hamrock, B. J., and Dowson, D., Isothermal Elastohydrodynamic Lubrication of Point Contacts, Part III—Fully Flooded Results. *J. Lubr. Technol.*, vol. 99, no. 2, pp. 264–275.

Lundberg, G., and Palmgren, A. (1947): Dynamic Capacity of Rolling Bearings. *Acta Polytech. Mech. Eng. Ser.*, vol. I, no. 3. (Ingeniors Vetenskaps Akadamien Handlingar, no. 196.)

Lundberg, G., and Palmgren, A. (1952): Dynamic Capacity of Roller Bearings. *Acta Polytech. Mech. Eng. Ser.*, vol. II, no. 4. (Ingeniors Vetenskaps Akadamien Handlingar, no. 210.)

Matt, R. J., and Gianotti, R. J. (1966): Performance of High Speed Ball Bearings With Jet Oil Lubrication. *Lubr. Eng.*, vol. 22, no. 8, pp. 316–326.

McCarthy, P. R. (1973): Greases. Interdisciplinary Approach to Liquid Lubricant Technology. P. M. Ku (ed.). *NASA Spec. Publ.* 318, pp. 137–185.

Meeks, C. R., and Ng, K. O. (1985a): The Dynamics of Ball Separators in Ball Bearings—Part I: Analysis. *ASLE Trans.*, vol. 28, no. 3, pp. 277–287.

Meeks, C. R., and Ng, K. O. (1985b): The Dynamics of Ball Separators in Ball Bearings—Part II: Results of Optimization Study. *ASLE Trans.*, vol. 28, no. 3, pp. 287–295.

Miyakawa, Y., Seki, K., and Yokoyama, M. (1972): Study on the Performance of Ball Bearings at High DN Values. *NAL–TR*–284, National Aerospace Lab., Tokyo, Japan. (*NASA TTF*–15017, 1973).

Palmgren, A. (1959): *Ball and Roller Bearing Engineering*, 3d ed. SKF Industries, Inc., Philadelphia.

Parker, R. J. (1980): Lubrication of Rolling Element Bearings. *Bearing Design—Historical Aspects, Present Technology and Future Problems*. W. J. Anderson (ed.). American Society of Mechanical Engineers, New York, pp. 87–110.

Parker, R. J., and Signer, H. R. (1978): Lubrication of High-Speed, Large Bore Tapered-Roller Bearings. *J. Lubr. Technol.*, vol. 100, no. 1, pp. 31–38.

Pirvics, J. (1980): Computerized Analysis and Design Methodology for Rolling Element Bearing Load Support Systems. *Bearing Design—Historical Aspects, Present Technology and Future Problems*. W. J. Anderson (ed.). American Society of Mechanical Engineers, New York, pp. 47–85.

Reynolds, O. (1876): On Rolling Friction. *Philos. Trans. Roy. Soc. London*, part 1, vol. 166, pp. 155–174.

Schuller, F. T. (1979): Operating Characteristics of a Large-Bore Roller Bearing to Speeds of 3×10^6 DN. *NASA Tech. Paper*-1413.

Scibbe, H. W. (1968): Bearings and Seals for Cryogenic Fluids. *SAE Pap.* 680550. (*NASA Tech. Memo*. X–52415.)

Signer, H., Bamberger, E. N., and Zaretsky, E. V. (1974): Parametric Study of the Lubrication of Thrust Loaded 120-mm Bore Ball Bearings to 3 Million DN. *J. Lubr. Technol.*, vol. 96, no. 3, pp. 515–524, 526.

Signer, H. R., and Schuller, F. T. (1982): Lubrication of 35-Millimeter-Bore Ball Bearings of Several Designs at Speeds to 2.5 Million DN. Problems in Bearings and Lubrication, *AGARD Conf. Proc.* 323, pp. 8–1 to 8–15.

Skurka, J. C. (1970): Elastohydrodynamic Lubrication of Roller Bearings. *J. Lubr. Technol.*, vol. 92, no. 2, pp. 281–291.

Stribeck, R. (1901): Kugellager fur beliebige Belastungen. *Z. Ver. Dtsch. Ing.*, vol. 45, pp. 73–125.

Tabor, D. (1955): The Mechanism of Rolling Friction. II. The Elastic Range. *Proc. Roy. Soc. London., Ser. A*, vol. 229, no. 1177, Apr. 21, pp. 198–220.

Tallian, T. E. (1967): On Competing Failure Modes in Rolling Contact. *Trans. ASLE*, vol. 10, pp. 418–439.

Tallian, T. E., Chiu, Y. P., and VanAmerongen, E. (1978): Prediction of Traction and Microgeometry Effects on Rolling Contact Fatigue Life. *J. Lubr. Technol.*, vol. 100, no. 2, pp. 156–166.

Todd, M. J., and Bentall, R. H. (1978): Lead Film Lubrication in Vacuum. Proceedings of 2nd ASLE International Conference on Solid Lubrication, pp. 148–157.

Walters, C. T. (1971): The Dynamics of Ball Bearings. *J. Lubr. Technol.*, vol. 93, no. 1, pp. 1–10.

Weibull, W. (1949): A Statistical Representation of Fatigue Failures in Solids. *Trans. Roy. Inst. Technol., Stockholm*, no. 27.

Zaretsky, E. V., Signer, H., and Bamberger, E. N. (1976): Operating Limitations of High-Speed Jet-Lubricated Ball Bearings. *J. Lubr. Technol.*, vol. 98, no. 1, pp. 32–39.

CHAPTER 25

ADDITIONAL ELASTOHYDRODYNAMIC LUBRICATION APPLICATIONS

In this chapter the film thickness equations for elliptical and rectangular conjunctions developed throughout the text are applied to specific machine elements, other than the rolling-element bearings covered in Chap. 24, to illustrate how the fluid film lubrication conditions can be analyzed. First, a simple involute gear is analyzed, and the film thickness predictions based on the elliptical conjunction equations developed in Chap. 22 are compared with those for a rectangular conjunction presented in Chap. 21. Next, the application to gears is extended to a simplified form of the continuously variable-speed drives being introduced into road vehicles and aircraft equipment. Then, the case of a railway wheel rolling on a wet or oily rail introduces an interesting and slightly more complicated problem in contact mechanics. Finally, the film thickness equations for highly deformable or "soft" materials are applied to synovial joints. The analysis of elastohydrodynamic lubrication in these load-bearing human joints is a more speculative example because understanding of their tribological behavior is incomplete.

25.1 INVOLUTE GEARS

It is well known that the contact between gear teeth at a distance $\bar{S}$ from the pitch line in a pair of involute gear wheels having radii r_a and r_b and a pressure angle $\bar{\psi}$ can be represented by two circular cylinders of radii $r_b \sin \bar{\psi} + \bar{S}$ and

$r_a \sin \bar{\psi} - \bar{S}$ rotating with the same angular velocities ω_a and ω_b as the wheels themselves. Indeed, this observation forms the basis of the two-disk machine that has been used so extensively and effectively in experimental studies of gear lubrication in general and of elastohydrodynamic lubrication in particular. The problem has been considered by Dowson and Higginson (1964, 1966). The present case demonstrates how the elastohydrodynamic minimum-film-thickness equation for elliptical conjunctions can be applied to a rectangular conjunction problem. The general geometry of the configuration is shown in Fig. 25.1.

Consider a pair of involute gear wheels having radii of 50 and 75 mm and a pressure angle of 20°. Let the angular velocity of the larger wheel be 210 rad/s ($\approx$ 2000 r/min); the width of the gear teeth, 15 mm; and the load transmitted between the teeth, 22,500 N. Let the essential properties of the lubricant and the wheel materials adopt the following values:

$$\eta_0 = 0.075 \text{ Pa} \cdot \text{s}$$

$$\xi = 2.2 \times 10^{-8} \text{ m}^2/\text{N}$$

$$E = 207 \text{ GPa}$$

$$\nu = 0.3$$

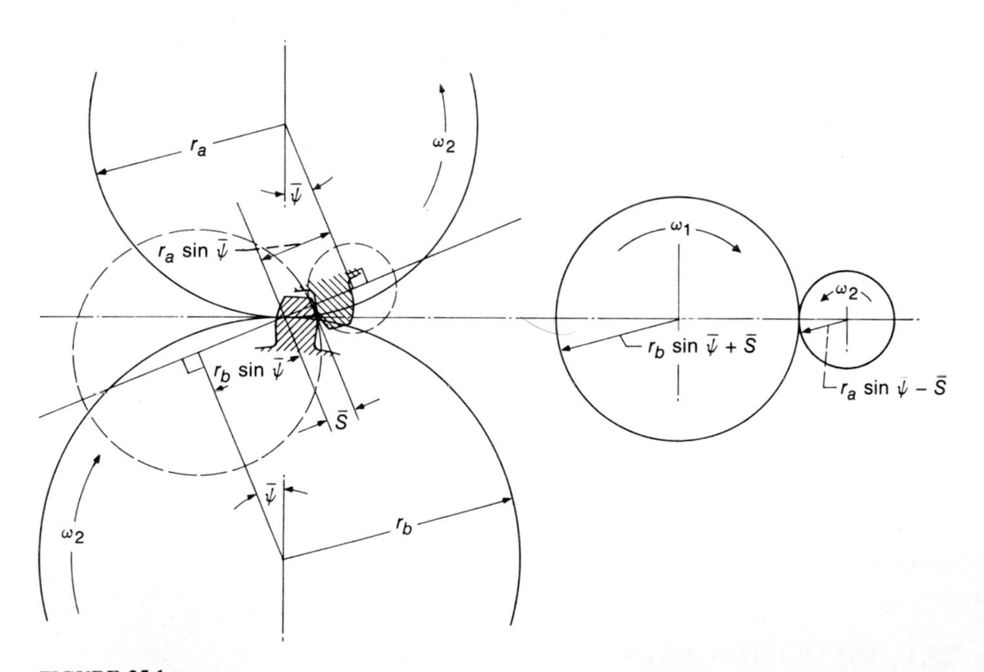

FIGURE 25.1
Involute spur gears and representation by equivalent cylinders. [*From Hamrock and Dowson (1981).*]

Calculate the minimum film thickness on the pitch line ($\bar{S} = 0$). Then,

$$r_b = 0.075 \text{ m}$$

$$r_a = 0.050 \text{ m}$$

$$\text{Gear rate} = \frac{r_b}{r_a} = \frac{\omega_a}{\omega_b} = 1.5$$

$$\text{Radii of cylinders} = r_b \sin \bar{\psi} + \bar{S} = r_b \sin \bar{\psi}$$

$$r_{ax} = 0.075 \sin 20° = 0.02565 \text{ m}$$

$$r_{ay} = \infty$$

and

$$r_a \sin \bar{\psi} - \bar{S} = r_a \sin \bar{\psi}$$

or

$$r_{bx} = 0.050 \sin 20° = 0.01710 \text{ m}$$

$$r_{by} = \infty$$

Hence,

$$\frac{1}{R_x} = \frac{1}{0.02565} + \frac{1}{0.01710} = 38.99 + 58.48$$

giving $R_x = 0.01026$ m.

$$\tilde{u} = \frac{\omega_b r_b \sin \bar{\psi} + \omega_a r_a \sin \bar{\psi}}{2} = \omega_b r_b \sin \bar{\psi} = (210)(0.02565) = 5.387 \text{ m/s}$$

$$E' = \frac{E}{1 - \nu^2} = 2.275 \times 10^{11} \text{ Pa}$$

Thus

$$U = \frac{\eta_0 \tilde{u}}{E' R_x} = \frac{(0.075)(5.387)}{(2.275 \times 10^{11})(0.01026)} = 1.731 \times 10^{-10}$$

$$G = \xi E' = 5005$$

$$W = \frac{w_z}{E' R_x^2} = \frac{22\,500}{(2.275 \times 10^{11})(0.01026)^2} = 9.395 \times 10^{-4}$$

and

$$k = \infty$$

Now

$$\tilde{H}_{e,\min} = 3.63 U^{0.68} G^{0.49} W^{-0.073} (1 - e^{-0.68k}) \tag{22.18}$$

$$\tilde{H}_{e,\min} = (3.63)(1.731 \times 10^{-10})^{0.68} (5005)^{0.49} (9.395 \times 10^{-4})^{-0.073} (1)$$

$$\tilde{H}_{e,\min} = (3.63)(2.302 \times 10^{-7})(64.97)(1.663)$$

$$\tilde{H}_{e,\min} = 0.903 \times 10^{-4}$$

The minimum film thickness is thus

$$\tilde{h}_{\min} = \tilde{H}_{e,\min} R_x = (0.903 \times 10^{-4})(0.01026) = 0.93\ \mu\text{m}$$

The rectangular conjunction formula for minimum elastohydrodynamic film thickness from Chap. 21 [Eq. (21.59)] is

$$\tilde{H}_{e,\min} = \frac{\tilde{h}_{\min}}{R_x} = 1.714(W')^{-0.128} U^{0.694} G^{0.568} \qquad (21.59)$$

where

$$W' = \frac{w_z}{E' R_x \ell} = \frac{22\,500}{(2.275 \times 10^{11})(0.01026)(0.015)} = 6.426 \times 10^{-4}$$

Hence,

$$\tilde{H}_{e,\min} = (1.174)(6.426 \times 10^{-4})^{-0.128} (1.731 \times 10^{-10})^{0.694} (5005)^{0.568}$$

$$= 0.638 \times 10^{-4}$$

$$\tilde{h}_{\min} = \tilde{H}_{e,\min} R_x = (0.638 \times 10^{-4})(0.01026) = 0.655\ \mu\text{m}$$

Just as for cylindrical roller bearings in Chap. 24, so here too the rectangular conjunction formulation is less than that for elliptical conjunctions. It is felt that the rectangular conjunction formulation is more accurate, since a much greater range of operating parameters was used in obtaining the results. Also the rectangular conjunction results agree better with the experimental work of Kunz and Winer (1977).

The dimensionless film parameter Λ is defined from Chap. 3 as

$$\Lambda = \frac{h_{\min}}{\left(R_{q,a}^2 + R_{q,b}^2\right)^{1/2}} = \frac{0.655}{\left[(0.3)^2 + (0.3)^2\right]^{1/2}} = \frac{0.655}{0.424} = 1.54 \qquad (3.22)$$

It is thus clear, from Chap. 24, that rolling-element bearings and also the pair of involute gears considered in this example should enjoy the benefits of some elastohydrodynamic minimum film thickness. However, Chap. 24 indicates that considerable improvement in gear life could be anticipated if Λ could be increased to about 3. In practical situations such an improvement would probably be achieved by increasing the lubricant absolute viscosity η_0 at $p = 0$ and constant temperature or by decreasing the rms surface roughness R_q of the gears.

25.2 CONTINUOUSLY VARIABLE-SPEED DRIVES

Interest in rolling friction drives for power transmission equipment dates back to the last century. A patent application for a variable-speed drive that used a

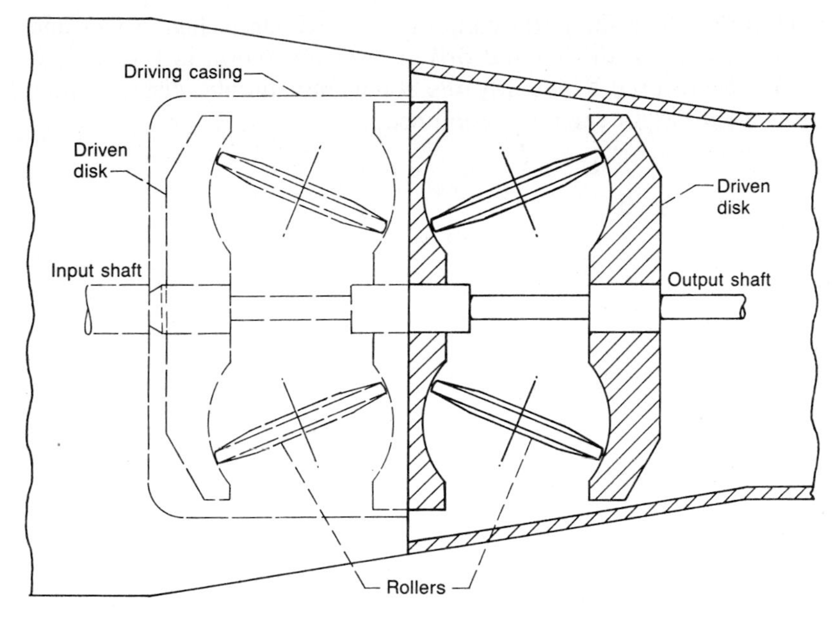

FIGURE 25.2
Continuously variable-speed drive. [*From Hamrock and Dowson (1981).*]

disk with toroidal surfaces and rollers which could be tilted to change the speed ratio was filed in 1899. Since that time, many different configurations have been designed, and several were included in a survey published by Cahn-Speyer (1957). A particular and more recent form of this type of drive, known as the "Perbury gear," has been described by Fellows et al. (1964).

Consider the double toroidal layout shown in Fig. 25.2 in which three symmetrically arranged rollers that can be tilted to change the ratio of input to output speeds are located on each side of the central disk. Power is delivered to the end disks, which either incorporate a mechanical end-loading device sensitive to input torque or, in an improved version, a hydraulic loading cylinder applying a load proportional to the sum of the input and output torques (i.e., proportional to the forces to be transmitted through the lubricant film). A roller control mechanism automatically ensures that the power transmitted through the unit is distributed equally between the six rollers. Such continuously variable-speed drives provide a smooth transmission of power, typically over a speed ratio range of about 5 : 1, in a jerk-free manner.

The life and efficiency of continuously variable-speed drives of this nature depend on the characteristics of the lubricating films between the rollers and the toroidal surfaces. The traction characteristics are beyond the scope of this book, but it is possible to show how the lubricant film thickness can be calculated.

The radii at the inner and outer conjunctions between a single roller and the toroidal surfaces on the driving and driven disks are shown in Fig. 25.3. If subscripts a and b are used to identify the roller and toroidal disks, respectively, it can be seen that for the inner conjunction

$$r_{ax} = r \qquad r_{ay} = \tilde{n}r$$

$$r_{bx} = \frac{\tilde{z}}{\sin\phi} = \frac{\tilde{Z} - r\sin\phi}{\sin\phi} \qquad r_{by} = -r$$

Hence,

$$\frac{1}{R_{x,i}} = \frac{1}{r_{ax}} + \frac{1}{r_{bx}} = \frac{1}{r} + \frac{\sin\phi}{\tilde{Z} - r\sin\phi}$$

giving $R_{x,i} = r[1 - (r/\tilde{Z})\sin\phi]$, and

$$\frac{1}{R_{y,i}} = \frac{1}{r_{ay}} + \frac{1}{r_{by}} = \frac{1}{\tilde{n}r} - \frac{1}{r}$$

giving $R_{y,i} = [\tilde{n}/(1 - \tilde{n})]r$. For the outer conjunction

$$r_{ax} = r \qquad r_{ay} = \tilde{n}r$$

$$r_{bx} = -\frac{\tilde{z}}{\sin\phi} = -\frac{\tilde{Z} + r\sin\phi}{\sin\phi} \qquad r_{by} = -r$$

Hence,

$$\frac{1}{R_{x,o}} = \frac{1}{r_{ax}} + \frac{1}{r_{bx}} = \frac{1}{r} - \frac{\sin\phi}{\tilde{Z} + r\sin\phi}$$

giving $R_{x,o} = r[1 + (r/\tilde{Z})\sin\phi]$, and

$$\frac{1}{R_{y,o}} = \frac{1}{r_{ay}} + \frac{1}{r_{by}} = \frac{1}{\tilde{n}r} - \frac{1}{r}$$

giving $R_{y,o} = [\tilde{n}/(1 - \tilde{n})]r$.

25.2.1 ELASTICITY CALCULATIONS. If the end load carried by each roller is w_z, the normal load at each of the inner and outer conjunctions $(w_z)_\phi$ is given by

$$(w_z)_\phi = \frac{w_z}{\cos\phi}$$

Consider a continuously variable-speed drive of the form shown in Figs. 25.2

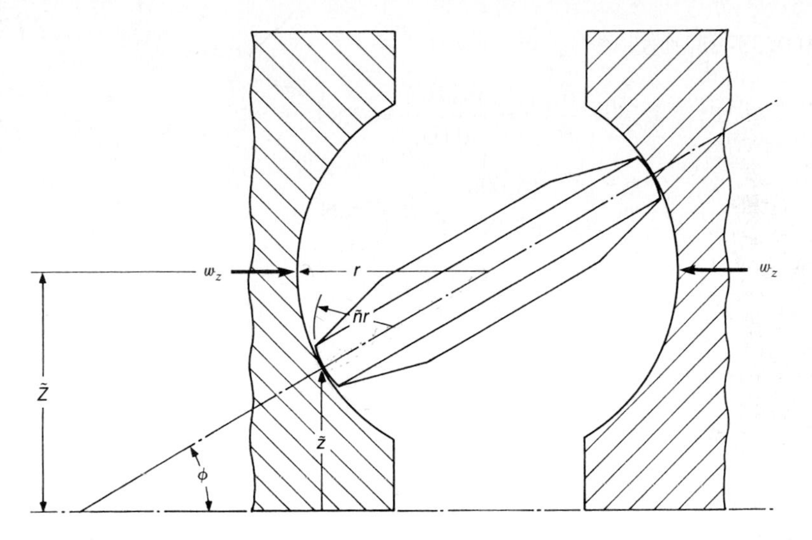

FIGURE 25.3
Geometry of crowned roller and toroidal surfaces in a continuously variable-speed drive. [*From Hamrock and Dowson (1981).*]

and 25.3 in which

$$w_z = 16{,}000 \text{ N} \qquad \tilde{n} = 0.7$$

$$\phi = 30^\circ \qquad E = 2.075 \times 10^{11} \text{ Pa}$$

$$r = 50 \text{ mm} \qquad \nu = 0.3$$

$$\tilde{Z} = 62.5 \text{ mm}$$

Then

$$E' = \frac{E}{1 - \nu^2} = \frac{2.075 \times 10^{11}}{0.91} = 2.28 \times 10^{11} \text{ Pa}$$

For the inner conjunction

$$R_{x,i} = 0.05(1 - 0.8 \sin 30^\circ) = 0.03 \text{ m}$$

$$R_{y,i} = \left(\frac{0.7}{1 - 0.7}\right)(0.05) = 0.1167 \text{ m}$$

$$\frac{1}{R} = \frac{1}{0.03} + \frac{1}{0.1167}$$

giving $R = 0.02387$ m.

$$\alpha_{r,i} = \frac{R_{y,i}}{R_{x,i}} = \frac{0.1167}{0.03} = 3.89$$

$$(w_z)_\phi = \frac{16{,}000}{\cos 30^\circ} = 18{,}475 \text{ N}$$

$$\bar{\mathscr{E}}_i = 1 + \frac{q_a}{\alpha_r} = 1.1467 \qquad (19.30)$$

$$\bar{\mathscr{F}}_i = \frac{\pi}{2} + q_a \ln \alpha_r = 2.3462 \qquad (19.32)$$

$$\bar{k}_i = \alpha_r^{2/\pi} = 2.375 \qquad (19.29)$$

$$D_y = 2\left[\frac{6\bar{k}^2\bar{\mathscr{E}}(w_z)_\phi R}{\pi E'}\right]^{1/3} = 5761\ \mu\text{m} \qquad (19.14)$$

$$D_x = 2\left[\frac{6\tilde{\mathscr{E}}(w_z)_\phi R}{\pi \bar{k} E'}\right]^{1/3} = 2426\ \mu\text{m} \qquad (19.15)$$

$$p_{\max} = \left[\frac{6(w_z)_\phi}{\pi D_x D_y}\right] = 2.52 \text{ GPa} \qquad (19.7)$$

The Hertzian contact zone is thus represented by a relatively large ellipse in which the ratio of major to minor axes is about 2.375.

For the outer conjunction

$$R_{x,o} = 0.05(1 + 0.8 \sin 30^\circ) = 0.07 \text{ m}$$

$$R_{y,o} = \left(\frac{0.7}{1 - 0.7}\right)(0.05) = 0.1167 \text{ m}$$

$$\frac{1}{R} = \frac{1}{0.07} + \frac{1}{0.1167}$$

giving $R = 0.0438$ m.

$$\alpha_{r,o} = \frac{R_{y,o}}{R_{x,o}} = \frac{0.1167}{0.07} = 1.667$$

$$(w_z)_\phi = \frac{16{,}000}{\cos 30^\circ} = 18{,}475 \text{ N}$$

$$\bar{\mathscr{E}} = 1 + \frac{q_a}{\alpha_r} = 1.3425 \qquad (19.30)$$

$$\bar{\mathscr{F}} = \frac{\pi}{2} + q_a \ln \alpha_r = 1.8624 \tag{19.32}$$

$$\bar{k} = 1.384 \tag{19.29}$$

$$D_y = 2\left[\frac{6\bar{k}^2\bar{\mathscr{E}}(w_z)_\phi R}{\pi E'}\right]^{1/3} = 5186\ \mu\text{m} \tag{19.14}$$

$$D_x = 2\left[\frac{6\bar{\mathscr{E}}(w_z)_\phi R}{\pi \bar{k} E'}\right]^{1/3} = 3747\ \mu\text{m} \tag{19.15}$$

$$p_{\max} = \left[\frac{6(w_z)_\phi}{\pi D_x D_y}\right] = 1.816\ \text{GPa} \tag{19.7}$$

The greater geometrical conformity at the outer conjunction provides a larger, less elongated Hertzian contact zone than that developed at the inner conjunction.

25.2.2 ELASTOHYDRODYNAMIC FILM THICKNESS CALCULATIONS. If it is assumed that the input disk rotates at 314 rad/s ($\approx$ 3000 r/min) and that the lubricant has an absolute viscosity η_0 of 0.0045 Pa · s at $p = 0$ and the effective operating temperature and a pressure-viscosity coefficient ξ of 2.2×10^{-8} m^2/N, then for pure rolling

$$U_i = \frac{\eta_0 \tilde{u}}{E' R_{x,i}} = \frac{(0.0045)(0.05)(314)}{(2.28 \times 10^{11})(0.03)} = 1.0329 \times 10^{-11}$$

$$U_o = \frac{\eta_0 \tilde{u}}{E' R_{x,o}} = \frac{(0.0045)(0.05)(314)}{(2.28 \times 10^{11})(0.07)} = 0.4427 \times 10^{-11}$$

$$G = \xi E' = 5016$$

$$W_i = \frac{(w_z)_\phi}{E'(R_{x,i})^2} = \frac{18{,}457}{(2.28 \times 10^{11})(0.03)^2} = 0.8995 \times 10^{-4}$$

$$W_o = \frac{(w_z)_\phi}{E'(R_{x,o})^2} = \frac{18{,}457}{(2.28 \times 10^{11})(0.07)^2} = 0.1652 \times 10^{-4}$$

The minimum film thickness in elliptical conjunctions under elastohydrodynamic conditions is given in dimensionless form by

$$\tilde{H}_{e,\min} = \frac{\tilde{h}_{\min}}{R_x} = 3.63 U^{0.68} G^{0.49} W^{-0.073} (1 - e^{-0.68k}) \tag{22.18}$$

Hence,

$$\left(\tilde{H}_{e,\min}\right)_i = (3.63)(1.0329 \times 10^{-11})^{0.68}(5016)^{0.49}(0.8995 \times 10^{-4})^{-0.073}$$
$$\times (1 - e^{-0.68\times 2.375})$$
$$= (3.63)(3.385 \times 10^{-8})(65.04)(1.9740)(0.7827)$$
$$= 1.235 \times 10^{-5}$$

and

$$\left(\tilde{h}_{\min}\right)_i = 0.371\ \mu\text{m}$$

Similarly,

$$\left(\tilde{H}_{e,\min}\right)_o = (3.63)(0.4427 \times 10^{-11})^{0.68}(5016)^{0.49}(0.1652 \times 10^{-4})^{-0.073}$$
$$\times (1 - e^{-0.68\times 1.384})$$
$$= (3.63)(1.9026 \times 10^{-8})(65.04)(2.2340)(0.6098)$$
$$= 0.6119 \times 10^{-5}$$

and

$$\left(\tilde{h}_{\min}\right)_o = 0.428\ \mu\text{m}$$

The power is thus transmitted through elastohydrodynamic films having minimum film thicknesses at the inner and outer conjunctions of 0.371 and 0.428 μm, respectively. These are quite acceptable film thicknesses, but for maximum life of the components and a dimensionless film parameter [Eq. (3.22)] of 3, it is clear that surface finishes comparable to those achieved in rolling-element bearings are required.

The large axial loads cause substantial, but not excessive, contact pressures of the order of 2.52 and 1.82 GPa on the inner and outer conjunctions, respectively. The Hertzian contact ellipses are quite large, having major axes 5761 and 5186 μm long at the inner and outer conjunctions. It has been assumed that pure rolling occurs, but spin losses will be inevitable in those conjunctions. The calculation of these losses, which probably account for most of the energy dissipation in such devices, requires a detailed knowledge of the lubricant rheology and the shape of the elastohydrodynamic film.

25.3 RAILWAY WHEELS ROLLING ON WET OR OILY RAILS

The construction of a modern track and locomotive driving wheel presented by Barwell (1974) is illustrated in Fig. 25.4. High-quality steel tires with a thickness of about 76 mm (3 in) are shrunk onto forged steel wheel centers. An important geometrical feature is the slight coning of the tires to facilitate automatic steering of the wheel sets as they roll along the track. This normally prevents contact between the wheel flanges and the rail. Studies of the Hertzian contact

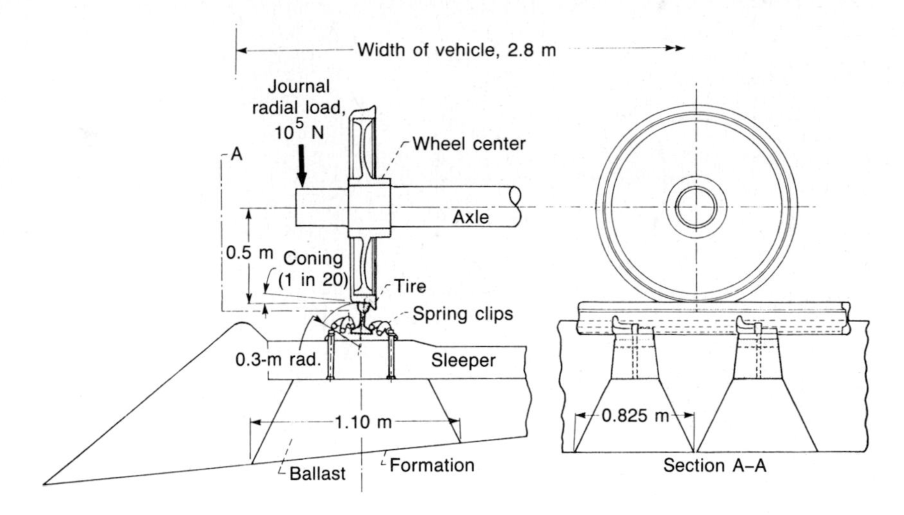

FIGURE 25.4
Construction of modern track and locomotive driving wheel. [*From Barwell (1974).*]

conditions for this situation have been presented by Barwell (1979) and the Engineering Sciences Data Unit (1978).

In this example it is assumed that the rail section transverse to the rolling direction has a radius of 0.3 m, the locomotive wheel has a radius of 0.5 m, and the coning angle is 2.86° ($\tan^{-1} 0.05$). The essential features of the Hertzian contact between the wheel and the rail are calculated for the following conditions:

Radial load carried by each wheel, N	10^5
Modulus of elasticity of wheel and rail E, Pa	2.07×10^{11}
Poisson's ratio ν	0.3

For these conditions

$$E' = \frac{E}{1-\nu^2} = \frac{2.07 \times 10^{11}}{0.91} = 2.2747 \times 10^{11} \text{ Pa}$$

25.3.1 INITIAL CALCULATION. If the coning of the tires is neglected and the contact is deemed to be equivalent to a wheel in the form of a circular cylinder of radius 0.5 m rolling on a track with a transverse radius of 0.3 m, as shown in

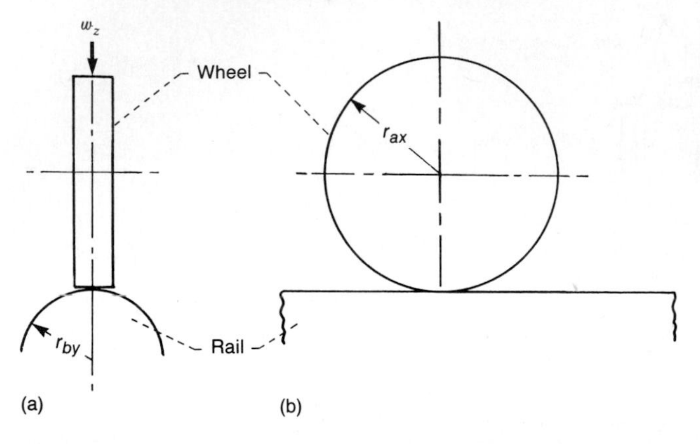

FIGURE 25.5
Contact between wheel and rail. [*From Hamrock and Dowson (1981).*]

Fig. 25.5,

$$r_{ax} = 0.5 \text{ m} \qquad r_{ay} = \infty$$

$$r_{bx} = \infty \qquad r_{by} = 0.3 \text{ m}$$

$$\frac{1}{R_y} = \frac{1}{r_{ay}} + \frac{1}{r_{by}} = \frac{1}{0.3 \text{ m}}$$

$$\therefore R_y = 0.3 \text{ m}$$

$$\frac{1}{R_x} = \frac{1}{r_{ax}} + \frac{1}{r_{bx}} = \frac{1}{0.5 \text{ m}}$$

$$\therefore R_x = 0.5 \text{ m}$$

$$\frac{1}{R} = \frac{1}{R_x} + \frac{1}{R_y} = \frac{1}{0.5} + \frac{1}{0.3} = 5.3333$$

$$R = 0.1875 \text{ m}$$

$$\alpha_r = \frac{R_y}{R_x} = \frac{0.3}{0.5} = 0.6$$

Since $\alpha_r < 1$, it follows that the semimajor axis of the elliptical Hertzian contact lies in the rolling direction. From Table 19.3 we get

$$\bar{\mathscr{E}} = 1 + q_a \alpha_r = 1.3425$$

$$\bar{\mathscr{F}} = \frac{\pi}{2} - q_a \ln \alpha_r = 1.8624$$

and

$$\bar{k} = \alpha_r^{2/\pi} = 0.7224$$

Hence

$$D_y = 2\left(\frac{6\bar{k}^2\bar{\mathscr{E}}w_z R}{\pi E'}\right)^{1/3}$$

$$= 2\left[\frac{(6)(0.7224)^2(1.3425)(10^5)(0.1875)}{(\pi)(2.2747 \times 10^{11})}\right]^{1/3} = 9.591 \text{ mm} \qquad (19.14)$$

$$D_x = 2\left(\frac{6\bar{\mathscr{E}}w_z R}{\pi \bar{k} E'}\right)^{1/3}$$

$$= 2\left[\frac{(6)(1.3425)(10^5)(0.1875)}{(\pi)(0.7224)(2.2747 \times 10^{11})}\right]^{1/3} = 13.277 \text{ mm} \qquad (19.15)$$

$$\delta = \bar{\mathscr{F}}\left[\frac{9}{2\bar{\mathscr{E}}R}\left(\frac{w_z}{\pi \bar{k} E'}\right)^2\right]^{1/3}$$

$$= 1.8624\left\{\frac{9}{(2)(1.3425)(0.1875)}\left[\frac{10^5}{(\pi)(0.7224)(2.2747 \times 10^{11})}\right]^2\right\}^{1/3}$$

$$= 0.1630 \text{ mm} \qquad (19.16)$$

and

$$p_{\max} = \frac{6w_z}{\pi D_x D_y} = \frac{(6)(10^5)}{(\pi)(0.01327)(0.009591)} = 1.501 \text{ GPa} \qquad (19.7)$$

The film thickness is calculated for two lubricants, one being water and the other a mineral oil, as a function of speed.

25.3.2 WATER. The viscosity of water varies little with pressure, and the fluid is generally assumed to be isoviscous. However, Hersey and Hopkins (1954) recorded a 7 percent increase in the viscosity of water over a pressure range of 1000 atm at 38°C (100°F), and this corresponds to a pressure-viscosity coefficient ξ of 6.68×10^{-10} m^2/N.

It is thus possible to adopt this value of ξ and to calculate the minimum film thickness from the equation used earlier in this chapter.

$$\tilde{H}_{e,\min} = \frac{\tilde{h}_{\min}}{R_x} = 3.63U^{0.68}G^{0.49}W^{-0.073}(1 - e^{-0.68k}) \qquad (22.18)$$

It will be assumed that the absolute viscosity of water η_0 at $p = 0$ and constant temperature is 0.001 $\mathrm{N \cdot s/m^2}$ and hence that

$$U = \frac{\eta_0 \tilde{u}}{E' R_x} = \frac{10^{-3}\tilde{u}}{(2.2747 \times 10^{11})(0.5)} = 8.7924\tilde{u} \times 10^{-15}$$

$$G = \xi E' = (6.68 \times 10^{-10})(2.2747 \times 10^{11}) = 152$$

$$W = \frac{w_z}{E' R_x^2} = \frac{10^5}{(2.2747 \times 10^{11})(0.5)^2} = 1.7585 \times 10^{-6}$$

With $\bar{k} = 0.7224$,

$$\tilde{H}_{e,\min} = \frac{\tilde{h}_{\min}}{R_x}$$

$$= (3.63)(8.7924\tilde{u} \times 10^{-15})^{0.68}(152)^{0.49}(1.7584 \times 10^{-6})^{-0.073}(1 - e^{-0.4912})$$

$$= (3.63)(2.7669 \times 10^{-10})(11.7247)(2.6309)(0.3881)(\tilde{u})^{0.68}$$

or

$$\left(\tilde{H}_{e,\min}\right)_{\text{water}} = 1.2024(\tilde{u})^{0.68} \times 10^{-8}$$

and

$$\left(\tilde{h}_{\min}\right)_{\text{water}} = 0.6012(\tilde{u})^{0.68} \times 10^{-8}\ \mathrm{m}$$

where $\tilde{u}$ in the preceding equations is in meters per second.

25.3.3 OIL. If the rail is covered by an oil film having a viscosity coefficient at absolute pressure η_0 of 0.1 Pa · s and a pressure-viscosity coefficient ξ of $2.2 \times 10^{-8}\ \mathrm{m^2/N}$,

$$U = \frac{\eta_0 \tilde{u}}{E' R_x} = \frac{0.1\tilde{u}}{(2.2747 \times 10^{11})(0.5)} = 8.7924\tilde{u} \times 10^{-13}$$

$$G = \xi E' = (2.2 \times 10^{-8})(2.2747 \times 10^{11}) = 5004$$

$$W = \frac{w_z}{E' R_x^2} = \frac{10^5}{(2.2747 \times 10^{11})(0.5)^2} = 1.7584 \times 10^{-6}$$

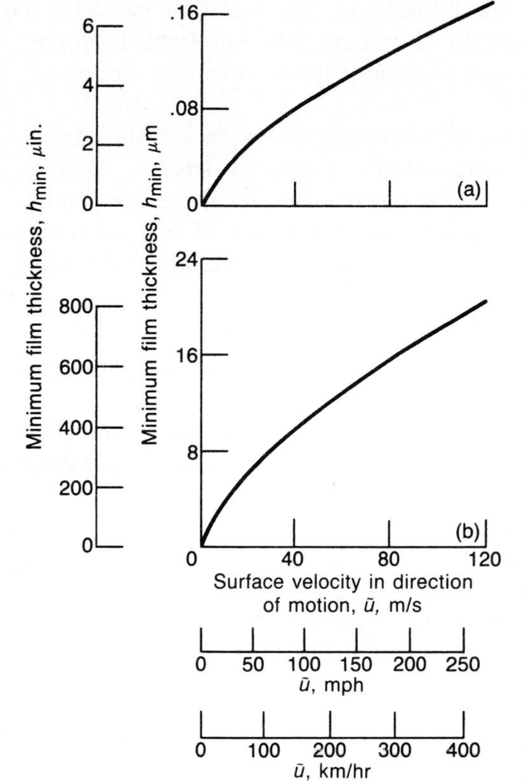

FIGURE 25.6
Variation of minimum film thickness with speed for wheel on rail. (a) Wet rail: absolute viscosity of water at $p = 0$ and constant temperature η_0, 0.001 Pa · s; viscosity-pressure coefficient ξ, 6.68 $\times 10^{-10}$ m^2/N. (b) oily rail: $\eta_0 = 0.1$ Pa · s; $\xi = 2.2 \times 10^{-8}$ m^2/N. [*From Hamrock and Dowson (1981).*]

Hence
$$\tilde{H}_{e,\min} = \frac{\tilde{h}_{\min}}{R_x}$$
$$= (3.63)(8.7924\tilde{u} \times 10^{-13})^{0.68}(5004)^{0.49}(1.7584 \times 10^{-6})^{-0.073} \times (1 - e^{-0.4912})$$
$$= (3.63)(6.3386 \times 10^{-9})(64.96)(2.6309)(0.3881)(\tilde{u})^{0.68}$$

or
$$\left(\tilde{H}_{e,\min}\right)_{\text{oil}} = 1.5261(\tilde{u})^{0.68} \times 10^{-6}$$

and
$$\left(\tilde{h}_{\min}\right)_{\text{oil}} = 0.7631(\tilde{u})^{0.68} \times 10^{-6} \text{ m}$$

where $\tilde{u}$ is in meters per second.

How $h_{\min}$ varies with locomotive speed in pure rolling is shown for both wet and oily rails in Fig. 25.6. It is clear from Fig. 25.6(a) that the elastohydrodynamic films developed by water alone are quite thin, and it is most unlikely that they will lead to effective hydrodynamic lubrication. The fact that moisture is known to influence traction quite markedly (Pritchard, 1981) probably strengthens the view that surface chemistry and boundary lubrication play important roles in wheel-rail friction.

For an oily track and a relatively high viscosity of 0.1 Pa · s, substantial film thicknesses are predicted for representative operating speeds. Clearly, traction will be reduced and skidding can be expected under these conditions. The results presented in Fig. 25.6(b) confirm the importance of a clean track. Alternative procedures for cleaning the rails in front of locomotive driving wheels are now being investigated in several countries.

25.4 SYNOVIAL JOINTS

To conclude this chapter on the applications of the elastohydrodynamic film thickness equations developed in the text, a human joint lubrication problem has been selected. This also allows the introduction of equations developed in Sec. 22.5 for the lubrication of elastic solids by isoviscous lubricants (soft EHL). It must be stated at the outset that film thickness calculations for synovial joints are more speculative than the other calculations presented in this chapter for engineering components. The properties of the materials, the magnitude of the applied load, and the motion and geometry of the contacting solids are all subject to debate. Indeed the lubrication mechanism of these remarkable bearings has yet to be resolved.

The essential features of a synovial joint are shown in Fig. 25.7. The bearing material is known as "articular cartilage" and the lubricant as "synovial

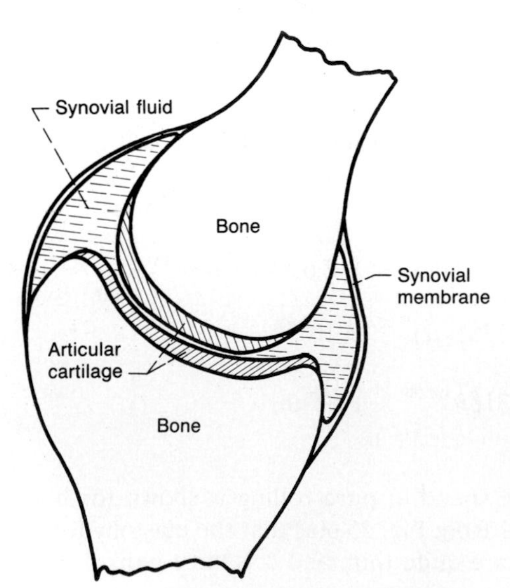

FIGURE 25.7
Illustration of synovial joint. [*From Hamrock and Dowson (1981).*]

fluid." The articular cartilage is a soft, porous material mounted on a relatively hard bone backing. The thickness of the cartilage in a synovial joint varies from joint to joint and also with age. In young, healthy subjects it may be several millimeters thick, but in elderly subjects it may be almost nonexistent in some regions. Although articular cartilage appears to be smooth, measurements suggest that the surface roughness R_a varies from about 1 to 3 μm. The effective modulus of elasticity E' varies with loading time and is difficult to specify with certainty. The various measurements which have been reported suggest that E' normally lies in the range 10^7 to 10^9 Pa.

Synovial fluid is a non-Newtonian fluid in which the effective viscosity falls markedly as the shear rate increases. The fluid's viscous properties appear to be governed by the hyaluronic acid, with the viscosity increasing almost linearly with acid concentration. It is also known that synovial fluid from normal, healthy joints is more viscous than pathological synovial fluid from both osteoarthritic and rheumatoid arthritic joints. At low shear rates the viscosity of normal synovial fluid is typically about 10 Pa · s; at the high shear rates encountered in joints, values between 10^{-3} and 10^{-2} Pa · s can be expected.

The effective radius of curvature of an equivalent sphere or cylinder near a plane is also difficult to determine but probably ranges from 0.1 to 1.0 m in the hip and from 0.02 to 0.1 m in the knee. Loads on the joints in the lower limb vary during walking, and the human bearings are thus subjected to dynamic conditions in which both loads and sliding speeds vary with time. When the leg swings freely in the walking cycle, the knee and the hip carry loads in the range zero to one times body weight; when the leg is in contact with the ground (stance phase), the peak loads range from three to seven times body weight.

The viscosity of synovial fluid varies negligibly with pressure over the pressure range encountered in a synovial joint. The articular cartilage deforms readily under physiological loads, and the lubrication regime can thus be regarded as isoviscous-elastic. The following expression was developed in Chap. 22 for the minimum film thickness under these conditions:

$$\tilde{H}_{e,\min} = 7.43U^{0.65}W^{-0.21}(1 - 0.85e^{-0.31k}) \qquad (22.22)$$

The hip joint is generally represented as a ball (femoral head)-and-socket (acetabulum) joint. In the present example it is assumed that the following quantities typify the peak loading periods of the stance phase in walking:

Radius of equivalent sphere near a plane R_x, m	1
Absolute viscosity of synovial fluid η_0, Pa s	2×10^{-3}
Effective elastic modulus E', Pa	10^7
Entraining mean velocity $\tilde{u} = (u_a + u_b)/2$, m/s	0.075
Applied load w_z, N	4500

Then,

$$U = \frac{\eta_0 \tilde{u}}{E' R_x} = \frac{(2 \times 10^{-3})(0.075)}{(10^7)(1)} = 1.5 \times 10^{-11}$$

$$W = \frac{w_z}{E' R_x^2} = \frac{4500}{(10^7)(1)} = 4.5 \times 10^{-4}$$

and

$$k = 1$$

Hence,

$$\begin{aligned} \tilde{H}_{e,\min} &= (7.43)(1.5 \times 10^{-11})^{0.65}(4.5 \times 10^{-4})^{-0.21}(1 - 0.85e^{-0.31}) \\ &= (7.43)(9.2142 \times 10^{-8})(5.0446)(0.3766) \\ &= 1.3006 \times 10^{-6} \end{aligned}$$

and

$$\tilde{h}_{\min} = 1.3\ \mu\text{m}$$

The calculation yields a film thickness similar to, but perhaps slightly less than, the roughness of articular cartilage. This suggests that prolonged exposure to the listed conditions could not conserve effective elastohydrodynamic lubrication. Dowson (1981) has discussed the question of lubrication in hip joints in some detail, and it emerges that if any hydrodynamic films are developed during the swing phase in walking, they might be preserved by a combination of entraining and squeeze-film action during the stance phase.

25.5 CLOSURE

The equation for minimum film thickness in rectangular and elliptical conjunctions under elastohydrodynamic conditions developed in Chaps. 21 and 22 was applied to involute gears in Sec. 25.1. The involute gear presents a rectangular conjunction and thus represents a limiting condition for applying the elliptical conjunction equation. The example nevertheless demonstrates the generality of the film thickness equations presented in Chap. 22 while allowing the accuracies of the formulas developed in Chap. 21 to be compared. The rectangular conjunction film thickness formula was found to be about 25 percent less than that for elliptical conjunctions. It is felt that the rectangular conjunction film formulation is more accurate, since a much greater range of operating parameters was used in obtaining the results. Also the rectangular conjunction results agree better with the experimental work of Kunz and Winer (1977). The dimensionless film parameter Λ, which appears to play an important role in determining the fatigue life of highly stressed, lubricated machine elements, was also given.

Continuously variable-speed drives like the Perbury gear, which present truly elliptical elastohydrodynamic conjunctions, are favored increasingly in mobile and stationary machinery. A representative elastohydrodynamic condition for this class of machinery was considered in Sec. 25.2 for power transmission equipment.

The possibility of elastohydrodynamic films of water or oil forming between locomotive wheels and rails was examined in Sec. 25.3. The important subject of traction on the railways is presently attracting considerable attention in various countries.

The final example, a synovial joint, introduced the equation developed in Sec. 22.5 for isoviscous-elastic lubrication regimes. This example is necessarily more speculative than others in this chapter owing to the varied and uncertain conditions encountered in human and animal joints. Other applications in the isoviscous-elastic regime include rubber tires on wet roads and elastomeric seals. This range of applications serves to demonstrate the utility of the film thickness equations developed in this text for elastohydrodynamic elliptical conjunctions. It is hoped that this study of elliptical conjunctions has extended in some measure the understanding of that recently recognized yet vitally important mode of lubrication in highly stressed machine elements known as elastohydrodynamic.

25.6 PROBLEMS

25.6.1 For a steel ball rolling on a steel plane, as shown in the following sketch, and lubricated with a mineral oil determine:

(*a*) Dimensions of the Hertzian contact zone D_x and D_y
(*b*) Maximum elastic deformation δ_{max}
(*c*) Maximum Hertzian pressure p_H
(*d*) Lubrication regime
(*e*) Minimum film thickness

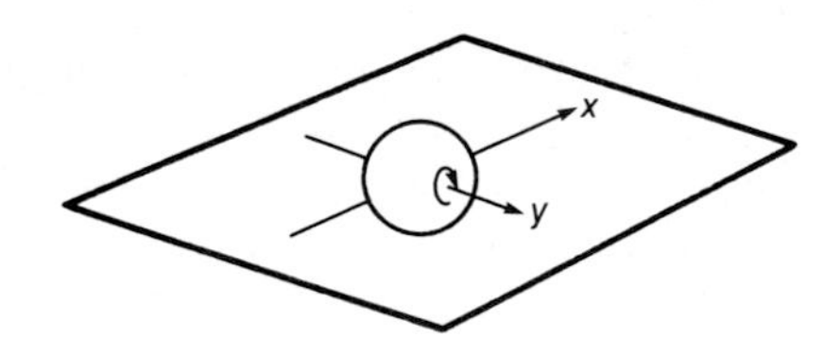

The steel ball has a diameter of 10 mm and weighs 0.04 N. The ball and the plane on which it rolls both have a modulus of elasticity of 2×10^{11} Pa and a Poisson's ratio of 0.3. Consider four loads: 0.04, 0.4, 4, and 40 N. The ball rolls over the flat surface with a mean velocity of 1 m/s in the presence of a lubricant having an absolute viscosity of 0.05 Pa · s and a pressure-viscosity coefficient of 2.0×10^{-8} m^2/N. Fully flooded, isothermal, and smooth surfaces are assumed.

25.6.2 A manufacturer has difficulty with persistent scuffing of the highly loaded cam shown in the sketch. One solution suggested is to eliminate metallic contact between the sliding faces. The cam has a surface finish of 6 μin cla and a minimum radius of 0.75 in at a distance of 1.7 in from the center of rotation. It is 0.8 in wide and subject to a load of 1250 lb. The cam rotates against a convex stationary follower that has a radius of 2.5 in, a surface finish of 7 μin CLA, and a sliding speed at the point of contact of approximately 1 ft/s. The contact zone is flooded with oil that has a viscosity of 0.86 P and a pressure-viscosity coefficient of 1.4×10^{-4} in^2/lb.

The material of both cam and follower has a modulus of elasticity of 30×10^6 lb/in^2 and a Poisson's ratio of 0.30.

The manufacturer states that the surface finishes are the best attainable and that the load and overall cam lift would be too costly to change. The cam's minimum radius of curvature is determined by the camshaft's rotational speed and is approximately proportional to the sliding speed at the contact point. Assuming a reasonably free hand with the other variables, list the courses of action possible, and state the advice you would give the manufacturer.

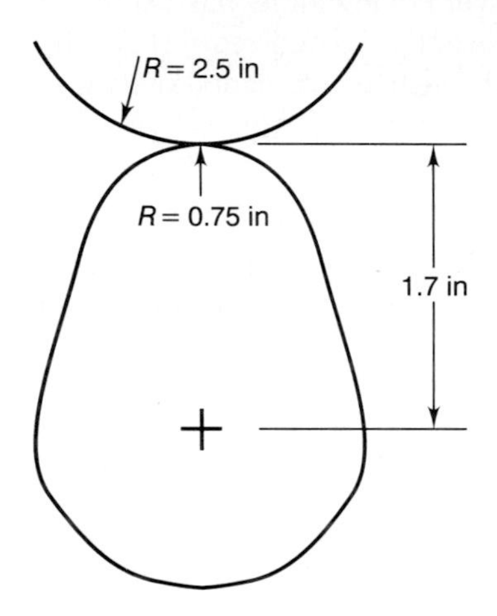

25.6.3 How would the oil film thickness vary along the contact for an involute gear tooth?

25.6.4 Two mating gears with involute teeth and a face width of 35 mm have radii of 60 and 180 mm, respectively and transmit 20 kW of power when the smaller gear rotates at 1440 r/min. If the pressure angle is 20°, suggest a suitable surface finish for the gears to avoid pitting. Assume

$$G(\text{materials parameter}) = 5000$$

$$E'(\text{effective modulus}) = 2.5 \times 10^{11}\ \text{N/m}^2$$

$$\eta(\text{oil viscosity}) = 0.015\ \text{kg/(m} \cdot \text{s)}$$

Is a surface finish of the order you calculate economically feasible?

25.7 REFERENCES

Barwell, F. T. (1974): The Tribology of Wheel on Rail. *Tribol. Int.*, vol. 7, no. 4, pp. 146–150.

Barwell, F. T. (1979): *Bearing Systems: Principles and Practice*. Oxford University Press, Oxford.

Cahn-Speyer, P. (1957): Mechanical Infinitely Variable Speed Drives. *Eng. Dig.* (*London*), vol. 18, no. 2, pp. 41–65.

Dowson, D. (1981): Lubrication of Joints: A. Natural Joints. *An Introduction to the Bio-Mechanics of Joints and Joint Replacements*. D. Dowson and V. Wright (eds.). Mechanical Engineering Publications, England, pp. 120–133.

Dowson, D., and Higginson, G. R. (1964): A Theory of Involute Gear Lubrication. *Gear Lubrication, Part 2—Theoretical Aspects of Gear Lubrication*, Proceedings of a Symposium Organized by the Mechanical Tests of Lubricants Panel of the Institute of Petroleum, Elsevier, London, pp. 8–15.

Dowson, D., and Higginson, G. R. (1966): *Elastohydrodynamic Lubrication, The Fundamentals of Roller and Gear Lubrication*. Pergamon, Oxford.

ESDU (1978): Contact Phenomena. I: Stresses, Deflections and Contact Dimensions for Normally Loaded Unlubricated Elastic Components. Engineering Sciences Data Unit, Item 78035, Institution of Mechanical Engineers, London.

Fellows, T. G., Dowson, D., Perry, F. G., and Plint, M. A. (1964): The Perbury Continuously Variable Ratio Transmission. *Advances in Automobile Engineering*, Part 2, N. A. Carter (ed.). Pergamon, Oxford, pp. 123–139.

Hamrock, B. J., and Dowson, D. (1981): *Ball Bearing Lubrication—The Elastohydrodynamics of Elliptical Contacts*. Wiley-Interscience, New York.

Hersey, M. D., and Hopkins, R. F. (1954): *Viscosity of Lubricants Under Pressure. Coordinated Data From Twelve Investigations*, ASME, New York.

Kunz, R. K., and Winer, W. O. (1977): Discussion on pp. 275–276 of Hamrock, B. J. and Dowson, D. Isothermal Elastohydrodynamic Lubrication of Point Contacts, Part III—Fully Flooded Results. *J. Lubr. Technol.*, vol. 99, no. 2, pp. 264–275.

Pritchard, C. (1981): Traction Between Rolling Steel Surfaces—A Survey of Railway and Laboratory Services. *Friction and Traction*. Proceedings of the Seventh Leeds-Lyon Symposium on Tribology. D. Dowson et al., (eds.). Mechanical Engineering Publications, London, pp. 197–206.

CHAPTER 26

NON-NEWTONIAN FLUID EFFECTS IN ELASTOHYDRODYNAMIC LUBRICATION

With the exception of Sec. 4.15 in which the limiting shear stress was introduced, the text up to this point has dealt strictly with the assumption that the fluid behaves in a Newtonian fashion. That is, the shear stress is linearly related to the shear strain rate by the fluid viscosity. In elastohydrodynamically lubricated conjunctions the lubricant experiences rapid and very large temperature changes, elongational viscosity, and, particularly in sliding contacts, high shear rates. As a result of high fluid film pressures and shear heating, the viscosity changes drastically within the lubricating film. In addition, the existence of very high shear rates may result in shear stresses that are well beyond the limiting shear stress of the lubricant. In elastohydrodynamically lubricated conjunctions the shear rates can range up to $10^7\ s^{-1}$. The maximum temperature within the conjunction can be as high as 500°C. Furthermore, the transit time required for the fluid to go from the inlet through the high-pressure zone to the outlet is between 10^{-3} and 10^{-4} s. The great severity of these conditions has called into question the normal assumption of Newtonian behavior.

26.1 FLUID RHEOLOGY MODELS

Hirst and Moore (1974) presented an experimental investigation of the dependence between shear stress and shear strain rate and proposed an Eyring type of relationship. Johnson and Tevaarwerk (1977) found the same type of relationship as did Hirst and Moore (1974). Houpert and Hamrock (1985), as well as Conry et al. (1987), incorporated Eyring's model (see Sec. 4.15) into the analysis of elastohydrodynamically lubricated line contacts by using a system approach to get a full numerical solution. It has been observed that the Eyring model does not exhibit limiting-shear-strength properties. In Eyring's model the rate at which shear stress increases with increasing shear strain rate is significantly smaller than the Newtonian linear rate at high shear strain rates. An upper limit to the shear stress increase can only be achieved for Eyring's model by resorting to a plastic transition.

The non-Newtonian property of limiting shear strength imposes that the lubricant does not experience shear stress exceeding this limit. The limiting shear strength is usually considered as an asymptotic limit. Should the fluid model, however, allow the shear stress to reach the limiting shear strength, slippage will occur either within the oil film or at the interface between the oil and one of the bounding surfaces. A fluid model that allows for such a slip phenomenon will be referred to as "a viscoplastic fluid model." The distinction between asymptotic and viscoplastic limiting-shear-strength behavior is an important one because the latter requires the replacement of the traditional velocity boundary conditions by stress boundary conditions. This means that there will be transitions in the type of boundary conditions imposed if the shear stress reaches the plastic limit inside the conjunction, a condition that may introduce singularities in the solution.

Bair and Winer (1979) and Gecim and Winer (1980) proposed alternative expressions for the relationship between shear stress and shear strain rate that do exhibit asymptotic-limiting-shear stress behavior. Both Bair and Winer (1979) and Gecim and Winer (1980) incorporated their models into the analysis of elastohydrodynamically lubricated line contacts by using a Grubin-like inlet analysis. In attempts to obtain a full numerical solution of the problem of elastohydrodynamically lubricated line contacts, Bair and Winer's model has been approximated into simple forms such as the linear model proposed by Trachman (1971), which was used in elastohydrodynamic lubrication by Wang and Zhang (1988) and Iivonen and Hamrock (1989), or the circular model proposed by Lee and Hamrock (1990a). Both the linear and the circular fluid models allow the establishment of a modified Reynolds equation by using an approach similar to the one that applies for a Newtonian fluid. The circular model is, however, not as severe a non-Newtonian fluid model as is the linear model, and it is therefore in better agreement with experimental observations. The modified Reynolds equation for the circular fluid model was solved by Lee and Hamrock (1990a) by using the fast system approach developed by Houpert and Hamrock (1986). This approach was later refined by Hsiao and Hamrock (1992) and used to investigate thermal effects in non-Newtonian elastohydrodynamic lubrication, which will be considered in the next chapter. The existence of the solution for a circular fluid model allows the most significant non-Newtonian feature to be incorporated into the analysis of elastohydrodynamic lubrication, namely, the limiting shear strength. It is, however, restrictive not to be able to model more freely the transition between Newtonian behavior at low shear strain rate and limiting-shear-strength behavior at high shear strain rate.

This chapter focuses on the circular fluid model for incorporating non-Newtonian fluid rheology effects into elastohydrodynamically lubricated conjunctions. The restriction of neglecting side leakage is assumed throughout the chapter. The steady-state and transient modified Reynolds equations as well as their appropriate stream functions are presented for the circular fluid model and applied to EHL and micro-EHL conditions. Results of pressure, film shape, streamlines, and shear stresses are presented. This chapter makes extensive use of the following technical papers: Shieh and Hamrock (1991), Lee and Hamrock (1990a, b), Elsharkawy and Hamrock (1991), and Myllerup et al. (1993).

26.2 FORMULATION OF FLUID RHEOLOGY MODELS

An alternative way of showing the non-Newtonian effect is to use the effective viscosity defined as

$$\eta_e = \frac{\tau}{s} \tag{26.1}$$

where τ is the shear stress and s is the shear strain rate $(\partial u/\partial z)$. This then defines the viscosity that is actually experienced within elastohydrodynamically lubricated conjunctions. Using this form, the non-Newtonian models described in Chap. 4 [Eqs. (4.29), (4.30), and (4.35)] can be written as

$$\frac{\eta_e}{\eta} = -\frac{\tau/\tau_L}{\ln(1 - \tau/\tau_L)} \tag{26.2}$$

$$\frac{\eta_e}{\eta} = \frac{\tau/\tau_L}{\tanh^{-1}(\tau/\tau_L)} \tag{26.3}$$

$$\frac{\eta_e}{\eta} = \frac{\tau/\tau_E}{\sinh(\tau/\tau_E)} \tag{26.4}$$

where τ_E is the shear stress at which the fluid first starts to behave nonlinearly when stress is plotted against shear strain rate.

Recently, a highly non-Newtonian fluid model was introduced by Iivonen and Hamrock (1989) that can be expressed as

$$\frac{\eta_e}{\eta} = 1 - \frac{\tau}{\tau_L} \tag{26.5}$$

This model is referred to in this chapter as "a straight-line model."

Another new non-Newtonian fluid rheology model developed by Lee and Hamrock (1990a) that shows considerable promise is

$$\frac{\eta_e}{\eta} = \left[1 - \left(\frac{\tau}{\tau_L}\right)^2\right]^{1/2} \tag{26.6}$$

This model is referred to in this chapter as "the circular model."

Elsharkawy and Hamrock (1991) proposed a general nonlinear viscous model in which the limiting shear strength has already been incorporated as

$$\frac{\partial u}{\partial z} = \frac{\tau}{\eta}\left[1 - \left(\frac{\tau}{\tau_L}\right)^n\right]^{-1/n} \tag{26.7}$$

where

$$\tau_L = \tau_{\text{atm}} + \gamma^* p \tag{26.8}$$

Equation (26.7) can be written as

$$\frac{\tau}{\tau_L} = \frac{\dfrac{\eta}{\tau_L}\dfrac{\partial u}{\partial z}}{\left[1 + \left(\dfrac{\eta}{\tau_L}\dfrac{\partial u}{\partial z}\right)^n\right]^{1/n}} \tag{26.9}$$

Equation (26.9) is plotted in Fig. 26.1 for different values of n. In Fig 26.1(a) the conventional way of describing the various models is used. Using the alternative representation given in Eq. (26.1) this non-Newtonian model can be expressed as

$$\eta_e = \tau\left(\frac{\partial u}{\partial z}\right)^{-1} = \eta\left[1 - \left(\frac{\tau}{\tau_L}\right)^n\right]^{1/n} \tag{26.10}$$

This relationship between shear stress and effective viscosity is shown in Fig. 26.1(b) for various values of the shape exponent n.

The main advantage of the general model is that all the other nonlinear viscous models with the limiting-shear-strength property are readily derived or closely approximated by appropriate choice of the exponent n. As shown in Fig. 26.1, $n = 1$ results in the linear model and $n = 2$ in the circular model. The general fluid model will, in fact, if incorporated into the analysis of elastohydrodynamic lubrication, allow a given relationship between shear strain rate and shear stress to also be simulated in the transition zone. In the general case of arbitrary n it is not possible to derive a modified Reynolds equation directly as was the case for $n = 1$ and $n = 2$.

In Fig. 26.1 the straight-line model is strongly non-Newtonian and thus probably overexaggerates the fluid effect. The straight-line model may be viewed as a limiting solution; the viscoplastic model is the other limit. An actual lubricant behaves somewhere between these two extremes. Table 26.1 shows the expression for the dimensionless effective viscosity and the shear strain rate for the various rheological models.

The circular model of Lee and Hamrock (1990a) and the $\tanh^{-1}$ model of Gecim and Winer (1980), shown in Fig. 26.1, behave more like a Newtonian model for dimensionless shear stresses less than 0.5 and then become highly non-Newtonian for dimensionless shear stresses greater than 0.5. This behavior is more like what one would expect for actual fluids used in elastohydrodynamically lubricated conjunctions. The major advantage that the circular model

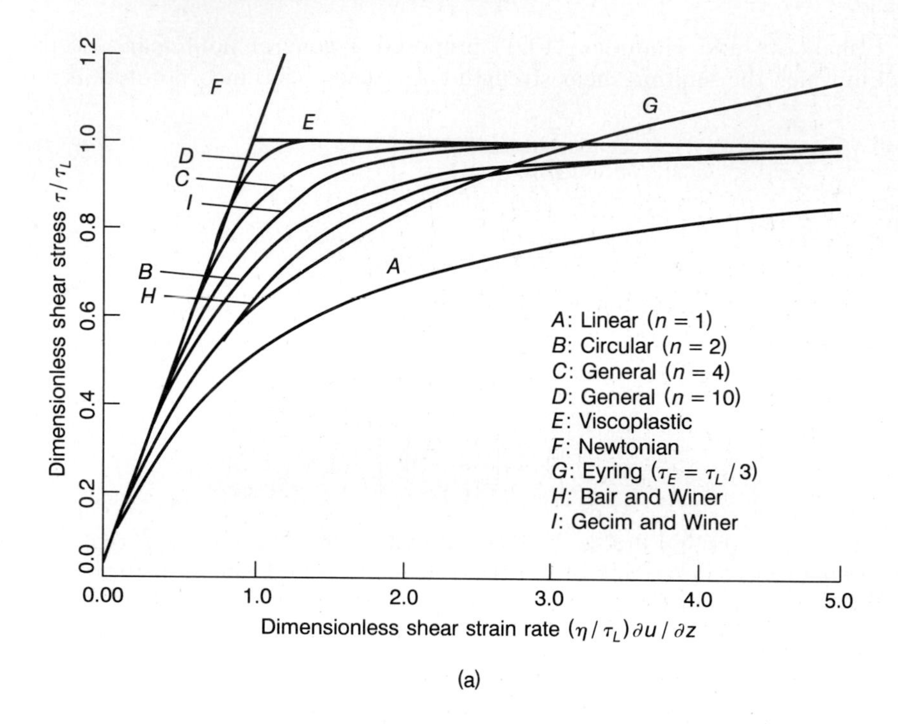

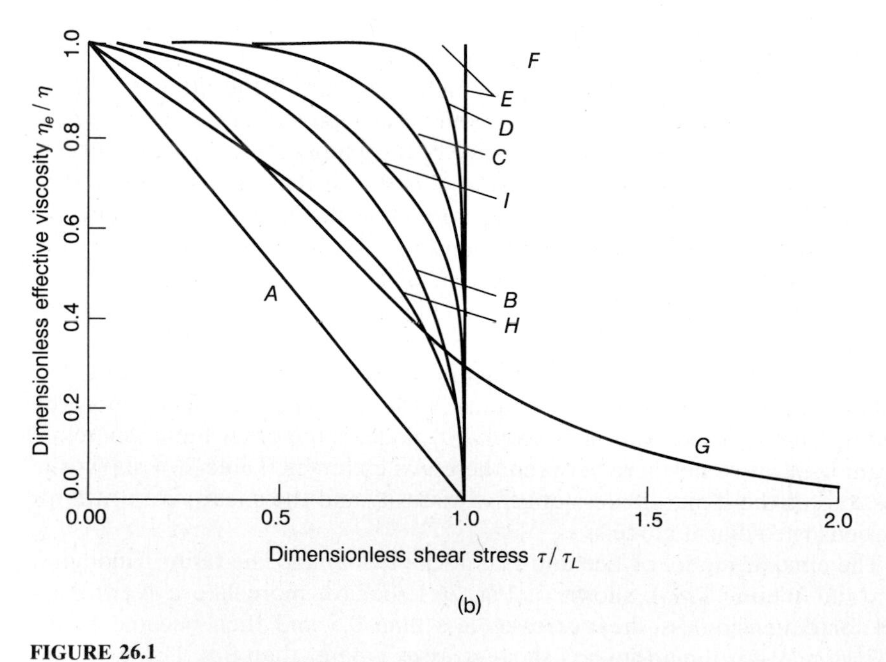

FIGURE 26.1
Non-Newtonian rheological models represented by (a) effect of shear strain rate on dimensionless shear stress and (b) effect of dimensionless shear stress on dimensionless effective viscosity. [*From Myllerup et al. (1993).*]

TABLE 26.1
Dimensionless effective viscosity for various non-Newtonian fluid models

Fluid model	Shear strain rate s	Dimensionless effective viscosity η_e/η
Newtonian	$\frac{\tau}{\eta}$	1
Eyring (1936) (sinh model)	$\frac{\tau_E}{\eta}\sinh\frac{\tau}{\tau_E}$	$\frac{\tau}{\tau_E\sinh(\tau/\tau_E)}$
Bair and Winer (1979) (ln model)	$\frac{\tau_L}{\eta}\ln(1-\bar{\tau})^{-1}$	$-\frac{\bar{\tau}}{\ln(1-\bar{\tau})}$
Gecim and Winer (1980) ($\tanh^{-1}$ model)	$\frac{\tau_L}{\eta}\tanh^{-1}\bar{\tau}$	$\frac{\bar{\tau}}{tanh^{-1}\bar{\tau}}$
Iivonen and Hamrock (1989) (straight-line model)	$\frac{\tau_L}{\eta}[(1-\bar{\tau})^{-1}-1]$	$1-\bar{\tau}$
Elsharkawy and Hamrock (1991) (general model)	$\frac{\tau}{\eta}[1-(\bar{\tau})^n]^{-1/n}$	$[1-(\bar{\tau})^n]^{1/n}$
Lee and Hamrock (1990a) (circular model)	$\frac{\tau}{\eta}[1-(\bar{\tau})^2]^{-1/2}$	$(1-\bar{\tau})^{1/2}$

offers over the $\tanh^{-1}$ model is that the analytical expressions can be obtained for the elements of the Jacobian by using the approach of Houpert and Hamrock (1986). Thus, the circular model is used in this chapter because it offers easy implementation of the numerical approach presented in Chap. 21, and yet the stress-strain relation agrees with the $\tanh^{-1}$ model of Gecim and Winer (1980) for which there is experimental evidence.

For the Eyring (1936), or sinh, model given in Eq. (26.4) and shown in Fig. 26.1 and Table 26.1, it was assumed that $\tau_E = \tau_L/3$. The Eyring (1936) model was found to have shear stress that increases monotonically with increasing strain rate as shown in Fig. 4.11. In Fig. 26.1(a) the dimensionless shear stress decreases monotonically with decreasing effective viscosity. The sinh model was used by Houpert and Hamrock (1985) in applying the elastohydrodynamic lubrication analysis to the problem of scuffing. Furthermore, the viscoplastic model shown in Fig. 26.1(a) was used by Jacobson and Hamrock (1984) in solving the elastohydrodynamic lubrication problem. That is, when the shear stress at the bearing surfaces exceeded the limiting shear stress, the shear stress was set equal to the limiting shear stress.

The rest of this chapter uses the circular non-Newtonian fluid model given in Eq. (26.6) and derives appropriate equations for the modified Reynolds equation, the effective viscosity, and the flow rate per unit width. These equations are compared with the corresponding equations for a Newtonian fluid model. Results obained from coupling the equilibrium equation while using the circular non-Newtonian fluid model and the elasticity equation are presented.

26.3 MODIFIED REYNOLDS EQUATION

The Reynolds equation given in Eq. (21.22) was obtained for linearly viscous lubricant behavior. This equation has to be modified in order to incorporate a non-Newtonian fluid model. Just as was done in Chap. 21, this modified Reynolds equation will assume an isothermal behavior and neglect side-leakage, transient, and squeeze effects. The equilibrium in the fluid can be written as

$$\frac{dp}{dx} = \frac{d\tau}{dz} \tag{26.11}$$

The film geometry and coordinates are shown in Fig. 26.2. Since the pressure p is not a function of z, integrating with respect to z and dividing by τ_L gives

$$\frac{\tau}{\tau_L} = \frac{z}{\tau_L}\frac{dp}{dx} + \frac{A_2}{\tau_L} \tag{26.12}$$

where A_2 is an integration constant. Applying the new coordinate system $(x, \bar{z})$ given in Fig. 26.2, where $\bar{z} = z - h_a$, to the equilibrium equation (26.12) gives

$$\bar{\tau} = \bar{Z}\tilde{T} + \bar{\tau}_a \tag{26.13}$$

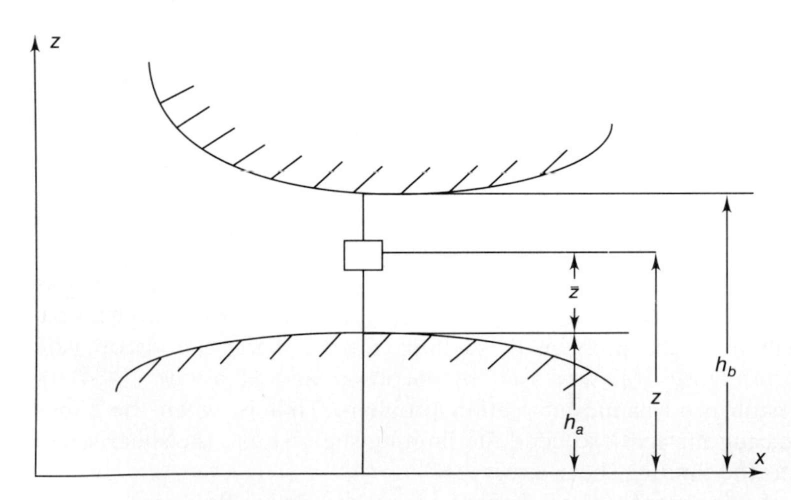

FIGURE 26.2
Film geometry and coordinates of conjunction. [*From Lee and Hamrock (1990a).*]

where

$$\bar{\tau}_a = \frac{\tau_a}{\tau_L} \qquad \bar{Z} = \frac{\bar{z}}{h} \qquad \tilde{T} = \frac{h}{\tau_L}\frac{dp}{dx} = \frac{2WH_r}{\pi\bar{\tau}_r}\frac{dP_r}{dX_r}$$

$$h = h_b - h_a \qquad \bar{\tau} = \frac{\tau}{\tau_L} \tag{26.14}$$

Note that $\bar{\tau}_a$ should just be viewed as an integration constant. Substituting Eqs. (26.1) and (26.6) into Eq. (26.13) gives

$$\frac{du}{d\bar{z}} = \frac{\tau_L}{\eta}\frac{\bar{Z}\tilde{T} + \bar{\tau}_a}{\left[1 - \left(\bar{Z}\tilde{T} + \bar{\tau}_a\right)^2\right]^{1/2}} \tag{26.15}$$

Integrating gives

$$\frac{u}{\tilde{u}} = -\frac{1}{S_c\tilde{T}}\left[1 - \left(\bar{Z}\tilde{T} + \bar{\tau}_a\right)^2\right]^{1/2} + A_4 \tag{26.16}$$

where

$$S_c = \frac{\eta\tilde{u}}{\tau_L h} = \frac{\pi U\bar{\eta}}{8WH_r\bar{\tau}_L} \tag{26.17}$$

The boundary conditions at surfaces a and b are

1. If $z = h_a$, $z = 0$, and $\bar{Z} = 0$, then $u = u_a$.
2. If $z = h_b$, $\bar{z} = h$, and $\bar{Z} = 1$, then $u = u_b$.

From boundary condition 1

$$A_4 = 1 - \frac{A_c}{2} + \frac{1}{S_c\tilde{T}}\left[1 - (\bar{\tau}_a)^2\right]^{1/2} \tag{26.18}$$

where the slide-roll ratio

$$A_c = \frac{2(u_b - u_a)}{u_b + u_a} \tag{26.19}$$

Figure 26.3 illustrates various cylindrical motions for several slide-roll ratios.

Substituting Eq. (26.18) into Eq. (26.16) gives

$$\frac{u}{\tilde{u}} = -\frac{1}{S_c\tilde{T}}\left[1 - \left(\bar{Z}\tilde{T} + \bar{\tau}_a\right)^2\right]^{1/2} + 1 - \frac{A_c}{2} + \frac{1}{S_c\tilde{T}}\left[1 - (\bar{\tau}_a)^2\right]^{1/2} \tag{26.20}$$

From boundary condition 2

$$A_cS_c\tilde{T} = \left[1 - (\bar{\tau}_a)^2\right]^{1/2} - \left[1 - \left(\tilde{T} + \bar{\tau}_a\right)^2\right]^{1/2} \tag{26.21}$$

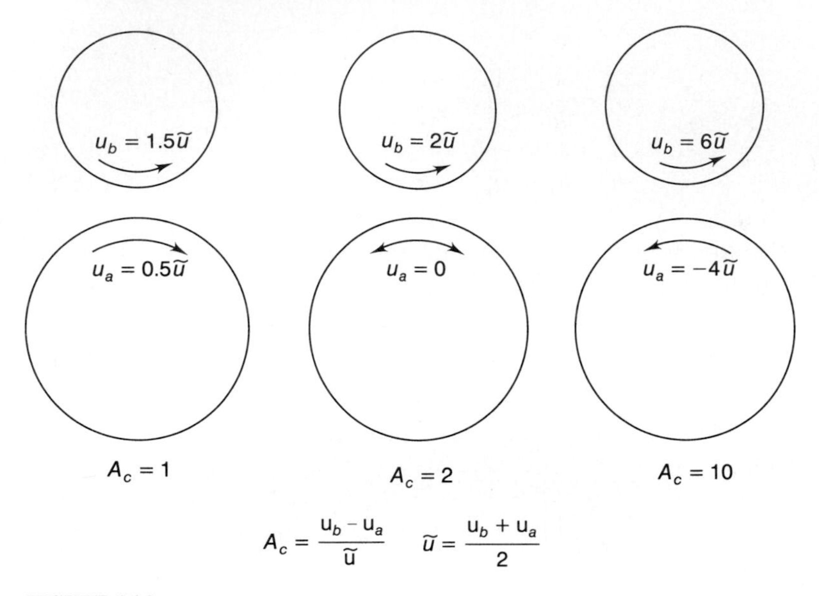

FIGURE 26.3
Illustration of cylindrical motion of surfaces for various slide-roll ratios.

Solving for $\bar{\tau}_a$ in this equation gives

$$\bar{\tau}_a = -\frac{\tilde{T}}{2} + A_c S_c \left[\frac{1}{1 + (A_c S_c)^2} - \frac{\tilde{T}^2}{4} \right]^{1/2} \tag{26.22}$$

This solution is only valid when

$$|\tilde{T}| \le \left[\frac{2}{(A_c S_c)^2 + 1} \right]^{1/2} \tag{26.23}$$

The range of $\tilde{T}$ that satisfies this inequality becomes smaller as $A_c S_c$ becomes larger. Recalling the definitions of $\tilde{T}$, A_c, and S_c, the implication of the preceding equations is that the pressure gradient is close to zero for high pressures and high slide-roll ratios.

Rewriting Eq. (26.20) while making use of Eq. (26.13) gives

$$u^* = \frac{u}{\tilde{u}} = 1 - \frac{A_c}{2} + \frac{1}{S_c \tilde{T}} \left(\sqrt{1 - \bar{\tau}_a^2} - \sqrt{1 - \bar{\tau}^2} \right) \tag{26.24}$$

From the continuity equation, considering compressibility but neglecting side leakage and the time-dependent term, the dimensionless stream function $\overline{\Phi}$ can be expressed as

$$\bar{\rho} H u^* = \frac{\partial \overline{\Phi}}{\partial \overline{Z}} \tag{26.25}$$

and

$$\bar{\rho} w^* = -\frac{\partial \overline{\Phi}}{\partial X_r} \tag{26.26}$$

where

$$\overline{Z} = \frac{z}{h} \qquad X_r = \frac{x}{D_x/2} \qquad W^* = \frac{wR}{\tilde{u}b} \qquad \overline{\Phi} = \frac{\phi R}{\rho_0 \tilde{u} b^2} \tag{26.27}$$

Recall that in Sec. 8.4.9 the stream function was developed for an incompressible Newtonian fluid when being applied to a fixed-incline slider bearing. However, this section considers a compressible non-Newtonian fluid. The dimensionless stream function of the non-Newtonian EHL line-contact problem can therefore be obtained by integrating Eq. (26.25) or

$$\begin{aligned}\overline{\Phi}(X_r, \overline{Z}) &= \int_0^{\overline{Z}} (\bar{\rho} H u^*)\, d\overline{Z} + f_4(X_r) \\ &= \bar{\rho} H \left[\left(1 - \frac{A_c}{2} + \frac{1}{S_c \tilde{T}} \sqrt{1 - \bar{\tau}_a^2} \right) \overline{Z} - \frac{1}{2 S_c \tilde{T}^2} \left(\bar{\tau} \sqrt{1 - \bar{\tau}^2} + \sin^{-1} \bar{\tau} \right) \right] \\ &\quad + f_4(X_r)\end{aligned} \tag{26.28}$$

Let $\overline{\Phi} = 0$ on the surface boundary $\overline{Z} = 0$; then,

$$f_4(X_r) = \frac{\bar{\rho} H}{2 S_c \tilde{T}^2} \left(\bar{\tau}_a \sqrt{1 - \bar{\tau}_a^2} + \sin^{-1} \bar{\tau}_a \right) \tag{26.29}$$

When $dp/dx \to 0$, this implies by definition that $\tilde{T} \to 0$. Observe from Eq. (26.20) that this results in 0/0. Applying Lhopital's rule gives

$$\left(\frac{u}{\tilde{u}} \right)_{\tilde{T} \to 0} = 1 - \frac{A_c}{2} + A_c \overline{Z} \tag{26.30}$$

The mass flow rate per unit width when $\tilde{T} \to 0$ is

$$(q')_{\tilde{T} \to 0} = \tilde{u} \rho_m h_m \int_0^1 \left(1 - \frac{A_c}{2} + A_c \overline{Z} \right) d\overline{Z} = \tilde{u} \rho_m h_m = \tilde{u} \rho_0 h_{\text{end}} \tag{26.31}$$

The mass flow rate per unit width must be the same at any location.

$$\rho h \int_0^1 u\, d\overline{Z} = \rho_0 h_{\text{end}} \tilde{u} \tag{26.32}$$

Making use of Eq. (27.20) gives

$$\rho_0 h_{\text{end}} = \rho h \left(1 + \frac{\bar{e} - \sin^{-1} \bar{e}}{2 S_c \tilde{T}^2} \right) \tag{26.33}$$

where

$$\bar{e} = \left(\tilde{T} + \bar{\tau}_a\right)\left(1 - \bar{\tau}_a^2\right)^{1/2} - \bar{\tau}_a\left[1 - \left(\tilde{T} + \bar{\tau}_a\right)^2\right]^{1/2} \tag{26.34}$$

Equation (26.33) is the Reynolds equation for the circular model. This equation can be written as

$$H_r^3 \frac{dP_r}{dX_r} + \bar{K}\bar{\eta}_c\left(\frac{H_{r,\text{end}}}{\bar{\rho}} - H_r\right) = 0 \tag{26.35}$$

where

$$\bar{\eta}_c = \frac{\tilde{T}^3\bar{\eta}/6}{\sin^{-1}\bar{e} - \bar{e}} \tag{26.36}$$

Note that $\bar{\tau}_L \to \infty$, $S_c \to 0$, $\tilde{T} \to 0$, and $\bar{\eta}_c = \bar{\eta}$. The solution to Eq. (26.35) is similar to that found in Chap. 21. Equation (26.35) is then the modified Reynolds equation for the circular model.

26.4 STEADY-STATE RESULTS

Smooth surfaces for two oils were studied by Lee and Hamrock (1990a). These were exactly the same fluids studied by Iivonen and Hamrock (1989). The properties of these oils are given in Table 26.2, where the values of the limiting-shear-strength proportionality constant γ^* and the solidification pressure p_s were taken from Höglund (1984) and the value of the shear stress at which the fluid first starts to behave nonlinearly when stress is plotted against shear strain rate $\bar{\tau}_E$ was previously used by Jacobson and Hamrock (1984). For all the results to be presented, the effective modulus of the solid materials was kept constant at $E' = 2.2 \times 10^{11}$ Pa. The results presented in Figs. 26.4(a) and (b) are for oil 2 of Table 26.2.

Figure 26.4(a) is the streamline pattern for a Newtonian EHL line contact where both surface boundaries have the same speed ($A_c = 0$). The lubricant used for the results presented in Fig. 26.4(a) was oil 2 of Table 26.2. The point S

TABLE 26.2
Properties of oils at 20°C
[From Lee and Hamrock (1990a)]

Property	Oil 1	Oil 2
Atmospheric viscosity η_0, Pa · s	0.0411	0.01326
Pressure-viscosity coefficient ξ, Pa^{-1}	2.276×10^{-8}	1.5816×10^{-8}
Shear strength proportionality constant γ^*	0.036	0.076
Maximum shear stress $\bar{\tau}_E$	9×10^{-5}	9×10^{-5}
Solidification pressure p_s, GPa	0.655	1.21

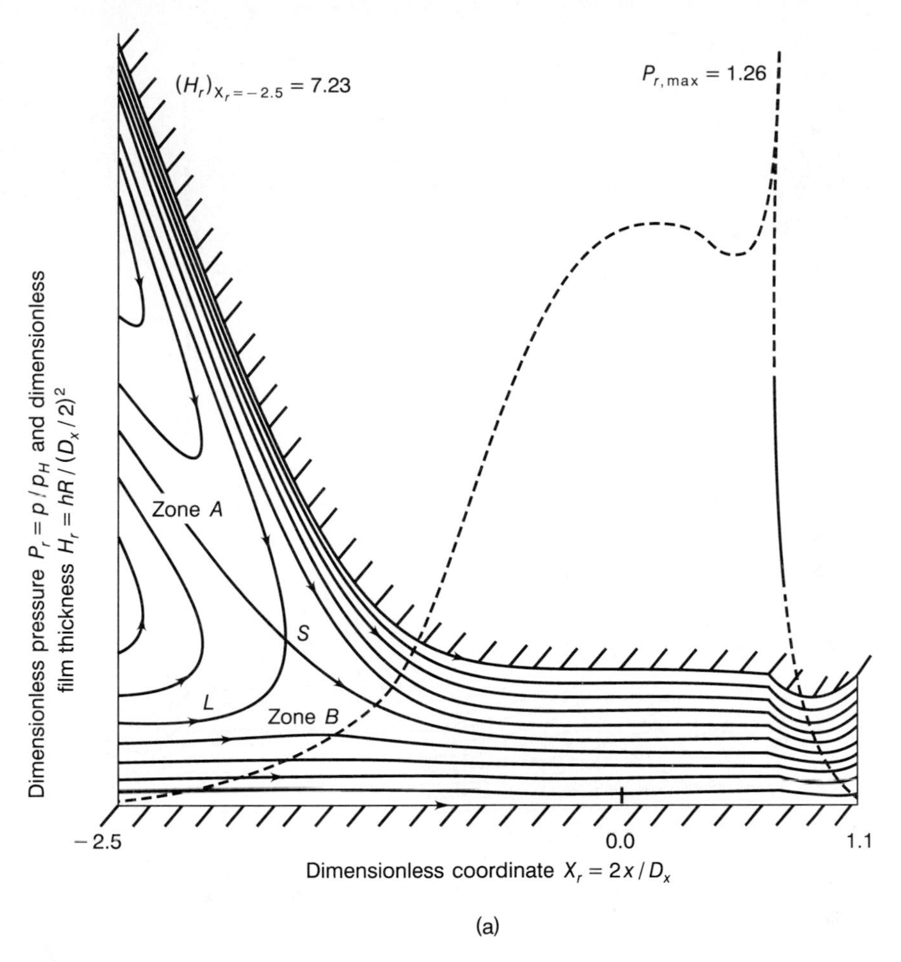

FIGURE 26.4
Pressure, film thickness, and streamline patterns in an EHL conjunction for oil 2 in (a) Newtonian and (b) non-Newtonian fluids. ($W' = 0.2 \times 10^{-4}$, $U = 1 \times 10^{-11}$, $G = 5007.2$, and $A_c = 0$.) [*From Shieh and Hamrock (1991).*]

in Fig. 26.4(a) is the stagnation point in the lubricated conjunction. The streamline L separates the whole lubricated conjunction into two regions: Zone A is the reverse-flow region, where two reverse-flow vortices are observed in this case; zone B is the pass-through region, where the lubricant can pass through the conjunction and has a lubricating effect. Thus, a large fraction of inlet lubricant will exhibit reverse flow in flooded EHL contacts.

The streamline pattern for a non-Newtonian line contact can be seen from Fig. 26.4(b). Comparison with Fig. 26.4(a) shows that the stagnation point S for a Newtonian fluid is closer to the inlet and that a greater fraction of inlet non-Newtonian lubricant is lubricating the conjunction. The reason is that the

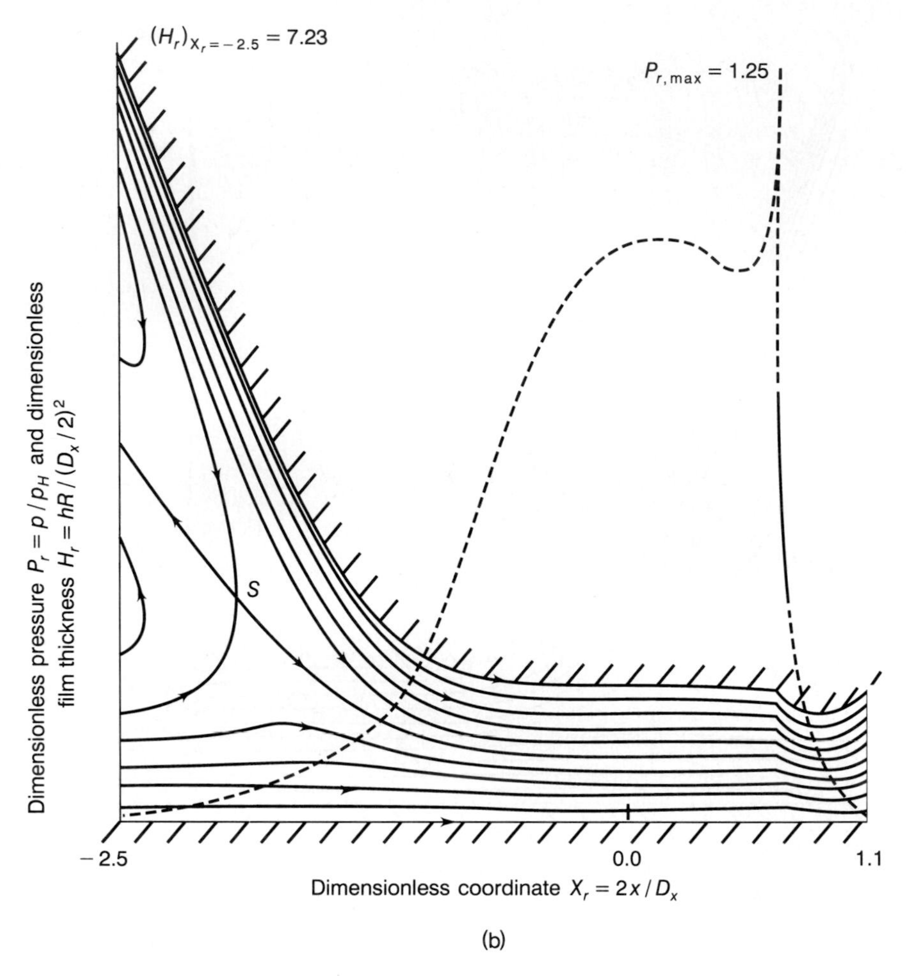

FIGURE 26.4 *Concluded.*

non-Newtonian effect makes the lubricant flow more easily under strong shear stresses.

The effects of various operating parameters, such as dimensionless speed parameter and slide-roll ratio, on the flow patterns in the lubricated conjunction for a non-Newtonian fluid are presented in Figs. 26.5 and 26.6. Figure 26.3 should be recalled to illustrate the physical situation for various slide-roll ratios. The effects of these parameters on suspected lubrication film breakdown were studied. The lubricant used for these studies was oil 2 of Table 26.2, for which the non-Newtonian properties can be obtained. Maintaining a fluid film of adequate magnitude is essential for the proper operation of lubricated machine elements. Examining the streamline pattern in the conjunction can lead to an understanding of how films collapse under certain operating conditions.

The effect of decreasing the dimensionless speed from 1×10^{-11} to 0.4×10^{-11} while holding the other operating parameters fixed at $W' = 0.6 \times 10^{-4}$, $G = 3480$, and $A_c = 0$ is shown in Fig. 26.5. As the speed decreases, the film thickness decreases and the stagnation point moves farther into the conjunction, thus resulting in more reverse flow. The movement of the stagnation point toward the conjunction as the speed decreases can only be observed with the aid of streamline patterns. This movement causes more lubricant reverse flow in the inlet region. It can be speculated from these results

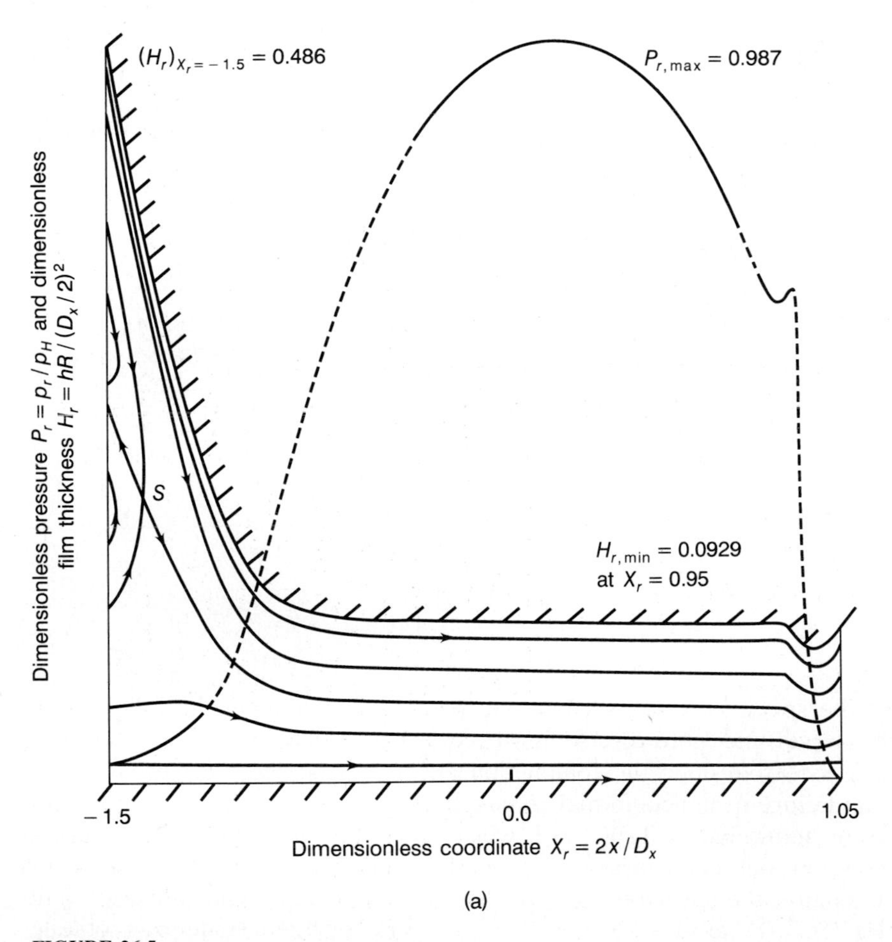

FIGURE 26.5
Pressure, film thickness, and streamline patterns in an EHL conjunction for oil 2 when the operating parameters are fixed at $W' = 0.6 \times 10^{-4}$, $G = 3480$, and $A_c = 0$. (a) $U = 1 \times 10^{-11}$; (b) $U = 0.4 \times 10^{-11}$. [*From Shieh and Hamrock, (1991).*]

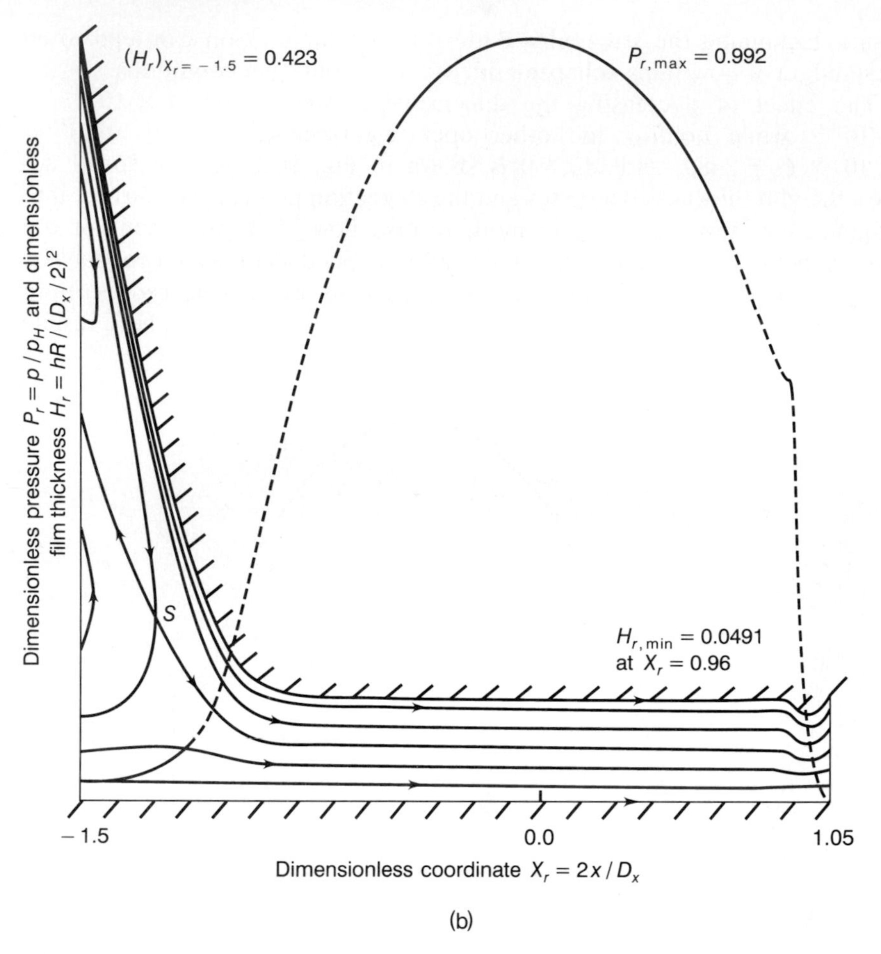

FIGURE 26.5 *Concluded.*

that as the speed becomes smaller, the pressure gradient in the inlet region becomes larger and more reverse flow occurs. This speculation implies that less lubricant passes through the conjunction and may describe the mechanism of fluid film failure in nonconformal conjunctions. These speed results along with others are quantified in Table 26.3, which shows the minimum film thickness and the stagnation point location for five slide-roll ratios.

Changing the slide-roll ratio A_c has the following physical effects: Note from Eq. (26.19) that $A_c = 0$ implies that $u_a = u_b$, or that cylinder a is rotating in a clockwise direction with the same magnitude as cylinder b while cylinder b is rotating in a counterclockwise direction. Also $A_c = 2$ implies that $u_a = 0$. Furthermore, $A_c = 10$ implies that $u_a = -2u_b/3$, or that cylinder a is rotating in a clockwise direction at a magnitude of $2u_b/3$ while cylinder b is also rotating in a clockwise direction. Finally, note that $A_c \to \infty$ implies that $u_a \to -u_b$, or that the cylinders are rotating with the same magnitude and in the same direction. This condition of $A_c \to \infty$ might be physically simulated by the motion between rollers in a cageless cylindrical roller bearing.

Figures 26.5(a) and 26.6 show the effect of changing the slide-roll ratio on the pressure, film thickness, and streamlines throughout the conjunction when the other operating parameters are held fixed at $W' = 0.6 \times 10^{-4}$, $U = 1 \times 10^{-11}$, and $G = 3480$. In going from $A_c = 0$ [Fig. 26.5(a)] to $A_c = 2$ [Fig. 26.6(*a*)] the pressure and the film thickness do not change appreciably but the streamlines do change. For $A_c = 2$, or a stationary lower surface, the stagnation point S moves to a lower surface. Figures 26.6(b) and (c) depict the pressure,

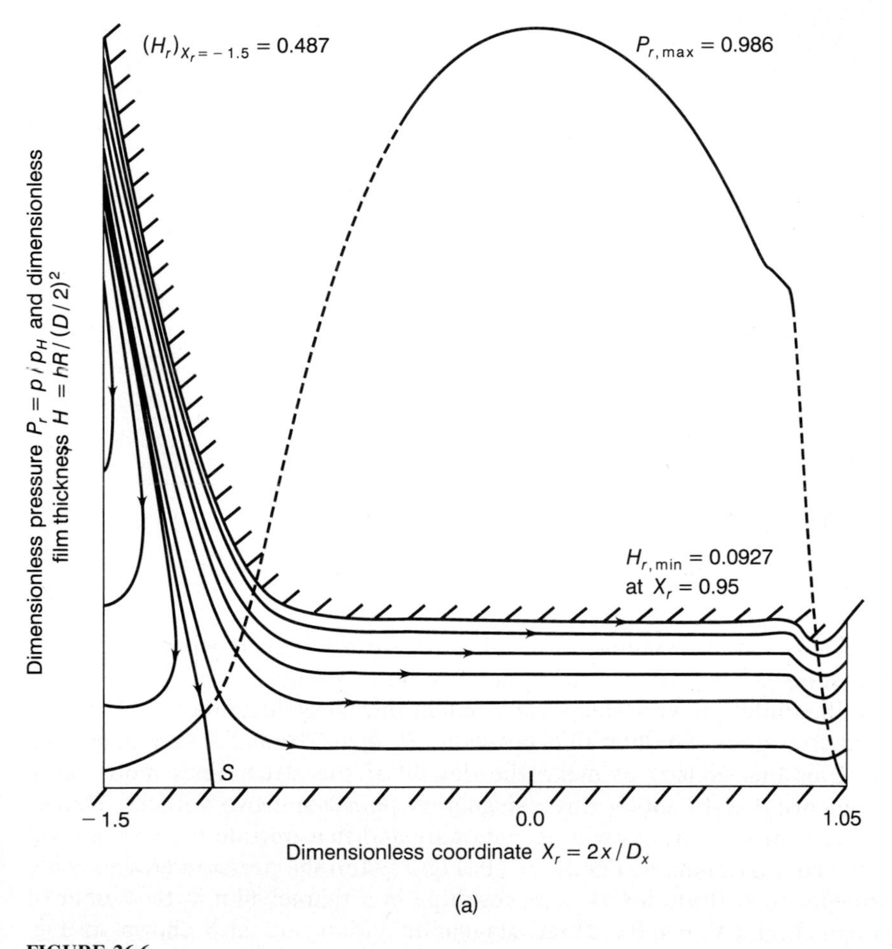

FIGURE 26.6
Pressure, film thickness, and streamline patterns in an EHL conjunction for oil 2 when the operating parameters are fixed at $W' = 0.6 \times 10^{-4}$, $U = 1 \times 10^{-11}$, and $G = 3480$. (a) $A_c = 2$; (b) $A_c = 5$; (c) $A_c = 10$. [*From Shieh and Hamrock (1991).*]

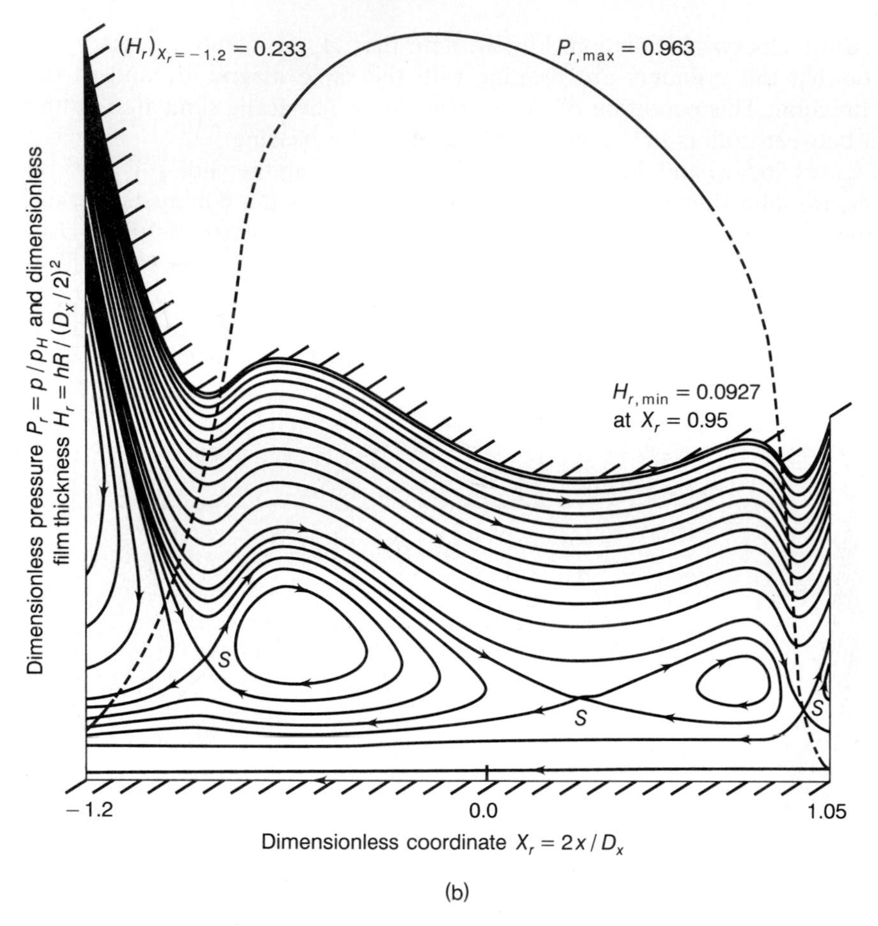

FIGURE 26.6 *Continued*

film thickness, and streamlines when the two surfaces are moving in opposite directions. For $A_c = 5$ [Fig. 26.6(b)] the pressure gradients in the inlet as well as those in the outlet are very steep, causing film thickness fluctuations at both the inlet and the outlet. The film thicknesses in Figs. 26.6(b) and (c) are magnified over that in Fig. 26.6(a) to make the details of the streamlines more easily visible. Figure 26.6(b) shows three stagnation points and two vortices. Hence, the lubricant at the vortices will be stationary and that outside the vortices will move in a circular manner. For $A_c = 10$ [Fig. 26.6(c)] the pressure gradients are even steeper than those for $A_c = 5$, resulting in a thinner film in the center of the conjunction ($X_r = 0.0$). Three stagnation points are also shown in Fig. 26.6(c), but the streamlines vary much more than those in Fig. 26.6(b). The results presented in Figs. 26.5(a) and 26.6 are quantified in Table 26.4. For $A_c = 10$ the film thickness decreases considerably from the $A_c = 5$ results,

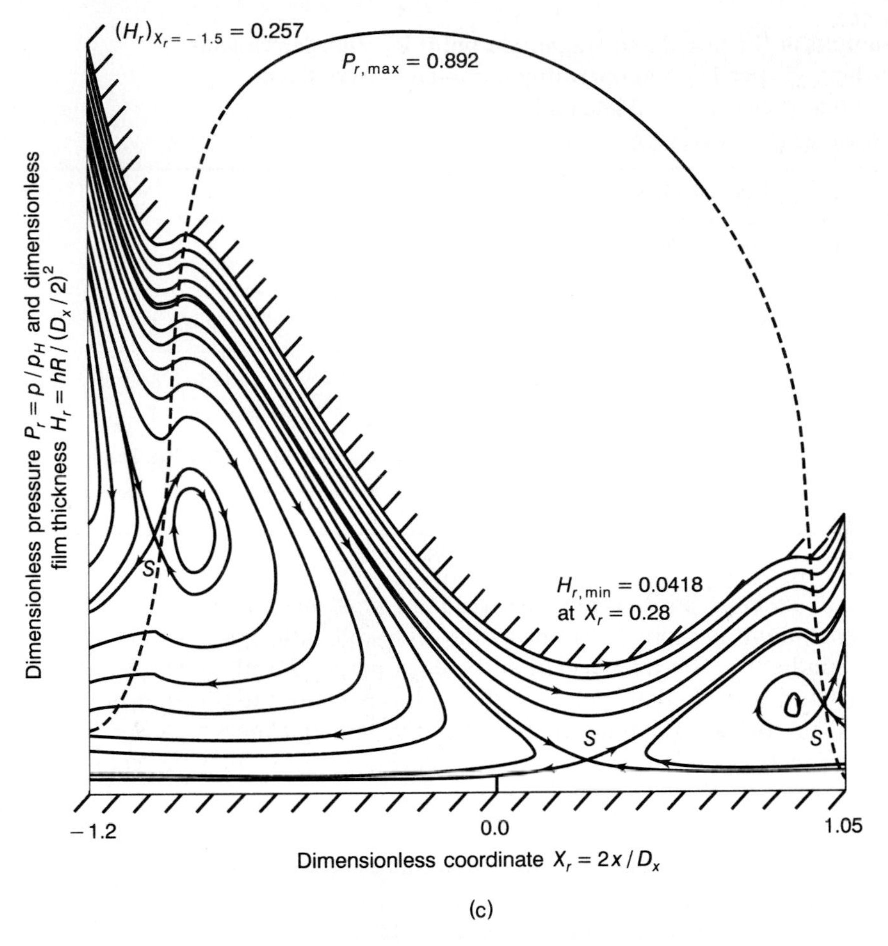

FIGURE 26.6 *Concluded.*

whereas in going from $A_c = 0$ to $A_c = 5$, very little change occurs. If the slide-roll ratio were to increase above 10, it is anticipated that further film collapse would take place. This drastic change in the film thickness is due to the non-Newtonian effect. If Newtonian lubricant is considered, the film shape and pressure profile will be the same for various slide-roll ratios while the mean velocity $\tilde{u}$ and other parameters are held fixed.

When the slide-roll ratio is larger than 2 (i.e., the lower surface is moving in the opposite direction to the upper surface), the shear stress at the upper surface is higher than that for slide-roll ratios less than 2 at the inlet region (around $X_r = 1.0$), and the non-Newtonian effect is more dominant. If the stress is near the limiting shear stress, "plastic" fluid flow occurs and the lubricant can more easily enter the conjunction. To satisfy the continuity

TABLE 26.3
Minimum film thickness and stagnation point location for various dimensionless speeds when operating parameters were fixed at $W' = 0.6 \times 10^{-4}$, $G = 3480$, and $A_c = 0$
[From Shieh and Hamrock (1991)]

Dimensionless speed parameter U, $\times 10^{-11}$	Dimensionless minimum film thickness for rectangular conjunctions $H_{r,\,min}$	Minimum film thickness ($R = 0.01$ m) h_{min}, μm	Dimensionless stagnation point location for rectangular coordinates $X_{r,\,s}$	Remarks
0.4	0.0491	0.075	−1.250	See Fig. 26.4(b)
0.6	0.0652	0.100	−1.302	
0.8	0.0796	0.121	−1.348	
1.0	0.0929	0.142	−1.385	See Fig. 26.4(a)
2.0	0.1491	0.277	−1.521	
4.0	0.2406	0.368	−1.709	

condition, a steeper positive pressure gradient must exist at the inlet region. This phenomenon is more apparent as the slide-roll ratio goes beyond 2. The pressure gradient at the inlet region becomes larger, and a nip is formed at the inlet which is similar to that formed at the outlet. Because the pressure profile is flatter in the conjunction than at the inlet, the film thickness is lower at the center conjunction region than at the inlet region in order that the continuity

TABLE 26.4
Minimum film thickness and stagnation point location for various slide-roll ratios when operating parameters were fixed at $W' = 0.6 \times 10^{-4}$, $U = 1 \times 10^{-11}$, and $G = 3480$
[From Shieh and Hamrock (1991)]

Slide-roll ratio A_c	Dimensionless minimum film thickness for rectangular conjunctions $H_{r\,min}$	Minimum film thickness ($R = 0.01$ m) h_{min}, μm	Dimensionless stagnation point location for rectangular coordinates $X_{r,\,s}$	Remarks
0	0.0929	0.142	−1.385	See Fig. 26.5(a)
1	0.0928		−1.354	
2	0.0927		−1.115	See Fig. 26.6(a)
5	0.0927		−0.854	See Fig. 26.6(b)
			0.297	
			0.962	
10	0.0418	0.064	−1.008	See Fig. 26.6(c)
			0.282	
			0.986	

equation be satisfied. Thus, two vortices are formed near the two nip cavities. Film failure may thus be accompanied by extremely steep pressures in the inlet so that the lubricant is unable to pass through the conjunction, and the film thickness is drastically reduced.

26.5 STATIONARY ASPERITY CONSIDERATION

In micro-EHL with a stationary surface irregularity in the lubricated conjunction, the modified Reynolds equation [Eq. (26.35)] and the stream function [Eq. (26.28)] are still valid. The only difference is the expression of the film thickness

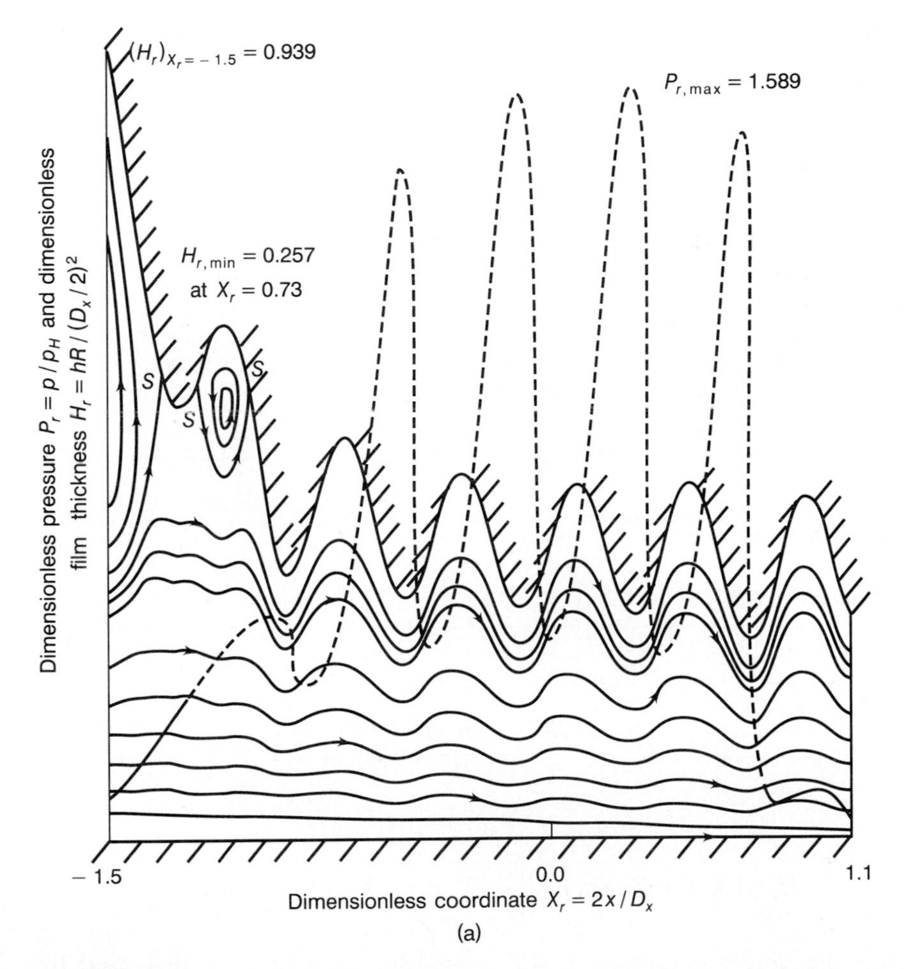

FIGURE 26.7
Pressure, film thickness, and streamline patterns in a micro-EHL conjunction for oil 2 when the operating parameters are fixed at $U = 1 \times 10^{-11}$, $G = 3480$, $A_c = -2$, $Z_m = 0.2$, and $X_{r,w} = 0.1$. (a) $W' = 0.2 \times 10^{-4}$; (b) $W' = 0.6 \times 10^{-4}$. [*From Shieh and Hamrock (1991)*.]

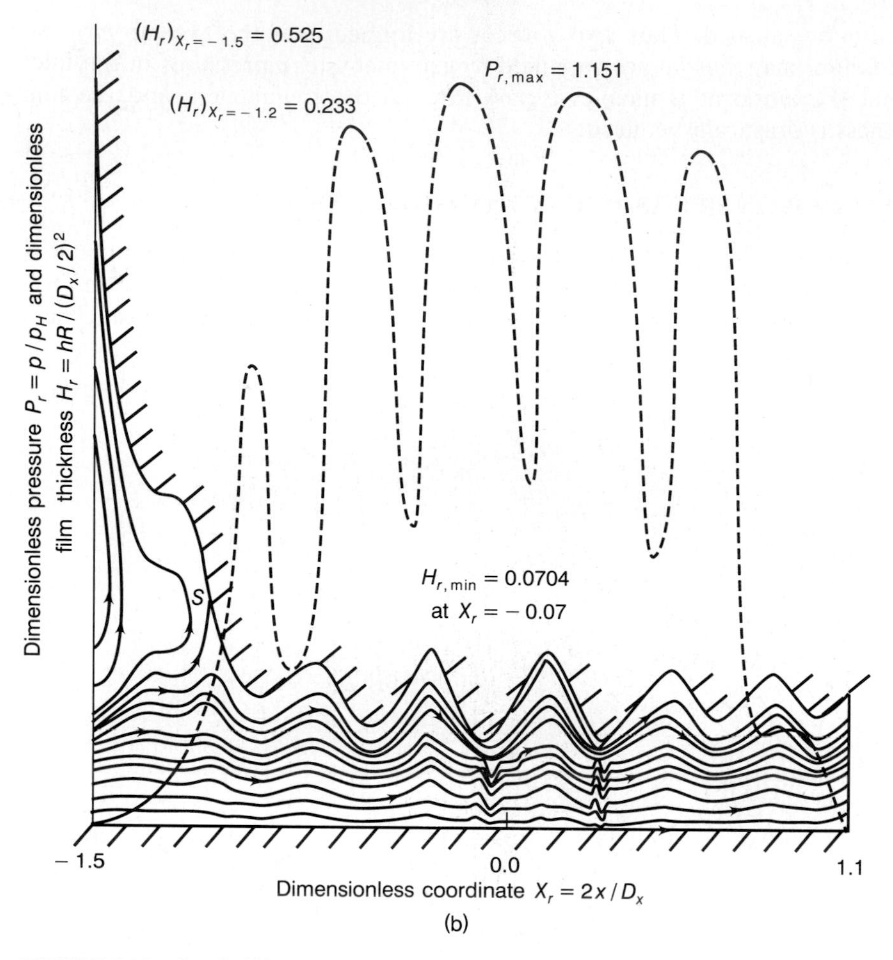

FIGURE 26.7 *Concluded.*

equation

$$H_r = H_{r,0} + \frac{X_r^2}{2} + \frac{4R_x\delta}{D_x^2} + H_{r,R}(X_r) \tag{26.37}$$

where

$$H_{r,R}(X_r) = Z_m \sin\left(\frac{\pi}{2X_{r,w}}X_r\right) \qquad Z_m = \frac{z_m R}{(D_x/2)^2}$$

and δ is the elastic deformation, which has the same form as that used by Houpert and Hamrock (1986). Thus, a sinusoidal surface waviness is assumed to simulate the stationary surface.

Figure 26.7 shows the pressure, film thickness, and streamlines in a microelastohydrodynamically lubricated conjunction. Note that for both parts

$A_c = -2$, which implies that the upper surface is stationary. Also the lubricant used for this figure was oil 2 from Table 26.2. Only the dimensionless load W' changes between Fig. 26.7(a) and (b). For the lighter load [Fig. 26.7(a)] a vortex occurs in the inlet; for the heavier load [Fig. 26.7(b)] no vortex occurs. That is, at light loads most of the lubricant conforms to the surface irregularity except in the inlet. The lubricant is trapped at the vortex, and if wear particles were to enter the inlet region close to the stationary surface, these particles would not flow out but would circulate in the groove. Increasing the load makes the amplitude of the surface irregularity flatter, as shown in Fig. 26.7(b). The vortex in the inlet disappears, but some tiny vortices are still observed adjacent to the groove in the conjunction.

Figure 26.8 shows the pressure, film thickness, and streamlines for a larger wavelength of surface roughness than that shown in Fig. 26.7(b). No vortex occurs in Fig. 26.8 either, but the streamlines conform more closely to the

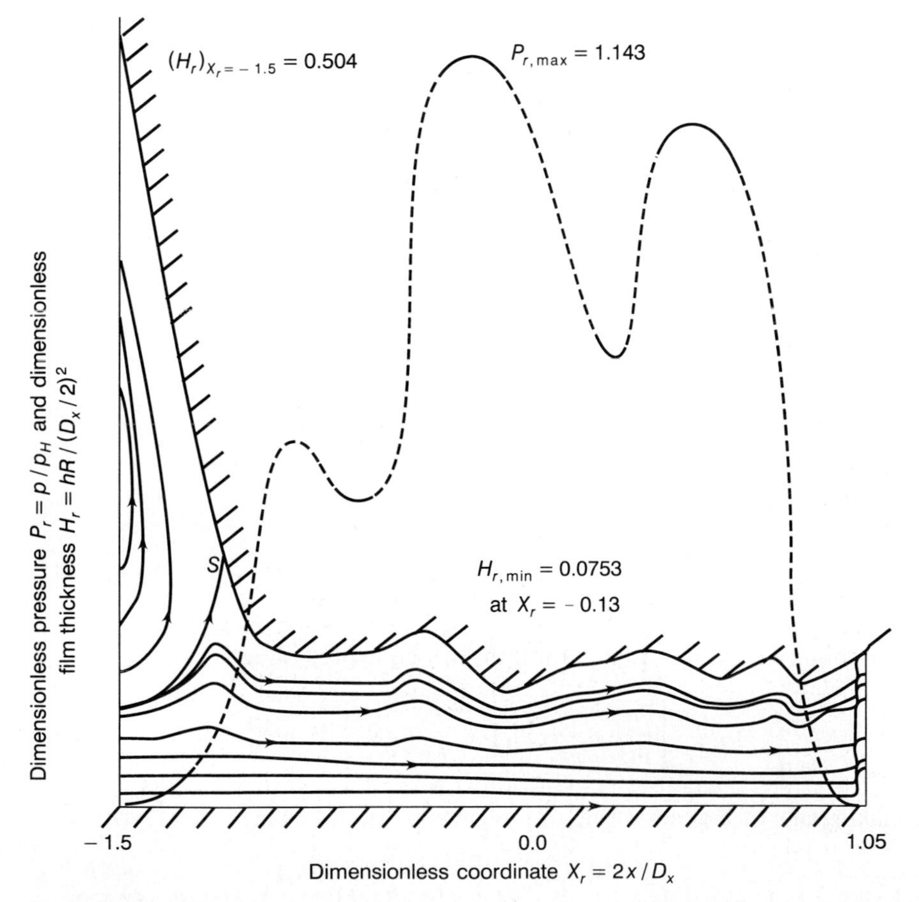

FIGURE 26.8
Pressure, film thickness, and streamline patterns in a micro-EHL conjunction for oil 2 when the operating patterns are fixed at $W' = 0.6 \times 10^{-4}$, $U = 1 \times 10^{-11}$, $G = 3480$, $A_c = -2$, $Z_m = 0.2$ and $X_{r,w} = 0.2$ [*From Shieh and Hamrock (1991).*]

surface texture. Thus, increasing the wavelength, which makes the surface profile close to a "smooth" assumption, may enable the lubricant to pass through the conjunction more easily. Note that for identical conditions, rough surfaces develop thinner films than smooth surfaces. Thus, surface contact caused by film breakdown is more likely to occur in micro-EHL than in EHL.

26.6 MODIFIED TRANSIENT REYNOLDS EQUATION

For the problem of a moving surface irregularity the modified Reynolds equation (26.35) is no longer valid. In this condition the transient effect must be considered. The modified transient Reynolds equation for line contacts can be derived from the reduced form of the Navier-Stokes and continuity equations. By the same procedure used in Sec. 26.2 the velocity component along the x axis becomes

$$\frac{u}{\tilde{u}} = 1 - \frac{A_c}{2} + \frac{1}{S_c \tilde{T}} \left[\sqrt{1 - \bar{\tau}_a^2} - \sqrt{1 - \left(\bar{Z}\tilde{T} + \bar{\tau}_a \right)^2} \right] \tag{26.38}$$

The mass flow rate per unit width in the x direction is defined as

$$q_x' = \int_{h_a}^{h_b} \rho u \, dz \tag{26.39}$$

Making use of Eq. (26.20) gives

$$q_x' = \rho h \tilde{u} - \left[1 + \frac{1}{2 S_c \tilde{T}^2} (e - \sin^{-1} e) \right] \tag{26.40}$$

or

$$q_x' = \rho h \tilde{u} - \frac{\bar{\eta}}{\bar{\eta}_c} \frac{\rho h^3}{12\eta} \frac{\partial p}{\partial x} \tag{26.41}$$

The continuity equation in integral form can be expressed as

$$\int_{h_a}^{h_b} \left[\frac{\partial \rho}{\partial t} + \frac{\partial}{\partial x}(\rho u) + \frac{\partial}{\partial z}(\rho w) \right] dz = 0 \tag{26.42}$$

By making use of a general rule of integration, the following can be written:

$$\int_{h_a}^{h_b} \frac{\partial}{\partial x} [f(x, z, t)] \, dz = \frac{\partial}{\partial x} \int_{h_a}^{h_b} f \, dz - f(x, h_b, t) \frac{\partial h_b}{\partial x} + f(x, h_a, t) \frac{\partial h_a}{\partial x} \tag{26.43}$$

Thus, the integrated continuity equation becomes

$$\frac{\partial}{\partial x}\int_{h_a}^{h_b}\rho u\,dz - \rho u_b\frac{\partial h_b}{\partial x} + \rho u_a\frac{\partial h_a}{\partial x} + \rho(w_b - w_a)$$

$$+\frac{\partial}{\partial t}(\rho h) - \rho\frac{\partial h_b}{\partial t} + \rho\frac{\partial h_a}{\partial t} = 0 \tag{26.44}$$

The surface velocities for both cylinders in a machine element are related to the translation in the z direction and the rotation about its own center. If the lower surface is assumed to be fixed and can only rotate, the surface velocities can be expressed as

$$w_a = u_a\frac{\partial h_a}{\partial x} \tag{26.45}$$

$$w_b = u_b\frac{\partial h_b}{\partial x} + \frac{\partial h}{\partial t} \tag{26.46}$$

Berthe and Godet (1974) indicate that these forms are only valid for the condition where the surface velocity envelopes and surface profiles coincide, or $h_a = 0$. When Eqs. (26.45) and (26.46) and the mass flow rate expression (26.41) are introduced into Eq. (26.44), the modified transient Reynolds equation becomes

$$\frac{\partial}{\partial x}\left(\rho h\tilde{u} - \frac{\bar{\eta}}{\bar{\eta}_c}\frac{\rho h^3}{12\eta}\frac{\partial p}{\partial x}\right) + \frac{\partial}{\partial t}(\rho h) = 0 \tag{26.47}$$

or in dimensionless form

$$\frac{\partial}{\partial X_r}\left(\frac{\bar{\rho}H_r^3}{12\bar{\eta}_c}\frac{\partial P_r}{\partial X_r}\right) = \bar{K}\frac{\partial\bar{\rho}H_r}{\partial X_r} + S^*\frac{\partial\bar{\rho}H_r}{\partial T'} \tag{26.48}$$

After integrating once, the integrated, dimensionless form of the transient Reynolds equation can be written as

$$H_r^3\frac{dP_r}{dX_r} = \bar{K}\bar{\eta}_c\left(\frac{H_{r,\mathrm{end}}}{\bar{\rho}} - H_r\right) + \frac{S^*\bar{\eta}_c}{\bar{\rho}}\int_X^{X_{\mathrm{end}}}\frac{\partial\bar{\rho}H_r}{\partial T'}\,dX_r = 0 \tag{26.49}$$

where the derivative of the transient term $\partial\bar{\rho}H_r/\partial T'$ is

$$\frac{\partial\bar{\rho}H_r}{\partial T'} = \frac{\partial\bar{\rho}H_{r,g}}{\partial T'} + \frac{\partial\bar{\rho}\delta}{\partial T'}$$

$$= \bar{\rho}_m\left(\frac{\partial H_{r,g}}{\partial T'}\right)_m + (H_{r,g})_m\frac{\bar{\rho}_m - \bar{\rho}_{m-1}}{\Delta T'} + \frac{(\bar{\rho}\delta)_m - (\bar{\rho}\delta)_{m-1}}{\Delta T'}$$

$$= -\bar{\rho}_m\left(\frac{\pi W_0}{8W}\right)_m + \bar{\rho}_m\frac{(H_{r,RI})_m - (H_{r,RI})_{m-1}}{\Delta T'} + \frac{\bar{\rho}_m H_{r,m}}{\Delta T'}$$

$$-\frac{\bar{\rho}_{m-1}}{\Delta T'}(H_{r,m} + \delta_m - \delta_{m-1}) \tag{26.50}$$

and

$$H_{r,g} = H_{r,0} + \frac{X_r^2}{2} + H_{r,R}(X) - H_{r,R}(0) \tag{26.51}$$

$$H_{r,RI} = H_{r,R}(X) - H_{r,R}(0) \tag{26.52}$$

The transient term can be expressed as

$$\begin{aligned}\int_{X_{r,i}}^{X_{r,\text{end}}} \frac{\partial \bar{\rho} H_r}{\partial T'} dX_r = & -\frac{1}{2} \sum_{j=i}^{N} \left(\bar{\rho}_{j,m} \frac{\pi W_0}{8W} \right) \Delta X_{r,j} \\ & + \frac{1}{2\,\Delta T'} \sum_{j=i}^{N} \left[(H_{r,j})_m + (H_{r,RI_j})_m - (H_{r,RI_j})_{m-1} \right] \bar{\rho}_{j,m} \Delta X_{r,j} \\ & - \frac{1}{2\,\Delta T'} \sum_{j=i}^{N} \left\{ \bar{\rho}_{j,m-1} \left[(H_{r,j})_m + (\delta_{r,j})_{m-1} - (\delta_{r,j})_m \right] \right\} \Delta X_{r,j}\end{aligned} \tag{26.53}$$

This transient term is almost the same as the last derivative from the smooth surface except that the rigid surface is moving.

26.7 CLOSURE

Non-Newtonian fluid rheology effects in elastohydrodynamic lubrication were considered in this chapter. Various non-Newtonian models were presented in terms of the effective viscosity, which defines the actual viscosity experienced in elastohydrodynamically lubricated conjunctions. The circular model behaves more like a Newtonian model for dimensionless shear stresses less than 0.5 and then becomes highly non-Newtonian for dimensionless shear stresses greater than 0.5. The circular model is believed to behave more like actual fluids and was the model used throughout the chapter.

The pressure, film thickness, and streamline patterns in EHL and micro-EHL conjunctions were presented. The non-Newtonian viscous behavior of the lubricant was taken into account. The circular non-Newtonian fluid model was used to derive the modified Reynolds equation and the stress function of the fluid. In micro-EHL problems a periodic surface waviness was assumed. The phenomenon of film collapse in the lubricated conjunction was studied from streamline contours. The following conclusions were drawn:

1. As the speed decreased, the film thickness decreased and the stagnation point moved farther into the conjunction. As a result, less of the inlet lubricant was able to pass through the conjunction.
2. When the slide-roll ratio A_e was between 0 and 2, the film thickness only changed slightly. But when A_c was larger than 2, a nip at the inlet (around $X_r = -1.0$), three stagnation points, and two vortices were observed in the

conjunction. These results are due to the non-Newtonian effect. It was also concluded that the film will eventually collapse at the central region.

3. In micro-EHL a vortex formed in the inlet groove if the load was low. Increasing the load or enlarging the wavelength of surface waviness decreased the amplitude of the surface irregularity and thus smoothed the flow pattern in the conjunction. The minimum film thickness in micro-EHL was less than that in EHL. Hence, film collapse is more likely to occur in micro-EHL than in EHL.

The procedure for solving the combined entraining and normal squeeze motion in EHL and micro-EHL line contacts was also presented in this chapter.

26.8 PROBLEM

26.8.1 Given the general non-Newtonian fluid model in Eq. (26.6), establish to the first decimal place the value of n in the general model that approximates the Bair and Winer (1979) model as well as the Gecim and Winer (1980) model.

26.9 REFERENCES

Bair, S., and Winer, W. O. (1979): A Rheological Model for Elastohydrodynamic Contacts Based in Primary Laboratory Data. *J. Lubr. Technol.*, vol. 101, no. 3, pp. 258–265.

Berthe, D., and Godet, M. (1974): A More General Form of Reynolds Equation—Application to Rough Surfaces. *Wear*, vol. 27, no. 3, pp. 345–357.

Conry, T. F., Wang, S., and Cusano, C. (1987): "A Reynolds-Eyring Equation for Elastohydrodynamic Lubrication in Line Contacts," *ASME J. of Tribol.*, vol. 109, pp. 648–658.

Elsharkawy, A. A., and Hamrock, B. J. (1991): Subsurface Stresses in Micro-EHL Line Contacts. *J. Tribol.*, vol. 113, no. 3, pp. 645–656.

Eyring, H. (1936): Viscosity, Plasticity, and Diffusion as Examples of Absolute Reaction Rates. *J. Chem. Phys.*, vol. 4, pp. 283–291.

Gecim, B., and Winer, W. O. (1980): Lubricant Limiting Shear Stress Effect on EHD Film Thickness. *J. Lubr. Technol.*, vol. 102, p. 213.

Hirst, W., and Moore, A. J. (1974): "Non-Newtonian Behavior in Elastohydrodynamic Lubrication," *Proc. Roy. Soc., London*, vol. 33, pp. 101–113.

Höglund, E. (1984): "Elastohydrodynamic Lubrication—Interferometric Measurements, Lubricant Rheology and Subsurface Stresses." Doctoral thesis 1984–32D. Luleå University of Technology, Sweden.

Houpert, L., and Hamrock B. J. (1985): Elastohydrodynamic Lubrication Calculations Used as a Tool to Study Scuffing. Proceedings of the Twelfth Leeds-Lyon Symposium on Tribology. Mechanical Engineering Publications, Bury St. Edmunds, Suffolk, England.

Houpert, L., and Hamrock, B. (1986): Fast Approach for Calculating Film Thickness and Pressure in Elastohydrodynamically Lubricated Contacts at High Loads. *J. Tribol.*, vol. 108, no. 3, pp. 411–420.

Hsiao, H. S., and Hamrock, B. J. (1992): "A Complete Solution for Thermal-Elastohydrodynamic Lubrication of Line Contacts Using the Circular Non-Newtonian Fluid Model," *ASME J. Tribol.*, vol. 114, no. 3, pp. 540–552.

Iivonen, H. T., and Hamrock, B. J. (1989): A Non-Newtonian Fluid Model for Elastohydrodynamic Lubrication of Rectangular Contacts. Proceedings of the Fifth International Congress on Tribology. Helsinki, Finland, June 1989, pp. 178–183.

Jacobson, B. O., and Hamrock, B. J. (1984): Non-Newtonian Fluid Model Incorporated Into Elastohydrodynamic Lubrication of Rectangular Contacts. *J. Tribol.*, vol. 106, no. 2, pp. 275–284.

Johnson, K. L., and Teraareverk, J. L. (1977): "Shear Behavior of EHD Oil Films," *Proc. Roy. Soc., London*, vol. 356, pp. 215–236.

Lee, R. T., and Hamrock, B. J. (1990a): A Circular Non-Newtonian Fluid Model: Part I—Used in Elastohydrodynamic Lubrication. *J. Tribol.*, vol. 112, no. 3, pp. 486–496.

Lee, R. T., and Hamrock, B. J. (1990b): A Circular Non-Newtonian Fluid Model: Part II—Used in Microelastohydrodynamic Lubrication. *J. Tribol.*, vol. 112, no. 3, pp. 497–505.

Myllerup, C. M., Elsharkawy, A. A., and Hamrock, B. J. (1993): Couette Dominance Used for Non-Newtonian Elastohydrodynamic Lubrication. J. Tribol. Paper 93–TRIB–5.

Shieh, J. A., and Hamrock, B. J. (1991): Film Collapse in EHL and Micro-EHL. *J. Tribol.*, vol. 113, no. 2, pp. 372–377.

Trachman, E. G. (1971): "The Rheological Effects of Friction in EHL," Ph.D. thesis, Northwestern University, Evanston, IL.

Wang, J., and Zhang, H. H. (1992): "A Higher Order Perturbational Approach in Lubricated EHL Contacts with Non-Newtonian Lubricant," *ASME J. of Tribol.*, vol. 114, pp. 95–99.

CHAPTER 27

THERMAL ELASTOHYDRODYNAMIC LUBRICATION

A complete solution is obtained in this chapter for elastohydrodynamically lubricated conjunctions in line contacts when the effects of temperature and the non-Newtonian characteristics of lubricants with limiting shear strength are considered. The fast approach developed in Chap. 21 is used to solve the thermal Reynolds equation by using the complete circular non-Newtonian fluid model and considering both velocity and stress boundary conditions. The reason and the occasion for incorporating stress boundary conditions in the circular model are discussed. A conservative form of the energy equation is developed by using the finite control volume approach. Analytical solutions for solid surface temperatures that take into consideration two-dimensional heat flow within the solids are used. A straightforward finite difference method called "successive overrelaxation by lines" is employed to solve the energy equation. Results of thermal effects on film shape, pressure profile, streamlines, and friction coefficient are presented.

27.1 HISTORICAL BACKGROUND

Elastohydrodynamic lubrication analysis that uses isothermal Newtonian fluid models has been well developed. The film shape and the pressure distribution have been reasonably well predicted. However, the traction could not be anticipated correctly. The need to incorporate the thermal effect, the non-Newtonian effect, or both was recognized long ago (e.g., by Cheng, 1965). Since then, many numerical analyses that considered either thermal effects on EHL with Newtonian lubricants or the effects of non-Newtonian lubricants on isothermal EHL conjunctions have been performed (e.g., by Dowson and Whitaker, 1965–66; Zhu and Wen, 1984; Ghosh and Hamrock, 1985; Sadeghi and Dow, 1987; and Lee and Hamrock, 1990). Some preliminary studies on both the effects of temperature rises and the characteristics of non-Newtonian fluids were conducted with various levels of numerical convergence difficulty (e.g., by Conry, 1981; Wang and Zhang, 1987; and Wang et al., 1991). With the recent discovery of the limiting shear strength of lubricants by Bair and Winer (1979), different rheological models have been proposed to characterize the behavior of the non-Newtonian fluids with limiting shear strength by Gecim and Winer (1980) and Lee and Hamrock (1990). More details about non-Newtonian fluid rheology models can be found in Sec. 26.1. The formulation and results presented in this chapter originate from Hsiao and Hamrock (1992), and to my knowledge, no other results have ever been published of an EHL analysis that incorporates the effects of two-dimensionally distributed temperature rises and the characteristics of non-Newtonian fluids with limiting shear strength.

The traditional thermal EHL models used the nonconservative form of the energy equation. This is believed to be the major cause of the difficulty in solving the numerical convergence problem that has been encountered, for example, by Cheng (1967), Conry (1981), Zhu and Wen (1984), and Wang et al. (1991). It meant that some delicate numerical procedures needed to be implemented to obtain a converged solution, especially when the reverse-flow zones were analyzed. The work presented in this chapter represents a conservative form of the energy equation that was solved by using a straightforward scheme of successive overrelaxation by lines, a popular finite difference method. No extra effort is needed to obtain a convergence solution for the entire conjunction including the inlet, Hertzian contact, and outlet zones.

In addition, the temperatures on the surfaces of the bounding solids have traditionally been approximated by using models based on the flash temperature concept proposed by Blok and mentioned by Archard (1958–59). According to this concept, one-dimensional heat flow perpendicular to the solid surface was assumed, and a high Peclet number of heat flow in the solid had to be imposed to keep the model valid. This constraint hinders analysis of a system with any bounding solid that is approaching its stationary state (e.g., with a slide-roll ratio near 2). In this work a general two-dimensional heat flow model is used to calculate the temperatures on the surfaces of the solids in contact. The analysis of a system with a slide-roll ratio approaching 2 becomes available.

In solving the Reynolds equation for EHL problems, Okamura (1982) introduced the Newton-Raphson scheme to solve the highly nonlinear system equation. Houpert and Hamrock (1986) successfully developed a systematic approach for solving the elastic deformation and Reynolds equations simultaneously. This is the well-known "fast approach" covered in Chap. 21. Lee and Hamrock (1990) improved the fast approach by solving for the outlet location instead of the dimensionless flow rate as one of the system unknowns. This improvement further automated the fast approach. In deriving the Jacobian factors for the new system unknown, however, Lee and Hamrock neglected the effect of pressure distribution on the deformation at the exit. This is amended in Hsiao and Hamrock (1992) and in this chapter to fully expedite the solution scheme.

Lee and Hamrock (1990) have successfully used the circular model in analyzing isothermal EHL problems. Their work established the general procedure of using the circular model. However, its implementation was not completed in two senses. First, the derivation of the Jacobian factors was not complete, especially the effects of the non-Newtonian viscosity variable on the dimensionless offset film thickness, one of the system unknowns. This incompleteness slowed down the convergence of the calculation and probably prevented a completely convergent solution. The formulation presented in this chapter as well as in Hsiao and Hamrock (1992) gives the derivation of the Jacobian factors and eliminates the potential difficulty of convergence. Second, as mentioned by Wilson and Huang (1990), stress boundary conditions may have to be substituted for the velocity boundary condition to continue the computation and obtain a converged solution. However, in contradiction to Wilson and Huang, the need to switch boundary conditions is not due to the circular model theory but to the inexact representation of a real number by a digital computer. When the product of the viscosity and the shear strain rate is greater than a certain amount, the value of the boundary shear stress calculated by the digital computer may reach or exceed the limiting shear strength even though the theoretical value of the shear stress is still less than the limiting shear strength. Numerical procedures are developed in the current work to remedy this shortcoming of digital computation.

Interest in film collapse in an EHL conjunction was raised by Shieh and Hamrock (1991). How temperature rises in the flow field affect film collapse has been pondered since then. This chapter attempts to answer this question. A major contribution to this chapter comes from Hsiao and Hamrock (1992).

27.2 THEORETICAL AND NUMERICAL SCHEMES

In solving a thermal EHL problem, two systems of equations that depend on each other must be solved (i.e., the thermal Reynolds equation and the energy equation). The coupling of these two systems of equations results from the pressure-temperature dependence of the lubricant properties. The resultant film thickness of the conjunction involves the elastic deformation of the bounding solids. The elastic deformation and the field pressure also depend on each other. A systematic approach has been developed by Houpert and Hamrock (1986) for solving for the elastic deformation and the field pressure simultaneously (see Chap. 21). The use of an asymptotic limiting-shear-stress model such as the circular model occasionally requires imposing various stress boundary conditions. Different Reynolds equations result from this consideration. Moreover, the resultant field temperature and the surface temperature of the bounding solids also depend on each other. A systematic approach developed by

Hsiao and Hamrock (1992) is used in this chapter to solve for the field temperature and the temperatures of the solid surfaces simultaneously. The following sections summarize the use of equations for the lubricant properties, the development of the thermal Reynolds equations that consider various boundary conditions, the derivation of a conservative form of the energy equation with a two-dimensional heat flow model for calculating solid surface temperatures, and the evolution of the systematic approach that solves for the field temperature and the surface temperature simultaneously. The numerical methods used to solve the individual equations and the coupling of the systems of equations are also described.

27.2.1 EQUATIONS FOR LUBRICANT PROPERTIES. The properties of any existing lubricant vary with pressure and temperature, some weakly, some strongly. Continuous functions are favored to describe these relationships in numerical models that simulate lubricant behavior. The following equations approximate some properties of lubricants as functions of pressure, temperature, or both. Those properties not mentioned here are considered to be constant.

First, the force density of a lubricant is represented as the multiplication of the force density at atmospheric conditions ρ_0; the dimensionless piezodensity $\bar{\rho}_p$, a dimensionless correction factor for force density that considers the pressure effect; and the dimensionless thermodensity $\bar{\rho}_T$, a dimensionless correction factor for force density that considers the temperature effect.

$$\rho = \rho_0 \bar{\rho}_p \bar{\rho}_T \tag{27.1}$$

where

$$\bar{\rho}_p = 1 + \frac{\alpha_{\rho,1} p}{1 + \alpha_{\rho,2} p} \tag{27.2}$$

$$\bar{\rho}_T = 1 - \beta_\rho \frac{t_m - t_{m,0}}{t_{m,0} - t_{m,R}} \tag{27.3}$$

and $\alpha_{\rho,1}, \alpha_{\rho,2}$ = piezodensity constants, m^2/N
p = field pressure, N/m^2
β_ρ = dimensionless thermodensity constant
$t_{m,R}$ = reference temperature used to convert field temperature and ambient temperature to absolute temperature scale, °C
$t_m, t_{m,0}$ = field temperature and ambient temperature of lubricant, respectively, °C

By denoting the average relative thermodensity across the film thickness as

$$\bar{\rho}_{T,\mathrm{avg}} = \frac{1}{h} \int_{h_a}^{h_b} \bar{\rho}_T \, dz = \int_0^1 \bar{\rho}_T \, dZ \tag{27.4}$$

where $h = h_b - h_a$ is the film thickness, the average force density of the lubricant across the film thickness can be written as

$$\rho_{\mathrm{avg}} = \rho_0 \bar{\rho}_{\mathrm{avg}} = \rho_0 \bar{\rho}_p \bar{\rho}_{T,\mathrm{avg}} \tag{27.5}$$

Richmond et al. (1984) indicated that the effect of extreme pressure on the thermal conductivity of a lubricant in EHL may not be negligible. This effect of pressure on thermal conductivity has been implemented in thermal EHL calculations by Wang et al. (1991). It is considered in this chapter in a similar manner. That is, the thermal conductivity of a lubricant is expressed as the multiplication of the thermal conductivity at atmospheric conditions K_0 and the dimensionless piezothermal conductivity $\overline{K}_b$.

$$K_b = K_0 \overline{K}_b \tag{27.6}$$

where

$$\overline{K}_b = 1 + \frac{\alpha_{\kappa,1} p}{1 + \alpha_{\kappa,2} p} \tag{27.7}$$

and $\alpha_{\kappa,1}$ and $\alpha_{\kappa,2}$ are the piezothermal conductivity constants. Because the thermal conductivity of a lubricant is assumed to be independent of temperature and the pressure is assumed to be constant across the film thickness, the thermal conductivity is constant across the film thickness.

For determining the pressure-temperature-viscosity effect the formula developed by Roelands et al. (1963) is adopted with modification to ensure that the viscosity reduces to the extreme-temperature viscosity η_{t_∞} as the extrapolated temperature approaches infinity. In short, the viscosity is denoted as the product of the viscosity at atmospheric conditions η_0, the dimensionless piezoviscosity $\overline{\eta}_p$, and the dimensionless thermoviscosity $\overline{\eta}_T$.

$$\eta = \eta_0 \overline{\eta}_p \overline{\eta}_T \tag{27.8}$$

where

$$\overline{\eta}_p = \left(\frac{\eta_0}{\eta_\infty} \right)^{(1+p/c_p)^{Z_1} - 1} \tag{27.9}$$

$$\overline{\eta}_T = \left(\frac{\eta_0 \overline{\eta}_p}{\eta_{t_\infty}} \right)^{[1+(t_m - t_{m,0})/(t_{m,0} - t_{m,R})]^{-s_0} - 1} \tag{27.10}$$

and η_∞ = pole viscosity corresponding to pole pressure c_p
Z_1 = dimensionless viscosity-pressure index
s_0 = dimensionless viscosity-temperature index

Although Z_1 may be pressure dependent, an approximate value is used for a given lubricant at a given ambient temperature and is considered to be a constant for the given problem. Furthermore, although the extreme-temperature viscosity may depend on the lubricant type and the pressure, a constant value, 4.5×10^{-4} Pa · s, is assumed.

Now the average viscosity across the film thickness at a longitudinal location in the flow field can be evaluated.

$$\eta_{\text{avg}} = \eta_0 \overline{\eta}_{\text{avg}} = \eta_0 \overline{\eta}_p \overline{\eta}_{T,\text{avg}} \tag{27.11}$$

where

$$\eta_{T,\text{avg}} = \frac{1}{h} \int_{h_a}^{h_b} \overline{\eta}_T \, dz = \int_0^1 \overline{\eta}_T \, dZ \tag{27.12}$$

Note that by defining the ambient piezoviscosity as

$$\eta_p = \eta_0 \bar{\eta}_p \tag{27.13}$$

the slope of $\ln \eta_p$ versus p as p approaches zero gage pressure gives the pressure-viscosity coefficient ξ. It was used by Barus (1893) to approximate the variation of viscosity with pressure. That is,

$$\xi = \left. \frac{\partial \ln \eta_p}{\partial p} \right|_{p \to 0} \tag{27.14}$$

By using Eqs. (27.9), (27.13), and (27.14), the relationship between ξ and Z_1 can be obtained; that is,

$$Z_1 = \frac{\xi c_p}{\ln(\eta_0/\eta_\infty)} \tag{27.15}$$

This relationship gives a quick way to calculate the viscosity-pressure index of a lubricant by knowing its pressure-viscosity coefficient, or vice versa.

Similarly, by defining the atmospheric thermoviscosity

$$\eta_{T,p_0} = \eta_0 \bar{\eta}_T \tag{27.16}$$

and the temperature-viscosity coefficient

$$\beta_\eta = - \left. \frac{\partial \ln \eta_{T,p_0}}{\partial t_m} \right|_{t_m \to t_{m,0}} \tag{27.17}$$

the relationship between the viscosity-temperature index and the temperature-viscosity coefficient can be derived.

$$s_0 = \frac{\beta_\eta (t_{m,0} - t_{m,R})}{\ln(\eta_0/\eta_{t_\infty})} \tag{27.18}$$

Again, this provides a quick way to obtain the value for the viscosity-temperature index by knowing the value of the temperature-viscosity coefficient, and vice versa.

The effect of pressure and temperature on the limiting shear strength of lubricants is also considered in this analysis. The limiting shear strength is denoted as the product of the limiting shear strength at atmospheric conditions $\tau_{L,0}$, the dimensionless pressure limiting shear strength $\bar{\tau}_{L,p}$, and the dimensionless temperature limiting shear strength $\bar{\tau}_{L,T}$

$$\tau_L = \tau_{L,0} \bar{\tau}_{L,p} \bar{\tau}_{L,T} \tag{27.19}$$

where

$$\bar{\tau}_{L,p} = 1 + \frac{\gamma p}{\tau_{L,0}} \tag{27.20}$$

$$\bar{\tau}_{L,T} = \exp\left[\beta_{\tau_L} \left(\frac{1}{t_m - t_{m,R}} - \frac{1}{t_{m,0} - t_{m,R}} \right) \right] \tag{27.21}$$

and γ^* is the pressure-limiting-shear-strength proportionality constant and β_{τ_L} is the temperature-limiting-shear-strength coefficient. Then the average limiting shear strength of the lubricant across the film thickness at a longitudinal location in the flow field can be evaluated.

$$\tau_{L,\text{avg}} = \tau_{L,0}\bar{\tau}_{L,\text{avg}} = \tau_{L,0}\bar{\tau}_{L,p}\bar{\tau}_{L,T,\text{avg}} \tag{27.22}$$

where

$$\bar{\tau}_{L,T,\text{avg}} = \frac{1}{h}\int_{h_a}^{h_b}\bar{\tau}_{L,T}\,dz = \int_0^1 \tau_{L,T}\,dZ \tag{27.23}$$

27.2.2 THERMAL REYNOLDS EQUATIONS CONSIDERING VELOCITY AND STRESS BOUNDARY CONDITIONS. The circular non-Newtonian fluid model introduced by Lee and Hamrock (1990) and covered in Chap. 26 is used as the basic rheology model in this thermal EHL analysis with the following assumptions:

1. Line contact is assumed (i.e., side leakage is neglected).
2. The film thickness is much smaller than the equivalent radius of the contact. This may further imply
   a. That the pressure gradient across the film thickness is negligible (that is, $\partial p/\partial z = 0$)
   b. That the velocity component in the z direction, which is the direction across the film thickness, is much smaller than that in the x direction, which is the main-stream direction (that is, $w \ll u$ and then $\partial w/\partial x \ll \partial u/\partial z$)
   c. That the shear stress acting on the yz plane is negligible
3. The inertia effect is negligible relative to the viscous effect.
4. The fluid flow and the heat flow are in steady state within the lubricant body.
5. The inlet conditions are ambient (i.e., the temperature is the given ambient temperature and the pressure is the zero gage pressure).
6. The pressure at the exit is the Reynolds boundary pressure (i.e., the pressure is the zero gage pressure and the gradient of the pressure is zero also). The thermal condition is adiabatic (and ambient) at a location infinitely far away from the exit. Practically, it is assumed to be adiabatic at the outlet, a location with a finite distance away from the exit.
7. The solid surface temperatures are calculated for the moving distributed heat sources that are generated by the temperature gradients of the lubricant adjacent to the liquid-solid interfaces. Only the start transient period is taken into account in lieu of the heat sink characteristic of the fluid beyond the outlet. The frictional heat generated at the liquid-solid interfaces is considered when the stress boundary condition is imposed.

On the basis of these assumptions, the circular non-Newtonian constitutive equation and the equation of force equilibrium can be written as

$$s = \frac{\partial u}{\partial z} = \frac{\tau}{\eta_{\text{avg}}}\left[1 - \left(\frac{\tau}{\tau_{L,\text{avg}}}\right)^2\right]^{-1/2} \tag{27.24}$$

and

$$\frac{dp}{dx} = \frac{\partial \tau}{\partial z} \tag{27.25}$$

where s = shear strain rate in x direction with respect to z direction
u = velocity component in x direction
τ = field shear stress acting on xy plane
$\eta_{\text{avg}}, \tau_{L,\text{avg}}$ = field viscosity and field limiting shear strength of lubricant, averaged across film thickness, respectively

By using the dimensionless coordinate system $Z = (z - h_a)/(h_b - h_a) = (z - h_a)/h$ and $X = 2x/D_x$, which were used by Lee and Hamrock (1990), denoting $\bar{\tau} = \tau/\tau_{L,\text{avg}}$, $W' = w'/E'R$, $h = h_b - h_a$, $H = 4hR/D_x^2$, and $P = p/p_H$, and then integrating Eq. (27.25) with respect to z once and making use of the boundary conditions $\bar{\tau} = \bar{\tau}_a$ at $Z = 0$ and $\bar{\tau} = \bar{\tau}_b$ at $Z = 1$, the shear stress equation at location x can be obtained.

$$\bar{\tau} = \bar{\tau}_a + \check{T}Z \tag{27.26}$$

where $\check{T} = \bar{\tau}_b - \bar{\tau}_a = 2W'H(dP/dX)/\pi\hat{\tau}_L$ is the difference of dimensionless boundary shear stresses and $\hat{\tau}_L = \tau_{L,\text{avg}}/E'$ is the average dimensionless limiting shear strength across the film thickness.

By letting $\tilde{u} = (u_a + u_b)/2$, $\bar{u} = u/\tilde{u}$, $D_x = 2R\sqrt{8W'/\pi}$, $\bar{S} = \pi U \bar{\eta}_{\text{avg}}/8W'H\hat{\tau}_L$, and $U = \eta_0\tilde{u}/E'R$, the dimensionless shear strain rate can be written as

$$\bar{s} = \frac{\partial \bar{u}}{\partial Z} = \frac{1}{\bar{S}}\frac{\bar{\tau}}{\sqrt{1 - \bar{\tau}^2}} \tag{27.27}$$

By integrating this equation and applying all relevant boundary conditions, the suitable field velocity equations can be obtained for all individual trans-film-thickness sections.

The normal boundary condition is velocity specified. Denoting the slide-roll ratio as $A_c = (u_b - u_a)/\tilde{u}$ gives

$$\bar{u} = \begin{cases} \bar{u}_a = 1 - \dfrac{A_c}{2} & \text{at } Z = 0 \\ \bar{u}_b = 1 + \dfrac{A_c}{2} & \text{at } Z = 1 \end{cases} \tag{27.28}$$

However, when the local limiting shear strength of the lubricant is approached or exceeded at any solid-liquid interface, the boundary condition must be treated as stress specified (i.e., the shear stress at this boundary must be set to be the local limiting shear strength of the lubricant and the velocity at this location becomes a dependent variable).

A special situation in the flow zone with velocity-specified boundaries is that the pressure gradient is zero at a trans-film-thickness section, such as the exit location. In this case the difference in boundary shear stresses vanishes (i.e., $\check{T} = 0$); therefore, the shear stresses and the shear strain rate are all constant. By integrating Eq. (27.27) and making use of Eq. (27.28) for the boundary conditions, the velocity equation for this special case becomes

$$\bar{u} = 1 - \frac{A_c}{2} + A_c Z \tag{27.29}$$

with

$$A_c = \frac{1}{\bar{S}} \frac{\bar{\tau}}{\sqrt{1 - \bar{\tau}^2}} \tag{27.30}$$

Solving this equation for the constant shear stress gives

$$\bar{\tau} = \frac{A_c \bar{S}}{\sqrt{(A_c \bar{S})^2 + 1}} \tag{27.31}$$

Integrating the product of density and velocity over the film thickness gives the total mass flow rate per unit length of the line-contact system. A general form of dimensionless total mass flow rate can be written as

$$Q_m = \bar{\rho}_p \bar{\rho}_{T,\text{avg}} H \int_0^1 \bar{u}\, dZ \tag{27.32}$$

Substituting Eq. (27.29) into Eq. (27.32) gives the mass flow rate formula for the trans-film-thickness sections where the pressure gradient is zero and the boundaries are velocity specified.

$$Q_m = \bar{\rho}_p \bar{\rho}_{T,\text{avg}} H \tag{27.33}$$

To satisfy the continuity equation, Q_m must be constant. By denoting $Q_{m,\text{end}} = \bar{\rho}_{T,\text{arg,end}} H_{\text{end}}$ as the dimensionless mass flow rate at the outlet section and equating all Q_m to $Q_{m,\text{end}}$ the thermal Reynolds equation for each trans-film-thickness section can be obtained. For the special case of velocity boundaries with a zero pressure gradient, the Reynolds equation becomes

$$\frac{\bar{\rho}_{T,\text{avg,end}} - H_{\text{end}}}{\bar{\rho}_p \bar{\rho}_{T,\text{avg}}} - H = 0 \tag{27.34}$$

Similarly, for those trans-film-thickness sections that have velocity boundaries and nonzero pressure gradients, the velocity and Reynolds equations can be obtained. That is, integrating Eq. (27.27) once while using Eq. (27.26) gives

$$\bar{u} = -\frac{1}{\bar{S}\check{T}} \sqrt{1 - (\bar{\tau}_a + \check{T} Z)^2} + f(X) \tag{27.35}$$

Making use of Eq. (27 28) as the boundary conditions results in the velocity equation

$$\bar{u} = 1 - \frac{A_c}{2} + \frac{1}{\bar{S}\check{T}}\left[\sqrt{1 - \bar{\tau}_a^2} - \sqrt{1 - \left(\bar{\tau}_a + \check{T}Z\right)^2}\right] \tag{27.36}$$

with

$$A_c = \frac{1}{\bar{S}\check{T}}\left[\sqrt{1 - \bar{\tau}_a^2} - \sqrt{1 - \left(\bar{\tau}_a + \check{T}\right)^2}\right]$$

or

$$\bar{\tau}_a = -\frac{\check{T}}{2} + A_c\bar{S}\sqrt{\frac{1}{\left(A_c\bar{S}\right)^2 + 1} - \frac{\check{T}^2}{4}} \tag{27.37}$$

Lee and Hamrock (1990) indicated the valid range of the difference between the boundary shear stresses as

$$|\check{T}| \le \frac{2}{\left(A_c\bar{S}\right)^2 + 1} \tag{27.38}$$

By making use of Eqs. (27.32) and (27.26) and applying the continuity equation, the Reynolds equation for the flow zone with velocity boundaries and nonzero pressure gradients can be written as

$$H^3\frac{dP}{dX} + \bar{K}\hat{\eta}_1\left(\frac{\bar{\rho}_{T,\,\text{avg, end}}H_{\text{end}}}{\bar{\rho}_p\bar{\rho}_{T,\,\text{avg}}} - H\right) = 0 \tag{27.39}$$

where $\bar{K}$ = dimensionless velocity-load synergistic parameter, $3\pi^2U/4W'^2$

$\hat{\eta}_1$ = dimensionless non-Newtonian viscosity variable of first kind, $\check{T}^3\bar{\eta}_{\text{avg}}/6\zeta_1$

$\bar{\eta}_{\text{avg}}$ = averaging dimensionless viscosity across film thickness, $\eta_{\text{avg}}/\eta_0 = \bar{\eta}_p\bar{\eta}_{T,\,\text{avg}}$

ζ_1 = dimensionless boundary shear stress variable of first kind, $\sin^{-1}\bar{\tau}_b - \sin^{-1}\bar{\tau}_a - \bar{\tau}_b\sqrt{1 - \bar{\tau}_a^2} + \bar{\tau}_a\sqrt{1 - \bar{\tau}_b^2}$

When a local shear stress at the surface of solid a reaches the limiting shear strength of the local lubricant while that at the surface of solid b is less than the limiting shear strength, the shear stress at the surface of solid a is set to the limiting shear strength with the sign of the shear stress, but the velocity-specified boundary condition remains at the other solid-liquid interface. Note that Eq. (27.35) is valid for all boundary conditions with nonzero pressure gradients. Applying this stress and the velocity boundary conditions to Eq. (27.35) gives the velocity equation for this second kind of flow zone.

$$\bar{u} = 1 + \frac{A_c}{2} - \frac{1}{\bar{S}\check{T}}\sqrt{1 - \left(\bar{\tau}_a + \check{T}Z\right)^2} \tag{27.40}$$

with $$-1 \le \bar{\tau}_a + \check{T} = \bar{\tau}_b \le 1 \qquad \text{and} \qquad |\bar{\tau}_a| = 1 \tag{27.41}$$

By making use of Eqs. (27.32) and (27.40) and applying the continuity equation, the Reynolds equation for the second kind of flow zone can be written as

$$H^3 \frac{dP}{dX} + \bar{K}\hat{\eta}_2 \left[\frac{\bar{\rho}_{T,\,\text{avg, end}} H_{\text{end}}}{\bar{\rho}_p \bar{\rho}_{T,\,\text{avg}}} - \left(1 + \frac{A_c}{2}\right) H \right] = 0 \tag{27.42}$$

where $\hat{\eta}_2$ = dimensionless non-Newtonian viscosity variable of second kind, $\check{T}^3 \bar{\eta}_{\text{avg}}/6\zeta_2$

ζ_2 = dimensionless boundary shear stress variable of second kind, $\sin^{-1}(\bar{\tau}_a + \check{T}) - \sin^{-1}\bar{\tau}_a + (\bar{\tau}_a - \check{T})\sqrt{1 - (\bar{\tau}_a + \check{T})^2}$

Similarly, when a local shear stress at the surface of solid b reaches the limiting shear strength of the local lubricant while that at the surface of solid a is less than the limiting shear strength, the shear stress at the surface of solid b is set to the limiting shear strength with the sign of the shear stress, but the velocity-specified condition remains at the other solid-liquid interface. Applying this boundary condition to Eq. (27.35) gives the velocity equation for this third kind of flow zone.

$$\bar{u} = 1 - \frac{A_c}{2} + \frac{1}{\bar{S}\check{T}} \left[\sqrt{1 - (\bar{\tau}_b - \check{T})^2} - \sqrt{1 - [\bar{\tau}_b - \check{T}(1 - Z)]^2} \right] \tag{27.43}$$

with $$-1 \le \bar{\tau}_b - \check{T} = \bar{\tau}_a \le 1 \qquad \text{and} \qquad |\bar{\tau}_b| = 1 \tag{27.44}$$

By making use of Eqs. (27.32) and (27.43) and applying the continuity equation, the Reynolds equation for the third kind of flow zone can be written as

$$H^3 \frac{dP}{dX} + \bar{K}\hat{\eta}_3 \left[\frac{\bar{\rho}_{T,\,\text{avg, end}} H_{\text{end}}}{\bar{\rho}_p \bar{\rho}_{T,\,\text{avg}}} - \left(1 - \frac{A_c}{2}\right) H \right] = 0. \tag{27.45}$$

where $\hat{\eta}_3$ = dimensionless non-Newtonian viscosity variable of third kind, $\check{T}^3 \bar{\eta}_{\text{avg}}/6\zeta_3$

ζ_3 = dimensionless boundary shear stress variable of third kind, $\sin^{-1}\bar{\tau}_b - \sin^{-1}(\bar{\tau}_b - \check{T}) - (\bar{\tau}_b + \check{T})\sqrt{1 - (\bar{\tau}_b - \check{T})^2}$

The severity of lubricant shearing may grow until both shear stresses at the local solid-liquid interfaces reach the limiting shear strength of the lubricant. Two situations may occur. One is that the shear stresses at two boundaries have the same sign, both positive or both negative. The other is that they have different signs, one positive and the other negative. When they have the same sign, it means that the difference in shear stresses at interfaces vanishes and the pressure gradient at this section must be zero. In this case, Eqs. (27.29) and

(27.34) become the velocity equation and the Reynolds equation of the fourth kind of flow zone, respectively.

When both shear stresses at the solid-liquid interfaces reach the local shear strength of the lubricant and they have different signs, the stress difference ($\check{T} = -2\bar{\tau}_a$) becomes a constant. The Reynolds equation and the velocity equation can be shown as

$$\frac{H}{\bar{S}} + 4\zeta_5\left[\frac{\bar{\rho}_{T,\,\mathrm{avg,\,end}}H_{\mathrm{end}}}{\bar{\rho}_p\bar{\rho}_{T,\,\mathrm{avg}}} - \bar{u}_{\mathrm{avg,\,fab}}H\right] = 0 \tag{27.46}$$

$$\bar{u} = \bar{u}_{\mathrm{avg,\,fab}} - \frac{1}{\bar{S}\check{T}}\sqrt{1 - \left(\bar{\tau}_a + \check{T}Z\right)^2} \tag{27.47}$$

27.3 NUMERICAL SOLUTION OF THERMAL EQUATIONS

The complete fast approach that was originated by Houpert and Hamrock (1986) (also given in Chap. 21), improved by Lee and Hamrock (1990) (also given in Chap. 26), and completed in Hsiao and Hamrock (1992) is used to solve the thermal Reynolds equations developed in the previous section. The original fast approach uses the Newton-Raphson scheme to solve the Newtonian isothermal Reynolds equation for the dimensionless mass flow rate $Q_{m,\,\mathrm{end}}$, the pressure distribution P, and the dimensionless offset film thickness H_0. The improved version uses the same scheme to solve for the pressure distribution and the dimensionless offset film thickness; however, it solves for the outlet location X_{end} instead of the dimensionless mass flow rate.

The improved fast approach does not carry out all relevant derivatives for the elements composing the Jacobian matrix, such as the effects of pressure distribution on the exit location. And although it extends the solution from that for a Newtonian fluid to that for a non-Newtonian fluid, some relevant derivatives are also neglected in this part of the extension, such as the effects of the dimensionless non-Newtonian viscosity variable $\hat{\eta}$ on the dimensionless offset film thickness.

The Hsiao and Hamrock (1992) work that is also used in this chapter completes all derivatives for all components of the Jacobian matrix and solves the problem fully automatically. The fast and complete convergence of solutions is therefore achieved. For a given set of operating parameters and material properties, the complete fast approach solves the circular non-Newtonian thermal Reynolds equations with a given temperature field and the predetermined or initially guessed pressure distribution, offset film thickness, exit location, and flow zone distribution with velocity or velocity and stress boundary conditions.

With the given temperature data the average dimensionless thermodensity $\bar{\rho}_{T,\,\mathrm{avg}}$, the average dimensionless thermoviscosity $\bar{\eta}_{T,\,\mathrm{avg}}$, and the average dimensionless temperature-limiting shear strength $\bar{\tau}_{L,T,\mathrm{avg}}$ can be calculated. With the given pressures the dimensionless piezodensity can be obtained. By knowing the offset film thickness, the exit location, and the pressures, the elastic deformation and the film shape of the contact can be calculated as

$$H_i = H_0 + \frac{X_i^2}{2} + \sum_{j=2}^{N} D_{ij} P_j \qquad i = 1, \ldots, N_{\mathrm{max}} \tag{27.48}$$

where subscripts i and j indicate the node number of the one-dimensional mesh system with $i = N + 1$ or $j = N + 1$ representing the exit node and D_{ij} are the influence coefficients that depict the effects of pressure distribution on the variational part of the elastic deformation. In addition to the influence coefficients, the two-node weighting factors a_{ij} for calculating the pressure gradients used in the Reynolds equations and the three-point Lagrange polynomial weighting factors C_j for calculating the reacting force can be obtained by using the formulas provided by Houpert and Hamrock (1986).

The number of system unknowns is $N + 1$, one for the exit location, one for the offset film thickness, and $N - 1$ for the discrete pressures distributed along the longitudinal contact direction. The $N - 1$ pressures exclude the inlet and exit pressures, which are set to zero. The $N + 1$ system equations required to solve the problem are taken from N Reynolds equations at N discrete nodes including the inlet and from another equation for the force balance between the applied load and the reacting contact pressures. In addition to the special Reynolds equation used for the nodes where the pressure gradient is zero, five kinds of Reynolds equation may be involved to compose the Jacobian matrix. The use of the alternative Reynolds equations depends on the severity of the shearing that the lubricant encounters. The system equations are

$$\begin{bmatrix} \frac{\partial f_1}{\partial X_{\mathrm{end}}} & \frac{\partial f_1}{\partial P_2} & \cdots & \frac{\partial f_1}{\partial P_j} & \cdots & \frac{\partial f_1}{\partial P_N} & \frac{\partial f_1}{\partial H_0} \\ \frac{\partial f_2}{\partial X_{\mathrm{end}}} & \frac{\partial f_2}{\partial P_2} & \cdots & \frac{\partial f_2}{\partial P_j} & \cdots & \frac{\partial f_2}{\partial P_N} & \frac{\partial f_2}{\partial H_0} \\ \cdots & \cdots & \cdots & \cdots & \cdots & \cdots & \cdots \\ \frac{\partial f_i}{\partial X_{\mathrm{end}}} & \frac{\partial f_i}{\partial P_2} & \cdots & \frac{\partial f_i}{\partial P_j} & \cdots & \frac{\partial f_i}{\partial P_N} & \frac{\partial f_i}{\partial H_0} \\ \cdots & \cdots & \cdots & \cdots & \cdots & \cdots & \cdots \\ \frac{\partial f_N}{\partial X_{\mathrm{end}}} & \frac{\partial f_N}{\partial P_2} & \cdots & \frac{\partial f_N}{\partial P_j} & \cdots & \frac{\partial f_N}{\partial P_N} & \frac{\partial f_N}{\partial H_0} \\ \frac{\partial f_{N+1}}{\partial X_{\mathrm{end}}} & \frac{\partial f_{N+1}}{\partial P_2} & \cdots & \frac{\partial f_{N+1}}{\partial P_j} & \cdots & \frac{\partial f_{N+1}}{\partial P_N} & \frac{\partial f_{N+1}}{\partial H_0} \end{bmatrix} \begin{bmatrix} \Delta X_{\mathrm{end}} \\ \Delta P_2 \\ \cdots \\ \Delta P_j \\ \cdots \\ \Delta P_N \\ \Delta H_0 \end{bmatrix} = \begin{bmatrix} -f_1 \\ -f_2 \\ \cdots \\ -f_i \\ \cdots \\ -f_N \\ -f_{N+1} \end{bmatrix} \tag{27.49}$$

where f_i, $i = 1, \ldots, N$, may be one of the following:

$$f_i = \frac{\bar{\rho}_{T,\text{avg,end}} H_{\text{end}}}{\bar{\rho}_{p,i} \bar{\rho}_{T,\text{avg},i}} - H_i = 0 \tag{27.50}$$

$$f_i = H_i^3 \left(\frac{dP}{dX} \right)_i + \bar{K} \hat{\eta}_{1,i} \left(\frac{\bar{\rho}_{T,\text{avg,end}} H_{\text{end}}}{\bar{\rho}_{p,i} \bar{\rho}_{T,\text{avg},i}} - H_i \right) = 0 \tag{27.51}$$

$$f_i = H_i^3 \left(\frac{dP}{dX} \right)_i + \bar{K} \hat{\eta}_{2,i} \left[\frac{\bar{\rho}_{T,\text{avg,end}} H_{\text{end}}}{\bar{\rho}_{p,i} \bar{\rho}_{T,\text{avg},i}} - \left(1 + \frac{A_c}{2} \right) H_i \right] = 0 \tag{27.52}$$

$$f_i = H_i^3 \left(\frac{dP}{dX} \right)_i + \bar{K} \hat{\eta}_{3,i} \left[\frac{\bar{\rho}_{T,\text{avg,end}} H_{\text{end}}}{\bar{\rho}_{p,i} \bar{\rho}_{T,\text{avg},i}} - \left(1 - \frac{A_c}{2} \right) H_i \right] = 0 \tag{27.53}$$

$$f_i = \frac{H_i}{\bar{S}_i} + 4\zeta_{5,i} \left[\frac{\bar{\rho}_{T,\text{avg,end}} H_{\text{end}}}{\bar{\rho}_{p,i} \bar{\rho}_{T,\text{avg},i}} - \bar{u}_{\text{avg,fab},i} H_i \right] = 0 \tag{27.54}$$

and

$$f_{N+1} = \sum_{j=2}^{N} C_j P_j - \frac{\pi}{2} = 0 \tag{27.55}$$

The complete derivation of all Jacobian factors can be performed accordingly.

A partial pivoting method is used to solve the system equation (27.49). The iterative Newton-Raphson scheme is applied until the fully converged solution (an error norm less than 1×10^{-6}) is obtained. In some situations, such as starting iteration with bad initial guesses or shear stresses reaching the limiting shear strength of the lubricant, the difference in boundary shear stresses $\check{T}$ may be out of range and result in a fatal numerical error during the calculation. Some numerical tricks need to be performed to get a converged solution. The basic idea of this treatment is to bring $\check{T}$ back to the limit of the range that it goes beyond. In other words, when $\check{T}$ is out of range at a trans-film-thickness section, a temporary remedy equation is substituted for the thermal Reynolds equation to correct $\check{T}$ by using the Newton-Raphson iterations. The remedy equations used for different situations that cause the ill-behaving $\check{T}$ are described in the following paragraph. The Jacobian factors for these substitutional Reynolds equations are lengthy analytical expressions, which are not shown here because of space limitation.

The situation most likely to happen is that the predetermined flow zone is of the first kind and the temporary result of $\check{T}$ is out of range; that is,

$$|\check{T}| > \frac{2}{(A_c \bar{S})^2 + 1}$$

It implies that the shear stress at either the surface of solid a or the surface of solid b is greater than the limiting shear strength of the lubricant at this location. In this case the false Reynolds equation used to correct $\check{T}$ is

$$f_i = \check{T}_i - \text{sign}\left(\check{T}_i\right)\frac{2}{\left(A_c\bar{S}_i\right)^2 + 1} = 0 \tag{27.56}$$

where sign is the function returning the sign of the argument.

The second situation is that the predetermined flow zone is of the second kind and the temporarily calculated shear stress at the surface of solid b is greater than the local limiting shear strength of the lubricant. Similarly, the third situation may happen when the predetermined flow zone is of the third kind and the calculated shear stress at the surface of solid a is greater than the local limiting shear strength of the lubricant. In these two cases the following equation is used to correct $\check{T}$.

$$f_i = \check{T}_i - \left[\text{sign}\left(\bar{\tau}_{b,i}\right) - \text{sign}\left(\bar{\tau}_{a,i}\right)\right] = 0 \tag{27.57}$$

One situation that is least likely to happen is that the pressure gradient is zero at a trans-film-thickness section and the predetermined shear stresses are less than the local limiting shear strength of the lubricant, but the Newton-Raphson method seems to predict that the shear stresses will go beyond the limiting shear strength. If this is the case, the temporary remedy equation is

$$f_i = A_c\bar{S}_i\sqrt{\frac{1}{\left(A_c\bar{S}_i\right)^2 + 1}} - \text{sign}\left(\bar{\tau}_{a,i}\right) = 0 \tag{27.58}$$

The use of these false Reynolds equations for iterations continues if the situations causing the ill-behaving $\check{T}$ remain. If the need for this remedial procedure persists and a converged solution (an error norm less than 1×10^{-6}) is obtained, the limiting shear strength is reached at the relevant locations. Thereafter, the velocity-specified boundary conditions at these locations is reset to the stress-specified boundary condition. A new iteration, which is based on the newly corrected boundary conditions, starts over again.

When the fully converged solution is obtained, the friction coefficient and the stream functions are then evaluated. The friction coefficient equation can be written as

$$\mu = \frac{1}{w'}\int_{-\infty}^{\infty}\tau_a\, dx \approx \sqrt{\frac{8}{\pi W'}}\int_{X_1}^{X_{N_{\max}}}\bar{\tau}_a\hat{\tau}_L\, dX \tag{27.59}$$

The stream functions for thermal EHL in line contact considering multiple boundary conditions can be derived by using the technique developed by Shieh and Hamrock (1991). That is, for compressible fluids, the stream functions are defined as $\bar{\rho}u = \partial\Phi/\partial z$ and $\bar{\rho}w = -\partial\Phi/\partial x$, where Φ is the stream function and w is the velocity component of the film thickness direction. The dimensionless stream function $\overline{\Phi}$ becomes

$$\bar{\rho}_p \bar{\rho}_{T,\text{avg}} H \bar{u} = \frac{\partial \overline{\Phi}}{\partial Z} \tag{27.60}$$

and

$$\bar{\rho}_p \bar{\rho}_{T,\text{avg}} \bar{w} \frac{2R}{D_x} = -\frac{\partial \overline{\Phi}}{\partial X} \tag{27.61}$$

where $\bar{w} = w/\tilde{u}$ and $\overline{\Phi} = 4\Phi R/\rho_0 \tilde{u} D_x^2$. Substituting the velocity equations (27.29), (27.36), (27.40), (27.43), or (27.47) into Eq. (27.60), integrating it, and setting $\overline{\Phi}(X,0) = 0$ give the stream functions for the flow zones with different boundary conditions.

First of all, using velocity equation (27.29) for the trans-film-thickness sections where the pressure gradient is zero gives

$$\overline{\Phi}(X,Z) = \overline{\Phi}(Z) = \bar{\rho}_p \bar{\rho}_{T,\text{avg}} H\left[\left(1 - \frac{A_c}{2}\right)Z + \frac{A_c}{2}Z^2\right] \tag{27.62}$$

For the flow zone of the first kind with velocity equation (27.36) and the flow zone of the third kind with velocity equation (27.43), we can obtain

$$\overline{\Phi}(X,Z) = \bar{\rho}_p \bar{\rho}_{T,\text{avg}} H\left[\left(1 - \frac{A_c}{2} + \frac{1}{\overline{S}\check{T}}\sqrt{1 - \bar{\tau}_a^2}\right)Z + \frac{1}{2\overline{S}\check{T}^2}\left(\bar{\tau}_a\sqrt{1-\bar{\tau}_a^2} + \sin^{-1}\bar{\tau}_a - \bar{\tau}\sqrt{1-\bar{\tau}^2} - \sin^{-1}\bar{\tau}\right)\right] \tag{27.63}$$

By using Eq. (27.40) for the flow zone of the second kind, the stream function becomes

$$\overline{\Phi}(X,Z) = \bar{\rho}_p \bar{\rho}_{T,\text{avg}} H\left[\left(1 + \frac{A_c}{2} + \frac{1}{\overline{S}\check{T}}\sqrt{1 - \bar{\tau}_b^2}\right)Z + \frac{1}{2\overline{S}\check{T}^2}\left(\sin^{-1}\bar{\tau}_a - \bar{\tau}\sqrt{1-\bar{\tau}^2} - \sin^{-1}\bar{\tau}\right)\right] \tag{27.64}$$

Finally, the stream function for flow zones with velocity equation (27.47) can be expressed as

$$\overline{\Phi}(X,Z) = \bar{\rho}_p \bar{\rho}_{T,\text{avg}} H\left[\bar{u}_{\text{avg, fab}} Z + \frac{1}{2\overline{S}\check{T}^2}\left(\sin^{-1}\bar{\tau}_a - \bar{\tau}\sqrt{1-\bar{\tau}^2} - \sin^{-1}\bar{\tau}\right)\right] \tag{27.65}$$

27.4 ENERGY EQUATION

The energy equation used in this chapter was developed by Hsiao and Hamrock (1992) by using the control volume approach. The first law of thermodynamics states that the rate of energy transfer by heat conducted into a control volume is balanced by the rate of energy transfer by work done by the control volume and the rate of the stored-energy change within the control volume. That is,

$$\frac{dQ_{cv}}{dt} = \frac{DE_{cv}}{Dt} + \frac{dW_{cv}}{dt} \tag{27.66}$$

The energy in the form of work W_{cv} involved in this study includes only the pressure work and the shear energy. It is called the "flow work."

From Fourier's law the rate of heat conduction for a control volume can be derived as

$$\frac{dQ_{cv}}{dt} = \iiint_{cv} \nabla \cdot K_f \nabla t_m \, d\forall = \iint_{cs} K_f \nabla t_m \cdot \mathbf{N} \, dA \tag{27.67}$$

For a steady flow the total rate of energy stored in a control volume can be expressed as

$$\frac{DE_{cv}}{Dt} = \iint_{cs} \left(\breve{u} + \frac{V^2}{2} + gz \right) \rho \mathbf{V} \cdot \mathbf{N} \, dA \tag{27.68}$$

By conducting flow work balance for a differential volume, using Taylor's expansion, and dropping the higher-order terms, the equation for the rate of energy transfer by flow work for a control volume becomes

$$\frac{dW_{cv}}{dt} = \iiint_{cv} -\nabla \cdot (\mathbf{T} \cdot \mathbf{V}) \, d\forall = \iint_{cs} -(\mathbf{T} \cdot \mathbf{V}) \cdot \mathbf{N} \, dA \tag{27.69}$$

where the stress tensor $\mathbf{T}$ can be expressed as

$$\mathbf{T} = \begin{bmatrix} \sigma_x & \tau_{xy} & \tau_{xz} \\ \tau_{yx} & \sigma_y & \tau_{yz} \\ \tau_{zx} & \tau_{zy} & \sigma_z \end{bmatrix} \tag{27.70}$$

Then for a steady flow the most general conservative form of the energy equation for a control volume becomes

$$\iint_{cs} K_f \nabla t_m \cdot \mathbf{N} \, dA = \iint_{cs} \left(\breve{u} + \frac{V^2}{2} + gz \right) \rho \mathbf{V} \cdot \mathbf{N} \, dA - \iint_{cs} (\mathbf{T} \cdot \mathbf{V}) \cdot \mathbf{N} \, dA \tag{27.71}$$

For a line-contact EHL flow field the gravity effect is negligible, and the flow work term can be further reduced to

$$\mathbf{T} \cdot \mathbf{V} = (\tau w - pu)\mathbf{i} + (\tau u - pw)\mathbf{k} \tag{27.72}$$

The flow in a line-contact EHL conjunction is two dimensional. The flow field can be divided into finite control volumes with a two-dimensional mesh comprising discrete grids. Each control volume encloses a grid (i, j). A system of equations can then be constructed by applying the energy equation (27.71) to all control volumes of the flow field. As long as the field pressures, velocities, and shear stresses are known, the temperatures at all grids can be estimated by solving the system of equations with a finite difference method.

The dimensionless system of equations used in this work has the following form:

$$C_b \bar{t}_{m,i,j-1} + C_c \bar{t}_{m,i,j} + C_t \bar{t}_{m,i,j+1} = C_\ell \bar{t}_{m,i-1,j} + C_r \bar{t}_{m,i+1,j} + \bar{E}_{\mathrm{KE}} + \bar{E}_{\mathrm{PW}} + \bar{E}_{\mathrm{SW}} \tag{27.73}$$

where

$$C_b = \bar{K}_{f,i} \frac{\Delta X_i}{\Delta Z_i} + \frac{P_{ef} Q_{m,b}}{2}$$

$$C_c = -\bar{K}_{f,\ell} \frac{\Delta Z_\ell}{\Delta X_\ell} - 2\bar{K}_{f,i} \frac{\Delta X_i}{\Delta Z_i} - \bar{K}_{f,r} \frac{\Delta Z_r}{\Delta X_r} + \frac{P_{ef}}{2}\left(Q_{m,\ell} + Q_{m,b} - Q_{m,r} - Q_{m,t}\right)$$

$$C_t = \bar{K}_{f,i} \frac{\Delta X_i}{\Delta Z_i} - \frac{P_{ef} Q_{m,t}}{2}$$

$$C_\ell = -\bar{K}_{f,\ell} \frac{\Delta Z_\ell}{\Delta X_\ell} - \frac{P_{ef} Q_{m,\ell}}{2}$$

$$C_r = -\bar{K}_{f,r} \frac{\Delta Z_r}{\Delta X_r} + \frac{P_{ef} Q_{m,r}}{2}$$

$$\bar{E}_{\mathrm{KE}} = \frac{F_{\mathrm{KE}}}{2}\left(\bar{V}_r^2 Q_{m,r} + \bar{V}_t^2 Q_{m,t} - \bar{V}_\ell^2 Q_{m,\ell}^2 - \bar{V}_b^2 Q_{m,b}\right)$$

$$\bar{E}_{\mathrm{PW}} = F_{\mathrm{WK}} \frac{p_H}{E'}\left[\frac{P_r Q_{m,r}}{\bar{\rho}_{\mathrm{avg},r}} - \frac{P_\ell Q_{m,\ell}}{\bar{\rho}_{\mathrm{avg},\ell}} + \frac{P_i}{\bar{\rho}_{\mathrm{avg},i}}\left(Q_{m,t} - Q_{m,b}\right)\right]$$

$$\bar{E}_{\mathrm{SW}} = F_{\mathrm{WK}} \hat{\tau}_{Li} \Delta X_i \left(\bar{\tau}_{i,j-1/2} \bar{u}_{i,j-1/2} - \bar{\tau}_{i,j+1/2} \bar{u}_{i,j+1/2}\right)$$

Note that the subscripts ℓ, r, b, and t denote the variables associated with the left, right, bottom, and top control surfaces, respectively. The last three terms of Eq. (27.73) depict the convection of kinetic energy, pressure work, and shear energy. The other terms are related to the heat conduction and the convection of internal energy.

Because analytical expressions for flow velocity across film thickness are available, the rate of mass flowing in the x direction and passing a control surface can be evaluated analytically. On the other hand, the rate of mass flowing in the z direction and passing a control surface can be obtained by applying the mass conservation law for a control volume and by using the condition of having solid boundaries.

The temperatures at the inlet of the flow field are assumed to be ambient, but the boundary condition at the outlet is taken to be adiabatic. On the other hand, the boundary conditions at the solid-liquid interfaces are far more complicated. The model traditionally used to calculate the boundary temperatures at the solid-liquid interfaces comes from the flash temperature concept proposed by Blok (1937). The flash temperature concept states that the moving heat source imposed on the solid surface is fast enough to be well approximated by assuming that the heat transferring into the solid is in the sole direction perpendicular to the solid surface. This model has been used by almost every tribologist to solve for the temperatures in EHL conjunctions. However, it has an important limitation. It cannot be used to evaluate the solid surface temperatures when the solid surface is nearly stationary. For example, it is invalid when the slide-roll ratio of the contact is near 2. A model considering two-dimensional heat flow in the bounding solid is used to improve the solutions.

The governing equation describing the two-dimensional transient heat flow in an infinite solid moving at a constant speed V in the x direction was given by Rosenthal (1946) as

$$\frac{\partial^2 t_m}{\partial x^2} + \frac{\partial^2 t_m}{\partial \zeta^2} - \frac{V}{\lambda}\frac{\partial t_m}{\partial x} = \frac{1}{\lambda}\frac{\partial t_m}{\partial t} \tag{27.74}$$

where $\lambda = \dfrac{K_f}{\rho^* C_p}$ = diffusivity, $\dfrac{m^2}{s}$

C_p = specific heat, J/kg K

The analytical solution of this equation with an instantaneous heat source discharging at any location in the moving solid at any instant can be obtained as shown by Tsai (1986).

In an EHL conjunction the situation of heat transfer at the solid-liquid interfaces may be simulated by considering that the heat sources are distributive and discharge continuously on the bounding solid surface as the solid passes by. That is, the heat sources distribute along the x axis at $x_1 \le x' \le x_{N_{\max}}$ and discharge at $0 \le t' \le t$. By neglecting the curvature effect, the bounding solid may be considered to be a semi-infinite solid. Then $\zeta = 0$ is set to be the coordinate of the solid surface in contact. By assuming that there is no heat loss through the surface of the solid, the surface temperature at t, x, and $\zeta = 0$ can be written as

$$t_{m,a/b} - t_{m,0} = \frac{1}{2\pi K_{f,a/b}} \int_{x_1}^{x_{N_{\max}}} \int_0^t \frac{\exp - \dfrac{\left[x - x' - u_{a/b}(t - t')\right]^2}{4\lambda_{a/b}(t - t')}}{t - t'} \dot{q}_{a/b}\, dt'\, dx' \tag{27.75}$$

In reality, the heat loss through the solid surface outside the range where the considered heat sources reside is not negligible. In fact, a large amount of heat is removed from the solid through the surface by means of cooling. This is a strong cyclic phenomenon. The quasi-stationary state can never be reached. Therefore, the surface temperature is calculated only for a finite value of time t.

In this work $t = (x_{N_{\max}} - x_1)/|u_a|$ or $t = (x_{N_{\max}} - x_1)/|u_b|$; whichever is less is assumed.

The discrete heat fluxes at the solid surfaces in contact can be evaluated once the discrete temperatures in the flow field are known. By assuming that the temperature distribution is of the second order, the heat fluxes directed into the solids at the location X_i can be written as

$$\dot{q}_{ai} = -K_{f,i}\frac{\partial t_m}{\partial z}\bigg|_{z=h_{ai}} = \frac{K_{f,i}}{2\,\Delta z}(-3t_{m,i,1} + 4t_{m,i,2} - t_{m,i,3}) \tag{27.76}$$

$$\dot{q}_{bi} = -K_{f,i}\frac{\partial t_m}{\partial z}\bigg|_{z=h_{bi}} = \frac{K_{f,i}}{2\,\Delta z}(-3t_{m,i,j_m+1} + 4t_{m,i,j_m} - t_{m,i,j_m-1}) \tag{27.77}$$

In integrating Eq. (27.75) the heat flux function between any two adjacent grid points is assumed to be linear. By doing so, the dimensionless solid surface temperature at the location X_i becomes

$$\bar{t}_{m,a/bi} = \check{K}_{a/b}\sum_{k=2}^{N_{\max}}\left[\bar{\dot{q}}_{a/bk-1}(O_{a1/b1i,k} - O_{a2/b2i,k}) + \bar{\dot{q}}_{a/bk}O_{a2/b2i,k}\right] \tag{27.78}$$

where

$$\check{K}_{a/b} = \frac{K_{f,0}R}{4\pi K_{f,a/b}b} \tag{27.79}$$

is the interface conductivity parameters,

$$O_{a1/b1i,k} = \int_{X_{k-1}}^{X_k}\int_0^t \frac{\exp -\dfrac{\left[b(X_i - X) - u_{a/b}(t-t')\right]^2}{4\lambda_{a/b}(t-t')}}{t-t'}\,dt'\,dX \tag{27.80}$$

is the thermal influence coefficient of the first kind, and

$$O_{a2/b2i,k} = \int_{X_{k-1}}^{X_k}\frac{X - X_{k-1}}{X_i - X_{k-1}}\int_0^t \frac{\exp -\dfrac{\left[b(X_i - X) - u_{a/b}(t-t')\right]^2}{4\lambda_{a/b}(t-t')}}{t-t'}\,dt'\,dX \tag{27.81}$$

is the thermal influence coefficient of the second kind.

By transforming the variables t' and X to $\bar{t}$ and $\bar{X}$ as

$$t' = \frac{t}{2}\bar{t} + \frac{t}{2} \tag{27.82}$$

and

$$X = \frac{X_k - X_{k-1}}{2}\bar{X} + \frac{X_k + X_{k-1}}{2} \tag{27.83}$$

the thermal influence coefficients can be rewritten as

$$O_{a1/b1i,k} = \frac{\bar{t}(\bar{X}_k - \bar{X}_{k-1})}{4} \iint_{-1}^{1} \frac{\exp - \dfrac{\left[b(\bar{X}_i - \bar{X}) - u_{a/b}(t - t')\right]^2}{4\lambda_{a/b}(t - t')}}{t - t'} \, d\bar{t}\, d\bar{X} \tag{27.84}$$

and

$$O_{a2/b2i,k} = \frac{\bar{t}}{4} \int_{-1}^{1} (\bar{X} - \bar{X}_{k-1}) \times \int_{-1}^{1} \frac{\exp - \dfrac{\left[b(\bar{X}_i - \bar{X}) - u_{a/b}(t - t')\right]^2}{4\lambda_{a/b}(t - t')}}{t - t'} \, d\bar{t}\, d\bar{X} \tag{27.85}$$

Equations (27.84) and (27.85) can be readily evaluated by using the Gauss-Legendre quadrature. In this chapter the eight-point Gauss-Legendre integration is used.

Because the dimensionless heat fluxes $\bar{q}_{ai}$ and $\bar{q}_{bi}$ depend on the flow field temperatures ($\bar{t}_{m,ai} = \bar{t}_{m,i,1}, \bar{t}_{m,i,2}, \bar{t}_{m,i,3}$) and ($\bar{t}_{m,bi} = \bar{t}_{m,i,j_m+1}, \bar{t}_{m,i,j_m}, \bar{t}_{m,i,j_m-1}$), respectively, the surface temperature equation (27.78) can be rewritten as

$$\bar{t}_{m,i,1} = \frac{1}{1 + 3O_{ai}\check{K}_a} \left\{ \sum_{k=2}^{N_{\max}} \left[\bar{q}_{ak-1}(O_{a1i,k} - O_{a2i,k}) + \bar{q}_{ak} O_{a2i,k}\right] - \bar{q}_{ai} O_{ai} + O_{ai}\left(4\bar{t}_{m,i,2} - \bar{t}_{m,i,3}\right) \right\} \tag{27.86}$$

and

$$\bar{t}_{m,i,j_m+1} = \frac{1}{1 + 3O_{bi}\check{K}_b} \left\{ \sum_{k=2}^{N_{\max}} \left[\bar{q}_{bk-1}(O_{b1i,k} - O_{b2i,k}) + \bar{q}_{bk} O_{b2i,k}\right] - \bar{q}_{bi} O_{bi} + O_{bi}\left(4\bar{t}_{m,i,j_m} - \bar{t}_{m,i,j_m-1}\right) \right\} \tag{27.87}$$

where

$$O_{ai} = O_{a2i,i} + O_{a1i,i+1} - O_{a2i,i+1} \tag{27.88}$$

and

$$O_{bi} = O_{b2i,i} + O_{b1i,i+1} - O_{b2i,i+1} \tag{27.89}$$

From Eqs. (27.73), (27.86), and (27.87), a tridiagonal system of equations can be set up for a specific location $\bar{X}_i$ to calculate the field temperatures $\bar{t}_{m,i,j}$ for $j = 2$ and $j = j_m$. The interface temperatures and the heat fluxes at this section are updated by using Eqs. (27.86), (27.87), (27.76), and (27.77) right after the field temperatures on this line are solved for. The computational scheme used in solving the energy equation is the procedure of successive overrelaxation by lines starting from $i = 2$ to $i = N_{\max}$. The criterion of convergence used in temperature calculations is 1×10^{-3} of Euclidean norm.

TABLE 27.1
Operating parameters for two cases

Parameters	Case 1	Case 2
Mean surface velocity in x direction $\tilde{u}$, m/s	0.59634	1,84843
Load per unit width w', N/m	0.147×10^6	0.147×10^6
Radius of curvature sum R, m	0.01114	0.01114
Ambient temperature $t_{m,0}$, °C	40	40

The overall computation procedure for coupling the thermal Reynolds equations and the energy equation starts from using the ambient temperature as the initial guess for the field temperature. The isothermal results, including film thickness, pressure, mass density, limiting shear strength, velocity, and shear stress, obtained from solving the Reynolds equations are used as the input information for solving the energy equation. The temperatures calculated from the energy equation are then used to update the temperature-related lubricant properties. The Reynolds equations are then solved again on the basis of the newly updated lubricant properties. The solution procedure continues until the resultant field temperature calculated by the energy equation converges to the preset criterion, 1×10^{-3} of Euclidean norm.

27.5 RESULTS AND DISCUSSION

Two series of computations, referred to as the "case-1" and "case-2" series, have been made to simulate. The operating conditions are listed in Tables 27.1 to 27.4. Table 27.1 shows the dimensional operating factors. Table 27.2 gives the properties of the bounding solids. The properties of the lubricants at ambient and atmospheric conditions are summarized in Table 27.3, while the pressure-temperature influencing factors on the properties of the lubricants are shown in Table 27.4. In addition, Table 27.5 gives the dimensionless operating parameters.

TABLE 27.2
Properties of bounding solids

Effective elastic modulus E', Pa	2.2×10^{11}
Mass density of solids ρ_a^* and ρ_b^*, kg/m^3	7850
Thermal conductivity of solids κ_a and κ_b, W/(m · K)	52
Specific heat of solids $c_{p,a}$ and $c_{p,b}$ J/(kg · K)	460

TABLE 27.3
Ambient-atmospheric properties of lubricant

Property	Case 1	Case 2
Mass density of lubricant at atmospheric conditions ρ_0^*, kg/m^3	866	866
Thermal conductivity of lubricant at atmospheric conditions $K_{f,0}$, $W/(m \cdot K)$	0.124	0.124
Specific heat of lubricant at atmospheric conditions $C_{p,0}$, $J/(kg \cdot K)$	2000	2000
Absolute viscosity at $p = 0$ and $t_m = 40°$ C, η_0, $Pa \cdot s$	0.04110	0.01326
Limiting shear strength at $p = 0$ and $t_m = 40°$ C, $\tau_{L,0}$, Pa	1.98×10^7	1.98×10^7

TABLE 27.4
Pressure-temperature influence factors on properties of lubricant

Factor	Case 1	Case 2
Viscosity-pressure index Z_1, dimensionless	0.689	0.580
Pressure-viscosity coefficient ξ, m^2/N	2.276×10^{-8}	1.582×10^{-8}
Viscosity-temperature index s_0, dimensionless	1.5	1.5
Temperature-viscosity coefficient β_η, 1/K	0.0216	0.0162
Limiting-shear-strength proportionality constant γ^*, Pa^{-1}	0.036	0.078
Temperature-limiting-shear-strength coefficient β_{τ_L}, 1/K	585	585
Piezodensity constant, first kind, $\alpha_{\rho,1}$, m^2/N	0.6×10^{-9}	0.6×10^{-9}
Piezodensity constant, second kind, $\alpha_{\rho,2}$, m^2/N	1.7×10^{-9}	1.7×10^{-9}
Thermodensity constant β_ρ, dimensionless	6.5×10^{-4}	6.5×10^{-4}
Piezothermal-conductivity constant, first kind, $\alpha_{\kappa,1}$, m^2/N	1.73×10^{-9}	1.73×10^{-9}
Piezothermal-conductivity constant, second kind, $\alpha_{\kappa,2}$, m^2/N	6.91×10^{-10}	6.91×10^{-10}

TABLE 27.5
Dimensionless operating parameters

Parameter	Case	
	1	2
Speed U	1.0×10^{-11}	1.0×10^{-11}
Load W	6.0×10^{-5}	6.0×10^{-5}
Materials G	5007	3480
Peclet number of lubricant P_{ef}	1147	3555
Kinetic energy parameter F_{KE}	0.652×10^{-3}	19.41×10^{-3}
Flow work parameter F_{WK}	0.466×10^{6}	1.433×10^{6}
Interface conductivity parameter $\check{K}_a$ and $\check{K}_b$	0.01535	0.01535

Note that case-2 series uses operating parameters identical to those that Shieh and Hamrock (1991) used in their isothermal study of non-Newtonian effects on film collapse. Using these parameters allows direct comparison between the thermal and isothermal results. The case-1 series uses operating parameters identical to those that Lee and Hamrock (1990) used in their study of circular non-Newtonian behavior in EHL conjunctions, in which the reach of the limiting shear strength at a low slide-roll ratio was encountered. This series of calculations demonstrates the numerical behavior of the asymptotically approached limiting-shear-strength model and shows the computational scheme that makes the convergent solution achievable.

Figures 27.1 to 27.6 give the results of case-2 computations. Figures 27.1(a) to (c) shows the three-dimensional temperature distributions for slide-roll ratios A_c of 1, 2, and 10. The temperature rises in the inlet zone are generally negligible relative to those in the Hertzian zone. In cases of pure rolling the maximum temperature rise in the whole conjunction is less than 1°C. The temperature distribution is completely flat in a scale such as that used in Fig. 27.1 and therefore is not shown. At low slide-roll ratios (that is, $A_c \leq 2$) the temperature distribution in the entire inlet zone is also flat. At greater slide-roll ratios, at which the solid surfaces are moving in opposite directions, the convective flows raise the downstream temperatures while maintaining low temperatures upstream in both the inlet and outlet zones.

Figure 27.2 gives a closer look at the temperature profiles at the lower solid-liquid interface ($Z = 0$), the midfilm ($Z = 0.5$), and the upper solid-liquid interface ($Z = 1$). It shows that at a lower slide-roll ratio [Fig. 27.2(a)] the temperature difference between the central film and the solid-liquid boundaries, as well as the magnitude of the temperature rises, is less than that at a higher slide-roll ratio. At low slide-roll ratios the midfilm temperature is higher than the solid surface temperatures in both the inlet and Hertzian contact zones, whereas at slide-roll ratios higher than 2 it is higher only at the central part of the Hertzian contact zone.

Comparing Fig. 27.3(b) with Fig. 27.3(a), which duplicates the results of Shieh and Hamrock (1991), shows the thermal effects on the formation of film shape, pressure profile, and streamlines at a slide-roll ratio of 10. With temperature effects not considered, Fig. 27.3(a) predicts the possibility of film collapse

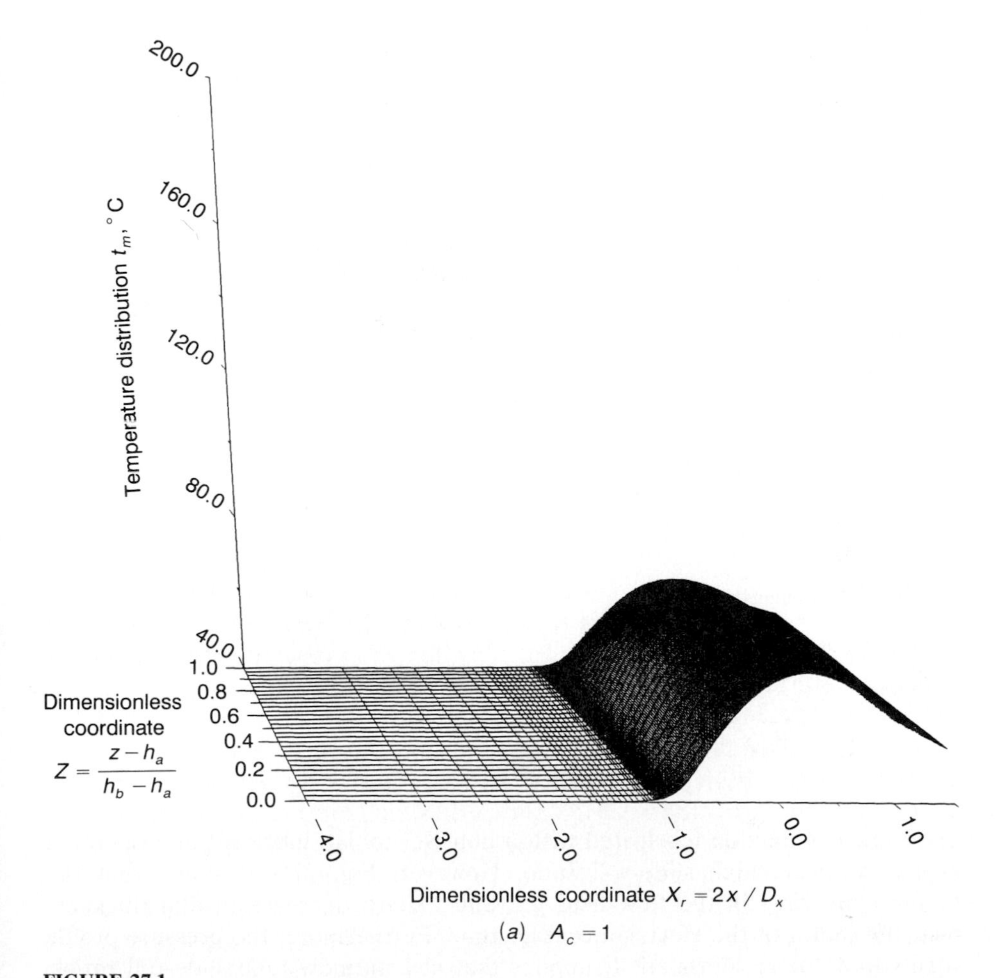

FIGURE 27.1
Temperature distribution for Case 2 at various slide-roll ratios of (a) $A_c = 1$, (b) $A_c = 2$, and $A_c = 10$. [*From Hsiao and Hamrock (1992).*]

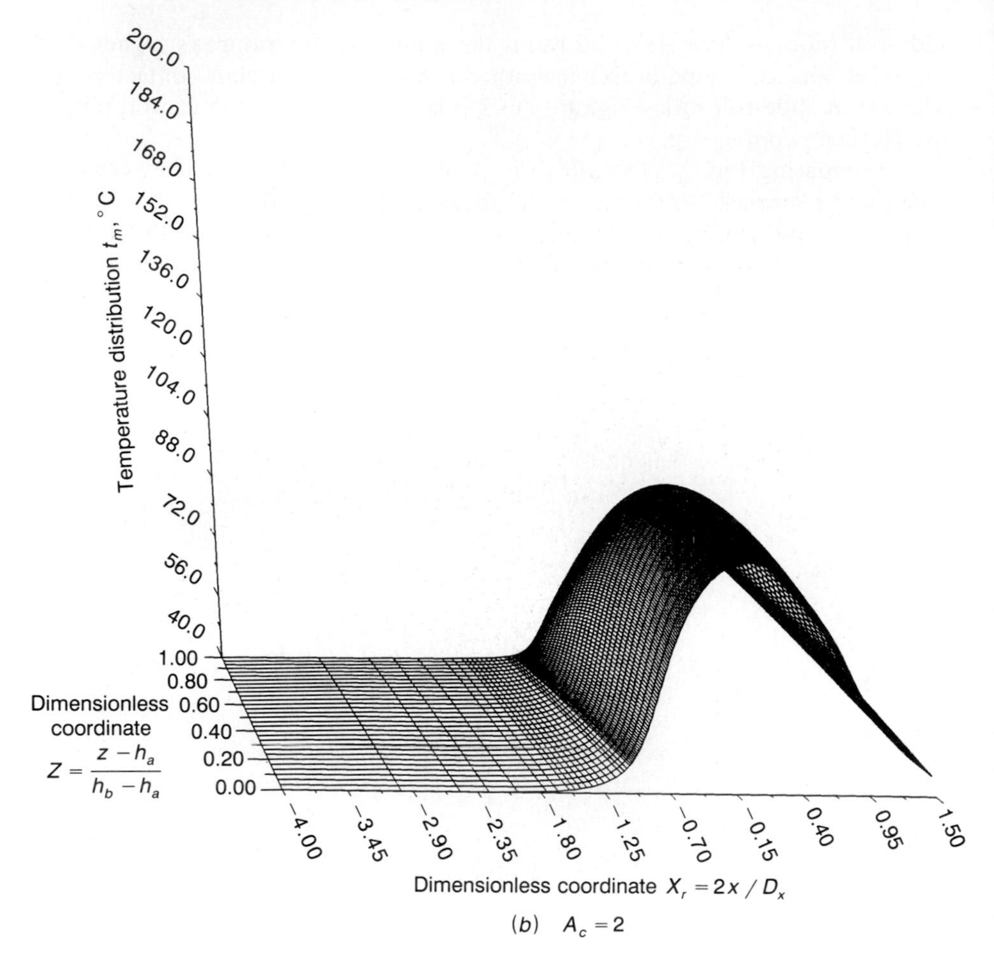

FIGURE 27.1 *Continued.*

in an EHL conjunction lubricated with a non-Newtonian lubricant and operated at an extremely high slide-roll ratio. However, Fig. 27.3(b) shows that the temperature rises in the flow field prevent a sharp decrease in film thickness near the center of the Hertzian contact zone. Furthermore, the pressure profile seems to be more Hertzian. It implies that at extremely high slide-roll ratios, two separate lubricant flows moving in opposite directions, instead of metal-to-metal contact, may develop in the Hertzian contact zone.

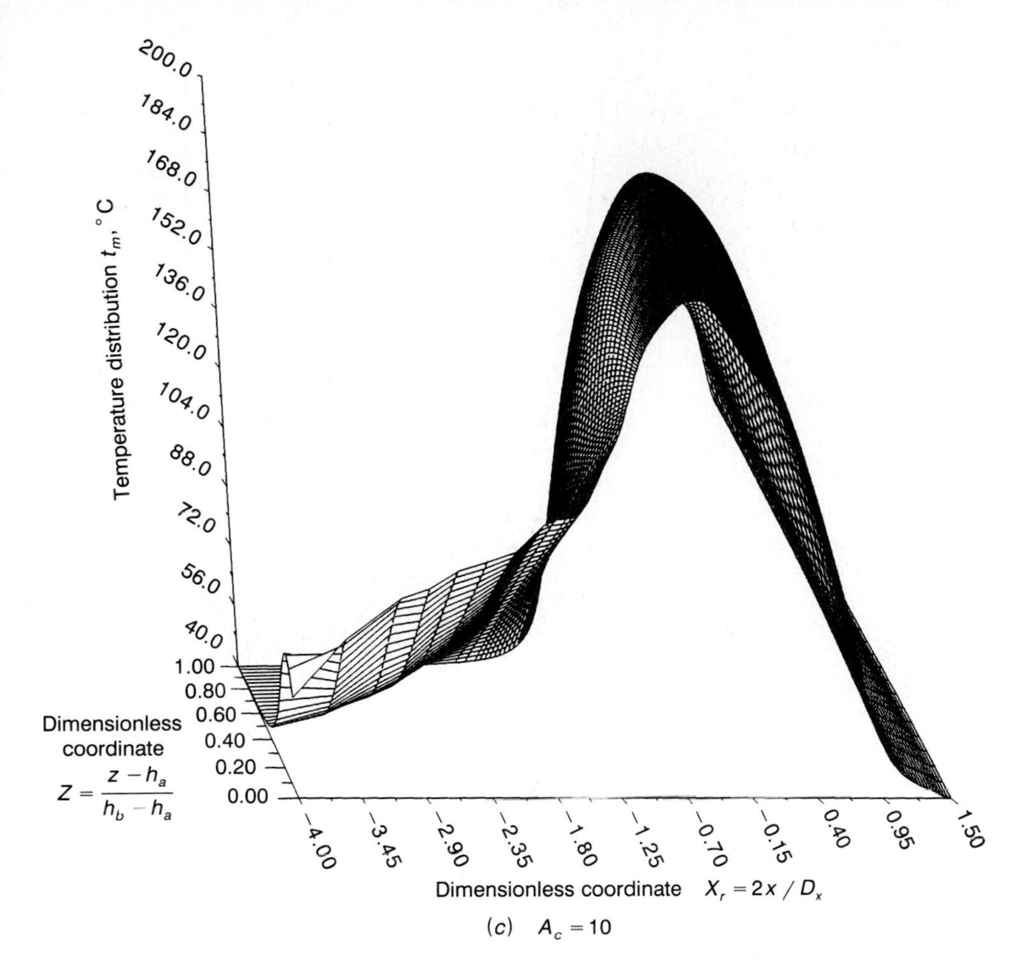

FIGURE 27.1 *Concluded.*

Figure 27.4 shows the viscosity profiles at slide-roll ratios of 0, 1, 2, and 10 for isothermal and thermal results. The viscosity profiles from the isothermal analysis reflect pressure distributions such as that shown in Fig. 27.3(a). It also shows that the viscosity profiles remain almost unchanged when the bounding solid surfaces move in the same direction or one of the surfaces is stationary. When the bounding surfaces start to move in opposite directions, the magnitude of viscosity (influenced by pressure) clearly decreases and the center of viscosity (pressure) shifts toward the starting point of the Hertzian contact zone.

On the other hand, the viscosity profiles from the thermal analysis indicate the synergistic effects of pressure and temperature. The decrease in viscosity with increasing slide-roll ratio is continuous and dramatic. The centers of these

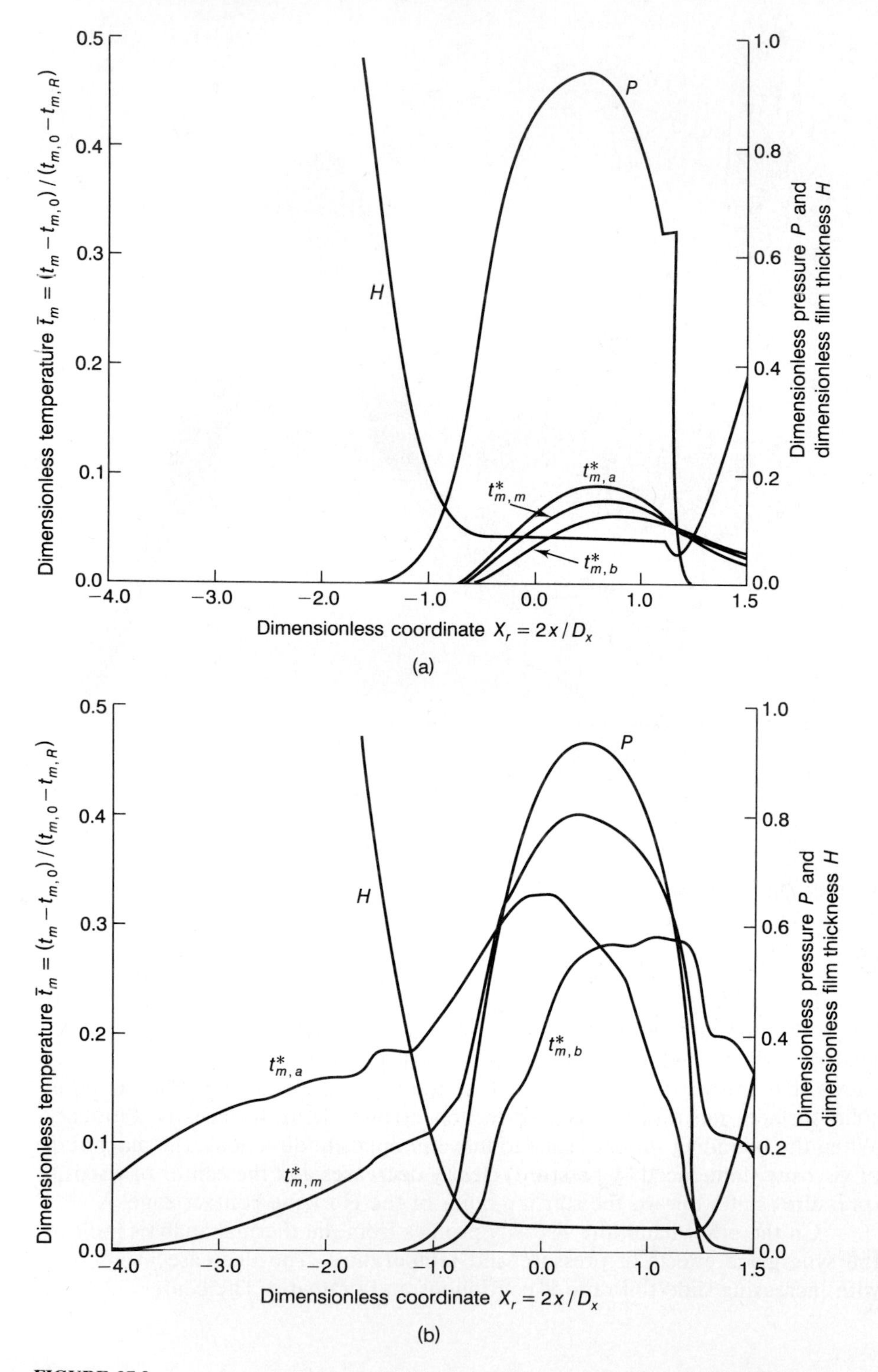

FIGURE 27.2
Surface and midfilm dimensionless temperatures as well as the film shape and pressure for case 2 and slide-roll ratios of (a) $A_c = 1.0$ and (b) $A_c = 10$. [*From Hsiao and Hamrock (1992).*]

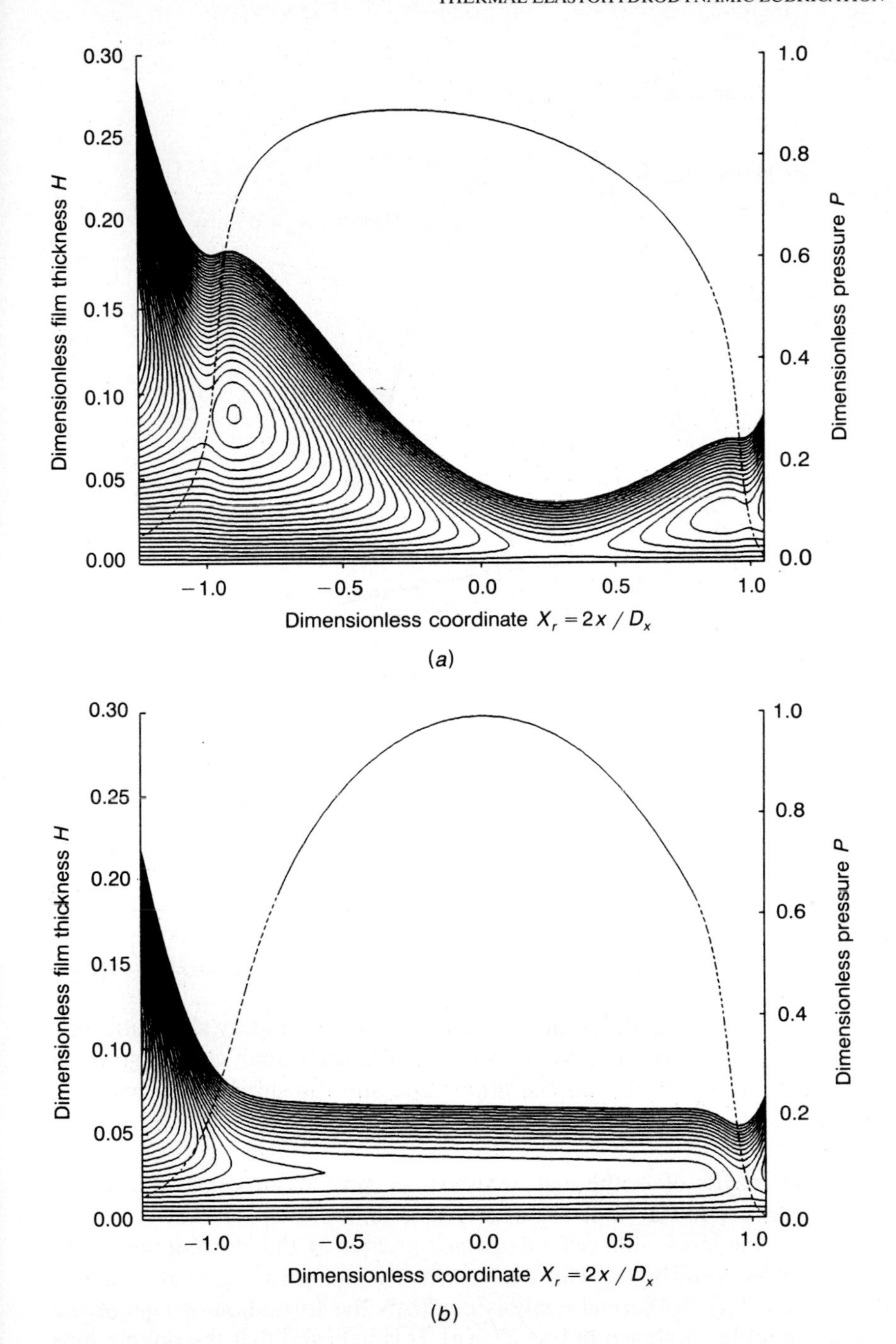

FIGURE 27.3
Streamlines as well as film shape and pressure profiles for case 2 at a slide-roll ratio of 10. (a) Isothermal results; (b) thermal results. [*From Hsiao and Hamrock (1992).*]

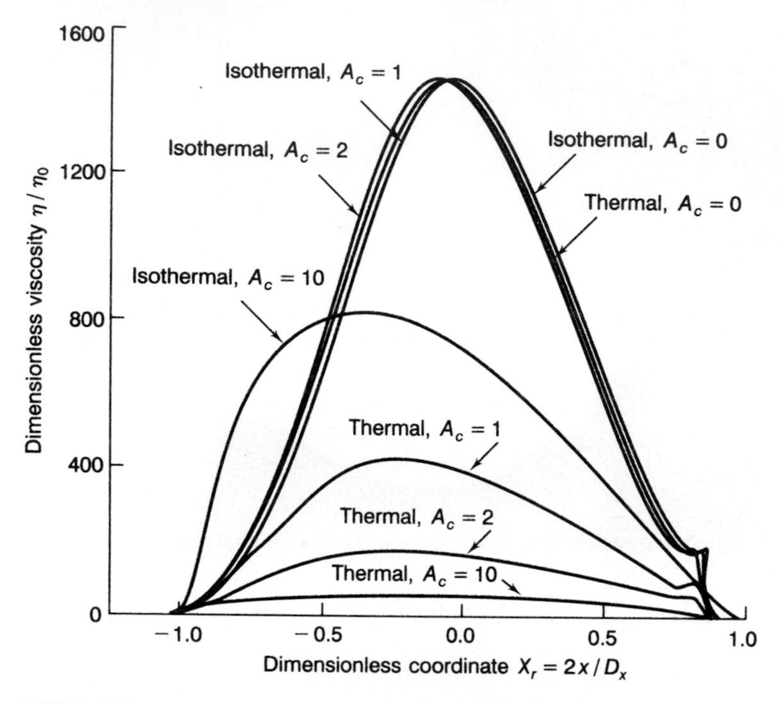

FIGURE 27.4
Viscosity profiles for case 2 at various slide-roll ratios for isothermal and thermal results. [*From Hsiao and Hamrock (1992).*]

profiles remain almost unchanged. This lack of change reflects the fact that the pressure profiles are more Hertzian and that the temperature profiles follow more or less the same shape as the pressure profiles in the Hertzian contact zone.

The profiles in $\check{T}$, the difference of dimensionless boundary shear stresses, for various slide-roll ratios from isothermal and thermal analyses are given in Fig. 27.5. The profile of $\check{T}$ depicts the film shape and the slope of the pressure profile as can be seen in the definition of the stress difference, $\check{T} = \bar{\tau}_b - \bar{\tau}_a = 2WH(dP/dX)/\pi\hat{\tau}_L$.

From the results of isothermal analyses the profiles of stress differences remain almost the same at slide-roll ratios less than 2. As the slide-roll ratio becomes larger, the stress difference becomes greater at the starting point and less at the exit location of the Hertzian contact zone, with the peaks shifting toward the ends. The isothermal analysis confirms the formation of nips in the film thickness profile, as shown in Fig. 27.3(a). It is believed that the double nips near the two ends create the circulating subflows within the Hertzian contact zone and prevent the lubricant from flowing through the conjunction.

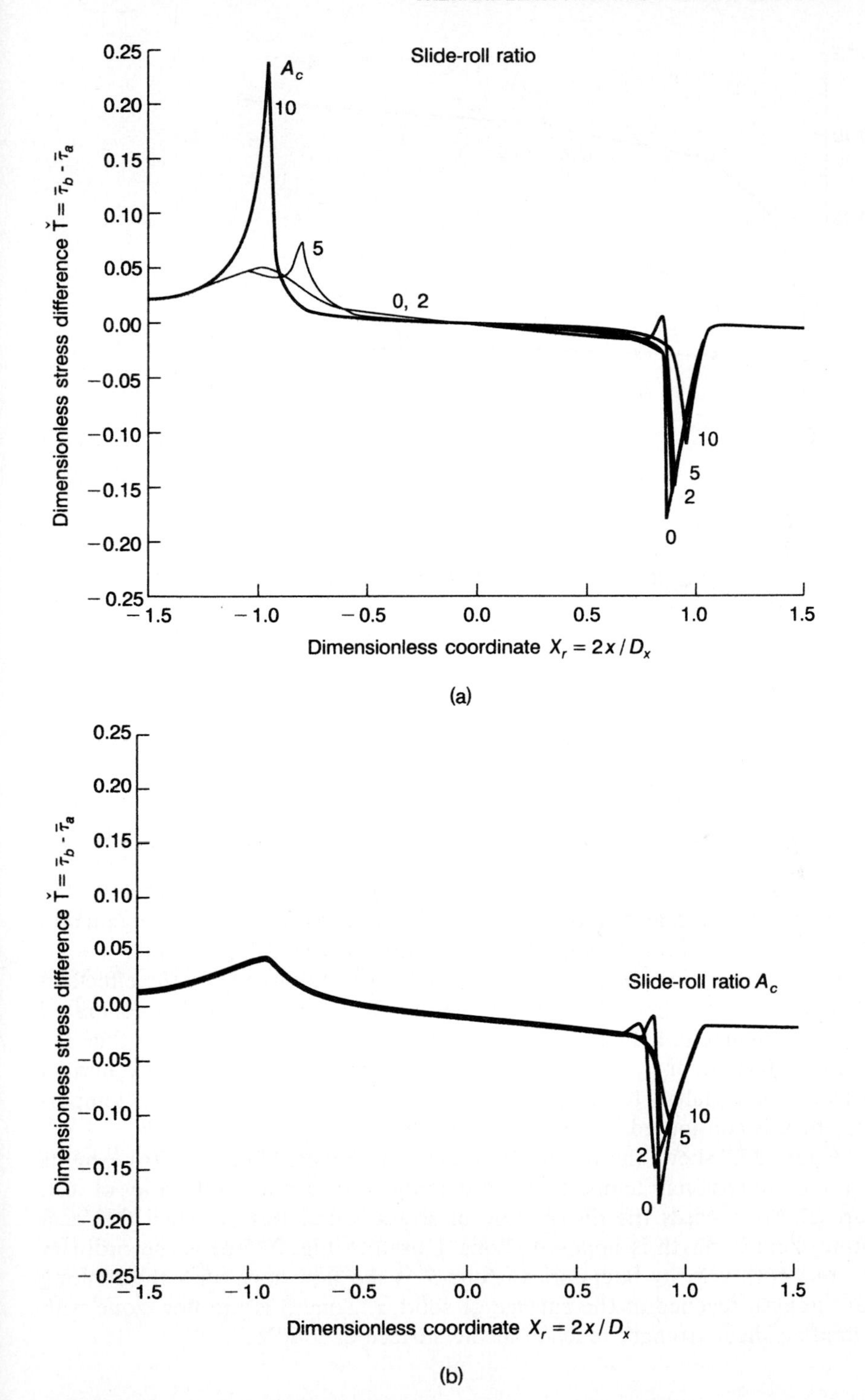

FIGURE 27.5
Stress-difference profiles for case 2 for various slide-roll ratios. (a) Isothermal results; (b) thermal results. [*From Hsiao and Hamrock (1992).*]

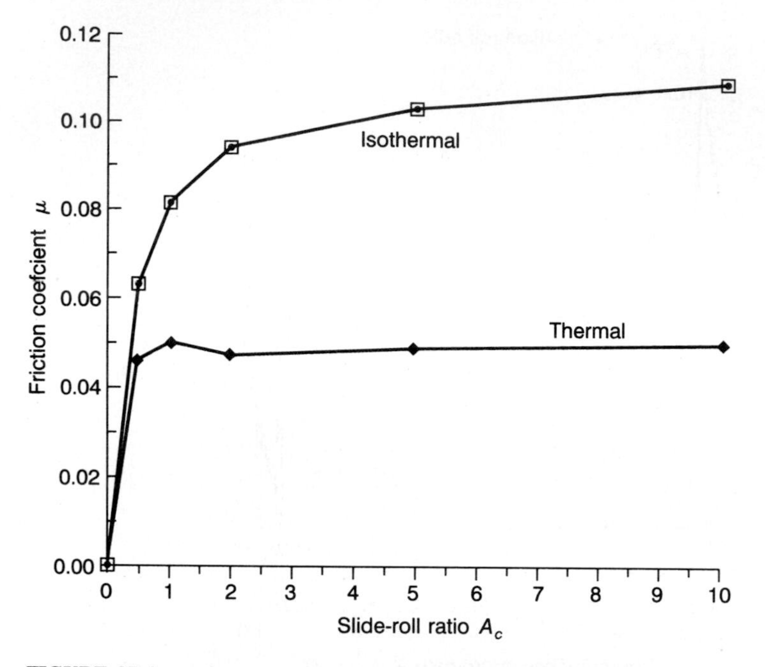

FIGURE 27.6
Effect of friction coefficient on slide-roll ratios for isothermal and thermal considerations. [*From Hsiao and Hamrock (1992).*]

However, the increase in stress differences at the starting point of the Hertzian contact zone means that more shear energy is created in this region. The increased shear energy increases the temperature by a significant amount and decreases the viscosity. The decrease in viscosity in turn decreases the pressure buildup. It may also prevent the formation of the nip in this region and allow more fluid to flow through the conjunction, as Fig. 27.3(b) shows.

One of the major objectives of the tribologist is to investigate the effects of both temperature and non-Newtonian characteristics on the traction generated in an EHL conjunction. Figure 27.6 gives one of the results, showing that the friction coefficient may be only one-half of that predicted by an isothermal non-Newtonian analysis for a high-slide-roll-ratio EHL conjunction if the temperature effect is considered.

Figure 27.7 shows the results of case-1 simulations. Figure 27.7(a) exhibits the three-dimensional temperature distribution at a slide-roll ratio of 0.1. Figure 27.7(b) records the distribution of stress boundaries at which the local limiting shear strength is imposed. Zone 1 used in Fig. 27.7(b) is the ordinary flow zone with velocity boundaries. Zone 2 is the flow zone with the limiting shear strength reached at the surface of solid *a*. Zone 3 is the flow zone with the limiting shear strength reached at the surface of solid *b*.

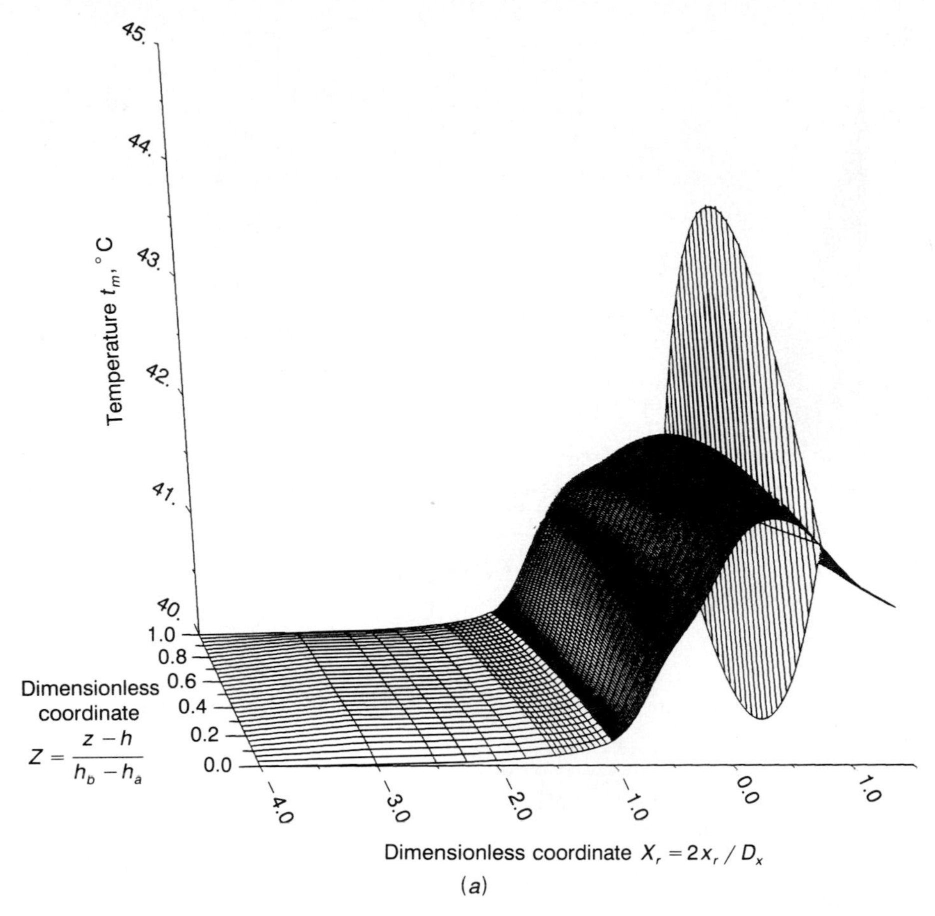

FIGURE 27.7
Results for case 1 at slide-roll ratio of 0.1. (a) Temperature distribution; (b) velocity contours indicating flow distribution; (c) surface and midfilm temperature as well as film thickness and pressure; (d) stress difference profile. [*From Hsiao and Hamrock (1992).*]

Figure 27.7(c) details the temperature profiles at the solid-liquid interfaces and the midfilm and shows the film shape and the pressure profile. Temperature spikes follow the pressure spike in the results of case-1 simulations, as depicted in Figs. 27.7(a) and (c). The causes of these temperatures spikes are not well understood, but the fluctuation of the boundary shear stress differences in these regions may explain them [Fig. 27.7(d)]. That is, when the stress difference sharply increases or decreases, the temperature increases or decreases in the same manner.

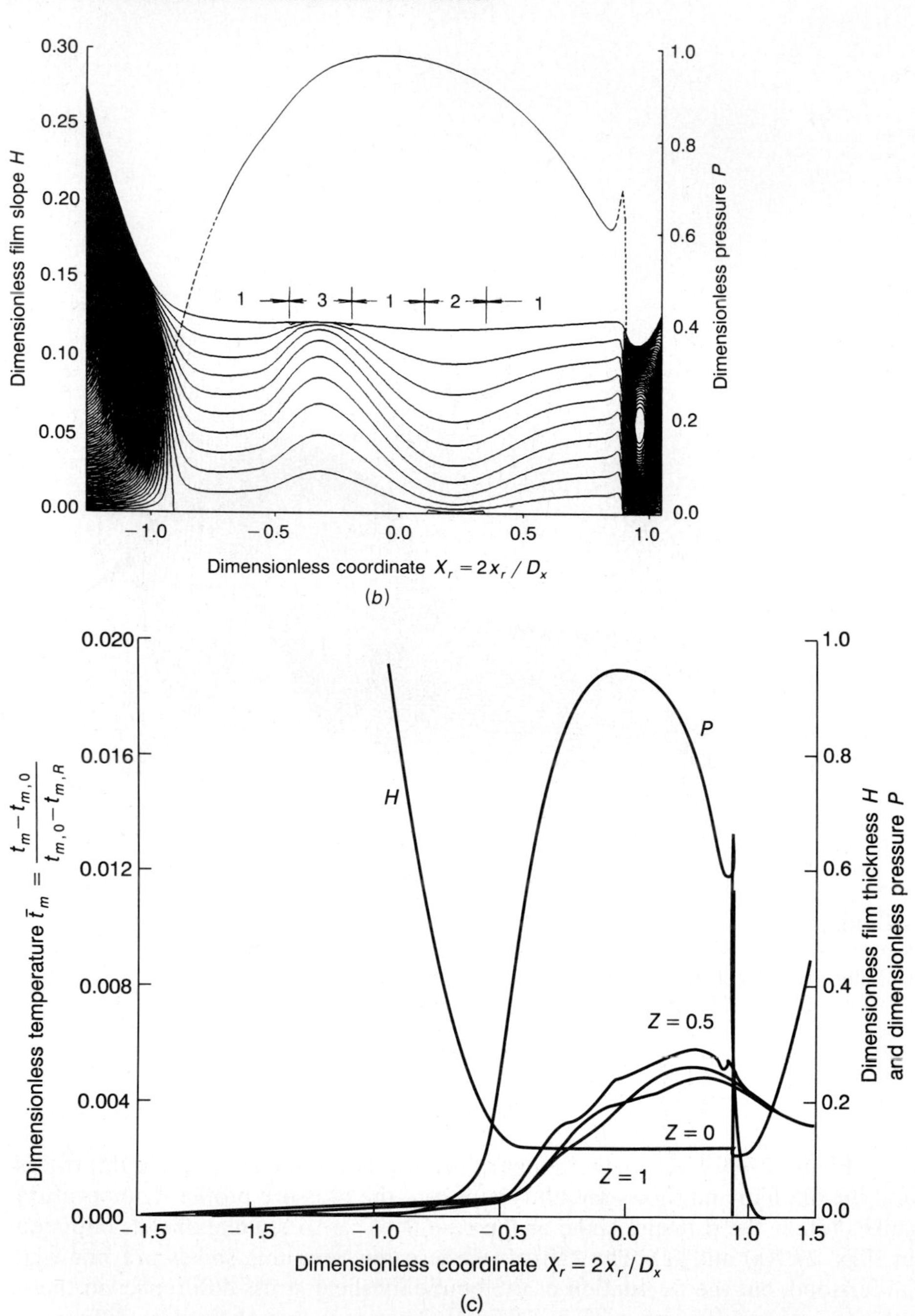

FIGURE 27.7 *Continued*

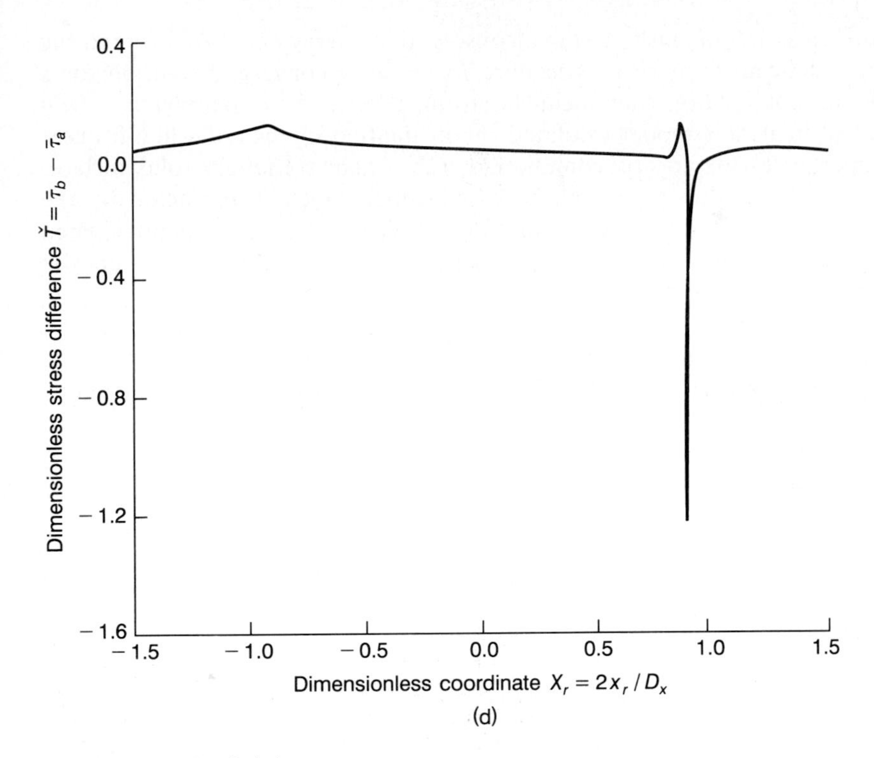

FIGURE 27.7 *Concluded*

27.6 CLOSURE

In this chapter the fast approach was completed and used to solve the thermal Reynolds equations for EHL conjunctions with circular non-Newtonian fluids. It fully automates and expedites the solution scheme.

The use of fictitious Reynolds equations ensured the validity of the solutions obtained and helped identify the flow zones with the boundary shear stresses that reached the local limiting shear strength of the lubricant. Imposing shear-stress boundary conditions enabled the simulation of an EHL conjunction lubricated by a circular non-Newtonian lubricant when the limiting shear strength of the lubricant is reached.

Using the thermal influence coefficients and embedding the solid-liquid interface temperatures into the discrete energy equations enabled the application of a straightforward successive-overrelaxation-by-lines finite difference method. This method solved for the field temperatures and the interface temperatures simultaneously. Using a conservative energy equation avoided the need for a delicate numerical procedure to obtain a converged solution for a thermal EHL conjunction that includes strong reverse flows. Applying a two-dimensional heat flow model validated the evaluation of the solid-liquid interface temperatures for an EHL conjunction with a near-stationary solid surface.

Film collapse due to a high slide-roll ratio, which is predicted by the isothermal circular non-Newtonian fluid model, does not seem to happen when the thermal effect is considered. Two separate streams flowing in opposite directions may develop instead.

The strong effect of temperature rises on the traction developed in an EHL conjunction was confirmed. When the thermal effect is considered at a high slide-roll ratio, the friction coefficient may be only one-half of that predicted by the isothermal model.

27.7 PROBLEMS

27.7.1 The moving face of a parallel-surface slider bearing has a velocity of u_0 and a temperature of $t_{m,0}$, while the stationary surface, which is at a distance h, has a temperature $t_{m,1}$. The lubricant is pressure fed so that a uniform pressure gradient $(-g)$ is established in the direction of motion. If it is assumed that steady-state conditions exist, the bearing is of infinite width, and the lubricant properties are invariant with temperature, show that the velocity profile across the film is parabolic. Also derive an expression for the temperature profile. Comment on the effect on energy dissipation and on the temperature profile if viscosity is taken to vary with temperature.

27.7.2 An incompressible, viscous fluid flows symmetrically in laminar flow along a circular-section tube of radius r_w. Flow is fully developed, and the velocity u at radius r is given by $u = 2v(1 - R^2)$, where v = mean velocity and $R = r/r_w$.

The temperature profile in the fluid is fully developed, with no temperature gradient in the axial direction. The inner surface of the tube is maintained at a constant temperature $t_{m,w}$. Show that the temperature t_m of the fluid at radius r is given by

$$t_m - t_{m,w} = \frac{\eta v^2}{K_f}(1 - R^4)$$

where K_f = thermal conductivity of fluid
η = viscosity of fluid

Explain the importance of the energy equation in the fundamental design of the bearings.

27.8 REFERENCES

Archard, J. F. (1958–59): The Temperature of Rubbing Surfaces. *Wear*, vol. 2, pp. 438–455.

Bair, S., and Winer, W. O. (1979): Shear Strength Measurements of Lubricants at High Pressure. *J. Lubr. Technol.*, vol. 101, pp. 251–257.

Barus, C. (1893): Isotherms, Isopiestics, and Isometrics Relative to Viscosity. *Am. J. Sci.*, vol. 45, pp. 87–96.

Blok, H. (1937): General Discussion on Lubrication. Institution of Mechanical Engineers, London. vol. 2, p. 222.

Cheng, H. S. (1965): A Refined Solution to the Thermal-Elastohydrodynamic Lubrication of Rolling and Sliding Cylinders. *ASLE Trans.*, vol. 8, pp. 397–410.

Cheng, H. S. (1967): "Calculation of Elastohydrodynamic Film Thickness in High Speed Rolling and Sliding Contacts." Tech. Rep. MTI–67 TR24. Mechanical Technology Incorporated.

Conry, T. F. (1981): Thermal Effects on Traction in EHD Lubrication. *J. Lubr. Technol.*, vol. 103, pp. 533–538.

Dowson, D., and Whitaker, A. V. (1965–66): A Numerical Procedure for the Solution of the Elastohydrodynamic Problem of Rolling and Sliding Contacts Lubricated by a Newtonian Fluid. *Proc. Inst. Mech. Eng., London*, vol. 180, no. 3B, pp. 57–71.

Gecim, B., and Winer, W. O. (1980): Lubricant Limiting Shear Stress Effect on EHD Film Thickness. *J. Lubr. Technol.*, vol. 102, pp. 213–221.

Ghosh, M. K., and Hamrock, B. J. (1985): Thermal Elastohydrodynamic Lubrication of Line Contracts. *ASLE Trans.*, vol. 28, no. 2, pp. 159–171.

Houpert, L. G., and Hamrock, B. J. (1986): Fast Approach for Calculating Film Thicknesses and Pressures in Elastohydrodynamically Lubricated Contacts at High Loads. *J. Tribol.*, vol. 108, pp. 411–420.

Hsiao, H. S., and Hamrock, B. J. (1992): A Complete Solution for Thermal-Elastohydrodynamic Lubrication of Line Contacts Using Circular Non-Newtonian Fluid Model. *J. Tribol.*, ASME paper 91-Trib-24.

Lee, R. T., and Hamrock, B. J. (1990): A Circular Non-Newtonian Fluid Model: Part I–Used in Elastohydrodynamic Lubrication. *J. Tribol.*, vol. 112, pp. 486–496.

Okamura, H. (1982): A Contribution to the Numerical Analysis of Isothermal Elastohydrodynamic Lubrication, *Tribology of Reciprocating Engines*. Proceedings of the Ninth Leeds-Lyon Symposium on Tribology. Butterworths, Guilford, England, pp. 313–320.

Richmond, J., Nilsson, O., and Sandberg, O. (1984): Thermal Properties of Some Lubricants Under High Pressure. *J. Appl. Phys.*, vol. 56, no. 7, pp. 2065–2067.

Roelands, C. J. A., Vlugter, J. G., and Waterman, H. I. (1963): The Viscosity-Temperature-Pressure Relationship of Lubricating Oils and Its Correlation With Chemical Constitution. *J. Basic Eng.*, pp. 601–610.

Rosenthal, D. (1946): The Theory of Moving Sources of Heat and Its Application to Metal Treatment. *ASME J.* pp. 849–866.

Sadeghi, F., and Dow, T. A. (1987): Thermal Effects in Rolling/Sliding Contacts: Part 2—Analysis of Thermal Effects in Fluid Film. *J. Tribol.*, vol. 109, pp. 512–518.

Shieh, J. A., and Hamrock, B. J. (1991): Film Collapse in EHL and Micro-EHL. *J. Tribol.*, vol. 113, pp. 372–377.

Tsai, C. L. (1986): "Welding Heat Transfer—Part I Conduction in Weldment." Class notes. Department of Welding Engineering, The Ohio State University.

Wang, S., Cusano, C., and Conry, T. F. (1991): Thermal Analysis of Elastohydrodynamic Lubrication of Line Contacts Using the Ree-Eyring Fluid Model. *J. Tribol.*, vol. 113, pp. 232–244.

Wang, S. H., and Zhang, H. H. (1987): Combined Effects of Thermal and Non-Newtonian Character of Lubricant on Pressure, Film Profile, Temperature Rise, and Shear Stress in EHL. *J. Tribol.*, vol. 109, pp. 666–670.

Wilson, W. R. D., and Huang, X. B. (1990): Discussion in a Circular Non-Newtonian Fluid Model: Part I—Used in Elastohydrodynamic Lubrication. Lee, R. T., and Hamrock, B. J. *J. Tribol.*, vol. 112, pp. 486–496.

Zhu, D., and Wen, S. Z. (1984): A Full Solution for the Thermalelastohydrodynamic Problem in Elliptical Contacts. *J. Tribol.*, vol. 106, pp. 246–254.

APPENDIX A

CALCULATION OF ELASTIC DEFORMATIONS

The details of evaluating Eq. (20.43) are presented in this appendix. The interval $[X_{\min}, X_{\text{end}}]$ can be divided into small intervals $[X_{j-1}, X_{j+1}]$ so that the deformation $\delta_{i,j}$ at node i is the sum of all the small elementary deformations $d\bar{\delta}_{i,j}$ calculated at node i and due to the pressure defined in the interval $[X_{j-1}, X_{j+1}]$.

$$\bar{\delta}_i = \sum_{j=2,4,\ldots}^{N-1} d\bar{\delta}_{i,j} \tag{A1}$$

In these small intervals, dP/dX' is assumed to vary linearly with X', and X' varies between X_{j-1} and X_j and between X_{j+1} and X_j as indicated in Fig. A. When these small intervals are used, the distance X_i to X_j, rather than simply X_i, has to be introduced.

$$d\bar{\delta}_{i,j} = -\frac{1}{2\pi}\int_{X_{j-1}\to X_j}^{X_{j+1}\to X_j} \frac{dP}{dX'}(X_i - X_j - X')\left[\ln(X_i - X_j - X')^2 - 2\right] dX' + \text{constant} \tag{A2}$$

The linear expression for dP/dX' reads

$$\frac{dP}{dX'} = (a_1 X' + a_2)P_{j-1} + (a_3 X' + a_4)P_j + (a_5 X' + a_6)P_{j+1} \tag{A3}$$

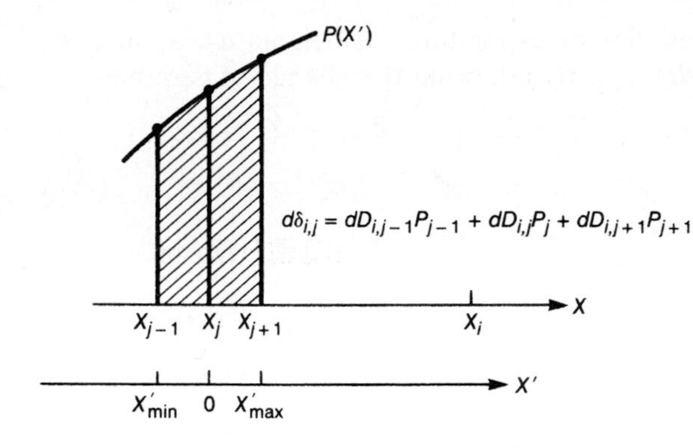

FIGURE A
Calculation of deformation $d\delta_{i,j}$ at node i due to pressure acting in interval $[X_{j-1}, X_{j+1}]$.

where
$$a_1 = \frac{2}{d_1} \tag{A4a}$$

$$a_2 = -\frac{(X_j + X_{j+1})}{d_1} + a_1 X_j \tag{A4b}$$

$$d_1 = (X_{j-1} - X_j)(X_{j-1} - X_{j+1}) \tag{A4c}$$

$$a_3 = \frac{2}{d_2} \tag{A4d}$$

$$a_4 = -\frac{(X_{j+1} + X_{j-1})}{d_2} + a_3 X_j \tag{A4e}$$

$$d_2 = (X_j - X_{j-1})(X_j - X_{j+1}) \tag{A4f}$$

$$a_5 = \frac{2}{d_3} \tag{A4g}$$

$$a_6 = -\frac{(X_j + X_{j-1})}{d_3} + a_5 X_j \tag{A4h}$$

$$d_3 = (X_{j+1} - X_{j-1})(X_{j+1} - X_j) \tag{A4i}$$

From the expression for dP/dX' it can be seen that $d\bar{\delta}_{i,j}$ can be expressed as

$$d\bar{\delta}_{i,j} = dD_{i,j-1}P_{j-1} + dD_{i,j}P_j + dD_{i,j+1}P_{j+1} + \text{constant} \tag{A5}$$

where $dD_{i,j}$ are the elementary influence coefficients calculated as

$$dD_{i,j-1} = -\frac{1}{2\pi}\int_{X_{j-1}\to X_j}^{X_{j+1}\to X_j} (a_1 X' + a_2)(X_i - X_j - X') \times \left[\ln (X_i - X_j - X')^2 - 2\right] dX' \tag{A6}$$

A similar relation that uses the corresponding coefficients a_3, a_4 and a_5, a_6 is used to define $dD_{i,j}$ and $dD_{i,j+1}$. By adopting the change of variables

$$Z = X_i - X_j - X' \qquad Z_{\min} = X_i - X_{j-1} \qquad Z_{\max} = X_i - X_{j+1}$$

$$b_2 = a_1(X_i - X_j) + a_2 \qquad dZ = -dX' \tag{A7}$$

Then $dD_{i,j-1}$ is calculated as

$$\begin{aligned} dD_{i,j-1} &= -\frac{1}{2\pi}\int_{Z_{\min}}^{Z_{\max}} -(-a_1 Z + b_2)Z(\ln Z^2 - 2)\,dZ \\ &= -\frac{1}{2\pi}\left(-b_2\int_{Z_{\min}}^{Z_{\max}} Z\ln Z\,dZ + a_1\int_{Z_{\min}}^{Z_{\max}} Z^2 \ln Z^2\,dZ + 2b_2 \right. \\ &\qquad \left. \times \left|\frac{Z^2}{2}\right|_{Z_{\min}}^{Z_{\max}} - 2a_1\left|\frac{Z^3}{3}\right|_{Z_{\min}}^{Z_{\max}}\right) \\ &= -\frac{1}{2\pi}\left| -2b_2\frac{Z^4}{4}(\ln Z^2 - 1) + 2a_1\frac{Z^3}{9}(\ln |Z|^3 - 1) \right. \\ &\qquad \left. + 2b_2\frac{Z^2}{2} - 2a_1\frac{Z^3}{3}\right|_{Z_{\min}}^{Z_{\max}} \end{aligned} \tag{A8}$$

Using the variables M, N, and K and rearranging gives

$$dD_{i,j-1} = -\frac{1}{2\pi}\left(a_1 K + a_2\frac{M}{2}\right) \tag{A9a}$$

$$dD_{i,j} = -\frac{1}{2\pi}\left(a_3 K + a_4\frac{M}{2}\right) \tag{A9b}$$

$$dD_{i,j+1} = -\frac{1}{2\pi}\left(a_5 K + a_6\frac{M}{2}\right) \tag{A9c}$$

where

$$M = Z_{\min}^2\left(\ln Z_{\min}^2 - 3\right) - Z_{\max}^2\left(\ln Z_{\max}^2 - 3\right) \tag{A10a}$$

$$N = Z_{\max}^3\left(\ln |Z_{\max}|^3 - 4\right) - Z_{\min}^3\left(\ln |Z_{\min}|^3 - 4\right) \tag{A10b}$$

$$K = M\frac{X_i - X_j}{2} + \frac{2N}{9} \tag{A10c}$$

There is no problem concerning the singularity occurring for $X_i - X_j$. When there is a singularity in the interval $[X_{i-1}, X_{i+1}]$, we can integrate along the two half-intervals $[X_{i-1}, X_i]$ and $[X_i, X_{i+1}]$ and use the relation

$$\lim_{Z\to 0}\left(Z^2 \ln Z^2\right) = \lim_{Z\to 0}\left(Z^3 \ln |Z|^3\right) = 0 \tag{A11}$$

which shows the relations (A9) to be also valid for $X_i = X_j$. Finally, the deformation $\bar{\delta}_1$ is obtained by summing all the elementary small deformations $d\bar{\delta}_{i,j}$

$$\bar{\delta}_i = \sum_{j=2,4}^{N-1} d\bar{\delta}_{i,j} \tag{A12}$$

$$\bar{\delta}_i = \sum_{j=1}^{N} D_{i,j}P_j - \frac{1}{4}\ln\left(R^2\frac{8W}{\pi}\right) \tag{A13}$$

where $D_{i,j} = dD_{i,j}$ if j is even. If j is odd ($j = 3$ in the following example), we calculate $D_{i,3}$ coming from the interval $[X_1, X_3]$ and add to it the value $dD_{i,j}$ coming from the interval $[X_3, X_5]$ to obtain the final value of $D_{i,3}$. The first and last values of $D_{i,j}$ are simply

$$D_{i,1} = dD_{i,1} \tag{A14a}$$

and

$$D_{i,N} = dD_{i,N} \tag{A14b}$$

Note now that $\sum_j D_{i,j}P_j$ is independent of the load. At high loads the dimensionless film thickness H_m becomes very small with respect to $\bar{\delta}_i$. However, by using an appropriate change of variable ($X = x/bc$), the maximum deformation $\bar{\delta}_m$ can be kept equal to H_m. Using the last change of variable gives

$$\bar{\delta}_i = c\sum_{j=1}^{N} D'_{i,j}P_j - \frac{1}{4}\ln\left(R^2\frac{8W}{\pi}c^2\right) \tag{A15}$$

where $D'_{i,j}$ are the new influence coefficients obtained with the new value of X. From the definition of c

$$c\sum D'_{i,j}P_j = H_m \tag{A16}$$

The maximum deformation $\bar{\delta}_m$ is close to the maximum Hertzian deformation $\bar{\delta}_H$, leading to

$$-\frac{1}{4}\ln\left(R^2\frac{8W}{\pi}\right) + \frac{1}{2}\ln 2 + \frac{1}{4} = H_m - \frac{1}{4}\ln\left(R^2\frac{8W}{\pi}c^2\right) \tag{A17}$$

The value of c can therefore be defined as

$$c = \frac{1}{2}\exp(2H_m - 0.5) \tag{A18}$$

This numerical "trick" was definitely helpful at high loads but was not used to obtain the results presented herein.

APPENDIX B

CORRECTIONS TO BE APPLIED TO WEIGHTING FACTORS DUE TO X

The term dP/dX is used both to calculate the influence coefficient $D_{i,j}$ and in the Reynolds equation. For the calculation of $D_{i,j}$, $(dP/dX)_i$ at node i is calculated by using three nodes.

$$\left(\frac{dP}{dX}\right)_i = a_{i,i-1}P_{i-1} + a_{i,i}P_i + a_{i,i+1}P_{i+1} \tag{B1}$$

where

$$a_{i,i-1} = \frac{X_i - X_{i+1}}{(X_{i-1} - X_i)(X_{i-1} - X_{i+1})} \tag{B2}$$

$$a_{i,i} = \frac{2X_i - X_{i+1} - X_{i-1}}{(X_i - X_{i-1})(X_i - X_{i+1})} \tag{B3}$$

$$a_{i,i+1} = \frac{X_i - X_{i-1}}{(X_{i+1} - X_{i-1})(X_{i+1} - X_i)} \tag{B4}$$

At the first and last nodes, X_1 and $X_{N_{max}}$, dP/dX is also defined by using three nodes (e.g., X_1, X_2, X_3 and X_{N-1}, X_{N-2}, $X_{N_{max}}$).

For the calculation of dP/dX in the Reynolds equations, two nodes may often be used if the numerical convergence is difficult to obtain with the three-node formula. For the two-node formula the weighting factors are

$$a_{i,i+1} = \frac{1}{X_{i+1} - X_i} \qquad a_{i,i} = -a_{i,i+1} \qquad a_{i,i-1} = 0 \tag{B5}$$

Minor corrections are also applied on the last values of $a_{n,j}$ to respect the boundary conditions at $X = X_{\text{end}}$. For $X = X_{\text{end}}$, $P = dP/dX = 0$. It is assumed that between X_N and X_{end}, $P(X)$ is described by a second-degree polynomial. Respecting the previously mentioned boundary condition gives

$$P = \frac{(X - X_{\text{end}})^2}{(X_N - X_{\text{end}})^2} P_N \tag{B6}$$

$$\frac{dP}{dX} = \frac{2(X - X_{\text{end}})}{(X_N - X_{\text{end}})^2} P_N \tag{B7}$$

For $X = X_N$, therefore,

$$\left(\frac{dP}{dX}\right)_N = \frac{2}{X_N - X_{\text{end}}} P_N \tag{B8}$$

which can also be expressed as

$$a_{n,n-1} = 0 \qquad a_{n,n} = \frac{2}{X_N - X_{\text{end}}} \qquad a_{n,n+1} = 0 \tag{B9}$$

The integral of the pressure between X_N and X_{end} lcads to

$$\int_{X_N}^{X_{\text{end}}} P\,dX = \frac{1}{3}(X_{\text{end}} - X_N)P_N = \Delta C_N P_N \tag{B10}$$

If N is odd, subtract from the value of C_N the value $dC_{N+1,N}$ coming from the interval $[X_N, X_{N+2}]$ and add ΔC_N to get the final value of C_N. If N is even, modify C_{N-1} and C_N. From C_{N-1} subtract $dC_{N,N-1}$ from the interval $[X_{N-1}, X_{N+1}]$ and add the weighting factor ΔC_{N-1} coming from the integration of P between X_{N-1} and X_N. Using the "trapeze" rule gives

$$\Delta C_{N-1} = \frac{X_N - X_{N-1}}{2} \tag{B11}$$

The value of C_N is finally defined as

$$C_N = \Delta C_{N-1} + \Delta C_N \tag{B12}$$

APPENDIX C

CALCULATION OF JACOBIAN FACTORS

The factors to be defined are

$$\frac{\partial f_i}{\partial(\bar{\rho}_m H_m)} \qquad \frac{\partial f_i}{\partial P_j} \qquad \frac{\partial f_i}{\partial H_0}$$

where

$$f_i = H_i^3\left(\frac{dP}{dX}\right)_i - \bar{K}\bar{\eta}_i\left(H_i - \frac{\bar{\rho}_m H_m}{\bar{\rho}_i}\right) \tag{C1}$$

$$H_i = H_0 + \frac{X_i^2}{2} + \sum_{j=1}^{N} D_{i,j}P_j \tag{C2}$$

and dP/dX, $\bar{\eta}_i$, and $\bar{\rho}_i$ are defined by Eqs. (21.22), (4.8) or (4.10), and (4.19). Before the final Jacobian factors are defined, we can define

$$\partial H_i/\partial H_0 = 1 \tag{C3a}$$

$$\partial H_i/\partial P_j = D_{i,j} \tag{C3b}$$

$$\partial \bar{\eta}_i/\partial P_j = \alpha p_H \bar{\eta}_i k_{i,j} \tag{C3c}$$

if Barus' viscosity [Eq. (4.8)] is used or

$$\frac{\partial \bar{\eta}_i}{\partial P_j} = 5.1 \times 10^{-9} p_H(\ln \eta_0 + 9.67)\left(1 + 5.1 \times 10^{-9} p_H P_i\right)^{Z-1} \bar{\eta}_i k_{i,j} \tag{C3d}$$

if Roelands' viscosity [Eq. (4.10)] is used, where $k_{i,j}$, the Kronecker symbol, equals 1 if $i = j$ and equals 0 if $i \neq j$.

$$\frac{\partial(1/\bar{\rho}_i)}{\partial P_j} = -\frac{0.6 \times 10^{-9} p_H}{1 + 2.3 \times 10^{-9} p_H P_i} k_{i,j} \tag{C3e}$$

$$\frac{\partial[(dP/dX)_i]}{\partial P_j} = a_{i,i-1} k_{i-1,j} + a_{i,i} k_{i,j} + a_{i,i+1} k_{i+1,j} \tag{C3f}$$

It is now easy to define the Jacobian factors.

$$\frac{\partial f_i}{\partial(\bar{\rho}_m H_m)} = \frac{\bar{K}\bar{\eta}_i}{\bar{\rho}_i} \tag{C4a}$$

$$\frac{\partial f_i}{\partial P_j} = 3H_i^2 \left(\frac{dP}{dX}\right)_i D_{i,j} + H_i^3 \frac{\partial[(dP/dX)_i]}{\partial P_j} - \bar{K}\frac{\partial \bar{\eta}_i}{\partial P_j}\left(H_i - \frac{\Delta\bar{\rho}_m H_m}{\bar{\rho}_i}\right) - \bar{K}\bar{\eta}_i \left[D_{i,j} - \bar{\rho}_m H_m \frac{\partial(1/\bar{\rho}_i)}{\partial P_j}\right] \tag{C4b}$$

and

$$\frac{\partial f_i}{\partial H_0} = 3H_i^2 \left(\frac{dP}{dX}\right)_i - \bar{K}\bar{\eta}_i \tag{C4c}$$

APPENDIX D

DEFINITION OF WEIGHTING FACTORS

The weighting factors C_j are defined by using the three-point Lagrange polynomial with a general mesh (nonconstant step). In the interval $[X_{j-1}, X_{j+1}]$ the pressure is described by a second-degree polynomial in X'.

$$P = \frac{X'(X' + X_j - X_{j+1})}{(X_{j-1} - X_j)(X_{j-1} - X_{j+1})} P_{j-1} + \frac{(X' + X_j - X_{j-1})(X' + X_j - X_{j+1})}{(X_j - X_{j-1})(X_j - X_{j+1})} P_j + \frac{(X' + X_j - X_{j-1})X'}{(X_{j+1} - X_{j-1})(X_{j+1} - X_j)} P_{j+1} \quad \text{(D1)}$$

We can now define the coefficients $dC_{j,j-1}$, $dC_{j,j}$, and $dC_{j,j+1}$ such that

$$\int_{X'_{\min}}^{X'_{\max}} P\, dX' = dC_{j,j-1}P_{j-1} + dC_{j,j}P_j + dC_{j,j+1}P_{j+1} \quad \text{(D2)}$$

where

$$X'_{\min} = X_{j-1} - X_j$$

and

$$X'_{max} = (X_{j+1} - X_j)$$

$$dC_{j,j-1} = \frac{1}{(X_{j-1} - X_j)(X_{j-1} - X_{j+1})} \left| \frac{(X')^3}{3} + (X_j - X_{j+1}) \frac{(X')^2}{2} \right|_{X'_{min}}^{X'_{max}} \tag{D3a}$$

$$dC_{j,j} = \frac{1}{(X_j - X_{j-1})(X_j - X_{j+1})} \left| \frac{(X')^3}{3} + (2X_j - X_{j-1}) \frac{(X')^2}{2} + (X_j - X_{j-1})(X_j - X_{j+1}) X' \right|_{X'_{min}}^{X'_{max}} \tag{D3b}$$

$$dC_{j,j+1} = \frac{1}{(X_{j+1} - X_{j-1})(X_{j+1} - X_j)} \times \left| \frac{(X')^3}{3} + (X_j - X_{j-1}) \frac{(X')^2}{2} \right|_{X'_{min}}^{X'_{max}} \tag{D3c}$$

These coefficients are calculated for $j = 2, 4, \ldots, N - 1$. When j is even, the coefficients C_j are finally defined as

$$C_j = dC_{j,j}$$

When j is odd ($j = 3$ in the following example), $dC_{2,3}$ is calculated corresponding to the interval $[X_1, X_3]$, and $dC_{4,3}$ coming from the interval $[X_3, X_5]$ is added to it to define C_j, or C_3 in this example. Minor corrections are also applied on the last values of C_j to respect the boundary condition for $X = X_{end}$, as shown in Appendix B.

INDEX

FUNDAMENTALS OF FLUID FILM LUBRICATION

McGraw-Hill Series in Mechanical Engineering

Consulting Editors

Jack P. Holman, Southern Methodist University
John R. Lloyd, Michigan State University

Anderson: *Modern Compressible Flow: With Historical Perspective*
Arora: *Introduction to Optimum Design*
Bray and Stanley: *Nondestructive Evaluation: A Tool for Design, Manufacturing, and Service*
Burton: *Introduction to Dynamic Systems Analysis*
Culp: *Principles of Energy Conversion*
Dally: *Packaging of Electronic Systems: A Mechanical Engineering Approach*
Dieter: *Engineering Design: A Materials and Processing Approach*
Eckert and Drake: *Analysis of Heat and Mass Transfer*
Edwards and McKee: *Fundamentals of Mechanical Component Design*
Gebhart: *Heat Conduction and Mass Diffusion*
Gibson: *Principles of Composite Material Mechanics*
Hamrock: *Fundamentals of Fluid Film Lubrication*
Heywood: *Internal Combustion Engine Fundamentals*
Hinze: *Turbulence*
Holman: *Experimental Methods for Engineers*
Howell and Buckius: *Fundamentals of Engineering Thermodynamics*
Hutton: *Applied Mechanical Vibrations*
Juvinall: *Engineering Considerations of Stress, Strain, and Strength*
Kane and Levinson: *Dynamics: Theory and Applications*
Kays and Crawford: *Convective Heat and Mass Transfer*
Kelly: *Fundamentals of Mechanical Vibrations*
Kimbrell: *Kinematics Analysis and Synthesis*
Martin: *Kinematics and Dynamics of Machines*
Modest: *Radiative Heat Transfer*
Norton: *Design of Machinery*
Phelan: *Fundamentals of Mechanical Design*
Raven: *Automatic Control Engineering*
Reddy: *An Introduction to the Finite Element Method*
Rosenberg and Karnopp: *Introduction to Physical Systems Dynamics*
Schlichting: *Boundary-Layer Theory*
Shames: *Mechanics of Fluids*
Sherman: *Viscous Flow*
Shigley: *Kinematic Analysis of Mechanisms*
Shigley and Mischke: *Mechanical Engineering Design*
Shigley and Uicker: *Theory of Machines and Mechanisms*
Stiffler: *Design with Microprocessors for Mechanical Engineers*
Stoecker and Jones: *Refrigeration and Air Conditioning*
Ullman: *The Mechanical Design Process*
Vanderplaats: *Numerical Optimization: Techniques for Engineering Design, with Applications*
White: *Viscous Fluid Flow*
Zeid: *CAD/CAM Theory and Practice*